Nonlinear Polymer Rheology

Nonlinear Polymer Rheology

Macroscopic Phenomenology and Molecular Foundation

Shi-Qing Wang
University of Akron, Ohio, US

WILEY

This edition first published 2018

Registered Office
John Wiley & Sons, Inc., 111 River Street, Hoboken, NJ 07030, USA

Editorial Office
111 River Street, Hoboken, NJ 07030, USA

For details of our global editorial offices, customer services, and more information about Wiley products visit us at www.wiley.com.

Wiley also publishes its books in a variety of electronic formats and by print-on-demand. Some content that appears in standard print versions of this book may not be available in other formats.

Library of Congress Cataloging-in-Publication Data

Names: Wang, Shi-Qing, author.
Title: Nonlinear polymer rheology : macroscopic phenomenology and molecular foundation / by Shi-Qing Wang.
Description: Hoboken, NJ : John Wiley & Sons, 2017. | Includes index. | Identifiers: LCCN 2017019098 (print) | LCCN 2017036798 (ebook) | ISBN 9781119029052 (pdf) | ISBN 9781119029045 (epub) | ISBN 9780470946985 (cloth)
Subjects: LCSH: Polymers–Rheology.
Classification: LCC TA455.P58 (ebook) | LCC TA455.P58 W355 2017 (print) | DDC 541/.2254–dc23
LC record available at https://lccn.loc.gov/2017019098

Cover design by Wiley

Set in 10/12pt WarnockPro by SPi Global, Chennai, India

10 9 8 7 6 5 4 3 2 1

Contents

Preface

Nonlinear Polymer Rheology explores the rich phenomenology of the mechanical behavior of polymer melts and concentrated polymer solutions. My main purpose is to expose the reader to the latest knowledge and understanding of the subject, developed in the past decade. This book explores and establishes a microscopic foundation that provides a coherent molecular-level interpretation for various nonlinear rheological behaviors. In absence of such a foundation, the book would not and could not have been written.

Covering nearly every aspect of the nonlinear rheological responses of entangled polymers, this book may be used as a textbook to introduce essential phenomenological information. The reader does not need to be an experienced researcher in the field of rheology. The book presents the subject in a self-contained manner, although familiarity with the literature on nonlinear polymer rheology would allow the reader to contrast different standpoints.

My approach to nonlinear polymer rheology places a great emphasis on understanding transient viscoelastic responses. While steady-flow behavior is also of interest, *Nonlinear Polymer Rheology* differs from other works in that it treats nonlinear responses as primary and linear responses as secondary. We aim to collect sufficient first-hand phenomenology before proposing theoretical concepts, although key concepts including yielding and disentanglement are utilized in as early as Chapters 6 and 7 without elaborative discussion. Since fresh viewpoints are required, the reader will recognize sharp contrasts with conventional knowledge and methodology.

Many excellent books have treated the subject of polymer rheology in a traditional way. The most classical literature is cited and discussed in books including those by Ferry,[1] Doi and Edwards,[2] Bird et al.,[3] Dealy and Larson,[4] Graessley,[5] and Phillies.[6] In general, these books do not have a sufficient discussion of nonlinear rheology that is based on a coherent gathering of key phenomenology. Limited by the available space, the present book omits discussion of the older literature before 2000 and only includes a few pertinent references since 2000.

Scientific inquiries develop in three stages: A. "Empirical," where we find out what happens; B. "Phenomenological," where we learn how it happens; C. "Theoretical," where we explain why it happens. For a complicated subject such as polymer rheology, it is not feasible to formulate a theory without first having sufficient and coherent phenomenological knowledge. *Nonlinear Polymer Rheology* acknowledges this logical sequence and strives to collect and establish the phenomenology before developing any theoretical treatment and formulation. However, these three types of research can and do often proceed interactively and interchangeably. For example, a particular theoretical idea or picture can prompt one to organize phenomenological information in a more coherent manner and to design additional experiments using hypothetical and unproven concepts.

The responses of polymeric liquids to large and rapid external deformations are challenging to depict and understand in molecular terms. The task is difficult because molecular behavior on nanometer scales dictates rheological properties characterized on macroscopic (millimeter) scales: There exists a gap of six orders of magnitude in length scales. Thus, we should begin by "listening to" what the experiments tell us, and first build a sound phenomenological base. Until we have a sense of the full picture, which could stem from an adequate analysis of the available phenomenology, it is challenging and risky to make theoretical simplifications, for example, in modeling such complex behavior as the response of polymer entanglement to sudden, fast, large external deformations.

Our goal is to arrive at a realistic physical picture for nonlinear polymer rheology. Having collected the essential phenomenology, the process of rational thinking must take us beyond empirical knowledge. To illustrate the merit of reason-guided thinking, let us consider, for example, the phenomenon of viscoelasticity. By definition, all viscoelastic materials are mechanically solid-like (elastic) when probed on short time scales. Viscoelastic liquids become completely viscous only on long time (relative to the material relaxation time) scales. Where does the "elasticity" come from in such a liquid? What is the structure of the material that produces the elasticity? How should we think about the microscopic origin of viscoelasticity?

Upon external deformation, it is clear from the established phenomenology that an initial elastic response must end, and a transition to flow must begin. In other words, all viscoelastic liquids undergo yielding when subjected to fast deformation. The transition to flow suggests that there exists some kind of a potential barrier on short time scales. The next question is what creates such a potential barrier. As soon as we ask questions like this, we can make progress toward an instructive understanding of viscoelasticity, instead of stopping at the level of phenomenological models including the Maxwell model and the Oldroyd[7] model.

This idea to associate the "elasticity" with a potential barrier for any viscoelastic materials is useful even for an external deformation rate $\mathcal{R}$ that is lower than the reciprocal of the dominant relaxation time τ, that is, when the Weissenberg number $Wi < 1$. In other words, a transition from elastic deformation to flow must occur even when the product $Wi = \mathcal{R}\tau$ is below unity. For $Wi \gg 1$, the initial elastic deformation can be rather remarkable, persisting up to many strain units in the case of entangled polymers. The termination of the elastic-dominant response apparently stems from a breakdown of some microscopic structure. It is an essential task of polymer rheology to identify and delineate the nature of potential barriers, in terms of intermolecular interactions, and show how the structural breakdown takes place.

Figuring out the nature of intermolecular interactions in entangled polymeric liquids under large deformation is a daunting task. Historically, the task has challenged the brightest minds in polymer science. An entangled polymer can be regarded as a physical network of Gaussian chains. Its stress response to startup shear reveals a finite cohesive strength of the network junctions. Maxwell was right[8]: Entangled polymer melts yield, just like ductile polymeric solids (e.g., glasses) do, under continuous external deformation. The rate dependence of the yielding response indicates that chain entanglements are dynamic and have finite lifetimes.

In 1979, Maxwell and Nguyen[9] described the stress overshoot of polystyrene melts upon startup shear by stating "the yielding behavior indicates that, as straining progresses, the structure of the melt is broken down, thereby permitting flow." In the same year, Doi and Edwards published Paper 4, completing their treatment of nonlinear response aspects in the tube model,[10–13] building on the appealing idea of reptation from de Gennes.[14] The tube model had a very different molecular interpretation of shear stress overshoot. Perhaps the tube model made the Maxwell and Nguyen's idea of yielding unnecessary and obsolete. Ever since 1979, theory, experiment, and interpretation of polymer rheology have developed on the presumption that the Doi–Edwards tube model encompasses the right physical picture and tells us how to understand the nonlinear rheological behavior of entangled polymers. The tube model paradigm provides a huge backdrop, against which this book discusses the same subject, polymer rheology, especially nonlinear rheology of entangled polymers.

It seems that modern scientific inquiries rarely follow the preaching of Karl Popper (1902–1994) concerning the objective of doing science, that is, (i) to develop falsifiable theories and (ii) to falsify existing theories with experiment. On the contrary, we prefer to work within an existing paradigm as described by Kuhn (1922–1996).[15] There is a tendency for one to do anything and everything to validate and defend a given theory instead of performing experiments aimed at falsifying it. The notion of scientific truth is often not established by objective criteria and logical rationalization in the Popperian sense but instead by the consensus of a scientific community. Consequently, for two reasons it may be difficult to carry out unconventional research: (i) We are predisposed to accept textbooks and literature results. (ii) Unconventional ideas can be inharmonious. Transformative knowledge find it hard to gain acceptance by the community, especially by people who are accustomed to the standard knowledge and approach. Fortunately, science is ultimately not an affair of democracy, dictated by popular vote. The state of our knowledge is not defined by the status quo.

Doing science often amounts to sorting out relationships between causes and effects. Depending on the level of description, causality can be confused or even reversed. Something taken as the cause at a coarser level may actually be an effect at a finer, deeper, or higher level. For polymer rheology, it is unnecessary to go to the quantum-mechanical level; but it is unacceptable to stay at the continuum mechanical level as the Maxwell model does when it is feasible to probe molecular origins. A short presentation has been posted at www.youtube.com/watch?v=2HDD51Mxu8U to discuss this matter of causality in *nonlinear polymer rheology*.

Our objective as well as the ultimate goal of Nonlinear Polymer Rheology is to raise and answer the following two questions to the best of our abilities: (i) Under macroscopic deformation, how and why do polymer chains deform in an entangled polymer solution or melt? (ii) When does chain deformation cease to increase and flow begins? In short, for entangled polymers, how does chain entanglement respond to external deformation? Intellectual independence is key to the search for answers. We are going to the places not because they were well lit; we are going to where the truth will be found.

This book (containing ca. 300 data-based figures) was motivated by and based on experiment, written to provide the reader with the comprehensive experimental information and conceptual discussions needed to develop a new, reliable, and realistic understanding of the nonlinear rheology of entangled polymers. As a fundamental step, one must accept that the interchain interactions in an entangled polymer system must be adequately treated to reflect the many-body nature. Intermolecular interactions due to chain uncrossability provide the initial potential barrier for elastic deformation during startup deformation. These interactions are of finite strength and are overcome when intrachain forces grow to a comparable magnitude, leading to force imbalance and yielding of the entanglement network, that is, chain disentanglement. As orientation, we include an introductory chapter to briefly summarize the characteristics of polymer rheology, including its definition and objectives, its history, tradition, culture, philosophy, and emergent new trends.

When I joined the faculty of Macromolecular Science and Engineering at Case Western Reserve University in the fall of 1989, I decided to work on polymer rheology although I was not trained in rheology during my doctoral study and did not know whether or not there would be any remaining important problems. From the early 1990s to 2006, before I developed my own systematic understanding of polymer rheology, I taught graduate courses at Case on polymer rheology based on the books of Ferry,[1] Bird,[3] and Macosko.[16] While teaching the subject according to these books, I actually had a lot of trouble in presenting the arguments and the discussions. Viscoelasticity seems rather abstract while shear thinning of polymeric liquids appears obvious. In my formative years, that is, the first 5 years at Case, I was lucky to work on an industrial project concerning extrusion of polyethylenes sponsored by BP Chemicals and thus had a chance to work on the incredibly important problem of polymer wall slip, into which P. G. de Gennes and F. Brochard had just developed some molecular insights. This research activity truly launched my career to work on polymer rheology. After 11 years at Case, I was recruited in 2000 to the graduate program on Polymer Science and Engineering at the University of Akron. Most of the contents in this book were developed at Akron over the past dozen years.

I have been very lucky to attract many talented students to come to Akron and join my lab. My graduate students (Prashant Tapadia, Pouyan E. Boukany, Sham Ravindranath, Yangyang Wang, Xin Li, Xiangyang Zhu, Shiwang Cheng, Gengxin Liu, Hao Sun, Panpan Lin, Xiaoxiao Li, Mengchen Wang, Jianning Liu, Xianggang Li and Zhichen Zhao) have kept me company. Consequently, I had never been alone on my journey to acquire new knowledge and a new fundamental understanding of polymer rheology. Without their coming to my research group, my scientific world would have been a rather empty place. Without their dedication to carry out all the essential experiments that led to the new worldview on polymer rheology, this book could not have been written. My understanding of nonlinear polymer rheology has changed because of their work. My knowledge about the subject has also increased because of Ed Laughlin who helped us construct many experimental apparatuses including the sliding-plate rheometer.

References

1 Ferry JD. Viscoelastic properties of polymers. John Wiley & Sons; 1980.
2 Doi M, Edwards SF. The theory of polymer dynamics. Oxford University Press; 1988.
3 Bird RB, Armstrong RC, Hassager O. Dynamics of polymeric liquids. In: Fluid mechanics. Vol. 1. John Wiley & Sons, Inc; New York, NY; 1987.
4 Dealy JM, Larson RG. Structure and rheology of molten polymers. Hanser, Munich; 2006.
5 Graessley WW. Polymeric liquids and networks: dynamics and rheology. Garland Science; 2008.
6 Phillies GDJ. Phenomenology of polymer solution dynamics. Cambridge University Press; 2011.
7 Oldroyd J. On the formulation of rheological equations of state. Proc Royal Soc London A: Math, Phys Eng Sci 1950;200(1063):523–541.
8 Maxwell B. The application of melt elasticity measurements to polymer processing. Polym Eng Sci. 1986;26(20):1405–1409.
9 Maxwell B, Nguyen M. Measurement of the elastic properties of polymer melts. Polym Eng Sci. 1979;19(16):1140–1150.
10 Doi M, Edwards S. Dynamics of concentrated polymer systems. Part 1 – Brownian motion in the equilibrium state. J Chem Soc, Faraday Trans 2: Mol Chem Phys. 1978;74:1789–1801.
11 Doi M, Edwards S. Dynamics of concentrated polymer systems. Part 2. – Molecular motion under flow. J Chem Soc, Faraday Trans 2: Mol Chem Phys. 1978;74:1802–1817.
12 Doi M, Edwards S. Dynamics of concentrated polymer systems. Part 3. – the constitutive equation. J Chem Soc, Faraday Trans 2: Mol Chem Phys. 1978;74:1818–1832.
13 Doi M, Edwards S. Dynamics of concentrated polymer systems. Part 4. – rheological properties. J Chem Soc, Faraday Trans 2: Mol Cheml Phys. 1979;75:38–54.
14 de Gennes P-G. Reptation of a polymer chain in the presence of fixed obstacles. J Chem Phys 1971;55:572–579.
15 Kuhn TS. The structure of scientific revolutions. 50th anniversary edition, University of Chicago Press; 2012.
16 Macosko CW. Rheology principles, measurements, and applications. Wiley-VCH, Inc, New York; 1994.

Acknowledgments

I am grateful to Eric Baer who included me in the Case-BP collaboration when I was a beginner, at the start of my career as a faculty, knowing little of polymer rheology. My thanks also go to Hatsuo Ishida for making his Monsanto capillary rheometer available to me. This piece of equipment was instrumental both for the Case-BP project and for my venture into nonlinear polymer rheology. A third colleague at Case, Alex Jamieson, also lent his helping hand, making his fluid rheometer available for my initial work. I am grateful to the Polymers program at the National Science Foundation for its continuous support that made our systematic investigations possible. I am so thankful for the inspiring company of Professor Ken Schweizer. With him I have had a countless number of e-mail exchanges to discuss the essence of nonlinear polymer rheology. I am grateful to his tireless moral support in my endeavors. His comments on several key chapters are also extremely valuable and helpful.

I would like to give special thanks to two people who directly helped me with the writing of the book. I am thankful to my current graduate student Jianning Liu who entered all the references for the book using Endnote, replotted several figures based on literature data, and helped me to make the Subject Index. I am infinitely indebted to Professor George Phillies who so kindly offered to copy-edit the book and raise questions about the presentation of various concepts. He made corrections and comments on every single page of the book. Because of George's commitment to good English writing, the book reads much more smoothly.... I cannot thank him enough for his friendly help. At the end, I wonder whether I have coauthored this book with George.

Finally, I would like to dedicate this book to my wife Ruth for her love and support. I gave up doing most house chores including cooking in the past 3 years (I used to do a little and always enjoy cooking) and lived like a parasite while writing this book. Over the past decade, I was able to work long hours toward the goal to figure out what is going on in nonlinear polymer rheology because Ruth selflessly supported this endeavor.

Introduction

In materials science and engineering, rheology is an important subject with a long and rich history. The first professional society on rheology and the first journal devoted to rheology both appeared in the United States of America in 1929. A comprehensive discourse of the history of rheology can be found in Tanner and Walters.[1] The specific subject of this book, nonlinear polymer rheology, also has an extensive history, and a brief description of its early history can be found in Chapter 1 of Rohn.[2] Polymer rheology can be regarded as one of the three major topics in polymer physics; the other two subjects concern with physics of polymers in crystalline and glassy states, respectively. In general, rheology is founded on the basis of macroscopic measurements. Challenges arise when we desire to figure out what microscopic physics is responsible for rheological behavior on macroscopic scales. Somehow polymer chains, with dimensions on the order of 10 nm, act either alone (dilute solutions) or collectively (concentrated solutions and melts) to produce many remarkable macroscopic phenomena. In this chapter, we will provide a brief summary of polymer rheology by highlighting its main ingredients and characteristics. We emphasize at the outset that identification of the molecular origin of mechanical stress during and after deformation is a leading task for polymer rheology. In the following discussions, some technical terms are inevitably used without a complete explanation, and some ideas are brought up without a complete background. The reader can use the Subject Index to obtain more rigorous elaborations in subsequent chapters.

1. Rheology: The Definition

Rheology is a study of how materials undergo deformation. This is a most common description of the task for rheology. By definition, solids have a finite limit of elastic deformability. Within the limit, that is, below a threshold of external deformation, a solid can return to its original state when the load (i.e., force) is removed. In other words, elastic deformation is essentially recoverable. Continuous external strain will ultimately cause a ductile solid to deform plastically without fracturing. When plastic deformation occurs, the sample cannot return its original shape or dimensions. Such irrecoverable or irreversible deformation is also known as flow.

In this book, we avoid using the two phrases deformation and flow interchangeably. While the meaning of flow is unambiguous, by "deformation" we imply strain that is not necessarily flow. In other words, "deformation" is clearly not synonymous with "flow". In fact, E.C. Bingham (1878–1945) also distinguished deformation from flow by stating "Rheology is the study of <u>deformation</u> and <u>flow</u> of matter".[3] We are discussing the careful usage of these two most frequently used words at this very beginning because there has been considerable confusion about the difference between the two. Often, flow is a heavily misused word in the rheology literature. Flow actually means something really explicit: irreversible deformation. Startup deformation of viscoelastic materials never starts with flow. Thus, we should not use the phrase "flow" to categorically refer to, for example, startup deformation as startup flow. This essential point will become clearer, hopefully, by the end of this introduction. Our more detailed evidence and reasoning will be presented later throughout the rest of the book.

Polymeric liquids are fluids of chain-like macromolecules that are not cross-linked to each other so that they may flow at long times by having their chains slide past one another. Experimental characterization indicates that polymeric liquids may respond like a solid when subject to sudden external deformation. At short times, they may be quite incapable of flowing. This simple observation turns out to provide some essential insights leading to a proper understanding of rheological behavior in viscoelastic liquids, particularly the entangled polymer solutions and melts that are the main objects of study in this book. It may be noteworthy that entangled polymeric liquids are uniquely capable of undergoing large elastic deformation in contrast to many other viscoelastic materials such as polymer glasses and colloidal dispersions.

In the past, the emphasis of experiment and theory was often placed on delineating the constitutive behavior in steady state. Shear thinning is one such characteristic, relating the steady shear stress to the shear rate. Moreover, rheometric measurements are frequently simplified to assume that the deformation field can be experimentally prescribed *a priori*. In this tradition, students of rheology rarely realize that the task of rheology is to separately and independently determine both the deformation field and the corresponding stress state.

2. Molecular Approach of Tube Model and Continuum-Mechanical Constitutive Modeling Versus a Phenomenology-Based Treatment

Polymeric materials are a unique class of matter in the sense that many of their physical properties depend explicitly on the chain length, which in the case of linear polymer chains is proportional to the total molecular weight M. Dynamic and rheological properties are particularly sensitive to M. Such experimental facts motivated scientists to build molecular models. Molecular modeling began as early as the 1940s, climaxing in the celebrated bead-spring theories of Rouse[4] and Zimm[5]. The long-awaited groundbreaking explanation of the scaling law $\eta_0 \sim M^3$ for melt viscosity came much later in 1971,[6] where we note that experimental data usually indicate a notably higher exponent than three. Doi and Edwards[7] took some significant steps to advance de Gennes' model of reptation. They asserted with sound reasoning that in an entangled polymer melt or solution, the emergent stress during either external shear or extension can be evaluated in terms of the intrachain retraction forces of a coarse-grained (test) chain confined to a fictitious tube. This Doi–Edwards (DE) model satisfactorily describes linear viscoelasticity of entangled polymer melts and solutions.[8–10] It also claims to describe and explain nonlinear rheological properties.[11] The subsequent three decades witnessed extensive efforts to improve the original de Gennes–Doi–Edwards model.[12–20] Experiments, too many to cite them here, seem to support the theoretical description.[1] However, it is actually not

1 The validation of the tube model usually adopts the following logic: Since the constitutive continuum description, extracted from the tube model, can describe such instabilities as shear banding, non-quiescent relaxation and necking instability in uniaxial extension it must have captured the correct molecular physics. Such reasoning overlooks the model degeneracy in science: Different models can produce similar data. A recent continuum-mechanics-level theoretical study on extensional necking most strikingly revealed such a degeneracy when Hoyle and Fielding showed in J. Rheol. 2016;60(6):1347-1375 that uniaxial extension is unstable against necking in several constitutive models, including a variant of the tube model and models that have no microscopic basis for chain entanglement. Thus, to declare that it could describe the macroscopic world of nonlinear polymer rheology, the tube model needs not incorporate the molecular physics that has to answer the question of whether an entangled polymer liquid should be treated as a junction-forming entanglement network and whether fast external deformation could weaken or destroy the network. This book suggests that we must move behind such a modest goal, collect more phenomenology, and search for a more useful and predictive theoretical framework by identifying the molecular origins for the key emergent phenomena including various macroscopic instabilities.

obvious that interchain interactions can be treated in such a smoothed-out manner. Moreover, admittedly, the tube model is inherently self-inconsistent.[21]

Treatments based on the tube description include the slip-link models.[22,23] These models aim to relate macroscopic rheological phenomena to molecular dynamics on nanometer scales, contrasting sharply with the constitutive continuum approach that preceded the tube model. James Clerk Maxwell (1831–1879) and Ludwig Boltzmann (1844–1906) explored phenomenological linear viscoelasticity at the continuum level. James Gardner Oldroyd[24] (1921–1982), among others, made well-known contributions to the field of constitutive modeling for nonlinear continuum behavior of viscoelastic fluids. Many textbooks present various constitutive descriptions without sufficient molecular ingredients.[25–31] According to these books, constitutive models contain physical parameters that may be determined from linear viscoelastic measurements. These parameters, such as the shear viscosity, are usually assumed to be constant, that is, fixed everywhere in space and time. The validity of assumptions of spatial and temporal uniformity has not been tested adequately in the past.

Separate from the continuum description and tube model for polymer rheology, there was a period of several decades in which the transient network models (TNMs) were the standard description, developed by Green and Tobolsky,[32] Lodge,[33] and Yamamoto.[34] To make the transient network models quantitative, for example, to describe shear thinning, the network junctions were assumed in an *ad hoc* manner to have a stress- or rate-dependent lifetime. Although both the tube model and the transient network model are phenomenological, there was an impression that the tube model is anything but *ad hoc* and thus far superior to the transient network model. In fact the tube model is also *ad hoc.*

There are two ways to explore what happens in experiment. In the first way, experiments are carried out because of an existing theory. For example, in the past decades, countless experimental studies have been made to vindicate the tube theory. Here the theory dictates how the phenomenology should be collected. The danger is that such an approach could be highly biased and lack basic objectivity. For example, up to 2004, nearly all rheological experiments had been analyzed on the basis of shear homogeneity because none of the extant theoretical approaches insisted *a priori* that the deformation field should be determined independently. In the second way, phenomenology is established without theoretical prejudice or favoritism. In this approach, the rheological behavior of entangled polymers is investigated in experiment without any presumptions about how the polymeric liquids should respond to various modes of external deformation. Instead of assuming homogeneous shear in rheometric measurements, experiments are performed to determine the state of deformation and stress field simultaneously and independently in real time. Such objective phenomenology must be collected first. Any reliable theoretical description of nonlinear rheology of entangled polymers only comes after the phenomenology. In this work, we follow the second approach. Neither the tube model nor the latest formulation of nonlinear rheology (presented in this book) has received explicit molecular-level confirmation. Thus, it would be unreasonable to only say that there is not yet any molecular-level substantiation of the new picture discussed in this book. The future molecular dynamics simulation and small angle neutron scattering measurements will tell which one is more realistic and useful.

3. Linear Versus Nonlinear Responses: Characterization Tool Versus Science of Rheology

The linear responses of viscoelastic liquids, such as entangled polymer solutions and melts, reflect and manifest the microstructure, which remains nearly intact during small external deformations. Through adequate molecular modeling, linear rheology thus becomes a characterization tool. Linear responses form one aspect of the subject. However, the core objective of polymer rheology has to do with something else, namely, studying nonlinear responses of entangled polymers to large fast deformation of different types. This book is devoted to exploring the physics behind the nonlinear rheological behavior of entangled polymeric liquids.

4. Shear Thinning, Stress Plateau, and Yielding

Nonassociating polymers, for example, most of the commodity polymers, including polyethylenes, polypropylenes, polystyrene, polybutadiene, and polyisoprene, with a global production rate of several hundred billion pounds per year, exhibit shear thinning as shown in Fig. 1(a). If steady shear can be reached, then the shear viscosity is found to decrease with increasing shear rate $\dot{\gamma}$. For monodisperse linear polymers, the shear thinning is steep: the shear viscosity η scales nearly reciprocally with the rate, that is, $\dot{\gamma}^{n-1}$ with $n \ll 1$. When we represent this steady-state behavior in terms of the shear rate as a function of the shear stress, we see a transition at a threshold stress level σ_c as shown in Fig. 1(b), where the nominal shear rate sharply rises. Beyond the terminal flow regime (where σ is linearly proportional to $\dot{\gamma}$), the stress level saturates with respect to increasing shear rate, a phenomenon that is known as stress plateau behavior. This stress plateau character is reminiscent of the yield-stress in materials such as ketchup, toothpaste, and yogurt that exhibit much higher fluidity only when a critical level of shear stress is exerted. The plateau width can span many orders of magnitude in the shear rate for well-entangled polymers and other yield-stress materials. Equivalently, the shear viscosity of these materials can drop by many decades within a narrow range of stress. Thus, the state of matter at low shear rates, in the terminal flow regime, must be rather different from the state of matter after yielding around σ_c. In other words, when the quasi-solid (i.e., the liquid with an enormous viscosity due to chain entanglement) turns into an inviscous state, the structure has undergone significant alternation. The change in viscosity must reflect a massive loss of entanglement.

5. Is There Always Homogeneous Deformation?

Let us consider simple shear generated by displacing two parallel plates in opposite directions. This is a standard protocol adopted in most studies of the shear rheology of entangled polymers. We have taken it for granted that homogeneous shear can be produced with such a boundary displacement, as depicted by Fig. 1.1(a), where the plates are taken to be separated by a distance H. The shear deformation is taken to be uniformly the same at all positions along the

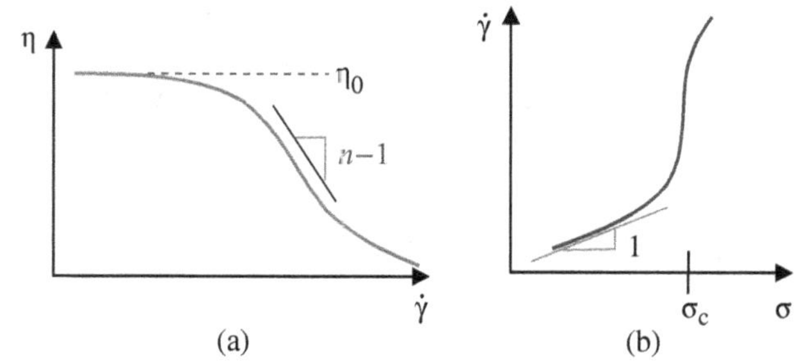

Figure 1 (a) Steady shear viscosity as a function of shear rate. (b) Steady shear rate as a function of shear stress σ, showing a stress plateau around σ_c.

gap. This assumption of homogeneous shear is not only out of convenience but also out of necessity. We could not perform any conventional rheometric measurements if a uniform shear field could not be prescribed by the imposition of a velocity V on one plate while holding the other plate stationary. Misleadingly, one could make a seemingly reasonable argument to assert that homogeneous shear should prevail since the steady shear stress should be constant along the gap direction, as discussed in Section 4.1.1. Traditionally, we tend to first think about the steady shear behavior, paying less attention to the fact that strongly viscoelastic liquids including entangled polymers are anything but a liquid when responding to sudden startup deformations. At times much shorter than the reptation time, they are transient solids and may experience solid-like breakup, for example, they undergo strain localization in spite of molecular diffusion. Thus, during any rheometric investigation of rheological behavior of entangled polymers, including startup shear and large stepwise shear, we should make efforts to verify the basic assumption that the shear rate is independent of position. Sound rheological measurements should involve experimental techniques that make *in situ* determination of the shear field.

The era of performing rheological studies based on an untested premise of homogeneous shear has come and gone. Since 2006, numerous studies from Akron[35–37] and other parts[38–40] of the world have recommended that the experimenters must determine the deformation field explicitly. In particular, particle-tracking velocimetry (PTV) has been applied in conjunction with commercial rheometric instruments to quantify shear banding during startup shear[41] and nonquiescent relaxation after stepwise shear both for well-entangled polymer solutions[37,42] and for melts.[43] It is found that localized yielding can lead to shear strain localization. Similarly, strain localization is a common feature in startup uniaxial extension of entangled melts. It is important to realize that the concepts of yielding and chain disentanglement remain valid even when there is no shear strain localization. The PTV observations of strain localization nonetheless most evidently demonstrate the nature of nonlinear rheological responses, leading to the new perspective that the entanglement network can yield and break up.

To reiterate, yielding can result in severe strain localization in well-entangled polymer solutions and melts. Thus, there is a breakdown in conventional rheometric analyses that are based on the textbook assumption of homogeneous shear. The demonstration of shear and tensile strain localization spells the end of conventional rheometry. In traditional rheology, the deformation field is chosen by the experimentalist and is assumed to be known *a priori*. In reality, in simple shear rheometry we can only control the speeds (0 and V) at the two parallel surfaces. We cannot dictate whether or not the velocity field is given by $v(y) = V(y/H)$, that is, we cannot assume that the strain field is homogeneous along the velocity gradient direction.

6. Rheology Versus Fluid Mechanics

The mission of rheology is entirely different from that of fluid mechanics. The aim of polymer rheology is to explore and plausibly derive constitutive relations from the principles of microscopic molecular physics. In contrast, fluid mechanics of polymeric liquids describes the state of deformation and field of stress by applying a given set of macroscopic constitutive relationships.

A detailed description of polymer processing requires fluid mechanical calculations based on constitutive equations. Such practice is reliable and realistic only when the constitutive behavior of polymeric liquids has been adequately captured in terms of molecular mechanisms. The subject of fluid mechanics generally requires us to formulate the stress and strain fields using tensorial representation. But the subject of rheology does not require tensor representations of stress and strain because we always specify the mode of deformation (e.g., shear or extension).

For two reasons this book does not include tensor analyses. First, many who would like to learn about the subject of polymer rheology often turn away upon opening the first two chapters of a typical book on rheology because they mistakenly think that the subject is highly mathematical and beyond their level of preparation. Second, no tensors are needed anywhere because we discuss simple shear and uniaxial extension separately. Once the mode of deformation is specified, all the mathematical treatments actually involve only scalar quantities. The present book thus departs from the tradition that invokes tensor analyses and matrix representations to introduce the subject of polymer rheology.

7. Emerging Trends

It is at least challenging if not risky to infer microscopic processes from macroscopic nonlinear responses of polymeric liquids to fast large deformation. We have arrived at a coherent molecular picture after a careful examination of the accumulated phenomenology. In the end, we are forced to suggest existence of an entropic barrier in Chapter 16 that provides the cohesion of the entanglement network against finite external deformation. Because of this cohesion barrier, after a sudden stepwise deformation of moderate magnitude chain deformation remains pinned down until reptative relaxation restores the chain conformation. This conclusion may be regarded as deductive and therefore speculative. To bridge the gap of six orders of magnitude in length scale we need to resort to microscopic observations such as neutron scattering measurements and computer simulations. As this book approached the finish line, we learned that new small-angle neutron scattering (SANS) measurements had recently been performed along with innovative model-independent analysis[44] that directly compares the SANS spectrum with the tube model prediction. During stress relaxation from stepwise extension of moderate magnitude, chains remain stretched on length scales larger than the averaged equilibrium tube segment according to the SANS spectrum even after 20 times Rouse time. Such measurements prompt us to suggest that chain deformation and stress arising from molecular deformation should be described at different length scales instead of a single scale of the equilibrium tube segment. The tube model assumed or made approximation that chain deformation and accompanying stress only needs to be characterized using a single length scale given by the equilibrium tube segment (or equilibrium entanglement strand). Consequently, the tube model may be highly unrealistic for nonlinear polymer rheology, as much of the experimental evidence presented in this book seems to indicate.

8. Summary

This book presents the subject in the order of phenomenology first and theory second. In contrast to other treatments in the literature that collect phenomenology based on the assumption that the deformation field could be prescribed *a priori*, this book shows new phenomenology obtained from the independent and separate characterization of the deformation field and the state of stress. To accomplish such a task, many inherent experimental limitations have been overcome. The emergent phenomenology motivates a physical picture of the nonlinear rheology of entangled polymers that is rather different from pictures associated with the tube theory. This phenomenology allows us to refocus on the two central questions in nonlinear rheology of entangled polymers: (a) Why do polymer chains deform elastically at the beginning in response to a macroscopic strain? (b) When do polymers cease to undergo further elastic deformation so that flow can commence. In the search for the answers, some new questions arise that are equally important to address: (c) What happens to chain entanglement when the elastic deformation is terminated at a considerable strain, for example, after a few strain units? (d) What is the role of melt elasticity in polymer rheology? This book describes the approaches taken to answer these foundational questions.

The book is divided into four parts: (a) Linear viscoelasticity and experimental methods; (b) Yielding – primary nonlinear response to ongoing deformation; (c) Decohesion and elastic yielding after large deformation; and (d) Emerging conceptual framework and beyond. The subject is discussed with minimal mathematical equations and without tensor analysis. The bulk of the book is a description of phenomenology, collected from state-of-the-art macroscopic rheometric measurements and particle-tracking velocimetric observations. The reader can find this phenomenological component in Parts 2 and 3 of the book. Part 4 explores the molecular foundation for nonlinear rheology of entangled polymers, based on the available phenomenological information described in Parts 2 and 3. Although the materials in Chapters 1–3 are not new, perhaps even too lengthy and simple for many readers, Chapters 1 and 2 provide the definitions and information needed for the developments in the subsequent Parts 2 through 4. Readers familiar with standard linear rheology may simply start reading from Chapter 4 or 6 and use Part 1 as a reference as needed. Conversely, for readers less fluent with the theoretical background, Chapters 1 and 2 are just a compilation of classical results from the established theories of polymer dynamics. Instead of going through the "derivations" in these chapters, the readers only need to go to these chapters for definitions of relevant concepts and quantities that appear in the reminder of the book. In short, the key chapters of the book are 6, 7, 9, 11, 12, 16, and 18, respectively, on *wall slip, yielding behavior, particle-tracking velocimetric observations of shear banding, various forms of strain localization in startup melt extension, elastic yielding phenomenon, the theoretical foundation, and troubles with existing frameworks* (a summary of different worldviews). To have a complete overview of the subject, it is better not to skip any chapter. For a short review of the new worldview described in this book, the reader may go to listen to a research seminar posted at https://www.youtube.com/watch?v=ffbhZYlkWcA.

In closing, a very specific comment needs to be made about how the content of this book is presented and discussed. To establish a correct and reliable conceptual foundation for the nonlinear rheology of entangled polymers, we must first describe the key phenomenology. This phenomenology will motivate the discussions of the theoretical concepts, leading to our molecular-network paradigm. To verify such a theoretical framework, we need to discuss further supporting phenomenology, including phenomenology that was uncovered as predictions of the new paradigm. In the chapters prior to Chapter 16 some phenomenology is presented and discussed using concepts that are only rigorously described in Chapters 15 and 16. These concepts include yielding, chain disentanglement, entropic barrier, and finite cohesion. These central phrases form the new language in which we discuss nonlinear polymer rheology. A reader may simply follow the discussion of the phenomenology, using the literal meanings of these terms. A reader may alternatively go to Chapter 16 to find more precise definitions and associated discussions for these terms as he or she reads through Parts 2 and 3.

Although this book focused on phenomenology derived from entangled polymers, many characteristics are expected to appear in other chain-like systems such as associative polymer solutions and melts as well as self-assembled aggregating systems including wormlike micelles. These systems must also yield upon large conformation. They could also undergo stress overshoot and various formations of strain location in either shear or extension or more complex configurations.

References

1 Tanner RI, Walters K. Rheology: an historical perspective. Elsevier, Amsterdam; 1998.
2 Rohn CL. Analytical polymer rheology: structure-processing-property relationships: Hanser Gardner Publications, Cincinnati, OH; 1995.
3 Tanner RI. Engineering rheology: Oxford University Press; 2000.
4 Rouse Jr, PE. A theory of the linear viscoelastic properties of dilute solutions of coiling polymers. J Chem Phys. 1953;21(7):1272–80.
5 Zimm BH. Dynamics of polymer molecules in dilute solution: viscoelasticity, flow birefringence and dielectric loss. J Chem Phys. 1956;24(2):269–78.
6 de Gennes P-G. Reptation of a polymer chain in the presence of fixed obstacles. J Chem Phys. 1971;55:572–9.
7 Doi M, Edwards SF. The theory of polymer dynamics. Oxford University Press, New York; 1988.
8 Doi M, Edwards S. Dynamics of concentrated polymer systems. Part 1.—Brownian motion in the equilibrium state. J Chem Soc Faraday Trans 2 Mol Chem Phys. 1978;74:1789–801.
9 Doi M, Edwards S. Dynamics of concentrated polymer systems. Part 2.—Molecular motion under flow. J Chem Soc Faraday Trans 2 Mol Chem Phys. 1978;74:1802–17.
10 Doi M, Edwards S. Dynamics of concentrated polymer systems. Part 3.—The constitutive equation. J Chem Soc Faraday Trans 2 Mol Chem Phys. 1978;74:1818–32.
11 Doi M, Edwards S. Dynamics of concentrated polymer systems. Part 4.—Rheological properties. J Chem Soc Faraday Trans 2 Mol Chem Phys. 1979;75:38–54.
12 Marrucci G, Grizzuti N. Fast flows of concentrated polymers-predictions of the tube model on chain stretching. Gazz Chim Ital. 1988;118(3):179–85.
13 Marrucci G. Dynamics of entanglements: a nonlinear model consistent with the Cox-Merz rule. J Non-Newtonian Fluid Mech. 1996;62(2–3):279–89.
14 Cates M, McLeish T, Marrucci G. The rheology of entangled polymers at very high shear rates. Europhys Lett. 1993;21(4):451.
15 Mead D, Larson R, Doi M. A molecular theory for fast flows of entangled polymers. Macromolecules. 1998;31(22):7895–914.
16 Watanabe H. Viscoelasticity and dynamics of entangled polymers. Prog Polym Sci. 1999;24(9):1253–403.
17 Milner ST, McLeish TCB, Likhtman AE. Microscopic theory of convective constraint release. J Rheol. 2001;45(2):539–63.
18 McLeish TCB. Tube theory of entangled polymer dynamics. Adv Phys. 2002;51(6):1379–527.
19 Graham RS, Likhtman AE, McLeish TC, Milner ST. Microscopic theory of linear, entangled polymer chains under rapid deformation including chain stretch and convective constraint release. J Rheol. 2003;47(5):1171–200.
20 Dealy JM, Larson RG. Structure and rheology of molten polymers. Hanser, Munich; 2006.
21 Lodge AS. Elastic recovery and polymer-polymer interactions. Rheol Acta. 1989;28(5):351–62.
22 Doi M, Takimoto J-I. Molecular modelling of entanglement. Philos Trans R Soc London, Ser A. 2003;361(1805):641–52.
23 Schieber JD, Neergaard J, Gupta S. A full-chain, temporary network model with sliplinks, chain-length fluctuations, chain connectivity and chain stretching. J Rheol. 2003;47(1):213–33.
24 Oldroyd J. On the formulation of rheological equations of state. Proc R Soc Lond A Math Phys Eng Sci. 1950;200(1063):523–41.

25 Noll W. On the continuity of the solid and the fluid states. J Ration Mech Anal. 1955;4:3–81.
26 Lodge AS. Elastic liquids: an introductory vector treatment of finite-strain polymer rheology. Academic Press, London; 1964.
27 Astarita G, Marrucci G. Principles of non-Newtonian fluid mechanics. McGraw-Hill Companies, New York; 1974.
28 Schowalter WR. Mechanics of non-Newtonian fluid. Pergamon, Oxford-Frankfurt; 1978.
29 Bird RB, Armstrong RC, Hassager O. Dynamics of polymeric liquids. Vol. 1. Wiley, New York; 1987.
30 Crochet MJ, Davies AR, Walters K. Numerical simulation of non-Newtonian flow. Elsevier, Amsterdam; 2012.
31 Larson RG. Constitutive equations for polymer melts and solutions. Butterworths, Boston; 1988.
32 Green M, Tobolsky A. A new approach to the theory of relaxing polymeric media. J Chem Phys. 1946;14(2):80–92.
33 Lodge A. A network theory of flow birefringence and stress in concentrated polymer solutions. Trans Faraday Soc. 1956;52:120–30.
34 Yamamoto M. The visco-elastic properties of network structure I. General formalism. J Phys Soc Jpn. 1956;11(4):413–21.
35 Tapadia P, Ravindranath S, Wang S-Q. Banding in entangled polymer fluids under oscillatory shearing. Phys Rev Lett. 2006;96(19):196001.
36 Tapadia P, Wang S-Q. Direct visualization of continuous simple shear in non-Newtonian polymeric fluids. Phys Rev Lett. 2006;96(1):016001.
37 Wang S-Q, Ravindranath S, Boukany P, Olechnowicz M, Quirk RP, Halasa A, et al. Nonquiescent relaxation in entangled polymer liquids after step shear. Phys Rev Lett. 2006;97(18):187801. A refined explanation for the observed elastic yielding was subsequently proposed. J Chem Phys. 2007;127, 064903.
38 Jaradat S, Harvey M, Waigh TA. Shear-banding in polyacrylamide solutions revealed via optical coherence tomography velocimetry. Soft Matter. 2012;8(46):11677–86.
39 Fang Y, Wang G, Tian N, Wang X, Zhu X, Lin P, et al. Shear inhomogeneity in poly(ethylene oxide) melts. J Rheol. 2011;55(5):939–49.
40 Hu YT, Wilen L, Philips A, Lips A. Is the constitutive relation for entangled polymers monotonic? J Rheol. 2007;51(2):275–95.
41 Boukany PE, Wang S-Q. Use of particle-tracking velocimetry and flow birefringence to study nonlinear flow behavior of entangled wormlike micellar solution: From wall slip, bulk disentanglement to chain scission. Macromolecules. 2008;41(4):1455–64.
42 Ravindranath S, Wang S-Q. What are the origins of stress relaxation behaviors in step shear of entangled polymer solutions? Macromolecules. 2007;40(22):8031–9.
43 Boukany PE, Wang S-Q, Wang X. Step shear of entangled linear polymer melts: new experimental evidence for elastic yielding. Macromolecules. 2009;42(16):6261–9.
44 Wang Z, Lam C, Chen W, Wang W, Liu J, Liu Y, et al. Fingerprinting molecular relaxation in deformed polymers. Phys Rev X. 2017;7:031003.

About the Companion Website

Don't forget to visit the companion website for this book:

www.wiley.com/go/nonlinearpolymerrheology

There you will find valuable material designed to enhance your learning, including:

- Videos, cited in Chapters 6, 9, 10, 12, 13, 16 and 19

Scan this QR code to visit the companion website.

Part I

Linear Viscoelasticity and Experimental Methods

Part of the physics on Earth is remarkably simple: There exists a subworld of linear response. The best-known example is Newton's second law: $\underline{F} = M\underline{a}$. In the field of mechanics, there are other famous examples of linear response, that is, Hooke's law for the deformation of elastic solids and Newton's law for the flow of viscous liquids. Here, the output, such as the degree of deformation or rate of deformation, is linearly proportional to the input, such as the imposed force, and vice versa.

It is conceptually interesting to note that viscoelastic materials including entangled polymeric liquids undergo yielding even in the limit of linear response. Since yielding implies removal of a constraining barrier, the structural state after yielding is different from the equilibrium state. In the linear response regime, the structural change is apparently insignificant. Indeed, all current theories that depict linear response assume that the structures of the liquids responsible for the linear viscoelasticity remain intact during a weak external deformation.

In this part of the book, we describe linear viscoelasticity of polymeric liquids, experimental characterization methods, and rheometric setups. Specifically, Chapter 1 provides a phenomenological discussion of linear viscoelasticity in terms of basic concepts and elementary models such as the Maxwell model, as well as a simple account of the rubber elasticity theory. Chapter 2 focuses on a theoretical description of various molecular models for polymer dynamics, for example, Rouse, Zimm, and reptation models as well as the packing and percolation models for polymer entanglements. Chapter 3 is devoted to a brief account of the various rheometric apparatuses and related methods that are applicable when deformation is homogeneous. Chapter 4 extends Chapter 3 by emphasizing that the objective of rheology is to characterize the mechanical responses *and* determine the deformation field. It introduces the particle-tracking velocimetric method that enables a comprehensive characterization of rheological behavior. Chapter 5 describes additional rheometric devices as well as apparatuses that involve more complicated geometries.

Nonlinear Polymer Rheology: Macroscopic Phenomenology and Molecular Foundation, First Edition. Shi-Qing Wang.

Companion Website: www.wiley.com/go/nonlinearpolymerrheology

1

Phenomenological Description of Linear Viscoelasticity

It has been known since James Clerk Maxwell's period that deformation and flow behavior of materials may not fall into two simple categories: ideally elastic solid and purely viscous liquid. Elastic solids were extensively studied by Robert Hooke (1635–1703). As a contemporary of Hooke, Isaac Newton (1643–1727) discovered the basic law for the motion of viscous liquids. Many materials are neither elastic solids nor viscous liquids. Their mechanical behavior exhibits time-dependent properties; that is, they appear solid-like on short time scales and liquid-like on long time scales. Such materials are clearly non-Newtonian. This viscoelastic behavior is actually a property of a wide range of structured materials, including colloids, liquid crystals, micelles, foams, gels, granular matter, emulsions and membranes, and, of course, polymers. Viscoelasticity arises from the existence of slow relaxation associated with material's structural change.

Linear viscoelasticity, by definition, depicts behavior that is phenomenologically simple: the mechanical resistance is linearly proportional to the degree of external perturbation. A linear response is possible when the external world does not cause the material to deviate so much away from its equilibrium state as to display obvious nonequilibrium properties. In the linear regime, the perturbation is so weak that the material response is uniform and homogeneous, making the phenomenology straightforward to interpret.

In the reminder of this chapter, we first discuss three protocols commonly used to probe the viscoelastic behavior of structured materials including polymers. Then we describe how the Boltzmann superposition principle and Maxwell model can be applied to describe the linear response behavior of viscoelastic materials. This chapter ends with a section on the classical network picture of rubber elasticity, showing the entropic origin of polymer viscoelasticity.

1.1 Basic Modes of Deformation

External deformation occurs in various forms such as simple shear, uniaxial extension, and compression. For each of these three types, there are at least three different ways to impose the deformation: (a) startup deformation where the boundaries of the body under deformation are suddenly displaced at some specified speed at $t = 0$; (b) stepwise deformation where a startup deformation is abruptly terminated at a specified magnitude of strain; and (c) oscillatory deformation where the moving boundary is made to undergo sinusoidal linear displacement. Only simple shear and uniaxial extension are extensively discussed in the book. Compression leading to squeezing or planar extension will first be discussed in Sections 5.3.2 and 5.3.3 and considered again in Sections 10.3 and 10.4 as well as in Section 13.5. In this chapter, we use simple shear to illustrate the three modes (a) to (c).

Nonlinear Polymer Rheology: Macroscopic Phenomenology and Molecular Foundation, First Edition. Shi-Qing Wang.

Companion Website: www.wiley.com/go/nonlinearpolymerrheology

1.1.1 Startup shear

Let us consider a sample of interest sandwiched between two parallel plates, as depicted in Fig. 1.1(a). To produce startup shear, the lower plate is held stationary, while the upper plate is made to suddenly displace in the X direction. We assume that the sample's upper boundary of area Σ will move by $X = Vt$ at time t when the upper plate is given a velocity V at time $t = 0$, as characterized in Fig. 1.1(b,c). Specifically, we consider no slip at the interfaces between the sample and two confining plates. During each time interval Δt, the upper plate moves by $\Delta x = V\Delta t$, producing an elemental shear strain of

$$\Delta\gamma = \Delta x/H \tag{1.1}$$

so that the shear rate is

$$\dot{\gamma} = \Delta\gamma/\Delta t \rightarrow V/H, \tag{1.2}$$

where $\Delta x/\Delta t|_{\Delta t\rightarrow 0} \rightarrow V$ by definition. This introduction of shear strain and shear rate rests on an important premise: The shear deformation is uniform along the thickness direction (Y axis) as depicted in Fig. 1.1(a). Assuming that the shear deformation is imposed at $t = 0$, with surface velocity V, then the accumulative external shear strain at time t is $\gamma(t) = \dot{\gamma}t$. As shown in Fig. 1.1(a), in response to shear deformation, a shear stress emerges – a force F is required to maintain shearing (Eq. (1.6)). In steady state, which is achieved when the stress responses no longer vary in time, we also need to argue about the validity to assume homogeneous shear. The illustration in Fig. 1.1(d) assumes a constant shear rate in every layer along the Y axis. We will defer a detailed discussion of the shear homogeneity assumption in Chapter 4.

From the onset, let us introduce a key yet elementary concept in rheology associated with the startup shear depicted in Fig. 1.1(a,b). Let us figure out how long it takes to produce a sizable deformation, for example, a shear strain of $\gamma_1 = 1.0$. At a displacement velocity V as shown in Fig. 1.1(d), the time $t_{1\text{exp}}$ it takes for the moving plate to travel a distance of H is given by

$$t_{1\,\text{exp}} = H/V = 1/\dot{\gamma}, \tag{1.3}$$

where the second equality follows from Eq. (1.2). This experimental time scale, $t_{1\text{exp}}$, is the time required to impose 100% shear deformation at given shear rate $\dot{\gamma}$. It is worth noting that Eq. (1.3) provides useful information only if homogeneous shear prevails. Two questions naturally arise: (a) When is shear deformation regarded slow enough to ensure linear response? (b) When does the assumption of shear homogeneity break down? We will answer (a) in Section 1.2.7 and defer (b) to Part 2.

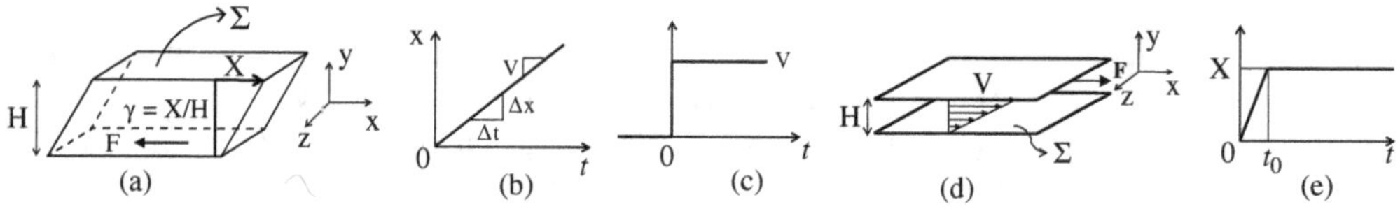

Figure 1.1 (a) Depiction of simple shear in three dimensions, showing two parallel surfaces at $y = 0$ (fixed) and H (displaced by X over time t). The force F required to hold the bottom surface stationary can be measured to define shear stress in Eq. (1.6) where Σ is the total area of the shearing surface. (b) Displacement X of the upper surface as a function of time at a constant shear rate V/H. (c) Startup shear shown by the step function of V versus time t. (d) Homogeneous continuous simple shear, produced by either the moving plate at speed V or a force F on the plate. (e) Step strain realized by the displacement of the moving surface by X over a period of t_0 for a gap distance of H, so that $\gamma = X/H$.

1.1.2 Step Strain and Shear Cessation from Steady State

Step strain is the simplest rheological experiment. As shown in Fig. 1.1(e), step strain is a variation of startup deformation, achieved by terminating the displacement in Fig. 1.1(a) after a certain amount of strain, $\gamma = X/H$. This protocol is commonly applied to characterize stress relaxation behavior after shear cessation of viscoelastic materials, including entangled polymeric liquids. Ideally, the step strain should be made to take place instantly. In the experimental reality, it always takes a finite period given by $t_0 = X/V = \gamma/(V/H)$. How short does t_0 need to be to avoid molecular relaxation during step strain? How small does γ need to be to ensure linear response? These questions are discussed in more detail in Section 1.2.7. Relaxation from steady shear is also a valuable protocol, which can be depicted in Fig. 1.1(e) by having $X/H \gg 1$ until steady state is reached. In summary, step strain tests are most often carried out to learn how stress relaxation takes place through molecular diffusion (in linear response regime) and how viscoelastic materials undergo structural changes due to large deformation, to which the entire Part 3 is devoted.

1.1.3 Dynamic or Oscillatory Shear

Oscillatory shear, also known as dynamic shear, is the most frequently applied method to probe viscoelasticity, particularly linear viscoelasticity. Instead of displacing the upper plate with a constant velocity, the top boundary moves according to a sine wave

$$x(t) = X_0 \sin \omega t \tag{1.4a}$$

with a velocity of $v(t) = V_0 \cos \omega t$, where $V_0 = X_0\omega$. In other words, the imposed strain $\gamma(t)$ is a sinusoidal function of time

$$\gamma(t) = \gamma_0 \sin \omega t, \tag{1.4b}$$

where ω is the oscillation frequency so that the period T is $2\pi/\omega$, and γ_0 is the amplitude of the oscillatory deformation: $\gamma_0 = X_0/H$. Assuming homogeneous shear, the time-dependent shear rate $\dot{\gamma}(t)$ is simply the time derivative of the right-hand side of Eq. (1.4b):

$$\dot{\gamma}(t) = \gamma_0\omega \cos \omega t \equiv \dot{\gamma}_0 \cos \omega t, \tag{1.5}$$

where the amplitude $\dot{\gamma}_0$ is given by the product of γ_0 and ω, that is, $\dot{\gamma}_0 = V_0/H$. Here, γ_0 and ω can be varied independently to explore viscoelastic responses.

1.2 Linear Responses

Both ideal elastic solids and viscous liquids are known to exhibit linear response as reviewed in the following paragraphs. Viscoelastic materials can also show linear responses, which may be characterized by the Maxwell and Voigt models. While a low magnitude of strain in stepwise deformation and oscillatory shear ensures linear response, sufficiently low rate of deformation or oscillation frequency is another way to make sure that the elastic structures remain intact during continuous or sinusoidal deformation.

Figure 1.1(a–d) explicitly depicts startup deformation in the displacement-controlled mode of simple shear. A liquid or solid resists the shear so that a finite horizontal force arises along the X axis. In the present case of simple shear, shear stress σ can be defined as the total resistance force on the bottom surface divided by the surface area Σ

$$\sigma_{xy} = F/\Sigma, \tag{1.6}$$

where σ_{xy} includes the conventional tensorial subscript "xy" to indicate that the force along the direction of shear, that is, along the X axis, is exerted on a plane whose normal is along the Y axis. Throughout this book, we will drop the subscript in σ_{xy}. The unit of stress is force per unit area, in SI units, N/m^2, which is Pascal (Pa), named after French mathematician Blaise Pascal (1623–1662).

1.2.1 Elastic Hookean Solids

Ideal "perfectly elastic" solids can undergo a finite amount of deformation instantaneously when a stress σ is suddenly applied. At small enough deformations, a linear relationship often exists between the elastic deformation and the stress σ, as studied by Hooke. Let us consider the case of simple shear as illustrated in Fig. 1.1(a). Hookean solids obey a linear relationship

$$\sigma(t) = G\gamma(t), \tag{1.7}$$

where G is the elastic modulus. In Fig. 1.2, G is the slope of the inclined straight line. Since the shear strain γ is dimensionless, G has the same dimension as that of stress. Thus, G may be expressed as energy per unit volume as well as force per unit area.

1.2.2 Viscous Newtonian Liquids

Liquids differ from solids in that deformation can proceed indefinitely. A viscous liquid undergoes flow instantly, whereas the flow of viscoelastic liquids (polymeric liquids being the primary example and focus of this book) is preceded by *elastic deformation* during startup deformation. Thus, it should be clear at the onset about the difference between *flow* and *deformation.* Throughout this book, deformation means something general, while flow has a very specific meaning, namely, *irrecoverable* deformation. In contrast to flow, elastic deformation is recoverable: an elastically deformed sample may return to its nondeformed state upon removing the external stress from the sample.

We begin by considering the simplest case of simple shear in viscous liquids. In terms of Fig. 1.1(d), Newton discovered for certain liquids such as water and milk that the force F required to make the upper plate move with velocity V is proportional to V. Since the total F is also proportional to the total area Σ of the top surface in Fig. 1.1(a), it is convenient to use the definition for shear stress given by Eq. (1.6). Newton's law for viscous liquids relates the shear stress σ to the shear rate $\dot{\gamma} = V/H$ depicted in Fig. 1.1(d) as

$$\sigma = \eta\dot{\gamma}. \tag{1.8}$$

Equation (1.8) also gives the basic definition for the viscosity η of liquids. For viscous Newtonian liquids, η is a constant, and Eq. (1.8) holds true at all times, as shown by the horizontal line in Fig. 1.2, starting at the moment when the shear is first applied. Newtonian liquids show *linear response*, as shown in Eq. (1.8), just as elastic solids show linear response, as shown by Eq. (1.7).

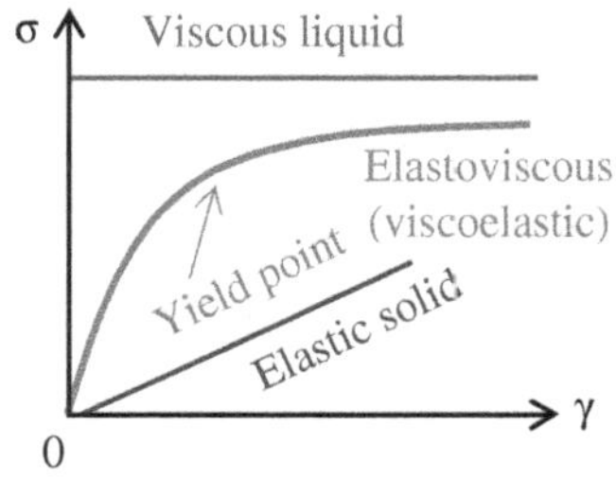

Figure 1.2 Sketches of various stress responses (liquid, solid, or viscoelastic liquid) during startup deformation in the linear response regime, that is, in the limit of $Wi \ll 1$, where the Weissenberg number Wi is defined in Section 1.2.7.

Equation (1.8) also explicitly indicates the unit of viscosity: Pa.s. In presenting Eq. (1.8), we assumed that the velocity field is uniform with a constant gradient as depicted in Fig. 1.1(d). While true for Newtonian liquids by definition, the assumption that the velocity gradient is a constant may be wrong for entangled polymeric liquids, except at very low shear rates, which are conditions to be treated in Section 1.2.7. Part 2 will consider whether the linear velocity profile of Fig. 1.1(d) remains true under all different shearing conditions in all systems. Section 4.1.1 discusses the assumption of homogeneous shear as a working principle for rheometric measurements. Chapter 9 shows when this assumption fails.

1.2.3 Viscoelastic Responses

1.2.3.1 Boltzmann Superposition Principle for Linear Response

Equations (1.7) and (1.8) are simplest examples of laws for linear response behavior, where the stress scales linearly with either strain or strain rate, respectively. By analogy with these equations, linear viscoelastic behavior can be generically described by applying the Boltzmann superposition principle, which generalizes the special cases of ideal viscous liquids and elastic solids. The specific phenomenological model for linear viscoelasticity was developed by James Clark Maxwell.

For viscoelastic materials, the concept of elastic modulus, for example, G in Eq. (1.7), is generalized to be time dependent: $G(t)$ is typically a decreasing function of time t. To illustrate viscoelasticity, let us consider a sudden step strain $\Delta\gamma(t_1)$ made over a vanishingly short period of Δt at time t_1. The sample may deform elastically and resist with a stress increment given by

$$\Delta\sigma(t) = G(t - t_1)\Delta\gamma(t_1), \quad \text{for } t > t_1. \tag{1.9}$$

Here, the relaxation modulus $G(t)$ replaces the elastic modulus G that appears in Eq. (1.7). Let us imagine imposing a series of small step strains $\Delta\gamma(t_i)$ at times t_i, $i = 1, 2, \ldots, n$, where t_1 can be taken to be the infinite past $(-\infty)$ and t_n defines the present time t. Ludwig Boltzmann (1844–1906) recognized that *for linear response* the stress $\sigma(t)$ produced by these consecutive steps of deformation should be additive, leading to[1,2]

$$\sigma(t) = \sum_{i=1}^{n} G(t - t_i)[\Delta\gamma(t_i)/\Delta t]\Delta t = \int_{-\infty}^{t} G(t - t')\dot{\gamma}(t')dt'. \tag{1.10a}$$

This formula, known as the Boltzmann superposition principle, offers a general relationship between the strain history and stress, in the linear response regime, for all viscoelastic materials. For startup shear at $t = 0$ with a constant rate $\dot{\gamma}$, Eq. (1.10a) can be rewritten to show the stress growth as

$$\sigma(t) = \dot{\gamma}\int_{0}^{t} G(s)ds \equiv \dot{\gamma}\eta(t). \tag{1.10b}$$

This stress increase is related to the relaxation modulus through the integration. The curve labeled "elastoviscous" in Fig. 1.2 illustrates the transition from the initial elastic deformation to the eventual flow state, a transition that we may call "voluntary yielding." In the steady state, the zero-shear viscosity is given by

$$\eta_0 = \frac{\sigma(\infty)}{\dot{\gamma}} = \int_{0}^{\infty} G(s)ds. \tag{1.10c}$$

It can be noted from Eq. (1.10c) that a viscosity can be given as a product of modulus and time.

1.2.3.2 General Material Functions in Oscillatory Shear

When an oscillatory external deformation, as described by Eqs. (1.4b) (1.5), is imposed onto a viscoelastic material, we can be certain in the linear response regime that the stress should also be oscillatory, given by

$$\sigma(t) \equiv \sigma_0(\omega)\sin[\omega t + \delta(\omega)] = \gamma_0(G' \sin \omega t + G'' \cos \omega t) \equiv G'(\omega)\gamma(t) + \eta'(\omega)\dot{\gamma}(t). \quad (1.11)$$

Here the introduction of storage and loss moduli $G' = (\sigma_0/\gamma_0)\cos\delta$ and $G'' = (\sigma_0/\gamma_0)\sin\delta$ arises naturally. By definition,

$$\tan\delta \equiv G''/G', \quad (1.12a)$$

and

$$\eta'(\omega) = G''(\omega)/\omega. \quad (1.12b)$$

By inserting Eq. (1.6) into Eq. (1.10a), we arrive at generic formulas relating the storage and loss moduli to the relaxation modulus G(*t*):

$$G'(\omega) = \omega\int_0^\infty G(s)\sin(\omega s)ds \quad (1.13a)$$

and

$$G''(\omega) = \omega\int_0^\infty G(s)\cos(\omega s)ds. \quad (1.13b)$$

These formal relationships imply that G′ and G″ provide the same information on the linear viscoelastic characteristics of the system that the relaxation modulus G(*t*) does. According to Eq. (1.11), the linear stress response to small-amplitude oscillatory shear (SAOS) is a combination of elastic deformation (the first term) and viscous flow (the second term).

1.2.3.3 Stress Relaxation from Step Strain or Steady-State Shear

Stress relaxation after step strain is one of few elementary experiments to probe viscoelastic behavior. Upon sudden imposition of a strain γ that occurs instantaneously at $t = 0$, that is, as t_0 in Fig. 1.1(e) approaches zero, we measure the residual stress $\sigma(t)$ for $t > t_0 \sim 0$. When γ is sufficiently small, linear response behavior is observed, so that for $t \geq 0$

$$\sigma(t) = G(t)\gamma. \quad (1.14)$$

The magnitude $\sigma(t)$ of the relaxing stress is linearly proportional to the imposed strain γ. The materials function G(*t*) is known as the relaxation modulus that first appeared in Eq. (1.10a). In an ideal experiment, the step strain takes place infinitely quickly. A realistic experiment can only impose a finite rate of deformation, that is, X/t_0 in Fig. 1.1(e) is finite rather than infinite. If there is a wide spectrum of relaxation times, then Eq. (1.14) may not hold true because fast relaxation modes may not contribute to G(*t*). Consequently, if there is a broad spectrum of relaxation times, different (nonideal) relaxation functions $G_{NI}(t)$ can be found, as though linear response were lost despite small step strains that do not alter the internal structure of the system. We will pick up this subtle issue in 1.2.5.2 after the introduction of the generalized Maxwell model in the following section.

1.2.4 Maxwell Model for Viscoelastic Liquids

Long before synthetic polymeric materials have been around, various natural materials were known to possess the duality of viscoelasticity. Molten cheese and chocolate can be viscoelastic.

Figure 1.3 (a) An abstract representation of the Maxwell element, made of a dashpot of viscosity η and spring of modulus G in series. (b) A shear setup made a layer of solid (or liquid) over a layer of liquid (or solid) with thickness H_s and H_l, which mimics the Maxwell model. A top surface displacement of X is a sum of the displacements x_l and x_s associated with the liquid and solid layers, respectively.

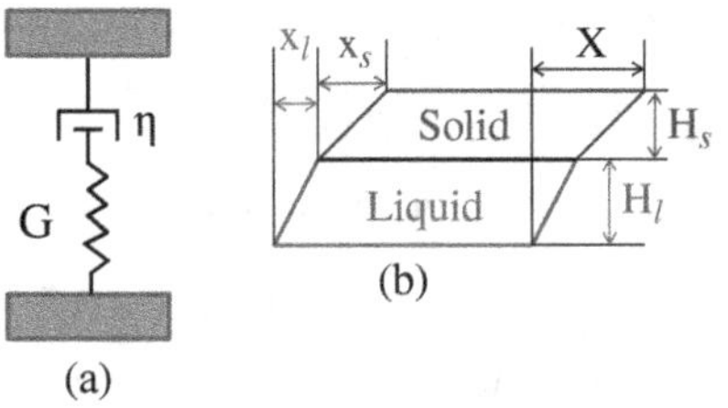

Maxwell (1831–1879) is the first to propose a *phenomenological* treatment of viscoelastic phenomena, famously known in the field as Maxwell model. Indeed, a piece of viscoelastic material can be modeled as an assembly of alternating solid and liquid layers. Conventionally, we simply assign a spring with Hookean elastic constant G and a dashpot with viscosity η to depict the Maxwell model,[2] as shown in Fig. 1.3(a).

Instead of considering a set of spring and dashpot, let us imagine placing a layer of elastic solid of modulus G on top of a liquid layer of viscosity η, as shown in Fig. 1.3(b), or vice versa. Suppose that the upper boundary is displaced with velocity V. Then, the solid layer elastically deforms according to

$$\sigma = G(x_s/H_s), \tag{1.15}$$

and the liquid layer flows as

$$\sigma = \eta(\dot{x}_l/H_l), \tag{1.16}$$

because the shear stresses are the same in the two layers. The time derivative of the total displacement X of the upper plate

$$X(t) = x_s + x_l \tag{1.17a}$$

is related to the overall shear rate $\dot{\gamma}$ as $\dot{X}/H \equiv V/H$. Upon differentiating both sides of Eqs. (1.15) and (1.17a), we find by adding both sides of the resulting Eqs. (1.15) and (1.16)

$$\sigma + \tau\dot{\sigma} = 2\eta\dot{\gamma}, \tag{1.18}$$

where the Maxwell relaxation time τ is given by

$$\tau = \eta/G, \tag{1.19}$$

which follows upon setting $H_l = H_s = H/2$ so that $\gamma = (\gamma_s + \gamma_l)/2$. In a standard exposition of the Maxwell model, the strain is linearly additive.

$$\gamma = \gamma_s + \gamma_l, \tag{1.17b}$$

that is, $X = \gamma H = x_s(H/H_s) + x_l(H/H_l)$ instead of Eq. (1.17a). In this case, the extra factor $H/H_l = 2$ would not appear on the right-hand side of Eq. (1.18). A construct analogous to Fig. 1.3(b) can be made for uniaxial extension. The same extra factor of 2 also shows up. Apart from this numerical discrepancy, construction of Fig. 1.3(b) offers a concrete way to contemplate viscoelasticity using a combination of purely viscous and purely elastic materials. Here in the Maxwell model, a relaxation time τ naturally emerges from a ratio of viscosity η to modulus G. The Maxwell model of Eq. (1.18) allows us to depict each of the three tests reviewed in Section 1.1.

1.2.4.1 Stress Relaxation from Step Strain

The Maxwell model allows us to depict the essence of a step strain test of magnitude $\gamma = X/H$, which is to evaluate stress relaxation. Since shear ceases at time t_0 as shown in Fig. 1.1(e) in such a test, we can set the right-hand side of Eq. (1.18) to zero for $t > t_0$ and solve this most common differential equation for the initial condition given by

$$\sigma(0) = G\gamma(H/H_s). \tag{1.20}$$

The solution is an exponential function:

$$\sigma(t) = \sigma(0)\exp(-t/\tau), \quad \text{for} \quad t > t_0. \tag{1.21a}$$

The "instantaneous" strain γ is imposed on the solid layer during a time scale $t_0 \ll \tau$; for the liquid layer to suffer the strain γ would correspond to emergence of a stress level larger than $\sigma(0)$ in Eq. (1.20) by a factor of $(\tau/t_0) \gg 1$, which could not feasibly occur. The condition of $t_0 \ll \tau$ translates to $\gamma/Wi \ll 1$, where the dimensionless parameter $Wi = \dot{\gamma}\tau$ is the Weissenberg number (to be formally introduced in Section 1.2.7). As long as $\gamma \ll 1$, the condition for instantaneous strain can be met for $Wi > 1$. During the step strain, the liquid layer experiences little flow: the amount of strain taking place in the liquid layer can be estimated to be $\gamma_l \sim \dot{\gamma}_l t_0 \sim \gamma(t_0/\tau) \ll \gamma$, where use is made of $\eta\dot{\gamma}_l \sim G\gamma$. Defining the relaxation modulus $G(t)$ as shown in Eq. (1.14), we have, according to Eqs. (1.20) and (1.21a),

$$G(t) = G\exp(-t/\tau), \tag{1.21b}$$

whereas the factor of (H/H_s) was omitted if we adopt the familiar convention, that is, using Eq. (1.17b) instead of Eq. (1.17a). Because according to Eq. (1.21a) the stress decreases to zero exponentially, the Maxwell model cannot describe viscoelastic solids. This should also be obvious from the construction of the model depicted in Fig. 1.3.

1.2.4.2 Startup Deformation

Taking the case of simple shear, according to the Maxwell model, viscoelastic materials must always initially undergo elastic deformation upon a startup shear for any values of *Wi*. This can be demonstrated by equating Eqs. (1.15) with (1.16) and expressing x_s in terms of x_l through Eq. (1.17a) to obtain the following equation for x_l:

$$x_l + \tau\dot{x}_l = X(t), \tag{1.22}$$

which is similar in the form to Eq. (1.18) for σ. It is straightforward to solve this elementary differential equation, given the initial condition $x_l(t=0)=0$, which states that the liquid layer suffers no deformation at $t=0$. We have

$$x_l(t) = X(t)\{1 + [\exp(-t/\tau) - 1]/(t/\tau)\}, \tag{1.23a}$$

where $X(t) = Vt$ as shown in Fig. 1.1(b). It is clear that as long as $t/\tau \ll 1$ we have, upon Taylor expansion, $x_l(t) \sim X(t)(t/2\tau)$, that is, $x_l/X \sim t/\tau \ll 1$, independent of $Wi = \dot{\gamma}\tau$. Numerically, we see from Fig. 1.4(a) that x_l is only 10% of the total displacement $x(t)$ at $t = 0.2\tau$. The condition of $x_l/x \ll 1$ implies the lack of flow and dominance of elastic deformation.

Accompanying the initial elastic deformation, stress rises from zero instead of having a finite value from the onset. It is easy to integrate Eq. (1.18) or to insert Eq. (1.21b) into Eq. (1.10a) and show both stress growth upon startup and stress relaxation after shear cessation, by setting $\dot{\gamma}(t)$ as $\dot{\gamma}$ for $t < t_1$ and 0 for $t > t_1$, respectively. Specifically, we have the stress growth given by

$$\sigma = \sigma_\infty[1 - \exp(-t/\tau)] \quad \text{for} \quad t < t_1 \tag{1.23b}$$

and the stress relaxation given by

$$\sigma = \sigma_\infty\{[\exp[-(t - t_1)/\tau] - \exp(-t/\tau)]\} \quad \text{for} \quad t > t_1, \tag{1.23c}$$

where $\sigma_\infty = \eta\dot{\gamma}$ is the steady shear stress. Experiment reveals that many types of viscoelastic liquids, including entangled polymer melts and solutions, show initial elastic response to startup

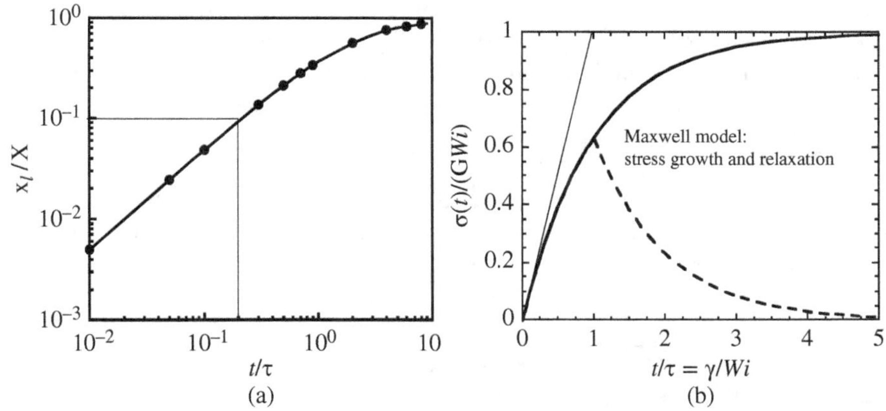

Figure 1.4 (a) Illustration of Eq. (1.23a). (b) Maxwell model's depiction of stress growth and relaxation (dashed curve) according to Eqs. (1.23b) and (1.23c), respectively.

shear, followed by flow at long times. This initial elastic strain γ grows with *Wi*. In the Maxwell model, the stress σ also initially grows with strain as shown by Eq. (1.23b) in Fig. 1.4(b). Thus, the Maxwell model qualitatively captures the initial rheological response of entangled polymers. On the other hand, being a model for linear response, the Maxwell model cannot anticipate and depict how the relation between stress and strain depends on *Wi*. According to Fig. 1.4(b), the initial stress growth associated with elastic deformation persists for a strain that is proportional to *Wi*. Viscoelastic fluids such as entangled polymers cannot remain elastic for a strain that linearly increases with *Wi*. Thus, the Maxwell model cannot describe fast deformation behavior at high strains.

The Maxwell model prescribes a smooth transition from elastic deformation to flow for any value of *Wi*. Thus, even Maxwell-like materials should show a sign of (voluntary) yielding. In the steady-shear limit, that is, $t/\tau \gg 1$, the solid component in the Maxwell element of Fig. 1.3(b) suffers a fixed amount of deformation $x_s(\infty) = X(\infty) - x_l(\infty) = V\tau$, corresponding to $\sigma_\infty = G(x_s/H_s) = (H/H_s)G(\tau\dot{\gamma}) \sim \eta\dot{\gamma}$. This depiction is reasonably close to reality for $Wi \ll 1$. But for $Wi > 1$, there is something missing in the Maxwell model: Rewriting Eq. (1.23b) as $\sigma = GWi[1 - \exp(-t/\tau)]$, we see σ/G becoming *Wi* at $t/\tau > 1$, that is, σ getting much larger than G for $Wi \gg 1$. This unphysical situation occurs because for $Wi \gg 1$ the Maxwell model is incapable of describing when and how elastic deformation is replaced by irrecoverable deformation, which is flow. To delineate the nature of the transition from elastic deformation to flow is the main focal point of this book. In solid mechanics, yielding refers to a material transitioning from elastic deformation to plastic deformation. If we do not distinguish between plastic deformation and flow and simply label them as irrecoverable deformation and if we can extend the concept of yielding to refer to a transition from elastic deformation to irrecoverable deformation, then we can legitimately use the phrase *yielding* to speak about the rheological behavior of any viscoelastic liquids, including entangled polymeric liquids.

1.2.4.3 Oscillatory (Dynamic) Shear

The stress response of an elastic solid to a strain sine wave is also a sine wave, while the stress response of a viscous liquid to a strain sine wave is a cosine wave. Viscoelastic materials will display a sinusoidal stress in steady state as indicated in Eq. (1.11), provided that γ_0 is low enough so that σ_0 is linearly proportional to γ_0. When the "phase lag" angle δ is vanishingly small, we see a solid-like response. Conversely, when δ is almost $\pi/2$, $\sigma(t)$ is nearly a cosine function and therefore in phase with the shear rate of Eq. (1.5), so the sample behaves like a liquid.

The Maxwell model has a specific quantification of how $\delta(\omega)$ and $\sigma_0(\omega)$ in Eq. (1.11) depend on ω. As shown in a standard textbook on viscoelasticity,[3] Eq. (1.18) can be readily solved using complex variable notation to obtain the functional forms of δ and σ_0. We may also perform the analysis in real variables. Inserting Eqs. (1.5) and (1.11) into Eq. (1.18), we have

$$\sigma_0(\omega)[\sin(\omega t + \delta) + \omega\tau\cos(\omega t + \delta)] = \gamma_0 G\omega\tau\cos\omega t, \tag{1.24}$$

where we ignore the extra factor of H/H_l from Eq. (1.18) to conform to the conventions used in the Maxwell model. Since there cannot be a term proportional to $\sin\omega t$ on the left-hand side of Eq. (1.24), we immediately find, by expanding the sine and cosine functions,

$$\tan\delta = 1/\omega\tau. \tag{1.25}$$

This equation implies $1 = \sin\delta\sqrt{1+(\omega\tau)^2}$ and $\omega\tau = \cos\delta\sqrt{1+(\omega\tau)^2}$, which can be, respectively, substituted into the left-hand side of Eq. (1.24) in front of each of the two terms giving a summation of $[\sin\delta\sin(\omega t+\delta) + \cos\delta\cos(\omega t+\delta)] = \cos\omega t$. This substitution allows us to determine σ_0 as

$$\sigma_0(\omega) = \frac{\gamma_0 G\omega\tau}{\sqrt{1+(\omega\tau)^2}} \equiv \gamma_0|G^*|, \tag{1.26}$$

where $G^* = G' + iG''$ is known as the complex modulus. According to the second equality of Eq. (1.11), the storage and loss moduli G' and G'' are simply related to σ_0 as

$$G' = [\sigma_0(\omega)/\gamma_0]\cos\delta = G(\omega\tau)^2/[1+(\omega\tau)^2] \tag{1.27a}$$

and

$$G'' = [\sigma_0(\omega)/\gamma_0]\sin\delta = G(\omega\tau)/[1+(\omega\tau)^2]. \tag{1.27b}$$

Here, the first equality holds true in general. The second two equalities represent a specific prediction of the Maxwell model. The specific predictions can also be derived by inserting Eq. (1.21b) into the generic expressions Eqs. (1.13a) and (1.13b).

Figure 1.5 depicts G' and G'' from Eqs. (1.27a) and (1.27b) as functions of ω. Depending on the imposed frequency ω, a Maxwellian material may behave like a solid displaying a rubbery plateau or a liquid showing a negligible G'. Such a material may be truly regarded as a solid as long as it is being mechanically examined on time scales much shorter than τ. Note that Eq. (1.11) gives the time-dependent stress in steady state. The initial transient response is not depicted by Eq. (1.11).

1.2.5 General Features of Viscoelastic Liquids

1.2.5.1 Generalized Maxwell Model

The Maxwell model offers a most useful and simplest phenomenological continuum-level description of viscoelastic liquids that possess a dominant relaxation time. In reality, a viscoelastic liquid may possess more than one elementary relaxation time. A generalized Maxwell

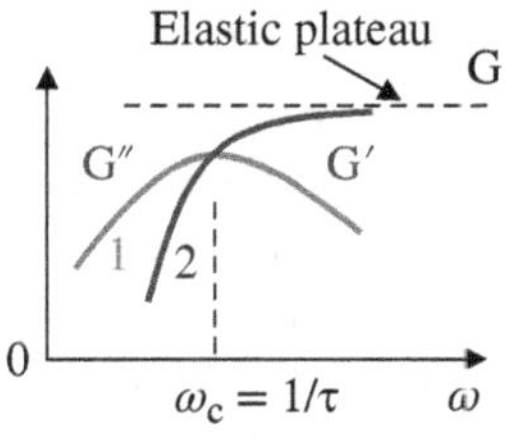

Figure 1.5 Schematic illustration on double-log of the Maxwell model description of the storage and loss moduli G'' and G'' as a function of the oscillation frequency ω. For $\omega\tau \ll 1$, $G' \sim \omega^2$ and $G'' \sim \omega$ and $G' = G''$ at $\omega\tau = 1$.

model[2] can be useful to describe linear viscoelastic processes in real systems. In particular, the storage and loss moduli G′, G″ in the generalized Maxwell model have the following forms:

$$G'(\omega) = \sum_{q=1}^{\infty} \frac{G_q(\omega\tau_q)^2}{1 + (\omega\tau_q)^2} \tag{1.28}$$

and

$$G''(\omega) = \sum_{q=1}^{\infty} \frac{G_q\omega\tau_q}{1 + (\omega\tau_q)^2}, \tag{1.29}$$

where $\{\tau_q\}$ gives a spectrum of relaxation times. Conventionally, $q = 1$ represents the longest relaxation time, with shorter time τ_q involving a larger number $q > 1$. Moreover, the relaxation modulus $G(t)$ is a summation of exponential functions

$$G(t) = \sum_{q=1}^{\infty} G_q \exp(-t/\tau_q). \tag{1.30}$$

Finally, the Newtonian viscosity η_0 of a generalized Maxwellian liquid is given by

$$\eta_0 = \sum_{q=1}^{\infty} G_q\tau_q, \tag{1.31}$$

which can be obtained by inserting Eq. (1.30) into the Boltzmann formula Eq. (1.10a). As we will see, most molecular models for polymer dynamics also share the same mathematical forms as given in Eqs. (1.28)–(1.31). Since the generalized Maxwell model is a phenomenological account, it cannot explain the molecular origin of the viscoelastic behavior, for example, it cannot determine how τ_q may depend on the chain length or molecular weight and how it varies with the mode index q and where the elasticity arises from.

1.2.5.2 Lack of Linear Response in Small Step Strain: A Dilemma

In realistic experiments, it takes a finite time to produce a step strain of magnitude γ_0. During t_0, fast relaxation modes compete with the imposed straining. The shorter the duration t_0, the less relaxation of the fast modes can take place. Step strain in real experiments amounts to shearing a sample for a very short time t_0 at high rate $\dot{\gamma}$ to a magnitude of $\gamma = \dot{\gamma}t_0$ before shear cessation at $t = 0$. Inserting this step strain condition and Eq. (1.30) into Eq. (1.10a), we have

$$\sigma(t) = \dot{\gamma}\sum_{q=1}^{\infty} G_q\tau_q[1 - \exp(-t_0/\tau_q)]\exp(-t/\tau_q), \text{ for } t \geq 0. \tag{1.32}$$

Thus, the relaxation modulus G_{NI} under the nonideal but realistic step strain condition follows from Eq. (1.32)

$$G_{NI}(t, \gamma_0) = \sigma(t)/\dot{\gamma}t_0 = \sum_{q=1}^{\infty} G_q \exp(-t/\tau_q)f(t_0/\tau_q), \tag{1.33}$$

where the function $f(x) = (1 - e^{-x})/x$ decreases monotonically from unity at $x = 0$ toward zero. Equation (1.33) differs from the ideal relaxation modulus $G(t)$ of Eq. (1.30) and is an explicit function of t_0/τ_q. When t_0/τ_q is not vanishingly small, the higher relaxation mode q makes a smaller contribution to the overall stress relaxation because f is unity only in the limit of $t_0/\tau_q = \gamma/(\dot{\gamma}\tau_q) \ll 1$. In other words, G_{NI} approaches Eq. (1.30) only when $t_0/\tau_q \ll 1$ for all the contributing modes indexed by q. Thus, when the relaxation dynamics are dictated by a series of time scales, G_{NI}, obtained from small step-strain with $\gamma \ll 1$, is explicitly dependent on

t_0/τ_q unless $\tau_q < 1$ for all values of q. Thus, in the presence of a broad relaxation spectrum as is found in the case of long-chain branched polymers and mixtures as well as melts with broadly polydisperse molecular weight distributions, without application of an extremely high rate, the relaxation modulus σ/γ could vary either with γ for a given rate $\dot{\gamma}$ or with the imposed rate $\dot{\gamma}$ for a fixed γ. Experimental measurements would yield the false impression that the sample always shows nonlinear response in stress relaxation, even though $\gamma \ll 1$, and Eq. (1.14) would not hold. Finite step strains take finite times to create, during which relaxation is also taking place. The time needed to produce a step-strain increases with the increasing strain γ and decreasing rate $\dot{\gamma}$. Corresponding to a larger γ and therefore larger t_0 at a given rate, with faster relaxation modes dropping out of the residual stress, the contributions of fast relaxation modes are smaller. Consequently, the stress relaxation from a larger strain γ_2 can be slower than that from a smaller strain $\gamma_1 < \gamma_2$ although both γ_1 and γ_2 are in the limit of linear response. When the imposed rate $\dot{\gamma}$ is smaller for a given γ the relaxation is slower. If the relaxation spectrum is wide, it is indeed an experimental challenge to obtain a unique relaxation modulus $G(t)$ such as that of Eq. (1.30). The nonideal relaxation modulus can also be discussed in a generic form. The leading correction to the ideal relaxation modulus $G(t)$ would be proportional to γ. A Taylor expansion of Eq. (1.33) produces $G_{NI}(t, \gamma) = G(t) + (\gamma/\dot{\gamma})\frac{\partial G(t)}{\partial t} + O(\gamma^2)$. The γ dependence would be negligible only if $G(t)$ does not change with t rapidly relative to the imposed rate $\dot{\gamma}$. Conversely, to recover linear response, we need to use a sufficiently small γ and high shear rate $\dot{\gamma}$ relative to the rate of change in $G(t)$. For amorphous polymers, this condition can always be reached by resorting to the time–temperature superposition principle (discussed in Section 2.4.1).

It is also instructive to consider stress relaxation after shear cessation at $t = 0$ from steady state. When $t_0/\tau_1 \gg 1$, Eq. (1.32) becomes

$$\sigma(t) = \dot{\gamma} \sum_{q=1}^{\infty} G_q \tau_q \exp(-t/\tau_q), \quad \text{for } t \geq 0. \tag{1.34}$$

Here two remarks are in order. First, the level of the relaxing stress from steady shear is linearly proportional to the imposed shear rate. This linearity arises because we invoked the Boltzmann formula Eq. (1.10a), which only describes linear response. Second, the stress decline described by Eq. (1.34) may deviate significantly from the decline following a step strain as seen in Eq. (1.30) or Eq. (1.32). Both declines involve the exponential decay factor $\exp(-t/\tau_q)$, but one is weighed by the component modulus G_q and the other is weighed by the component viscosity $\eta_q = G_q \tau_q$. Since τ_q drops quickly with the index q, the relaxing stress from steady-state shear receives less contribution from the fast relaxation modes. Therefore, the stress relaxation would be slower after cessation of a steady shear (Eq. (1.34)) than after a step strain (Eq. (1.30)).

1.2.6 Kelvin–Voigt Model for Viscoelastic Solids

By construction, the Maxwell model describes viscoelastic *liquids*. It certainly cannot reveal the behavior of viscoelastic *solids*. The simplest continuum mechanics model for viscoelastic solids was invented by William T. Kelvin (1824–1907) and Woldemar Voigt (1850–1919). Instead of placing the viscous and elastic elements in series, they arranged them in parallel as shown in Fig. 1.6(a,b).[2] The equation of motion in the Kelvin–Voigt model is much more straightforward to derive because each component contributes additively to the total stress, namely

$$\sigma = G\gamma + \eta\dot{\gamma}. \tag{1.35}$$

Figure 1.6 (a) The Kelvin–Voigt element with a spring of modulus G and dashpot of viscosity η in parallel. (b) Experimental version of the Kelvin–Voigt model made of a solid block adjacent to a liquid block in parallel.

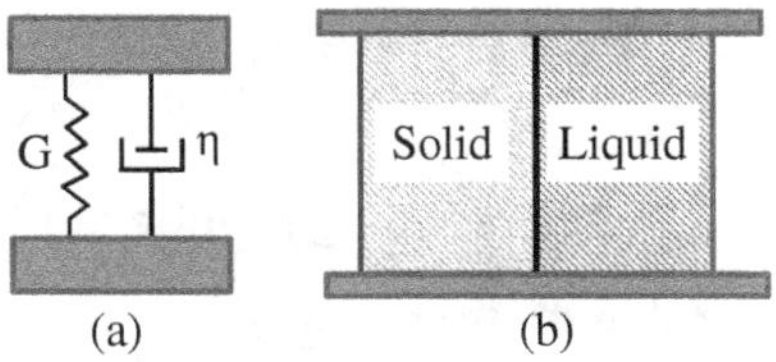

In the following section, we give two examples where the Kelvin–Voigt model provides some crude account of the simplest phenomenology.

1.2.6.1 Creep Experiment

A creep test applies a constant stress σ_0 starting at time $t = 0$. The Kelvin–Voigt model has a specific prediction about how the strain changes with time under σ_0. It is straightforward to solve Eq. (1.34) to find the exponential solution as

$$\gamma(t) = \gamma(\infty)[1 - \exp(-t/\tau)], \tag{1.36}$$

where the limiting strain is given by $\gamma(\infty) = \sigma_0/G$ and the relaxation time τ is the same as that defined in the Maxwell model $\tau = \eta/G$. This expression for τ is inescapable because it is the only combination of η and G to yield a quantity with the unit of time. It is important to note that the strain cannot instantly reach a finite level in such materials. The initial rate depends on the imposed stress level σ_0. Specifically, the initial rate is given by $\dot{\gamma}(0) = \gamma(\infty)/\tau = \sigma_0/\eta$. Creep is commonly applied to obtain linear viscoelastic information. Nonlinear response to creep in simple shear is far more interesting, as shown in Chapter 7 in the example of entangled polymer solutions. Creep in uniaxial extension is less straightforward to implement in the nonlinear response regime because the cross-sectional area of the sample shrinks considerably. To keep the Cauchy stress constant, the total applied tensile force needs to decrease in proportion to the shrinkage of the transverse dimensions.

1.2.6.2 Strain Recovery in Stress-Free State

The Kelvin–Voigt model is clearly the simplest that can describe elastic recoil. After a viscoelastic material has suffered a step strain or has reached the steady shear flow, what happens if we withdraw the external stress that has emerged from the applied deformation? An elastic solid, upon a step shear, would *instantly* return to its original position, that is, completely recoil to its undeformed state, so long as the step shear was small enough so that there was no permanent structural change in the solid. However, when the material is viscoelastic, the recoil cannot occur instantaneously because there would be an infinitely high shear rate and correspondingly an infinitely high shear stress (Eq. (1.8)). If the viscous component is lower, the recoil is expected to take a shorter time. Actually, Eq. (1.35) gives a crude depiction of how elastic-driven strain recovery takes place. Setting $\sigma = 0$ in Eq. (1.35), we arrive at a solution of the form

$$\gamma(t) = \gamma(0)\exp(-t/\tau), \tag{1.37}$$

where $\gamma(0)$ is the strain corresponding to the moment when the system is set stress free. $\tau = \eta/G$ dictates the time scale for the recoil. Being viscoelastic, the recoil is never instantaneous and approaches completion at long times. Some viscoelastic solids do not fully return to the original position in such a test, contrary to Eq. (1.37). Being a simplest linear model, the Kelvin–Voigt model evidently does not encompass yielding behavior and cannot depict irrecoverable deformation that may occur in the real world of rheology of viscoelastic materials. Thus, similar to the Maxwell model, the Kelvin–Voigt model cannot depict nonlinear rheology: By construction, the solid block in Fig. 1.6(b) never breaks down and is always available to produce complete recoil. This is plausible provided the applied stress is low so that the resulting deformation is small.

1.2.7 Weissenberg Number and Yielding during Linear Response

For step strain generated at a sufficiently high rate, linear response is guaranteed by a sufficiently small magnitude of the amplitude $\gamma_0 < \gamma_{LR}$, so that the internal structure such as entanglement network remains intact, except for cases with broad relaxation spectrum (Section 1.2.5.2). For entangled polymer solutions and melts, γ_{LR} just needs to be below about 0.7 or so.[4,5] Assuming that the dominant relaxation time is τ, step strain is realized in experiment when $t_0 \ll \tau$ so that there is little relaxation during the step strain. This condition actually states $\dot{\gamma}\tau \gg \gamma$. Thus, when $\gamma \sim 1$, we will need $\dot{\gamma}\tau \gg 1$ so that the step strain may be regarded instantaneous. Nonlinear responses arise in a step strain when the generated stress $\sigma(t)$ is no longer linearly proportional to γ, which happens when γ exceeds a critical value. For small step strain we only consider step strain that is fast enough, so that we can ignore the complications revealed in Section 1.2.5.2.

For nonlinear response to occur in startup shear (Fig. 1.1(a–d)), the applied rate needs to be high so that $t_{1\text{exp}}$ of Eq. (1.3) is considerably shorter than τ. We introduce the Weissenberg number

$$Wi \equiv \tau / t_{1\text{exp}} \equiv \dot{\gamma}\tau \tag{1.38a}$$

named after Karl Weissenberg (1893–1976). The Weissenberg number Wi denotes the amount of strain γ elapsed during the relaxation time τ. Thus, under the condition of $Wi \gg 1$, nonlinear rheological responses can be expected. The onset of nonlinearity is characterized by the critical condition of

$$Wi_c = \tau\dot{\gamma}_c \approx 1, \tag{1.38b}$$

corresponding to $t_{1\text{exp}} = \tau$. At $Wi \gg 1$, a significant amount of affine deformation may take place first.[1] We note that Wi has first appeared in Section 1.2.4.1 and the preceding paragraph as well as in the caption of Fig. 1.2. When startup deformation involves $Wi \gg 1$, the central question is how much elastic (nearly affine) deformation takes place before the flow fully develops. In other words, does the initial elastic deformation last as many strain units as the upper bound given by Wi? What terminates the elastic deformation and launches the flow? To address such questions is the central task in polymer rheology. This book is devoted to investigating such basic questions for entangled polymeric liquids, starting in Chapter 6.

The transformation from initial elastic deformation to subsequent irreversible deformation (flow) is a natural nonlinear response to startup deformation of either shear or extension. Such a transformation is conceptually analogous to yielding in large deformations of solids. For $Wi \ll 1$, the stress response can be described by the viscoelastic curve in Fig. 1.2 for entangled polymers: The initial deformation is elastic in nature and may be even recoverable. However, the response at long times (i.e., in steady state) is flow, approaching the liquid horizontal line of Fig. 1.2. We label the total response as "elastoviscous" to indicate that it starts elastically

1 In polymer rheology, affine deformation is a key concept. We can regard it to originate from the formulation of the rubber elasticity theory, to be briefly described in Section 1.3. Affine deformation is a theoretical idea used to bridge the many orders-of-magnitude gap in length scale when relating macroscopic mechanical behavior to microscopic processes. For example, the concept of affine deformation is called for when we desire to learn how a polymer is responding on molecular scales to the imposed macroscopic deformation. Thus, we cannot speak of "affine deformation" without specifying on the length scale beyond which the deformation is affine. For a cross-linked rubber or an entangled melt, affine deformation usually means that on a length scale equal to or larger than the subchain between cross-links or between neighboring entanglements the deformation can be specified to be identical to that occurring macroscopically. Throughout the book, the phase "affine" assumes the meaning provided here.

and then turns viscous. Here, the transformation from elastic deformation to flow should be regarded as *yielding* albeit structural changes may be small. It is worth noting that the Maxwell model captures this feature rather well as seen in the similarity between Fig. 1.4(b) and the elastoviscous curve of Fig. 1.2.

1.3 Classical Rubber Elasticity Theory

This chapter has focused on phenomenological accounts of linear viscoelasticity, treating viscoelastic materials as a continuum during macroscopic deformation. The following account of rubber elasticity serves as a logical bridge to the next chapter on molecular theories of chain dynamics because it gives an example of a polymer being treated from a molecular view point.

Polymeric materials are uniquely capable of undergoing large recoverable deformations and displaying significant elasticity. In polymer physics, molecular modeling began with an account of rubber elasticity in terms of a phantom network of Gaussian chains. This classical theory of Guth and James[6–9] actually provided a first theoretical framework for modern polymer physics. Given the importance of the elastic character in large deformation behavior of entangled polymers, we choose to first discuss the statistical mechanical theory[9] for rubber elasticity.

1.3.1 Chain Conformational Entropy and Elastic Force

Boltzmann quantified the notion of entropy S as being proportional to the log of the number ϖ of states accessible to a system. Let us first consider the simplest case of a single Gaussian chain with end-to-end distance vector **R**. The chain possesses a Boltzmann entropy S given by

$$S = k_B ln\varpi\{\mathbf{R}\}$$

where k_B is the Boltzmann constant. A Gaussian chain has the property that the probability distribution function P{**R**} is given by a Gaussian function

$$\mathrm{P}\{\mathbf{R}\} \sim \exp(-3\langle \mathbf{R}^2\rangle/2Nl_K^2), \tag{1.39a}$$

where the ensemble-averaged end-to-end distance squared is proportional to the chain length N, as

$$Nl_K^2 = \langle \mathbf{R}^2\rangle \equiv \mathrm{R}^2 \tag{1.39b}$$

Here l_K is the Kuhn length and the number N of Kuhn segments per chain is frequently termed "chain length." In terms of the characteristic ratio C_∞ and bond length l of the repeated unit in the backbone, for simplicity, we will use the following approximate relation:[10]

$$l_K \simeq C_\infty l \tag{1.39c}$$

instead of $l_K = (C_\infty + 1)l$. Since ϖ should be proportional to P, we can write down a Helmholtz (1821–1894) free energy A for the system as

$$\mathrm{A}\{\mathbf{R}\} = \mathrm{U} - \mathrm{TS} = \mathrm{U} + (3\mathbf{R}^2/2Nl_K^2)k_B\mathrm{T}, \tag{1.40}$$

where U represents internal energy and T the absolute temperature. Then, the first law of thermodynamics,

$$\mathrm{T}dS = \mathrm{dU} - \mathrm{d}W, \tag{1.41}$$

allows us to show that

$$\mathrm{dA} = -S\mathrm{dT} - \mathrm{Pd}V + \boldsymbol{f}\cdot\mathrm{d}\mathbf{R}, \tag{1.42}$$

where $dW = -PdV + \boldsymbol{f} \cdot d\mathbf{R}$ is the work done on the system. Under isothermal condition, increasing the end-to-end distance can result in a change in the chain's Helmholtz free energy even in the absence of any volume change. The basic operation of calculus permits Eq. (1.42) to be rewritten to show that the entropic force $\boldsymbol{f}$ is given by[11]

$$\boldsymbol{f} = \left(\vec{x}\frac{\partial}{\partial R_x} + \vec{y}\frac{\partial}{\partial R_y} + \vec{z}\frac{\partial}{\partial R_z}\right)_{T,V} A = H\mathbf{R}, \tag{1.43}$$

where the Hookean constant H is defined as

$$H = 3k_B T/\langle \mathbf{R}^2 \rangle. \tag{1.44}$$

Equation (1.44) suggests that a longer chain is softer because its H would be smaller. Force measurements[12,13] on single chains have been carried out to verify the simple relationship of Eq. (1.43).

1.3.2 Network Elasticity and Stress–Strain Relation

An elastomeric network was first proposed to model rubber elasticity.[6–9] Consider a slab-shaped sample of volume ΣH such as that depicted in Fig. 1.1a. Suppose that it contains a total of n active (load-bearing) strands whose average end-to-end distance is R. A simple analysis can be carried under the assumption that these strands are Gaussian chains and act independently of one another.[9] The total Helmholtz free energy associated with this system is, by analogy with Eq. (1.40), given by

$$A = U + k_B T \sum_{i=1}^{n} \left(\frac{3\mathbf{R}_i^2}{2R^2}\right), \tag{1.45}$$

where all chains are assumed to be independent of one another so that the conformational entropy associated with them is simply additive. The Helmholtz free energy $A(\gamma)$ in the presence of a simple shear deformation γ (as depicted in Fig. 1.1a) can be readily evaluated. Under the affine deformation assumption, the x-component of the end-to-end distance vector **R** is displaced according to

$$R_{ix} \rightarrow R_{ix} + \gamma R_{iy} \tag{1.46}$$

while the other two components remain unchanged during simple shear, as shown in Fig. 1.7. Thus, upon simple-shear deformation, the Helmholtz free energy is given by

$$A(\gamma) = A(0) + \left(\frac{3k_B T}{R^2}\right) \sum_{i=1}^{n} \gamma^2 R_{iy}^2 = A(0) + (n k_B T/2)\gamma^2, \tag{1.47}$$

where $A(0)$ is the free energy in the absence of deformation, the first equality follows because $\sum_{i=1}^{n}(R_{ix}R_{iy}) = 0$ for isotropic Gaussian chains and the second equality arises from $\langle R^2 \rangle = \langle R_{ix}^2 + R_{iy}^2 + R_{iz}^2 \rangle = 3\sum_{i=1}^{n} R_{iy}^2/n$ since the chains are statistically isotropic and the average given by the summation is the same as the ensemble average in the limit of $n \rightarrow \infty$.

It is worth mentioning that an expression for uniaxial extension, which is depicted in Fig. 1.8, can be similarly derived in the affine deformation limit as[9]

$$A(\lambda) = A(0) + (n k_B T/2)(\lambda^2 + 2/\lambda), \tag{1.48}$$

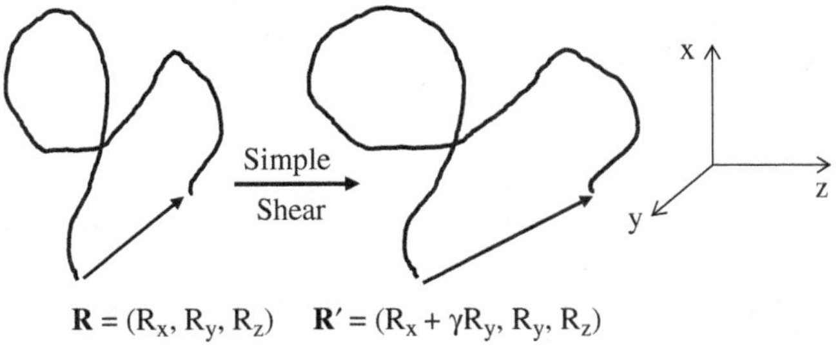

Figure 1.7 Illustration of transformation in the end-to-end distance vector from **R** to **R′** during affine simple shear, where $\gamma = X/H$ (Fig. 1.1(a)).

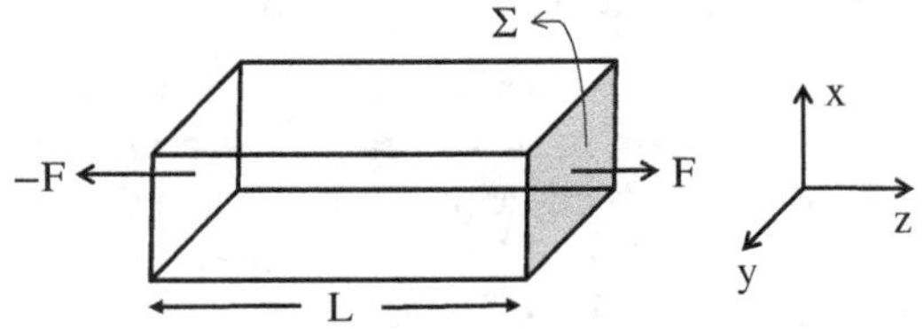

Figure 1.8 Illustration of uniaxial extension along Z-axis, involving a tensile force F over the shaded cross-sectional area Σ for a specimen of length L.

where λ is the stretching ratio of final length L to initial length L_0. Here, we skip the derivation because it is widely available in many textbooks.[3] In contrast, few books describe the case of simple shear deformation.

The total elastic force F is given by the variation of the free energy A with the displacement $X = \gamma H$ (Fig. 1.1(a)), that is,

$$F = \left(\frac{\partial \Delta A}{\partial X}\right)_{T,V} = \frac{\partial \Delta A}{\partial \gamma}\frac{\partial \gamma}{\partial x} = n k_B T\gamma/H, \tag{1.49}$$

so that the shear stress $\sigma = F/\Sigma$ has the following form

$$\sigma = G\gamma, \tag{1.50}$$

where G can be identified as the shear modulus G given by

$$G = (n/\Sigma H)k_B T = \rho_p k_B T, \tag{1.51a}$$

with ρ_p being the number of active strands per unit volume. Denoting the physical volume of a network strand as $\nu_\times$, then Eq. (1.51a) can be rewritten as

$$G = k_B T/\nu_\times = k_B T/pR_\times^2, \tag{1.51b}$$

where the subscript × stands for "cross-linking" and the second equality follows from the fact that $\nu_\times$ is proportional to the chain length of an active strand, that is, to $R_\times^2$. Here, a new parameter p, known as the packing length,[10,14] is used to quantify the relation between $\nu_\times$ and coil size $R_\times$. More discussion on this concept is provided in Section 2.2.3.1.

This Gaussian-chain network model prescribes that the elastic modulus G is simply proportional to the number density of chains as shown by Eq. (1.51a). We can actually arrive at Eq. (1.51a) by dimensional analysis. It is straightforward to argue, given the dimension of G being energy per unit volume, that G should be $k_B T$ divided by a specific volume, that is, the physical volume of the strand, $\nu_\times$, because these are the only two quantities having the dimensions of energy and volume, respectively.

For uniaxial extension, the following expression relating the engineering stress $\sigma_{engr} = F/\Sigma_0$ (Σ_0 being the initial value of the cross-sectional area) to $\lambda = L/L_0$ can be readily obtained from Eq. (1.48)

$$\sigma_{engr} = G(\lambda - 1/\lambda^2), \tag{1.52}$$

where the elastic modulus G is given by Eq. (1.51a,b). It is necessary to emphasize that Eqs. (1.50) and (1.52) for simple shear and uniaxial extension, respectively, are two key formulas. They provide the reference for the subsequent discussion of deformation of uncross-linked, entangled melts and solutions. In passing, it is useful to note that Eqs. (1.50) and (1.52) can also be written in terms of the areal density of network strands and the entropic force per strand. This is explained in the following sections.

1.3.3 Alternative Expression in terms of Retraction Force and Areal Strand Density

The preceding statistical mechanical account of rubber elasticity within the framework of a Gaussian chain network traces the origin of elasticity to chain conformation entropy. It is a linear-response description and does not deal with the question of what happens at sufficiently large deformations. We can imagine applying the same concept to describe a Maxwellian *liquid* as long as the relevant experimental time scale defined by Eq. (1.3) is much shorter than τ, that is, for $Wi \gg 1$. In fact, many polymeric liquids should exhibit strong elasticity as depicted in Eqs. (1.50) and (1.52) as long as the deformation occurs within a time scale significantly shorter than their internal relaxation times. Once again, we reach the point where we wonder what molecular parameters determine the relaxation times in polymeric materials, and whether and why these times become shorter under fast large deformation. This is the subject for the rest of the book.

Available textbooks rarely present an alternative, more intuitive derivation or interpretation of Eq. (1.50) and Eq. (1.52). From the standpoint of mechanics, we can describe the cross-linked Gaussian chain network as shown in Fig. 1.9, taking the example of simple shear. The shear stress by definition arises simply from the elastic retraction forces $f_{retract}$ of those active strands whose areal number density $\psi_\times$ is related to the average end-to-end distance $R_\times$ (coil size) of a strand between neighboring cross-links. The magnitude of the retraction force can be estimated from Eqs. (1.43) and (1.44) as

$$f^* = k_B T/R_\times \tag{1.53}$$

so that the magnitude of $f_{retract}$ for a given shear strain is given by

$$f_{retract} = \gamma f^*. \tag{1.54}$$

In terms of $f_{retract}$ and $\psi_\times$, we can straightforwardly express the shear stress as

$$\sigma = \psi_\times f_{retract}. \tag{1.55}$$

By identification between Eq. (1.55) and Eq. (1.50) and knowing G from Eq. (1.51b), it follows that

$$\psi_\times \approx 1/pR_\times, \tag{1.56a}$$

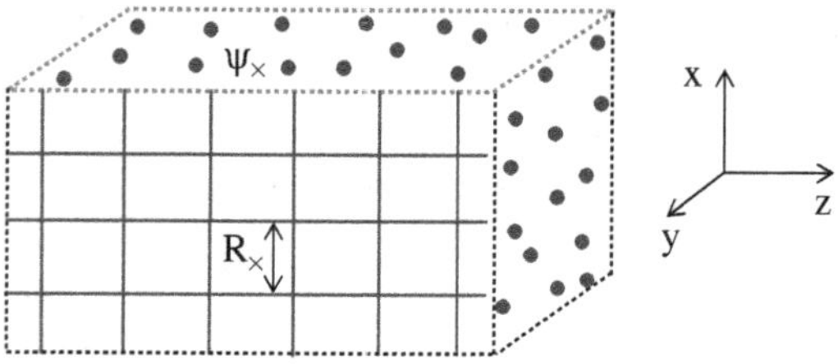

Figure 1.9 Microscopic view of a chain network where the basic building block has a coil size of R. In the case of simple shear, the areal density of strands is denoted by $\psi_\times$ of Eq. (1.56b), which would be unchanged in the presence of cross-linking.

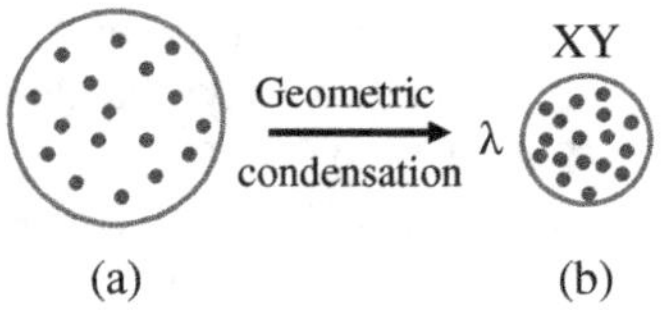

Figure 1.10 Illustration of the concept of geometric condensation that emerges during uniaxial extension where strand areal density in the XY plane increases from the initial ψ_x to $\lambda\psi_x$ upon extension by a factor of λ. A detailed conceptual and mathematical description of the geometric condensation concept can be found in the Appendix in Chapter 8.

and

$$G = \psi_\times f^*. \tag{1.56b}$$

Equation (1.56a) is a basic property of Gaussian chains in solvent-free melt state and can be independently derived as shown in Section 2.2.3.1. Equation (1.56a) is actually identical in form to that of Eq. (2.31). Moreover, the resemblance between Eq. (1.56b) and Eq. (2.30c) is also clear. Unlike Eq. (1.50), Eq. (1.55) clearly correlates the stress versus strain relation to the network structure. The coil size $R_\times$ of the strand controls the level of the entropic stress according to (1.53), and the areal density of such strands, $\psi_\times$ of Eq. (1.56a), as a prefactor, defines the intensity of the stress.

Similarly, Eq. (1.52) for uniaxial extension can be rewritten in terms of the areal density $\psi_\times$ in the undeformed state and the elastic retraction force $f_{retract}$, based on the image depicted in Fig. 1.10 where the effect of geometric condensation emerges from the cross-sectional areal shrinkage, via a factor given by the stretching ratio λ. Specifically, we have

$$\sigma = (\lambda\psi_\times)f_{retract}, \tag{1.57}$$

where the elastic force per strand is given by

$$f_{retract} = f^*(\lambda - 1/\lambda^2), \tag{1.58}$$

with f^* given in Eq. (1.53). Equation (1.57) can be rewritten in terms of the engineering stress as

$$\sigma_{engr} = \sigma/\lambda = \psi_\times f_{retract}. \tag{1.59}$$

The combination of Eq. (1.58) and Eq. (1.59), along with Eq. (1.56a) for G, and Eq. (1.53) for f^*, recovers Eq. (1.52).

This chapter aimed to provide the basic definitions in linear viscoelasticity and rheology, such as stress, strain, viscosity, and modulus. Through the introduction of the Maxwell model, we give the first example on how the material relaxation time is related to the material's viscosity and elastic modulus. While most of the information can be found in existing textbooks, the discussion in Section 1.2.5.2 appears to be original. At the end of this chapter, we also presented the classical theoretical treatment of rubber elasticity since no discussion of chain dynamics is necessary in Section 1.3.2. Section 1.3.3 is also nonstandard, where we showed how the expressions in the rubber elasticity theory can be cast in terms of intrachain retraction force $f_{retract}$ and the areal density $\psi_\times$ of cross-linked subchains.

References

1 Ferry JD. Viscoelastic properties of polymers. John Wiley & Sons; 1980.
2 Tschoegl NW. The phenomenological theory of linear viscoelastic behavior: an introduction. Springer-Verlag; 1989.
3 Shaw MT, MacKnight WJ. Introduction to polymer viscoelasticity. John Wiley & Sons; 2005.
4 Cheng S, Lu Y, Liu G, Wang S-Q. Finite cohesion due to chain entanglement in polymer melts. Soft Matter. 2016;12(14):3340–51.

5 Wang S-Q, Ravindranath S, Wang Y, Boukany P. New theoretical considerations in polymer rheology: Elastic breakdown of chain entanglement network. J Chem Phys. 2007;127(6):064903.
6 James HM, Guth E. Theory of the elastic properties of rubber. J Chem Phys. 1943;11(10):455–81.
7 James HM, Guth E. Simple presentation of network theory of rubber, with a discussion of other theories. J Polym Sci. 1949;4(2):153–82.
8 Treloar L. The elasticity and related properties of rubbers. Rep Prog Phys. 1973;36(7):755.
9 Treloar LRG. The physics of rubber elasticity. Oxford University Press, USA; 1975.
10 Wang S-Q. On chain statistics and entanglement of flexible linear polymer melts. Macromolecules. 2007;40(24):8684–94.
11 De Gennes P-G. Scaling concepts in polymer physics. Cornell University Press; 1979.
12 Smith SB, Finzi L, Bustamante C. Direct mechanical measurements of the elasticity of single DNA molecules by using magnetic beads. Science. 1992;258(5085):1122–6.
13 Rief M, Oesterhelt F, Heymann B, Gaub HE. Single molecule force spectroscopy on polysaccharides by atomic force microscopy. Science. 1997;275(5304):1295–7.
14 Fetters L, Lohse D, Richter D, Witten T, Zirkel A. Connection between polymer molecular weight, density, chain dimensions, and melt viscoelastic properties. Macromolecules. 1994;27(17):4639–47.

2

Molecular Characterization in Linear Viscoelastic Regime

We must go beyond the phenomenological description of linear viscoelasticity (LVE) to inquire about how the relaxation time depends on molecular parameters. In this chapter, we will start with a review of the molecular dynamic theories due to Kirkwood, Rouse, and Zimm for dilute polymer solutions. Rheological features of chain entanglement will be presented next, before describing transient network models and molecular-level models for polymer entanglement. Then we give a basic description of the reptation idea and the tube theory for well-entangled polymeric liquids before ending the chapter with a brief discussion of the time–temperature superposition principles and related topics.

To move beyond the Maxwell model, we need to have molecular-level theories to express the relaxation times in terms of molecular parameters including molecular weight and chain architecture. The Maxwell model gives us a hint that an expression for viscosity η may not be as formidable to derive as is an expression for η of a simple liquid. Equation (1.31) suggests that the Newtonian η_0 is related to a spectrum of relaxation times, weighted by the corresponding moduli. The molecular theories of polymer dynamics and rheology are also based on the same philosophy, gathering intractable unknowns into a few phenomenological parameters. For example, dynamics of polymer solutions are expressed to be related to the solvent viscosity η_s, which is treated as an experimental input.

2.1 Dilute Limit

2.1.1 Viscosity of Einstein Suspensions

Albert Einstein (1879–1955) first showed how a liquid viscosity η_s can be altered upon addition of a small number of solid spherical particles.[1,2] In the dilute limit, it does not matter whether or not the particles perform strong Brownian movements. The specific viscosity, which is the ratio of the viscosity increment due to the particles to the solvent viscosity η_s, was found by Einstein to be proportional to the volume fraction ϕ of solid particles:

$$[\eta(\phi) - \eta_s]/\eta_s = 2.5\phi. \tag{2.1}$$

This result simply indicates that the effect of particles is additive as expected in the dilute limit. It turns out that, apart from some subtle corrections, viscosity of a dilute solution made of linear flexible polymer chains is similarly related to the solvent viscosity and to the pervaded volume fraction of polymer chains.

Nonlinear Polymer Rheology: Macroscopic Phenomenology and Molecular Foundation, First Edition. Shi-Qing Wang.

Companion Website: www.wiley.com/go/nonlinearpolymerrheology

2.1.2 Kirkwood–Riseman Model

Kirkwood and Riseman[3] had the first right idea about how to derive the intrinsic viscosity $[\eta]$ of a dilute polymer solution. It was clear to them that intrachain hydrodynamic interactions are important. Hydrodynamic interactions cause the motions of different segments of a flexible linear chain to be hydrodynamically correlated. The movement of a segment perturbs the velocity field around any other segment in the chain. Since a modern description of the Kirkwood–Riseman theory is available in the literature,[4] we only cite the final Kirkwood–Riseman expression for $[\eta]$

$$[\eta] = \lim_{c\to 0}[(\eta - \eta_s)/\eta_s c] = \Phi_0 N_a \langle \mathbf{R}^2\rangle^{3/2}/M, \tag{2.2}$$

where the prefactor Φ_0 is 0.477 in the infinitely long chain limit and N_a is Avogadro's constant equal to $6.022 \times 10^{23}\ \mathrm{mol}^{-1}$. The number density of polymer chains, ρ_p, is related to the polymer mass concentration c as

$$\rho_p = (c/M)N_a, \tag{2.3}$$

where we adopt the notation already seen in Eq. (1.47). Equation (2.2) actually has the same form as the Einstein formula Eq. (2.1) because the pervaded volume of a single chain is on the order of $\langle \mathbf{R}^2\rangle^{3/2}$. Plugging Eq. (2.3) into Eq. (2.2) allows us to rewrite the latter as

$$\eta_{sp} \equiv (\eta - \eta_s)/\eta_s - \rho_p\langle \mathbf{R}^2\rangle^{3/2} \sim \phi, \tag{2.4}$$

where ϕ is the volume fraction pervaded by molecular coils. As a consequence of hydrodynamic interactions that were accounted for by Kirkwood–Riseman, a linear flexible polymer as a Gaussian coil behaves nearly like an impenetrable sphere. Equivalently, the solvent within the space occupied by the coil is essentially immobile and moves along with the chain as the chain diffuses so that a dilute polymer solution behaves like a dilute suspension of spherical particles. The Kirkwood–Riseman model has been significantly extended by Phillies and students as recently reviewed by Phillies.[5]

2.1.3 Zimm Model

The Kirkwood–Riseman model does not depict time-dependent viscoelastic behavior. Zimm[6] extended the Rouse model[7] to obtain a full dynamic theory for dilute polymer solutions. The Rouse and Zimm models are based on bead-spring models, in which a flexible linear chain is modeled as a collection of N frictional *beads* of size l_K (conveniently taken to be the Kuhn length) and friction coefficient ζ, connected by elastic *springs* with Hookean spring constants equal to $3k_B T/l_K^2$. A cartoon of the model appears in Fig. 2.1. In a bead-spring model, both the viscous interactions of a chain with the solvent and chain elasticity are accounted for. Unlike the Rouse model, the Zimm model takes hydrodynamic interactions into account and is thus able to depict correctly the viscoelastic properties of dilute polymer solutions. In particular, it shows that the viscosity increment due to the polymer has the same form as Eq. (1.31) where Zimm found G to be independent of the mode index q and is of the form given in Eq. (1.51a) with ρ_p of Eq. (2.3). The Zimm relaxation times τ_q are given by

$$\tau_q = \tau_1/q^{3\nu}, \quad q = 1, 2, 3, \ldots \tag{2.5}$$

with longest relaxation time τ_1

$$\tau_1 \equiv \tau_Z = \eta_s R^3/(3\pi)^{1/2} k_B T. \tag{2.6}$$

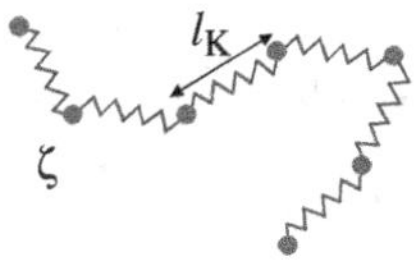

Figure 2.1 Illustration of a subchain in a bead-spring model.

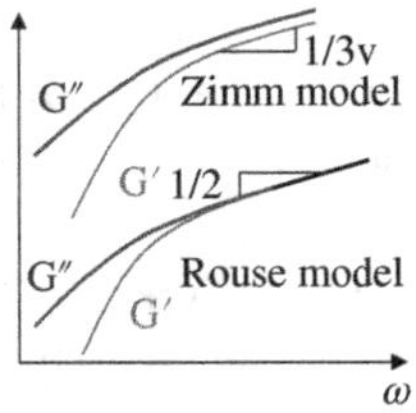

Figure 2.2 Storage and loss moduli G′ and G″ according to the Rouse and Zimm models where the Flory exponent ν is 1/2 for theta solutions and 3/5 for good solvents.

The original Zimm calculation only considers polymers near the theta (θ) condition so that $\nu = 1/2$ in Eq. (2.5). For notational simplicity, we denote the average chain end-to-end distance as R since Chapter 1. Similarly, the chain's radius of gyration will be denoted as $R_g = 6^{-1/2}R$. The expressions in Eqs. (2.5) and (2.6) are general. They include the case of polymers in good solvents, for which the Flory exponent $\nu = 3/5$.

Formulas (2.5) and (2.6) could also be derived from dimensional analysis on requiring that τ_1 must be a function of η_s, coil size R, and thermal energy k_BT. The only combination of these variables that produces a quantity with units of time is that given in Eq. (2.6). Dimensional analysis does not, of course, predict the numerical prefactor. The molecular weight dependence of τ_1 is apparently in agreement with experimental data.[8]

It is clear that the specific viscosity η_{sp} must have the same form as that appearing in Eq. (2.4). Actually, for a θ solvent, the Zimm model[9] yields Eq. (2.2) for the intrinsic viscosity [η], with $\Phi_0 = 0.425$. The more powerful part of the theoretical prediction is the storage and loss moduli $G'(\omega)$ and $G''(\omega)$, which have the same form as given in Eqs. (1.28) and (1.29) of the generalized Maxwell model.[8,9] It can be shown numerically that G″ stays above G′ at all values of ω, with a common limiting slope given by $1/3\nu$ for both good and theta solvents, respectively. These predictions are as sketched in Fig. 2.2, in agreement with experiment. In contrast, the Rouse model (presented below in 2.1.4) predicts that G′ merges with G″ at high frequencies, as shown in Fig. 2.2.

There is one anomaly that was not addressed in the original Zimm model and was subsequently studied in detail by Schrag and coworkers.[10] In some systems, the solution viscosity is smaller than the viscosity of the pure solvent. From an empirical viewpoint, the anomalous behavior of a dilute polymer solution having a smaller viscosity is rather straightforward to explain.[11] The behavior occurs only when the polymer has a lower glass transition temperature T_g than the solvent does. In this case, dissolution of polymer chains in the solvent amounts to plasticization of the solvent.

2.1.4 Rouse Bead-Spring Model

Although the Rouse model was intended to describe single-chain dynamics in dilute polymer solutions, its omission of hydrodynamic interaction (HI) ensures its failure. Interestingly, there are two cases where neglect of HI is not only acceptable but also reasonable. It is believed that the intrachain and interchain hydrodynamic interactions are effectively screened in polymer melts. Such screening is actually not much different from the screening of excluded volume interactions in concentrated solutions and melts. Indeed, in melts and concentrated solutions, chains mutually interpenetrate. As a consequence, a given chain no longer behaves hydrodynamically like a solid sphere. The frictional interactions of a linear chain with the solvent are additive, proportional to the chain length, namely, N. When the solvent is polymeric, even dilute solution was speculated by Daoud and de Gennes to become Rouse like:[12] When the solvent chain length is sufficiently long, it can screen the intrachain HI within an individual long chain, making it fully exposed to the viscous liquid medium so that the Rouse model becomes applicable.[13] Thus, the Rouse model turns out to be useful in describing dynamics of dense polymer systems. However, more recent research has raised questions about the validity of the screening concept.[5] In the following section, we will discuss a few basic concepts before depicting the Rouse dynamics.

2.1.4.1 Stokes Law of Frictional Force of a Solid Sphere (Bead)

A proper description of the bead-spring model requires an introduction of a basic concept about the viscous interaction between a solid object and a liquid medium. George G. Stokes (1819–1903) discovered that the force F required to make a spherical body of radius R travel at velocity V relative to its surrounding medium of viscosity η is proportional to V. Specifically, the Stokes' law states

$$F = \zeta V \equiv (6\pi\eta R)V. \tag{2.7}$$

Here, the friction coefficient ζ can actually be argued to be proportional to ηR on the basis of a dimensional analysis: ζ should only depend on η and R, and their product is the only combination to have the same unit as $[\zeta] = [F/V] = \text{mass/time}$.

2.1.4.2 Brownian Motion and Stokes–Einstein Formula for Solid Particles

A key transport phenomenon is Brownian motion. A sufficiently small particle can be seen to execute violent and rapid random movements in a liquid medium, as first found by Robert Brown (1773–1858) in his studies of pollen and other particles in water. Such random motion obeys a simple rule, as discovered by Einstein.[14] As a particle wanders around, it diffuses from its initial position $\boldsymbol{r}(0)$ at $t = 0$ to another position $\boldsymbol{r}(t)$ at time t. On the average, the distance between these two points is related to the elapsed time t by

$$\left\langle [\boldsymbol{r}(t) - \boldsymbol{r}(0)]^2 \right\rangle = 6Dt, \tag{2.8}$$

where the proportionality constant D is known as diffusion constant.

Einstein deduced how to compute D in terms of the particle size R and medium viscosity η. Recognizing that the origin of Brownian motion is thermal agitation, we can apply dimensional analysis to show that the unit of D, namely, $[(\text{length})^2]/[\text{time}]$, should be related to a unique combination of the thermal energy $k_B T$ and the friction constant ζ. Actually, the Einstein–Stokes formula for the diffusion constant of a spherical particle relates D and ζ in a reciprocal manner, namely

$$D = k_B T/\zeta = k_B T/6\pi\eta R. \tag{2.9}$$

Equation (2.9) is a powerful formula. For example, it indicates how long it takes for a particle to travel a distance comparable to its own size R. We may recognize such a time as a characteristic time scale τ for the Brown motion of the particle. According to a combination of Eqs. (2.8) and (2.9), the time taken by a particle to diffuse a distance comparable to its own size R is

$$\tau \sim R^2/D \sim \eta R^3/k_B T, \tag{2.10}$$

which has the same form as that seen in Eq. (2.6) and thus supports the idea that a flexible linear polymer chain can be regarded as a solid sphere of radius R in dilute solutions.

2.1.4.3 Equations of Motion and Rouse Relaxation Time τ_R

The subject of polymer dynamics is an extension of classical statistical mechanics. Rouse's treatment of polymers models one chain immersed in a continuum liquid. This view may be justifiable for a dilute solution. On one hand, since hydrodynamic interactions are important, the simplification in the Rouse model to omit hydrodynamic interactions renders the Rouse model unusable. On the other hand, since hydrodynamic interactions die out in polymer melts, the Rouse scheme may be borrowed to treat chain dynamics of polymer melts. In retrospect, the Rouse model provides a starting point to develop a mean-field approach to polymer melt dynamics, treating the surrounding of a test chain as a continuum. The equations of motion for the test chain of N beads (i.e., Kuhn segments) are based on application of Newton's second law and Stokes law to each frictional bead (the ith) connected to two springs:

$$m\frac{\partial^2 \boldsymbol{r}_i}{\partial t^2} = \boldsymbol{F}_i^{el} - \zeta\frac{\partial \boldsymbol{r}_i}{\partial t} + f_i(t), \quad i = 1, 2, \ldots, N, \tag{2.11}$$

where f_i is a random force related to the thermal agitation. Under the condition of constant temperature and volume, the elastic $\boldsymbol{F}_i^{el}$ can be evaluated from the Helmholtz free energy A of the test chain

$$\mathrm{A} = \mathrm{U} - \mathrm{TS} = \mathrm{U} + (H_1/2)\sum_{i=1}^{N} (\boldsymbol{r}_{i+1} - \boldsymbol{r}_i)^2, \tag{2.12}$$

as the spatial derivatives

$$\boldsymbol{F}_i^{\mathrm{el}} = \left(\vec{x}\frac{\partial}{\partial r_{ix}} + \vec{y}\frac{\partial}{\partial r_{iy}} + \vec{z}\frac{\partial}{\partial r_{iz}}\right)_{\mathrm{T,V}} \mathrm{A} = H_1(\boldsymbol{r}_{i+1} + \boldsymbol{r}_{i-1} - 2\boldsymbol{r}_i).$$

Here we have the Hookean constant $H_1 = 3k_\mathrm{B}\mathrm{T}/l_\mathrm{K}^2$, analogous to H in Eq. (1.44), and the second equality follows from the fact that the internal energy U should not change with the coordinates of the beads. Upon setting the left-hand side of Eq. (2.11) to zero, Eq. (2.11) is a Langevin equation. The rationale for neglecting the acceleration term is that the viscous force rapidly damps out any inertia. The time scale associated with the inertia, given by $\tau_\mathrm{v} = m/\zeta$, can be shown to be negligibly short relative to the Brownian time scale τ_0, given below in Eq. (2.17). Rouse solved these coupled differential equations by expressing the spatial vector $\boldsymbol{r}_i$ in terms of normal coordinates $\{\theta_q\}$

$$\boldsymbol{r}_i = \sum_{q=1}^{N} \boldsymbol{\theta}_q \cos[i(\pi q/N)], \tag{2.13}$$

and inserting this transformation into Eq. (2.11). All the differential equations for $\boldsymbol{\theta}_q$ are decoupled unlike Eq. (2.11) for $\boldsymbol{r}_i$,[9]

$$\tau_q(\mathrm{d}\theta_q/\mathrm{d}t) + \theta_q = \boldsymbol{\phi}(t), \tag{2.14}$$

where $\boldsymbol{\phi}$ is related to the random force. The relaxation times naturally emerge for the normal modes

$$\tau_q = (\zeta l_\mathrm{K}^2/3k_\mathrm{B}\mathrm{T})/[2\sin(\pi q/2N)]^2, \quad q = 1, 2, 3, \ldots, N. \tag{2.15}$$

For long chains, namely, $N \gg 1$, the longest relaxation time τ_1 is given by

$$\tau_1 \equiv \tau_\mathrm{R} = \tau_0 N^2, \tag{2.16a}$$

where the elemental time scale τ_0 is associated with the chain segment represented by the bead

$$\tau_0 = (\zeta l_K^2 / 3\pi^2 k_B T). \tag{2.17}$$

In the modern notation, τ_R is referred to as the Rouse (relaxation) time, dependent quadratically on the chain length N as shown in Eq. (2.16a). The time scales for the higher modes have the following simplified form for $N \gg q$ according to Eq. (2.15)

$$\tau_q = \tau_1 / q^2. \tag{2.18}$$

2.1.4.4 Rouse Dynamics for Unentangled Melts

The Rouse description of chain dynamics did not work for dilute linear polymer solutions but eventually found application in polymer melts. If we wanted a mean-field single-chain approach to treat polymer dynamics of unentangled melts, we could invoke the Rouse model because it may be the only single-chain treatment that we could formulate. Any alternative has yet to be developed sufficiently to offer specific, quantitative improvement.[15–17] Although a single-chain mean-field approach might be too simple and naïve, we have no alternative.

According to the Rouse model[7,9] the Newtonian viscosity η_0, commonly known as zero-shear viscosity, also has the same form as Maxwell model formula Eq. (1.31), involving a summation over the product of relaxation times $\{\tau_q\}$ and moduli $\{G_q\}$. It turns out[9] that $G_q = G_R$ is independent of the normal mode index q. Along the same line of reasoning given in the paragraph beneath Eq. (1.47), G_R has the dimension of [energy]/[volume]. It should thus involve $k_B T$ and a molecular volume such as the physical volume occupied by a chain:

$$\nu_M \equiv M/\rho N_a = pR^2. \tag{2.19}$$

Here, ρ is the mass density of a polymer melt and p is known as the packing length.[18,19] Equation (2.19) may also be said to provide the definition for p. In terms of the packing length, we can argue based on dimensional analysis that

$$G_R \approx k_B T/\nu_M \equiv k_B T/pR^2 \equiv \rho RT/M, \tag{2.20}$$

where $R = k_B N_a$ is the gas constant. This Rouse modulus G_R is inversely proportional to the molecular weight or chain length. Insertion of Eq. (2.20) and (2.16–2.18) into Eq. (1.31) produces

$$\eta_R = G_R \sum_{q=1}^{N} \tau_q = (\rho RT/M)\tau_1 \sum_{q=1}^{N} q^{-2} \equiv \zeta N/18p, \tag{2.21}$$

where $\sum_{q=1}^{N} q^{-2} = \frac{\pi^2}{6}$ for $N \gg 1$. Thus, a Rouse melt has a zero-shear viscosity η_0 that is essentially proportional to the total friction constant ζN, and thus to chain length N. Here, ζ is a coarse-grained phenomenological parameter from the bead-spring model, whereas p is a molecular parameter that has precise values for various polymers.[18]

The difference between various polymers is reflected in the coarse-grained phenomenological parameter ζ, which is the friction coefficient assigned to each Rouse bead in the bead-spring model. The packing length p in Eq. (2.19) is a molecular length related to the chain thickness and is typically between 0.1 and 1 nm.[18] Equation (2.21) may be viewed as a generalized Stokes formula, treating a Rouse chain as a Stokes object: For a linear chain of thickness p, the total friction constant $\Xi = \zeta N$ is related to the medium viscosity η_R as $\Xi = 18p\eta_R$. This form may be contrasted with the familiar formula $\zeta = 6\pi R\eta_0$ for a sphere of radius R in a medium of viscosity η_0.

The packing length p is thus the relevant length scale in the formula linking the friction of a Rouse chain to the viscosity of the Rouse melt.

Significant extensions of the Rouse theory have recently been made to describe the dynamics of unentangled melts.[20–22] The work explored the concept of viscoelastic hydrodynamic interactions. A detailed discussion of these developments is well beyond the scope of this book.

2.1.5 Relationship between Diffusion and Relaxation Time

On molecular time and distance scales, thermal agitation is very strong, causing rapid Brownian motion and fast chain relaxation. As discussed in Eq. (2.8), the Brownian motion can be quantified by a diffusion constant D. Diffusion and relaxation have the same origin in Rouse melts; the two processes are related. The ensemble-averaged time needed for a polymer chain to diffuse through a distance of its own coil size R is comparable to the overall chain relaxation time τ_R, namely

$$D_R \sim R^2/\tau_R. \quad (2.22)$$

This result could also simply be obtained from the argument that D_R has dimensions of $[\text{length}]^2/[\text{time}]$ in accordance with the fundamental definition of D in Eq. (2.8). In the case of Rouse melts, insertion of Eq. (2.16a) into Eq. (2.22) yields

$$D_R \sim k_B T/\zeta N, \quad (2.23)$$

which is Einstein–Stokes formula Eq. (2.9), given the total friction coefficient of a chain is $\Xi = \zeta N$. The diffusion constant D_R for a Rouse chain is thus inversely proportional to the chain length. Conversely, by making an argument that D_R of a Rouse chain is given by the Einstein–Stokes formula of Eq. (2.23), the dimensional relationship of Eq. (2.22) immediately yields an expression

$$\tau_R \sim \Xi R^2/k_B T, \quad (2.16b)$$

for the Rouse relaxation time. Equation (2.16b) has the same form as that of Eq. (2.17) because $\Xi = N\zeta$ is the friction coefficient for the entire chain of size R. Finally, a combination of Eq. (2.21) and Eq. (2.23) gives

$$p \sim k_B T/18\eta_R D_R,$$

which may be used (if so desired) to measure p from measurements of η_R and D_R at some convenient temperatures well above the glass transition temperature.

In the case of the Zimm model for dilute solutions, insertion of Eq. (2.6) for τ_Z into dimensional form $D_Z \sim R^2/\tau_Z$ yields

$$D_Z \sim k_B T/\eta_s R. \quad (2.24)$$

This equation is in the same form as the Einstein–Stokes formula Eq. (2.9) and suggests that the Gaussian chain is equivalent to a sphere of size R. In contrast to Rouse chains, flexible linear chains in dilute solution diffuse with diffusion constant that decreases with N as $D_Z \sim N^{-\nu}$, with ν equal to 1/2 in a θ solvent and up to 3/5 for a chain in a good solvent. The conceptually correct quantity to appear in Eq. (2.24) is actually a hydrodynamic radius R_H, instead of R. In the limit that the coil truly behaves hydrodynamically like a hard impenetrable sphere, Eq. (2.24) may be valid. This nondraining limit is approachable when the chain length is exceedingly long. As discussed in Section 2.3.4, chain diffusion and stress relaxation are rather different processes and may not be simply related in entangled polymer melts and concentrated solutions.

2.2 Entangled State

The Rouse model was accepted as a description of polymer melts perhaps because it predicts a linear dependence of the zero-shear melt viscosity η_0 on the molecular weight M, in agreement with experiment on a variety of linear polymers. Figure 2.3 illustrates the linear variation below a critical molecular weight M_c. Thus, a chain can perhaps indeed freely execute Rouse dynamics in a sea of the same chains so long as $M < M_c$. The experiment shows that the Rouse model is not applicable for chains having $M > M_c$. This result is not very surprising given that linear chain-like macromolecules are random coils, and thus, chains in melts must strongly interpenetrate one another, more strongly the longer the chains. Given our knowledge about the mechanical behavior of chemically cross-linked polymer melts, it is reasonable to expect melts and concentrated solutions of high-molecular weight to be rubber like, at least temporarily. In the following section, we review the well-known phenomenology in the linear response regime, which suggests that melts and concentrated solutions of high-molecular weight are indeed transiently elastic. Somehow topological interchain interactions not only cause the zero-shear viscosity to depend much more strongly on M than it does in Rouse melts, as shown in Fig. 2.3, but also lead to the formation of a chain network. This cross-linking like effect is often referred to as chain entanglement.

2.2.1 Phenomenological Evidence of chain Entanglement

A system (melt or concentrated solution) of linear chains produces rubbery elasticity if the chains are chemically cross-linked. The classical account given in Section 1.3 suggests entropic origin for the observed rubber elasticity of chemically cross-linked chains. Let us consider the same system except that no cross-linking is present. If the chains are long enough, namely, the polymer's molecular weight is sufficiently high, transient elasticity appears.

There are at least four additional ways to demonstrate how elasticity arises in non-cross-linked polymeric liquids. Drawing analogy with vulcanized rubbers, the display of transient elasticity may be said to arise from intermolecular chain uncrossability. We may say that chain uncrossability produces *entanglement*.

2.2.1.1 Elastic Recovery Phenomenon

A classical classroom demonstration used by every polymer rheologist is to roll a high-molecular-weight (MW) poly(dimethyl siloxane) melt into a ball and throw it to the floor rapidly. It bounces up upon contact with the floor surface like a rubber ball although it may not be quite as bouncy. The initial response is elastic even though the PDMS chains are not cross-linked. This effect can be more precisely depicted by considering an elastic recovery experiment as follows. Let us impose a sudden simple shear γ or extension $(L - L_0)/L_0$ to a polymer melt or solution as shown in Fig. 2.4(a,b), which may be realized by either applying a given stress for a period or displacing one surface for a distance of γH or $(L - L_0)$. If the

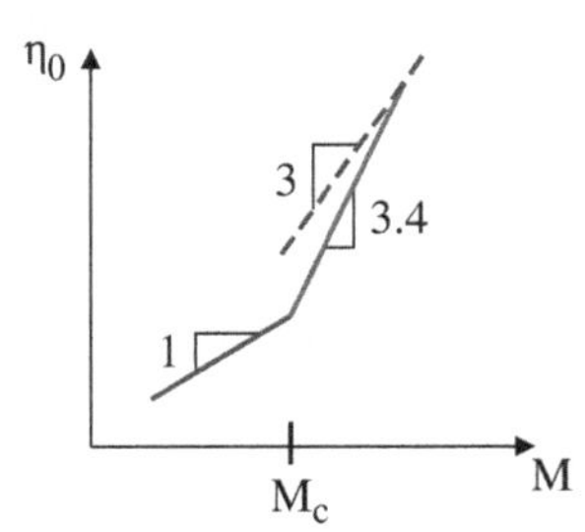

Figure 2.3 Zero-shear viscosity η_0 versus molecular weight M on a log–log plot. Here, the critical molecular weight M_c separates Rouse melts (after iso-free volume corrections) from entangled melts. The dashed line showing exponent 3 is the idealized scaling behavior, to be discussed in Section 2.3.2. Due to finite chain length effects, η_0 is always smaller than its ideal value.

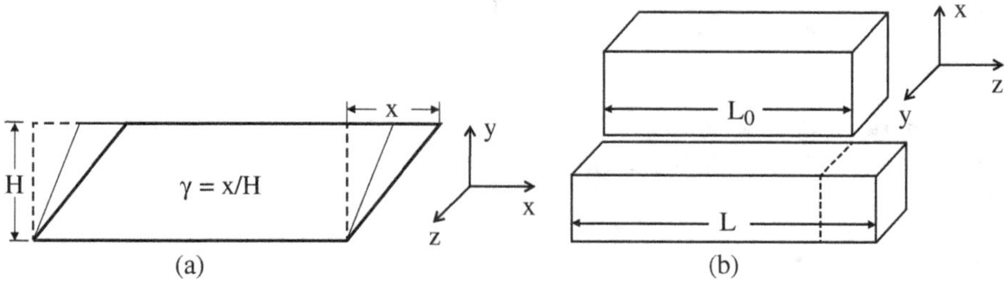

Figure 2.4 Illustration of elastic recovery under (a) simple shear and (b) uniaxial extension where the dashed lines are used to show the equilibrium positions. Full recoil corresponds to the deformed sample returning to the shapes indicated by the dashed lines. The thin lines reveal the positions of the samples after (partial) elastic recoil in (a).

deformation is produced on a time scale much shorter than the longest (i.e., terminal) relaxation time τ of the polymer, the sample should not undergo flow during the deformation. When the moving surface or end is suddenly released so that no external forces are acting on the sample, an entangled polymeric fluid could return to its initial shape as depicted by the dashed lines, exhibiting complete elastic recoil. This return toward its initial position is possible because the sample is initially incapable of flow as we demonstrated in Section 1.2.3 for any viscoelastic Maxwellian materials.

Under other conditions, namely, when the total strain can be viewed to consist of a combination of elastic deformation and flow, the recovery will be partial, namely, the sample only partially recoils, as indicated by the inclined thin lines in Fig. 2.4(a). The ability of a non-cross-linked polymer liquid to resist irrecoverable deformation (flow), and to display elastic response, is evidence that the coiled chains in the liquid are unable to relax their deformed states by passing around one another on time scales appreciably shorter than τ. This inability is said to be due to chain entanglement. Partial or complete elastic recoil supports the idea that there is "chain entanglement".

2.2.1.2 Rubbery Plateau in Creep Compliance

The ability of entangled polymer solutions and melts to display elasticity also appears in a classical creep experiment where a constant small stress is applied to a sample, either in simple shear or in uniaxial extension. At sufficiently high-molecular weights, the sample reacts to the applied force by reaching, within a very short period, a fixed level of deformation. This deformation level remains little changed (i.e., the strain reaches a plateau) until sufficient time elapses to allow creep. The sample starts to flow when $t \sim \tau$, as schematically depicted in Fig. 2.5. For actual experimental data, see Fig. 16.8. We can also regard the experiment as a way to reveal a transformation from elastic deformation to flow, showing how given sufficient time the system yields, allowing molecular diffusion to unlock the entanglement network. A creep test introduces a quantity known as *creep compliance* $J(t)$. The creep compliance is well defined in the linear response regime; it is proportional to the strain $\gamma(t)$ arising from a given applied stress σ_0, namely

$$J(t) = \gamma(t)/\sigma_0. \tag{2.25}$$

In the linear regime, the strain is linearly proportional to the applied stress σ_0; therefore, J in Eq. (2.25) is independent of σ_0. J_e^0 in Fig. 2.5 is the *steady-state compliance*; it reflects a saturated level of elasticity present during steady shear. Upon removal of an applied shear stress σ_0, the sample under the stress-free condition recoils by an amount $\sigma_0 J_e^0$, as sketched in Fig. 2.5. The elastic recoil by $\sigma_0 J_e^0$ is a clear sign of rubber elasticity in a flowing system. The presence

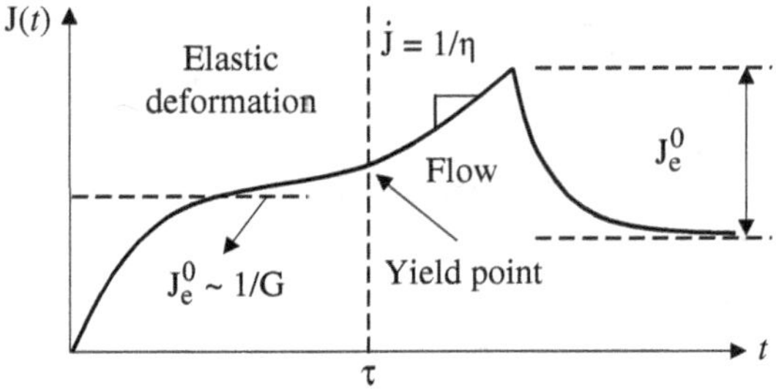

Figure 2.5 Time-dependent compliance of a viscoelastic material during and after creep in the linear response regime, where the elastic J_e^0 is comparable to the recoverable compliance and approximately given by the reciprocal elastic modulus G. Chain entanglement produces the creep plateau before flow that begins beyond the time marked by the vertical dashed line. So the response to the imposed stress in such a creep test is initially elastic and eventually viscous; the transition occurs at the vertical dashed line.

of elastic recoil in an entangled polymeric liquid indicates that the chains are interconnected and have formed an entanglement network as seen in Fig. 2.9. Elastic recoil is circumstantial evidence for chain entanglement. In passing, we note that at high stresses, yielding can occur much earlier than τ, as discussed in Section 7.4.

2.2.1.3 Stress Relaxation

The most familiar indication of chain entanglement comes from a stress relaxation experiment, namely, a step strain test that has been introduced in Section 1.1.2. Schematically, in the linear response regime, the relaxation modulus $G(t)$ of an entangled polymeric fluid shows a rubbery plateau, reflecting a very long-lived stress, as sketched in Fig. 2.6. When τ is long and easily determined experimentally, it is rather impressive that a polymeric liquid can sustain a great amount of stress over times $t < \tau$. Only solids are capable of maintaining a finite stress indefinitely without *ongoing* deformation. Thus, the lack of stress decline for a period of τ is truly noteworthy and can only occur if the material is solid like.

2.2.1.4 Elastic Plateau in Storage Modulus G′

The most popular way to reveal the presence of chain entanglement is to conduct small amplitude oscillatory shear (SAOS) frequency sweep measurements. Under proper experimental conditions, non-cross-linked monodisperse high-molecular-weight polymers display frequency dependences of the storage and loss moduli G' and G'' that are nearly identical to those depicted by the generalized Maxwell model at low frequencies, as sketched in Fig. 2.7. By analogy with the Maxwell model, we can regard the reciprocal of the first crossover frequency ω_c as the terminal relaxation time τ as indicated in Fig. 2.7. At frequencies above

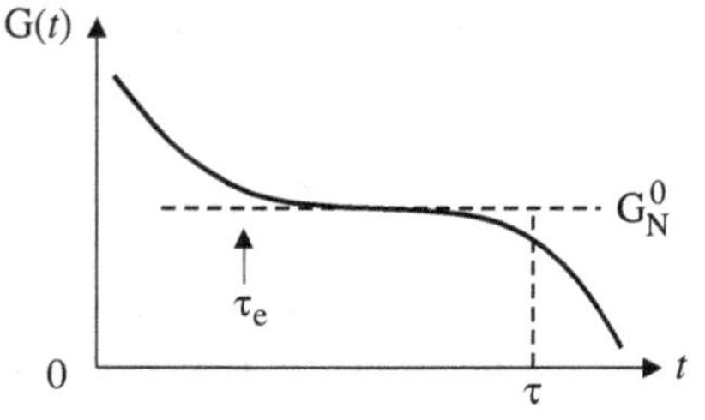

Figure 2.6 Relaxation modulus $G(t)$ of an entangled polymer melt as a function of time t, showing an elastic plateau modulus G_N^0 and two characteristic times, the disengagement time τ and tube confinement time τ_e.

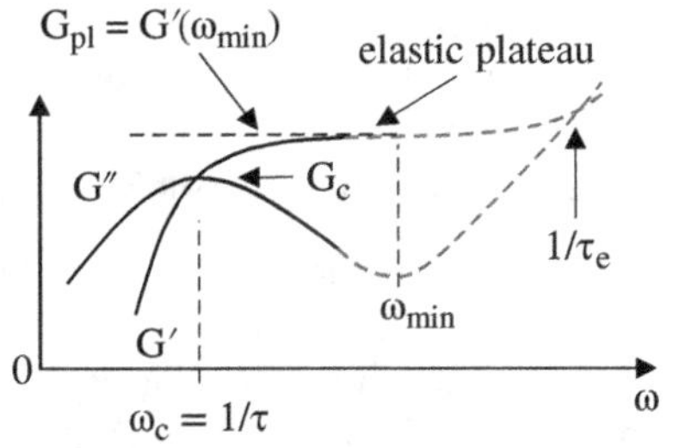

Figure 2.7 Storage and loss moduli G′ and G″ as a function of the oscillation frequency, typically obtained from frequency sweep of small-amplitude oscillatory shear. For monodisperse entangled solutions or melts, the crossover frequency ω_c can be taken as the reciprocal terminal relaxation time or reptation time or disengagement time. The magnitude G_c at ω_c is about a factor of 3.5 smaller than G_{pl} for monodisperse well-entangled polymer solutions and melts. The Maxwell model (solid lines) may capture the behavior at low frequencies. The complete curves are depicted by adding the dashed lines.

ω_c, the sample behaves as if it is a solid with elastic plateau modulus G_{pl}. This emergence of elasticity is attributed to chain entanglement. A real entangled polymer melt with a narrow molecular weight distribution has a high-frequency response beyond the Maxwell curves as shown by dashed lines of Fig. 2.7. The second crossover frequency where G′ and G″ cross is conventionally labeled as $1/\tau_{e,}$ whose definition will become clear in Section 2.3.

Chain entanglement makes polymeric liquids respond like a vulcanized rubber and produces the four leading different rheological responses in an entangled polymer. The four features are illustrated in Figs. 2.4–2.7: (a) elastic recoil from a deformed state, (b) elastic deformation under controlled stress before onset of flow, (c) stress plateau during relaxation from a deformed state, (d) a wide range of frequencies over which an elastic modulus emerges in oscillatory shear. The most realistic way to envisage an entangled polymer melt made of linear flexible chains is to look at a bowl of crystal noodles photographed in Fig. 2.8. Basically, the inability of long chains to pass around or through one another forces each chain to remain in its initial position for a long period until the chains move away from one another. In the absence of any external deformation or when the external deformation is small either in rate or in magnitude, chains are not forced to pass by one another; the chains move around only because of their Brownian motions. Intuitively, the longer the chain, the more time it takes to diffuse. Before any significant diffusion can take place, namely, on time scales much shorter than τ, chains are locally trapped as though chemical cross-links were present. Consequently, melts of long-chain polymers are capable of displaying elastic responses as depicted in Figs. 2.4–2.7. While this effect of chain entanglement is understandable by analogy with rubber elasticity, it is less obvious how entanglement (or molecular jamming) produces the strong molecular weight dependence of the relaxation time τ and viscosity η, as shown in Fig. 2.3. Thus, apart from the question of what chain entanglement is, there are questions including why the chain entanglement produces these rheological consequences as shown in Figs. 2.3–2.7. Many such elusive questions lingered around for decades[23–26] until P. G. de Gennes (1932–2007) came up with a simple

Figure 2.8 A snapshot of an entangled melt, magnified 1 million times – a photo of a bowl of crystal noodle.

explanation known as the idea of chain reptation in a tube.[27] Prior to de Gennes, the effect of chain entanglement on polymer dynamics had only been modeled with limited success.

2.2.2 Transient Network Models

Extending the success of the classical rubber elasticity theory, the network picture may be applied to depict non-cross-linked systems provided the junctions are allowed to have a finite lifetime τ as shown in Fig. 2.9. Here, each entanglement strand is modeled as a Gaussian subchain with an end-to-end distance l_{ent}. This end-to-end distance is conventionally known as an entanglement length. In a transient network model, the "cross-links" are perceived to renew on a phenomenological time scale. The first attempt by Green and Tobolsky[28] and subsequent improvements by Lodge[29] assumed that the relaxation time τ is independent of the external conditions. With this assumption, behaviors similar to those of the Maxwell model and generalized Maxwell model were accounted for. However, these models only predict linear viscoelastic properties. To describe any nonlinear behavior, including the decrease in viscosity with increasing shear rate (shear thinning), one must depict how the junction lifetime τ depends on deformation of the strands between junctions. Yamamoto considered such a scenario, driven by a desire to describe shear thinning.[30,31] The merit of these transient network models lies in their recognition that polymers of high molecular weight form a molecular network (i.e., entanglement network). However, like the Maxwell model, these network models cannot express τ as a function of molecular parameters and do not describe the molecular origin of chain entanglement.

The Maxwell relaxation time τ of Eq. (1.19) can also arise from consideration of the transient network picture of Fig. 2.9. During shear at a rate $\dot{\gamma}$, a transient network can only undergo an amount of elastic shear deformation given approximately by $\gamma_{el} = \dot{\gamma}\tau$, resulting in a shear stress level given by $\sigma = G\gamma_{el}$ (Eq. (1.7)), where G is an elastic modulus. The combination of the two expressions ($\gamma_{el} = \dot{\gamma}\tau$ and $\sigma = G\gamma_{el}$) amounts to $\sigma \simeq G\tau\dot{\gamma}$ whose identification with the Newton's law of Eq. (1.8) leads to $\eta = G\tau$, which is Eq. (1.19). To describe nonlinear responses such as shear thinning, one assumes in an *ad hoc* manner that τ_{eff} is a decreasing function of $\dot{\gamma}$ so that

$$\eta(\dot{\gamma}) = G\tau_{eff}(\dot{\gamma})$$

is a decreasing function of $\dot{\gamma}$. Thus, generalizing Eq. (1.8), we have

$$\sigma = \eta(\dot{\gamma})\dot{\gamma} < \eta_0\dot{\gamma},$$

where $\eta_0 = \eta(\dot{\gamma} \to 0)$ is the zero-shear viscosity. There are also modern attempts to construct network models[32–37] such as slip-link models[33,35,36] to describe entangled polymer solutions and melts.

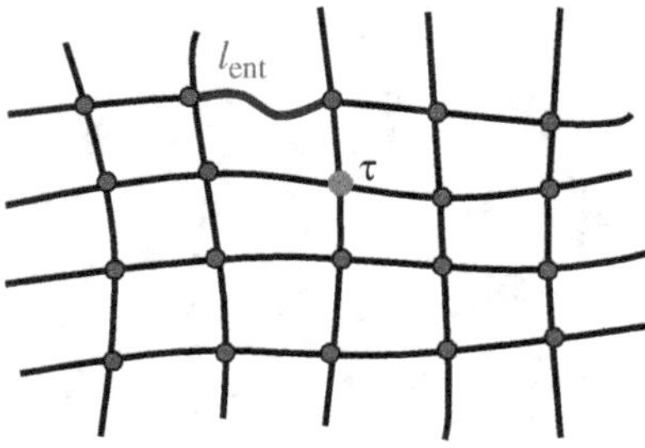

Figure 2.9 Transient network picture showing a mesh size l_{ent} and effective junctions with an average lifetime τ.

2.2.3 Models Depicting Onset of Chain Entanglement

Where does entanglement coupling come from in melts? It arises from the fractal property of a Gaussian chain: As shown in Eq. (1.39b), the mass of a random-walk (Gaussian) chain, which is linearly proportional to the chain length N, scales quadratically with its linear dimension (coil size R). Since the fractal dimension of a chain in a melt is two, and the chains in a melt do not collapsed into space-filling objects, many chains may interpenetrate into one another. Thus, in the pervaded volume (on the order of R^3) of a given long chain, there must be many other chains. Conceivably, all the chains within the pervaded volume of a given chain are topologically trapped and dynamically jammed. In the following section, we review two leading models proposed to visualize chain entanglement.

2.2.3.1 Packing Model

It is reasonable to think that chain entanglement arises from the fact that different chains must share the same space and interpenetrate. The longer the chains are, the more chains are in the same pervaded volume $V_{perv} \sim (4\pi/3)R^3$, making it more difficult for any given chain to move about without dragging other chains with it. When is a chain sufficiently long to get so "entrapped"? It should depend on when chains are sufficiently interpenetrated. Thus, it should be feasible to characterize the onset of chain entanglement.

In the mid-1980s, several groups independently arrived at a similar way to think about the onset of the entangled state. This approach is today known as the packing model for chain entanglement. Rault,[38] Heymans,[39] Lin,[40] Kavassalis, and Noolandi[41] all had similar ideas. The concept is that chain entanglement occurs in a monodisperse melt of molecular weight M when enough interchain topological intertwining builds up with increasing M. As the pervaded volume

$$V_{perv}(n) = (4\pi/3)R_g^3 \tag{2.26}$$

increases with the chain length or number n of backbone bonds, the number Q of chains required to fill it up increases according to

$$Q(n) = V_{perv}/v_M = (2\pi/6^{1/2}9)R/p, \tag{2.27}$$

where the physical volume v_M of the chain is given by Eq. (2.19). The end-to-end distance R of a flexible linear chain can be approximately related to the number of backbone bonds n through the characteristic ratio $C_n = R^2/nl^2$, which accounts for the deviation of real chains from the ideal flexible linear Gaussian chains. Since $C_n \to C_\infty$ for $n \gg 1$, we can estimate $R = \sqrt{C_\infty nl} = \sqrt{6}R_g$ in the long chain limit, where l is the backbone length.

For Gaussian chains, $Q \sim M^{1/2}$. Sufficient topological interchain interference occurs to cause entanglement when Q increases to a critical value Q_e. The conjecture of the packing model is that Q_e is a relatively universal constant independent of chemical details. Therefore, as molecular weight M increases, at the onset of chain entanglement all linear flexible polymers have about the same value of Q_e. Setting Q to Q_e at $M = M_e$ or $n = n_e$, we rewrite Eq. (2.27) as

$$l_{ent}/p \sim Q_e = Q(n_e) = \text{constant for most linear flexible polymers.} \tag{2.28}$$

The coil size at the entanglement molecular weight M_e, namely, the end-to-end distance at n_e, is given in Eq. (1.39b) by setting N to N_e:

$$R(n_e) \equiv l_{ent} = \sqrt{N_e l_K} = \sqrt{C_\infty n_e l}. \tag{2.29}$$

This coil size l_{ent} is known as the entanglement spacing. Equation (2.29) assumes that this subchain is still long enough to be Gaussian and shows that there are N_e Kuhn segments in such an entanglement strand. In conclusion, the packing conjecture (Eq. (2.28)) leads to the result that l_{ent} is proportional to p.

Since Q_e in Eq. (2.28) is a universal constant, Eq. (2.28) shows that the coil size l_{ent} at the entanglement molecular weight M_e is larger for a polymer with greater packing length p. According to the definition of p given in Eq. (2.19), p depicts the chain thickness.[22] Thus, bulkier chains need to be larger for entanglement to occur. Since most common linear chains are comparably flexible and have similar coil size at a given n,[22] Eq. (2.28) asserts that n_e is larger for bulkier chains, namely, chains with larger p. A great number of different polymers indeed appear to obey Eq. (2.28) as shown in Fig. 2.10(a,b). For these figures, the packing length p was evaluated according to Eq. (2.19), while the entanglement spacing l_{ent} was estimated from the rheological determination of the (elastic) melt plateau modulus of the polymer melts

$$G_N^0 \approx \rho RT/M_e = k_B T/l_{ent}^2 p. \tag{2.30a}$$

The equality follows from Eq. (2.19) and the gas constant $R = N_a k_B$. Equation (2.30a) resembles Eq. (2.20) and can be derived by dimensional analysis in the same manner: $G_N^0 \sim k_B T/\nu_e$ where the physical volume of an entanglement strand is $\nu_e = p(l_{ent})^2$. The good correlations in Fig. 2.10(a,b) confirm the conjecture of the packing model in Eq. (2.28): $l_{ent} \sim p$. This conjecture can be inserted into Eq. (2.30a) to produce

$$G_N^0 \approx k_B T/p^3 \tag{2.30b}$$

for all linear flexible polymer melts. Equation (2.30b) is another way to present the experimental data in support of the packing model. In passing, we note that the result given in Eq. (2.28) has also been obtained by Schweizer and Szamel based on the polymer-mode-coupling theory.[43,44] Thus, the packing conjecture appears to have sound theoretical underpinning.

The concepts encompassed by the packing model also allow us to express the state of entanglement in an alternative way. We can define a number of entanglement strands per unit area, ψ_{ent}, in equilibrium. The quantity ψ_{ent} can be expressed in terms of $Q(n_e)$ of Eq. (2.28). Let us consider taking a cross-sectional cutting through the center of the pervaded volume of a single strand. Over an area of πl_{ent}^2, the cross-section intercepts Q_e strands. Therefore, we have the following relationship

$$\psi_{ent} = Q_e/\pi l_{ent}^2 \sim 1/p l_{ent}. \tag{2.31}$$

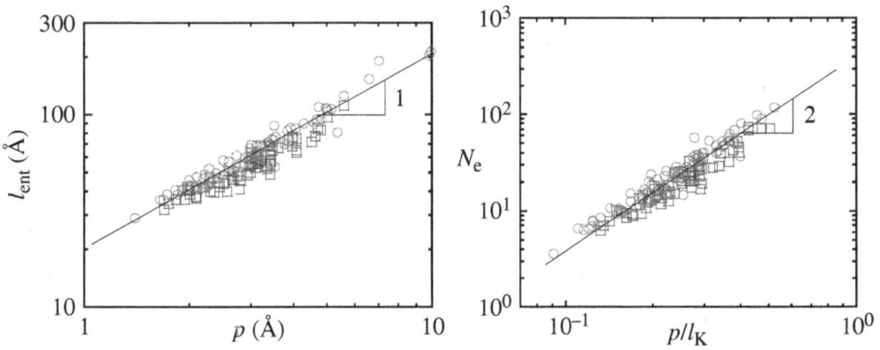

Figure 2.10 (a) Linear dependence of entanglement spacing l_{ent} on packing length p. (b) Quadratic dependence of N_e on p/l_K based on the same data as those in (a). *Source:* Data taken from Fetters *et al.* 1994 [18]; Wang 2007 [19] & Fetters *et al.* 2007 [42].

The conjecture of the packing model, expressed in Eq. (2.28), implies $\psi_{ent} \sim 1/p^2$. Moreover, in terms of ψ_{ent}, we can reexpress the plateau modulus Eq. (2.30a) as

$$G_N^0 = \psi_{ent} f_{ent}, \tag{2.30c}$$

which resembles Eq. (1.56b). Here, we have defined a new quantity f_{ent}, a force that we[45] term the entanglement force or cohesion force, given by

$$f_{ent} = k_B T / l_{ent}. \tag{2.30d}$$

The entanglement force is analogous to the retraction force of Eq. (1.53). This definition refers to an intact entanglement network with mesh size l_{ent}. The preceding treatment of chain entanglement provides the basis for the alternative treatment of Section1.3.3 that leads to Eq. (1.59) for rubber elasticity. We note that Eq. (2.31) is of the same form as Eq. (1.56a).

Figure 2.11 summarizes the scaling behavior of Eq. (2.30b) and provides a quantitative illustration of the magnitude of the areal density ψ_{ent} and its scaling dependence on p. Specifically, ψ_{ent} is in a range from 0.2 to 2 nm^{-2} for most of the two dozen polymers considered by Ref. [46].

The formula in Eq. (2.27) is valid for any Gaussian chain of size R in the melt state. Upon diluting the melt with a solvent or a low-molecular-weight oligomer of the same polymer at concentration ϕ where the flexible linear chains are still Gaussian, the number Q of chains of size R in the pervaded volume of a given chain simply drops by a factor ϕ, so that

$$Q_{sol} \sim \phi R(\phi)/p. \tag{2.32}$$

Similarly, the areal density of such chains is given by

$$\psi(\phi) \sim Q_{sol}/[R(\phi)]^2 \sim \phi/pR(\phi). \tag{2.33}$$

Introducing the entanglement spacing in a solution of concentration ϕ as $l_{ent}(\phi)$, an expression similar to Eq. (2.31) characterizes the entanglement network of an entangled solution as

$$\psi_{ent}(\phi) \sim \phi/pl_{ent}(\phi). \tag{2.34}$$

We also have a modified expression for the entanglement force as

$$f_{ent}(\phi) \sim k_B T/l_{ent}(\phi). \tag{2.35}$$

This force is weaker because it involves a long entanglement strand.

In terms of Eqs. (2.34) and (2.35), we have a further relationship to express the reduction of the plateau modulus by the dilution, as

$$G_{pl}(\phi) \sim \psi_{ent}(\phi) f_{ent}(\phi) \sim k_B T\phi/p[l_{ent}(\phi)]^2 \tag{2.36a}$$

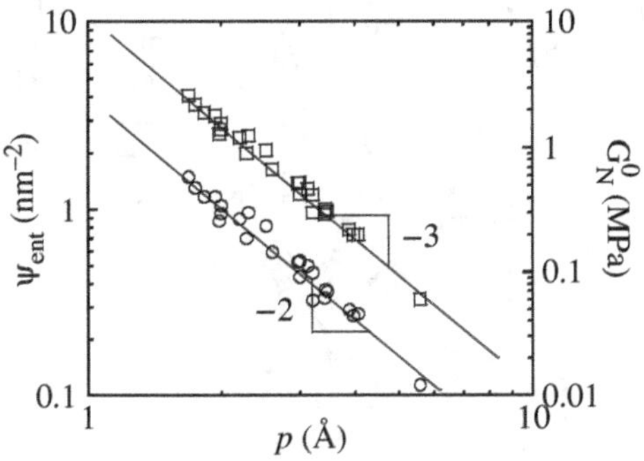

Figure 2.11 Dependence of the areal density ψ_{ent} of entanglement strands in Eq. (2.13) on the packing length p as well as that of G_N^0 (right Y axis) for various polymers. *Source:* Data from Wang 2007 [19].

on one hand, and

$$G_{pl}(\phi) \sim \phi\rho RT/M_e(\phi) \quad (2.36b)$$

on the other hand. Experimental measurements of the plateau modulus reveal[46,47]

$$G_{pl}(\phi) = G_{pl}(\phi = 1)\phi^{(2.1\sim2.3)}, \quad (2.36c)$$

where $G_{pl}(\phi = 1) \equiv G_N^0$ is melt plateau modulus. The scaling behavior in Eq. (2.36c) approximately agrees with the result $l_{ent} \sim \phi^{-2/3}$ from a scaling argument.[48] Given the experimental uncertainty regarding the exponent in Eq. (2.36c), we often use 2.2–2.3 instead of $(1 + 4/3) = 7/3$ to estimate the plateau modulus of concentrated solutions from the plateau modulus of the melt. Therefore, by comparison between Eq. (2.36a,b) and Eq. (2.36c), we have the following scaling relations for the entanglement spacing $l_{ent}(\phi)$ and entanglement molecular weight $M_e(\phi)$

$$l_{ent}(\phi) = l_{ent}(\phi = 1)\phi^{-(0.6\sim0.65)}, \quad (2.37a)$$

and

$$M_e(\phi) = M_e(\phi = 1)\phi^{-(1.2\sim1.3)}. \quad (2.37b)$$

2.2.3.2 Percolation Model

Starting from a completely different standpoint, Wool[49] proposed a percolation model to depict the onset of chain entanglement. The model arises by considering mechanical consequences of entanglement. Entanglement would arise to bear a load when a chain is able to make a couple of returns across a flat surface as shown in Fig. 2.12. To proceed, we first introduce a relevant molecular parameter s, in terms of which and of the backbone length l the physical volume v_1 of a repeated unit is given by $v_1 = v_M/n = sl$, where use is made of Eq. (2.19). Specifically, we have[19]

$$s = v_M/nl = pl_K, \quad (2.38a)$$

which is an effective molecular cross-sectional area of the chain backbone. Here the Kuhn length l_K in Eq. (1.39c) does not vary greatly among a number of common polymers.[19]

It is straightforward to estimate that the number q of times a Gaussian chain intercepts a plane is, on average,

$$q(n) = \pi R_g^2/sQ. \quad (2.39a)$$

where Q was defined in Eq. (2.27) as the number of chains filling the volume pervaded by a single chain. For Gaussian chains, $qQ \sim R_g^2 \sim M$. Therefore, similar to Q, q also scales like $M^{1/2}$. Inserting Eq. (2.27) for Q into Eq. (2.39a), we can write

$$q(n) = 6^{1/2}(3/4)Rp/s. \quad (2.39b)$$

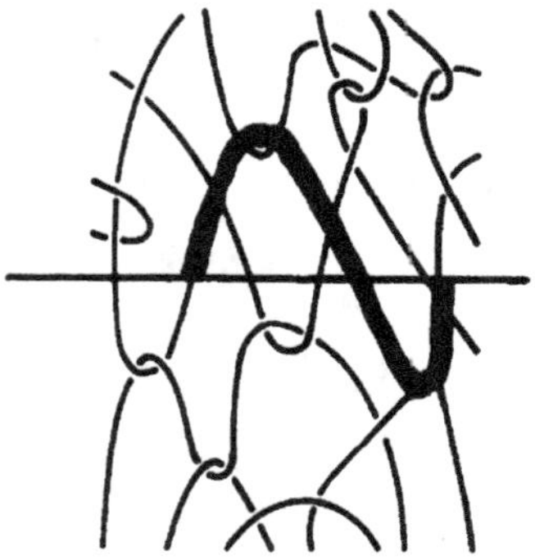

Figure 2.12 Wool's depiction of chain entanglement in his percolation model. It shows that a minimum chain length is as long as being able to return to a flat surface (denoted by the horizontal line) three times. *Source:* Wool 1993 [49]. Reproduced with permission of American Chemical Society.

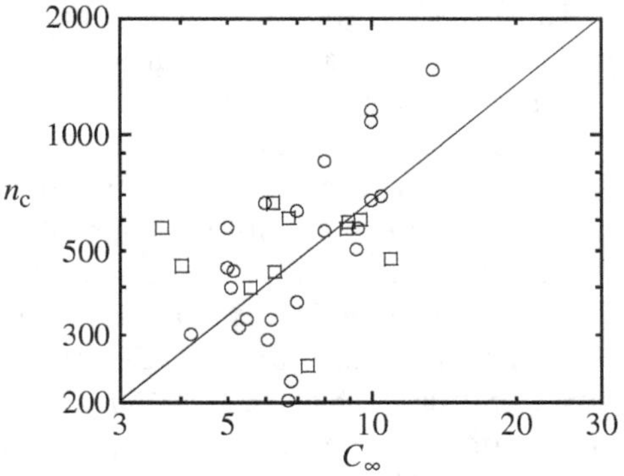

Figure 2.13 The number n_c of backbone lengths corresponding to a subchain of molecular weight M_c. *Source:* Data taken from Wang 2007 [19]; Aharoni 1986 [50] & Fetters 1999 [51].

The percolation model proposes that a linear flexible polymer is entangled when it intercepts a plane three times, namely, when $q(n_c) = 3$, to form entanglement points with other chains as shown in Fig. 2.12. This condition of $q(n_c) \sim 3$ for Eq. (2.39b) prescribes a threshold chain size l_c for entanglement

$$l_c \equiv R(n_c) = (8/3)^{1/2}(s/p) \sim l_K. \tag{2.38b}$$

The percolation model suggests that the coil size of an entanglement strand varies linearly with the Kuhn length l_K instead of varying with the packing length p. So l_c depends on the chain flexibility of the polymer, given by l_K, rather than the chain thickness given by p. The percolation model does not give a good description of how M_e, obtained from G_N^0 according to Eq. (2.30a), varies with p among the different linear polymers,[19] so we use M_c defined in Fig. 2.3 to see whether M_c is given in terms of the molecular characteristics per Eq. (2.38b). Figure 2.13 shows a poor fit of the limited data to the prediction of $M_c \sim n_c \sim C_\infty$ from Eq. (2.38b). Thus, M_c in Fig. 2.3 does not involve the critical condition that the percolation model aims at describing.

In our judgment, the percolation model shown in Fig. 2.12 may be an appealing depiction of chain entanglement during large deformation. The hairpin concept embedded in the percolation model is probably more relevant when we try to evaluate the onset of chain disentanglement under large deformations. Such a speculation has yet to be investigated by systematic experiments.

2.3 Molecular-Level Descriptions of Entanglement Dynamics

Chain entanglement is presumably responsible for the five leading features observed experimentally as sketched in Figs. 2.3–2.7. Only Fig. 2.3 speaks about the molecular weight dependence of a leading rheological property, namely, zero-shear viscosity η_0. One explanation for the strong molecular weight dependence of η_0 emerged[27] as early as 1971 soon after Berry and Fox[52] collected most available data. In the following section, we will briefly present the 1971 idea of reptation from de Gennes, the tube model of Doi–Edwards based on the image of chains moving in fixed tubes and finally a more first-principles-based polymer-mode-coupling theory of Schweizer for chain dynamics in dense systems.

2.3.1 Reptation Idea of de Gennes

By now, the reptation idea of de Gennes[27] is familiar to all working in the fields of polymer science and rheology. A careful and extensive discussion of reptation can be found in the book of

Doi and Edwards.[9] Thus, we can be brief and less rigorous in our presentation of the reptation picture. In a conceptual breakthrough, de Gennes proposed that the chain dynamics of dense polymeric systems can be adequately described by a single-chain mean-field framework. While stress relaxation may indeed involve the collective dynamics of multiple chains, self-diffusion of long chains in a melt or concentrated solution is treated as a single-chain process. It is in the treatment of polymer self-diffusion that the reptation idea found the most convincing support. By regarding the test chain to be confined to move along its own contour in a tube of length L, de Gennes cleverly simplified the many-body system to a single-chain problem. The effect of all surrounding chains is to restrict large-scale transverse motions by forming a tube. Inside this imaginary tube, the test chain would diffuse "freely," but only parallel to the tube, like a Rouse chain with diffusion constant D_R, given by Eq. (2.23). In the language of the reptation picture, the diffusion coefficient is known as the curvilinear diffusion constant $D_c = D_R$. For the center of mass of the chain of interest to diffuse a distance on the order of the chain's own coil size R, the chain would need to move a distance L along its contour in the tube as shown in Fig. 2.14.

Dimensional analysis immediately predicts a time τ_d required for the test chain's center of mass to diffuse a distance R

$$\tau_d \sim L^2/D_R, \tag{2.40}$$

where the subscript "d" is customarily used to denote this disengagement process. τ_d should never be regarded as designating a disentanglement time in quiescence because the chain is always entangled in the absence of external deformation. This distinction between disengagement and disentanglement will become clearer when the concept of disentanglement is discussed in detail in subsequent chapters. Since the contour length L is proportional to the chain length N, it must be proportional to R^2 where the "prefactor" would be another length scale. This unknown length is the so-called tube diameter or entanglement spacing l_{ent}, which appears in the following expression to make L be proportional to N

$$L l_{ent} = R^2 = N l_K^2. \tag{2.41}$$

Since the tube diameter is the chain length of an entanglement strand, we use l_{ent}, following the notation introduced in Section 2.2.3.1. Substitution of Eq. (2.41) for L into Eq. (2.40) produces the famous result

$$\tau_d \sim \zeta N^3 l_K^4 / l_{ent}^2 k_B T, \tag{2.42}$$

where use is made of Eq. (2.23) for $D_c = D_R$. According to the Doi–Edwards tube model,[53–56] the missing prefactor is $1/\pi^2$. Thus, de Gennes explained that the time scales involved in diffusion, stress relaxation, and zero-shear viscosity are the same and are expected to scale with the cube of the chain length. The model describes the limit in which the tube is regarded as immobile, which is perhaps close to reality as $N \to \infty$.

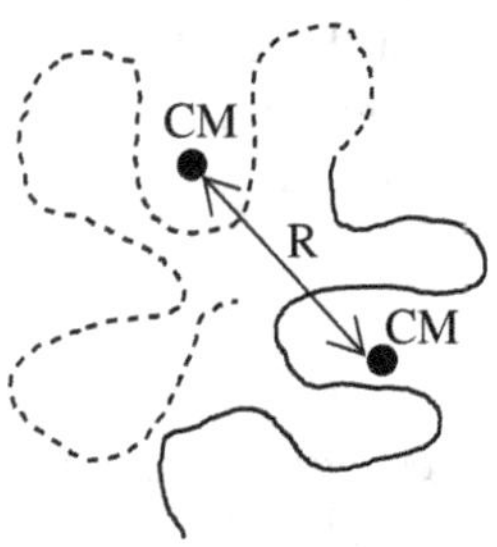

Figure 2.14 Chain movement from the position given by the solid line to that indicated by the dashed line amounts to diffusion of chain's center-of-mass by a distance comparable to its size R.

It is useful to write τ_d in terms of the longest Rouse relaxation time τ_R of Eqs. (2.16a) and (2.17), as

$$\tau_d = 3(N/N_e)\tau_R. \tag{2.43}$$

where the number of Kuhn segments between entanglement points, N_e, is related to the tube diameter as shown in Eq. (2.29). It is also useful to rewrite Eq. (2.43) as

$$\tau_d = 3\tau_e(N/N_e)^3 \tag{2.44}$$

where the tube constraint time τ_e is defined as

$$\tau_e = \tau_0 N_e^2 = \tau_d(N_e) \tag{2.45}$$

with τ_0 given by Eq. (2.17). In the tube model of Doi and Edwards, the tube restriction time τ_e shows up in small amplitude oscillatory shear, as shown in Fig. 2.7.

In the literature, the level of chain entanglement is commonly characterized by the number of entanglement points per chain

$$Z = \mathrm{L}/l_{ent} = (N/N_e), \tag{2.46}$$

where the second equality follows from use of Eq. (2.41). There is obviously little tube constraint when L is comparable to l_{ent}, and τ_d in Eq. (2.43) reduces to the Rouse relaxation time τ_R as Z approaches unity.

It is widely known that the asymptotic exponent 3 in Eq. (2.42) does not match the experimental observation. The experimental data usually appear to have an exponent varying between 3.0 and 3.6 over a wide range of molecular weight. The discrepancy between the de Gennes exponent of 3 and experimental value about 3.4 is well understood to arise from the crude assumption that the tube could be regarded fixed. When $\mathrm{Z} \to \infty$, the experimental data are proposed to approach the Gennes limit.

In the asymptotic limit of $\mathrm{Z} \to \infty$, there is a consensus that the scaling exponent should be 3. Only in this limit may we apply dimensional arguments to relate τ_d to the self-diffusion constant $\mathrm{D_s}$ of a test chain, giving

$$\mathrm{D_s} \sim \mathrm{R}^2/\tau_d \sim \mathrm{D_R}/\mathrm{Z} \sim N^{-2}, \quad \text{for } \mathrm{Z} \to \infty \tag{2.47}$$

where $\mathrm{D_R}$ is given by Eq. (2.23) and $\mathrm{D_s} \sim \mathrm{D_R}/\mathrm{Z}$ stems from expressing τ_d in terms of Eq. (2.40) along with Eq. (2.46) for Z. Phillies cited[57] Ref. [58] to note that many models can produce the scaling law Eq. (2.47). Indeed, the polymer-mode-coupling theory of Schweizer and coworkers[59,60] is one such alternative. Because Z is finite in real systems, the experimental data on $\mathrm{D_s}$ tend to reveal a more negative exponent[61,62] than −2. Convergence to the scaling form of Eq. (2.47) occurs at a sufficiently large value of Z. We will return for more discussion on the relationship between relaxation and diffusion processes in Section 2.3.4.

2.3.2 Tube Model of Doi and Edwards

While the notion of tube confinement was first introduced by Edwards,[63] it was de Gennes who figured out how a primitive chain would execute dynamic motions under the tube constraint. The reptation concept did not get widely known and accepted until Doi and Edwards published a series of four papers[53–56] in 1978 and 1979 that formulated a full molecular-level depiction of polymer viscoelasticity within the conceptual framework of reptation.

Assuming the existence of chain entanglement and modeling it in terms of tube confinement, the Doi–Edwards (DE) tube model introduces elastic plateau modulus $\mathrm{G_N^0}$ for melts as a phenomenological parameter. Regarding an entangled melt as the network depicted by Fig. 2.9, $\mathrm{G_N^0}$

should be given in terms of the number of strands per unit volume, according to the dimensional analysis leading to Eq. (2.30a). Following the long tradition of rubber elasticity theory, the Doi–Edwards model employs an Edwards tube to describe intermolecular interactions and computes the stress only from the intramolecular forces that are entropic and elastic in origin. In this construction, the residual stress is proportional to the fraction $\psi(t)$ of the primitive chain still inside the deformed tube, so that at long times (i.e., $t > \tau_e$) the relaxation modulus $G_{rept}(t)$ is proportional to the end-to-end vector correlation function $\psi(t)$[11]

$$G_{rept}(t) = G_N^0 \psi(t),\ \psi(t) = \sum_{q;odd} \frac{8}{q^2\pi^2} \exp(-q^2 t/\tau_d), \quad \text{for } t > \tau_e. \tag{2.48a}$$

The DE model for G(t) has the same mathematical structure as the generalized Maxwell model. The relaxation time spectrum is given by

$$\tau_q = \tau_d/q^2, \quad q = 1, 2, 3, \ldots \tag{2.49}$$

The longest relaxation time, namely, the reptation time τ_d is given by Eqs. (2.17) and (2.43–2.45). Comparison between Eq. (1.30) and Eq. (2.48a) yields

$$G_q = (8/\pi^2 q^2) G_N^0, \quad \text{for } q = 1, 3, 5, \ldots \tag{2.48b}$$

and

$$G_q = 0, \quad \text{for } q = 2, 4, 6, \ldots \tag{2.48c}$$

Insertion of Eq. (2.48a–c) into Eq. (1.10) produces explicit expressions for G′ and G″, which are identical in form to Eqs. (1.28) and (1.29), respectively. In particular, it can be shown numerically in Fig. 2.15 that the crossover frequency ω_c is approximately identical to $1/\tau_d$. In other words, for monodisperse melts, the DE model identifies the crossover ω_c to be approximately the reciprocal of the reptation time τ_d. Throughout this book, we mostly deal with monodisperse polymers and therefore take the reciprocal of ω_c as the terminal relaxation time τ. The symbols τ and τ_d will be used interchangeably with the understanding that the tube model prescribes an accurate account of τ in terms of τ_d.

On the other hand, at ω_c, $G_c = 0.40 G_N^0$, which shows a small numerical discrepancy with the empirical relation that $G_c = G_N^0/3.56$. Experimental data based on monodisperse melts show

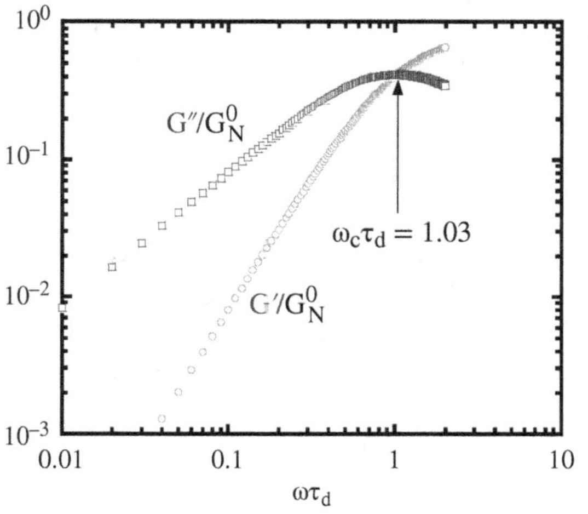

Figure 2.15 Normalized storage and loss moduli according to the tube model given by Eq. (2.48a).

that the two curves of G′ and G″ meet at ω_c, where G″ shows a maximum. Therefore, the tube model captures such linear viscoelastic data very well.

Based on the expression Eq. (2.48a) for G_{rept} and Eq. (2.30a) for G_N^0, the tube model produces via Eq. (1.10c) the following expression for the zero-shear viscosity η_0:[9]

$$\eta_0 = \frac{\pi^2}{12} G_N^0 \tau_d \approx \frac{\zeta N^3}{36 N_e^2 p}, \tag{2.50a}$$

where the second expression follows use of Eqs. (2.17) and (2.43) for τ_d. Equation (2.50a) can be rewritten in the form of Eq. (2.21) if we assign an effective friction factor ζ_{eff} to replace ζ in Eq. (2.21). Specifically, we have $\eta_0 = \zeta_{eff} N/18p$ with $\zeta_{eff} \sim \zeta Z^2/2$. Equating Eq. (2.50a) with Eq. (2.21) permits us to determine the location of the kink in Fig. 2.3 as

$$N_c = \sqrt{2} N_e. \tag{2.51}$$

Conventionally, N_c is typically taken to be $2N_e$.

On the other hand, the discrepancy in the scaling exponent for η_0 between the reptation value of 3 and the experimental value of about 3.4 is well known and has been taken most seriously by both proponents and adversaries of the tube model. The consensus among experts on the tube model is that this deviation arises from "tube erosion" at the end chains, known as contour length fluctuations.[53–56,64] However, experimental data suggests that constraint release, arising because the "tube" is made of other identical mobile chains, plays an important role in accelerating chain relaxation and producing a smaller η_0 than the ideal scaling of Eq. (2.50a) as indicated in Fig. 2.3.

In summary, the Doi–Edwards model not only provides a detailed description of all linear viscoelastic characteristics of entangled melts, in terms of the relaxation modulus of Eq. (2.48a). As to be presented in Chapter 16, the model also claims to describe nonlinear responses of entangled polymers under large rapid deformation. After several iterations of improvement, the tube model has become the standard theoretical description for nonlinear polymer rheology.[65] We defer to Chapter 16 our assessment of its ability to depict nonlinear rheological properties of entangled polymers. It suffices to mention that the tube model, being a single-chain approach, was not constructed in a self-consistent manner and is unable to anticipate a host of nonlinear phenomena that are obviously multichain cooperative in nature.

2.3.3 Polymer-Mode-Coupling Theory of Schweizer

As a heuristic model to visualize chain dynamics in entangled polymers, the reptation model is appealing. However, severe shortcomings are extensively noted by Phillies[57] when comparing with various experimental data. We appropriately note that indeed the picture advocated by the reptation model is not the only scenario.[58] According to Schweizer and coworkers,[66] "entanglements arise due to time and space correlations of the excluded volume forces exerted by the surrounding matrix on a tagged connected macromolecule." As a consequence of entanglement, linear chains cannot easily go around each other when they become very long. Mathematically, it indeed looks formidable to account for the dynamic effects of entanglement using first principles, force-based, time-dependent statistical mechanics. A relatively early attempt in this direction was initiated nearly 30 years ago by Schweizer and coworkers.[59,60,67] It was based on the microscopic mode-coupling-theory ideas that were initially developed to describe dynamical caging and emergent slow relaxation and viscoelasticity in glass-forming simple liquids. This so-called polymer-mode-coupling (PMC) theory[66] assumes, for simplicity and tractability reasons, an isotropic motion perspective. This assumption is applicable in an ensemble-averaged "effective medium" sense for a globally isotropic fluid of identical polymer molecules, but it is not necessarily literally true at the level of trajectories of individual polymer chains per the foundational idea of the reptation-tube model. The key physics in PMC theory for capturing emergent entangled dynamics is the structural de Gennes long-range correlation hole, corresponding to long-range interchain correlations between pairs of segments on different coils that extends to the radius-of-gyration scale. This effect is a universal consequence of chain connectivity, random walk conformation, and intermolecular excluded volume interactions that becomes relevant when polymers are sufficiently long and highly interpenetrating. Physically, the effect is envisioned in the PMC theory approach to induce space-time dynamical correlations between all polymer chains in a spherical region of space of diameter defined by macromolecular size. PMC theory predicts a crossover from Rouse to entangled behavior with increasing chain length. It makes qualitative scaling and quantitative predictions for all single-chain time correlation and response functions, the non-Fickian segmental and center-of-mass mean square displacements, relaxation times, the self-diffusion constant, viscosity, and plateau shear modulus, which are in good accordance with experiment.

In the asymptotic limit where Z of Eq. (2.46) goes to infinity, the PMC theory produces the same scaling results for the self-diffusion constant D_s, zero-shear viscosity η_0 and terminal relaxation time τ_d in Eqs. (2.47), (2.50a), and (2.43), respectively,[59,60] as does the reptation-tube model.[53–56] Moreover, dynamical fluctuation effects important for long but finite length chains, which reduce the anisotropic character of polymer motion, are also captured including the analogs of contour length fluctuations and matrix constraint release. The relative importance of the latter effects is predicted to be different for self-diffusion and chain relaxation.

Importantly, the apparent successes of microscopic PMC theory should not be taken as definitive evidence against the phenomenological reptation-tube model since the isotropic motion picture adopted is an assumption, not a prediction. A virtue of the force-based PMC approach is that it relates interchain interactions, thermodynamics, structure, and slow dynamics. Hence, it can be naturally generalized to treat entangled dynamics in more complex polymeric systems such as blends, microdomain-forming block copolymer solutions and melts, liquids of fractal macromolecular objects, and cross-linked gels.[66] But how to generalize such force-based methods of modern time-dependent statistical mechanics to treat *anisotropic* dynamics in entangled polymer liquids, in equilibrium let alone under nonlinear rheological conditions, has been a large outstanding challenge. Significant progress in this direction has been very recently achieved, as briefly sketched in Section 16.8.

2.3.4 Self-diffusion Constant versus Zero-shear Viscosity

Two leading properties of entangled polymers are the self-diffusion constant D_s and the zero-shear viscosity η_0. Are they dictated by the same physics? In the asymptotic limit of $Z = M/M_e$ approaching infinity, pure reptation is dominant so that Eq. (2.50a) has no correction. Thus, we can argue $D_s \sim R^2/\tau_d \sim R^2/\eta_0 \sim M^{-2}$. For finite values of M/M_e, there are deviations from these ideal scaling laws as illustrated by the comparison between the dashed and solid lines in Figs. 2.3 and 2.16. For example, D_s crosses over from a dependence on N^{-1} in the unentangled regime, given in Eq. (2.23), to a dependence N^{-a} for $a > 2$ in the entangled regime.[61,68,69] Experimental data in Fig. 2.17(a) show in solid symbols that D_s scales approximately as $M^{-2.4}$ as indicated by the line with slope −0.4. Thus, superficially,[62,71] according to Eq. (2.47), the nonideal scaling of D_s appeared to have the same physical origin as that of η_0. But actually the physics governing diffusion is different from that controlling stress relaxation time and shear viscosity. For example, at finite M/M_e, self-diffusion would

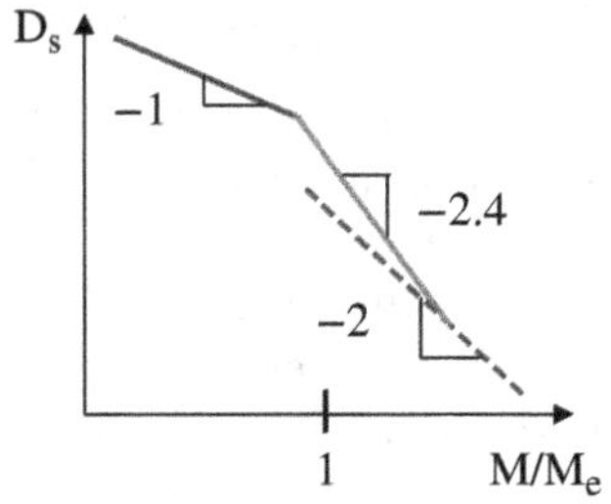

Figure 2.16 Self-diffusion constant D_s as a function of the reduced molecular weight M/M_e on double-logarithmic scales. The dashed line showing exponent −2 is the idealized scaling behavior. Due to the finite chain length effects, D_s is higher than the ideal values. For illustrative purpose, $M/M_e = 1$ is used to denote the borderline between Rouse chain and entangled chain diffusions.

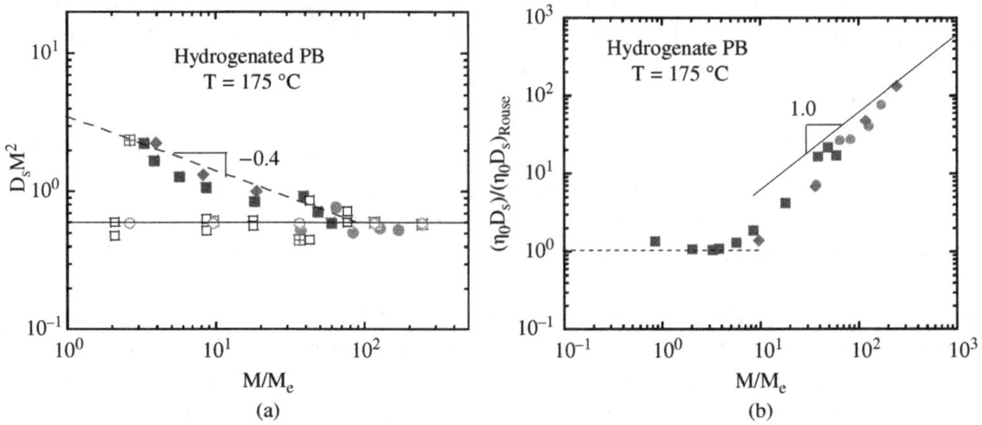

Figure 2.17 (a) Scaling behavior of the product of self-diffusion coefficient D_s and molecular weight M squared for hydrogenated polybutadiene at 175 °C, on double-logarithmic scales in the unit of g^2 (cm^2/s) mol^{-2}. The open symbols represent trace-diffusion and confirm the asymptotic scaling behavior. (b) The product of zero-shear viscosity η_0 and diffusion coefficient D_s, normalized by that for a Rouse melt, for hydrogenated polybutadiene at 175 °C. The dependence of the product is noticeably strongly than indicated by the slope of unity (straight line). *Source:* Wang 2003 [70]. Reproduced with permission of Wiley.

not speed up due to contour length fluctuations (CLFs) as the relaxation time does. On the other hand, the constraint release (CR) should speed up the self-diffusion, and its effect can be readily examined by the so-called trace diffusion measurements, where the diffusion takes place in a matrix of ultra-high molecular weight. In confirmation with the statement that self-diffusion should be little affected by CLF, the open symbols in Fig. 2.17(a) show[70,72] the trace-self-diffusion constant $D_{ts} \sim M^{-2}$. Moreover, the self-diffusion constant D_s converges to D_{ts} beyond $M/M_e \sim 30$, indicating that CR effect is diminishing at high M/M_e.

Unlike the diffusion measurements, it is difficult to perform rheometric measurements that would discern whether CLF or CR is more important to produce the deviation for η_0 from the ideal scaling of M^3. We cannot measure selectively the relaxation time associated with a trace amount of chains of molecular weight M in a matrix of much higher molecular weight $M^* > M$. One could perhaps apply rheo-optics[73] or rheodielectric[74,75] method to monitor the relaxation process of labeled chains in a matrix of nonlabeled chains. In any event, the finite-size effect, namely, Z being finite instead of infinite, prevents us from applying the scaling argument for any interconnection concerning the M dependences of D_s and η_0, as given in Eq. (2.47). Figure 2.17(b) confirms that the product of $D_s\eta_0$ indeed scales with M more strongly than linearly, implying the failure of Eq. (2.47) since $\eta_0 \sim G_N^0\tau_d$, η_0 obeys the same scaling as τ_d does.

2.3.5 Entangled Solutions

Entangled solutions are important in their own right. Many industrial processes including fiber spinning involve entangled solutions rather than melts. The entanglement networks of polymer solutions are looser and softer than the networks of melts. They can be good model systems to learn about nonlinear viscoelasticity of entangled polymers. For well-entangled solutions with polymer concentration at a volume fraction of ϕ, their plateau modulus G_{pl} is related to that of corresponding melts according to Eq. (2.36b,c). Moreover, the ϕ dependence of $\tau_d(\phi)$ partially stems from the increased molecular weight between entanglement points associated with the swollen tube diameter $l_{ent}(\phi)$.

According to Eq. (2.19), Gaussian chains satisfy $M_e(\phi) \sim [l_{ent}(\phi)]^2$. We therefore have in Eq. (2.37a,b) expressions for the concentration dependence of both quantities. We can generalize Eq. (2.42) using Eq. (2.37a) to find the concentration dependence for the reptation time in solutions and mixtures.

$$\tau_d(\phi) \sim \phi^{1.2\sim1.3}\tau_d(\phi = 1). \tag{2.52a}$$

The prefactor depends on the viscosity of the solvent; the concentration dependence is more complicated if the polymer "plasticizes" the solvent or the solvent "plasticizes" the polymer.[72] We note that the reptation time of solutions may have different temperature dependence than that of the pure melt.

When the molecular weight of the short-chain component M_s is comparable to M_e in a binary mixture made of long and short chains of the same polymer, the prefactor in Eq. (2.52a) for $\tau_d(\phi)$ is on the order of unity. Because of the tube dilation, it takes a longer time for a test chain to encounter the constraint due to surrounding chains so that τ_e of Eq. (2.44) changes to

$$\tau_{e(solution)} = \tau_e[l_{ent}(\phi)/l_{ent}(\phi = 1)]^4. \tag{2.52b}$$

With M_s increasing beyond M_e, $\tau_d(\phi)$ is upper bonded by $\tau_d(\phi = 1)$, which is the relaxation time of the pure long chain melt. Moreover, $\tau_{e(solution)}$ no longer follows Eq. (2.52b) and develops strong dependence[72] on M_s, for example, M_s^3. Indeed, as shown in Fig. 2.18(a), the experimental data[72] all fall between these two limits, indicating that the first theoretical treatments of such binary mixtures either over-[76] or underestimate[77] the terminal relaxation time τ_d. Note that the

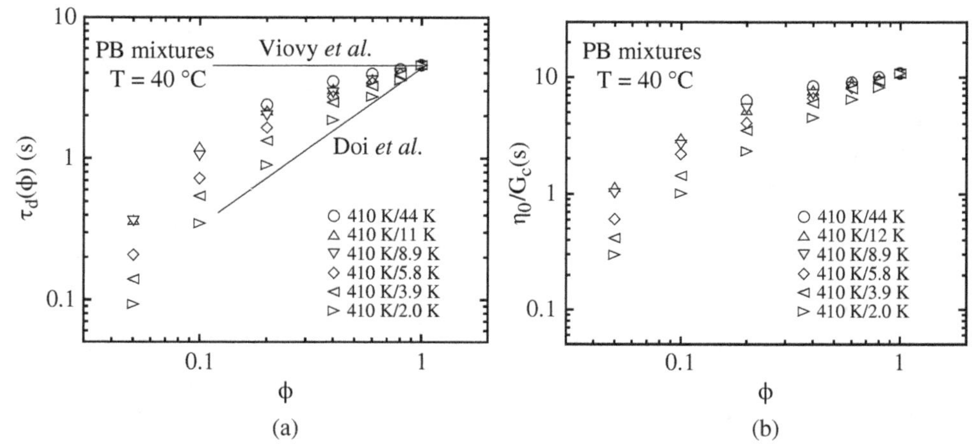

Figure 2.18 (a) Disengagement time τ_d as a function of concentration ϕ for different binary mixtures involving a common "parent" 1,4-polybutadiene of M = 410 kg/mol and various polybutadienes of lower molecular weights. (b) Terminal relaxation time given by η_0/G_c for the same set of binary mixtures as those in (a). *Source:* Wang *et al.* 2003 [72]. Reproduced with permission of American Chemical Society.

zero-shear viscosity of such mixtures or solutions is simply related to the terminal relaxation time $\tau_d(\phi)$ by

$$\eta_0(\phi) \approx G_{pl}(\phi)\tau_d(\phi), \tag{2.50b}$$

which is analogous to Eq. (2.50a) for entangled melts. As shown in Fig. 2.18(b), the experimental data confirms the validity of Eq. (2.50b), where G_c is defined in Fig. 2.7, proportional to G_{pl}.

The investigation of polymer dynamics of binary mixtures has interesting and important implications. The usefulness of working with such binary mixtures will become clear in Part 2. When entangled polymer solutions are binary mixtures, as shown in Chapters 6 and 9, there is unprecedented freedom to dictate whether entangled polymers suffer wall slip or undergo bulk shear strain localization.

2.4 Temperature Dependence

The temperature dependence of viscoelasticity in polymeric liquids is a most important aspect of polymer rheology. Efficient, effective polymer processing in manufacturing of all plastic and rubber products depends on reliable knowledge of how polymer dynamics change with temperature, above the melting temperature T_m for semicrystalline polymers and above the glass transition temperature T_g for noncrystalline polymers.

2.4.1 Time–Temperature Equivalence

Essentially all dynamic theories of polymer solutions and melts are based on a bead-spring type picture, including those of Rouse, Zimm, Doi–Edwards, and Schweizer. Moreover, except for the PMC theory, these molecular models assume that stresses arise from bonded intrachain forces. The temperature dependence of the chain dynamics is phenomenologically modeled by a friction coefficient ζ associated with the beads. Although the concept of ζ was first introduced through Stokes' law Eq. (2.7), we cannot in the case of pure melt sensibly relate ζ of a bead on the chain to the macroscopic melt viscosity. We need to assume phenomenologically that ζ contains all the information about how the segmental dynamics, characterized by the elemental time scale τ_0 of Eq. (2.17), depend on temperature. In all of these molecular theories, the terminal chain relaxation time thus has the same temperature as the segmental time. Equations (2.42)–(2.45) of the tube model provide an example.

The dynamics of polymeric liquids slow down upon cooling. Polymers have a diverging viscosity near the glass transition temperature T_g but can flow easily at temperatures well above T_g. Classically, we attribute this temperature dependence to the strong temperature dependence of $\zeta(T)$. Rheologists have fully exploited this characteristic of polymers. For polymer solutions and melts whose (entanglement) structures remain unaltered when the temperature is changed, fast dynamics can be studied experimentally by going to low temperatures. Conversely, slow dynamics can be accessed by raising the temperature. When polymers undergo a glass transition during cooling, the density change with temperature is weaker. A phenomenological description based on the concept of free volume has been proposed to depict how polymer dynamics change with temperature.

Focusing on the overall relaxation behavior of polymers, Williams, Landel, and Ferry (WLF)[78] found, based on experiment that many polymers follow a universal formula for the temperature dependence of their dynamics. Taking T_0 as a reference temperature, the ratio a_T of the overall relaxation time $\tau(T)$ to the relaxation time at T_0, which according to most molecular theories is the same as $\zeta(T)/\zeta(T_0)$, is given by

$$log\, a_T = log[\tau(T)/\tau(T_0)] = log[\zeta(T)/\zeta(T_0)] = -C_{10}(T - T_0)/(C_{20} + T - T_0), \tag{2.53}$$

where the WLF constants C_{10} and C_{20} depend on the choice of T_0 and are different for different polymers. When the glass transition temperature T_g is chosen as the reference temperature, replacing T_0 in Eq. (2.53) with T_g, the corresponding WLF constants C_1 and C_2 are nearly universal in their values,[8] typically with C_1 between 13 and 20, and C_2 between 40 and 60. Here, C_1 and C_2 are C_{10} and C_{20} if T_0 is set equal to T_g. The two sets of $\{C_{10},C_{20}\}$ and $\{C_1,C_2\}$ are interrelated[47,79–89] according to

$$C_2 = C_{20} - (T_0 - T_g), \tag{2.54a}$$

and

$$C_1 = C_{10}C_{20}/C_2, \tag{2.56b}$$

which correspond to the Table 11-II in Ref. [8] that lists values of C_{10} and C_{20} for a number of polymer systems at various temperatures T_0.

When the relaxation spectrum, embedded in G' and G'' from small-amplitude oscillatory shear measurements, depends on temperature in accordance with the WLF description so that τ in Fig. 2.7 obeys the WLF equation of Eq. (2.53), we have time–temperature equivalence. Polymer dynamics at any temperature well above T_g can be systematically related to the dynamics at any other temperature. This principle is commonly applied to access dynamics that are too fast or too slow to lie within the experimental measurement window at one temperature. A convenient frequency range for ω is between 100 and 0.01 rad/s in a typical rheometric instrument. Commercial rheometers usually do not operate reliably beyond 100 rad/s. Although lower frequencies than 0.01 rad/s are accessible, the measurements take a time that grows linearly with $1/\omega$.

Suppose that we are interested in linear viscoelastic behavior, namely, in obtaining Fig. 2.7 at a reference temperature T_{ref}, where $\omega_c = 1/\tau(T_{ref})$ would be 10^{-4} rad/s for this hypothetical sample. Instead of conducting small-amplitude oscillatory shear measurements at frequencies around 10^{-4} rad/s that may take $2\pi \times 10^4$ s, namely, about 17 h, we can resort to the principle of time–temperature equivalence and heat the sample to a higher temperature $T_1 > T_{ref}$ to make small-amplitude oscillatory shear measurements. If $\tau(T_1)$ would be shorter by three orders of magnitude, namely, around 10 s, corresponding to $\omega_c \sim 0.1$ rad/s, the SAOS measurement would finish in a time as short as a minute. Conversely, if a sample's relaxation time τ at T_0 is 1/500 s, it is inaccessible for a rheometric instrument with an upper frequency limit of 100 s^{-1}. In this case, we cool down the sample to slow its chain dynamics, so that a frequency sweep measurement can cover the needed range of time scales, for example, to reveal the crossover behavior at ω_c. A full set of G' and G'' curves may well span more than 10 decades in frequency. They are usually constructed by performing frequency sweep in small-amplitude oscillatory shear measurements at many different temperatures above and below T_{ref}. The resulting master curves as sketched in Fig. 2.7 arise from shifting the data (obtained at different temperatures) to superimpose them into continuous curves for G' and G''. A WLF shift factor a_T as a function of temperature arises from such a construction. The construction of the master curves for $G'(\omega)$ and $G''(\omega)$ is known as application of the time–temperature superposition principle. We will illustrate concretely in Section 17.3.1 how to evaluate a_T in the application of the time–temperature superposition for small-amplitude oscillatory shear.

2.4.2 Thermo-rheological Complexity

There are two types of thermorheological complexities involving different physical origins. When two miscible polymers A and B are blended together, the chain dynamics associated with A and with B may still have different temperature dependences, despite A and B having intimate molecular-level mixing. The glass transition temperature of each is influenced by the other. Depending on the molecular weight and composition of the two polymers, when the temperature is changed, the separate dynamics associated with chain A and chain B can shift differently, resulting in a breakdown of the time–temperature superposition principle.

When G' and G'' and the loss tangent $\tan\delta = G''/G'$ are measured over a wide enough range of frequency ω at a single temperature, it is found for some pure polymer melts that the curve of $\tan\delta(\omega)$ versus ω is different at different temperatures. Such observations clearly suggest that dynamics at different lengths and time scales do not depend on temperature in the same way. The origin of this second type of thermorheological complexity has received a theoretical explanation[90] after many heuristic arguments in the literature.

The failure of time–temperature superposition principle has important implications. At the least, it suggests that bead-spring model-based molecular theories miss some essential physics. The failure of time–temperature superposition typically occurs at low temperatures. Thus, at temperatures sufficiently above the glass transition temperature, molecular theories such as Rouse–Zimm and reptation models may still be regarded as a good approximation.

Questions concerning the validity of the time–temperature superposition principle are conventionally discussed for linear viscoelastic characteristics.[79–89] Rarely does the literature discuss any failure of TTS for nonlinear rheological properties.[91] If we focus our attention only on the steady flow behavior of polymers, it can be argued that time–temperature superposition is always valid. The reason is straightforward: In the steady state, rheological behavior is controlled by terminal dynamics that already average over local, segmental dynamics. In Section 17.3 of Chapter 17, we will discuss cases in which time–temperature superposition fails in a dramatic manner, involving nonlinear transient rheological responses of entangled polymer solutions and melts to startup shear and extension.

2.4.3 Segmental Friction and Terminal Relaxation Dynamics

It is remarkable that most theoretical treatments of polymer dynamics, such as those of Rouse, Zimm, de Gennes, Edwards, Doi, and others, are able to address how polymer dynamics depend on parameters such as molecular weight and concentration while bypassing difficult questions concerning the microscopic origin of the temperature dependence of chain dynamics. Apparently at the length and time scales involving an entire chain in either solution or melt, we can treat all polymers in a universal way, leaving out all the chemical details that render one polymer glassy (e.g., polystyrene) and another (e.g., 1,4-polybutadiene) far above T_g at room temperature. As indicated in the preceding subsection, on local (the so-called segmental) scales, perhaps involving a handful of monomers, the dynamics may be more temperature responsive. Well above the glass transition temperature nonbonded interactions between physically nearby nonbonded monomers are obviously short lived and determine an effective segmental friction ζ. However, we know phenomenologically that all segments find it more difficult to move when the temperature is lowered toward T_g. This entrapment or caging effect is diluted over longer length scales. Consequently, as proposed by Sokolov and Schweizer,[90] the terminal chain dynamics may indeed have a weaker temperature dependence than do the segmental dynamics. The success of the phenomenological approach in polymer rheology lies with questions that can be modeled and explored in a generic and universal and often phenomenological way, leaving out the challenge of accounting for chemical specificity that must be addressed for polymer glass transition.

This chapter introduced the standard yet necessary information regarding how modern microscopic theories describe chain dynamics in the two limits of dilute solutions (Section 2.1) and well-entangled melts or solutions (Section 2.3). The phenomenological evidence for entanglement was reviewed in Section 2.3.1 to highlight the fact that considerable rubber-elasticity-like behavior occurs in the linear response regime. Since available textbooks on polymer physics and rheology do not introduce the packing model, Section 2.2.3 presented the phenomenological models that derive the onset condition for polymer entanglement. In the packing model, the average entanglement strand length l_{ent} is argued to linearly scale with the packing length p; the elastic plateau modulus G_{pl} is inversely proportional to p^3; G_{pl} can also be expressed in terms of the cohesion force f_{ent} and areal density ψ_{ent} of entanglement strands. Finally, temperature dependence of chain dynamics was discussed in Section 2.4 in terms of the time–temperature superposition principle and observed failure of this principle.

References

1 Einstein A. A new determination of molecular dimensions. Ann Phys. 1906;19(2):289–306.
2 Mark H, Simha R. Viscosity and molecular structure. Nature. 1940;145:571–3.
3 Kirkwood JG, Riseman J. The intrinsic viscosities and diffusion constants of flexible macromolecules in solution. J Chem Phys. 1948;16(6):565–73.
4 Yamakawa H. Modern theory of polymer solutions. Harper & Row; 1971.
5 Phillies GDJ. The hydrodynamic scaling model for the dynamics of non-dilute polymer solutions: A comprehensive review, arXiv.org: 1606.09302. 2016.
6 Zimm BH. Dynamics of polymer molecules in dilute solution: viscoelasticity, flow birefringence and dielectric loss. J Chem Phys. 1956;24(2):269–78.
7 Rouse Jr PE. A theory of the linear viscoelastic properties of dilute solutions of coiling polymers. J Chem Phys. 1953;21(7):1272–80.
8 Ferry JD. Viscoelastic properties of polymers. John Wiley & Sons; 1980.
9 Doi M, Edwards SF. The theory of polymer dynamics. Oxford University Press; 1988.
10 Schrag JL, Stokich TM, Strand DA, Merchak PA, Landry CJT, Radtke DR, et al. Proceedings of the international discussion meetings on relaxations in complex systems local modification of solvent dynamics by polymeric solutes. J Non-Cryst Solids. 1991;131:537–43.
11 Yang X, Wang S-Q. Exploring the origin of negative viscosity change in polymer solutions. Macromolecules. 2000;33(8):3128–33.
12 Daoud M, De Gennes P. Some remarks on the dynamics of polymer melts. J Polym Sci Part B: Polym Phys. 1979;17(11):1971–81.
13 Wang S, Elkasabi Y, Wang S-Q. Rheological study of chain dynamics in dilute binary polymer mixtures. Macromolecules. 2005;38(1):125–33.
14 Einstein A. On the movement of small particles suspended in stationary liquids required by the molecular-kinetic theory of heat. Ann Phys. 1905;17:549–60.
15 Guenza M. Cooperative dynamics in unentangled polymer fluids. Phys Rev Lett. 2001;88(2):025901.
16 Guenza M. Cooperative dynamics in semiflexibile unentangled polymer fluids. J Chem Phys. 2003;119(14):7568–78.
17 Debnath P, Guenza M. Cooperative dynamics in polymer melts from the unentangled to the entangled regime. Philos Mag. 2008;88(33–35):4131–6.
18 Fetters L, Lohse D, Richter D, Witten T, Zirkel A. Connection between polymer molecular weight, density, chain dimensions, and melt viscoelastic properties. Macromolecules. 1994;27(17):4639–47.
19 Wang S-Q. On chain statistics and entanglement of flexible linear polymer melts. Macromolecules. 2007;40(24):8684–94.

20 Farago J, Semenov A, Meyer H, Wittmer J, Johner A, Baschnagel J. Mode-coupling approach to polymer diffusion in an unentangled melt. I. The effect of density fluctuations. Phys Rev E. 2012;85(5):051806.

21 Farago J, Meyer H, Baschnagel J, Semenov A. Mode-coupling approach to polymer diffusion in an unentangled melt. II. The effect of viscoelastic hydrodynamic interactions. Phys Rev E. 2012;85(5):051807.

22 Semenov A, Farago J, Meyer H. Length-scale dependent relaxation shear modulus and viscoelastic hydrodynamic interactions in polymer liquids. J Chem Phys. 2012;136(24):244905.

23 Bueche F. Viscosity, self-diffusion, and allied effects in solid polymers. J Chem Phys. 1952;20(12):1959–64.

24 Bueche F. Physical properties of polymers. Interscience Publishers, Inc., New York; 1962.

25 Graessley WW. The entanglement concept in polymer rheology. The entanglement concept in polymer rheology. Springer, Berlin Heidelberg; 1974. p. 1–179.

26 Graessley WW. Entangled linear, branched and network polymer systems – molecular theories. Adv Polym Sci. 1982;47:67–117.

27 de Gennes P-G. Reptation of a polymer chain in the presence of fixed obstacles. J Chem Phys. 1971;55:572–9.

28 Green M, Tobolsky A. A new approach to the theory of relaxing polymeric media. J Chem Phys. 1946;14(2):80–92.

29 Lodge AS. Constitutive equations from molecular network theories for polymer solutions. Rheol Acta. 1968;7(4):379–92.

30 Yamamoto M. The visco-elastic properties of network structure I. General formalism. J Phys Soc Jpn. 1956;11(4):413–21.

31 Yamamoto M. The visco-elastic properties of network structure II. Structural viscosity. J Phys Soc Jpn. 1957;12(10):1148–58.

32 Tanaka F, Edwards S. Viscoelastic properties of physically crosslinked networks: Part 3. Time-dependent phenomena. J Nonnewton Fluid Mech. 1992;43(2–3):289–309.

33 Hua CC, Schieber JD. Segment connectivity, chain-length breathing, segmental stretch, and constraint release in reptation models. I. Theory and single-step strain predictions. J Chem Phys. 1998;109(22):10018–27.

34 Masubuchi Y, Takimoto J-I, Koyama K, Ianniruberto G, Marrucci G, Greco F. Brownian simulations of a network of reptating primitive chains. J Chem Phys. 2001;115(9):4387–94.

35 Doi M, Takimoto J-i. Molecular modelling of entanglement. Philos Trans R Soc London, Ser A. 2003;361(1805):641–52.

36 Schieber JD, Neergaard J, Gupta S. A full-chain, temporary network model with sliplinks, chain-length fluctuations, chain connectivity and chain stretching. J Rheol. 2003;47(1):213–33.

37 Yaoita T, Isaki T, Masubuchi Y, Watanabe H, Ianniruberto G, Greco F, et al. Statics, linear, and nonlinear dynamics of entangled polystyrene melts simulated through the primitive chain network model. J Chem Phys. 2008;128(15):154901.

38 Rault J. Conjecture on the distance between entanglements in polymer melts. C R Acad Sci. 1985;300(10):433–6.

39 Heymans N. Diffuse versus point entanglements—homopolymers and blends. J Mater Sci. 1986;21(6):1919–26.

40 Lin Y. Number of entanglement strands per cubed tube diameter, a fundamental aspect of topological universality in polymer viscoelasticity. Macromolecules. 1987;20(12):3080–3.

41 Kavassalis TA, Noolandi J. A new theory of entanglements and dynamics in dense polymer systems. Macromolecules. 1988;21(9):2869–79.

42 Fetters L, Lohse D, Colby R. Chain dimensions and entanglement spacings. Physical properties of polymers handbook. Springer; 2007. p. 447–54.
43 Schweizer KS, Szamel G. Crossover to entangled dynamics in polymer solutions and melts. J Chem Phys. 1995;103(5):1934–45.
44 Schweizer KS, Szamel G. Microscopic theory of concentration scaling in entangled polymer solutions and melts. Macromolecules. 1995;28(22):7543–8.
45 Wang S-Q, Ravindranath S, Wang Y, Boukany P. New theoretical considerations in polymer rheology: elastic breakdown of chain entanglement network. J Chem Phys. 2007;127(6):064903.
46 Raju V, Menezes E, Marin G, Graessley W, Fetters L. Concentration and molecular weight dependence of viscoelastic properties in linear and star polymers. Macromolecules. 1981;14(6):1668–76.
47 Yang X, Wang S-Q, Ishida H. A solution approach to component dynamics of A/B miscible blends. 1. Tube dilation, reptation, and segmental friction of polymer A. Macromolecules. 1999;32(8):2638–45.
48 Colby RH, Rubinstein M. Two-parameter scaling for polymers in θ solvents. Macromolecules. 1990;23(10):2753–7.
49 Wool RP. Polymer entanglements. Macromolecules. 1993;26(7):1564–9.
50 Aharoni SM. Correlations between chain parameters and the plateau modulus of polymers. Macromolecules. 1986;19(2):426–34.
51 Fetters LJ, Lohse DJ, Milner ST, Graessley WW. Packing length influence in linear polymer melts on the entanglement, critical, and reptation molecular weights. Macromolecules. 1999;32(20):6847–51.
52 Berry GC, Fox TG. The viscosity of polymers and their concentrated solutions. Fortschritte der Hochpolymeren-Forschung. Springer; 1968. p. 261–357.
53 Doi M, Edwards S. Dynamics of concentrated polymer systems. Part 1.—Brownian motion in the equilibrium state. J Chem Soc, Faraday Trans 2: Mol Chem Phys. 1978;74:1789–801.
54 Doi M, Edwards S. Dynamics of concentrated polymer systems. Part 2.—Molecular motion under flow. J Chem Soc, Faraday Trans 2: Mol Chem Phys. 1978;74:1802–17.
55 Doi M, Edwards S. Dynamics of concentrated polymer systems. Part 3.—The constitutive equation, J Chem Soc, Faraday Trans 2: Mol Chem Phys. 1978;74:1818–32.
56 Doi M, Edwards S. Dynamics of concentrated polymer systems. Part 4.—Rheological properties. J Chem Soc, Faraday Trans 2: Mol Chem Phys. 1979;75:38–54.
57 Phillies GDJ. Phenomenology of polymer solution dynamics. Cambridge University Press; 2011.
58 Skolnick J, Kolinski A. Dynamics of dense polymer systems: computer simulations and analytic theories. Adv Chem Phys. 1990;77:223–78.
59 Fuchs M, Schweizer KS. Polymer-mode-coupling theory of finite-size-fluctuation effects in entangled solutions, melts, and gels. 1. General formulation and predictions. Macromolecules. 1997;30(17):5133–55.
60 Fuchs M, Schweizer KS. Polymer-mode-coupling theory of finite-size-fluctuation effects in entangled solutions, melts, and gels. 2. Comparison with experiment. Macromolecules. 1997;30(17):5156–71.
61 Tao H, Lodge TP, von Meerwall ED. Diffusivity and viscosity of concentrated hydrogenated polybutadiene solutions. Macromolecules. 2000;33(5):1747–58.
62 Lodge TP. Reconciliation of the molecular weight dependence of diffusion and viscosity in entangled polymers. Phys Rev Lett. 1999;83(16):3218–21.
63 Edwards S. The statistical mechanics of polymerized material. Proc Phys Soc. 1967;92(1):9.

64 Doi M. Explanation for the 3.4 power law of viscosity of polymeric liquids on the basis of the tube model. J Polym Sci Part C Polym Lett. 1981;19(5):265–73.
65 Graham RS, Likhtman AE, McLeish TC, Milner ST. Microscopic theory of linear, entangled polymer chains under rapid deformation including chain stretch and convective constraint release. J Rheol. 2003;47(5):1171–200.
66 Schweizer KS, Fuchs M, Szamel G, Guenza M, Tang H. Polymer-mode-coupling theory of the slow dynamics of entangled macromolecular fluids. Macromol Theory Simul. 1997;6(6):1037–117.
67 Schweizer KS. Microscopic theory of the dynamics of polymeric liquids - general formulation of a mode mode-coupling approach. J Chem Phys. 1989;91(9):5802–21.
68 Wheeler LM, Lodge TP. Tracer diffusion of linear polystyrenes in dilute, semidilute, and concentrated polyvinyl methyl-ether) solutions. Macromolecules. 1989;22(8):3399–408.
69 Von Seggern J, Klotz S, Cantow H. Reptation and constraint release in linear polymer melts: an experimental study. Macromolecules. 1991;24(11):3300–3.
70 Wang SQ. Chain dynamics in entangled polymers: diffusion versus rheology and their comparison. J Polym Sci, Part B: Polym Phys. 2003;41(14):1589–604.
71 Rubinstein M, Colby RH. Polymer Physics: OUP Oxford; 2003. p. 364 and Fig. 9.2.
72 Wang S, Wang S-Q, Halasa A, Hsu W-L. Relaxation dynamics in mixtures of long and short chains: Tube dilation and impeded curvilinear diffusion. Macromolecules. 2003;36(14):5355–71.
73 Fuller GG. Optical rheometry of complex fluids. Oxford University Press; 1995.
74 Matsumiya Y, Watanabe H, Inoue T, Osaki K, Yao M-L. Rheo-dielectric behavior of oligostyrene and polyisoprene. Macromolecules. 1998;31(22):7973–5.
75 Hiroshi W, Yumi M, Tadashi I. Rheo-dielectrics in oligomeric and polymeric fluids: a review of recent findings. J Phys: Condens Matter. 2003;15(11):S909.
76 Viovy JL, Rubinstein M, Colby RH. Constraint release in polymer melts: Tube reorganization versus tube dilation. Macromolecules. 1991;24(12):3587–96.
77 Doi M, Graessley W, Helfand E, Pearson D. Dynamics of polymers in polydisperse melts. Macromolecules. 1987;20(8):1900–6.
78 Williams ML, Landel RF, Ferry JD. The temperature dependence of relaxation mechanisms in amorphous polymers and other glass-forming liquids. J Am Chem Soc. 1955;77(14):3701–7.
79 Plazek DJ. Temperature dependence of the viscoelastic behavior of polystyrene. J Phys Chem. 1965;69(10):3480–7.
80 Plazek DJ, O'Rourke VM. Viscoelastic behavior of low molecular weight polystyrene. J Polym Sci Part B: Polym Phys. 1971;9(2):209–43.
81 Plazek DJ, Chelko AJ. Temperature dependence of the steady state recoverable compliance of amorphous polymers. Polymer. 1977;18(1):15–8.
82 Plazek DJ. The temperature dependence of the viscoelastic behavior of poly (vinyl acetate). Polym J. 1980;12(1):43–53.
83 Plazek DJ. The temperature dependence of the viscoelastic softening and terminal dispersions of linear amorphous polymers. J Polym Sci Part B: Polym Phys. 1982;20(4):729–42.
84 Ngai K, Plazek D. Relation of internal rotational isomerism barriers to the flow activation energy of entangled polymer melts in the high-temperature Arrhenius region. J Polym Sci, Polym Phys Ed. 1985;23(10):2159–80.
85 Ngai K, Plazek D. A quantitative explanation of the difference in the temperature dependences of the viscoelastic softening and terminal dispersions of linear amorphous polymers. J Polym Sci, Part B: Polym Phys. 1986;24(3):619–32.

86 Plazek D, Zheng X, Ngai K. Viscoelastic properties of amorphous polymers. I. Different temperature dependences of segmental relaxation and terminal dispersion. Macromolecules. 1992;25(19):4920–4.

87 Plazek D, Chay I-C, Ngai K, Roland C. Viscoelastic properties of polymers. 4. Thermorheological complexity of the softening dispersion in polyisobutylene. Macromolecules. 1995;28(19):6432–6.

88 Plazek DJ. 1995 Bingham Medal Address: Oh, thermorheological simplicity, wherefore art thou? J Rheol. 1996;40(6):987–1014.

89 Ngai KL, Plazek DJ, Rendell RW. Some examples of possible descriptions of dynamic properties of polymers by means of the coupling model. Rheol Acta. 1997;36(3):307–19.

90 Sokolov AP, Schweizer KS. Resolving the mystery of the chain friction mechanism in polymer liquids. Phys Rev Lett. 2009;102(24):248301.

91 Dealy J, Plazek D. Time-temperature superposition—a users guide. Rheol Bull. 2009;78(2):16–31.

3

Experimental Methods

Polymer rheology is an experimental science. Experiment must guide theoretical development because polymeric liquids, especially entangled polymers, are complicated many-body systems. Obtaining crucial information requires effective experimental methods that impose desired modes of deformation and flow. We review conventional experimental apparatuses that may or may not be commercially available. At the end, we discuss the merit of combining rheometry with *in situ* structural characterization techniques.

3.1 Shear Rheometry

Simple shear deformation can be made in a variety of devices. There are two common types of instruments. The first type is more commonly used and involves the displacement of the boundaries that confine the sample. Moving boundaries is the preferred setup to probe the constitutive behavior of viscoelastic materials. Since there exists a uniform shear stress field across the sample thickness in simple shear, it is conventionally assumed that the deformation and flow fields are known *a priori*, that is, without any knowledge of the constitutive properties of the material under study. Here by "constitutive," we refer to knowledge about how the local shear stress is related to the local deformation or flow field. Starting from Chapter 4, we will discuss how this assumption may have to change. On the other hand, a pressure-driven capillary or channel flow device imposes an inherent stress gradient, making it impossible to delineate *a priori* the deformation field without detailed information of the constitutive properties of the sample. In anticipation of what is to come in subsequent chapters, we must emphasize that the following review of shear rheometry is based on a fundamental assumption that wall slip does not occur and that shear deformation is always homogeneous, as depicted in Fig. 1.1(a,d). Unfortunately, for well-entangled polymeric liquids, this assumption hardly ever applies. In simple shear, the interfacial failure known as wall slip is a primary response to large deformation. This leading case of inhomogeneous shear is discussed in Chapter 6.

3.1.1 Shear by Linear Displacement

For investigation of transient rheological behaviors under finite deformations, it may be most convenient to produce a simple shear by linear displacement either at constant speed or by driving with a constant force. Depending on requirements of the application, different geometries may be considered.

The simplest geometry involves displacing one plate over a second stationary plate, the moving and stationary plates being parallel, as shown in Fig. 3.1(a,b). Here the driving mechanism can be a step-motor to control the displacement or an air cylinder to exert a constant total force.

Nonlinear Polymer Rheology: Macroscopic Phenomenology and Molecular Foundation, First Edition. Shi-Qing Wang.

Companion Website: www.wiley.com/go/nonlinearpolymerrheology

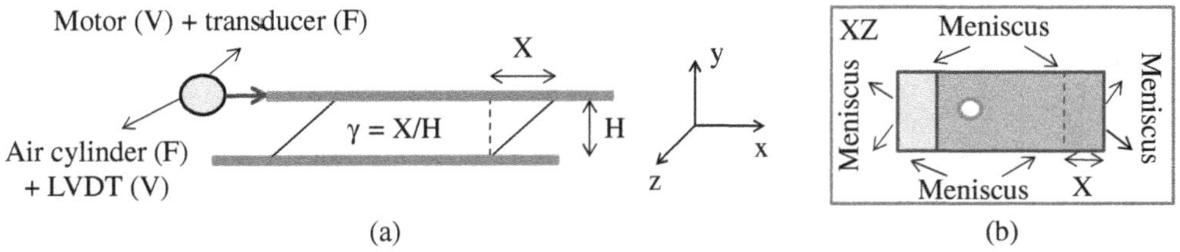

Figure 3.1 (a) Schematic illustration of a shear cell (with gap distance H) with linear displacement of the upper plate, either by a step-motor with a force transducer, or by a constant force (air cylinder) along with a linear-variable-differential transformer. (b) XZ view of the parallel-plate shear cell where a load cell can be installed to measure the local shear stress at the wall, as done in the literature.[1]

In the first case, a load cell can be employed to capture the time dependence of the resulting shear force during and after shear. In the second scenario of constant force, a linear voltage differential transformer can be attached to track the displacement during creep. Video recording would be another convenient way to evaluate the displacement as a function of creep time. Theoretically, one can think of the size of the plates to be infinite and ignore edge effects. Shear deformation between the parallel plates is known as planar Couette flow. But we shall avoid using the term "flow" because a viscoelastic sample does not immediately undergo flow upon such sudden startup deformation as depicted in Fig. 1.1. See Section 1.2.4.2 for a previous discussion based on the Maxwell model.

In reality, a sample is of finite size, bounded by the meniscus. The amount of shear deformation is limited by the aspect ratio of the sample, that is, $L/H \gg 1$. Given these limitations, such an apparatus is not commercially available. One obvious difficulty is that at large strain the meniscus can be severely distorted. For highly viscoelastic polymeric liquids, the edge fracture can be so severe at large strain that the sample boundary on the XZ plane can turn from the initial rectangular shape to a totally irregular one, making rheometric measurements unreliable. Nevertheless, the effectiveness and usefulness of such equipment to allow examination of polymer melts during and after large but finite shear deformation, have been well recognized, especially when combined with a particle-tracking velocimetric method to be described in Section 5.2. Until commercial rheometric companies manufacture proper instruments to meet the need for studying entangled melts under large deformation, sliding-plate rheometers as sketched in Fig. 3.1(a,b) can be custom-made in individual research labs. Their value cannot be underestimated because often the most revealing information comes from the transient responses to sudden deformation, and the sliding plate is perfectly suited for finite shear. Sliding plate rheometers[2] are particularly simple and useful for studying stepwise shear that may involve only a few strain units of shear.

3.1.2 Shear in Rotational Device

The obvious limitation of linear-displacement shear device is overcome in rotational rheometric instruments. Since the sample is sheared circularly in the tangential direction, the beginning and ending are closed onto each other. All commercial shear rheometers employ rotational motion. However, their ability to produce steady-state behavior still depends on the type of viscoelastic materials and the experimental condition because they do not completely eliminate all meniscuses.

3.1.2.1 Cone-Plate Assembly

The most widely used and known rheometric device comprises a cone and corresponding disk, whose side view is shown in Fig. 3.2. In the small gap approximation, the shear stress σ produced

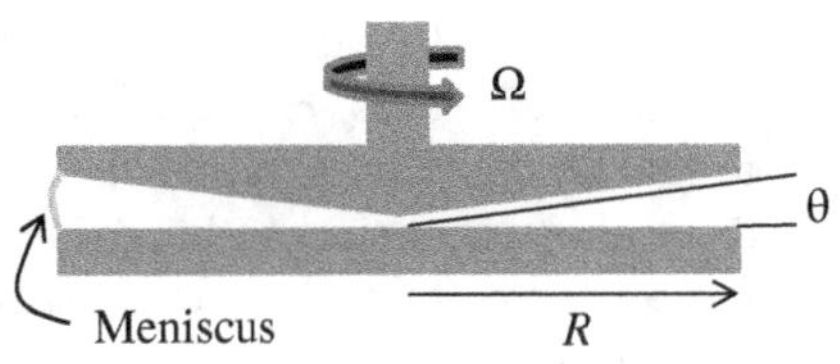

Figure 3.2 Illustration of a cone-plate rotational shear apparatus in terms of the plate radius R and cone angle θ. The stationary fixture can also be the cone instead of the plate. Also indicated is the meniscus, which in this system is a free surface surrounding the gap in the cone-plate arrangement.

within the gap can be regarded as being uniform at different heights and as being independent of the radial position r because the shear rate is perceived to be constant in different layers and radii. In particular, the constant (apparent) shear rate is given by $\dot{\gamma} = \Omega/\theta$, where Ω is the angular velocity, and θ rarely exceeds 0.1 rad or 5.7° in practice. This feature of constant shear stress across the sample thickness is the basic reason why the cone-plate combination is so popular. It is straightforward to show that the radial-independent shear stress σ corresponds to a total torque T

$$T = \int_0^R dr\,(2\pi r\sigma)r = \frac{2\pi}{3}\sigma R^3. \tag{3.1}$$

Obviously, by adjusting the cone radius R we can broaden the measurement range, given that the available range of T is fixed in any given rheometer. This device also allows the normal (vertical) force F_N to be measured with a force transducer. The first normal stress difference N_1 can be shown[3] to be related to F_N as $N_1 = 2F_N/\pi R^2$.

There are two severe problems for any commercial rotational rheometer. First, the upper bound of T is typically 0.2 Nm, which is hardly high enough for examination of entangled melts having elastic modulus G_{pl} at a level of 1 MPa. According to Eq. (3.1), the sample size would have to be as small as $R \sim 3.3$ mm to accommodate a shear stress level of 1 MPa. Often, the normal force range of a rheometer (20–50 N) is exceeded before the torque limit is reached. For entangled polymers, the normal force originates from the elastic resistance to chain deformation.

The second problem arises from the meniscus, as indicated in Fig. 3.2. Under continuous shear deformation, at rates considerably larger than the overall relaxation rate, that is, for $Wi \gg 1$, meniscus tends to become unstable and irregular for entangled melts and solutions of high concentrations. Such behavior is known as edge fracture.[4] As edge fracture progresses, there can be less and less contact between the shearing surfaces and the sample, making it difficult to obtain steady-state measurements. During severe edge fracture, the torque keeps dropping over time. Instead of confronting this experimental challenge, it is customary to bypass this difficulty and resort to capillary flow rheometry to make measurements at high shear rates. Some solutions are available to improve the reliability of rheometric measurements for nonlinear rheological properties. We will discuss them in Chapter 5.

3.1.2.2 Parallel Disks

The setup made of two parallel circular disks is often known as parallel-plate fixture, replacing the cone in Fig. 3.2 with a second identical disk. Because the apparent shear rate linearly increases from zero at the center to a maximum at the rim according to $\dot{\gamma}(r) = r\Omega/H$, the shear stress varies radially. The relation between the torque and stress is unknown unless the constitutive properties of the sample are known. In the limit of Newtonian liquids, $\sigma(r)$ is linear in r, that is, $\sigma(r) = (r/R)\sigma(R)$, the integral in Eq. (3.1) yields the following simple relationship:

$$T = (\pi/2)R^3\sigma(R). \tag{3.2}$$

The parallel-disk setup may not be desirable as a convenient geometry for studying nonlinear rheological responses. However, it can be a useful shearing device that naturally generates a shear rate gradient in the radial direction. We could think of using it to establish a combinatorial method, allowing the smoothly varying shear rate to imprint a spatial variation of properties or structures for samples that are sensitive to the shear rate.

One important use of the parallel-disk configuration is to probe linear responses such as small-amplitude oscillatory shear. The configuration is particularly suitable for stiff samples that may be time consuming to load into a cone-plate shear cell for small-amplitude oscillatory shear measurements. When small-amplitude oscillatory shear needs to be carried out at different temperatures, once again parallel disks provide the desirable setup because of the ease with which the sample gap is adjusted during temperature change. Last but not least, with parallel disks the sample thickness need not be the same from one sample loading to the next, although it is common to have H between 0.5 and 1 mm. For reasons that have remained elusive, for viscoelastic materials such as polymeric liquids, the measured torque tends to depend on the gap distance H, specifically, be smaller when H is below 0.5 mm.

Available commercial rotational rheometers have 2000 g cm or 200 mNm as their torque limit. Therefore, with $T_{max} = 0.2$ Nm, and for a typical value $R = 1.25$ cm, according to Eq. (3.1) σ_{max} is ca. 5×10^4 Pa. This value is far smaller than the elastic plateau modulus of most common polymer melts. Consequently, commercial rotational rheometers are hardly suitable for studying nonlinear rheological behavior of entangled polymer melts.

3.1.2.3 Circular Couette Apparatus

Maurice Couette (1858–1943) considered the deformation of a liquid filling the gap between two concentric cylinders, as shown in Fig. 3.3. In the limit that the cylinders are infinitely long and the liquid is Newtonian, the Navier–Stokes equation reduces to a single-variable differential equation for the tangential velocity $u(r)$ in the cylindrical coordinate system

$$\frac{d^2u}{dr^2} + \frac{1}{r}\frac{du}{dr} - \frac{u}{r^2} = 0. \tag{3.3}$$

The G.I. Taylor's (1886–1975) solution of Eq. (3.3) is

$$u(r) = -C_1 r + C_2/r, \tag{3.4a}$$

where the integration constants can be determined from the no-slip boundary condition, that is, $u(R + \Delta) = 0$ and $u(\text{R}) = \Omega R$. The constants are

$$C_1 = \Omega\alpha/(1 - \alpha) \text{ and } C_2 = \Omega R^2/(1 - \alpha), \tag{3.4b}$$

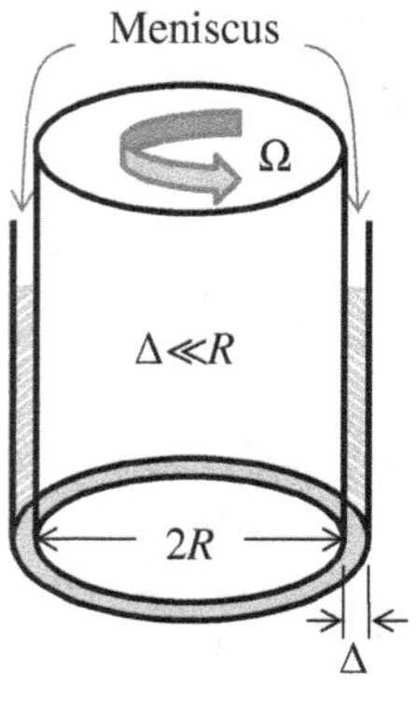

Figure 3.3 Illustration of a Couette device, made of inner cylinder of radius R concentrically placed inside a cylindrical cup with radius $(R + \Delta)$, where the inner cylinder is shown to rotate at an angular velocity Ω. Depending on its design to connect to the step-motor, the outer cylinder can be rotational instead of the inner cylinder.

where $\alpha = [R/(R+\Delta)]^2 < 1$ and Ω is the angular velocity. Because of the difference in the diameters R and $(R+\Delta)$ of the inner and outer cylinders, there is a variation in the velocity gradient across the gap. The ratio of the velocity gradients at the inner and outer cylinder walls is given by

$$\frac{du/dr|_{r=R}}{du/dr|_{r=R+\Delta}} = (1+\alpha)/2\alpha, \tag{3.5}$$

which approaches unity as Δ/R vanishes.

Using a cup as the outer cylinder, the Couette cell is commonly used to examine rheological behavior of less viscoelastic materials including dilute and semidilute polymer solutions. Commercial rotational rheometers are often supplied with a Couette device along with cone-plate and parallel-disk setups. However, the Couette device has the same intrinsic limitation as cone-plate, that is, the presence of a meniscus as shown in Fig. 3.3. The meniscus can turn unstable for viscoelastic materials during rotation at apparent shear rates $\dot{\gamma} \cong \Omega R/\Delta > 1/\tau$, although the edge failure mechanism is not the same as the mechanism encountered in cone-plate configurations. Finally, it is worth noting that this apparatus does not allow normal force to be measured readily.

3.1.3 Pressure-Driven Apparatus

Rheometers based on pressure-driving mechanisms have a special place in rheometry because they are closely related to systems encountered in the processing of materials. For example, pressure-driven rheometers can imitate extrusion, as illustrated in Fig. 3.4. Pressure-driven rheometers have found a wide range of applications, particularly in dealing with nonlinear or strong flow conditions, where rotational rheometry suffers from meniscus fracture. However, they should not be regarded as rigorous rheometric instruments because the deformation field inside the pipe is unknown *a priori*. The pressure available at the entry of the pipe, often known as "die," can be generated in two different modes, so that a capillary rheometer can be called either pressure-controlled or piston-speed-controlled. In the former, a given pressure is exerted onto the piston, forcing the material to enter the die and exit from the die end. In the latter, the piston is driven at a constant speed, which also leads to a buildup of pressure at the die entry. Depending on the design of the capillary rheometer, the information related to the pressure can be determined either at the die entry, by installing a pressure transducer at the bottom of the barrel, or through a load cell connected to the piston. For all the data in Chapters 6 and 10, the pressure refers to the piston pressure. For simplicity, we do not distinguish the two methods when presenting analytical expressions in this chapter and Chapter 6.

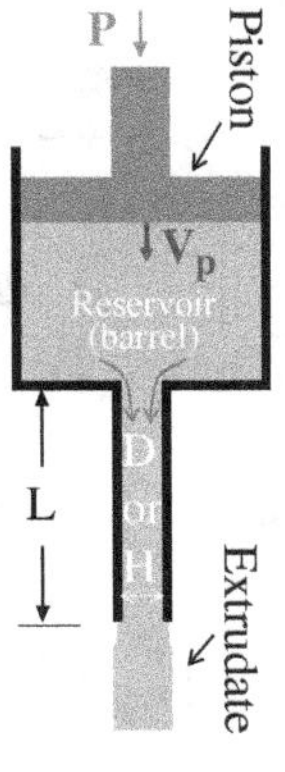

Figure 3.4 Pressure-driven shear through extrusion, by either displacement of the piston at speed V_p or pressure P, where the die can be either capillary or slit, as shown in Fig. 3.5.

3.1.3.1 Capillary Die

Materials forced to move through a circular pipe is known as Poiseuille [Jean Poiseuille (1797–1869)] flow if the liquid flows like a Newtonian liquid. In the field of rheology, where the liquid may be viscoelastic and may have a rate-dependent viscosity, pipe flow is known as capillary flow. In both cases, the ideal condition is dealt with, that is, steady-state flow behavior is considered in a pipe whose length is so long that no end effects are accounted for. In steady state, by ignoring the finite length effects, we can write down a generic expression relating the driving pressure gradient dP/dz to the shear stress σ at any radial position r by solving the force balance equation

$$\nabla \cdot \overleftrightarrow{\sigma} + \nabla P = 0 \qquad (3.6)$$

in cylindrical coordinates, as shown in Fig. 3.5(a). Integration of Eq. (3.6) yields

$$\sigma(r) \equiv \sigma_{rz} = -(r/2)dP/dz. \qquad (3.7a)$$

This equation holds for a material with any constitutive relationship, showing that the shear stress grows radially in a linear fashion. On the other hand, without explicit constitutive information about the material under study, Eq. (3.7a) cannot be solved. Without a solution to Eq. (3.7a), the velocity field would not be known, and thus no unambiguous constitutive relationship could be determined from capillary rheometry.

Capillary rheometry is still very useful as an approximate tool when we model an unknown material as a Newtonian liquid. We can learn a great deal about what capillary rheometry does by examining the simplest Newtonian case. Inserting Eq. (1.8a) $\sigma = \eta_0 dV_z/dr$ into Eq. (3.7a) and assuming that the velocity is only along the Z direction and dependent only on r, for Newtonian liquids we can straightforwardly integrate the resulting differential equation

$$\eta_0 dV_z/dr = -(r/2)|dP/dz|, \qquad (3.8)$$

and obtain

$$V(r) = V_{max}(1 - r^2/R^2). \qquad (3.9)$$

Here the subscript z is dropped for notational simplicity. The maximum velocity V_{max} takes place at the central line, $r = 0$, where the gradient of the velocity field is zero. The maximum shear rate is at the wall, given by

$$dV/dr|_{r=R} \equiv \dot{\gamma}_w = -4V_{max}/D \equiv \sigma_w/\eta_0, \qquad (3.10)$$

where the minus sign indicates that the velocity decreases with increasing r.

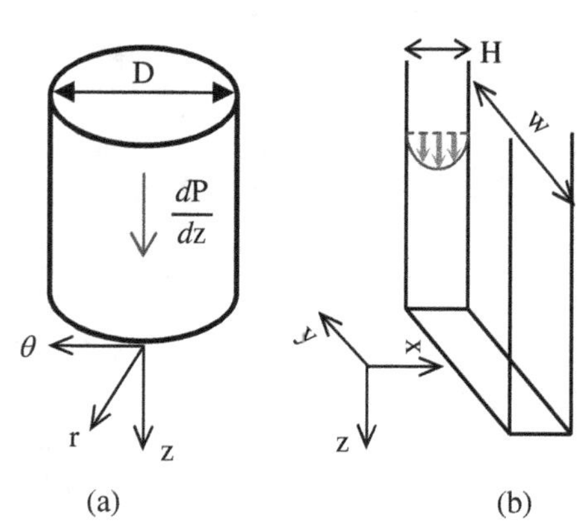

Figure 3.5 (a) Capillary die with diameter D and (b) slit die with width w and thickness H.

The shear stress at the wall σ_w is given by setting $r = D/2$ in Eq. (3.7a),

$$\sigma_w = P(D/4L). \tag{3.7b}$$

Here we have taken $|dP/dz| = P/L$, with P being the pressure at the die entry and L being the die length, upon neglecting the end effects. In other words, for a given pressure at the entry, the maximum shear stress depends on the aspect ratio L/D of the die. In terms of P, V_{max} in Eq. (3.9) is given by

$$V_{max} = (PD^2/16L\eta_0), \tag{3.11}$$

which is related to the volumetric flow rate Q_f, which is the amount of fluid coming out of the die per unit time. It is straightforward to show

$$Q_f = \int_0^R 2\pi r V(r) dr = (\pi/2)R^2 V_{max} = (\pi D^3/32)\dot{\gamma}_w, \tag{3.12a}$$

where the shear rate $\dot{\gamma}_w$ at the wall is defined in Eq. (3.10). Note that Q_f scales cubically with D. Eq. (3.12a) can be used to define an average speed $\overline{V}$ of the material in the die as

$$(1/4)\,\pi D^2\overline{V} \equiv Q_f, \quad \text{i.e.,} \quad \overline{V} = V_{max}/2 = D\dot{\gamma}_w/8. \tag{3.12b}$$

Although Eq. (3.12b) is only valid for Newtonian liquids, the dimensional form or the functional dependence of Q_f on the die diameter and wall shear rate is useful to note for any liquid.

For viscoelastic materials of unknown constitutive characteristics, the velocity profile inside the die cannot be calculated. For example, to calculate Q_f according to Eq. (3.12a), we need to have the full functional form of $V(r)$. This means that we do not even have a straightforward relationship (such as Eq. (3.12a)) between throughput Q_f and wall shear rate $\dot{\gamma}_w$. However, through multiple runs, the wall shear rate can be deduced from the relationship between Q_f and P. Specifically, the Weissenberg–Rabinowitsch equation estimates the wall shear rate $\dot{\gamma}_{w(n\text{-}Newt)}$ in terms of how Q_f varies with the wall shear stress σ_w as[3]

$$\dot{\gamma}_{w(n\text{-}Newt)} = \frac{\dot{\gamma}_w}{4}\left(3 + \frac{d \ln Q_f}{d \ln \sigma_w}\right), \tag{3.12c}$$

where $\dot{\gamma}_w$ is the Newtonian result, given by Eq. (3.12a). Nevertheless, as pointed out in Appendix 11.A of Chapter 11, the utility of capillary rheometry is limited for exploration of the nonlinear bulk constitutive behavior of viscoelastic materials.

3.1.3.2 Channel Slit

In this section, we consider liquid flow and solid deformation in a channel die (Fig. 3.5(b)). The subject of elastic deformation in pressure-driven die is rarely treated in the literature, but it is clearly pertinent when considering the initial response of viscoelastic fluids to start up extrusion.

Viscous Flow Given the rectangular-shaped cross section, we can solve Eq. (3.6) in Cartesian coordinates by integration to obtain

$$\sigma \equiv \sigma_{xz} = -x dP/dz \tag{3.13}$$

in the limit where the channel width w is much greater than H so that we can ignore the boundary effects at the side walls at $y = \pm w/2$. Integration of Eq. (3.13) by inserting the Newton's law, we get

$$V(x) = V_{max}[1 - (2x/H)^2], \tag{3.14}$$

where V_{max} is given by

$$V_{max} = (H/2)^2(P/2L\eta_0). \tag{3.15}$$

The volumetric flow rate Q_f can be evaluated in a way similar to Eq. (3.12a) as

$$Q_f = \int_{-H/2}^{H/2} wV(x)dx = (2wH/3)V_{max} = (wH^2/6)\dot{\gamma}_w, \tag{3.16a}$$

where the shear rate $\dot{\gamma}_w$ at the wall is related to the wall shear stress σ_w as σ_w/η_0. Here $\sigma_w = HP/2L$ is given by Eq. (3.13) at $x = H/2$. In the preceding discussion, the 3D flow problem of channel flow is simplified to a 1D calculation, as seen in Eqs. (3.13), (3.14), and (3.16a) for Newtonian liquids. These results remain useful as an approximate way to visualize viscoelastic fluids undergoing channel flow. Experimentally speaking, for us to ignore the side wall and the entry effects, we must consider high aspect ratios of $w/H \gg 1$ and $L/H \gg 1$. Finally, we can define an average speed, similar to Eq. (3.12b), as

$$\overline{V} = (2/3)V_{max} = H\dot{\gamma}_w/6, \tag{3.16b}$$

showing that the pertinent length scale in channel flow is H.

Elastic Deformation For entangled polymeric liquids, one often sought theoretical treatment and experimental characterization of steady-state flow and paid little attention to the initial state at the onset of extrusion. It should be clear for viscoelastic materials in either pressure- or speed-controlled extrusion that a flow state cannot be established instantly. Upon exertion of a constant piston speed or pressure in a large barrel, a highly viscoelastic liquid such as an entangled polymer is expected to undergo elastic deformation until the point of yield. Let us consider prefilling a channel with the material and allowing the material in the channel to relax indefinitely. As shown in Eq. (3.13), the shear stress inside the channel is related to the pressure gradient $P/L = dP/dz$ independent of the constitutive characteristics of the material. On applying a pressure gradient P/L at time $t = 0$ and treating the sample to be a Hookean solid with a plateau modulus G_{pl}, we can obtain the response by solving Eq. (3.13). In the simplest case of linear solids obeying Hooke's law, Eq. (1.7), we have

$$\sigma \equiv \sigma_{xz} = (P/L)x = -G_{pl}\partial Z/\partial x. \tag{3.17}$$

The solution of Eq. (3.17) describes the elastic deformation of the sample inside the channel. Assuming that the sample adheres to the channel wall, that is, $Z(x = \pm H/2) = 0$, we see, by integrating Eq. (3.17), how the material at different values of x, is displaced. The displacement is approximately instantaneous for ideally pure elastic deformation. Integration of Eq. (3.17) produces

$$Z(x) = (H/2)[1 - (2x/H)^2](\sigma_w/2G_{pl}), \tag{3.18}$$

which is a parabolic function of x, analogous in form to Eq. (3.14). The maximum strain γ_w occurs next to the wall: $\gamma_w = \partial Z/\partial x|_{x=\pm H/2} = \sigma_w/\sigma_{pl}$, where the maximum shear stress $\sigma_w = P(H/2L)$ is at the two walls ($x = \pm H/2$). This analysis shows that the initial elastic deformation would not be large unless σ_w exceeds the elastic plateau modulus G_{pl}.

It is equally interesting to contemplate a controlled-speed scenario. Suppose a piston speed V_{pist} is exerted at time $t = 0$. The volumetric flow rate is fixed at $Q_f = (\pi/4)D_b^2V_p$ for $t > 0$. The quasielastic material is squeezed into the channel in the manner depicted in Eq. (3.18).

The amount of material Ω_f that has entered the channel is related to the pressure buildup at that moment

$$\Omega_f = 2w \int_0^{H/2} z(x)dx = (wH^2/6)(\sigma_w/G_{pl}), \tag{3.19}$$

which increases linearly in time according to $\frac{d\Omega_f}{dt} = Q_f$. Thus, the pressure at the die entrance should initially grow linearly in time. The rate of the increase is constant and given by differentiation of Eq. (3.19), that is, by

$$\frac{d\sigma_w}{dt} = G_{pl}(6Q_f/wH^2), \tag{3.20}$$

showing the rate of the pressure (stress) buildup depends on the normalized magnitude of the throughput, in the parenthesis. Taking $\sigma_w = G_{pl}\gamma_w$, Eq. (3.20) amounts to showing $d\gamma_w/dt = (6Q_f/wH^2)$, which is identical in form to Eq. (3.16a). However, here the rate of strain $d\gamma_w/dt$ depicts the rate of elastic deformation, while $\dot{\gamma}_w$ designates Newtonian flow in the channel. The maximum strain occurs at the channel walls and grows with time as $\gamma_w = \sigma_w/G_{pl} = (6t/wH^2)Q_f$. Obviously, the duration of such initial elasticity does not increase indefinitely. The analysis does not show how yielding takes place at high strains.

3.2 Extensional Rheometry

An equally important component of rheometry concerns extensional deformation. It is typically more demanding to produce well-defined extensional rheological conditions than it is to produce well-defined shear deformations. For materials of low viscosity, extensional flow can be approximately generated by either injecting the material onto a flat wall or having two opposing jets facing each other. TA Instruments had at one point commercialized an extensional flow rheometer based on the latter flow pattern. Another well-known design, so-called four-roll mill, employs four counter-rotating cylinders.[5–8] In this book, we focus on apparatuses that deal with the other extreme, that is, materials that are sufficiently stiff and have reasonable good mechanical integrity on the experimental time scale. In this case, the analysis of extensional deformation is straightforward since the sample can be treated essentially like a cross-linked rubber and mounted onto an Instron stretcher that pulls the two ends of a strip in opposite directions.

3.2.1 Basic Definitions of Strain and Stress

Let us consider stretching a specimen so that its ends move in opposite directions with a relative speed of $V(t)$ that can be a function of time. The sample experiences an elemental strain given by

$$\Delta\varepsilon = \Delta L/L \tag{3.21}$$

over an interval of $\Delta t = \Delta L/V$. These forms are analogous to Eq. (1.1) for simple shear. Here $L(t)$ is the total length of the sample at time t. The stretch ratio λ is related to the Hencky (Heinrich Hencky (1885–1951)) strain ε through an integration of Eq. (3.21) with respect to the differential change of the length L.

$$\varepsilon = ln\lambda \equiv ln(L/L_0), \tag{3.22}$$

where L_0 is the initial length of the sample. The corresponding Hencky strain rate is given by

$$\dot{\varepsilon} = \lim_{\Delta t \to 0} \frac{\Delta \varepsilon}{\Delta t} = V/L. \tag{3.23}$$

Conversely, the relationship between the stretching ratio λ and Hencky strain (rate) can be expressed by rewriting Eq. (3.22) as

$$\frac{L}{L_0} = \lambda(t) = \exp\left(\int_0^t \dot{\varepsilon}(t)dt \right) \equiv \exp(\varepsilon). \tag{3.24}$$

In a simple case where $\dot{\varepsilon}$ is constant, $\lambda = L(t)/L_0$ increases exponentially with time t.

When the degree of stretching is significant, which can be the case for rubbery polymeric liquids, the cross-sectional area A shrinks continuously during extension. It is then necessary to introduce and distinguish the engineering from true (Cauchy) stresses, that is, σ_{engr} from σ_E. Viscoelastic polymers resist extension by creating a tensile force F (see Fig. 1.8). F has two different forms when written as an intensive quantity, either as

$$\sigma_{engr}(t) = F(t)/A_0 \tag{3.25}$$

or

$$\sigma_E(t) = F(t)/A(t) = \sigma_{engr}(t)\lambda, \tag{3.26}$$

where A is given by

$$A = A_0/\lambda = A_0 \exp(-\varepsilon). \tag{3.27}$$

$A(\varepsilon)$ decreases exponentially with the Hencky strain ε. We note that σ_E differs from σ_{engr} in proportion to the stretching ratio λ due to the change in cross-sectional area that corresponds to the change in length. In extensional as well as shear deformation and flow of polymeric liquids the volume of the sample is conserved, so that an increase in length by λ is accompanied by a reduction in cross section by the same factor. Whether σ_{engr} or σ_E should be used to depict uniaxial extension depends on the nature of the material deformation. It makes sense to employ the true extensional stress σ_E if steady-state flow can be finally produced. For well-entangled polymeric liquids, steady-state extensional flow is only possible at a Hencky strain rate $\dot{\varepsilon}$ below the overall relaxation rate τ, that is, for the Weissenberg number $Wi = \dot{\varepsilon}\tau \ll 1$, as discussed in detail in Chapter 11. Any further discussion is only meaningful after a description of available rheometric apparatuses for studying uniaxial extension behavior of viscoelastic materials including polymer melts.

3.2.2 Three Types of Devices

When a viscoelastic material is sufficiently stiff and possesses a sufficiently long relaxation time, it is relatively straightforward to subject it to well-defined uniaxial extensional tests. We briefly review two apparatuses that are adequate to produce rheometric information during startup extension of well-entangled polymer melts. When the sample has a level of viscoelasticity too large for a four-roll mill device[8] but too small for Instron-like device, a third type of stretcher also mentioned in this section might be of some value.

3.2.2.1 Instron Stretcher

The simplest way to make a uniaxial extension of a sample with strong mechanical integrity is to use an Instron tensile tester (www.Instron.com). The instrument has two clamps to grab the two ends of a stripe-like specimen and to move them apart at a relative velocity V. If V is a constant V_0 in time, then the Hencky rate is not a constant but given by $\dot{\varepsilon}(t) = V_0/(L_0 + V_0 t)$. The sample can be also stretched at a constant Hencky rate, in which case $V(t)$ increases exponentially with time. Instron is popular for stretching solids including polymer glasses and vulcanized rubbers where the extension does not require large extensional rate.

The deformation of viscoelastic materials such as entangled polymeric liquids is highly rate sensitive. It is therefore desirable to maintain a constant Hencky strain rate $\dot{\varepsilon}$ during stretching. For a setup such as an Instron tester, this means that the crosshead needs to move exponentially faster as time goes on: $L = L_0 \exp(\dot{\varepsilon}t)$ or $V = dL/dt = \dot{\varepsilon}L$. Although challenging, it is indeed possible to achieve a constant $\dot{\varepsilon}$.[9–13] The more severe limitation for an Instron tester to observe extensional rheology of polymer melts is instrumentational: if we desire to reach a Hencky strain of $\varepsilon = 4$, we have to reach $L = 55\ L_0$, which exceeds 1 m for L_0 as shorter as 2 cm.

It is often crucial to maintain a constant extensional strain rate (i.e., Hencky rate) since viscoelasticity is usually best characterized on fixed time scales. At a constant rate of stretching, that is, having a constant Hencky strain rate, the following alternative paths to probe time-dependent responses are better.

3.2.2.2 Meissner-Like Sentmanat Extensional Rheometer

According to Eq. (3.23), a simple way to produce a constant Hencky rate $\dot{\varepsilon}$ during stretching is to extend a specimen having a fixed length L with a constant speed V. Such a specimen becomes thinner exponentially fast according to Eq. (3.27). The first-generation design employed four rotating clamps to grab the two ends of the specimen.[14] A subsequent improvement replaced the rotating clamps with conveyor belts and had the device commercialized by TA Instruments as an extensional rheometer.[15] Since gravity can cause a sample to sag, the apparatus is typically rather tricky to operate.

A further improvement in early 2000s made it possible to utilize (commercial) existing rotational rheometers, greatly extending the availability of uniaxial extensional rheometry to a much wider community. This new design is made of two counter-rotating cylinders,[16–18] between which a specimen stretches upon attachment of the two ends to the two rotating drums. The sample length L to fit into the Sentmanat extension rheometer (SER) is smaller than 2 cm. In contrast to Meissner's design, the stripe-shaped specimen stands vertically to minimize sagging due to gravity. Nevertheless, the specimen still needs to be stiff enough so that it may not creep significantly under its own weight during the measurement. The first-generation SER has a severe limitation; specimen can only wrap around the drums for one turn. This limitation amounts to a limiting Hencky strain ε_H of ca. 4. Interestingly and perhaps surprisingly, most entangled melts cannot sustain such a high strain without encountering cohesive failure after yielding. We will return to the important subject of yielding under uniaxial extension in Section 7.5.

3.2.2.3 Filament Stretching Rheometer

A third type of apparatus, known as a filament stretching rheometer (FSR), has been popularized by a few workers studying concentrated solutions and melts.[19–21] Placing a polymeric liquid between two parallel plates whose separation is increased in an Instron-type apparatus, the method uses a feedback loop to achieve a given Hencky rate at one (middle) point of the specimen by adjusting the relative speed of the two separating ends. The merits and drawbacks of FSR will be discussed in detail in Section 11.4 and Appendix 11.A.

3.3 *In Situ* Rheostructural Methods

For rheometric measurements to offer structural information about the degree of chain deformation and state of chain entanglement, one must have specific and realistic molecular models. *In situ* structural characterization during and after deformation/flow can be used to delineate the state of deformation and molecular responses independent of any particular model. Such an approach might potentially also help identify required features for theoretical modeling, since it independently reveals complementary knowledge about the molecular state. Collectively, we term all such experimental techniques *rheostructural* characterization. By performing rheostructural experiments, we may not only better understand the microscopic origin of rheological behavior but also help determine structure–property relationships emerging from polymer processing.

In principle, many experimental characterization methods for quiescent measurements could be adopted to capture structural changes during deformation. Nonpolarized light transmission allows turbidity to be monitored, which could be due to incipient crystallization.[22–25] Polarized light transmission observes optical birefringence, which can be related to chain orientation.[26] Light, X-ray, and neutron scattering measurements can be made across in different planes, most commonly by incident along Y direction in the case of simple shear depicted in Figs. 1.1(a) and 1.2. When beam of radiation (light, X-ray, neutron) is incident along Y direction, the resulting information is an average over the sample thickness. This integration across the sample thickness masks whether a structural change is homogeneous along the Y direction, where the shear stress can be taken to be homogeneous. Spectroscopic techniques such as NMR, fluorescence, infrared, Raman, and dielectric spectroscopy may be employed to examine a sample during deformation or flow. Below we focus on the optical flow birefringence and only briefly mention other rheostructural methods that have been implemented and yielded additional information beyond what is available from rheometry.

3.3.1 Flow Birefringence

Chain orientation occurs during deformation and flow when the thermal agitation cannot cause polymer chains to recover their preferred isotropic orientational distribution fast enough relative to the rate of external deformation. Since polymer chains are optically anisotropic, that is, have different polarizability along and perpendicular to their backbones, overall chain orientation produces macroscopic optical anisotropy, which can be quantified by light transmission experiments. In simple shear as depicted in Fig. 1.1(a) and Fig. 1.2, the ensemble-averaged shape of a chain takes an ellipsoidal form, with its principal axis making an extinction angle χ with respect to the X axis. Light with polarization parallel to this principal axis sees a refractive index n_e, and light with polarization perpendicular to this axis sees a different refractive index n_o.

According to standard electromagnetism theory, the optical field of nonpolarized incident light with wavelength λ in the Z direction (in an isotropic medium of refractive index n) is given by the amplitude $A = E_0 sin(kz - \omega t)$, where $k = (2\pi/\lambda)n$. Let us consider sending this beam along the Z axis through an optical system made of a sample under shear sandwiched by a polarizer and an analyzer as sketched in Fig. 3.6.

For a case where simple shear is imposed on a sample so that optical anisotropy is produced at an angle (known as extinction angle) χ as shown in Fig. 3.6, the polarized light (after passing through the polarizer) travels in the sample as a combination of two polarized beams with polarization vectors along and perpendicular to the principal optical axis, which lies at an angle χ with X axis. The amplitudes of these beams are given, respectively, by

$$A_e = E_0 cos\chi\, sin(k_e w - \omega t), \tag{3.28a}$$

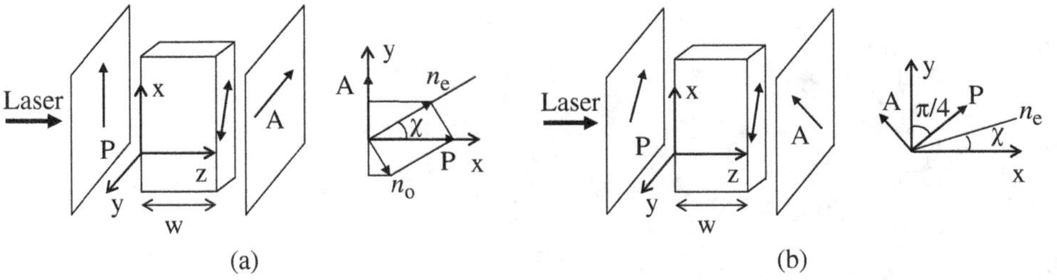

Figure 3.6 Rheo-optical setup for simple shear to measure strain-induced birefringence with the polarization P of the incident laser either (a) parallel to or (b) at 45° with the shearing direction (X axis). The analyzer A is always perpendicular to P.

and

$$A_o = E_0 sin\,\chi sin(k_o w - \omega t), \tag{3.28b}$$

where $k_{e,o} = (2\pi/\lambda)n_{e,o}$. At the analyzer, the projection along Y axis is given by

$$A_T = A_e sin\,\chi - A_o cos\,\chi. \tag{3.29}$$

The transmitted light intensity I is then given by

$$I \sim (A_T)^2 = I_0 sin^2 2\chi sin^2(\delta/2), \tag{3.30}$$

where use is made of $sin\,\alpha - sin\,\beta = 2sin[(\alpha - \beta)/2]cos[(\alpha + \beta)/2]$ and $2sin\,\chi\,cos\,\chi = sin\,2\chi$, and the retardance δ is related to the birefringence Δn by

$$\delta = |k_e - k_o|w = (2\pi w/\lambda)\Delta n. \tag{3.31}$$

The retardance grows linearly with the sample thickness w. Depending on the magnitude of $\Delta n = |n_e - n_o|$ and sample thickness w, the light intensity will go through zeros and maxima at $\delta/2 = 0, \pi, 2\pi, \ldots$, and $\pi/2, 3\pi/2, \ldots$, respectively. This scheme (a) of having the polarizer along the shearing direction (i.e., X axis) is effective when the extinction angle χ is not far from its unperturbed value of 45°.

When the extinction angle χ is expected to be small, it is more effective to have the polarizer and analyzer oriented 45° to the shearing direction as shown in Fig. 3.6. This amounts to replacing χ in Eq. (3.30) with $\pi/4 - \chi$, so that Eq. (3.30) becomes

$$I \sim (A_T)^2 = I_0 cos^2 2\chi sin^2(\delta/2), \tag{3.32}$$

which has maximum intensity at $\chi = 0$. Uniaxial extension along X axis adopts this configuration where χ stays at zero. Obviously, when χ is changing in time from 45° toward 0°, it is desirable to rotate the polarizer and analyzer accordingly. A more advanced treatment is available from a monograph by Fuller.[26]

According to Eq. (3.31), the retardance $\delta(\lambda)$ depends on the light wavelength λ. Therefore, if white light is sent through the system, the intensity of the light after the analyzer is different for different wavelengths, creating colors instead of whiteness at different values of the differential optical path length $w\Delta n$, as shown by the Michel-Levy chart (Fig. 3.7). There is thus advantage of using white light source to perform birefringence observations because the differentiation in color permits illustration of spatial changes in birefringence.

3.3.1.1 Stress Optical Rule

Rheo-optical measurements have been crucial in creating support for the foundation of modern molecular-level dynamic theories. Most analytical, molecular theories for chain dynamics evaluate stresses from bonded intrachain interactions. It is challenging for the tube model[27] to

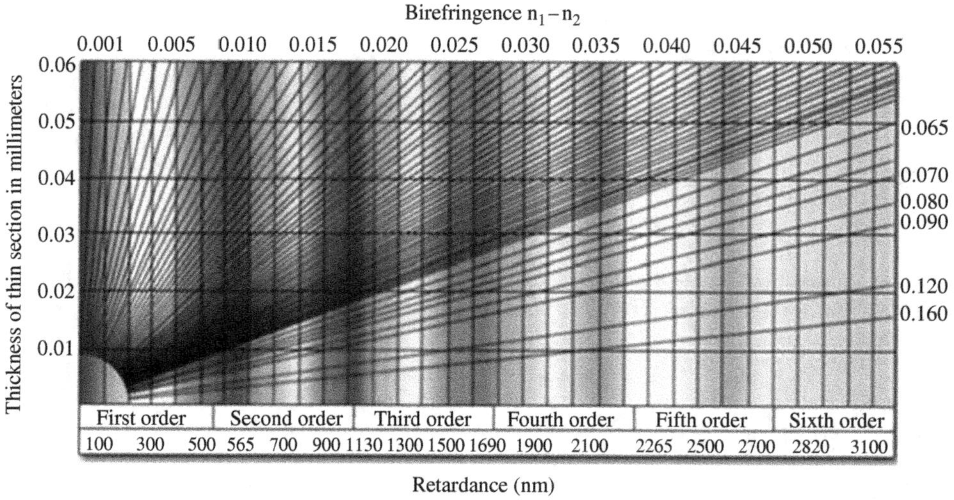

Figure 3.7 Michel-Levy birefringence chart from http://www.olympusmicro.com/primer/techniques/polarized/michel.html. *Source:* Reproduced with permission of Olympus America Inc. (*See color plate section for the color representation of this figure.*)

include nonbonded interactions in the evaluation of mechanical stresses. On the other hand, the idea of including only bonded forces in the calculation is not just a matter of convenience. Certainly, in some limit, especially the linear response regime, the level of stress indeed appears to be linearly proportional to the degree of chain deformation as measured by the birefringence. In the case of simple shear, if the stress is linearly proportional to the birefringence, then we have the following expression relating shear stress σ to birefringence Δn

$$\sigma = (\Delta n/2C)sin2\chi, \tag{3.33}$$

where C is the stress-optic coefficient.[3] N_1, the first normal stress difference, is

$$N_1 = \sigma_{xx} - \sigma_{yy} = (\Delta n/C)cos2\chi. \tag{3.34}$$

N_1 can be approximately measured in a cone-plate apparatus from the normal force F according to $N_1 = 2F/\pi R^2$. Many flow birefringence studies can be found in the monograph by Janeschitz-Kriegl.[28]

3.3.1.2 Breakdown of Stress-Optical Rule

The *stress-optical rule* can be most straightforwardly tested for cross-linked polymeric systems where it is conventional to associate the stress with the network deformation. Ordinarily, the *stress-optical rule* holds for such elastomeric materials. In polymer rheology, the empirical rule usually holds.[28] The theoretical explanation usually involves a demonstration that the optical anisotropic tensor is proportional to the stress tensor, which is taken to be entirely intrachain in origin.[27] Under two obvious conditions, the *stress–stress rule* is expected to be invalid. First, when the glass transition is approached, the stress optical rule fails.[29] As the temperature is cooled toward T, nonbonded forces become dominant. A great deal of stress may correspond to a negligible amount of chain deformation, and therefore a negligible amount of birefringence. It is then no longer instructive to correlate the stress level with any level of optical anisotropy. The stress optical rule is therefore perhaps not so informative in investigations of polymer rheology near the glass transition temperature. Second, when large affine deformation occurs to cause high chain (non-Gaussian) stretching where the chain orientation saturates with further stress buildup.[30,31]

3.3.2 Scattering (X-Ray, Light, Neutron)

Possible effects of deformation and flow to induce structural changes in polymeric materials are often examined with scattering methods. For example, shear could cause demixing or mixing in polymer blends, shifting of the order–disorder transition temperature in block copolymers, lowering the crystallization temperature in semicrystalline polymers, and so on. Scattering techniques can be effective as an *in situ* characterization tool. However, we omit any discussion of the scattering methods because it will not be applied in the rest of the book.

3.3.3 Spectroscopy (NMR, Fluorescence, IR, Raman, Dielectric)

Microscopic structural changes could occur during deformation and flow. It would be desirable to detect such changes by resorting to spectroscopic methods. Many dielectric measurements have been carried out during continuous deformation of polyisoprene-based solutions.[32–34] Such experiments offer information on the state of chain deformation. NMR imaging has been applied to probe velocity field in a simple shear apparatus.[35,36] Fluorescence spectroscopy has been employed to visualize chain dynamics of fluorescently labeled DNA molecules in solutions.[37–40] Combining *in situ* structural characterization with rheometric measurements is a very effective approach necessary for future development in the field, but no further mention of it will be made in this book.

3.3.4 Microrheology and Microscopic Force Probes

Microrheology[41,42] is a popular approach, often applied to probe linear response behaviors of complex fluids. Recently, such a methodology has been extended[43–46] to investigate nonlinear rheological behavior under large deformation with considerable success. It can be hoped that the method will undergo significant further development to combine with single-molecule (DNA) observations. The new method can potentially detect chain disentanglement under large deformation, locally generated by the force-probe bead.

3.4 Advanced Rheometric Methods

The basic modes of shear deformation are startup, stepwise, and oscillatory shear including large-amplitude oscillatory shear. It is often helpful to combine these modes to gain additional information. We call such options "advanced methods" only because they are not used more frequently in rheological studies, though they should be.

3.4.1 Superposition of Small-Amplitude Oscillatory Shear and Small Step Strain during Steady Continuous Shear

It is well established (p. 527 of Ref. [47]) that small-amplitude oscillatory shear can be superimposed onto ongoing steady shear to provide insight about how steady flow has altered the viscoelastic structure of materials such as entangled polymers. Modern rheometers can readily perform this combination: At least it is straightforward to impose small-amplitude oscillatory

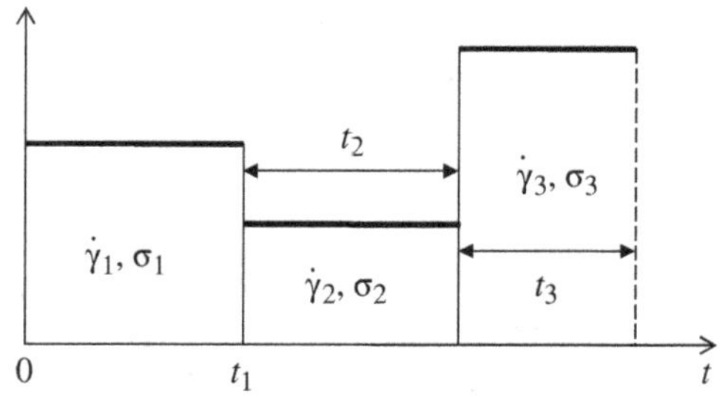

Figure 3.8 Depiction of a general protocol involving either rate or stress switching or a combination of rate and stress switching.

shear in the same shearing direction as the steady shear. We will illustrate the application of this method in Section 7.3.1. A more sophisticated setup is needed to superimpose transverse small-amplitude oscillatory shear onto steady shear. According to the literature,[48] the transverse superimposition reveals similar information as does the superimposed small-amplitude oscillatory shear parallel to the underlying shear.

The idea of the superposition can be extended to contemplate adding a small step strain made at a rate much larger than the underlying steady shear rate, and then determining how the additional stress buildup relaxes during the steady shear. Section 7.3.2 describes how such an experiment demonstrates the nature of steady shear in an entangled polymer solution.

3.4.2 Rate or Stress Switching Multistep Platform

Several characteristic rheological features reflect the state of molecular deformation, either in response to a sudden startup deformation or in the form of elastic recoil after a large deformation. These features will be described in detail in Chapter 7. After the first shear in either rate- or stress-controlled mode, we can interrogate the response by making a rate or stress switch at various later times. With the choice $t_2 = 0$, Fig. 3.8 illustrates a rate or stress hike at t_1. With $\dot{\gamma}_2 = 0$ or $\sigma_2 = 0$ and $t_2 \to \infty$, we can examine stress relaxation or elastic recoil. We can also ask what happens after a certain period (t_2) of relaxation if we then impose a new startup shear, whose $\dot{\gamma}_3$ or σ_3 may be the same as $\dot{\gamma}_1$ or σ_1, or different. The same protocol is valid for uniaxial extension so long as the extension preserves homogeneous deformation without strain localization.

3.5 Conclusion

In this chapter, we have reviewed most of the common rheometric methods for rheological studies that have been applied in linear response and nonlinear regimes. Shear rheometry is discussed in Section 3.1, and extensional rheometry is discussed in Section 3.2. This chapter also briefly discussed birefringence due to deformation-induced optical anisotropy as well as less standard rheological protocols. Conventional rheometry may be inadequate when large deformation becomes inhomogeneous. For well-entangled polymeric liquids, there is no guarantee that homogeneous deformation will always prevail. Specifically, it is worth noting that in many past decades the assumption of shear homogeneity in cone-plate and parallel-disk assemblies was rarely questioned. In the following chapter, we will discuss the argument leading to the assumption of homogeneous shear and consider how to set up experimental systems that can verify this prevailing assumption in modern rheometry.

References

1 Hatzikiriakos SG, Dealy JM. Wall slip of molten high-density polyethylene. 1. Sliding plate rheometer studies. J Rheol 1991;35(4):497–523.
2 Sodeifian G. Non-linear rheology of polymer melts: constitutive equations, rheological properties of polymer blends, shear flow, sliding plate rheometers: LAP Lambert Acad Publ, Germany; 2011, ISBN: 9783845405698.
3 Macosko CW. Rheology principles, measurements, and applications. Wiley-VCH, Inc., New York; 1994.
4 Larson RG. Instabilities in viscoelastic flows. Rheol Acta. 1992;31(3):213–63.
5 Bentley B, Leal L. Computer-controlled four-roll mill for investigations of particle and drop dynamics in two-dimensional linear shear flows. J Fluid Mech. 1986;167:219–40.
6 Bentley B, Leal L. An experimental investigation of drop deformation and breakup in steady, two-dimensional linear flows. J Fluid Mech. 1986;167:241–83.
7 Stone H, Bentley B, Leal L. Experimental study of transient effects in the breakup of viscous drops. J Fluid Mech. 1986;173(1):131–58.
8 Lagnado R, Leal LG. Visualization of three-dimensional flow in a four-roll mill. Exp Fluids. 1990;9(1–2):25–32.
9 Vinogradov G. Viscoelasticity and fracture phenomenon in uniaxial extension of high-molecular linear polymers. Rheol Acta. 1975;14(10):942–54.
10 Vinogradov G, Malkin AY, Volosevitch V, editors. Some fundamental problems in viscoelastic behavior of polymers in shear and extension. J Appl Polym Sci Appl Polym Symp; 1975.
11 Vinogradov G, Volosevitch V, Shatalov V, Yudin V. Flow, high-elastic (recoverable) deformation, and rupture of uncured high molecular weight linear polymers in uniaxial extension. J Polym Sci Polym Phys Ed. 1975;13(9):1721–35.
12 Vinogradov G. Ultimate regimes of deformation of linear flexible chain fluid polymers. Polymer. 1977;18(12):1275–85.
13 Vinogradov G, Malkin AY. Rheology of Polymers (Mir, Moscow, 1980). Translated from the Russian edition; 1977.
14 Meissner J. Development of a universal extensional rheometer for the uniaxial extension of polymer melts. Trans Soc Rheol. 1972;16(3):405–20.
15 Meissner J, Hostettler J. A new elongational rheometer for polymer melts and other highly viscoelastic liquids. Rheol Acta. 1994;33(1):1–21.
16 Sentmanat ML. Dual windup drum extensional rheometer. US patent US6691569 B1; 2004.
17 Sentmanat ML. Miniature universal testing platform: from extensional melt rheology to solid-state deformation behavior. Rheol Acta. 2004;43(6):657–69.
18 Sentmanat M, Wang BN, McKinley GH. Measuring the transient extensional rheology of polyethylene melts using the SER universal testing platform. J Rheol. 2005;49(3):585–606.
19 Sridhar T, Tirtaatmadja V, Nguyen D, Gupta R. Measurement of extensional viscosity of polymer solutions. J Nonnewton Fluid Mech. 1991;40(3):271–80.
20 Bach A, Rasmussen HK, Hassager O. Extensional viscosity for polymer melts measured in the filament stretching rheometer. J Rheol. 2003;47(2):429–41.
21 Tirtaatmadja V, Sridhar T. A filament stretching device for measurement of extensional viscosity. J Rheol. 1993;37(6):1081–102.

22 Kumaraswamy G, Issaian AM, Kornfield JA. Shear-enhanced crystallization in isotactic polypropylene. 1. Correspondence between in situ rheo-optics and ex situ structure determination. Macromolecules. 1999;32(22):7537–47.
23 Kumaraswamy G, Verma R, Issaian A, Wang P, Kornfield J, Yeh F, et al. Shear-enhanced crystallization in isotactic polypropylenePart 2. Analysis of the formation of the oriented "skin". Polymer. 2000;41(25):8931–40.
24 Kornfield JA, Kumaraswamy G, Issaian AM. Recent advances in understanding flow effects on polymer crystallization. Ind Eng Chem Res. 2002;41(25):6383–92.
25 Kumaraswamy G, Kornfield JA, Yeh F, Hsiao BS. Shear-enhanced crystallization in isotactic polypropylene. 3. Evidence for a kinetic pathway to nucleation. Macromolecules. 2002;35(5):1762–9.
26 Fuller GG. Optical rheometry of complex fluids. Oxford University Press, New York; 1995.
27 Doi M, Edwards SF. The theory of polymer dynamics. Oxford University Press, Oxford; 1988.
28 Janeschitz-Kriegl H. Polymer melt rheology and flow birefringence. Springer Science & Business Media; 2012.
29 Muller R, Pesce JJ. Stress optical behavior near the t(g) and melt flow-induced anisotropy in amorphous polymers. Polymer. 1994;35(4):734–9.
30 Matsumoto T, Bogue DC. Stress birefringence in amorphous polymers under nonisothermal conditions. J Polym Sci Part B Polym Phys. 1977;15(9):1663–74.
31 Muller R, Froelich D. New extensional rheometer for elongational viscosity and flow birefringence measurements – some results on polystyrene melts. Polymer. 1985;26(10):1477–82.
32 Watanabe H, Ishida S, Matsumiya Y. Rheodielectric behavior of entangled cis-polyisoprene under fast shear. Macromolecules. 2002;35(23):8802–18.
33 Watanabe H, Matsumiya Y, Inoue T. Dielectric and viscoelastic study of entanglement dynamics: a review of recent findings. Macromol Symp. 2005;228:51–70.
34 Horio K, Uneyama T, Matsumiya Y, Masubuchi Y, Watanabe H. Rheo-dielectric responses of entangled cis-polyisoprene under uniform steady shear and LAOS. Macromolecules. 2014;47(1):246–55.
35 Callaghan P, Cates M, Rofe C, Smeulders J. A study of the "spurt effect" in wormlike micelles using nuclear magnetic resonance microscopy. J Phys II. 1996;6(3):375–93.
36 Callaghan PT. Rheo NMR and shear banding. Rheol Acta. 2008;47(3):243–55.
37 Teixeira RE, Babcock HP, Shaqfeh ES, Chu S. Shear thinning and tumbling dynamics of single polymers in the flow-gradient plane. Macromolecules. 2005;38(2):581–92.
38 Teixeira RE, Dambal AK, Richter DH, Shaqfeh ES, Chu S. The individualistic dynamics of entangled DNA in solution. Macromolecules. 2007;40(7):2461–76.
39 Schroeder CM, Teixeira RE, Shaqfeh ESG, Chu S. Dynamics of DNA in the flow-gradient plane of steady shear flow: observations and simulations. Macromolecules. 2005;38(5):1967–78.
40 Schroeder CM, Shaqfeh ESG, Chu S. Effect of hydrodynamic interactions on DNA dynamics in extensional flow: simulation and single molecule experiment. Macromolecules. 2004;37(24):9242–56.
41 Mason TG, Ganesan K, vanZanten JH, Wirtz D, Kuo SC. Particle tracking microrheology of complex fluids. Phys Rev Lett. 1997;79(17):3282–5.
42 Crocker JC, Valentine MT, Weeks ER, Gisler T, Kaplan PD, Yodh AG, et al. Two-point microrheology of inhomogeneous soft materials. Phys Rev Lett. 2000;85(4):888–91.

43 Chapman CD, Robertson-Anderson RM. Nonlinear microrheology reveals entanglement-driven molecular-level viscoelasticity of concentrated DNA. Phys Rev Lett. 2014;113(9):5.

44 Li YF, Hsiao KW, Brockman CA, Yates DY, Robertson-Anderson RM, Kornfield JA, et al. When ends meet: circular DNA stretches differently in elongational flows. Macromolecules. 2015;48(16):5997–6001.

45 Falzone TT, Robertson-Anderson RM. Active entanglement-tracking microrheology directly couples macromolecular deformations to nonlinear microscale force response of entangled actin. ACS Macro Lett. 2015;4(11):1194–9.

46 Gurmessa B, Fitzpatrick R, Falzone TT, Robertson-Anderson RM. Entanglement density tunes microscale nonlinear response of entangled actin. Macromolecules. 2016;49(10):3948–55.

47 Ferry JD. Viscoelastic properties of polymers. John Wiley & Sons, New York; 1980.

48 Simmons J. Dynamic modulus of polyisobutylene solutions in superposed steady shear flow. Rheol Acta. 1968;7(2):184–8.

4

Characterization of Deformation Field Using Different Methods

Following a long tradition in fluid mechanics, researchers have pursued viscoelastic measurements by designing shear rheometers with the assumption that one surface displacing relative to the other at any speed V would ensure uniform shear. This assumption is illustrated in Fig. 4.1(a). The figure shows a single XY-plane at a fixed z. Every point on the vertical dashed line at $x = x_0$ and time $t = 0$ displaces according to $x = x_0 + (y/H)Vt$. Experimental rheology is founded on this assumption, which is thought to be necessary in order to explore the constitutive behavior of viscoelastic materials. Without the assumption of homogeneous shear, the task of establishing a quantitative relationship among shear stress, strain, and rate of strain is formidable. In this chapter, we begin by presenting a problematic argument that simple shear apparatus should always produce homogeneous shear. After a quick review of yielding in Bingham fluids in Section 4.2 and mention of polydisperse polymers as a candidate to have homogeneous shear in Section 4.3, particle-tracking velocimetry is described in Section 4.4. Important discoveries have been made because of this particle-tracking velocimetric method, which can be applied to geometries including simple shear in Chapter 9, squeezing and entry deformation, both in Chapter 10.

4.1 Basic Features in Simple Shear

4.1.1 Working Principle for Strain-Controlled Rheometry: Homogeneous Shear

It is often assumed that we know *a priori* and can control the deformation and flow fields in displacement-driven shear apparatuses, such as those described in Sections 3.1.1 and 3.1.2. The homogeneous shear depicted in Fig. 1.1(a), 1.2, or 4.1(a) is a necessary condition for conventional design and operation of rheometers. Is this condition just a convenient assumption? Do we really know *a priori* whether shear homogeneity prevails? If we neglect transient physical processes leading to the steady state and conjecture that there is one-to-one correspondence between steady-shear stress σ and local shear rate $\dot{\gamma}$, then it is straightforward to show that homogeneous shear is expected. The argument goes as follows. Let us consider a layer of area Σ and thickness Δ in the sample interior as shown in Fig. 4.1(b). The Newtonian law $F = Ma$ for this layer can be written as

$$\rho \Sigma \Delta \frac{\partial V}{\partial t} = \Sigma[\sigma(y + \Delta) - \sigma(y)]. \tag{4.1a}$$

Equation (4.1a) turns into the equation of motion for the continuum for simple shear when we take $\Delta \rightarrow 0$:

$$\rho \frac{\partial V}{\partial t} = \frac{\partial \sigma}{\partial y}. \tag{4.1b}$$

Nonlinear Polymer Rheology: Macroscopic Phenomenology and Molecular Foundation, First Edition. Shi-Qing Wang.

Companion Website: www.wiley.com/go/nonlinearpolymerrheology

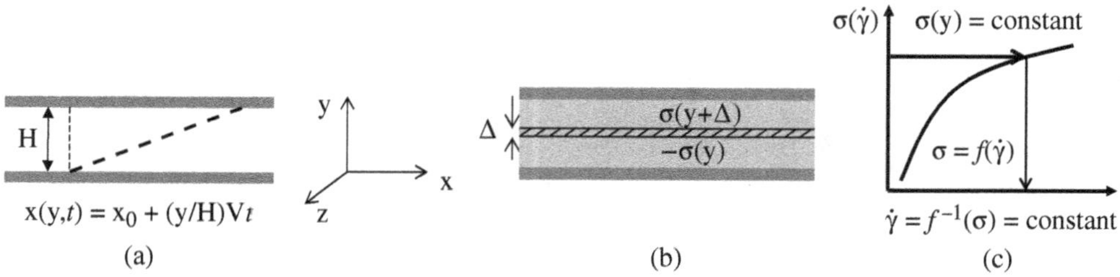

Figure 4.1 (a) Sketch of simple shear based on homogeneous deformation assumption. (b) Illustration of an arbitrarily chosen layer of thickness Δ in the interior of the shear cell that experiences zero net force after the inertia effect vanishes. (c) "Flow curve" or constitutive curve showing one-to-one correspondence between shear stress σ and shear rate $\dot{\gamma}$, so that the relation is invertible, rendering support for the existence of a constant shear rate across the gap.

In steady state, where the shear field is no longer time varying, $\frac{\partial \sigma}{\partial y} = 0$ or $\sigma(y) = \text{constant}$, independent of y. The constant shear stress across the gap does not imply that the shear rate is constant, namely, $\dot{\gamma}(y) = \text{constant}$, unless we assume that there is one-to-one correspondence relating the stress to the rate as $\sigma = f[\dot{\gamma}(y)]$ in steady state, which is invertible to yield $\dot{\gamma}(y) = f^{-1}[\sigma(y)]$, as illustrated in Fig. 4.1(c). When the slope of the curve in Fig. 4.1(c) approaches zero, it is obviously questionable whether it is still reliable to assume this invertibility.

In the field of polymer rheology, the relationship between initial response and steady state has not been well explored. The preceding analysis says nothing about shear homogeneity during transient responses. Can any shear inhomogeneity take place? What would be the cause? Can inhomogeneity persist to the steady state in a metastable manner without violating a monotonic relation between σ and $\dot{\gamma}$ in the steady state? In the area of polymer processing, it is not obvious that the processed materials ever get to reach steady state. Usually the deformation does not reach steady state. Extrusion instabilities such as sharkskin-like surface roughness and gross melt fracture (briefly discussed in Chapter 19) have little to do with the steady-state flow behavior and much to do with the material elastic response at the die exit or entry. Thus, the analysis in the previous paragraph may in general be inapplicable.

We need to recognize that shear homogeneity cannot be assumed for nonlinear rheological responses of viscoelastic materials. In a displacement-driven shear apparatus, only the boundary movement is experimentally controlled. The movement of any layer in the sample interior cannot be determined *a priori* in the absence of any knowledge of the constitutive properties of the material under study. In other words, in studying viscoelastic materials, the experimenter cannot be sure that the simple shear device illustrated in Fig. 3.1(a) creates a uniform strain field so that every layer experiences the same strain γ.

It is useful at this junction to recall our discussion of rheometry in Chapter 3. We assumed that only displacement-driven devices could generate deformation fields in a controlled way *a priori*, independent of the constitutive properties of viscoelastic materials. This assumption is not quite true: As the subsequent chapters will discuss, it turns out that under a variety of rheological conditions shear homogeneity, Fig. 4.1(a–c), cannot be guaranteed to take place. Consequently, the working principle in rheology may collapse.

4.1.2 Stress-Controlled Shear

There is one basic reason why the "rate controlled" mode is the conventional or preferred method to produce shear deformation and probe constitutive behavior in rheometric instruments: By dictating the displacement speed V as shown in Figs. 1.1(a) and 4.1, we hope that this boundary movement produces a constant shear rate of V/H across the sample thickness so that constitutive properties can be directly determined by measuring the shear stress σ. Theoretical modeling (e.g., the tube model) is also conventionally formulated by considering a given system under simple shear and evaluating the stress response. Consequently, stress-controlled (creep) mode has rarely been applied to probe nonlinear rheological responses of entangled polymeric liquids.

It turns out that shearing in controlled-stress mode is as valuable as shearing in controlled-rated mode as an experimental protocol for determination of rheological behavior of viscoelastic materials. The controlled-stress mode is actually rather distinct from the rate-controlled shear mode: We do not prescribe an experimental time scale, as discussed in Eq. (1.3) for rate-controlled shear. The rheological property of the material determines how the material responds to the applied stress, and on what the time scale its structure evolves in time. There is no fundamental difficulty in producing shear deformation at a constant shear stress. In Section 7.4, we show that the creep test actually offers new insights into the nature of shear thinning and polymer disentanglement. Specifically, the test determines in real time how the shear viscosity decreases with time at a constant applied shear stress. The shear rate evolves in time, even though the applied stress is fixed. In the creep mode, however, it is also unknown *a priori* whether homogeneous shear would prevail. Analogous to the argument made in the preceding section, in creep, homogeneous shear should prevail in the steady state: Since the shear stress is the same in every layer, the corresponding shear rate should also be the same, namely, there should be a uniform shear rate across the sample thickness. This inference turns out to be nothing more than a sensible speculation. The validity of the speculation is determined by whether or not the system can quickly reorganize in a globally homogeneous manner in response to the imposed stress.

There is another motivation to explore nonlinear rheological behavior in the creep mode. For well-entangled polymers, especially monodisperse solutions and melts, apart from their capacity to undergo massive wall slip, stress plateau is a commonly known feature. In a stress plateau, the steady-shear stress saturates with increasing shear rate, as sketched in Fig. 7.23. This stress plateau resembles a transition in which the controlling variable is shear stress. It is interesting to examine this transition using stress-controlled shear. Using controlled stress to investigate nonlinear rheological responses of entangled polymers is actually quite unconventional; its use has led to the discovery of the entanglement–disentanglement transition (EDT), detailed in Chapter 7.

4.2 Yield Stress in Bingham-Type (Yield-Stress) Fluids

The perception of homogeneous shear has been around since the beginning of rheology. Let us look as an example at the first paper in the Journal of Rheology, Volume 1 and Issue 1 by Markus Reiner, who treated the problem of plastic flow of Bingham fluids. Even for yield-stress materials that follow the simple Bingham law

$$\dot{\gamma} = 0, \quad \text{for } \sigma < \sigma_y, \tag{4.2a}$$

and

$$\dot{\gamma} = (\sigma - \sigma_y)/\eta, \quad \text{for } \sigma > \sigma_y, \tag{4.2b}$$

it was perceived that simple shear flow could take place at the *constant* shear rate given in Eq. (4.2b). Here, η is the viscosity upon yielding. The Bingham description assumes that the state after yield is structurally uniform, possessing a common viscosity η for at all values of y along the direction of the sample thickness (cf. Fig. 4.1(a)). For Eq. (4.2b) to be true, the structure of the fluid must collapse homogeneously at the yield stress. For yield-stress materials, known as Bingham fluids it seems mathematically problematic to use Eq. (4.2a) to depict the state of deformation below the yield stress, which implies an infinite viscosity before yielding.[1] In reality, the drop in viscosity at σ_y is of a finite magnitude. There is no mystery[2] concerning the behavior of yield-stress materials. The characterization should focus on the sharp drop of the nominal viscosity over a narrow range of the applied stress rather than on whether the material can creep or not by any measure below the yield stress.

Real yield-stress materials may possess a range of cohesive strengths corresponding to the different levels of structural buildup. Thus, beyond the yield stress, the system viscosity η could drop far more steeply than linearly with increasing stress, namely, the system after yield could be strongly shear thinning. The Bingham formula does not depict the rich physics and may be replaced by the following expressions to depict steady-state behavior

$$\dot{\gamma} = \sigma^{B}, \quad \text{for } \sigma < \sigma_y, \tag{4.3a}$$

$$\dot{\gamma} - \dot{\gamma}_0 = (\sigma - \sigma_y)^{A}, \quad \text{for } \sigma_y + \Delta > \sigma > \sigma_y, \tag{4.3b}$$

$$\dot{\gamma} - \dot{\gamma}_1 = (\sigma - \sigma_y - \Delta)^{C}, \quad \text{for } \sigma \geq \sigma_y + \Delta, \tag{4.3c}$$

where all the prefactors are omitted for notational simplicity, so that $\dot{\gamma}_0 = \sigma_y^B$ and $\dot{\gamma}_1 = \dot{\gamma}_0 + \Delta^A$. Typically, a yielding transition takes place over a narrow range of stress as indicated in Eq. (4.3b) and Figure 4.2 where strong shear thinning takes place with exponent $A \gg 1$. Away from the transition region either below or above yielding, the relationship between stress and shear rate can be much closer to linearity, namely, the exponents B and C are much smaller than A. As illustrated in Fig. 4.2, the yielding phenomenon revealed in a controlled-stress experiment may be viewed as a transition. The Bingham expression seems both inaccurate and incomplete. Rheological behavior of many systems can fall into this category of the so-called yield-stress materials, including well-entangled polymeric liquids with monodisperse molecular weight distributions. The extreme shear thinning behavior of monodisperse entangled polymers is precisely what Eq. (4.3b) depicts; their terminal flow regimes obey Eq. (4.3a) with $B = 1$. A phenomenological description similar to that given in Eqs. (4.3a)–(4.3c) assumes homogeneous shear in the steady state: Because the stress σ is position independent, the shear rate $\dot{\gamma}$ must also be a constant everywhere along the gap as argued in Section 4.1.1. However, the argument could be incorrect, as noted above. When inhomogeneity is present, Eqs. (4.3a)–(4.3c) do not provide a complete description of the rheological states.

For highly viscoelastic materials that exhibit yield stress, it is perhaps a bold assumption that such materials transform in a uniform manner during yielding transition. Does the yielding, namely, the *transition from elastic deformation to irrecoverable flow*, occur in a spatially homogeneous manner in materials such as dense suspensions, strongly entangled polymers, gels, and granular materials? Obviously, we need to develop effective experimental methods to determine whether or not the assumption of homogeneous deformation and flow, which is a cornerstone

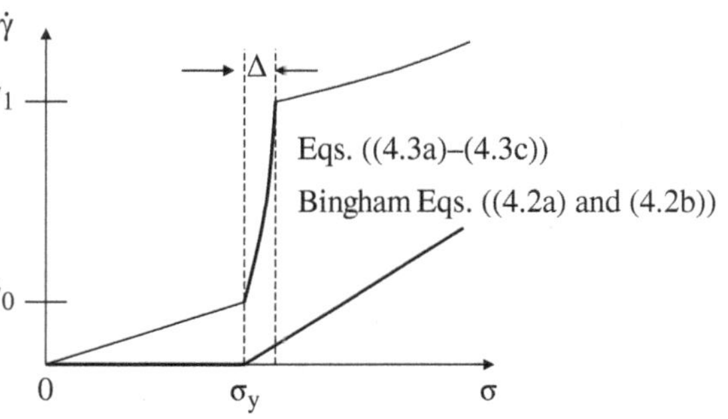

Figure 4.2 Comparison between the Bingham equation and the alternative expression given in Eqs. (4.3a)–(4.3c), where the rates are merely the nominal rates.

of the modern rheometry, is indeed valid. Before proceeding to Section 4.4 on the method of particle-tracking velocimetry for transparent materials, we discuss cases in which homogeneous shear should prevail.

4.3 Cases of Homogeneous Shear

Well-entangled monodisperse polymeric liquids exhibit a wide stress plateau in steady shear, beyond the terminal flow regime, namely, for Weissenberg number $Wi > 1$. Equivalently speaking, we expect a sharp decrease in the apparent shear viscosity over a narrow range of shear stress as shown in Fig. 7.2(a) as well as in Fig. 7.23. We may wonder whether these liquids can transform uniformly in space from the resting (initial) state of high viscosity to a state of much reduced viscosity. The high zero-shear viscosity $\eta_0 = G_{pl}\tau$ originates from existence of a long relaxation time τ due to chain entanglement, as explained by the reptation-based tube model briefly reviewed in Section 2.3.2. For polymeric liquids that are not entangled or only weakly entangled, the stress plateau is narrow and not so "flat," so that there may be a one-to-one correspondence between shear stress and rate.

When the exponent A in Eq. (4.3b) is much larger than unity, a very small change in σ results in a huge change in $\dot{\gamma}$, creating the possibility that during steady shear different layers across the sample thickness may have different values of $\dot{\gamma}$. Polymeric liquids that have broad molecular weight distributions tend to show exponent A not greatly less than unity and, therefore, do not possess a significant stress plateau. Thus, homogeneous shear likely prevails in strongly polydisperse polymer samples. Different local rates at different positions along the sample thickness direction (Y-axis, which is the direction of the velocity gradient) would involve different shear stresses in the absence of a stress plateau. Since the shear stress is the same along the Y-axis in steady state, only a single shear rate is expected across the thickness. Thus, shear homogeneity can be expected for sufficiently polydisperse polymer.

In the absence of any reliable theoretical understanding, we cannot be sure when shear inhomogeneity will take place. Therefore, it is imperative that effective methods be developed and applied to measure *in situ* deformation and flow fields during conventional rheometric measurements. In the following section, we review one such method that has been successfully implemented to unravel the phenomenology of nonlinear polymer rheology.

4.4 Particle-Tracking Velocimetry (PTV)

Several experimental methods have been applied to determine the velocity profile inside a shear rheometer. Among them,[3] particle-tracking velocimetry (PTV) is the simplest[4] and least expensive to set up. When set up properly, PTV affords good temporal resolution, allowing measurements of the time-dependent velocity field across the gap of a shear cell.[5] Because of its unique advantages over other methods, we provide a relatively thorough review of the PTV method. Reference [3] reviewed other means to evaluate the shear strain field in simple shear.

There are several choices for the tracking particles that can be dispersed into a sample. Silver-coated hollow glass spheres (HGSs) ranging from 5 to 30 μm are a convenient choice if the spatial resolution needs not be higher than 50 μm. The conventional rheometric setup involves a sample thickness on the order of 1 mm. Thus, these hollow glass spheres are well suited for PTV observations.

When examining a sample sheared in a much smaller gap, for example, about 50 μm, confocal microscopy is required. Micrometer-sized fluorescent particles are good tracking particles.

Quantum dots (QDs) are nanosized particles, bright upon excitation and have been used for tracking. Use of QD eliminates any concern that solid particles could have produced invasive effects on the structural continuum. For our purpose, such a concern is unnecessary because the volume fraction of the PTV tracking particles is typically a few hundred parts per million.

4.4.1 Simple Shear

There are strong reasons to examine deformation fields and confirm the validity of the homogeneous shear assumption, even in such a simple shear device as a cone-plate assembly. If shear were not homogeneous, rheometric measurements would not be a straightforward characterization tool. Let us take the simplest case of viscosity measurement according to Eq. (1.8a) along with Fig. 1.1(d). Both shear stress σ and shear rate $\dot{\gamma}$ in the gap must be determined in order to measure η. If $\dot{\gamma}$ cannot be prescribed according to $\dot{\gamma} = V/H$, as depicted in Fig. 1.1(d), the conventional simple rheometric determination of σ would be inadequate. The success of conventional rheometry depends on the validity of the assumption of shear homogeneity. Shear homogeneity is assumed not just for convenience but because it is necessary. Without the shear homogeneity condition, rheometric measurements do not provide enough information for analysis.

4.4.1.1 Velocities in XZ-Plane

A conventional description of simple shear designates X as the shearing direction, Y as the direction of the velocity gradient, and Z as the vorticity axis, as shown in Fig. 4.1(a). It is rather straightforward to place a video CCD camera above a sample looking down onto the XZ plane. Such an arrangement has been implemented before in the literature.[6] The drawback of such a setup is twofold: (a) this configuration cannot track particle motions in different XZ planes at the same time; (b) the microscopic objective lens attached to a CCD camera has poor vertical resolution, making it difficult to make accurate velocity measurements of layers spaced along Y-axis. Thus, the XZ-PTV method is essentially limited to monitoring motions of particles near the sample/wall interface or making crude characterizations of the steady-state velocity profile, in cases where we can take the time to scan different layers across the sample thickness. The XZ-PTV protocol has its time resolution limited by how fast the CCD camera can be moved along Y-axis to scan different layers in a short period.

Vertical resolution along the velocity gradient direction can be improved using confocal microscopy. However, the cost of a confocal microscope is orders of magnitude higher than that of a conventional microscope lens. On the other hand, with confocal microscopy, high spatial resolution can be achieved using this configuration with an objective lens as high as 60× or 100×. In this case, the gap distance needs to be much smaller than 1 mm because of the short working distances for such high-magnification lens. Moreover, if the gap distance H needs to be as small as 50 μm, the XZ-PTV configuration with confocal microscope is the only choice.

Shearing at a gap of H = 50 μm is an attractive scenario because the edge instability can no longer affect the determination of shear strain field in rotational rheometry that is 3 mm away from the rim, as shown in Fig. 4.3. On the other hand, because of the much reduced gap, wall slip can be a dominant feature, which is a subject of great importance, to be discussed in Chapter 6. In Chapter 9, we discuss how to suppress wall slip and observe shear strain localization free of edge effects in such an unconventionally low gap distance.

4.4.1.2 Deformation Field in XY Plane

It is actually rather straightforward to accomplish particle-tracking velocimetric (PTV) observations in the XY-plane for both rotational and linearly displaced shear apparatuses, as shown

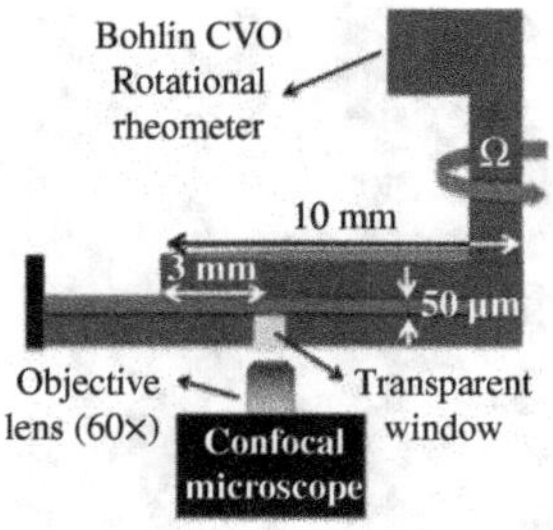

Figure 4.3 A rheomicroscopic setup involving a rotational rheometer such as Bohlin CVO over a confocal microscope, where the maximum velocity could be 500 μm/s, corresponding to a shear rate of 10 s^{-1}. *Source:* Boukany *et al.* 2010 [7]. Reproduced with permission of American Physical Society.

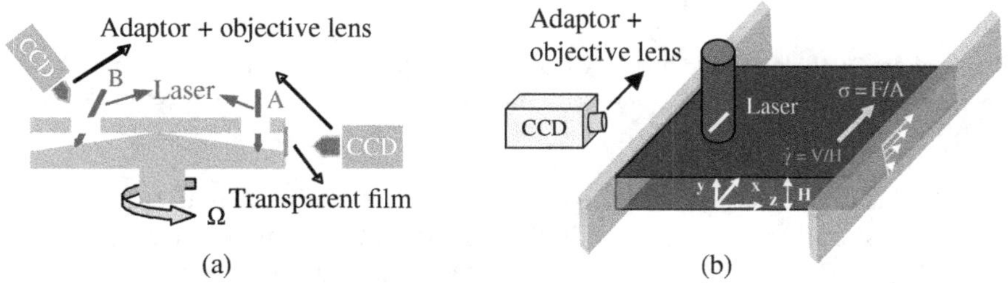

Figure 4.4 (a) Illustration of a cone-plate apparatus coupled to particle-tracking velocimetry (PTV). *Source:* Wang 2007 [4]. Reproduced with permission of Wiley. (b) PTV setup for a sliding-plate shear cell. *Source:* Boukany and Wang 2007 [8]. Reproduced with permission of American Institute of Physics. (*See color plate section for the color representation of this figure.*)

in Fig. 4.4(a,b), respectively. Although Fig. 4.4(a) is based on rotational shearing that should be appropriately described using cylindrical coordinates, the PTV observation is so localized that the plane made of the shearing and velocity gradient axes is conveniently denoted as "XY" plane. The XY-PTV method monitors the deformation field across the sample thickness (i.e., along the velocity gradient direction) with excellent temporal resolution because every video frame records the positions of all particles in the XY plane. In this experiment, a laser beam is incident along the Y-axis and the movements of individual particles are video-recorded for postmeasurement analysis. A crucial aspect of the XY-PTV is the design of the illuminating laser beam. Since (2–3)× objective lenses have rather poor vertical (Z) resolution, it is necessary to ensure that the incident laser beam is a sheet of light whose cross section is a thin line desirably of thickness below 0.1 mm, as indicated in Fig. 4.4(b). In other words, by having a thin sheet of laser beam, the difficulty to have high Z resolution is overcome. In scheme B in Fig. 4.4(a), multiple imaging can be minimized if the illuminating beam has a thin sheet-like cross section.

When the PTV is based on a low-speed video camera with 30 frames per second (fps) and exposure time of $\Delta t = 1/60$ s, significant streaks can show up as the illuminated particles draw bright lines in Fig. 4.5. Such a feature can be used to evaluate the local velocity. The appearance of streaks sets up the upper limit for the low-speed PTV technique. If we limit the length of streak to 50 μm, then the upper speed limit of a tracking particle is $V^* = 50\,\mu m/(1/60)s = 3\,mm/s$. For a conventional gap of H = 1 mm, severe streak cannot be avoided for any shear rate larger than $V/H = 3\,s^{-1}$. In practice, a typical PTV study, based on a conventional rotational rheometer, does not involve shear rate larger than 5–10 s^{-1}.

$$V^*(PTV) < 5\text{–}10\,mm/s. \tag{4.4}$$

For example, Fig. 4.5 is PTV images at 10 s^{-1} where the length of the streak reveals the particle speed V. On the other hand, there is no difficulty to replace a low-speed CCD camera with a high-speed camera although higher laser intensity will typically be required because the high-speed cameras tend to be less light sensitive.

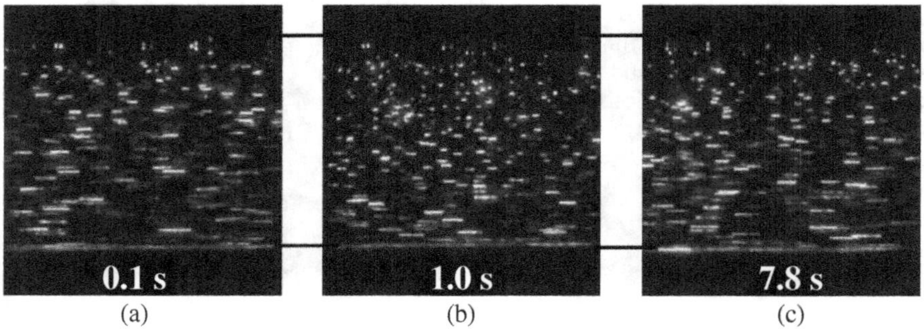

Figure 4.5 Three frames from PTV video recording of the illuminated tracking particles at various times during startup at a nominal shear rate of $10\,s^{-1}$. *Source:* Tapadia and Wang 2006 [9]. Reproduced with permission of American Physical Society.

The XY-PTV method is imperfect in the sense that it can only examine a small region in the sheared sample, namely, about 2 mm by 1.5 mm in the XY plane. Fortunately, the deformation to be examined with the XY-PTV is typically laminar and has translational invariance, and the shear field is largely the same at different planes specified by different Z values.

4.4.2 Channel Flow

The PTV method is not limited to simple shear devices where shear is produced by the displacement of the boundaries. Particle-tracking velocimetry can be readily applied to interrogate the deformation field in pressure-driven apparatuses. Such applications of the PTV method turn out to be rather valuable because they enable us to visualize how entangled polymers respond at the die entrance and in the die land and extend our knowledge about whether and when spatial structural heterogeneity develops. The shear stress is typically smoothly varying in the die entry, so in the absence of shear strain localization, the velocity field should also vary continuously in space. If the velocity field becomes discontinuous under some extrusion conditions, PTV can tell us how it happens, as shown in Section 10.2. In such geometric setups, conventional variables such as the pressure at the die entry and die throughput are clearly insufficient for a complete depiction of rheological responses, when strain localization is present.

Having pointed out the necessity and importance to carry out PTV observations in pressure-driven apparatuses, we describe one such PTV setup from the literature. To allow laser illumination, one side of the channel is made of glass as shown in Fig. 4.6. One of the narrow sides of the rectangular channel die is also made of glass, through which a CCD camera peeks along the Y-axis. By opening two optical windows, the extrusion behavior at

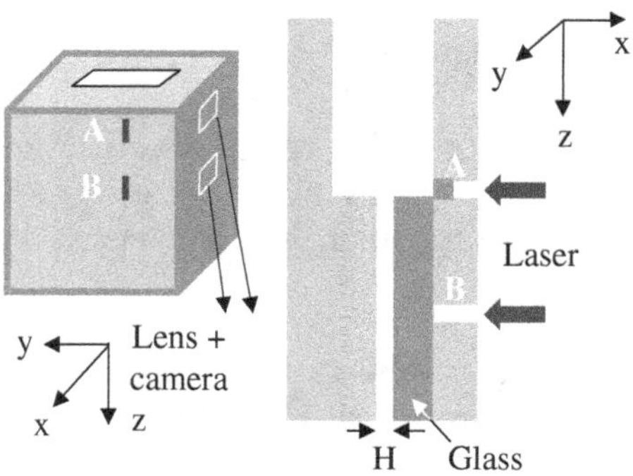

Figure 4.6 Particle-tracking velocimetric setup for observation of strain localization at the slit die entry and inland. Here the laser passes along X-axis through thin openings to illuminate a thin plane of PTV particles in the sample. The objective lens along with a video camera is placed along the Y-axis.

the die entrance and in the die land can be separately examined; the deformation field can be video-recorded in real time. Depending on whether the material under investigation is an entangled melt or solution of low elastic plateau modulus, the setup in Fig. 4.6 refers to either a capillary melt rheometer[10] or a customer-built channel-flow device driven by a syringe pump.[11] Both cases are discussed in some detail in Section 10.2.

4.4.3 Other Geometries

The value of implementing PTV is so obvious that it should be applied to many cases. In this book, we illustrate two more cases, discussed first in Section 5.3 and subsequently in Chapter 10. When squeezing a sample confined laterally so that it has only one direction in which to move, PTV observations make such setups valuable as a rheometric apparatus. Insightful information can be obtained about the highly nonlinear rheological responses of entangled polymer melts.

4.5 Single-Molecule Imaging Velocimetry

Particle-tracking velocimetry reviewed in the preceding sections was developed as an accessory to commercial rotational shear rheometers.[5] While it is adequate to determine the deformation field in real time, it lacks the spatial resolution needed to provide direct information on molecular behavior during simple shear in entangled solutions including wormlike micelle and DNA solutions.[12] While synthetic polymer chains are typically too small for any time-resolved visualization technology to image during deformation, DNA (deoxyribonucleic acid) molecules can be big enough to be spatially resolved with a conventional confocal fluorescence microscope. More importantly, aqueous DNA solutions can be rather entangled at just 1% or 2% concentration. They are in every way rheologically indistinguishable and thus appear equivalent to entangled polymer solutions and melts.[7,13–17] Furthermore, DNA molecules can readily be stained with fluorescent dye molecules to make them individually visible under microscope. When a DNA of sufficiently high-molecular weight is extended, it is many micrometers long. Equipped with a 60× objective lens, such conformational changes can be spatial resolved. Thus, as applied to DNA solutions, single-molecule imaging velocimetry has become a reality and has the potential to delineate how polymer entanglement responds to large deformations. An example of applying SMIV will be presented in Chapter 6, examining wall slip based on the configuration shown in Fig. 4.3.

4.6 Other Visualization Methods

While PTV and PIV (particle imaging velocimetry) are effective and straightforward techniques, other methods have been applied to determine velocity profiles. These alternative methods include NMR imaging velocimetry,[18–20] ultrasonic speckle velocimetry (USV),[21–23] photon correlation spectroscopy (dynamic light scattering, DLS),[24–28] and laser Doppler velocimetry (LDV).[29–34] Comparisons between these techniques have been discussed in the literature.[3] It suffices to mention that NMR and USV can be applied to turbid systems where optical methods (PTV, DLS, and LDV) may not work well. NMR uniquely does not require the incorporation of tracer particles into the system. However, in general, introduction of a minute number of particles into the system does not complicate the interpretation of PTV measurements.

References

1 Barnes HA. The yield stress—a review or 'παντα ρει'—everything flows? J Nonnewton Fluid Mech. 1999;81(1):133–78.
2 Barnes H, Walters K. The yield stress myth? Rheol Acta. 1985;24(4):323–6.
3 Manneville S. Recent experimental probes of shear banding. Rheol Acta. 2008;47(3):301–18.
4 Wang SQ. A coherent description of nonlinear flow behavior of entangled polymers as related to processing and numerical simulations. Macromol Mater Eng. 2007;292(1):15–22.
5 Wang S-Q. Rheometer allowing direct visualization of continuous simple shear in non-Newtonian fluid. US Patent 7,908, 905; 2011.
6 Archer L, Larson R, Chen Y-L. Direct measurements of slip in sheared polymer solutions. J Fluid Mech. 1995;301:133–51.
7 Boukany PE, Hemminger O, Wang S-Q, Lee L. Molecular imaging of slip in entangled DNA solution. Phys Rev Lett. 2010;105(2):027802.
8 Boukany PE, Wang S-Q. A correlation between velocity profile and molecular weight distribution in sheared entangled polymer solutions. J Rheol. 2007;51(2):217–33.
9 Tapadia P, Wang SQ. Direct visualization of continuous simple shear in non-Newtonian polymeric fluids. Phys Rev Lett. 2006;96(1):016001.
10 Zhu X, Yang W, Wang S-Q. Exploring shear yielding and strain localization at the die entry during extrusion of entangled melts. J Rheol. 2012;57(1):349–64.
11 Hemminger OL, Boukany PE, Wang S-Q, Lee L. Flow pattern and molecular visualization of DNA solutions through a 4: 1 planar micro-contraction. J Non-Newtonian Fluid Mech. 2010;165(23):1613–24.
12 Boukany PE, Wang S-Q. Use of particle-tracking velocimetry and flow birefringence to study nonlinear flow behavior of entangled wormlike micellar solution: From wall slip, bulk disentanglement to chain scission. Macromolecules. 2008;41(4):1455–64.
13 Boukany PE, Hu YT, Wang S-Q. Observations of wall slip and shear banding in an entangled DNA solution. Macromolecules. 2008;41(7):2644–50.
14 Hemminger OL, Pouyan EB, Wang S-Q, Fan L, Lee LJ, editors. Visualization of Individual DNA Chains In Entangled Solutions Using High-Speed Confocal Microscopy. The 2008 Annual Meeting; 2008.
15 Boukany PE, Wang S-Q. Shear banding or not in entangled DNA solutions depending on the level of entanglement. J Rheol. 2009;53(1):73–83.
16 Boukany PE, Wang S-Q. Exploring the transition from wall slip to bulk shearing banding in well-entangled DNA solutions. Soft Matter. 2009;5(4):780–9.
17 Boukany PE, Wang S-Q. Shear banding or not in entangled DNA solutions. Macromolecules. 2010;43(17):6950–2.
18 Fischer E, Callaghan P. Is a birefringence band a shear band? Europhys Lett. 2000;50(6):803.
19 Holmes W, Lopez-Gonzalez M, Callaghan P. Fluctuations in shear-banded flow seen by NMR velocimetry. Europhys Lett. 2003;64(2):274.
20 Callaghan PT. Rheo NMR and shear banding. Rheol Acta. 2008;47(3):243–55.
21 Manneville S, Bécu L, Colin A. High-frequency ultrasonic speckle velocimetry in sheared complex fluids. Eur Phys J Appl Phys. 2004;28(3):361–73.
22 Manneville S, Salmon J-B, Bécu L, Colin A, Molino F. Inhomogeneous flows in sheared complex fluids. Rheol Acta. 2004;43(5):408–16.
23 Manneville S, Salmon J-B, Colin A. A spatio-temporal study of rheo-oscillations in a sheared lamellar phase using ultrasound. Eur Phys J E Soft Matter. 2004;13(2):197–212.
24 Fuller G, Rallison J, Schmidt R, Leal L. The measurement of velocity gradients in laminar flow by homodyne light-scattering spectroscopy. J Fluid Mech. 1980;100(03):555–75.

25 Salmon J-B, Bécu L, Manneville S, Colin A. Towards local rheology of emulsions under Couette flow using dynamic light scattering. Eur Phys J E Soft Matter. 2003;10(3):209–21.
26 Salmon J-B, Colin A, Manneville S, Molino F. Velocity profiles in shear-banding wormlike micelles. Phys Rev Lett. 2003;90(22):228303.
27 Di Leonardo R, Ianni F, Ruocco G. Flow between rotating finite disks with a closed end condition studied by heterodyne photon-correlation. J Fluid Mech. 2005;525:27–36.
28 Salmon J-B, Manneville S, Colin A. Shear banding in a lyotropic lamellar phase. I. Time-averaged velocity profiles. Phys Rev E. 2003;68(5):051503.
29 Wassner E, Schmidt M, Münstedt H. Entry flow of a low-density-polyethylene melt into a slit die: an experimental study by laser-Doppler velocimetry. J Rheol. 1999;43(6):1339–53.
30 Münstedt H, Schmidt M, Wassner E. Stick and slip phenomena during extrusion of polyethylene melts as investigated by laser-Doppler velocimetry. J Rheol. 2000;44(2):413–27.
31 Münstedt H, Schwetz M, Heindl M, Schmidt M. Influence of molecular structure on secondary flow of polyolefin melts as investigated by laser-Doppler velocimetry. Rheol Acta. 2001;40(4):384–94.
32 Schwetz M, Münstedt H, Heindl M, Merten A. Investigations on the temperature dependence of the die entrance flow of various long-chain branched polyethylenes using laser-Doppler velocimetry. J Rheol. 2002;46(4):797–815.
33 Hertel D, Münstedt H. Dependence of the secondary flow of a low-density polyethylene on processing parameters as investigated by laser-Doppler velocimetry. J Nonnewton Fluid Mech. 2008;153(2):73–81.
34 Schuberth S, Münstedt H. Simultaneous measurements of velocity and stress distributions in polyisobutylenes using laser-Doppler velocimetry and flow induced birefringence. Rheol Acta. 2008;47(1):111–9.

5

Improved and Other Rheometric Apparatuses

When describing the various rheometric setups in Chapter 3, we omitted any mention of major experimental difficulties. It is fair to say that significant progress in the field of nonlinear rheology of strongly entangled polymer solutions and melts requires us to first confront the severe experimental limitations that make rheological quantification unreliable. The dominant experimental challenges stem from finite size effects. In the following two sections, we describe how to either minimize or eliminate the influence of edge instabilities on rheometric characterization of nonlinear rheological behavior. Because edge fracture is so pronounced in the large deformation of entangled melts in rotational shear devices including cone-plate and parallel-plate setups, no commercial rotational rheometer is designed to accommodate rheometric measurements of melts under the condition of $Wi > 1$. Specifically, existing rheometers have a rather low range of torque T and normal force F_N tolerance. Specifically, these rheometers typically have $T_{max} = 0.2$ N m and $F_{N(max)} = 20–50$ N.

Melts have shear stress at large deformation on the order of 1 MPa with the normal stress in excess of 1 MPa. When the fixture, for example, a cone plate, adopts reduced dimensions, for example, $R < 5$ mm in order to stay below the instrumental limits, the influence of any edge effects is greatly amplified by the size reduction. The commercially available shear rheometric instruments are not suitable to study nonlinear rheological responses of entangled melts. The exceptions are the capillary rheometers. Unfortunately, capillary rheometry is not suitable for establishment of constitutive behaviors because the deformation field inside the die is unknown *a priori* – there is a stress gradient, as shown in Eq. (3.7a). Appendix 11.A of Chapter 11 comments further on the limitation of capillary rheometry. In contrast, boundary-displacement-driven shear apparatuses such as cone plate prescribe a constant shear stress across the gap. Consequently, as discussed in Section 4.1, it can be argued that the shear field may be homogeneous as depicted in Fig. 1.1(a,d). Although the assumption of shear homogeneity needs to be verified on a case-by-case basis by particle-tracking velocimetry (PTV) observations for instance, displacement-controlled shear configuration does offer a much simpler scenario than capillary rheometry and permits an objective investigation of the possibility of strain localization.

Our knowledge of nonlinear rheological behavior of entangled polymers will largely be based on experiments on entangled polymer solutions involving commercial rotational rheometry. For entangled solutions, the finite-size effect associated with meniscus failure can be dealt with. As for entangled melts, specially built shear devices[1,2] have been employed effectively to study certain aspects of nonlinear rheological responses. Sliding plate rheometers can be used to produce stepwise shear, without encountering complications arising from the meniscus instability. In Section 5.1, we describe a slight variation of the sliding plate rheometer, which has been successfully used to study interfacial stick–slip transition in Section 6.2.2.

Nonlinear Polymer Rheology: Macroscopic Phenomenology and Molecular Foundation, First Edition. Shi-Qing Wang.

Companion Website: www.wiley.com/go/nonlinearpolymerrheology

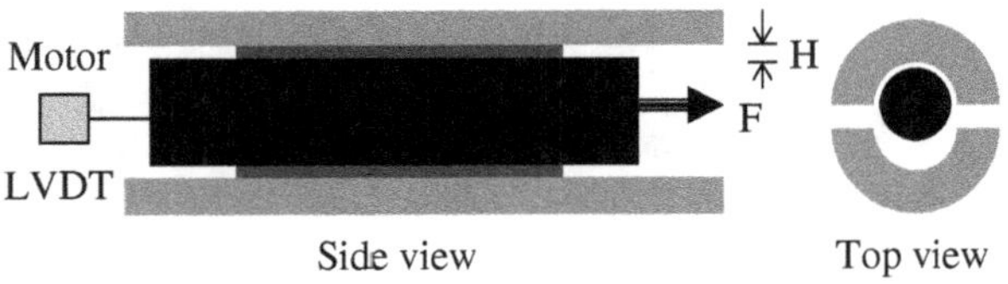

Figure 5.1 Depiction of customer-made device for linear displacement, comprising an inner cylindrical rod and an outer cylinder that are two half cylinders. *Source:* Boukany *et al.* 2006 [4]. Reproduced with permission of American Institute of Physics.

5.1 Linearly Displaced Cocylinder Sliding for Simple Shear

There is no commercially available rheometric apparatus that uses linear boundary displacement. Unlike the rotational rheometers, such a device cannot produce an unlimited amount of strain. The range of operation of a sliding plate rheometer is limited by the inherent meniscus instability at the perimeter. Here we consider a more stable configuration that significantly reduces the amount of meniscus.[3] Let us imagine wrapping a sheet of sample around an inner cylinder that is placed inside a larger second cylinder as shown in Fig. 5.1, where the outer cylinder is shown to be made of two halves. In contrast to the "free-standing" sliding plate scenario shown in Fig. 3.1(b), the meniscus is eliminated on two sides. The remaining meniscuses at the front and back are much more stable against large deformations. This apparatus[5,6] thus allows significantly larger strains to be exerted on the sample in comparison to the sliding plate shear cell. Once again, in the same small-gap approximation, simple shear can be produced by displacing the inner (or equivalently outer) cylinder with a constant velocity V or force F. For a finite ratio of the gap Δ to the inner radius R, there is a small shear stress gradient: the ratio of the shear stress at the outer surface to that on the inner surface should be approximately $(1 + \Delta/R)^{-1}$ because the area under shear at the inner surface is smaller, yet the total force on the inner surface should be the same as that on the outer surface. In practice, it is easy to attain Δ/R as small as a few percent.

This apparatus is convenient for sluggish well-entangled melts or concentrated solutions that have sufficient structural integrity on feasible experimental scales. To ease sample loading, the outer cylinder is made of two semicylinders as shown in Fig. 5.1. It is most effective and inexpensive to operate in the constant force mode using an air cylinder to drive the inner rod. If there is one-to-one relationship between stress and shear rate, this device may allow steady-state properties to be probed. In the constant force mode, we can use such a coaxial-shear device to characterize the wall slip behavior of entangled polymer solutions and melts, as shown in Section 6.2.2.

5.2 Cone-Partitioned Plate (CPP) for Rotational Shear

Viscoelastic materials are expected to suffer from a variety of mechanical or structural instabilities whenever the deformation rate is significantly larger than the dominant relaxation rate. These material instabilities are unwelcome in rheometry and make the study of nonlinear rheological behavior challenging to carry out. Because all apparatuses for simple shear are necessarily of finite dimensions, a meniscus is inevitably part of the measurement system. The meniscus is the free surface at the rim of the sample, as shown in Fig. 3.2. During continuous shear in rotational rheometry at $Wi > 1$ using such as cone-plate or parallel disk apparatus, the meniscus for well-entangled polymeric liquids may become unstable. It then becomes challenging to attain a steady state. The conventional explanation identifies the emergence of normal stresses as the cause of edge instability and fracture.[7] Such understanding is based on a continuum mechanical analysis[8] and therefore does not explore the molecular origin of edge failure of entangled polymeric liquids. The confusion about causality has led Inn *et al.*[9] and Sui and McKenna[10] to incorrectly suggest that edge instabilities are the origin of the entanglement–disentanglement transition phenomenon discussed in Section 7.4.

The difficulty associated with edge instability is so severe that it really threatens the use of cone-plate assembly in rheometric studies of nonlinear rheological behavior. However, capillary flow rheometry is not a desirable alternative if the purpose is to probe the constitutive behavior: The lack of a spatially uniform shear stress in a capillary or channel flow device makes it impossible to quantify the constitutive relationship between shear stress and velocity gradient away from the polymer/wall interface (cf. Appendix 11.A).

In principle, rheometric measurements can be made reliably without complications from meniscus instability, by adopting a clever design proposed long ago by Meissner *et al.*[11] and popularized by Schweizer.[12–14] Fig. 5.2 depicts a cone-partitioned plate (CPP) setup. For this design to function well, the torque transducer must be connected to the stationary side. Torque-controlled rotational rheometers typically apply rotational motion and measure the corresponding torque from the same shaft and are therefore not suitable for implementation of cone-partitioned plate methods. A strain-controlled rotational rheometer, available commercially from TA Instruments, is suitable to accommodate CPP. In principle, we can make $R_2/R_1 \gg 1$ so that the sample under shear between the cone and inner disk remains intact and unaffected by meniscus fracture even in the long time limit. In practice, the full potential of the CPP is still limited by the low tolerance of normal force (2 kg). As analyzed in the following text, CPP is not suitable for entangled melts of high elastic modulus. On the other hand, the CPP makes it possible to explore thoroughly nonlinear rheological behavior of softer viscoelastic materials including entangled polymer solutions, thus greatly extending the capability of available rheometric methods.

The key idea here is to realize radial decoupling by making two pieces, a disk and a ring as shown in Fig. 5.2(b) to replace the traditional single disk. This partition occurs on the stationary side. Only the torque produced by deformation within the inner disk is measured in the rheometer. The sample typically does not escape from the small clearance between the disk and ring unless there is a great deal of normal force. Normally, the clearance is filled with the same sample. In the presence of a significant normal force, the gap between the plate and ring can be covered up as shown in Fig. 5.2(a).

We cannot overstate the value of CPP. It has not only improved the reliability and accuracy of rotational rheometric measurements of entangled polymer solutions during continuous shear but it has also made it possible, for the first time, to identify nonlinear rheological responses

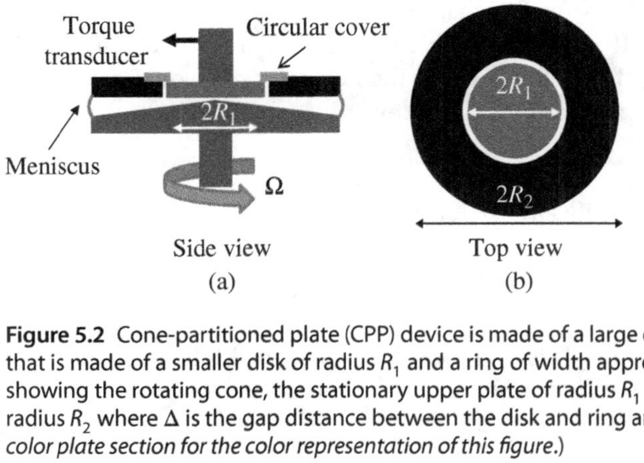

Figure 5.2 Cone-partitioned plate (CPP) device is made of a large cone (of radius R_2) against a partitioned plate that is made of a smaller disk of radius R_1 and a ring of width approximately equal to $(R_2 - R_1)$. (a) Side view showing the rotating cone, the stationary upper plate of radius R_1 and a ring of inner radius $R_1 + \Delta$ and outer radius R_2 where Δ is the gap distance between the disk and ring and (b) top view of the partitioned plate. (*See color plate section for the color representation of this figure.*)

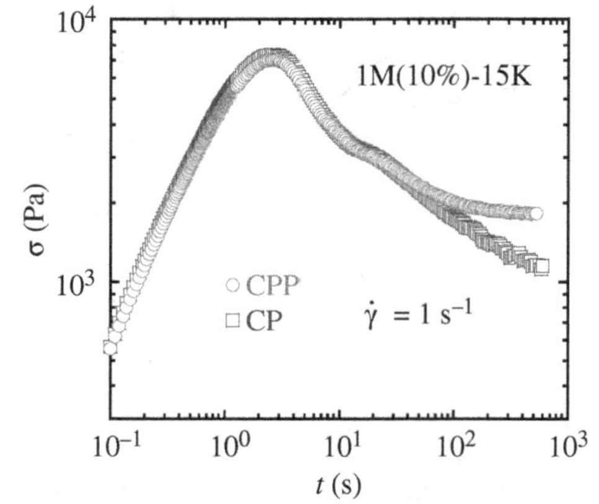

Figure 5.3 Comparison of shear stress growth upon startup shear at a shear rate of $1.0\ s^{-1}$ between C/P and CPP device. *Source:* Ravindranath and Wang 2008 [15]. Reproduced with permission of American Institute of Physics.

of entangled polymer solutions with confidence and assurance. The effectiveness of CPP is demonstrated in Fig. 5.3 by its comparison with the conventional CP setup. Figure 5.3 shows that a well-entangled polybutadiene solution studied without CPP suffers from edge fracture. The accumulation of the edge effect causes the shear stress to decline "precipitously" with increasing time. The shear stress should have leveled off after 100 s. Instead, with a conventional cone-plate setup, we see continuing stress decrease due to edge instability or fracture. An increasing fraction of the sample ceases to participate in the rotational shear.

It is challenging in a commercial rotational device to overcome the meniscus instability that occurs during nonlinear rheological response. Nonlinear rheological measurements mean that shear and normal stress (σ and N_1) readily increase beyond the level set by plateau modulus G_N^0 of entangled melts. Because G_N^0 is large, the cone-plate fixture must have a rather small diameter. As seen in Fig. 7.8(a,b), the maximum shear stress σ_y reaches G_N^0 in the weakly sheared regime when the yield strain γ_y is around two. Let us consider two polymer melts, polystyrene (PS) and 1,4-polyisoprene (PI) with G_N^0 equal to 0.2 and 0.6 MPa, respectively. Given that the upper limits of the torque T and the normal force F_N are, respectively, 0.2 N m and 20 N for most commercial rheometers, we can only use a CP with radius equal to 7.8 or 5.4 mm for PS and PI, respectively. According to Eq. (3.1), we can then make some modest nonlinear rheological measurements. In this regime, we estimate N_1 as $N_1 \sim G_N^0\gamma^2$, growing rapidly with the imposed strain γ. Even if we just take N_1 to equal G_N^0, we would find

$$F_N(CP; R) = (\pi R^2/2)N_1 \tag{5.1}$$

to reach 19 N for PS and to be around 27 N for PI, where CP stands for "cone-plate." This is a scenario in the absence of edge effect. If we have to implement cone-partitioned plate (CPP) to combat edge effects, adding a ring of larger radius R2, the normal force would grow even greater. In spherical coordinates, the normal stress is given by

$$\sigma_{\theta\theta}(r) = (N_1 + 2N_2)ln(r/R_2) + N_2 + p_a, \tag{5.2}$$

where p_a is the ambient temperature. Specifically, we have

$$F_N(CPP; R_1, R_2) = -2\pi \int_0^{R_1} r\sigma_{\theta\theta}dr + p_a\pi R_1^2 = \pi R_1^2[N_1/2 + (N_1 + 2N_2)ln(R_2/R_1)], \tag{5.3}$$

which reduces to Eq. (5.1) when $R_1 = R_2 = R$. This result was first shown by Meissner *et al.*[11] The present exercise reveals that if we want to protect the rheometric measurements from influence of edge instability by adopting cone-partitioned plate with, say, $R_2 = 2R_1$, the normal force is no longer that for a cone-plate of radius R_1. Since $ln\,2 = 0.69$ and N_1 typically far exceeds N_2, we have more than doubled the normal force if $R_2 = 2R_1$:

$$F_N(CPP; R_1, R_2 = 2R_1) \cong 2.4F_N(CP; R_1). \tag{5.4}$$

$F_N(CPP;\ R_1,\ 2R_1)$ is fortunately considerably smaller than $F_N(CP;\ 2R_1) = 4F(CP;R_1)$ because $F_N(CP; R)$ grows quadratically with R while $F_N(CPP; R_1, R_2)$ only grows logarithmically with the outer radius R_2. Nevertheless, it is impractical to apply CPP to investigate strongly nonlinear rheological behavior in entangled melts.[13,14]

5.3 Other Forms of Large Deformation

For entangled polymer melts, to avoid edge effect, rheological studies of nonlinear shear behavior are limited to a finite amount of strain. It can be shown that sliding-plate rheometry is suitable to produce simple shear for up to 10 strain units, which is sufficient to study stress overshoot response to startup shear and stress relaxation behavior after large stepwise shear, to be described in Sections 7.2.1.2 and 12.2.2, respectively. Since well-entangled melts may undergo massive wall slip, it is actually difficult to investigate *bulk* rheological responses of strongly entangled melts. Causing melts to adhere to shearing walls presents a formidable experimental challenge. Serrated or cleated surfaces[16] can hardly prevent apparent wall slip. For some styrene–butadiene rubbers, an effective method to fix them onto metallic walls involves an application of the readily available superglue. Results in Sections 7.2.1.2 and 12.2.2 were obtained using this primitive method.

We have no means to probe bulk rheological responses of entangled polymer melts in *steady* simple shear. Fortunately, the steady shear behavior of polymer melts is usually not important in the development of guiding principles for polymer processing. For example, many practical situations in polymer processing never involve steady shear. In the following section, we enumerate a few cases of melt deformation that do not involve simple shear and avoid edge instability.

5.3.1 Deformation at Converging Die Entry

Let us consider capillary extrusion. The quiescent polymer in the barrel, as sketched in Fig. 3.4, is forced to enter into a narrow die, in either pressure-controlled or piston-speed-controlled mode. Severe shear deformation, mixed with extensional stretching can occur at die entry. The effect of this converging deformation relaxes during capillary "flow" through the die. The residence time in the die is given on average by $t_{resid} \sim L/\overline{V} = (8L/D)/\dot{\gamma}_w$, where the equality follows from Eq. (3.12b). The time t_{resid} may not be long enough for the capillary flow to reach steady state. Measured in terms of the terminal relaxation time τ, $t_{resid}/\tau = (8L/D)/Wi$, which is smaller than unity for $Wi > (8L/D)$. It is well known that the effect of deformation typically takes a time longer than τ to relax in entangled melts.

Broadly speaking, we should treat the severe deformation at the die entrance as an interesting and important form of high shear deformation. There is no free surface and thus no meniscus effect at the die entrance: the system is completely closed. Coupled with PTV observations (cf. Fig. 4.6) deformation at die entry can be examined to reveal nonlinear responses of entangled melts, as shown in Section 10.2. The PTV method yields important information that is not possible to acquire from commercial rheometers.

5.3.2 One-Dimensional Squeezing

Another way to produce significant deformation without any influence from meniscus instability is a squeezing deformation. Unlike the conventional "squeezing flow"[17] where the sample is squeezed away radially, we consider one-dimensional squeezing as shown in Fig. 5.4. Neglecting inertia, the equation of motion for the shear stress is

$$\frac{\partial \sigma_{zx}}{\partial z} + \frac{\partial P'}{\partial x} = 0, \tag{5.5}$$

where P′ includes σ_{xx}, and use is made of $\frac{\partial \sigma_{yx}}{\partial y} = 0$. In the limit of $h/w \ll 1$, the state of stress can be approximately viewed as having no y dependence. The same kind of approximation to ignore "finite size effects" was made to obtain the simplified result of Eq. (3.13). On the other hand, the incompressibility condition

$$\nabla \cdot \vec{v} = 0, \text{ i.e., } \frac{\partial v_x}{\partial x} + \frac{\partial v_z}{\partial z} = 0, \tag{5.6}$$

holds true for polymeric liquids, where the velocity has no y dependence (because $h/w \ll 1$) and $v_y = 0$. Since v_z is not a function of x (unless $|x| \sim L/2$), its derivative, namely, the second term in Eq. (5.6), cannot be a function of x either. Therefore, the first term of Eq. (5.6) must also be independent of x. This implies that v_x must be linear in x. We can assume the following form for v_x

$$v_x = x f(z) \tag{5.7}$$

so that Eq. (5.6) becomes

$$\frac{\partial v_z}{\partial z} = -f(z), \tag{5.8}$$

where the function f is to be determined from Eq. (5.5) based on the constitutive relation between σ_{zx} and velocity gradient field.

Similar to capillary rheometry, without knowing the constitutive equation, we cannot describe the deformation field. So let us consider the simplest case. It is straightforward to work out the results for a Newtonian liquid. Inserting the Newton's law

$$\sigma_{zx} = -\eta \frac{\partial v_x}{\partial z} \tag{5.9}$$

into Eq. (5.5) allows us to see that P′ must be quadratic in x, since the first term of Eq. (5.5) is linear in x, given the linear dependence of v_x on x in Eq. (5.7). An even decreasing function of x for P′ makes sense because the sample should be symmetric with respect to the middle plane at x = 0. Thus, assuming

$$P'(x) = B[(L/2)^2 - x^2], \tag{5.10}$$

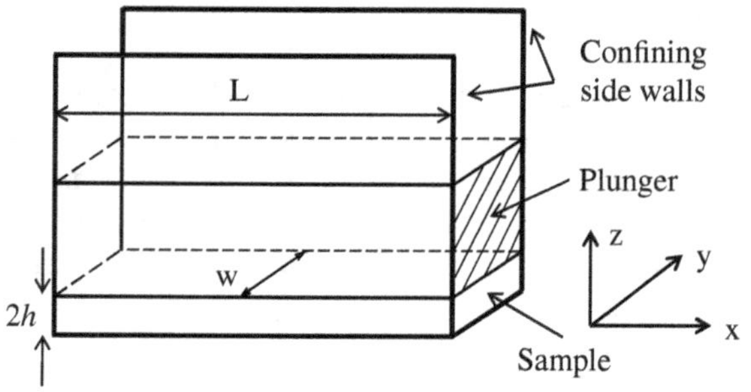

Figure 5.4 One-dimensional squeeze setup, made of a trench with width w and length L and a matching plunger that is moving downward along the Z-axis, where the sample thickness is 2h. The bottom of the trench is stationary. If the plunger is pushed down, the sample is expelled from the ends of the trench at x = ±L/2.

we can solve for the velocity fields in terms of f.

Specifically, the combination of Eqs. (5.5), (5.7), and (5.9) shows

$$\frac{\partial^2 f}{\partial z^2} = -\frac{2B}{\eta} \tag{5.11}$$

which is integrated to produce the solution

$$f(z) = (B/\eta)(h^2 - z^2). \tag{5.12}$$

This solution satisfies the no-slip boundary condition of $v_x = xf(z = \pm h) = 0$ and indicates

$$v_x(x, z) = (B/\eta)x(h^2 - z^2). \tag{5.13a}$$

Integrating Eq. (5.8) with Eq. (5.12) as input, we have

$$v_z(z) = -(B/\eta)h^3[(z/h) - (1/3)(z/h)^3] + C, \tag{5.13b}$$

where the two unknown constants B and C can be determined from the boundary conditions

$$v_z(z = h) = -2\dot{h}; v_z(z = -h) = 0. \tag{5.14}$$

Here $2\dot{h}(>0)$ denotes the speed with which the sample's thickness shrinks. Specifically, we have

$$B = (3/2)\eta(\dot{h}/h^3) \tag{5.15a}$$

and

$$C = -\dot{h}. \tag{5.15b}$$

Upon inserting B and C into Eqs. (5.13a) and (5.13b), we obtained the final results

$$v_x(x, z) = (3/2)x(\dot{h}/h)[1 - (z/h)^2] \tag{5.16}$$

and

$$v_z(z) = -(3/2)\dot{h}[z/h - (1/3)(z/h)^3] - \dot{h} \tag{5.17}$$

respectively. The expression Eq. (5.16) readily shows its meaning: the sample has zero horizontal speed at the center, $x = 0$, as well as at $z = \pm h$ because of the no-slip boundary condition and attains maximum speed at $z = 0$ (middle plane); the sample picks up speed linearly as x increases toward the edges of the apparatus, that is, there is considerable extensional deformation. Equation (5.17) is best discussed in terms of a plot shown in Fig. 5.5, which evidently reveals the expected symmetry with respect to $z = 0$. It is interesting to note that although the vertical speed v_z is half, at the middle plane, of that at the top surface, its decrease from $2\dot{h}$ to 0 is nonlinear.

By inserting Eq. (5.15a) into (5.10), we derive an expression for the pressure $P'(x)$ due to the squeezing action

$$P'(x) = (3/2)\eta(\dot{h}/h)[(L/2h)^2 - (x/h)^2], \tag{5.18}$$

which shows a maximum at the middle ($x = 0$). For a large aspect ratio $L/h \gg 1$, $P_{max} = P'(x = 0)$, measured in unit of viscous stress $\eta(\dot{h}/h)$, increases quadratically with L/h. The squeezing force F_z is given, after integration of $P'(x)$ in Eq. (5.18) from $x = -L/2$ to $x = L/2$, by

$$F_z = wL\eta(\dot{h}/h)(L/2h)^2. \tag{5.19}$$

An averaged pressure can be defined as

$$\overline{P} = F_z/wL = \eta(\dot{h}/h)(L/2h)^2. \tag{5.20}$$

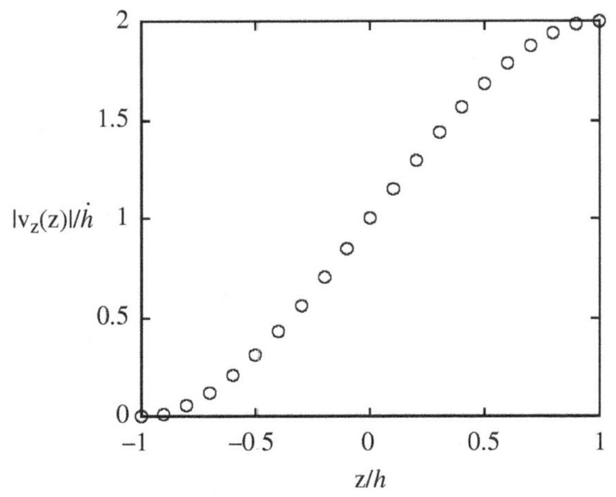

Figure 5.5 Plot of the z-component velocity profile depicted in Eq. (5.17).

As expected, Eq. (5.20) looks similar in form to the known result[18] for squeezing of a Newtonian liquid sandwiched between two circular disks of radius R: $\overline{P}_{disk} = (3/8)\eta(\dot{h}/h)(R/h)^2$.

Finally, we evaluate the shear stress by inserting the derivative of v_x (with respect to z) into Eq. (5.9), obtaining

$$\sigma_{zx} = 3\eta(\dot{h}/h)(xz/h^2) = -\eta\dot{\gamma}(x, z). \tag{5.21}$$

where $\dot{\gamma}$ is given by differentiating Eq. (5.16) for $v_x(x, z)$ with respect to z. Equation (5.21) shows that σ_{zx} increases linearly with x and z, respectively, and attains a maximum at the two walls $z = \pm h/2$

$$|\sigma_w| = 3\eta(\dot{h}/h)(|x|/h). \tag{5.22}$$

In all three Eqs. (5.20)–(5.22), h depends explicitly on time as $h(t) = h_0 - \dot{h}t$. Both Eqs. (5.21) and (5.22) show that the shear stress is zero at the center (x = 0) and the middle plane (z = 0) and maximum at the ends marked by $x = \pm L/2$. The shear stress actually diverges at the ends in proportion to the aspect ratio of L/h. The stress level is proportional to $\eta(\dot{h}/h)$, amplified by a factor of the aspect ratio L/h. Although the present treatment is for Newtonian liquids, the results allow us to recognize that the critical shear stress for wall slip of well-entangled polymer melts may be readily reached at $|x|/h \gg 1$. It is conceivable for an entangled melt that under squeezing the middle section of the sample does not undergo wall slip while the outer sections do so.

The factor $(\dot{h}/h)$ in Eqs. (5.21) and (5.22) can be regarded as a squeezing rate. This form reminds us of the Hencky rate introduced for uniaxial extension and compression. To see the resemblance, we define the squeezing ratio as

$$\lambda = h_0/h \tag{5.23}$$

and "Hencky strain" as

$$\varepsilon_{sqz} = ln\lambda \tag{5.24}$$

Then, we have

$$\dot{h}/h = d(ln\lambda)/dt = \dot{\varepsilon}_{sqz}. \tag{5.25}$$

According to (5.16), the Hencky strain rate in the squeezing is proportional to the rate given in Eq. (5.25). Comparing with the shear rate in Eq. (5.21), the Hencky rate is smaller by a factor of L/h. Thus, under the boundary condition of no-slip, the squeezing is dominantly shear.

Although such a setup cannot replace the standard rheometer described in Chapter 3, it has its utility. In particular, when combined with particle-tracking velocimetry, it can be used to reveal strongly nonlinear rheological responses including shear strain localization without complications from meniscus instabilities. By construction, any observation of the deformation field inside this squeezing setup is free of anything happening at $x = \pm L/2$. Actual PTV observations on entangled melts during squeezing are presented in Section 10.3.

5.3.3 Planar Extension

For well-entangled polymers, it is always necessary to apply particle-tracking velocimetry or a comparable technique to reliably determine the strain field during rheometric measurements. In the case of entangled melts, because of the high normal force, the shearing fixture on a commercial rotational rheometer has to be so small that it is infeasible to perform particle-tracking velocimetric measurements. Thus, we have to resort to other modes of deformation to learn about nonlinear rheological behavior in entangled melts. Conversely, the application of particle-tracking velocimetry has made it much more interesting to examine nonlinear rheological responses of entangled melts in unconventional geometries. A third example of such uncommon setup for melt deformation is planar extension, also known as pure shear. Such deformation can be produced using the same configuration shown in Fig. 5.4, where one dimension is fixed at w and the thickness reduction translates to lengthening of the sample along the X-axis. Unlike the one-dimensional squeezing, planar extension requires application of an effective lubricant at the confining surfaces ($z = \pm h$) to remove the no-slip boundary condition and to avoid shear deformation. Using particle-tracking velocimetry, we have an opportunity to determine whether and how any strain localization could take place upon planar extension of entangled melts. This demonstration is the subject of Section 10.4.

In the presence of wall slip caused either by shear stress or lubrication, characterized by slip velocity V_s at $x = \pm L/2$, Eq. (5.11) can be integrated to produce the solution

$$f(z) = (B/\eta)(h^2 - z^2) + 2V_s/L. \tag{5.26}$$

This solution satisfies the slip boundary condition of $v_x(x = \pm L/2) = (L/2)f(z = \pm h) = \pm V_s$. Integrating Eq. (5.8) with Eq. (5.26) as input, we have

$$v_z(z) = -(B/\eta)h^3[(z/h) - (1/3)(z/h)^3] + C - (2z/L)V_s, \tag{5.27}$$

where the two unknown constants B and C can be determined from the boundary conditions

$$v_z(z = h) = -2\dot{h}; v_z(z = -h) = 0. \tag{5.28}$$

Here $2\dot{h}(>0)$ denotes the speed with which the sample's thickness shrinks. Specifically, we have

$$B = (3/2)\eta(\dot{h}/h - 2V_s/L)/h^2 \tag{5.29a}$$

and

$$C = -\dot{h}. \tag{5.29b}$$

Upon inserting B and C into Eqs. (5.26) and (5.27), and Eq. (5.26) into Eq. (5.7), we obtained the final results

$$v_x(x, z) = (3/2)x[(\dot{h}/h) - 2V_s/L][1 - (z/h)^2] + 2xV_s/L \tag{5.30}$$

and

$$v_z(z) = -(3/2)(\dot{h}/h - 2V_s/L)[z - (1/3)h(z/h)^3] - \dot{h} - (2z/L)V_s. \tag{5.31}$$

The volumetric flow rate coming out of the *two* ends is given by the following integration

$$Q_f = 2w \times 2\int_0^h dz v_x(x = L/2, z) = wL(2\dot{h}) = \frac{d\Omega_f}{dt}. \tag{5.32}$$

Clearly, the volume change per unit time is $wLv_z(h)$ where $v_z(h)$ is the relative speed with which the two surfaces approach each other.

In the limit of perfect lubrication, we have "plug flow" and Eqs. (5.30) and (5.31) become

$$v_x(x, z) = 2xV_s/L \tag{5.33a}$$

and

$$v_z(z) = -\dot{h}(1 + z/h), \tag{5.33b}$$

where v_x no longer varies with z, implying there is no shear deformation. The amount of material exiting from the two opening ends is at a rate of

$$Q_f = 2 \times w(2h)V_s = 4whV_s. \tag{5.34}$$

Equating Eqs. (5.32)–(5.34) yields the condition for complete lubrication, given by

$$\dot{h}/h = 2V_s/L. \tag{5.35}$$

When the left-hand side of Eq. (5.35) is kept constant, we achieve a constant extensional rate $2V_s/L$. In other words, the squeezing speed controls the extensional rate. To reiterate, with perfect lubrication, insertion of Eq. (5.35) into Eq. (5.30) shows that the sample has the same velocity as given by Eq. (5.33a) at every layer along the thickness direction z. Similarly, B = 0 under the condition of Eq. (5.35) so that the pressure no longer varies along X-axis. Moreover, the shear stress from Eq. (5.9) is zero because v_x is no longer z dependent. In this limit, we reach the condition of planar extension. When Eq. (5.35) holds, namely, in planar extension, there is no shear, and the pressure $\overline{P}$ decreases by the amplification factor $(L/2h)^2$, from that in Eq. (5.20) to

$$\overline{P} \approx 3(\dot{h}/h)\eta, \tag{5.36}$$

where 3η is the extensional viscosity. In general, in the presence of finite wall slip ($2V_s/L < \dot{h}/h$), the bulk shear rate and shear stress at the wall are both reduced from their respective values valid in the absence of wall slip.

5.4 Conclusion

Large shear deformation is particularly challenging to impose on strongly entangled polymers. The complication usually comes from the intrinsic finite size effects. For example, shear in rotational devices cannot avoid failure of the meniscus, yet rotational configuration is necessary to allow continuous buildup of the shear strain. The meniscus instability is so severe for entangled melts that commercial rheometry does not offer the capability to handle nonlinear rheology of polymer melts, which would require much larger torque and normal force tolerance than the ranges available in current rheometers. Chapter 5 describes a few improvements for simple shear configuration and indicates other shear-like configurations that are commonly encountered and therefore useful to consider. Specifically, cone-partitioned-plate apparatus in Section 5.2 is particularly valuable as a means to remove effects of edge instability on certain rheometric measurements. In each of the settings described in Sections 5.2, 5.3.1 through 5.3.3, it is desirable to apply particle-tracking velocimetry (depicted in Chapter 4) and determine to what extent wall slip takes place and when bulk deformation becomes strong as the applied shear rate increases.

By concluding Chapter 5, we have reached the end of Part 1 and laid the foundation for a thorough investigation of all aspects of nonlinear rheological responses of entangled polymer solutions and melts. While much of the content of Part 1 is standard, available from existing textbooks on polymer rheology and dynamics, some description is highly uncommon, including Sections 1.2.5.2, 1.3.3, 2.2.3, Chapter 4, and Chapter 5. In particular, Section 2.2.3 on phenomenological models for entanglement should have been discussed in books on Polymer Physics that were published after 1990s. Because of their importance, we presented the packing model thoroughly along with the percolation model. Chapters 4 and 5 are not the usual materials in available books because the discussion is based heavily on new methods that can be applied to investigate strain localization. This book makes a thorough treatment of the subject of strain localization while existing books hardly dwell on the challenges.

References

1 Boukany PE, Wang S-Q. Exploring origins of interfacial yielding and wall slip in entangled linear melts during shear or after shear cessation. Macromolecules. 2009;42(6):2222–8.

2 Boukany PE, Wang S-Q, Wang X. Step shear of entangled linear polymer melts: new experimental evidence for elastic yielding. Macromolecules. 2009;42(16):6261–9.

3 Myers A, Faucher J. A modified pochettino viscometer for the measurement of the viscoelastic properties of high polymer melts. Trans Soc Rheol. 1968;12(1):183–93.

4 Boukany PE, Tapadia P, Wang S-Q. Interfacial stick–slip transition in simple shear of entangled melts. J Rheol. 2006;50:641.

5 Segel M. Über eine Methode zur Bestimmung der inneren Reibung fester Körper. Phys Z 1903;4:493.

6 Pochettino A. Su le proprietà dei gorpi plastici. Nuovo Cimento. 1914;8(1):77–108.

7 Larson RG. Instabilities in viscoelastic flows. Rheol Acta. 1992;31(3):213–63.

8 Tanner R, Keentok M. Shear fracture in cone-plate rheometry. J Rheol. 1983;27(1):47–57.

9 Inn YW, Wissbrun KF, Denn MM. Effect of edge fracture on constant torque rheometry of entangled polymer solutions. Macromolecules. 2005;38(22):9385–8.

10 Sui C, McKenna GB. Instability of entangled polymers in cone and plate rheometry. Rheol Acta. 2007;46(6):877–88.

11 Meissner J, Garbella R, Hostettler J. Measuring normal stress differences in polymer melt shear flow. J Rheol. 1989;33(6):843–64.

12 Schweizer T. Measurement of the first and second normal stress differences in a polystyrene melt with a cone and partitioned plate tool. Rheol Acta. 2002;41(4):337–44.

13 Schweizer T. Comparing cone-partitioned plate and cone-standard plate shear rheometry of a polystyrene melt. J Rheol. 2003;47(4):1071–85.

14 Schweizer T, Schmidheiny W. A cone-partitioned plate rheometer cell with three partitions (CPP3) to determine shear stress and both normal stress differences for small quantities of polymeric fluids. J Rheol. 2013;57(3):841–56.

15 Ravindranath S, Wang SQ. Steady state measurements in stress plateau region of entangled polymer solutions: controlled-rate and controlled-stress modes. J Rheol. 2008;52(4):957–80.
16 Nickerson CS, Kornfield JA. A "cleat" geometry for suppressing wall slip. J Rheol. 2005;49(4):865–74.
17 Macosko CW. Rheology principles, measurements, and applications. Wiley-VCH, Inc., New York; 1994.
18 Bird RB, Armstrong RC, Hassager O. Dynamics of polymeric liquids. Vol. 1, Fluid mechanics, John Wiley & Sons, New York. 1987.

Part II

Yielding – Primary Nonlinear Responses to Ongoing Deformation

Nonlinear rheological behavior arises when the mechanical response is no longer linearly proportional to the magnitude of the external deformation produced by either an imposed stress or a boundary displacement. We cease to observe linear response when the underlying structure of the material is altered by the external deformation. The physical world would have been extraordinarily simple if linear response prevailed under all conditions. In reality, the opposite is true: the world is full of interesting systems that show nonlinear rheological responses.

Entangled polymeric liquids are a major class of materials that are nonlinearly viscoelastic. They show interesting behavior even in the regime of linear response, to be discussed in Section 7.1.1. During startup shear with large Weissenberg number *Wi*, the shear stress rises linearly with increasing time before "strain softening" sets in. A shear stress maximum then appears on a time scale that can be much shorter than the reptation (i.e., terminal) time τ. The maximum is also termed the *overshoot*. After the maximum, the shear stress declines, signaling the end of elastic deformation. Bryce Maxwell (1911–2001) appropriately called[1,2] the peak stress at the overshoot the "yield stress." Writing in 1986, he borrowed this familiar term from engineering mechanics; however, in 1979, Doi and Edwards were already completing their construction of the tube model for nonlinear rheology of entangled polymers[3] and had also offered an explanation for the stress overshoot.

From 1979 until the mid-2000s, the idea of yielding in nonlinear rheology of polymers was hardly ever mentioned.[4] During this period of 25 years, the field of nonlinear polymer rheology depended on the tube model to provide molecular pictures and interpretations. Since 2006, there has been ample phenomenological evidence[5] to suggest that nonlinear rheology might be very different from the tube model description.[6] Considerable new information has since been introduced, urging us to think differently about microscopic structural changes during the nonlinear response of entangled polymer solutions and melts.

We find the concept of yielding[1] to provide deep insights to understand nonlinear rheological properties of entangled polymeric liquids. Yielding prompts us to explore the scaling

1 In the literature of materials mechanics, the nomenclature *yielding* typically refers to a transition from elastic deformation to plastic deformation or irrecoverable deformation. Until the mid-2000s, yield had never been invoked in the literature on rheology of entangled solutions and melts with the exception of Maxwell and Nguyen. We find it concise and convenient to invoke this phrase to refer to the termination of the dominantly elastic state during startup deformation of entangled polymer solutions and melts. An alternative phrase could be *softening*. However, softening or any other words we can think of hardly reflects the essence of the nonlinear responses of entangled polymers.

Nonlinear Polymer Rheology: Macroscopic Phenomenology and Molecular Foundation, First Edition. Shi-Qing Wang.

Companion Website: www.wiley.com/go/nonlinearpolymerrheology

characteristics of stress overshoot and allows us to discuss shear and extensional rheology on an equal footing. Moreover, in terms of yielding, it is straightforward to understand the true meaning of "strain hardening" in startup uniaxial extension, as discussed in Chapter 8. It appears that various types of strain localization, in simple shear, uniaxial extension, and more complicated geometries, as discussed in Chapters 9–11, are all due to localized yielding of the entanglement network.

Beginning with Chapters 6 and ending in Chapter 14, we present a comprehensive description of the key rheological phenomenology concerning nonlinear responses of entangled polymers. Since many of these phenomena were discovered as predictions[5] of our new conceptual framework, concepts such as disentanglement, yielding, and molecular forces are inevitably mentioned in the exposition of the rich phenomenology. All essential features, including the transition from no-slip to wall slip, stress overshoot during startup deformation and subsequent shear banding, and disentanglement transition under creep, are explained in Chapter 16. The reader is referred to Chapter 16 for more precise discussions of disentanglement and yielding.

Our discussion of the phenomenology is divided into two parts, in order to emphasize the significant differences between yielding during deformation and elastic yielding after a large deformation. Specifically, Part 2 explores nonlinear responses to ongoing external deformation and identifies the characteristics of yielding, that is, the transition from initial elastic deformation to subsequent flow. Part 3 discusses how elastic deformation causes yielding after cessation of external deformation. Throughout Part 2 and Part 3, yielding is a phenomenological concept to be invoked without any detailed discussion of the corresponding molecular picture or theory, which is deferred to Chapter 16. The notion of yielding is a thread that provides a unifying language to describe all nonlinear rheological phenomena within a common framework.

Part 2 begins with a thorough presentation on wall slip. Wall slip is rarely treated in textbooks on polymer rheology, but it is fundamental for a basic understanding of nonlinear (entangled) polymer rheology. No proper understanding of the constitutive bulk behavior of entangled polymers can be obtained experimentally without an appropriate treatment of wall slip behavior. The emergence of wall slip involves the same physics (chain disentanglement) that is involved in yielding in the bulk. In fact, the onset of wall slip is a sign of interfacial yielding. Chapter 6 provides a description of the latest knowledge and understanding on wall slip. We omit a discussion of those previous references by other workers that have already been reviewed in Refs. [7–9].

References

1 Maxwell B, Nguyen M. Measurement of the elastic properties of polymer melts. Polym Eng Sci. 1979;19(16):1140–50.

2 Maxwell B. The application of melt elasticity measurements to polymer processing. Polym Eng Sci. 1986;26(20):1405–9.

3 Doi M, Edwards S. Dynamics of concentrated polymer systems. Part 4.—Rheological properties. J Chem Soc Faraday Trans 2 Mol Chem Phys. 1979;75:38–54.

4 Tapadia P, Wang S-Q. Nonlinear flow behavior of entangled polymer solutions: yieldlike entanglement–disentanglement transition. Macromolecules. 2004;37(24):9083–95.

5 Wang S-Q. Nonlinear rheology of entangled polymers at turning point. Soft Matter. 2015;11(8):1454–8.

6 Wang SQ. The tip of iceberg in nonlinear polymer rheology: entangled liquids are "solids". J Polym Sci Part B Polym Phys 2008;46(24):2660–5.
7 Wang S-Q. Molecular transitions and dynamics at polymer/wall interfaces: Origins of flow instabilities and wall slip. In: Granick S, Binder K, de Gennes PG, Giannelis EP, Grest GS, Hervet H, et al., editors. Polymers in confined environments. Berlin, Heidelberg: Springer Berlin Heidelberg; 1999. pp. 227–75.
8 Denn MM. Extrusion instabilities and wall slip. Annu Rev Fluid Mech. 2001;33(1):265–87.
9 Hatzikiriakos SG, Migler KB. Polymer processing instabilities: control and understanding. New York: CRC Press; 2004.

6

Wall Slip – Interfacial Chain Disentanglement

Wall slip is a leading example of strain localization during the nonlinear response of entangled polymeric liquids to high shear. The transition from no-slip to slip upon startup shear is best understood as yielding at the polymer–wall interface. Thus, our discussion of nonlinear responses begins with a systematic exploration of wall slip.

Entangled liquids made of linear chains or aggregates (e.g., wormlike micelles) are uniquely capable of undergoing significant wall slip. Historically speaking, the search for violations of the no-slip boundary condition started in the time of Claude-Louis Navier (1785–1836) and George Gabriel Stokes (1819–1903). Navier was the first to contemplate the possibility of wall slip for fluids. For small molecule liquids, it is simply impossible to find evidence of slip on macroscopic length. Polymeric liquids such as entangled melts appeared to display wall slip.[1–8] However, before de Gennes' analysis,[9] we had no clue as to why high-molecular-weight polymers should display considerable wall slip. In contrast to the Mooney analysis[5,10] of wall slip, de Gennes' account not only pinpointed the molecular origin of wall slip but also showed how to evaluate the degree of wall slip in terms of Navier's notion of the extrapolation length b. Many studies[11–20] have since been carried out on wall slip.[1–8] The purpose of this chapter is to summarize our knowledge of the subject and not to review past studies of other workers. The results below represent comprehensively the current state of scientific understanding of wall slip.

This chapter is organized as follows. Section 6.1 describes basic quantities such as slip velocity and extrapolation length that are used to characterize the level of wall slip. Importantly, the concept of limiting slip velocity V_s^* is introduced in terms of materials parameters, defined at the condition of complete wall slip that involves a threshold shear stress. Section 6.2 is devoted to the topic of stick–slip transition where the value of V_s^* is listed for several polymer melts. Section 6.3 emphasizes the notion of interfacial yielding by describing how wall slip emerges from the initial condition of no-slip using experimental data. Section 6.4 addresses the key link between this chapter and the rest of the book, especially Chapters 7 and 9. This section discusses in Section 6.4.1 the critical condition when rheological responses change from wall slip to bulk nonlinear shear deformation as well as how shear thinning reduces the wall slip correction to the shear field. It is also shown that rheometric measurements may not show any gap dependence during wall slip when slip is only partial rather than complete. Moreover, a sharp transition is predicted to occur, taking the system from wall slip dominant to strong bulk shearing, as the gap distance is varied and the applied nominal shear rate is at a threshold value. Section 6.5 reviews the molecular evidence for wall slip based on the literature data. Section 6.6 shows that the boundary condition can be unstable, for example, oscillatory, under certain shearing conditions.

Nonlinear Polymer Rheology: Macroscopic Phenomenology and Molecular Foundation, First Edition. Shi-Qing Wang.

Companion Website: www.wiley.com/go/nonlinearpolymerrheology

6.1 Basic Notions of Wall Slip in Steady Shear

Let us consider two steady-shear scenarios, as depicted in Fig. 6.1(b,c). Here during steady shear a solid wall loses its ability to grip an entangled polymer liquid, either due to disentanglement or desorption. The fully entangling interface seen in Fig. 6.1(a) is the reference state. When there is sufficient shear stress on weakly adsorbing surfaces, chain desorption can take place, so that the shearing surfaces no longer have the molecular "hooks," that is, tethered (physically adsorbed) chains, to transmit the boundary displacement or shear stress to the bulk of the entanglement network. This desorption state is illustrated by Fig. 6.1(c). It is also possible in the case of strong polymer/wall adsorption that the monolayer of adsorbed chains reaches a condition of disengagement as illustrated in Fig. 6.1(b) in which the chains at the surface remain attached to the surface but disentangle from the chains in the bulk. The interfacial layer of diminishing chain entanglement has a thickness a_i or a_i'. In the case of Fig. 6.1(c), de Gennes assumed a_i' to be a monomeric scale.[9] For the situation of Fig. 6.1(b), a_i may be comparable to the coil size of an entanglement strand, that is, l_{ent}.

The detailed mechanism for such disentanglement will be discussed in Chapter 16. Until Chapter 16, we use the phrase "disentanglement" in the literal sense depicted in Fig. 6.1(b) in contrast to the reference state of Fig. 6.1(a). The consequences of chain desorption or disentanglement are similar: In the absence of chain entanglement at the monolayer between the solid surface and bulk polymer, we encounter a much reduced interfacial viscosity η_i that reflects only the viscous interaction of the polymer chains with the surfaces. It is convenient to discuss wall slip first in the presence of steady shear flow and then in Section 6.3 describe interfacial yielding signaling a transition from no-slip to slip upon startup shear.

6.1.1 Slip Velocity V_s and Navier–de Gennes Extrapolation Length *b*

de Gennes proposed that the slip velocity V_s present at the polymer/wall interface should be proportional to shear stress on the wall

$$V_s = \sigma/\beta, \tag{6.1a}$$

as first proposed by Navier.[21] The simple linear relation of Eq. (6.1a) can be regarded as the assumed constitutive equation for the polymer/wall interface. Other constitutive equations may exist, but this linear relationship is the simplest and often holds true. All discussion in this book on slip is based on Eq. (6.1a). From Eq. (6.1a), de Gennes rediscovered the concept of extrapolation length *b* first identified by Navier. This Navier–de Gennes extrapolation length *b* has its abstract meaning and definition illustrated in Fig. 6.2(a).

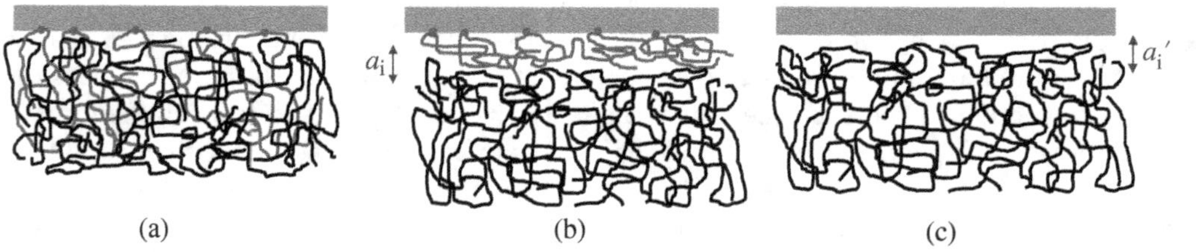

Figure 6.1 Different states at various polymer/wall interfaces: (a) in quiescence with strong polymer adsorption or under weak shear (corresponding either to slow shear rate or to low shear stress) where the depiction is somewhat misleading: the adsorbed chains are actually still inter-penetrating into the unbound chains (see Fig. 16.6 for a more realistic illustration of the interfacial layer) and unbound chains are disentangling in the sense that there is significant sliding between adsorbed and free chains; (b) full disentanglement between adsorbed chains and unbound (bulk) chains during complete wall slip; (c) chain desorption in the presence of shear at weakly adsorbing interfaces. *(See color plate section for the color representation of this figure.)*

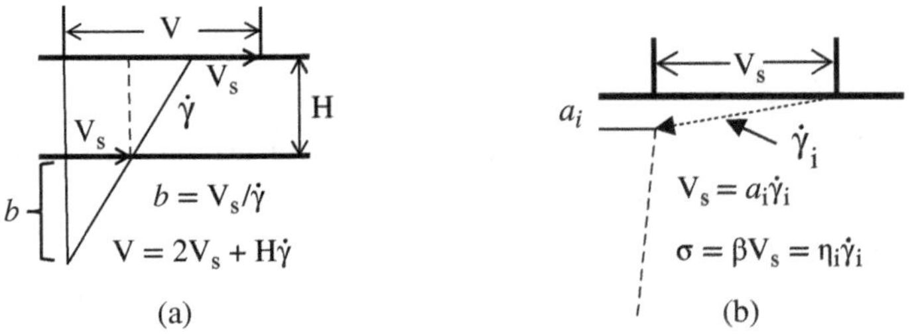

Figure 6.2 (a) Schematic illustration of wall slip in simple steady shear, produced by the displacement of the upper surface with velocity V while fixing the lower surface at zero speed. The effective shear rate $\dot{\gamma}$ in the bulk is smaller than the apparent rate V/H because of wall slip, characterized by either the slip velocity V_s or the extrapolation length *b*. Here *b* is termed the *extrapolation length* because it is not a physical dimension of or length scale in the system. For example, it is not the thickness of the interfacial slip layer. Rather, it is defined by extrapolating along the bulk velocity gradient to where the velocity would be zero outside the sample of thickness H as shown. (b) Interfacial layer of thickness a_i characterized by presence of high shear rate $\dot{\gamma}_i$ and low viscosity η_i, where the long-dashed line shows the bulk shear rate.

Figure 6.2(a) depicts a symmetric case with equal amount of wall slip on both surfaces. According to Figure 6.2(a), we have the following kinematic relation between *b* and V_s

$$b = V_s/\dot{\gamma} = \eta/\beta, \tag{6.2}$$

where the second equality follows from Eqs. (6.1a) and (1.8). de Gennes further suggested that the interfacial friction factor β is related to an interfacial viscosity η_i and the interfacial layer thickness a_i (Fig. 6.2(b)) by

$$\beta = \eta_i/a_i. \tag{6.3}$$

Equation (6.3) is obtained by dimensional analysis and can be derived straightforwardly from Eq. (6.1a), where the kinematic definition of V_s is given by, as shown in Fig. 6.2(b)

$$V_s = \dot{\gamma}_i a_i \tag{6.1b}$$

and

$$\sigma = \eta_i\dot{\gamma}_i. \tag{6.1c}$$

Here $\dot{\gamma}_i$ is the shear rate in the interfacial layer. Conversely, Eq. (6.1a) along with Eq. (6.3) follows from inserting Eq. (6.1b) into Eq. (6.1c). We note that Eq. (6.1c) is a statement of stress continuity, that is, the shear stress σ is constant along the Y-axis, its value at the wall is the same as its value in the bulk.

Equations (6.2) and (6.3) depict how *b* is dependent on the viscosity ratio η/η_i and the interfacial layer thickness a_i. The expressions in Eqs. (6.1a)–(6.1c), (6.2), and (6.3) are generic; they are not restricted to the ideal wall slip condition that slip involves only one monolayer (i.e., $a_i = l_{ent}$). When a_i is vanishingly small relative to H but involves multiples of the characteristic molecular scale (l_{ent}), Fig. 6.2(a) can be used to describe *apparent wall* slip. It is necessary to distinguish apparent wall slip from true wall slip, a term that we reserve for the cases shown in Fig. 6.1(b,c), where the wall slip involves an interfacial layer whose thickness is the size of a subcoil, for example, that of an entanglement strand.

6.1.2 Correction of Shear Field due to Wall Slip

Wall slip can make conventional rheometric measurements unreliable for strongly entangled polymers. Take the simple case shown in Fig. 6.2(a). Assuming that there is slip velocity V_s, we can relate V_s to the applied V as

$$V_s/V = (b/H)/(1 + 2b/H), \tag{6.4a}$$

which follows immediately from Fig. 6.2(a). If there is an equal amount of wall slip on both walls, the actual bulk shear rate $\dot{\gamma}$ is related to the apparent rate V/H by

$$\dot{\gamma}/(V/H) = 1/(1 + 2b/H) < 1, \tag{6.4b}$$

or

$$\dot{\gamma}/(V/H) = (1-2V_s/V) < 1. \tag{6.4c}$$

The correction due to slip is negligible when $b/H \ll 1$ or $V_s/V \ll 1$. For well-entangled polymeric liquids, wall slip can be severe for a conventional gap distance H on the order of 1 mm. b/H can be comparable to or even much larger than unity, leading to significant wall-slip corrections to rheometric measurements.

When wall slip (or shear banding as described in Chapter 9) takes place, it is necessary to differentiate the apparent rate of shear from the true bulk shear rate $\dot{\gamma}$. Correspondingly, it is necessary to define an apparent Weissenberg number

$$Wi_{app} = (V/H)\tau, \tag{6.5}$$

which differs from the original definition of Weissenberg number (Eq. (1.38a)) that characterizes shear rate in the bulk. (Henceforth, Wi always refers to the bulk Weissenberg number.) When $Wi_{app} > 1$, wall slip tends to occur in entangled polymer liquids. Specifically, experiments indicate that wall slip begins when the bulk rate $\dot{\gamma}$ approaches a critical level satisfying the condition given in Eq. (1.38b). Whether or not wall slip results in a measurable change in the bulk shear field depends on the value of Wi_{app}. The extrapolation length b increases with increasing Wi_{app} and reaches its maximum b_{max} before the bulk of the system undergoes shear with $Wi > 1$. Experimentally, many well-entangled polymer liquids have $b_{max} \gg H \sim 1$ mm. Therefore, wall slip can be a severe form of strain localization in simple shear of entangled polymers.

For a given $Wi_{app} > 1$, it is interesting to determine whether or not the measured shear stress σ_{slip}

$$\sigma_{slip} = \eta(\dot{\gamma})\dot{\gamma} \tag{6.6}$$

depends on H. For a given Wi_{app}, Eq. (6.4b) shows that the bulk rate $\dot{\gamma}$ would vary with b/H, implying that σ_{slip} would depend explicitly on b/H according to Eq. (6.6). It is customarily assumed that wall slip is negligible if the rheometric measurement does not depend on H. This understanding is erroneous: When varying H while keeping Wi_{app} fixed, we need to examine whether the slip velocity V_s would stay constant, as discussed in Section 6.4.4.

6.1.3 Complete Slip and Maximum Value for *b*

The preceding discussions depicted wall slip in its generic forms in terms of Fig. 6.2. This section quantifies the complete slip in terms of the linear viscoelastic characteristics of entangled polymer liquids. Complete wall slip occurs, as shown in Fig. 6.1(b), when full disentanglement takes place at the polymer/wall interface, leading to a stick–slip transition at a critical shear stress σ_c (see Section 6.2). Upon this stick–slip transition, β plunges to its minimum β_{min}. For an entangled melt with entanglement molecular weight M_e, the viscosity η_i in the interfacial layer has a lower bound of $\eta(M_e) = \eta_e$, which denotes the viscosity of a Rouse melt whose molecular weight is M_e. Because of the complete disentanglement η_i drops to η_e. For entangled solutions, η_i may be as low as the solvent viscosity η_s. Thus, we have

$$\beta_{min} \sim \eta(M_e)/l_{ent} \quad \text{for entangled melts} \tag{6.7a}$$

or

$$\beta_{min} \sim \eta_s/l_{ent}(\phi) \quad \text{for entangled solutions/mixtures} \tag{6.7b}$$

Here, the thickness a_i of the interfacial slip layer in Eq. (6.3) is taken to be the entanglement spacing l_{ent} given by Eqs. (2.29) and (2.37a), respectively. The approximation of replacing η_i with η_s is relatively reliable for $\phi \ll 1$, although polymers having high glass transition temperatures (e.g., polystyrene) tend to deplasticize inviscous small-molecule solvents. When the solutions are polymer mixtures made of high (M_h)- and low (M_l)-molecular-weight components of the same species, at volume fractions of ϕ and $(1-\phi)$, respectively, it is nontrivial to estimate η_i. Depending on whether M_l is smaller or larger than M_e, η_i is expected to show opposite trends. Upon full disentanglement, η_i is the viscosity of a mixture of two components having $\eta(M_l) = \eta_s$ and $\eta(M_e)$, respectively. Thus, if the "solvent" has $M_l > M_e$, we can expect $\eta_i < \eta_s$, otherwise for $M_l < M_e$, we should have $\eta_i > \eta_s$. For $\phi \ll 1$, Eq. (6.7b) is a good guide.

Experiments reveal that the critical stress σ_c for the stick–slip transition is a fraction of the elastic plateau modulus G_{pl}. Since they are comparable in magnitude, we simply write

$$\sigma_c \sim G_{pl}, \tag{6.8}$$

so that the full disentanglement produces a limiting slip velocity

$$V_s^* = \sigma_c/\beta_{min} \sim (l_{ent}/\tau_e) \quad \text{for entangled melts.} \tag{6.9a}$$

Here use was made of Eq. (6.7a) and $\eta_e \approx G_{pl}\tau(M_e)$ where $\tau_e = \tau(M_e)$ is defined in Equation (2.45). Equation (6.9a) shows how the local time and length scales (τ_e and l_{ent}) in combination determine the limiting interfacial slip velocity at σ_c. Equation (6.9a) shows that V_s^* is independent of molecular weight. This is expected since characteristics of a fully disentangled state do not depend on molecular weight. Similarly, we have

$$V_s^* = \sigma_c/\beta_{min} \sim [G_{pl}(\phi)/\eta_s]l_{ent}(\phi), \quad \text{for entangled solutions/mixtures,} \tag{6.10}$$

where the elastic plateau modulus $G_{pl}(\phi)$ is given by Eq. (2.36c). When a solution is prepared with a highly inviscous solvent so that $\eta_s \ll \eta_e$, the value of V_s^* can be greater than that of the corresponding melt. Since V_s^* in Eqs. (6.9a) and (6.10) is defined at σ_c, its magnitude is given in terms of basic viscoelastic characteristics and structural properties of the entangled polymers.

For walls with strong polymer adsorption, there is no wall slip for $Wi_{app} < 1$. However, wall slip typically shows up as soon as Wi_{app} increases beyond unity and develops progressively in magnitude. In rate-controlled shear, as long as the applied V in Fig. 6.2(a) does not exceed V_s^*, the magnitude of wall slip will continue to increase. Therefore, over a wide range, with increasing V, V_s increases proportionally, so that the shear rate in the bulk does not have to increase and the shear stress shows a plateau. This shear stress plateau is comparable to the critical shear stress σ_c at which the stick–slip transition occurs in the controlled-stress mode. Consequently, the bulk shear rate $\dot{\gamma}$ remains constant around $1/\tau$ (and $Wi \sim 1$) until complete wall slip takes place, with β reaching β_{min}: The condition of $\sigma_c = \eta\dot{\gamma}_c \sim G_{pl}$ is the same as $\dot{\gamma}_c \sim 1/\tau$ because $\eta \approx G_{pl}\tau$. Complete wall slip happens when V equals $2V_s^* + \dot{\gamma}_c H$. Under this condition of stress plateau or at the stick–slip transition at σ_c, the extrapolation length b reaches a maximum, given by

$$b_{max} = V_s^*/\dot{\gamma}_c \approx \tau V_s^*. \tag{6.11}$$

According to its definition in Eq. (6.2), b_{max} occurs upon full interfacial chain disentanglement and when the bulk shear rate $\dot{\gamma}$ is also at the minimum necessary to assure complete slip, namely at $\dot{\gamma}_c \sim 1/\tau$. Equation (6.11) is valid for both entangled polymer solutions and melts.

By inserting Eq. (6.9a) and Eq. (6.10) into Eq. (6.11) and acknowledging $\tau/\tau_e = \eta/\eta_e$, we obtain the following scaling relations,[22]

$$b_{max} \approx [\eta(M)/\eta(M_e)]l_{ent} \approx (M/M_e)^{(3.2-3.4)}l_{ent}, \quad \text{for melts} \tag{6.12}$$

and

$$b_{max} \approx [\eta(\phi)/\eta_s] l_{ent}(\phi), \quad \text{for solutions} \tag{6.13}$$

where use is made of $\eta(\phi) \approx G_{pl}(\phi)\tau$. In Eq. (6.12), for monodisperse melts, we invoked an empirical scaling relation $\eta(M)/\eta(M_e) = (M/M_e)^{(3.2-3.4)}$ instead of $\eta(M)/\eta(M_e) = (M/M_e)^3$ from Eq. (2.50a) that may apply only in the limit of extremely large M/M_e. Since M_e for polystyrene and PMMA is as high as 15 kg/mol, in practice (when M is less than 200 kg/mol), these two melts have values for *b*, according to Eq. (6.12), that are much smaller than 1 mm. Therefore wall slip is hardly observable or relevant for PS and PMMA in industrial processing.

6.2 Stick–Slip Transition in Controlled-Stress Mode

Wall slip has been observed to occur in capillary extrusion since 1958, in the form of a spurt transition under the condition of controlled pressure. The physical origin of the spurt phenomenon was debated for a long time since its first observation. de Gennes' account[9,24] provided a starting point to rationalize the ability of polymer melts to undergo significant wall slip, allowed us to clarify the interfacial nature of spurt flow, which involves a stick–slip transition. The Navier–de Gennes extrapolation length *b* gives an adequate quantification of the degree of wall slip associated with the transition,[22] in contrast to the Mooney analysis[10] that extracts the magnitude of the slip velocity V_s from the data.

6.2.1 Stick–Slip Transition in Capillary Extrusion

6.2.1.1 Analytical Description

For interfaces where polymer adsorption is strong, as shown in Fig. 6.1(a), controlled shear stress experiments can reveal a discontinuous transition due to chain disentanglement at the polymer/wall interface, as depicted in Fig. 6.1(b). The cohesion at the polymer/solid interface can be overcome by a sufficiently high shear stress that produces complete disentanglement between the adsorbed chain and chains in the bulk.

Experimental data show that for monodisperse melts[6–8,25] and solutions[26] spurt flow occurs at a critical wall stress level where the polymer liquids behave Newtonian like – the shear stress is linearly proportional to the shear rate. Under the critical wall shear stress σ_c, or critical pressure $P_c = \sigma_c(4D/L)$, the stick–slip transition leads to a velocity field change from the parabolic profile given by Eq. (3.9) to a combination of plug flow and Poiseuille flow, as shown in Fig. 6.3. Correspondingly, the volumetric throughput Q_f jumps from Q_c, which is the throughput just before the spurt, to

$$Q_s = Q_c + (\pi D^2/4)V_s^*. \tag{6.14a}$$

Specifically, we have the terminal flow characterized by

$$Q_c = (\pi D^3/32)\dot{\gamma}_c. \tag{6.14b}$$

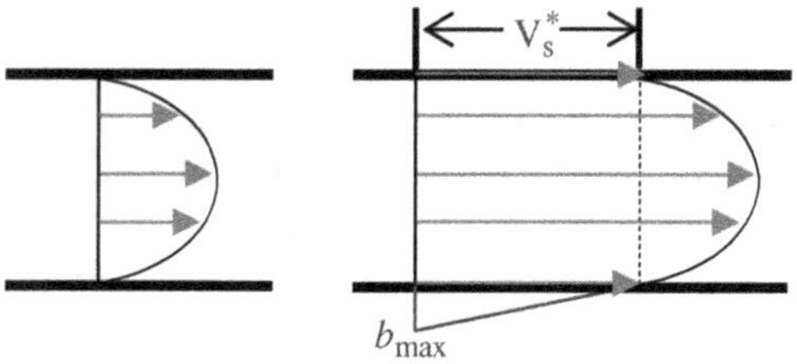

Figure 6.3 Velocity profiles in the cases of a capillary or a slit (channel) die extrusion before and after the stick–slip transition. Because the stress level is the same at the transition, the velocity profiles in the bulk are identical.

Here, the critical shear rate is related to the critical wall stress σ_c through the Newtonian law

$$\dot{\gamma}_c \approx \sigma_c/\eta_0. \tag{6.14c}$$

Equation (6.14b) is an expression for Poiseuille flow of Newtonian liquids, as shown in Section 3.1.3.1. Expressions from Eqs. (6.14a) to (6.14c) hold true for the stick–slip transition because the transition takes place within the Newtonian (terminal flow) regime. In terms of the local wall shear rate, defined as $\dot{\gamma}_s = 32Q_s/\pi D^3$, Eq. (6.14a) can be rewritten as

$$\dot{\gamma}_s = \dot{\gamma}_c + 8V_s^*/D \tag{6.15}$$

Both Eqs. (6.14a) and (6.15) can be regarded as the Mooney formula that evaluates the slip velocity V_s according to the D dependence of the throughput Q_s or $\dot{\gamma}_s$. For example, in the tradition of Mooney, the increase in the shear rate is plotted as a function of the reciprocal die diameter 1/D so that the slope is $8V_s^*$.

To show how such a stick–slip transition experiment quantifies a polymer's ability to undergo wall slip, we rewrite Eq. (6.14a). At the transition, the jump in the throughput or apparent shear rate at the wall is accurately captured by the ratio

$$Q_s/Q_c = \dot{\gamma}_s/\dot{\gamma}_c = 1 + 8b_{max}/D, \tag{6.14d}$$

which indicates that the jump in the throughput is greater for a smaller capillary diameter D. The D dependence is by itself sufficient to reveal the interfacial origin of the stick–slip transition. Here, b_{max} is the quantity discussed in Section 6.1.3 because the stick–slip transition involves the state of complete wall slip at shear stress σ_c.

6.2.1.2 Experimental Data

The earliest observation of the spurt flow was rather revealing. In contrast, more recent capillary rheometric studies[16,17,27,28] of wall slip were based on contemporary speed-controlled rheometers and thus could not quantitatively evaluate the characteristics of the stick–slip transition. We defer any further discussion of speed-controlled capillary rheometry to Section 10.1. Focusing on pressure-controlled extrusion and controlled-stress simple shear, we first reproduced the 1958 measurements in Fig. 6.4.[4] The 1958 experiment not only discovered the remarkable ability of high-density polyethylene (HDPE) to undergo spurt but also indicated that spurt is a first order transition with a significant hysteretic character. The hysteresis loop can be characterized in terms of the critical stress σ_c (or pressure P_c) for spurt and the critical stress $\sigma_h(P_h)$ below which the capillary extrusion returns to no-slip behavior, indicated by the circles in Fig. 6.4. The hysteresis actually reveals the interfacial nature of the spurt[4]: In capillary extrusion, fresh material keeps being fed into the extrusion die and being extruded. Only the interface is not

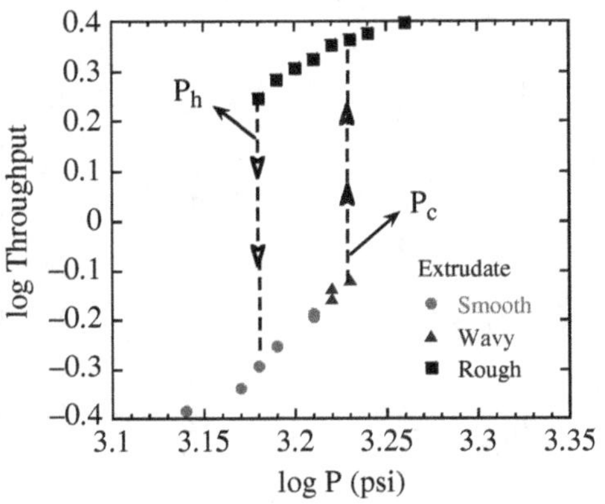

Figure 6.4 Spurt flow of a linear polyethylene from pressure-controlled capillary rheometry at 190 °C, based on a die of diameter D = 0.5 mm and length L = 4.5 mm. As indicated in the original paper, the extrudate is *wavy* (denoted by the triangles) just below the spurt transition at P_c and turns *rough* on the slip branch upon the transition, as indicated by the squares. *Source:* Data from Bagley *et al.* (1958) [4].

"extruded out," allowing the system to remain in the state B without returning to the state A as the applied pressure is reduced. The interfacial state is displaying hysteresis.

The appearance of extrudate distortion, as mentioned in Fig. 6.4 and observed by Tordella,[1–3] has caused considerable confusion, distracting attention away from the interfacial origin of the spurt phenomenon. Subsequent three decades of research aimed at understanding the nature of this spurt flow,[1–3,6–8] ended with Denn's declaring in his 1990 review[29] that spurt flow was a "paradox." For example, extensive data[6–8] from Vinogradov's laboratory indicates that the critical stress for spurt is similar to the rubbery plateau modulus $G_{pl,}$ as shown in Fig. 6.5. This correlation caused Vinogradov and coworkers to speculate that spurt flow was a constitutive behavior of the polymer melt, since G_{pl} is a bulk constitutive property of polymer melts. Calling spurt flow a "fluid-to-rubber" transition, Vinogradov *et al.* specifically stated[6] that the "phenomena of the discontinuous increase in volume flow rate" are "due to the induced transition of the polymer from the fluid to the high-elastic state at larger deformation rates." In reality, the flow rate in the bulk is still in the terminal flow regime, as indicated by Eq. (1.38b). In a later publication, Vinogradov and coworkers explored a plausible correlation between spurt and observations of melt rupture near the wall.[30]

The confusion about the nature of the stick–slip transition cleared up in the mid-1990s, when highly *polydisperse* HDPE resins were examined in the style of Bagley[4] in a series of pressure-controlled capillary rheometric studies,[31–33] of which Fig. 6.6 shows an example. Because of the strong polydispersity in the molecular weight distribution, above the stick–slip transition, the extrudate was smooth, dismissing the suspicion that the spurt originates from a constitutive instability. A more refined series of studies ensued, based on monodisperse 1,4-polybutadiene (PB) melts[25,34] and solutions.[26] These data, as shown in Fig. 6.7, resemble

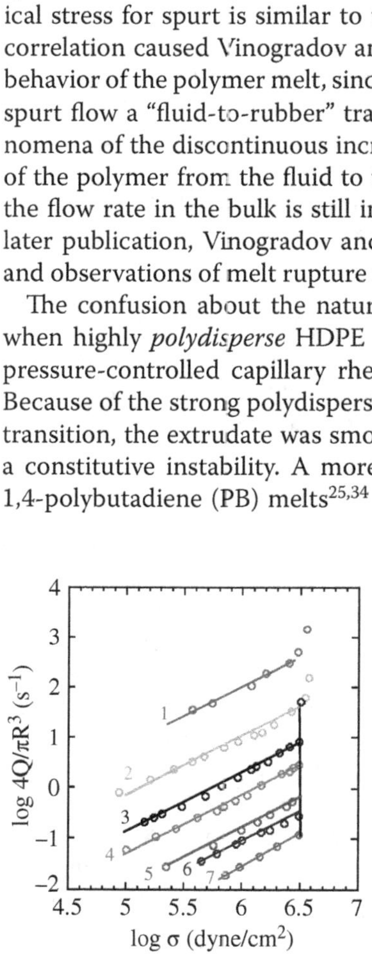

Figure 6.5 Volumetric output versus applied pressure (shear stress at the wall) for a series of relatively monodisperse polybutadiene melts, from capillary extrusion with a die of diameter D = 0.96 mm and length L = 24.8 mm. The numbers from 1 to 8 label the molecular weights of 38, 68, 102, 151, 204, 240, 320, 580 kg/mol. *Source:* Data from Vinogradov *et al.* (1972) [6].

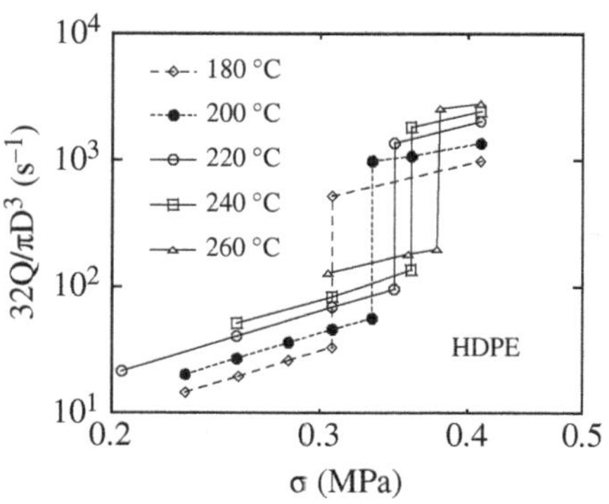

Figure 6.6 Stick–slip transition of a high-density polyethylene at five temperatures, expressed in terms of the die output versus the wall shear stress, obtained with a Monsanto (pressure-controlled) capillary rheometer. The critical stress for the transition is seen to increase with rising temperature. In the original literature, this temperature dependence of σ_c was invoked to justify the conclusion that the stick–slip transition is not caused by chain desorption. *Source:* Data from Wang and Drda (1996) [31].

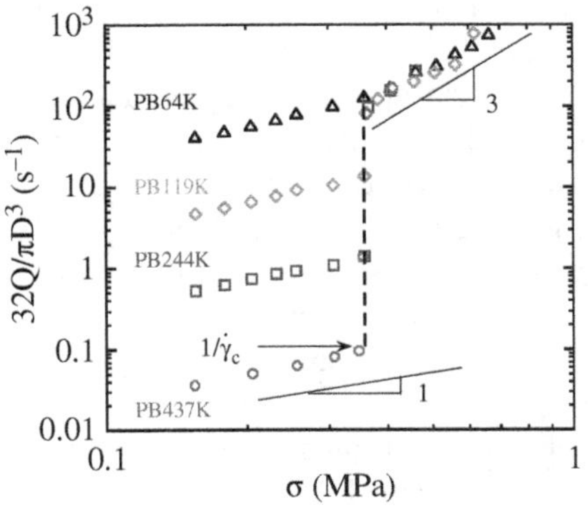

Figure 6.7 Stick–slip transition of 1,4-polybutadiene melts of four different molecular weights, $M_w = 64, 119, 244$, and 437 kg/mol, respectively, using a Monsanto capillary rheometer. *Source:* Data from Yang *et al.* 1998 [25].

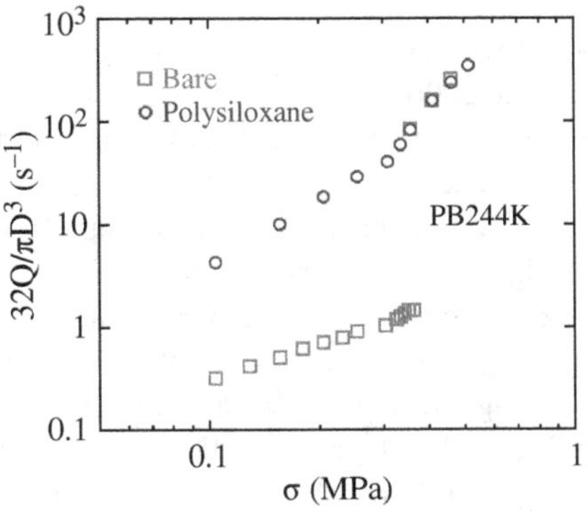

Figure 6.8 Removal of the stick–slip transition by a surface coating that induces desorption at high shear stress at the capillary wall. *Source:* Data from Yang *et al.* 1998 [25].

data from Vinogradov and coworkers in the sense that there is a unique critical wall stress for the stick–slip transition independent of molecular weight that ranges from 64 to 437 kg/mol. However, contrary to the conclusions of Vinogradov and coworkers the spurt behavior was shown to be entirely interfacial, as shown in Fig. 6.8. Had spurt been due to bulk behavior during capillary flow, coating the die wall would not have removed the spurt character. However, coating the die wall did indeed remove spurt behavior. Moreover, the magnitude of the stick–slip transition is indeed predicated by Eq. (6.12): As shown in Ref. [25], b_{max} determined from Eq. (6.14d) scales with M with an exponent of circa 3.5, in agreement with Eq. (6.12).

There are some noticeable differences in the characteristics of the stick–slip transition between high density polyethylene (HDPE) and 1,4-polybutadiene. The stick–slip transition occurs at the same critical stress σ_c for 1,4-PB melts of different molecular weights, whereas HDPE melts show an increased σ_c for HDPE of lower molecular weight.[32] Moreover, the ratio of σ_c/G_N^0 is much lower for HDPE than for 1,4-PB. These differences might indicate a lack of adequate polymer adsorption by the HDPE and a dependence of adsorption on the molecular weight distribution for HDPE.

6.2.2 Stick–Slip Transition in Simple Shear

Since the shear stress in a capillary die grows linearly from zero at the center line to the maximum at the die wall, as shown in Eq. (3.7a), it is interesting to question whether the stick–slip transition would occur in a non-pressure-driven apparatus, that is, in displacement-driven simple shear. The data from the literature[35] confirm that the stick–slip transition also takes place at a critical stress σ_c whose magnitude is a significant fraction of G_{pl}, just like the case of the pressure-driven stick–slip transition. At the stick–slip transition, the apparent shear rate jumps from $\dot{\gamma}_c \sim 1/\tau$ to $\dot{\gamma}_s$, given by

$$\dot{\gamma}_s/\dot{\gamma}_c = V^*/V_c = 1 + 2b_{max}/H, \tag{6.16}$$

where $V^* = 2V_s^* + V_c$ and $V_c = \dot{\gamma}_c H$. Although the stick–slip transition has been reported using capillary rheometry since 1958, the first observation of the stick–slip transition in simple shear arrived[35] as late as 2006, as shown in Fig. 6.9(a,b). Analogous to Fig. 6.7,

$\sigma_c \sim 0.2$ MPa $< G_{pl} = 1$ MPa is independent of the molecular weight of these 1,4-polybutadiene melts. Moreover, Fig. 6.9(b) confirms the scaling relation between b_{max} and the M, as shown by Eq. (6.12). The 2006 study was accomplished using the special apparatus depicted in Section 5.1.

Figure 6.10 reveals the interfacial nature of the stick–slip transition. The inset shows that the magnitude of the jump in the apparent shear rate depends on the gap distance, in agreement with Eq. (6.16). Simple shear and capillary extrusion produce similar stick–slip transitions, as shown in Fig. 6.11. We note in passing that b_{max} evaluated from the inset of Fig. 6.10 is smaller than that given in Fig. 6.9(b), suggesting small uncertainties in the identification of the stick–slip transition magnitude and possible variation of the surface condition from the capillary die wall to the shearing surfaces of the simple shear device.

When $2b_{max}/H \gg 1$ or $8b_{max}/D \gg 1$, a polymer of different molecular weights has different values for b_{max} but should reach a common limiting shear rate $\dot{\gamma}_s$ at the stick–slip transition. According to Eqs. (6.14d) and (6.16), we have

$$\dot{\gamma}_s = 32Q_s/\pi D^3 \sim 8V_s^*/D \quad \text{for } b_{max}/D \gg 1, \tag{6.17a}$$

and

$$\dot{\gamma}_s \sim 2V_s^*/H, \quad \text{for } b_{max}/H \gg 1, \tag{6.17b}$$

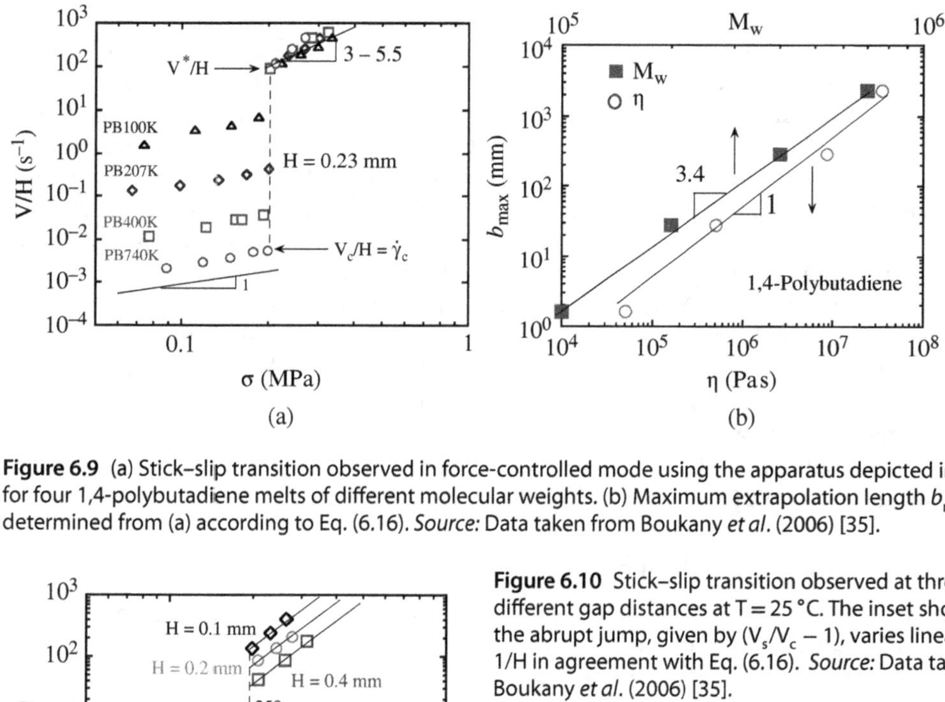

Figure 6.9 (a) Stick–slip transition observed in force-controlled mode using the apparatus depicted in Fig. 5.1, for four 1,4-polybutadiene melts of different molecular weights. (b) Maximum extrapolation length b_{max} determined from (a) according to Eq. (6.16). *Source:* Data taken from Boukany *et al.* (2006) [35].

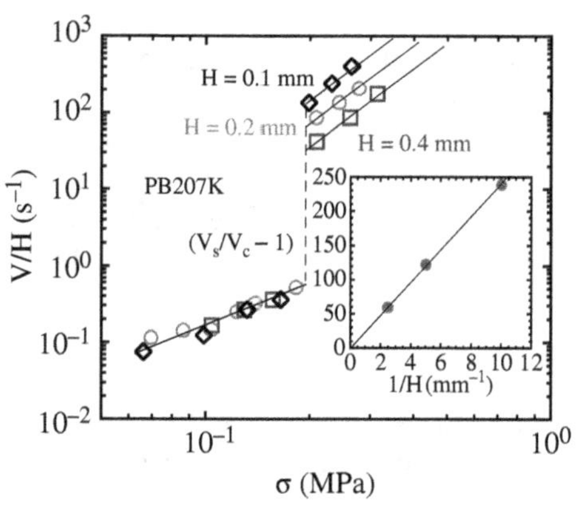

Figure 6.10 Stick–slip transition observed at three different gap distances at T = 25 °C. The inset shows that the abrupt jump, given by $(V_s/V_c - 1)$, varies linearly with 1/H in agreement with Eq. (6.16). *Source:* Data taken from Boukany *et al.* (2006) [35].

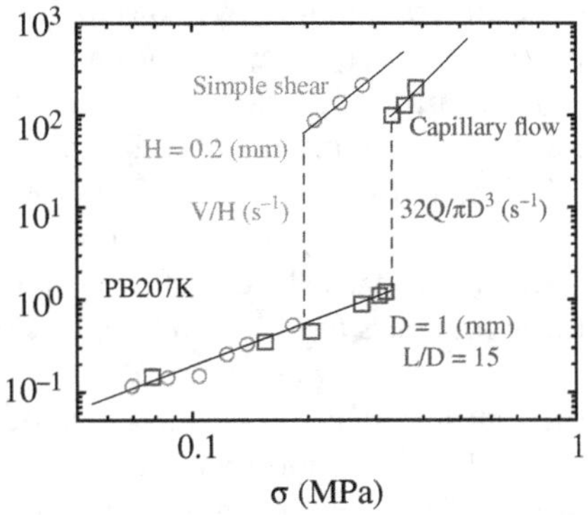

Figure 6.11 Comparison of the stick–slip transition between capillary extrusion and simple shear of the same sample (PB207K) in terms of the wall stress σ. *Source:* Data taken from Boukany *et al.* (2006) [35].

where use is made of Eq. (6.11) and $\tau\dot{\gamma}_c \sim 1$. The convergence of data upon the stick–slip transition in Figs. 6.7 and 6.9(a) confirms Eqs. (6.17a) and (6.17b) and shows that V_s^* is indeed independent of the polymer molecular weight, as predicted by Eq. (6.9a).

Comparison of the stick–slip transition characteristics between pressure-driven capillary rheometry and displacement-controlled simple shear for a given melt reveals an interesting level of complexity. Figure 6.11 shows that the critical shear stress at the wall for the stick–slip transition is noticeably higher in capillary extrusion. This difference suggests the presence of pressure loss at the die entry, which is well known and is briefly described in Section 10.1.2.2. It should be possible to test this effect by using capillary dies with the same die entry angle but different L/D. Currently, such data are unavailable.

Finally, we note that chain architecture appears to affect the ultimate slip velocity V_s^*. Figure 6.12 shows that the slip branches of linear and star 1,4-PB melts do not overlap, as observed both with capillary extrusion and with parallel-plate shear. This difference of nearly one decade in the magnitude of the stick–slip transition is instructive, indicating that a star polymer has different structure and dynamics after chain disentanglement. According to Eq. (6.9a), τ_e is greater for star-like melts than for linear melts.

6.2.3 Limiting Slip Velocity V_s^* for Different Polymer Melts

The comparison of Eq. (6.9a) for V_s^* with Eq. (6.12) for b_{max} reveals several interesting points. First, the extrapolation length b_{max} is independent of temperature and thus suitable as a parameter to evaluate the intrinsic capacity for wall slip among different polymer melts.

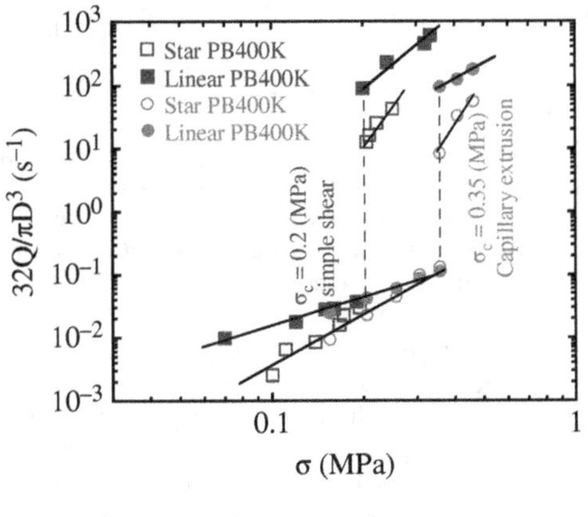

Figure 6.12 Stick–slip transition of linear and star-like PB melts in simple shear (H = 0.2 mm) and capillary extrusion (D = 1 mm) at T = 25 °C. *Source:* Data taken from Boukany *et al.* (2006) [35].

In contrast, the Mooney slip velocity V_s^* is greater at a higher temperature because of the faster dynamics. But greater V_s^* does not mean a greater correction by wall slip. Second, V_s^* may vary significantly from one polymer to another, making some polymers more suitable than others for direct visualization of wall slip by various techniques including the PTV method.

Third, we can estimate V_s^* from the characteristics such as b_{max} associated with the stick–slip transition. When particle-tracking velocimetric observations are available, V_s^* at the stick–slip transition can also be directly measured, as shown in Fig. 6.13 for polyisoprene (PI). Such direct measurements turn out to be consistent with the rheometric determination of the stick–slip transition; they yield $V_s^* \approx 0.27$ mm/s for 1,4-polyisoprene (PI) at room temperature. From the measurement of V_s^* and knowledge of l_{ent}, we can estimate τ_e by using Eq. (6.9a). Let us take a 1,4-polyisoprene melt of $M_w = 550$ kg/mol as an example.[36] From the published value of $l_{ent} \approx 6$ nm for PI melts[37] and the value for V_s^* in Table 6.1, we find $\tau_e \sim 2.2 \times 10^{-5}$ s. This magnitude is consistent with an estimate from the empirical formula $\tau/\tau_e \approx (M_w/M_e)^{3.3}$, which gives $\tau_e \approx 1.7 \times 10^{-5}$ s, given $\tau = 50$ s for this polyisoprene[36] and $M_e = 6$ kg/mol. Similarly, for

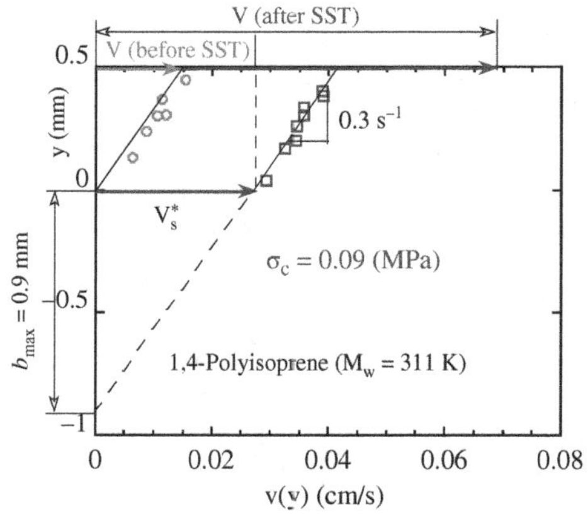

Figure 6.13 PTV determination of the velocity profile in a sliding-plate setup (cf. Fig. 4.4(b)) to reveal wall slip at the stick–slip transition. *Source:* Data taken from Boukany et al. (2006) [35].

Table 6.1 Characterization of various entangled melts studied with PTV.

Samples	References	Temperature	V_s^* (mm/s)
HDPE	[31]	200 °C	200
1,4-PB	[25]	Room temp.	10
1,4-PI	[35,36]	Room temp.	0.27
SBR241[a)]	[38]	70 °C	1.3
SBR241[b)]	[38]	Room temp.	0.05
SBR250[c)]	Unpublished	Room temp.	0.0093

a) SBR is acronym for styrene butadiene rubber. Being a random copolymer of styrene and butadiene that can have a varying level of vinyl content, this particular SBR's microstructure is characterized as having 21% styrene and 39.9% vinyl.

b) At room temperature (23–25 °C), SBR241 is slower than at 70 °C by a factor of 24, according to the small-amplitude oscillatory shear (SAOS) data in Ref. [38]. V_s^* is evaluated from the data at 70 °C by division by 24.

c) This SBR250 has 20% styrene and 54.7% vinyl. Therefore it is expected to have a higher T_g and smaller V_s^*, which is evaluated according to unpublished capillary rheometric data on stick–slip transition, made at University of Akron by Dr. Xianggang Li.

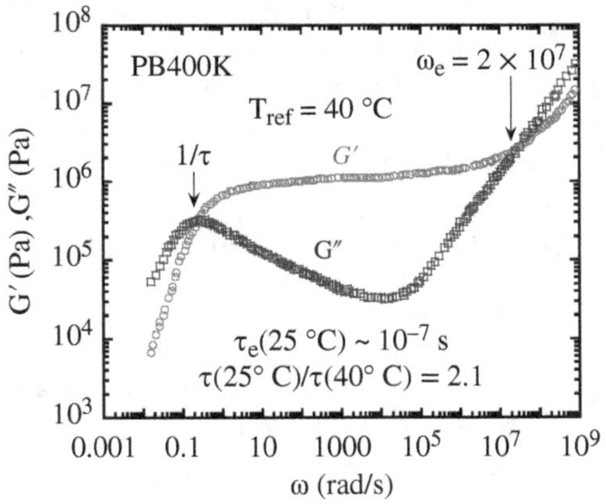

Figure 6.14 Storage and loss moduli G′ and G″ of PB400K melt at reference temperature $T_{ref} = 40$ °C. According to the Williams–Landel–Ferry shift factor, the chain dynamics are slower by a factor of 2.1 at 25 °C. *Source:* Data taken from Wang *et al.* (2003) [39].

1,4-polybutadiene melts, Fig. 6.9(a) shows that $2V_s^*/H \sim 100\ s^{-1}$, which gives $V_s^* = 11$ mm/s for $H = 0.23$ mm. Given $l_{ent} \approx 4.4$ nm,[37] Eq. (6.9a) gives $\tau_e \approx 4.4 \times 10^{-7}$ s. The value of τ_e for 1,4 PB is also available in the literature. For example, Fig. 6.14 shows small-amplitude oscillatory shear (SAOS) data at 40 °C for a similar 1,4-PB(400 K) melt. If we identify τ_e as $1/\omega_e$, then $\tau_e \sim 10^{-7}$ s at 25 °C, which is comparable to the estimate of τ_e from the properties such as V_s^* of the stick–slip transition.

The preceding confirmation of the relationship between V_s^* and τ_e is a worthwhile exercise because it rules out that the initial suggestion by de Gennes[9] that the interfacial viscosity η_i was monomeric and therefore given by $\eta_i = \eta_e/N_e^2$. In the absence of polymer desorption, we would have estimated a much greater value for τ_e than those values given in the previous paragraph that would be inconsistent with independent measurements of τ_e. Specifically, replacing Eq. (6.7a) for β_{min} with $\beta'_{min} = \eta_i/l_K = \beta_{min}(N_e)^{-3/2}$, we would have $\tau_e = [l_{ent}/V_s^*](N_e)^{3/2}$ instead of l_{ent}/V_s^*.

In order to obtain the full curves in Fig. 6.14, the time–temperature superposition principle was applied to construct the master curves from small amplitude oscillatory shear (SAOS) measurements at a series of low temperatures down to − 100 °C. Thus, Fig. 6.14 shows the traditional method to estimate the value of τ_e at 25 °C from SAOS tests. In contrast, the estimate Eq. (6.9a) based on the characteristics of the stick–slip transition may sometimes turn out to be convenient. Conversely, in the absence of any quantification of the stick–slip transition, V_s^* can be estimated from the rheometric SAOS determination of τ_e and available values for l_{ent}. In Table 6.1, we tabulate V_s^* for different melts, from the available literature data on their stick–slip transitions in capillary rheometry and sliding plate rheometry, estimated according to Eqs. (6.17a) and (6.17b).

To understand why there is such a wide range of V_s^* as shown in Table 6.1, we rewrite Eq. (6.9a) as

$$V_s^* \approx l_{ent}/\tau_e = (l_K/\tau_0)N_e^{-3/2} \sim (l_K/\tau_0)(l_K/p)^3. \qquad (6.9b)$$

Here the elemental time τ_0 is given in Eq. (2.17) and is assumed to characterize the temperature dependence of polymer dynamics in all molecular models and N_e the number of Kuhn segments of length l_K in an entanglement strand. The last expression follows from Eq. (2.28) of the packing model, p being the packing length. Equation (6.9b) shows that V_s^* scales with p as p^{-3} in units of (l_K/τ_0). This p dependence is expected from Eq. (6.9a) for $V_s^* \sim G_{pl}/\beta_{min}$ because $G_{pl} \sim p^{-3}$ according to Eq. (2.30b) and β_{min} of Eq. (6.7a) scales as $k_B T\tau_0/(l_K)^4$ and thus has no p dependence.

If the state of chain desorption changes with temperature T, $V_s^*(T)$ may have an additional temperature dependence, so that V_s^* may not show the same temperature dependence as that of $1/\tau_e$, which is obtained from small amplitude oscillatory shear measurements of G′ and G″. For polymer melts that show a breakdown of time–temperature superposition on time scales between τ_e and τ, an accurate characterization of the stick–slip transition on strongly adsorbing surfaces at different temperatures could be a means to elucidate the failure of time–temperature superposition. No such investigations have been reported in the literature.

6.2.4 Characteristics of Interfacial Slip Layer

In the absence of surface contamination that could lead to chain desorption at shear stresses much lower than G_N^0, the data reviewed in Sections 6.2.1 and 6.2.2 indicate that the SST takes place at σ_c given by Eq. (6.8). This information has been used to derive Eqs. (6.9a) and (6.10) for the slip velocity at the complete slip. These results allow us to show that the shear rate in the entanglement-free interfacial slip layer is given by either

$$\dot{\gamma}_i^* = V_s^*/l_{ent} \sim 1/\tau_e, \quad \text{for entangled melts} \tag{6.18a}$$

or

$$\dot{\gamma}_i^* = V_s^*/l_{ent}(\phi) = G_{pl}(\phi)/\eta_s, \quad \text{for entangled solutions.} \tag{6.18b}$$

In the simple case where the solution is based on a solvent of the same polymer with molecular weight comparable to M_e, we can replace η_s in Eq. (6.18b) with η_e and take a ratio of Eq. (6.18b) to Eq. (6.18a)

$$\frac{\dot{\gamma}_i^*(\text{solution with } M_e)}{\dot{\gamma}_i^*(\text{melt})} = \frac{G_{pl}(\phi)}{G_{pl}(\phi = 1)} \tag{6.18c}$$

whose scaling dependence on the concentration ϕ is shown by Eq. (2.36c). Given a comparable viscosity $\eta_i = \eta_e$ in the slip layer, the smaller shear rate of the slip layer in the solution stems from the fact that the stress level is on the order of $G_{pl}(\phi)$, lower than that of the melt given by $G_{pl}(\phi = 1)$.

6.3 Wall Slip during Startup Shear – Interfacial Yielding

The treatment in the preceding sections only addressed wall slip in steady state. Massive wall slip, that is, $b_{max}/H \gg 1$, occurs in steady shear because of the great disparity between the state of entanglement in the slip layer and the state of entanglement in the bulk, characterized by the enormous viscosity ratio in Eqs. (6.12) and (6.13) for b_{max}. So far, our description avoids the question of how wall slip arises and replaces the state of no-slip. In the absence of external deformation, the wall/polymer interface may be depicted as shown by Fig. 6.1(a). Here, the unbound free chains would remain entangled with the adsorbed chains on a time scale comparable with the terminal relaxation time τ. When the applied shear rate is low, so that $Wi = \dot{\gamma}\tau$ is below unity in the bulk, Brownian motion is fast enough to maintain the topological coupling between adsorbed and free chains. For $Wi_{app} < 1$, the adsorbed chains would remain entangled with the bulk chains; no wall slip occurs in the presence of adequate polymer adsorption. In this section, we discuss how we should understand the emergence of wall slip, when a sudden startup shear is suddenly applied with $Wi_{app} > 1$.

Because of melt compression in the barrel, capillary rheometry operated at a constant piston speed is not a convenient protocol to characterize at various shear rates how wall slip emerges upon startup shear.[16,17,41–45] In the literature, wall slip has been studied in rate-controlled simple shear,[19,46–48] usually for steady state. It is rare to find publications that demonstrate how wall slip comes into existence. In this context, the work of Laun[46] was pioneering because it showed the appearance of wall slip only after a considerable amount of bulk deformation had occurred following a startup shear.

6.3.1 Theoretical Discussions

As early as 1979, de Gennes proposed[9] an explanation for how massive wall slip is possible for entangled polymers. Under some limiting interfacial conditions, for example, by considering an interfacial state where there is on the average no more than one adsorbed chain in an area of the coil size R, Brochard- and de Gennes[24] developed a treatment for envisioning interfacial wall slip in steady shear. Similar scaling arguments[49–52] and other more detailed modeling[53,54] have been presented to describe the magnitude of the wall slip in terms of the shearing conditions. Most of these studies explored the consequences of having different values of V_s at the polymer/wall interface, by assuming that the slip velocity preexists at the interface. For example, for a scaling argument to apply, Brochard and de Gennes considered the limit of low surface coverage where interfacial disentanglement can be expected upon exposing tethered chains to a sufficiently large V_s. These publications only offered information on the wall slip characteristics in the steady state, while clearly the equally important and more interesting questions are how and when wall slip arises during startup shear.

Since the state of chain adsorption for a given pair of polymer and wall is nontrivial to depict, the simplest theoretical modeling of wall slip assumes full polymer adsorption, as shown in Fig. 6.1(a). Upon a startup shear with $Wi_{app} > 1$, in the presence of the significant polymer adsorption that typically occurs on metallic (high surface energy) walls, an entangled polymeric liquid initially undergoes elastic shear deformation until the shear stress reaches a threshold level. Further shearing is expected to produce a transition from no-slip to slip at the interface between the sample and the shearing surfaces. This change should naturally be called *interfacial yielding*.[36] Whether or not a change of the interfacial state from entanglement to disentanglement can cause stress overshoot (mentioned in the introductory paragraphs of Part 2) depends on the ratio b/H. As shown in Chapter 7, a nonmonotonic relation between shear stress and shear strain (known as overshoot) can occur in the absence of wall slip during startup shear. In general, rheometric information alone may not be sufficient to discern whether or not measurable interfacial yielding has taken place.

Because entangled melts are usually too stiff for commercial rotational rheometers to study, it has been an experimental challenge to examine melt wall slip in simple shear. Before presenting two examples that avoid or overcome the stiffness limitation in different ways, we first illustrate the concept of wall slip in the presence of elastic bulk deformation.

The subject of nascent wall slip during startup shear is actually rather different from wall slip in steady shear where both the interface and the bulk are in flow state. The molecular mechanism for interfacial yielding will receive a further account in Section 16.5. At this point, we just need to develop a sense of how interfacial slip happens and what consequences it has. Upon startup shear at V/H, significant elastic deformation takes place for well-entangled polymeric liquids as long as $Wi_{app} \gg 1$. Shear stress monotonically increases with increasing strain. Interfacial disentanglement then takes place between the adsorbed chains and the bulk chains, upon force imbalance as described in Section 16.5, leading to interfacial slip. The magnitude of the wall slip is determined by the slip velocity V_s that is required to transmit the shear stress from the wall to the bulk. At the time slip commences, the bulk is still undergoing elastic deformation – had slip not occurred the system could have been undergoing further elastic deformation. This means that we cannot describe the stress level using Eq. (6.6), which is only valid in steady shear. On the other hand, the effect of wall slip can still be expressed in terms of a velocity ratio V_s/V.

When σ increases to a level comparable to G_{pl}, we have

$$V_s/V = (\sigma_c/\beta)/V \sim [(G_{pl}\tau/\beta)/H]/Wi_{app} = (b_0/H)(Wi_{app})^{-1}. \tag{6.19}$$

Equation (6.19) follows from Eq. (6.1a), with Wi_{app} given by Eq. (6.5), while b_0 is an extrapolation length defined by

$$b_0 = G_{pl}\tau/\beta \approx \eta_0/\beta. \tag{6.20}$$

The abovementioned expression resembles the general expression for b in Eq. (6.2) and becomes the same as Eq. (6.2) if we take η in Eq. (6.2) to be close to the zero-shear viscosity η_0. In Eq. (6.20), the second equality arises from the Maxwellian relation $\eta_0 \approx G_{pl}\tau$ for the zero-shear viscosity. Unless interfacial yielding produces complete wall slip with β approaching β_{min} as given in Eq. (6.7a) or (6.7b), b_0 is smaller than b_{max} of Eq. (6.12) or (6.13). Unlike the expression of Eq. (6.4a) for steady shear, V_s is transient, reflecting the polymer velocity a monolayer away from the shearing surface. The ratio V_s/V can therefore exceed unity if the interfacial yielding leads to bulk elastic recoil. Equation (6.19) shows that $V_s/V > 1$ occurs when $b_0/H > Wi_{app}$. The data presented in Fig. 6.15(a) shows that $V_s/V > 1$ can indeed take place transiently. In steady flow, the condition of $V_s/V > 1$ can never be true. Conversely, according to Eq. (6.19), wall slip does not affect the startup shear if $b_0/H \ll 1$ or if $Wi_{app} \gg b_0/H$.

It is necessary to emphasize that the limiting slip velocity V_s^* depicted in Sections 6.1.3 and 6.2.3 is the velocity found at the level of the stress plateau during steady shear. In other words, V_s^* is not the maximum slip velocity. According to Eq. (6.1a), in the presence of a higher stress σ than the plateau stress σ_c, we have

$$V_s = V_s^*(\sigma/\sigma_c) > V_s^*, \tag{6.21}$$

where V_s^* is given by either Eq. (6.9a) or (6.10). V_s can grow beyond V_s^* when $\sigma > \sigma_c$. However, we should note that at higher stresses Eq. (6.1a) no longer holds and V_s may vary quadratically with σ.

6.3.2 Experimental Data

The slip velocity V_s can be very large in aqueous entangled solutions where the solvent is water because η_s is as low as 10^{-3} Pa s. Thus, a well-entangled DNA (deoxyribonucleic acid) solution

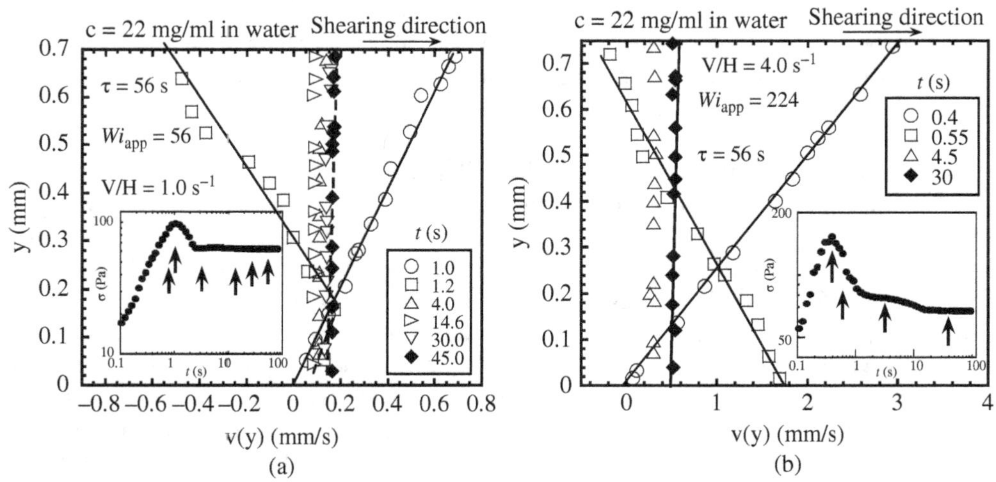

Figure 6.15 Velocity profiles at different times during startup shear with apparent rate V/H (a) 1 s^{-1} and (b) 4 s^{-1} for an entangled DNA solution. The insets show the corresponding stress versus time curves, with stress rising to a maximum (overshoot) and then decreasing to its steady-state value. *Source:* Data taken from Boukany *et al.* (2009) [55].

is an ideal candidate to demonstrate interfacial yielding during startup shear. The experimental system can be examined with a commercial shear rheometer, equipped with a PTV system, as shown in Fig. 4.4(a). For example, a water-based DNA solution at a concentration of c = 22 mg/ml has an estimated $b_{max} > 700$ mm. A conventional PTV setup can readily observe wall slip in this solution.

Figure 6.15(a,b) shows how startup shear at V/H = 1 and 4 s^{-1} produces interfacial yielding.[55] Before the stress overshoots at $\gamma = 1.0$ and 1.6, respectively, the shear field is free of any wall slip correction. Between 1 and 1.2 s (for V/H = 1 s^{-1}) and between 0.4 and 0.55 s (for V/H = 4 s^{-1}), massive wall slip takes place, leading to enormous sample recoil, as shown by the squares in both Fig. 6.15(a,b). At longer times, elastic recoil ceases and wall slip dominates the response as shown by the diamonds that show more wall slip on the moving surfaces.

To observe interfacial yielding in entangled melts, we have to resort to a home-built shearing apparatus. A common design is a sliding parallel-plate assembly coupled to PTV device, as shown in Fig. 4.4(b). After establishing excellent polymer adsorption on the metallic surfaces, which may take hours to days of contact between a polymer melt and the shearing surfaces, there is no measurable wall slip during startup shear until the applied strain exceeds unity, at which time the shear stress reaches a level of the order comparable with the melt plateau modulus G_N^0. In general, the interfacial interaction between the polymer and the walls is realized through chain entanglement between the adsorbed chains and bulk chains, as shown in Fig. 6.1(a). The emergence of wall slip is a transformation of the interfacial state from full entanglement to disentanglement between the adsorbed chains and the chains in the bulk, that is, a transformation from Fig. 6.1(a) to Fig. 6.1(b). As discussed in the following section, many intermediate states separate Fig. 6.1(a) from Fig. 6.1(b). These partially disentangled states lead to partial slip.

Figure 6.16(a) examines the rheological response of a melt. It shows the time dependence of the stress with shear starting at $t = 0$ at apparent shear rate V/H = 4 s^{-1} and being terminated at $t = 0.5$ s. Up to a strain of 1.5 ($t = 0.38$ s), there is no wall slip, as shown Fig. 6.16(b). After a further 0.05 s (at $t = 0.43$ s), apparent wall slip is observed. Within the spatial resolution of the PTV, we see an interfacial velocity $V_i \sim 1$ mm/s, which is larger than V_s^* in Table 6.1 by a

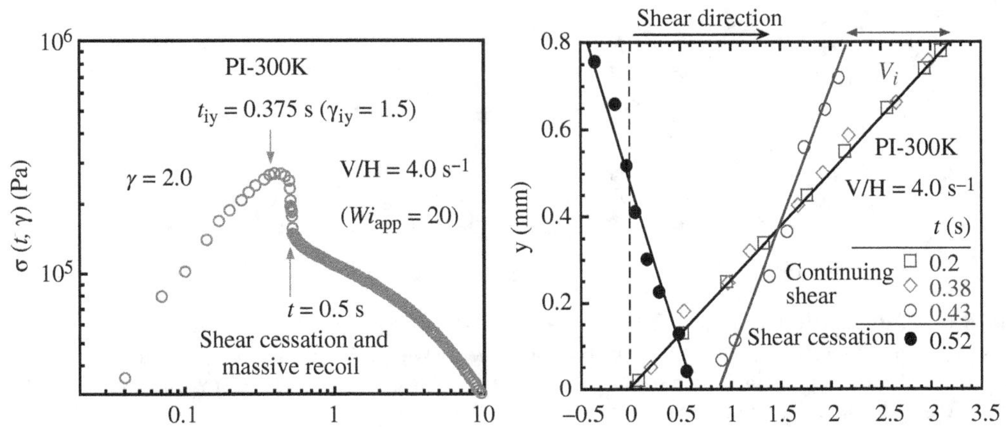

Figure 6.16 (a) Stress responses upon a startup shear at V/H = 4 s^{-1}. The startup shear is terminated at 0.5 s. The stress peaks at t_{iy}, before the shear is terminated, and then has a steep drop after 0.5 s. (b) Corresponding particle-tracking velocimetric observations reveal interfacial failure at 0.43 s during shear and recoil-like motions after the shear ceases. *Source:* Data taken from Boukany and Wang (2009) [36].

factor of 4. Because this magnitude of 1 mm/s is considerably larger than V_s^*, the interfacial disentanglement may have involved several monolayers so that a_i is greater than l_{ent}. For an impression of how such interfacial yielding takes place, see a movie provided with the book.[1]

6.4 Relationship between Slip and Bulk Shear Deformation

6.4.1 Transition from Wall Slip to Bulk Nonlinear Response: Theoretical Analysis

Rheological studies of entangled polymers under large shear deformation cannot proceed without a clear understanding of various aspects of wall slip behavior. Wall slip is the first sign of inhomogeneous shear and is inevitably measurable for well-entangled polymeric liquids. Wall slip can change the shear field to a degree that prevents the bulk from being sheared into the nonlinear response regime. There is no wall slip for $Wi_{app} < 1$, as depicted in Fig. 6.17(a). When Wi_{app} of Eq. (6.5) first exceeds unity, stress overshoot upon startup shear may be due to wall slip. In steady state, the bulk may undergo terminal flow with Wi_c of Eq. (1.38b), as shown in Fig. 6.17(b), so long as Wi_{app} has not exceeded a threshold. When $b_{max}/H \gg 1$, this threshold is large. There would be a wide range of Wi_{app} over which the magnitude of the wall slip grows progressively with increasing Wi_{app}. Consequently, entangled polymers capable of massive wall slip will exhibit a wide stress plateau, as recently reported.[56]

With increasing Wi_{app}, the steady shear stress should stay flat at a level high enough to retain wall slip until a complete wall slip takes place. This assertion is equivalent to the statement that before wall slip reaches its maximum, that is, before b grows to b_{max} through the decrease in η_i in Eq. (6.3), the bulk can remain sheared at $Wi_c \sim 1$ as shown in Fig. 6.17(b,c). Before Wi_{app} reaches the threshold value $Wi_{ws\text{-}bnl}$ (to be defined in Eq. (6.23)), there cannot be complete wall slip in steady state. Specifically, Fig. 6.17(b) shows

$$Wi_{app} \equiv (V/H)\tau = Wi_c(1 + 2b/H). \tag{6.22}$$

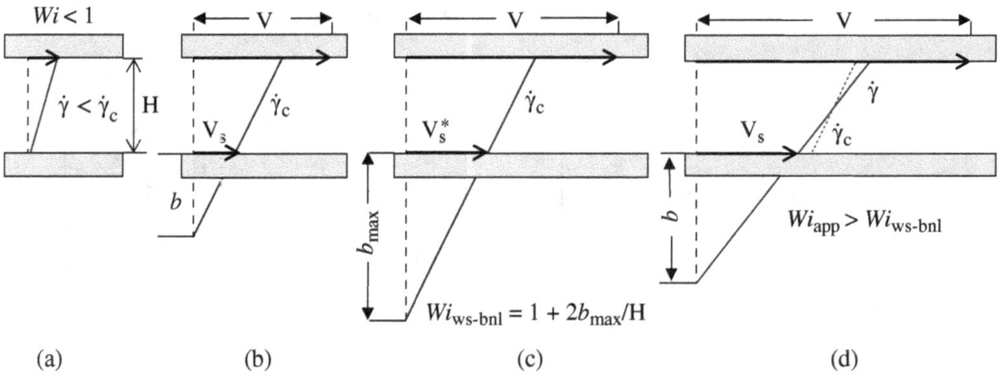

Figure 6.17 Illustration of simple shear (a) in absence of wall slip, which is usually guaranteed for $Wi_{app} < 1$, (b) in presence of measurable wall slip in the stress plateau between $Wi_{app} > 1$ and $Wi_{app} < Wi_{ws\text{-}bnl}$, (c) at the maximum of b, that is, the upper bound of Wi_{app} where the bulk shear rate still locks in at $\dot{\gamma}_c$ with $Wi_c \sim 1$, and (d) when the bulk undergoes strong shear, that is, $\dot{\gamma} > \dot{\gamma}_c$ or $Wi > 1$.

1 Video 6.1: Interfacial yielding of linear polyisoprene melt with M = 500 kg/mol, sheared in sliding-plate rheometer at $V/H = 1\,s^{-1}$ at room temperature, made at University of Akron.

This relationship is valid until complete wall slip takes place. Equation (6.22) states that b/H grows linearly with Wi_{app}, since Wi_c is a constant around unity. This equation is the essence of the stress-plateau conjecture in the window ranging from zero wall slip to complete slip.

Whenever wall slip occurs in steady state, the bulk deformation satisfies $Wi \sim 1$, and the stress plateau prevails as long as b has not grown to b_{max}. In this regime, $b = V_s/\dot{\gamma}_c \sim V_s\tau$. Since b grows linearly with Wi_{app} as shown by Eq. (6.22), V_s also increases linearly with Wi_{app}. Our conclusion is that with increasing Wi_{app} b is proportional to V_s and is reminiscent of the Brochard–de Gennes description[24] of the marginal slip regime where $b \sim V_s$. Thus, our assumption, inspired by the character of the stick–slip transition and expressed in Fig. 6.17(b,c), is consistent with the theoretical treatment of Brochard–de Gennes, although their theoretical study was not constructed to address the condition of full polymer adsorption.

In this regime, the slip velocity V_s increases linearly with the applied velocity V of the moving plate (cf. Figure 6.2(a)). The progressive increase in the slip velocity continues to V_s^* when Wi_{app} reaches a threshold value given by

$$Wi_{ws\text{-}bnl} = Wi_c(1 + 2b_{max}/H), \tag{6.23}$$

which is the condition shown in Fig. 6.17(c). We note that the condition (Eq. (6.23)) was first mentioned in Ref. [55] and subsequently elaborated in Ref. [57] where Wi_c was taken as unity. When $Wi_{app} > Wi_{ws\text{-}bnl}$, we necessarily have $Wi > 1$, as shown in Fig. 6.17(d), because the complete wall slip is no longer sufficient to "absorb" the magnitude of the boundary speed V. Here the subscript "ws-bnl" stands for "transition from *w*all *s*lip to *b*ulk *n*on*l*inear response." In Fig. 6.17(d) for steady shear, we have assumed for simplicity that with $Wi_{app} > Wi_{ws\text{-}bnl}$ V_s increases further, the bulk shear rate $\dot{\gamma}$ exceeds $\dot{\gamma}_c$, and the shear stress σ grows beyond σ_c. However, b decreases from its maximum because V_s does not increase as strongly as $\dot{\gamma}$ does. According to Eq. (6.21), V_s is expected to scale as $V_s \sim \sigma \sim \dot{\gamma}^n$ with $n < 1$ because of shear thinning (cf. Eq. (7.1)). Thus, for $Wi_{app} > Wi_{ws\text{-}bnl}$, b is expected to decrease from its maximum at $Wi_{ws\text{-}bnl}$, as shown below in Eq. (6.27).

Given Fig. 6.17(a–d), the more intuitive alternative expression for the condition to produce nonlinear responses in the bulk is for the applied velocity V to satisfy

$$V > V_{ws\text{-}bnl} = H\dot{\gamma} + 2V_s^* > 2V_s^*. \tag{6.24a}$$

The plate velocity V needs to be at least twice the limiting slip velocity V_s^* (taking the case of equal wall slip on both shearing surfaces for simplicity) before $Wi > 1$ can take place. When Eq. (6.24a) is met, both the slip velocity V_s and bulk shear rate $\dot{\gamma}$ can grow with increasing V, beyond V_s^* and $\dot{\gamma}_c \sim 1/\tau$, respectively, as shown in Fig. 6.17(d). In contrast, the extrapolation length b may decrease from its maximum, attained at σ_c either at the stick–slip transition in controlled-stress shear or at $Wi_{ws\text{-}bnl}$ in rate-controlled shear in steady state. To conclude the present discussion, we note that the ratio of Wi_{app} to $Wi_{ws\text{-}bnl}$ is given by

$$Wi_{app}/Wi_{ws\text{-}bnl} = (V/V_c)/(2V_s^*/V_c + 1), \tag{6.24b}$$

where

$$V_c = H\dot{\gamma}_c \sim H/\tau \tag{6.24c}$$

is the magnitude of the relative velocity of two shearing surfaces over a gap distance of H needed to produce a critical shear rate $\dot{\gamma}_c \sim 1/\tau$.

In closing of this section, we note that according to Eq. (6.13), in entangled polymer solutions or mixtures, the magnitude of wall slip increases with the viscosity ratio $\eta(\phi)/\eta_s$. At a given concentration and for a given molecular weight, η of a well-entangled "solution" is usually not linearly proportional to η_s. Therefore, we can suppress b_{max} in Eq. (6.13) by elevating the value of η_s, for example, using polymeric solvents.

6.4.2 Experimental Evidence of Stress Plateau Associated with Wall Slip

In a simple shear experiment, the structural continuity of the entanglement network, and the translational symmetry, is broken by the two shearing surfaces. The first entanglement layers adjacent to the two surfaces are rather different from the layers in the bulk. As a consequence, the loss of entanglement preferentially takes place at the polymer/wall interfaces. Wall slip due to interfacial chain disentanglement can persist in the steady state, even if the steady-state shear stress is lower than the interfacial yield stress. Such behavior is consistent with the fact that wall slip is hysteretic. Deferring a detailed discussion of the molecular mechanism for interfacial chain disentanglement to Chapter 16, we observe the equilibrium interface shown in Fig. 6.1(a). At the beginning of startup simple shear with $Wi_{app} > 1$ or $\gg 1$, the entanglement between the adsorbed chains and the bulk free chains is effective, so that the boundary displacement produces a simple shear strain field where the bulk (entanglement network) undergoes dominantly elastic deformation. Chain deformation ceases in the first entanglement layers when the entanglements can no longer sustain the growing entropic retraction force associated with the chain deformation. The adsorbed chains presumably pull out from the sea of the bulk entangling chains, giving rise to a slip layer that is less entangled with the bulk chains and has a reduced viscosity. The limiting case of full disentanglement is depicted in Fig. 6.1(b).

6.4.2.1 A Case Based on Entangled DNA Solutions

Discussion in 6.4.1 asserts that for $Wi_{app} < Wi_{ws\text{-}bnl}$ the shear stress locks in at a level corresponding to $Wi \sim 1$, as illustrated in Fig. 6.17(b,c). *In situ* rheometric and particle-tracking velocimetric experiments[55] based on well-entangled DNA-water solutions have confirmed this expectation. Specifically, the rheometric measurements in Fig. 6.18 confirm that after the overshoot there is convergence to steady stress σ_{ss} that is indeed insensitive to V/H. As shown in Fig. 6.19, the peak shear stress shows a weak dependence on the applied rate V/H. A line of 1/3 is indicated as a reference. The peak strain γ_{iy} scales similarly with V/H as shown in Fig. 6.20. Here $Wi_{ws\text{-}bnl}$ is estimated to be as high as 1470, so wall slip should be the leading character for most of the applied rates in Fig. 6.18. Reference [55] estimated V_s^* to be 13 mm/s, which is greater

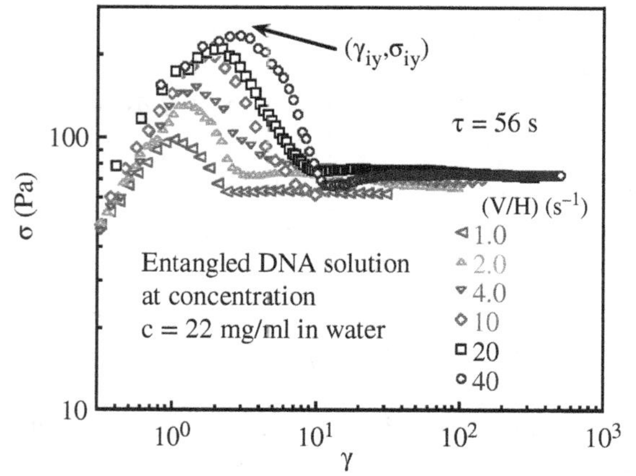

Figure 6.18 Stress plateau at a level nearly independent of the applied rate for a entangled DNA solution. *Source:* Data taken from Boukany and Wang (2009) [55].

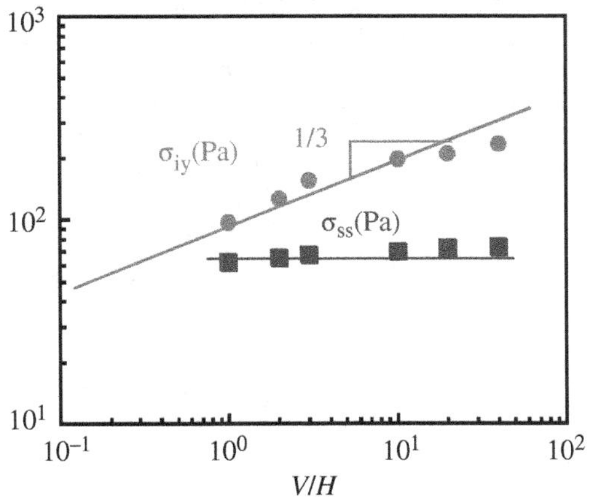

Figure 6.19 Interfacial yield stress σ_{iy} and steady-state stress σ_{ss} as a function of shear rate V/H according to Fig. 6.18.

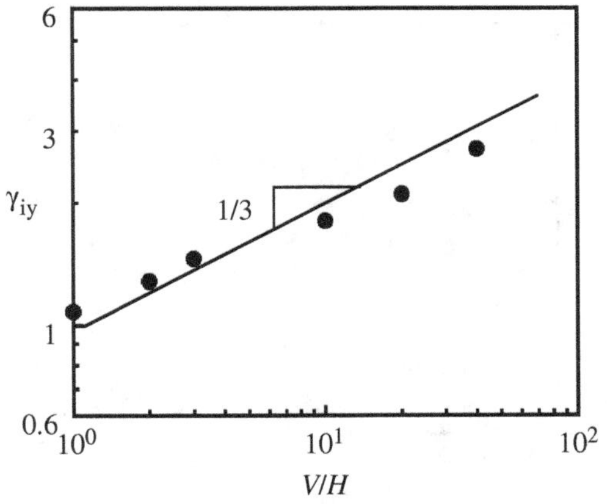

Figure 6.20 Interfacial yield strain γ_{iy} versus shear rate V/H according to Fig. 6.18.

than the limiting speed $V^* = 5$ mm/s in Eq. (5.3) for the low-speed particle-tracking velocimetric observations. Thus, for this particular solution we cannot examine the transition from wall slip to bulk nonlinear responses. It is shown by the measurements given in Fig. 6.15(b) that at $Wi_{app} = 224 < Wi_{ws\text{-}bnl} = 1470$ there was only wall slip, as expected. Figure 6.15(a,b) shows that V_s essentially grows linearly with Wi_{app}, increasing from 0.5 m/s at $Wi_{app} = 56$ to 2.4 mm/s at $Wi_{app} = 224$.

6.4.2.2 Entangled Polybutadiene Solutions in Small Gap Distance H ~ 50 μm

When the gap distance H is reduced from the conventional 1 mm, down to 50 μm, the same sample sees its b_{max}/H increased by a factor of 20. Reducing H is an effective way to amplify the correction due to wall slip because wall slip produces a greater correction in a smaller dimension apparatus. Table 6.2 gives linear viscoelastic characteristics of an entangled 1,4-polybutadiene (PB) solution from the small amplitude oscillatory shear measurements shown in Fig. 6.21. The characteristic parameters for wall slip, estimated from such linear viscoelastic properties, are also given. For example, Eq. (6.11) can be used to estimate V_s^* from b_{max} whose values are obtained using Eq. (6.13). Based on Section 6.4.1, the PB1M(13%)-1.5K solution, made of 87% 1,4-PB of M = 10 g/mol and 13% 1,4-PB of M = 1.5 kg/mol, has a significant capacity for wall slip, in the sense that for Wi_{app} in the range from 1 to 240 the response in steady shear is wall slip dominant. This blend is an ideal system to be used as a test of the analysis made in Section 6.4.1.

Table 6.2 Linear viscoelastic properties and slip characteristics.

Sample	G_{pl} (kPa)	τ (s)	η_s (Pa s)	η (kPa s)	$l_{ent}(\phi)$ (nm)	V_s^* (μm/s)	b_{max} (mm)	$2b_{max}$/H
1M(13%)-1.5K	9.7	36	0.7	280	15	167	6.0	240
1M(13%)-10K	9.4	68	14	390	15	6.2	0.42	17

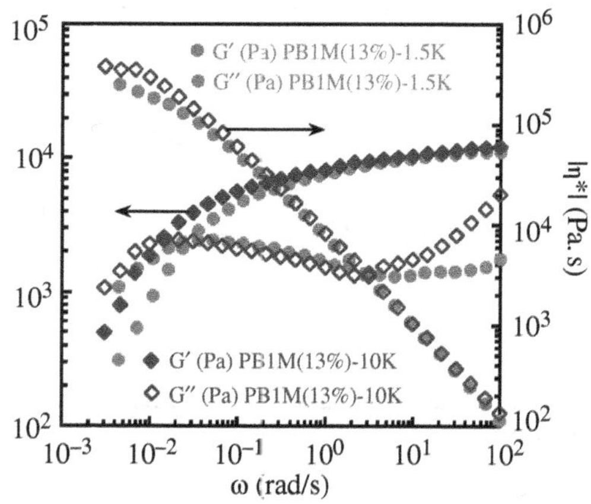

Figure 6.21 Linear viscoelastic behavior of PB1M(13%)-1.5K and PB1M(13%)-10K, as revealed by small amplitude oscillatory shear measurements. *Source:* Boukany *et al.* [58], http://pubs.rsc.org/-/content/articlehtml/2015/sm/c5sm01429h. Used under CC BY 3.0, https://creativecommons.org/licenses/by/3.0/.

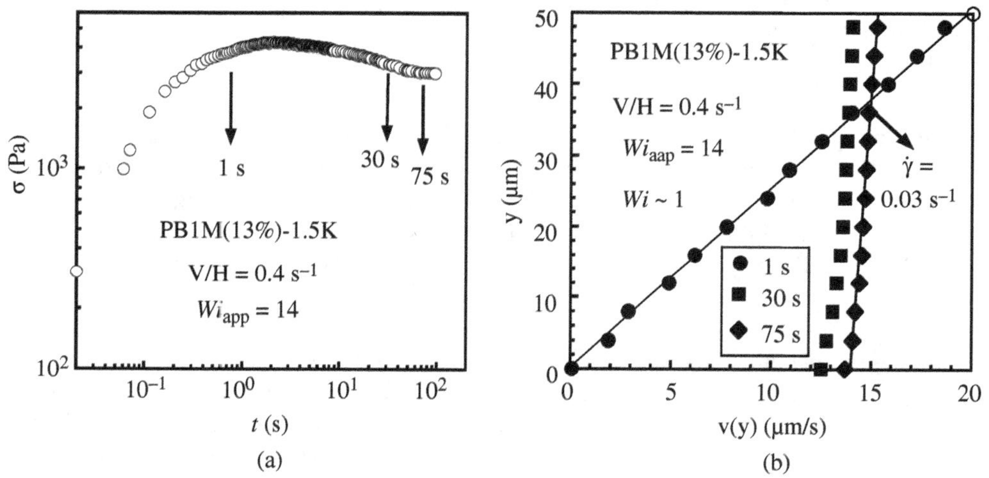

Figure 6.22 Rheometric and particle-tracking velocimetric measurements during startup shear of PB1M(13%)-1.5K, based on a 20-mm-diameter parallel-disk shear cell having a 50-μm gap. (a) Shear stress response to a startup shear of 0.4 s^{-1} ($Wi_{app} = 14$). (b) Evolution of the velocity profile at different times after applying the startup shear ($t = 1$, 30, and 75 s). Before the shear stress maximum is reached, the velocity profile is linear across the gap. At long times, massive slip emerges. *Source:* Boukany *et al.* [58], http://pubs.rsc.org/-/content/articlehtml/2015/sm/c5sm01429h. Used under CC BY 3.0, https://creativecommons.org/licenses/by/3.0/.

We begin with a moderate shear rate to impose a startup shear with $Wi_{app} = 14$. Figure 6.22(a) reveals a weak stress overshoot. Here the stress decline after the overshoot can convincingly be demonstrated to be due to interfacial yielding, that is, wall slip. Indeed, prior to the stress peak, shear homogeneity prevails, as shown in Fig. 6.22(b). The long-time behavior is simply wall slip. Since $Wi_{app} = 14$ is so far away from $Wi_{ws\text{-}bnl} \sim 240$, the magnitude of wall slip is modest: $V_s = 14$ μm/s is far less than V_s^*, which we estimate to be ca. $b_{max}/\tau = 167$ μm/s. The PTV data in Fig. 6.22(b) not only directly measures V_s but also determines the bulk shear rate to be around 0.03 s^{-1}, which is close to the reciprocal of τ. With increasing Wi_{app}, we expect V_s to grow monotonically. Figure 6.23(a) shows the stress responses for three larger values of Wi_{app}. As also seen in Fig. 6.18, the steady stress converges to about the same level at all shear rates. PTV observations reveal nothing other than massive wall slip, as shown in Fig. 6.23(b).

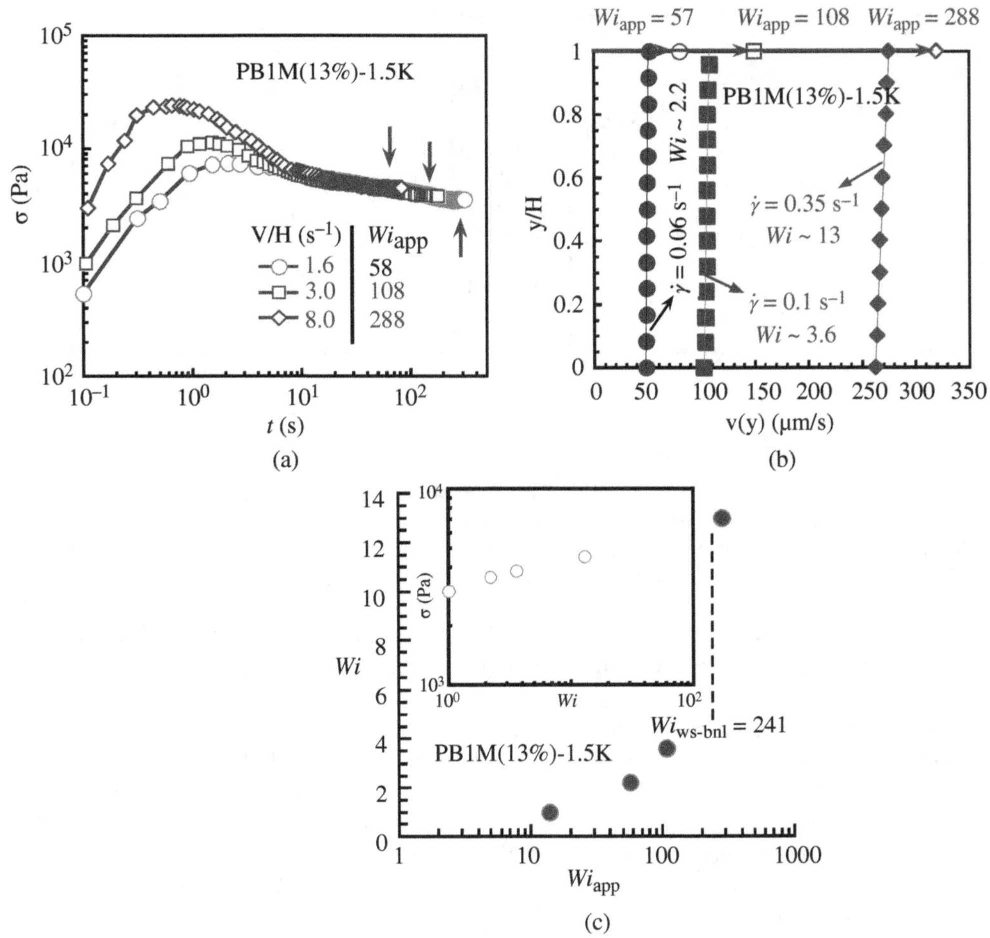

Figure 6.23 (a) Stress response upon startup shear at shear rates 1.6, 3.0, and 8.0 s^{-1} respectively, where the vertical arrows indicate the moments when the particle-tracking velocimetric observations were made. (b) Velocity profiles at the indicated times and Weissenberg numbers. As also shown in Fig. 6.22(b), the profiles are completely plug-flow like due to the massive wall slip. While the startup shear at $Wi_{app} = 57$ and 108 involved H = 50 μm, the startup test with $Wi_{app} = 288$ was made for H = 40 μm. (c) Plot of the bulk Weissenberg number *Wi* versus the apparent Weissenberg number Wi_{app}, based on the information from Figs. 6.22(b) and 6.23(b), where the vertical dashed line marks the onset of strong nonlinear responses of the bulk. The inset shows how the long-time stresses vary with *Wi*, measured at moments indicated by the vertical arrows in Figs. 6.22(a) and 6.23(a). *Source:* Boukany *et al.* [58], http://pubs.rsc.org/-/content/articlehtml/2015/sm/c5sm01429h. Used under CC BY 3.0, https://creativecommons.org/licenses/by/3.0/.

In the slip dominated regime, with $Wi_{app} < Wi_{ws\text{-}bnl} = 240$, the bulk deformation is indeed mild as shown in Fig. 6.23(c) where *Wi* jumps up upon crossing the borderline. The weak increase of *Wi* with Wi_{app} shows that the stress plateau conjecture is only approximate. The appreciable increase of σ with *Wi* shown in the inset of Fig. 6.23(c) may be due to the fact that neither σ nor *Wi* is evaluated in steady state.

The less wall-slip capable solution of 1M(13%)-10K can still exhibit plug-like-flow because of the massive effect of wall slip at such a low gap distance. Figure 6.24(a) shows the weak stress overshoot at $Wi_{app} = 20$. The PTV measurement in Fig. 6.24(b) reveals that wall slip is complete, as expected (because Wi_{app} is larger than $Wi_{ws\text{-}bnl}$). We estimate V_s^* as $b_{max}/\tau = 6.2$ μm/s.

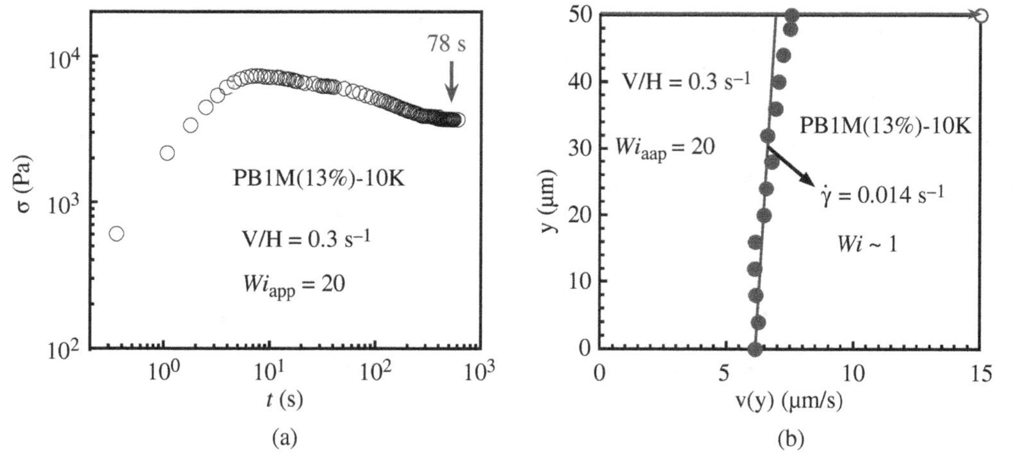

Figure 6.24 (a) Stress response to startup shear at an apparent rate 0.3 s^{-1} ($Wi_{app} = 20$). (b) The velocity profile showing wall slip for PB1M(13%)-10K where $Wi \sim Wi_c \sim 1$, obtained from the 20-mm-diameter parallel-disk shear cell with separation of 50 μm. *Source:* Boukany *et al.* [58], http://pubs.rsc.org/-/content/articlehtml/2015/sm/c5sm01429h. Used under CC BY 3.0, https://creativecommons.org/licenses/by/3.0/.

In Fig. 6.24(b) the apparent slip velocity V_s on the upper moving surface slightly exceeds 6.2 μm/s while the bulk liquid remains in the Newtonian terminal flow regime, with $Wi \sim 1$.

6.4.2.3 Verification of Theoretical Relation by Experiment

The theoretical analyses presented in Section 6.4.1 should not be regarded as a microscopic theory. They are not even rooted in a molecular model. The analyses were based on the three basic characteristics of wall slip. (a) Experiments reveal that wall slip takes place at a critical stress σ_c when examined in controlled-stress mode. (b) At σ_c, the bulk deformation is still within the terminal regime, that is, $Wi \sim Wi_c = 1$. (c) Associated with the stick–slip transition is a narrow window of hysteresis as shown in Fig. 6.4. Consequently, wall slip can be sustained at $\sigma_h < \sigma_c$. In rate-controlled mode, the same law is obeyed, that is, the lower bound of shear stress for wall slip is σ_h, as illustrated in Fig. 6.25(a). The stress plateau is not entirely flat, but instead looks like the dashed line in Fig. 6.25(a). Correspondingly, in this window between $Wi_c = 1$ and $Wi_{ws\text{-}benl}$, as the shear stress increases from σ_h to σ_c, Wi is not fixed at $Wi_c = 1$. In consequence, Eq. (6.23) is only a zeroth order account. Figure 6.23(c) already indicated that the stress plateau condition is an approximation to simplify the analysis. However, as long as $b/H \gg 1$, V_s should increase roughly linearly with V, as expected from Eq. (6.22).

The condition depicted by Eq. (6.22) can be relaxed. A more general expression than Eq. (6.22) is

$$Wi_{app} = Wi(1 + 2b/H). \tag{6.25}$$

According to Eq. (6.25), when $b_{max}/H \gg 1$, deep in the wall slip regime, it is always true that $Wi \ll Wi_{app}$. Eq. (6.25) can be rewritten as

$$V_s = (Wi_{app} - Wi)(H/2\tau). \tag{6.26}$$

Thus, as long as $Wi_{app} \gg Wi$, Eq. (6.26) shows that $2V_s/V_c \sim Wi_{app}$, where V_c is defined in Eq. (6.24c). The small anticipated rise of Wi above $Wi_c = 1$ does not affect the linear increase of V_s with Wi_{app}. Combining the PTV data in Figs. 6.22(b), 6.23(b), and 6.24(b), along with the data in Ref. [58] (not reproduced here), Fig. 6.25(b) shows that V_s indeed increases linearly with Wi_{app} over a wide range of Wi_{app} from 14 to 288 for PB1M(13%)-1.5K although Wi is shown

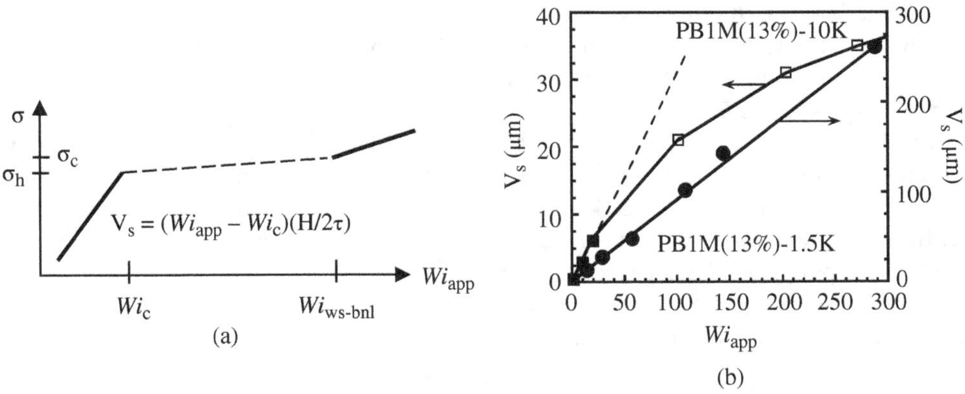

Figure 6.25 (a) Illustration of the stress plateau: over a wide range of applied shear rates, with Wi_{app} changing from $Wi_c \sim 1$ to $Wi_{ws\text{-}bnl}$, the shear stress only varies in the narrow range between σ_h and σ_c. (b) Dependence of the slip velocity on Wi_{app} for the two PB solutions. *Source:* Boukany *et al.* [58], http://pubs.rsc.org/-/content/articlehtml/2015/sm/c5sm01429h. Used under CC BY 3.0, https://creativecommons.org/licenses/by/3.0/.

in Fig. 6.23(c) to appreciably increase with Wi_{app}. Even for PB1M(13%)-10K, there is a narrow linear regime.

Let us first focus on 1M(13%)-10K. The slope of the dashed straight line in Fig. 6.25(b) is indeed comparable to (H/2τ) in Eq. (6.26). The deviation beyond $V_s^* = 6.2$ μm/s from this linear dependence, that is, the third (filled square) data point in Fig. 6.25(b), indicates entry into the regime of $Wi_{app} > Wi_{ws\text{-}bnl}$ where Eq. (6.26) no longer holds. As will be shown in Chapter 9, for the three open squares, one encounters both shear banding and apparent wall slip, from which V_s is determined. Since V_s in open squares is significantly larger than $V_s^* \sim 6$ μm/s, the interfacial slip layer is most likely of thickness multiples of l_{ent}. At $Wi_{app} > Wi_{ws\text{-}bnl}$, what is observed is apparent wall slip rather than ideal wall slip.

The case of 1M(13%)-1.5K can be depicted as ideal wall slip. The asymmetric wall slip indicates that the two shearing surfaces have different conditions. Specifically, although V_s on the bottom plate is sufficiently high and in fact larger than the estimated $V_s^* = 167$ mm, the slip velocity on the upper moving plate is far smaller. Perhaps unknown surface contamination on the bottom plate has produced excessive slip on the bottom plate to save the system from undergoing full slip on the top plate. Consistently, the long-time shear stress presented in the inset of Fig. 6.23(c) is rather low, lower than half of G_{pl}.

6.4.3 Influence of Shear Thinning on Slip

For strongly shear thinning polymers, for example, monodisperse entangled polymer solutions and melts, V_s^* hardly increases beyond the values estimated in Section 6.1.3, because slightly above σ_c there is a well-known stress plateau, as depicted in Fig. 7.2(a,b) of the following Chapter 7. As Wi_{app} increases beyond $Wi_{ws\text{-}bnl}$, Wi grows linearly with Wi_{app}. According to Eq. (6.21), V_s of an ideal slip will not rise significantly above V_s^* so long as the shear stress σ remains in its plateau region.

After complete wall slip is produced at $Wi_{ws\text{-}bnl}$ of Eq. (6.23), a further increase in Wi_{app} causes Wi to grow, as explained in Fig. 6.17(d) and Eq. (6.25). On the other hand, the extrapolation length b will start to decrease from it maximum value because of the shear thinning property of entangled polymers. By definition, we have

$$b = V_s/\dot{\gamma} = b_{max}(\sigma/\sigma_c)/Wi \sim b_{max} Wi^{-(1-n)}, \tag{6.27}$$

where use is made of Eq. (6.21) for V_s and Eq. (6.11) for b_{max} as well as $\sigma/\sigma_c \sim (Wi)^n$, which is essentially Eq. (7.1), to be presented in Chapter 7. The origin for the decrease of b with Wi or Wi_{app} is shear thinning. An alternative way to show the Wi dependence of b is to use Eq. (6.2) to show that $b \sim \eta \sim (Wi)^{-(1-n)}$. For steady shear produced in controlled-rate mode, we can crudely describe how b changes with Wi_{app} in the diagram shown in Fig. 6.26(a). The function $b(Wi_{app})$ is nonmonotonic, while the function $V_s(Wi_{app})$ is monotonic. In the controlled-stress mode, b attains its maximum at the stick–slip transition at σ_c and decreases as the flow enters the shear thinning regime, as shown in Fig. 6.26(b). Because the shear thinning exponent $n \ll 1$, for monodisperse polymers b changes much more sharply with σ than with Wi_{app}.

Substitution of Eq. (6.27) into Eq. (6.25) allows us to examine how nonlinear bulk shear takes place, as Wi changes with Wi_{app} according to

$$Wi + (2b_{max}/H)(Wi)^n = Wi_{app}. \tag{6.28}$$

In the limit $Wi_{app} \gg 2b_{max}/H$, Wi grows linearly with Wi_{app}, as shown in Fig. 6.27, which is plotted according to Eq. (6.28) by choosing $n = 0.1$, $2b_{max}/H = 10$ and 50, respectively, for example. Wall slip cannot save an entangled polymer from suffering strong shearing in the bulk, because, as seen in Fig. 6.27 at large Wi_{app}, the increase of Wi with Wi_{app} is as strong as linear.

6.4.4 Gap Dependence and Independence

Now we are ready to answer the questions raised at the end of Section 6.1.2. If the gap distance H is doubled at a fixed value of $Wi_{app} = (V/H)\tau$, does the steady shear stress vary with H? The conjecture of a stress plateau, discussed in Section 6.4.1, can apply here to predict or anticipate the consequence of gap distance change. For a fixed value of V/H in the range that satisfies

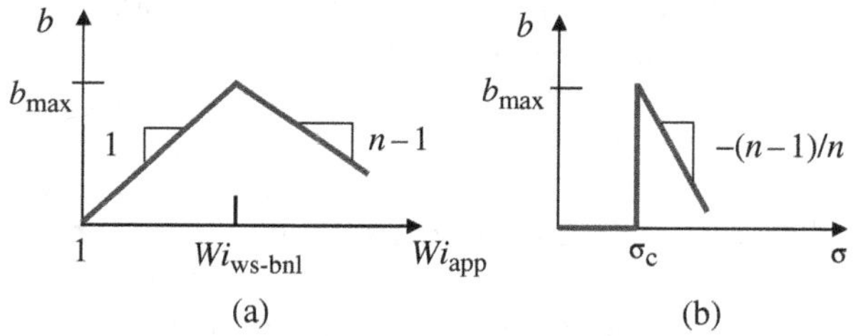

Figure 6.26 (a) Dependence of the extrapolation length b on the apparent Weissenberg number Wi_{app} during rate-controlled steady shear of entangled polymeric liquids. (b) Dependence of b on the applied shear stress σ, with a stick–slip transition at σ_c.

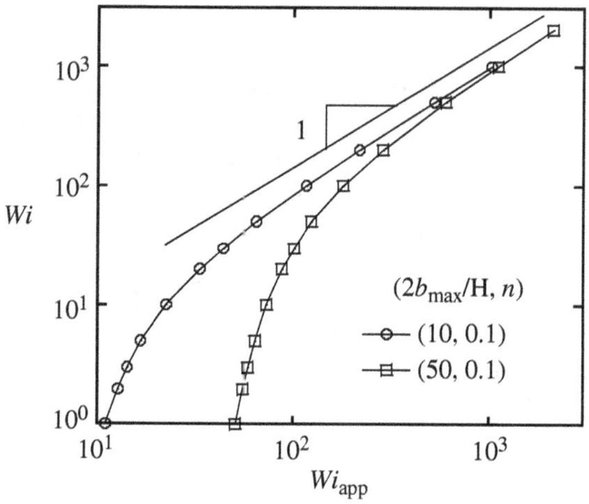

Figure 6.27 Plot of Eq. (6.28) for bulk Weissenberg number Wi versus Wi_{app}.

$1 < Wi_{app} < Wi_{ws\text{-}bnl}$ of Eq. (6.23), when H doubles, so does V by definition. Upon changing H, if the bulk shear rate $\dot{\gamma}$ is to stay constant at $\dot{\gamma}_c$, then Eq. (6.4b) dictates that b/H must remain constant, namely, b changes linearly with H. Similarly, V_s changes linearly with H. Figure 6.28(a,c) illustrates the effect of changing H: When V doubles upon doubling H to keep V/H constant, the only way to avoid larger bulk shear rate is for V_s to increase accordingly. When V and H are both doubled, V_s must also double to keep the same $\dot{\gamma}_c$ in the bulk. This trend is expected to continue to occur until V_s reaches V_s^*, and a hypothetical situation of Fig. 6.28(b) would not occur.

The prediction depicted in Fig. 6.28(c) is a consequence of the stress plateau conjecture, which has been confirmed in Figs. 6.18, 6.19, and 6.23(a) when Wi_{app} is not high enough to produce complete wall slip. Changing the gap distance is another interesting test of the conjecture. Figure 6.29 shows that indeed an increase in V does not result in any measurable change in the shear stress when H changes by a factor of three. Lack of H dependence can be used as verification that the complete slip has not yet taken place. Thus, for entangled

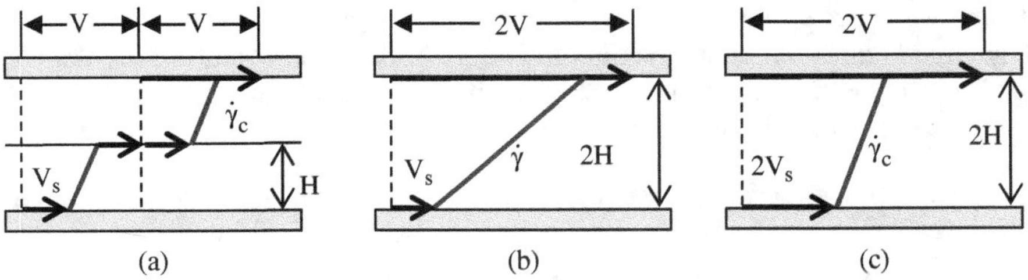

Figure 6.28 Illustration of the effect of doubling the gap distance in the marginal slip regime: Upon increasing the gap distance from H to 2H as shown in (a), if the slip velocity would stay the same, then shear rate in the bulk would double as shown in (b), resulting in a higher shear stress level. Such gap dependence of shear stress does not occur in the marginal state because the slip velocity V_s also doubles, so that the bulk shear rate and stress would remain unchanged as shown in (c). This behavior is a property of the marginal state: Multiple values for V_s, including its upper bound V_s^*, exist at a common shear stress σ_c.

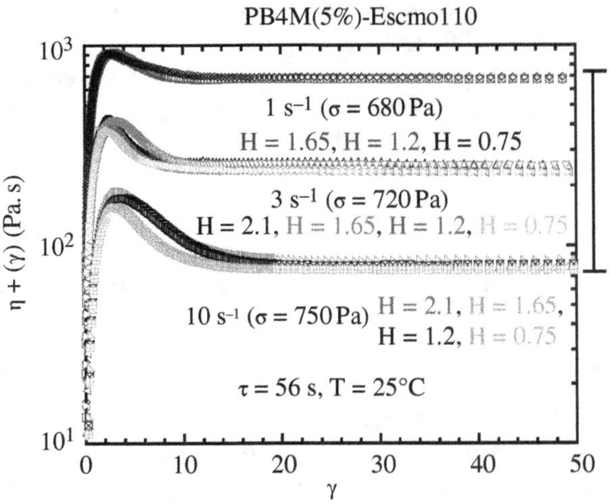

Figure 6.29 Transient shear viscosity η as a function of strain at different gap distances for three shear rates for a 5% 1,4-polybutadiene solution. The polybutadiene has an ultrahigh molecular weight 4×10^6 g/mol. The solvent is Escmo110 (from Kraton). With a decade of change in the applied shear rate, the resulting steady-shear stress is nearly the same because over the shear rate change from 1 to 10 s^{-1} the shear viscosity also drops nearly a decade as indicated by the vertical bar. This stress plateau and the extreme shear thinning are caused by the massive wall slip in steady state. It reveals the existence of a marginal slip state depicted in Fig. 6.25(a). *Source*: Unpublished data taken by Mengchen Wang at University of Akron.

polymers examined under the condition $H < H_c$, the absence of H dependence of rheometric measurements should not be taken to indicate that there is no wall slip. H dependence shows up only when $Wi_{app} > Wi_{ws\text{-}bnl}$ of Eq. (6.23).

Since V_s increases with H, increasing H at a fixed Wi_{app} is equivalent to increasing V. Thus, the change from Fig. 6.17(b) to Fig. 6.17(c) is equivalent to the change from Fig. 6.28(a) to Fig. 6.28(c). Such a peculiar conclusion occurs because the parameter $Wi_{ws\text{-}bnl}$ defined in Eq. (6.23) explicitly changes with H: When H changes, $Wi_{ws\text{-}bnl}$ is a moving target for Wi_{app} to approach. Thus, as H changes to make $Wi_{ws\text{-}bnl}$ change, the value of Wi_{app} relative to $Wi_{ws\text{-}bnl}$ changes. To reiterate, as H increases, $Wi_{ws\text{-}bnl}$ diminishes. Starting from H_i, as H increases, V_s will eventually reach V_s^* at a critical value of H, which we call H_c. In terms of $Wi_{app}(H_i)$, H_c is determined by

$$H_c = \frac{2b_{max}}{Wi_{app}(H_i) - Wi_c}, \tag{6.29}$$

which follows from setting b to b_{max} at $H = H_c$ in Eq. (6.22). If $Wi_{app}(H_i) = Wi_c$, then by definition even an infinitely large H does not allow the bulk rate to go beyond the condition of Wi_c. For $2b_{max}/H_i \gg 1$, at $Wi_{app}(H_i) = 2$, H_c is as high as $2b_{max}$, which is much higher than H_i because $2b_{max} = H_i(2b_{max}/H_i) \gg H_i$. Conversely, we have

$$H_c = H_i \frac{(2b_{max}/H_i)}{Wi_{app}}, \quad \text{for } Wi_{app} \gg Wi_c \sim 1. \tag{6.30}$$

In the example of Fig. 6.29, where the PB solution has b_{max} around 418 mm,[2] for $H_i = 0.75$ mm, we have $H_c = 15$ and 1.5 mm for the shear rates of 1 ($Wi_{app} = 56$) and 10 ($Wi_{app} = 560$) respectively. Thus, in principle, at H = 2.1 mm and 10 s^{-1} in Fig. 6.29, we have reached the condition for the bulk to undergo shear with $Wi > Wi_c \sim 1$. However, since the shear stress plateau is a characteristic of the bulk rheological response, any increase of the bulk shear rate beyond $\dot{\gamma}_c$ may not result in an appreciable increase in the steady stress.

The preceding analysis suggests that at fixed $Wi_{app} < Wi_{ws\text{-}bnl}$ the bulk shear rate does not increase with increasing H until H_c, at which, $Wi_{ws\text{-}bnl}$ has dropped to meet Wi_{app}. Now we can consider the interesting effect of changing H from a different standpoint. Let us start by apply $Wi_{app}(H_i)$ at H_i that is already as high as $Wi_{ws\text{-}bnl}(H_i)$. When H increases from H_i while keeping $Wi_{app}(H)$ constant at $Wi_{ws\text{-}bnl}(H_i)$, how does the bulk shear rate $\dot{\gamma}$ increase beyond $\dot{\gamma}_c$? The value of $\dot{\gamma}$ or $Wi = \dot{\gamma}\tau$ can be straightforwardly estimated from the condition $Wi_{app}(H) = Wi_{ws\text{-}bnl}(H_i)$, that is,

$$Wi(2b/H + 1) = Wi_c(2b_{max}/H_i + 1). \tag{6.31}$$

On inserting Eq. (6.27) for b into Eq. (6.31) and taking $Wi_c = 1$, we obtain an equation relating Wi to H/H_i as

$$Wi + (2b_{max}/H_i)(Wi)^n(H_i/H) = 1 + (2b_{max}/H_i). \tag{6.32}$$

For $b_{max}/H_i \gg 1$, Wi grows sharply with H/H_i (>1) as expected based on the physical consideration that $V_s \sim \sigma \sim (Wi)^n$ does not increase strongly with Wi for $n \ll 1$. Eq. (6.33) can be numerically solved to show how Wi varies with H/H_i. The relationship is illustrated in Fig. 6.30 for two values of $2b_{max}/H_i$ and $n = 0.1$. The rather sharp transition from terminal flow ($Wi = 1$) to high values of Wi occurs in a very narrow range of gap sizes. For example by doubling the gap distance, Wi jumps to half of its limiting value, which is close to $Wi_{ws\text{-}bnl}$. It would be interesting for future work to study this interesting transition by measuring the velocity profile as a

2 For a detailed description of the sample, see the MS thesis by Mengchen Wang, University of Akron.

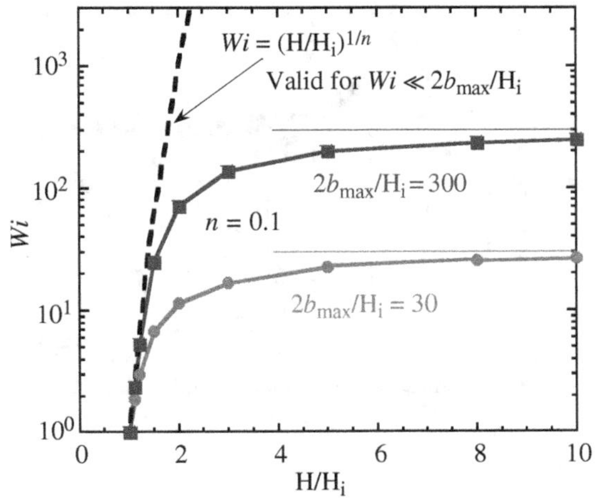

Figure 6.30 Bulk Weissenberg number *Wi* as a function of the gap distance H/H_i according to Eq. (6.32). In the limit $Wi \ll 2b_{max}/H$, Eq. (6.32) simply reduces to $(Wi)^n(H_i/H) = 1$, as shown by the dashed line which is also computed for $n = 0.1$. The two horizontal lines indicate the limiting values of 30 and 300 respectively for *Wi*.

function of the gap distance at $Wi = Wi_{ws\text{-}bnl}$. As shown in Fig. 6.30, the transition is larger for a higher value of $2b_{max}/H_m$. It is interesting to note that the steep transition-like dependence of *Wi* on H/H_i in Fig. 6.30 stems from the strong shear thinning characteristics, that is, $n \ll 1$. If $n = 1$, in the limit $2b_{max}/H \gg 1$, *Wi* can only grow as strongly as linearly with H/H_i.

6.5 Molecular Evidence of Disentanglement during Wall Slip

Up to now, all the discussions of the wall slip phenomena are based on macroscopic rheometric and particle-tracking velocimetric measurements. The theoretical analysis, pioneered by de Gennes,[9] actually convinces us of the possibility of observing massive corrections to the shear field on macroscopic scales due to ideal wall slip. The observed rheological features reveal presence of wall slip. The ideal slip supposedly involves disentanglement at the entangled-polymer/wall interface, which is truly a molecular event taking place within a monolayer. It is desirable for some experiment to provide relevant or explicit evidence to support the disentanglement picture suggested by Yang *et al.*[25] Pioneering studies by Shaqfeh and coworkers[59,60] have attempted to bridge between bulk rheology and molecular behavior in both dilute solutions and entangled solutions. One study[59] made direct visualization of individual DNA molecules in a weakly entangled solution during startup shear and during stepwise shear. Inspired by such earlier efforts, Boukany and coworkers[61] carried out *in situ* single-molecule imaging velocimetry (SMIV) to search for evidence of interfacial chain disentanglement leading to wall slip.

The investigation of Boukany *et al.* retained the advantage of the particle-tracking velocimetric observations by placing a commercial rotational rheometer over the stage of a confocal microscope, so that rheometric measurements were made along with the *in situ* SMIV observations. Because the required spatial resolution was high, SMIV employs a different PTV configuration as shown in Fig. 4.3. Figure 6.31(a) shows the stress overshoot and decline due to wall slip upon startup shear. During shear, the confocal microscope focuses onto the bottom stationary plate to track the fluorescence-stained DNA molecules. Before the overshoot, the DNA molecules are bright spots in their coiled conformations. The adsorbed DNA became disentangled from the bulk DNA chains beyond the stress overshoot. At $t \sim 5$ s, the absorbed chains began to become elongated and stretched in the shearing direction. Correspondingly, the SMIV detected significant wall slip as shown in the inset of Fig. 6.31(a). Here the stained

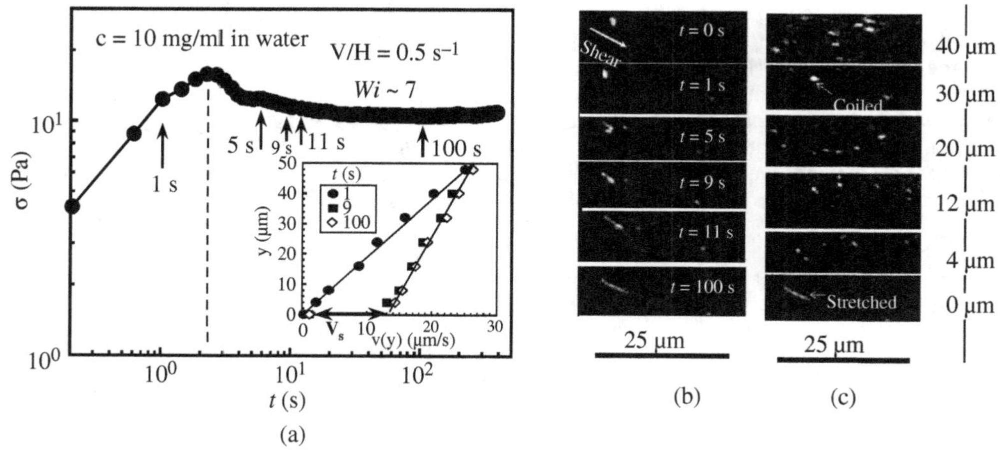

Figure 6.31 (a) Stress versus time of an aqueous entangled DNA solution upon startup shear at an apparent rate 0.5 s^{-1} corresponding to $Wi \sim 7$, where the inset indicates no slip prior to the stress overshoot and slip afterward according to the method depicted in Fig. 4.3 in Chapter 4. (b) Conformations of individual DNA molecules in the slip layer at different times during the startup shear depicted in (a), where the shearing direction is indicated with the arrow. (c) Conformations of DNA molecules across the gap in steady state that show coiled state except for the one at 0 μm from the shearing wall. *Source:* Data taken from Boukany *et al.* (2010) [61].

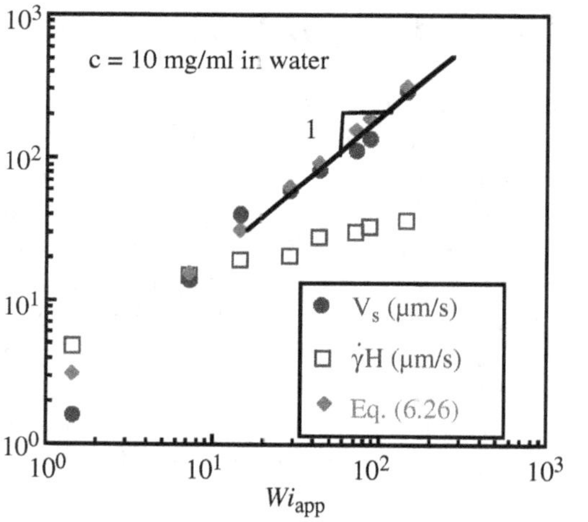

Figure 6.32 Slip velocity V_s as well as $\dot{\gamma}H$, which is the velocity increase from the stationary plate to the moving place due to the bulk shear, along with the depiction by Eq. (6.26), showing linear variation of V_s with Wi_{app}. *Source:* Data taken from Boukany *et al.* (2010) [61].

DNA, at the level of 0.01% of all DNA chains, acts as the tracking particles. SMIV also records change in the chain conformations. Figure 6.31(b) shows the progressive coil-stretch transition of the adsorbed DNA during startup shear. In steady state ($t = 100$ s and beyond), the magnitude of the wall slip reaches a steady value. Figure 6.31(c) shows the DNA conformations at the different positions along the sample thickness. The DNA molecules are only disentangled in the first monolayer; they remain coiled everywhere else.

Single-molecule imaging velocimetry (SMIV) allows the degree of wall slip to be measured as a function of the applied shear rate. Figure 6.32 shows that V_s increases linearly with Wi_{app}, consistent with Eq. (6.26). Also plotted in Fig. 6.32 is the maximum velocity $\dot{\gamma}H$ associated with the bulk shear. The bulk shear rate $\dot{\gamma}$ is rather insensitive to Wi_{app} and considerably smaller than V_s/H. In Fig. 6.32, when the circles stay well above the squares, $Wi_{app} \gg Wi$ so that according to Eq. (6.26) V_s should scale linearly with Wi_{app}. For $H = 60$ μm and $\tau = 14$ s, the prediction of

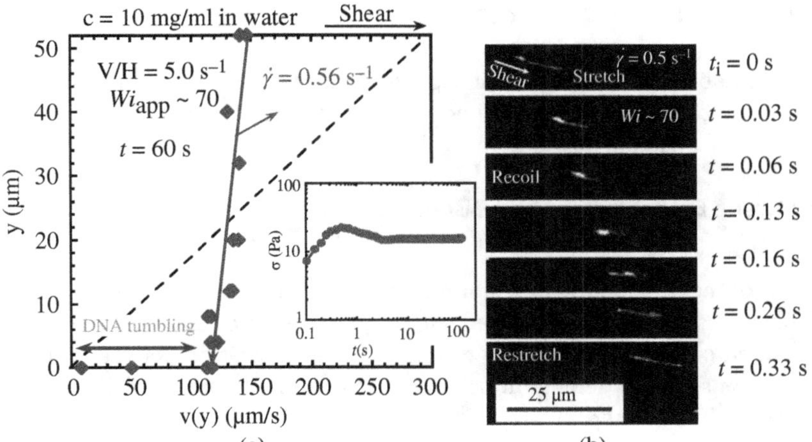

Figure 6.33 (a) Steady state velocity field at V/H = 5 s^{-1} with the inset showing the stress versus time curve. Here the bulk shear rate 0.56 s^{-1} is only one tenth of the nominal rate 5 s^{-1}. (b) Tumbling of DNA molecule at the wall during steady shear with considerable slip at the same time as the moment examined by SMIV (cf. Fig. 4.3). *Source:* Data taken from Boukany *et al.* (2010) [61].

Eq. (6.26) is given by the diamonds in Fig. 6.32. For $Wi_{app} \gg 1$, the experimental data in circles are in good agreement with the diamonds.

SMIV also reveals some evidence for chain disentanglement. For example, apart from the considerable slip indicated by the high speed of the DNA molecules just next to the stationary solid surface, as shown in Fig. 6.33(a), there is oscillatory conformational change in these molecules. As shown in Fig. 6.33(b), DNA molecules in the slip layer stayed stretched most of the time and spent a small fraction of the time tumbling, at a relatively high $Wi_{app} = 70$. This interesting tumbling behavior can take place only if the DNA molecule in the slip layer is free of entanglement. At V/H = 5 s^{-1}, the DNA molecule takes less than 0.05 s to tumble but spends a long time (ca. 1 s) remaining stretched and nontumbling. DNA tumbling has previously been observed only in dilute nonentangled solutions.[60] The observed tumbling in entangled solutions confirms that the disentanglement state is not so different from that of free chains in dilute solution.

Both experiment[60] and simulation[62] indicate that the tumbling time scales with the underlying shear rate with a negative exponent are close to −3/4. Here let us simply take the reciprocal of $t_{tumble} \sim 0.05$ s as the local shear rate, that is, $\dot{\gamma}_i \sim 20\,s^{-1}$, in the interfacial layer. On the other hand, SMIV data in Fig. 6.33(a) reveal a slip velocity $V_s = 115$ μm/s. If we use Eq. (6.1b) and Fig. 6.2(b) to infer that the thickness of the slip layer is $a_i = V_s/\dot{\gamma}_i = 115/20 = 5.75$ μm, we would find that the thickness of the slip layer is nearly ten times the coil size R_g of the DNA molecules. Two comments are therefore worth making. First, ideal wall slip should involve an a_i that is the size of the entanglement spacing l_{ent}, which is around 0.2 μm as listed in Table 9.2. $a_i \sim 6$ μm is significantly larger than $R_g = 0.65$ μm, so the phenomena reported in Ref. [61] plausibly involved shear banding rather than ideal wall slip. Second, since the shear stress level is $\sigma \sim 16$ Pa according to the inset of Fig. 6.33(a), we can estimate the interfacial viscosity to be $\eta_i = \sigma/\dot{\gamma}_i \sim 0.8$ Pa s, which is larger than η_s of water by a factor of 10^3. This extremely large value for η_i suggests that the interfacial layer has only partially disentangled, so that η_i is about two orders of magnitude smaller than the bulk shear viscosity $\eta \sim 100$ Pa s and three order of magnitude larger than $\eta_s = 0.9$ cp of water. These conclusions are based on the estimated $\dot{\gamma}_i$ from the tumbling characteristics.

6.6 Uncertainties in Boundary Condition

Because of their significant ability to undergo wall slip, when the applied rate is in a window, not high enough to ensure stable wall slip but high enough to produce some wall slip, entangled polymers typically experience unstable flow due to oscillatory boundary conditions.

6.6.1 Oscillations between Entanglement and Disentanglement Under Constant Speed

The well-known example of the oscillation of the boundary condition is related to the so-called pressure oscillation during piston-speed-controlled capillary extrusion of entangled polymer melts.[45] Here the finite melt compressibility helps to regulate the alternation between no-slip and slip boundary conditions. Some details on such an oscillation are provided in Section 10.1.2.1 as well as Sections 19.3.1.1 and 19.3.1.2. In a certain range of Wi_{app} there can be a quasi-periodic stick and slip oscillation, even under simple shear, as shown in a movie of startup shear of a polybutadiene mixture, which is based on particle-tracking velocimetric observations.[3] Systematic studies of boundary condition oscillation do not appear to exist in the current literature.

6.6.2 Oscillations between Stick and Slip under Constant Pressure

When the magnitude of the stick–slip transition is extremely high due to an extraordinary level of entanglement, an oscillation between stick and slip boundary condition can occur even in pressure-controlled capillary extrusion.[34] For a cylindrical barrel that has a relatively small diameter D_b and is filled to a considerable height, upon the stick–slip transition at P, the jump in throughput Q_s amounts to strong capillary flow in the barrel that amounts to a sizable level of pressure loss P′. As a consequence, the pressure available to drive the capillary flow in the die, given by $[P - P'(Q_s)]$, can drop below P_h that defines the size of the hysteresis loop, of which Fig. 6.4 shows an example. When the capillary flow in the die returns to the no-slip branch at a sharply reduced throughput Q_c, the diminished flow in the barrel no longer consumes much pressure, freeing the total pressure P to drive the capillary flow in the die to undergo stick–slip transition again. The cycle perpetuates at frequencies that keep changing as the melt in the barrel empties over time. Although the physics controlling this oscillation is different from that observed in the constant-piston speed mode (as shown in Section 10.1.2.1), the oscillation in both cases is due to the alternation between stick and slip boundary conditions for the capillary wall.

6.7 Conclusion

We have here made a comprehensive treatment of the subject of wall slip, not only because little discussion can be found in existing books on polymer rheology, but because wall slip is foundational for treating the central theme of this book, that is, nonlinear bulk rheology of entangled polymers. Due to the limited space here, we did not review other theoretical and experimental investigations. The reader can go to review papers of Leger[63] and Denn[64] to read how they and others[16,17,52] have since the 1990s approached the topic of wall slip. Finally, we

3 Video 6.2: Stick–slip oscillation of a linear polybutadiene mixture (70% PB600K and 30% PB44K) at V/H = 0.3 ($1\ s^{-1}$) at room temperature, made by Yangyang Wang at University of Akron.

chose not to include any account of the theoretical works by de Gennes, Brochard, and other workers. The reader should go directly to the original publications.

We have treated the subject of wall slip from our perspectives in an extremely thorough manner so that this Chapter can provide a sound basic starting point for our subsequent discussion of the main theme of the book. A comprehensive understanding of wall slip is a prerequisite for a reliable investigation of the nonlinear rheological behavior of entangled polymers. Advances in research on nonlinear polymer rheology depend on the in-depth knowledge of wall slip. Specifically, after introduction of de Gennes' extrapolation length *b* in Section 6.1, interfacial stick–slip transition was examined in Section 6.2 for both pressure-driven capillary extrusion and force-driven non-pressure driven simple shear. The concept of complete wall slip at the stick–slip transition was described in terms of the limiting slip velocity V_s^* at $\sigma_c \sim G_{pl}$ (melt plateau modulus). It is shown that the extrapolation length *b* attains its maximum value upon the transition. Particle-tracking velocimetric measurements reveal how wall slip arises after sufficient straining as an interfacial yielding phenomenon in Section 6.3. Based on both theoretical analysis and experimental evidence, Section 6.4 was devoted to a crucial discussion of how wall slip prevents the bulk from undergoing strong shear until a critical value of nominal Weissenberg number Wi_{app} is reached. It was also shown in Section 6.4 how shear thinning causes the wall slip correction to diminish.

In passing, we emphasize that the concept of *b* can also be applied to describe strain localization away from the wall in the bulk. In fact, as will be shown in Sections 9.1.2.4 and 12.3.2, the magnitude of *b* relative to the characteristic dimension H determines whether shear banding and nonquiescent stress relaxation can take place.

References

1 Tordella J. Fracture in the extrusion of amorphous polymers through capillaries. J Appl Phys. 1956;27(5):454–8.

2 Tordella JP. Unstable flow of molten polymers: a second site of melt fracture. J Appl Polym Sci. 1963;7(1):215–29.

3 Tordella JP. Unstable flow of molten polymers. In: Eirich FR, editor. Rheology: theory and applications. Vol. 5. New York: Academic Press; 1969. pp. 57–92.

4 Bagley EB, Cabott, IM, West, D. C. Discontinuity in the flow curve of polyethylene. J Appl Phys. 1958;29:109–10.

5 Blyler L, Hart A. Capillary flow instability of ethylene polymer melts. Polym Eng Sci. 1970;10(4):193–203.

6 Vinogradov G, Yanovskii YG, Borisenkova E, Yarlykov B, Berezhnaya G. Viscoelastic properties and flow of narrow distribution polybutadienes and polyisoprenes. J Polym Sci Part B Polym Phys. 1972;10(6):1061–84.

7 Vinogradov GV, Beknazarov A, Malkin AI. Rheology of polymers: viscoelasticity and flow of polymers. New York: Springer-Verlag; 1980.

8 Vinogradov G, Protasov V, Dreval V. The rheological behavior of flexible-chain polymers in the region of high shear rates and stresses, the critical process of spurting, and supercritical conditions of their movement at T > Tg. Rheol Acta. 1984;23(1):46–61.

9 De Gennes P. Ecoulement viscosimetriques de polymers enchevetres. CR Acad Sci 1979;288:219–20.

10 Mooney M. Explicit formulas for slip and fluidity. J Rheol. 1931;2(2):210–22.

11 Kraynik A, Schowalter W. Slip at the wall and extrudate roughness with aqueous solutions of polyvinyl alcohol and sodium borate. J Rheol. 1981;25(1):95–114.

12 Atwood B, Schowalter W. Measurements of slip at the wall during flow of high-density polyethylene through a rectangular conduit. Rheol Acta. 1989;28(2):134–46.
13 Lim FJ, Schowalter W. Wall slip of narrow molecular weight distribution polybutadienes. J Rheol. 1989;33(8):1359–82.
14 Hill DA, Hasegawa T, Denn MM. On the apparent relation between adhesive failure and melt fracture. J Rheol. 1990;34(6):891–918.
15 Piau J, El Kissi N, Tremblay B. Influence of upstream instabilities and wall slip on melt fracture and sharkskin phenomena during silicones extrusion through orifice dies. J Nonnewton Fluid Mech. 1990;34(2):145–80.
16 Hatzikiriakos SG, Dealy JM. Role of slip and fracture in the oscillating flow of HDPE in a capillary. J Rheol. 1992;36(5):845–84.
17 Hatzikiriakos SG, Dealy JM. Wall slip of molten high density polyethylenes. II. Capillary rheometer studies. J Rheol. 1992;36(4):703–41.
18 Piau J, El Kissi N. Measurement and modelling of friction in polymer melts during macroscopic slip at the wall. J Nonnewton Fluid Mech. 1994;54:121–42.
19 Migler K, Hervet H, Leger L. Slip transition of a polymer melt under shear stress. Phys Rev Lett. 1993;70(3):287.
20 Migler K, Massey G, Hervet I, Leger L. The slip transition at the polymer-solid interface. J Phys Condens Matter. 1994;6(23A):A301.
21 Navier M. Memoire Sur les lois du mouvement des fluides. In: Didot F, editor. Mémoires de l'Académie des Sciences de l'Institut Impérial de France. Vol. 6. Académie des sciences (France), Paris, France; 1823. pp. 389–440. Cited on Goldstein S. Modern developments in fluid dynamics: an account of theory and experiment relating to boundary layers, turbulent motion and wakes. London: Oxford University Press; 1938. 676–80 p.
22 Wang S-Q. Molecular transitions and dynamics at polymer/wall interfaces: Origins of flow instabilities and wall slip. In: Granick S, Binder K, de Gennes PG, Giannelis EP, Grest GS, Hervet H, et al., editors. Polymers in confined environments. Berlin, Heidelberg: Springer Berlin Heidelberg; 1999. pp. 227–75.
23 Doi M, Edwards S. Dynamics of concentrated polymer systems. Part 4.—Rheological properties. J Chem Soc Faraday Trans 2 Mol Chem Phys. 1979;75:38–54.
24 Brochard F, De Gennes PG. Shear-dependent slippage at a polymer/solid interface. Langmuir. 1992;8(12):3033–7.
25 Yang X, Wang S-Q, Halasa A, Ishida H. Fast flow behavior of highly entangled monodisperse polymers 1. Interfacial stick–slip transition of polybutadiene melts. Rheol Acta. 1998;37(5):415–23.
26 Plucktaveesak N, Wang S, Halasa A. Interfacial flow behavior of highly entangled polybutadiene solutions. Macromolecules. 1999;32(9):3045–50.
27 Piau JM, Elkissi N, Toussaint F, Mezghani A. Distortions of polymer melt extrudates and their elimination using slippery surfaces. Rheol Acta. 1995;34(1):40–57.
28 Elkissi N, Piau JM. The different capillary-flow regimes of entangled polydimethylsiloxane polymers – macroscopic slip at the wall, hysteresis and cork flow. J Non-Newtonian Fluid Mech. 1990;37(1):55–94.
29 Denn MM. Issues in viscoelastic fluid mechanics. Annu Rev Fluid Mech. 1990;22(1):13–32.
30 Vinogradov G, Isayev A, Mustafaev D, Podolsky YY. Polarization-optical investigation of polymers in fluid and high-elastic states under oscillatory deformation. J Appl Polym Sci. 1978;22(3):665–77.
31 Wang S-Q, Drda PA. Superfluid-like stick–slip transition in capillary flow of linear polyethylene melts. 1. General features. Macromolecules. 1996;29(7):2627–32.

32 Wang S-Q, Drda PA. Stick–slip transition in capillary flow of polyethylene. 2. Molecular weight dependence and low-temperature anomaly. Macromolecules. 1996;29(11):4115–9.

33 Wang S-Q, Drda PA. Stick–slip transition in capillary flow of linear polyethylene: 3. Surface conditions. Rheol Acta 1997;36(2):128–34.

34 Yang X, Wang S-Q, Halasa A, Ishida H. Fast flow behavior of highly entangled monodisperse polymers 2. Barrel correction, hysteresis and self-oscillation in capillary flow of PB. Rheol Acta. 1998;37(5):424–9.

35 Boukany PE, Tapadia P, Wang S-Q. Interfacial stick–slip transition in simple shear of entangled melts. J Rheol 2006;50:641.

36 Boukany PE, Wang S-Q. Exploring origins of interfacial yielding and wall slip in entangled linear melts during shear or after shear cessation. Macromolecules. 2009;42(6):2222–8.

37 Fetters L, Lohse D, Richter D, Witten T, Zirkel A. Connection between polymer molecular weight, density, chain dimensions, and melt viscoelastic properties. Macromolecules. 1994;27(17):4639–47.

38 Zhu X, Yang W, Wang SQ. Exploring shear yielding and strain localization at the die entry during extrusion of entangled melts. J Rheol. 2013;57(1):349–64.

39 Wang S, Wang S-Q, Halasa A, Hsu W-L. Relaxation dynamics in mixtures of long and short chains: tube dilation and impeded curvilinear diffusion. Macromolecules. 2003;36(14):5355–71.

40 Tapadia P, Wang S-Q. Nonlinear flow behavior of entangled polymer solutions: yieldlike entanglement–disentanglement transition. Macromolecules. 2004;37(24):9083–95.

41 Lupton JM, Regester HW. Melt flow of polyethylene at high rates. Polym Eng Sci. 1965;5(4):235–45.

42 Vinogradov GV, Ivanova LI. Viscous properties of polymer melts and elastomers exemplified by ethylene-propylene copolymer. Rheol Acta. 1967;6(3):209–22.

43 Ramamurthy A. Wall slip in viscous fluids and influence of materials of construction. J Rheol. 1986;30(2):337–57.

44 Kalika DS, Denn MM. Wall slip and extrudate distortion in linear low-density polyethylene. J Rheol. 1987;31(8):815–34.

45 Pearson JR. Mechanics of polymer processing. Netherlands: Springer Science & Business Media; 1985.

46 Laun H. Elastic properties of polyethylene melts at high shear rates with respect to extrusion. Rheol Acta. 1982;21(4–5):464–9.

47 Archer L, Larson R, Chen Y-L. Direct measurements of slip in sheared polymer solutions. J Fluid Mech. 1995;301:133–51.

48 Reimers MJ, Dealy JM. Sliding plate rheometer studies of concentrated polystyrene solutions: nonlinear viscoelasticity and wall slip of two high molecular weight polymers in tricresyl phosphate. J Rheol. 1998;42(3):527–48.

49 Adjari A, Brochard-Wyart F, de Gennes P-G, Leibler L, Viovy J-L, Rubinstein M. Slippage of an entangled polymer melt on a grafted surface. Phys A Stat Mech Appl. 1994;204(1–4):17–39.

50 Brochard-Wyart F, De Gennes P-G, Hervert H, Redon C. Wetting and slippage of polymer melts on semi-ideal surfaces. Langmuir. 1994;10(5):1566–72.

51 Brochard-Wyart F, Gay C, De Gennes P-G. Slippage of polymer melts on grafted surfaces. Macromolecules. 1996;29(1):377–82.

52 Mhetar V, Archer L. Slip in entangled polymer solutions. Macromolecules. 1998;31(19):6639–49.

53 Joshi YM, Lele AK, Mashelkar R. Molecular model for wall slip: role of convective constraint release. Macromolecules. 2001;34(10):3412–20.
54 Joshi YM, Lele AK. Dynamics of end-tethered chains at high surface coverage. J Rheol. 2002;46(2):427–53.
55 Boukany PE, Wang S-Q. Exploring the transition from wall slip to bulk shearing banding in well-entangled DNA solutions. Soft Matter. 2009;5(4):780–9.
56 Sabzevari SM, Cohen I, Wood-Adams PM. Wall slip of bidisperse linear polymer melts. Macromolecules. 2014;47(9):3154–3160.
57 Wang S-Q, Ravindranath S, Boukany P. Homogeneous shear, wall slip, and shear banding of entangled polymeric liquids in simple-shear rheometry: a roadmap of nonlinear rheology. Macromolecules. 2011;44(2):183–90.
58 Boukany PE, Wang S-Q, Ravindranath S, Lee LJ. Shear banding in entangled polymers in the micron scale gap: a confocal-rheoscopic study. Soft Matter. 2015;11(41):8058–68.
59 Teixeira RE, Dambal AK, Richter DH, Shaqfeh ES, Chu S. The individualistic dynamics of entangled DNA in solution. Macromolecules. 2007;40(7):2461–76.
60 Teixeira RE, Babcock HP, Shaqfeh ES, Chu S. Shear thinning and tumbling dynamics of single polymers in the flow-gradient plane. Macromolecules. 2005;38(2):581–92.
61 Boukany PE, Hemminger O, Wang S-Q, Lee L. Molecular imaging of slip in entangled DNA solution. Phys Rev Lett. 2010;105(2):027802.
62 Saha Dalal I, Albaugh A, Hoda N, Larson RG. Tumbling and deformation of isolated polymer chains in shearing flow. Macromolecules. 2012;45(23):9493–9.
63 Leger L, Raphael E, Hervet H. Surface-anchored polymer chains: their role in adhesion and friction. In: Granick S, editor. Polymers in confined environments. Advances in polymer science. Vol. 138 Berlin: Springer; 1999. pp. 185–225.
64 Denn MM. Extrusion instabilities and wall slip. Annu Rev Fluid Mech. 2001;33(1):265–87.

7

Yielding during Startup Deformation: From Elastic Deformation to Flow

Polymer rheology historically started by exploring constitutive equation descriptions[1–4] derived from the principle of material objectivity, also known as principle of material frame indifference. Much attention is given to steady flow behavior such as shear thinning, and little attention has been paid to the idea of Maxwell,[5] proposed in 1979, that stress overshoot upon startup shear should be recognized as a signature of *yielding*. In 1979, the tube model,[6–8] upon extending its initial domain of applicability to nonlinear rheology,[9] prescribed a different scenario for the shear and engineering stress overshoots observed in startup simple shear and uniaxial extension, respectively. In contrast to the constitutive equation approach, the tube model attempted to relate nonlinear rheological responses of polymeric liquids to molecular processes. However, by construction, the tube model is unable to delineate whether, how, and why entangled polymers may undergo yielding because the theoretical description is not self-consistently formulated and does not prescribe any means to characterize structural changes of the entanglement network. The tube model assumes the existence of entanglement, instead of showing how entanglement may arise when the chain length is long enough. Against this historical background, we examine in this chapter various rheological characteristics in some detail , including stress overshoot upon startup shear, steady shear behavior, and nonlinear rheological responses of entangled melts to startup extension.

Concerning the notion of yielding, we have already made some generic remarks in the Preface and Introduction. We should reiterate. (a) All viscoelastic materials undergo yielding during large external deformation and (b) the structure responsible for the elastic component of the viscoelasticity must suffer breakdown under fast continuous deformation. All viscoelastic materials initially respond to startup external deformation elastically (at least partially) before turning viscous or plastic. Therefore, it is more suitable to call this vast class of materials "elastoviscous" instead of "viscoelastic." The elasticity can emerge, characteristically at short times, because of a potential barrier. Specifically, behind any observable viscoelasticity, there must be a transient solid-like structure: There is always a dynamic "caging" constraint to localize the constituents in viscoelastic materials. Like all other "elastoviscous" materials, entangled polymeric liquids must derive their ability to exhibit time-dependent elasticity from entropic barriers that localize every chain on short time scales. A detailed discussion of this concept is deferred to Chapter 16.

In the absence of any large external deformation, the state of entanglement renews after each chain diffuses away from its initial location in a reptation time τ_d. It is hardly difficult to imagine that large deformation changes the state of chain entanglement at large values of the Weissenberg number Wi ($\gg$1). The entanglement cannot withstand *affine* deformation for strains as high as $Wi \gg 1$ without breaking apart. It seems that chain disentanglement and ensuing yielding of the network is inevitable before the deformation time reaches the dominant relaxation time (e.g., τ_d). Detailed and more rigorous theoretical discussions of concepts such

Nonlinear Polymer Rheology: Macroscopic Phenomenology and Molecular Foundation, First Edition. Shi-Qing Wang.

Companion Website: www.wiley.com/go/nonlinearpolymerrheology

as yielding and disentanglement can be found in Chapter 16. For now and in all subsequent chapters leading to Chapter 16, we will employ these phrases and expect the reader to take their literal meanings.

The adoption of the concept of yielding is beneficial. It reminds us that we must pay attention to how and why the initial elastic response is eventually replaced by irrecoverable processes and explain what structure has changed during yielding. For this reason, we persistently use this nomenclature in this book, even though yielding rarely appeared in the literature on the rheology of entangled polymeric liquids before the mid-2000s.[10,11]

We will demonstrate in this chapter that upon a startup deformation well-entangled polymer solutions and melts first undergo elastic deformation. But the initial elastic deformation cannot last as long as the terminal relaxation time τ. If the deformation rate is much larger than the reciprocal of τ, namely, Wi of Eq. (1.38a) is much greater than one and if the elastic deformation lasts as long as τ, the shear stress would grow monotonically till the elapsed strain reaches $Wi \gg 1$. We will see that elastic deformation terminates at a strain γ_y much smaller than Wi. In particular, for $Wi > 1$ or $\gg 1$, shear stress shows a peak or pronounced maximum to reveal the onset of yielding. The concept of yielding in polymer rheology has allowed us to establish a common framework to describe the nonlinear responses of entangled polymers to both shear and extensional deformation on equal footing. As shown in Chapter 8, this new worldview permits us to identify the special characteristics of entangled melts in extensional rheology.

Specifically, this chapter is organized as follows. Section 7.1 discusses why yielding also occurs at $Wi < 1$. In this section, we also briefly discuss characteristics of shear thinning in well-entangled polymers. Section 7.2 focuses on stress overshoot from startup shear in both low and high rate regimes and demonstrates the structural meaning of the overshoot through elastic recoil tests and rate-switching experiments. Section 7.3 discloses interesting characteristics of steady shear including the effective relaxation time as a function of the applied shear rate. Section 7.4 emphasizes the insightful entanglement–disentanglement transition as a demonstration of decohesion of the entanglement network by the external shear stress. Section 7.5 shows that fast uniaxial extension also drives entangled polymer melts to yield. Since the cross-sectional area of a specimen under extension keeps decreasing, the signature for yielding, namely, transition from elastic deformation dominant to irrecoverable deformation dominant, is the emergence of engineering stress maximum.

7.1 Yielding at $Wi < 1$ and Steady Shear Thinning at $Wi > 1$

7.1.1 Elastic Deformation and Yielding for $Wi < 1$

When the deformation rate is small so that the nominal Weissenberg number Wi_{app} is well below unity, an entangled polymer solution or melt should exhibit linear response, and homogenous deformation is expected to prevail. Whenever there is homogeneous deformation, namely, in the absence of wall slip and shear strain localization (cf. Chapter 9), we have $Wi_{app} = Wi$, and for notational simplicity, we will avoid the notation Wi_{app} in favor of Wi. Since the topics in this chapter are all discussed under the idealized scenario of homogeneous deformation, only Wi is utilized here. Correspondingly, the bulk shear rate $\dot{\gamma}$ is the nominal rate in this chapter. In the case of startup continuous shear, the linear response means that the steady shear stress is linearly proportional to Wi, namely, the shear viscosity is independent of Wi. This regime is readily accessible for polymers with narrow molecular weight distribution. The condition $Wi \ll 1$ assures linear response because the state of chain entanglement will be left intact. However, even in this least interesting case of the so-called terminal flow, there are important points to observe. First, even though $Wi < 1$, the initial response to startup deformation is elastic and not flow, which we already demonstrated in Chapter 1 using the Maxwell model. Second, Hookean-like monotonic stress growth ($\sigma = G_{pl} \times \dot{\gamma}t$) is expected to last for a period as long as the terminal relaxation (reptation) time τ (τ_d), after which an entangled polymeric liquid undergoes flow. In the traditional representations of transient viscosity or shear stress, namely, in either Fig. 7.1(a or b), dashed curves are characteristic for $Wi < 1$. While there is no stress overshoot, the evidence for yielding is still apparent, with the monotonic stress rise being replaced at longer times by a Newtonian-like unchanging stress. This behavior is "voluntary" or unforced yielding: The barrier that exists to allow the initial elastic deformation is removed over time by the molecular diffusion. Specifically, on the time scale of τ, the microstructure (responsible for the observed macroscopic viscoelasticity) gets reconstructed in the presence of ongoing shear, resulting in a state of entanglement that no longer constitutes a potential barrier at the underlying shear rate. It is in this sense that the state of polymer entanglement in steady shear should be different from that of the equilibrium state, even for $Wi < 1$. We expect the deviation from the equilibrium state to grow with increasing Wi.

For completeness, we have also drawn in Fig. 7.1(a,b) the characteristic nonmonotonic behavior for $Wi > 1$.[13–20] As indicated, the "strain softening," namely, weaker than linear increase of σ with time t, occurs before the stress maximum. Throughout the book, this phrase in quotation marks refers to such a weaker-than-linear stress increase. This departure can be viewed as reflecting partial yielding. Figure 7.1(c) presents an actual set of experimental data from the literature. Section 7.2 will be devoted to a detailed discussion on the significance of the overshoot.

7.1.2 Steady Shear Rheology: Shear Thinning

Regarding nonlinear rheological responses of polymeric liquids, earlier studies paid much greater attention to the steady shear relationship between the stress σ^{ss} and the shear rate $\dot{\gamma}$ and much less attention to the transient responses depicted in Fig. 7.1(a)–(b). In steady shear, the well-known behavior is shear thinning, associated with the stress plateau behavior at $Wi > 1$,

$$\sigma^{ss} \sim \dot{\gamma}^n, \tag{7.1}$$

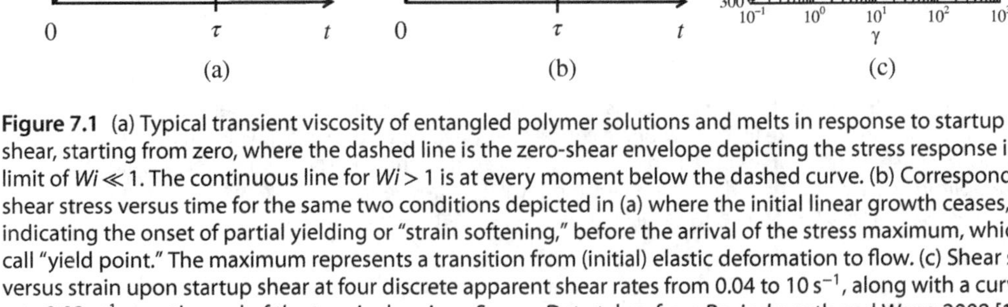

Figure 7.1 (a) Typical transient viscosity of entangled polymer solutions and melts in response to startup shear, starting from zero, where the dashed line is the zero-shear envelope depicting the stress response in the limit of $Wi \ll 1$. The continuous line for $Wi > 1$ is at every moment below the dashed curve. (b) Corresponding shear stress versus time for the same two conditions depicted in (a) where the initial linear growth ceases, indicating the onset of partial yielding or "strain softening," before the arrival of the stress maximum, which we call "yield point." The maximum represents a transition from (initial) elastic deformation to flow. (c) Shear stress versus strain upon startup shear at four discrete apparent shear rates from 0.04 to 10 s^{-1}, along with a curve at rate 0.02 s^{-1} near the end of the terminal region. *Source:* Data taken from Ravindranath and Wang 2008 [12].

with $n \ll 1$. For monodisperse polymers with sufficient chain entanglement, Eq. (7.1) holds for decades of the applied rate, as reviewed in many textbooks on polymer rheology and dynamics, for example, by Shaw[21] and Phillies.[22] Equivalently, the non-Newtonian viscosity η decreases rapidly over a very narrow range of shear stress, as sketched in Fig. 7.2(a). Specifically, the shear viscosity η drops sharply with respect to the shear stress σ as

$$\eta = \sigma^{ss}/\dot{\gamma} \sim (\sigma^{ss})^{-(1-n)/n}. \tag{7.2}$$

The sharp decrease occurs near a critical stress level σ_c, which can be determined accurately for a monodisperse polymer and which is comparable in magnitude to the elastic plateau modulus G_{pl}. Because $n \ll 1$, the exponent $(1-n)/n \gg 1$. The schematic plot in Fig. 7.2(a) also allows us to introduce a critical shear rate $\dot{\gamma}_c = \sigma_c/\eta_0 = G_{pl}/\eta_0 \sim 1/\tau$, corresponding to the onset of shear thinning at the critical Weissenberg number $Wi_c \sim 1$, a condition that was presented in Eq. (1.38b).

For a sample with a narrow molecular weight distribution, the power law exponent n can be smaller than 0.1, corresponding to a significant rise in the apparent shear rate and a sharp drop of the shear viscosity η near a threshold stress σ_c. Let us take some actual experimental data as an example. In Fig. 7.2(b), as the shear rate increases from 0.04 to 10 s^{-1}, which is an increase by a factor of 250, the shear stress only grows by a factor of 1.5. The data in Fig. 7.2(b) show that the stress plateau behavior occurs at a stress not far below G_{pl}. Thus, Fig. 7.2(b) resembles the response of a yield-stress material that undergoes yielding beyond a critical shear stress.

In the power law regime indicated by Eq. (7.1), capillary rheometry shows the pressure at the die entry P is related to the volumetric output Q_f as $P \sim Q_f^n$ so that Eq. (3.12c) can be simplified. More generally, by neglecting the entry effect (cf. Chapter 10), the formulas in Section 3.1.3 show how shear stress and shear rate at the wall can be calculated from P and Q_f in steady state. However, we should not be misled to conclude that capillary rheometry provides

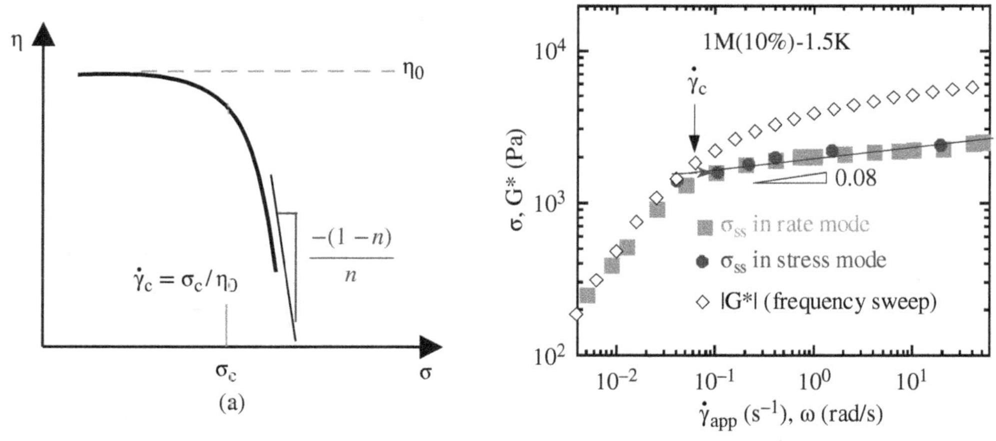

Figure 7.2 (a) Illustration of steady shear viscosity of monodisperse (entangled) polymer melts as a function of stress showing a transition-like feature around a critical stress level σ_c where η steeply declines with σ. (b) Stress versus apparent shear rate curves obtained in both controlled-rate and controlled-stress modes, along the small-amplitude oscillatory shear data $|G^*|$ versus the oscillation frequency ω, where the shear thinning exponent is as small as 0.08. *Source:* Data taken from Ravindranath and Wang 2008 [12].

adequate steady shear information on nonlinear rheological behavior of entangled polymers. More comments are provided in Appendix 11.A of Chapter 11.

7.2 Stress Overshoot in Fast Startup Shear

Entangled polymeric liquids (solutions, melts, or mixtures) are strongly viscoelastic. As such, on short time scales, they are transient solids and possess appreciable elasticity. Their transient rheological behavior is full of interesting characteristics. Startup deformation is the most commonly applied method for exploring rheological responses of polymeric liquids. It can be implemented in both simple shear and uniaxial extension. We first describe simple shear and then turn in Section 7.5 to extensional deformation. Under the condition $Wi > 1$, the elastic deformation is significant during startup shear. Unlike a Newtonian liquid, during simple shear, the shear stress σ of entangled polymers grows from zero to a high level, showing an initial solid-like response of the system. Had flow, namely, irrecoverable deformation, taken place at the beginning of startup, there would have been a finite stress as shown by the horizontal line in Fig. 1.2. The initial Hooke-like response, given by $\sigma(t) = G_{pl}\,\gamma$, can be rewritten as $\eta^+ = \sigma(t)/\dot{\gamma} = G_{pl}t$ for all shear rates, as sketched in Fig. 7.1(a). As a first guess, we could speculate that the elastic deformation would last as long as τ, after which molecular diffusion to "dissolve" the barrier associated with the elastic structure, as seen for $Wi < 1$. This speculation leads to the prediction that σ increases monotonically until a shear strain of Wi is reached, before leveling off at a level of $G_{pl}\,Wi$. In reality, the response is far more interesting and intricate. In particular, shear stress overshoot emerges for $Wi > 1$. Shear stress overshoot is a leading character of nonlinear rheological responses in polymeric materials. For entangled polymers, the overshoot occurs at a strain larger than unity. By analogy with the mechanical behavior of ductile polymeric solids, we should regard the overshoot as an indication of yielding. Bryce Maxwell was the first to advocate such a view.[5] Since his proposal and prior to the recent systematic investigations,[23–26] however, there has been only a brief mention of the idea in Matsuoka's book.[27]

Specifically, while it takes a time $t_{flow} \sim \tau$ to reach the flow state for $Wi < 1$, for $Wi \gg 1$, the transition from elastic deformation to flow occurs at times much shorter than τ. Therefore, η^+ deviates downward from the linear increase at a much earlier time, as depicted in Fig. 7.1(a), and the stress builds up much faster, as shown in Fig. 7.1(b). The actual experimental data in Fig. 7.1(c) show that the stress $\sigma(t)$ goes through a peak as a function of the elapsed strain $\gamma = \dot{\gamma}t$. The initial stress versus strain curves show a linear relationship, involving the same elastic modulus (the slope), roughly independent of shear rate. This Hookean feature informs us that entangled polymers are transient solids. For viscoelastic materials such as entangled polymer fluids, flow does not initially occur upon a startup shear, even though there is shearing motion due to the moving upper surface. The velocity field indicates that shear deformation is taking place at a finite rate and does not mean that the sample is initially undergoing flow (irrecoverable deformation).

Having discussed the significance of the initial stress growth, we turn our attention to a striking feature, the stress maximum known as overshoot[13] as observed for $Wi > 1$. Overshoot typically occurs at a strain γ_y that is either smaller or much smaller than Wi, indicating that the initial elastic shear deformation does not persist for as long as the equilibrium relaxation time τ. When does the initial elastic deformation cease and flow begin? The stress signal gives us a clue for startup shear: elastic deformation ceases when the shear stress is no longer increasing with the increasing strain. The drop in shear stress at strains and times beyond the maximum implies an end to the elastic deformation. It is thus reasonable to regard stress overshoot as signature of yielding, namely, the end of elastic deformation and beginning of nonelastic (irrecoverable) deformation. We term the shear stress maximum the *yield point*, following Maxwell.[28]

7.2.1 Scaling Characteristics of Shear Stress Overshoot

As discussed in Section 7.1, for $Wi < 1$, there is no stress overshoot. Nevertheless, the response still transforms from elastic deformation to flow. For $Wi > 1$, the characteristics of stress overshoot can be conveniently described in terms of the so-called Rouse–Weissenberg number, given by the product of the deformation rate and Rouse relaxation time τ_R from Eq. (2.17):

$$Wi_R = \dot{\gamma}\tau_R \text{ for simple shear or } \dot{\varepsilon}\tau_R \text{ for uniaxial extension.} \tag{7.3}$$

Nonlinear responses can be classified according to whether $Wi_R < 1$ or $Wi_R > 1$. In the following section, we describe features of these two regimes, which we, respectively, call the viscoelastic deformation and elastic deformation regimes. To properly describe the universal behavior during startup shear, seven different entangled polybutadiene (PB) solutions and melts are examined. Their linear viscoelastic properties at room temperature are listed in Table 7.1. All the startup experiments in Section 7.2 involve measurements at room temperature. Pioneering studies[15–17] in the past have examined polymer solutions with fewer entanglements per chain and in some cases might not have yet adequately suppressed wall slip.

To ensure that the overshoot reflects the bulk rheological response of entangled polymers, the capacity of the system to undergo measurable wall slip is suppressed in these solutions since entangled polybutadiene melts are employed as the "solvent." Thus, these solutions are in fact binary PB mixtures and have rather small values for b_{max}/H. This strategy seems essential and has been effectively employed to examine the intrinsic rheological behavior arising from bulk shear. For the styrene–butadiene rubber (SBR) listed in Table 7.1, superglue is used to adhere SBR melts onto the shearing plates and to prevent wall slip.

Table 7.1 Characterization of seven entangled PB solutions and SBR melts at room temperature.

Samples	M_w (kg/mol)	M_w/M_n	$M_e(\phi)$	G_{pl} (kPa)	$Z = M_w/M_e(\phi)$	τ (s)	τ_R (s)[a)]
0.7M(10%)-9K	700	1.02	26	5.6–6.3	27	18	0.23
1M(10%)-15K	1100	1.05	27	5.6–6.3	40	71	0.60
1M(15%)-15K	1100	1.05	17	14.6	64	83	0.42
SBR100K	94.8	1.05	4	550	24	25	1.1
SBR170K	174	1.07	3.3	670	53	120	2.3
SBR250K*	250	1.04	3.3	680	76	310	4.1
SBR500K	497	1.19	3.1	720	160	2100	13

a) We have corrected the values for Rouse time τ_R of the PB solutions, dividing the published values in Ref. [12] by a factor of 3 so that $\tau_R = \tau/3Z$. This new choice for τ_R allows the onset of the scaling behavior as shown in Fig. 7.8(b) to take place at $Wi_R = 1$. The same values for τ_R of SBR melts as in Ref. [24] given by τ/Z, are kept here because the scaling behavior starts at $Wi_R = 1$ when using this relation as shown in Fig. 7.8(b). When there is uncertainty in assigning values for τ_R, we can define τ_R according to the transitional feature in figures such as Fig. 7.8(b). Appendix 7.A discusses four different methods to evaluate the Rouse time.

7.2.1.1 Viscoelastic Regime ($Wi_R < 1$)

The stress response to startup shear produced at $Wi > 1$ is signified by the stress overshoot with a peak (yield) stress σ_y. Figure 7.3 shows a typical example in this regime based on a well-entangled polybutadiene solution.[23] As Wi changes by a factor of 10 both the yield stress σ_y and the strain γ_y change notably. We also note that the initial slope is close to G_{pl} up to a strain $\gamma_R = Wi_R < 1$, at which the elapsed time is τ_R. Beyond this characteristic time scale, the slope deviates downward from G_{pl}. We defer detailed theoretical interpretations for the various features to Chapter 16. Here, it suffices to say that the yield stress, namely, the stress maximum, is a measure of the cohesive strength of the entanglement network.

7.2.1.2 Elastic Deformation (Scaling) Regime ($Wi_R > 1$)

The characteristic of the stress overshoot can be depicted in terms of the coordinates (σ_y, γ_y) at the yield point. For $Wi_R > 1$, as the applied shear rate increases, the yield stress σ_y and yield strain γ_y develop stronger dependences on $\dot{\gamma}$. Figure 7.4 shows representative stress–strain curves in this elastic deformation regime. Several features are noteworthy. (a) Unlike the viscoelastic deformation regime, the initial slope matches the rubbery plateau modulus G_{pl} and persists to shear strains larger than unity. (b) The coordinates of the yield point have a simple property:

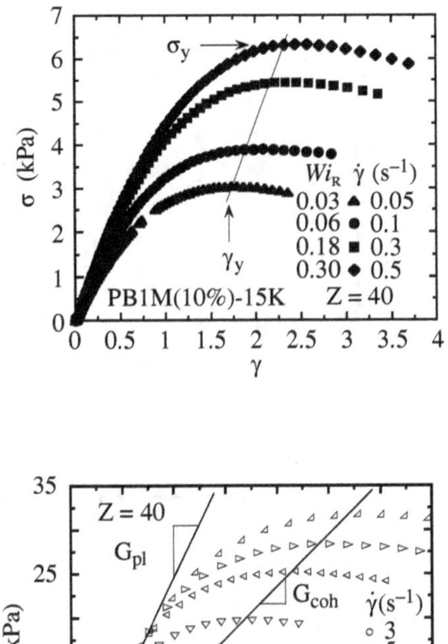

Figure 7.3 Shear stress overshoot as a function of the elapsed strain for $Wi_R < 1$, for a 10% polybutadiene solution with $Z = 40$ entanglements per chain. *Source:* Data taken from Ravindranath and Wang 2008 [23].

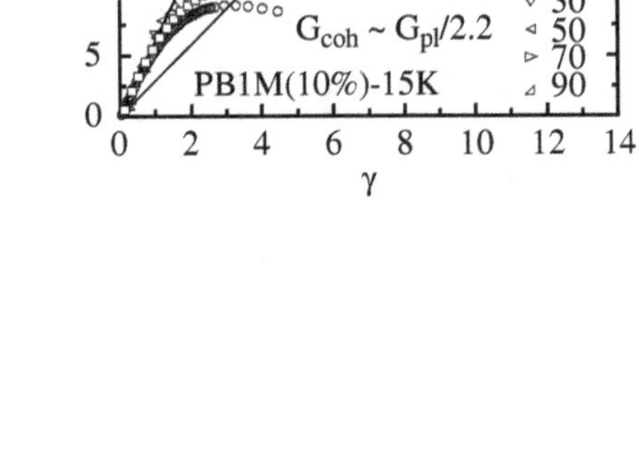

Figure 7.4 Shear stress as a function of the elapsed strain for $Wi_R > 1$, where two lines of slopes G_{pl} and G_{coh} are drawn to, respectively, indicate the initial perfect elastic response and a common modulus $G_{coh} = G_{pl}/2.2$ shared by all the curves at the stress maxima. *Source:* Data taken from Ravindranath and Wang 2008 [23].

the shear stress maximum σ_y is related to the yield strain γ_y linearly, namely,

$$\sigma_y = G_{coh}\gamma_y, \tag{7.4a}$$

where this cohesive modulus G_{coh} is comparable to the plateau modulus G_{pl},

$$G_{coh} \cong G_{pl}/2.2. \tag{7.4b}$$

Here G_{coh} characterizes the cohesion of the entanglement network at the onset of the global yielding at σ_y. (c) Stress–strain curves obtained at $Wi_R > 1$ all have the same functional form and can be made to collapse onto each other with a proper normalization in variables, as shown in Fig. 7.5. A crude way of interpreting Eq. (7.4b) is to say that global yielding takes place when the entanglement network loses more than half of its strands.

Figure 7.6 shows the near-master curves for stress against strain for two different PB solutions including both the viscoelastic deformation and elastic deformation regimes. It is noteworthy that the stress–strain curves for $Wi_R < 1$ show "strain softening" similar to that observed with $Wi_R > 1$. The appearance of strain softening at large strain suggests that the path to global yielding of the entanglement network is similar regardless of whether the yielding happens through one mechanism or another. The initial linear growth is nothing else but the rubber elasticity formula for the stress–strain curve: $\sigma = G_{pl}\gamma$, Eq. (1.50). Partial yielding begins at σ equal to about half of σ_y, as shown in Fig. 7.6.

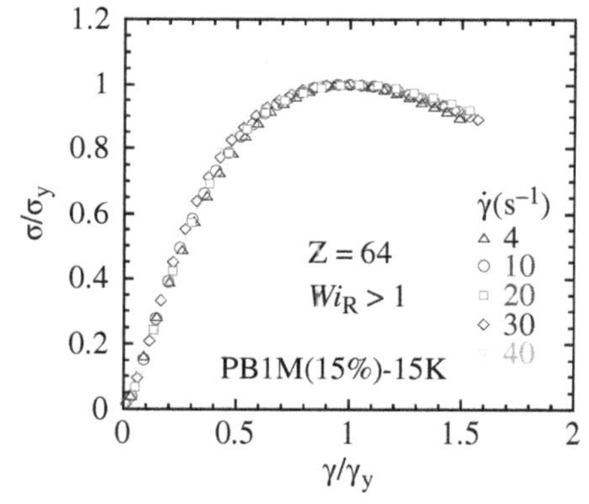

Figure 7.5 Collapse of all the curves from Fig. 7.4 upon normalizing the strain and stress with the coordinates of (γ_y, σ_y). *Source:* Data taken from Ravindranath and Wang 2008 [23].

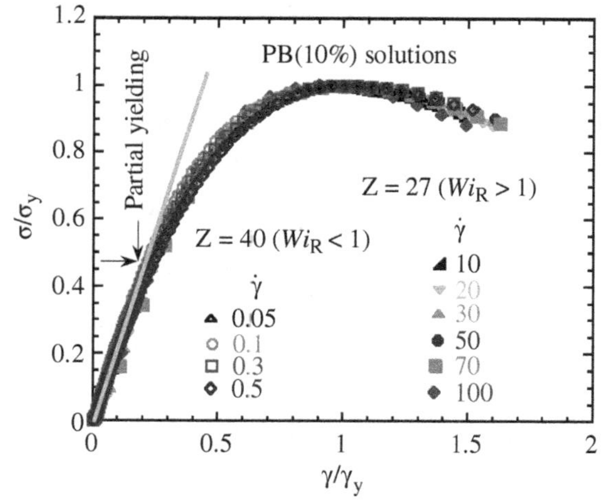

Figure 7.6 Master curves based on the normalized stress versus strain curves from two different PB solution in different regimes. *Source:* Data taken from Ravindranath and Wang 2008 [23].

According to the literature, entangled melts show the same yielding behavior as these PB solutions. Thus, the characteristics of yielding appear to be universal, independent of the concentration ϕ and molecular weight M, so long as ϕ and M are sufficiently large that the chains are highly entangled. The stress versus strain curves for a styrene–butadiene melt (Fig. 7.7) were obtained using a custom-made parallel-sliding-plate rheometer. Within the experimental uncertainty, the data on the melts display the same relationships seen in Fig. 7.4 for entangled PB solutions.

The universality of Eq. (7.4a) is further demonstrated in Fig. 7.8(a), which includes both entangled solutions and melts having different degrees of chain entanglement. Another way to illustrate the commonality between melts and solutions is to show how the yield strain γ_y and therefore σ_y via Eq. (7.4a) change with the applied shear rate, or Wi_R. Figure 7.8(b) summarizes the scaling behavior for both entangled solutions and melts. The figure confirms that Wi_R is a controlling variable dictating the nature of the nonlinear rheological response to startup shear. Specifically, for $Wi_R > 1$, the data in Fig. 7.8(b) support the scaling law

$$\gamma_y = \gamma_c (Wi_R)^{1/3}, \tag{7.5}$$

where the onset strain γ_c is between 2.0 and 3.0. A subsequent study[29] in the literature also found the same 1/3 scaling law.

Figure 7.7 Shear stress overshoot as a function of the elapsed strain for an SBR500K melt at $Wi_R > 1$, where two slopes are defined in the same way as in Fig. 7.4. *Source:* Data taken from Boukany *et al.* 2009 [24].

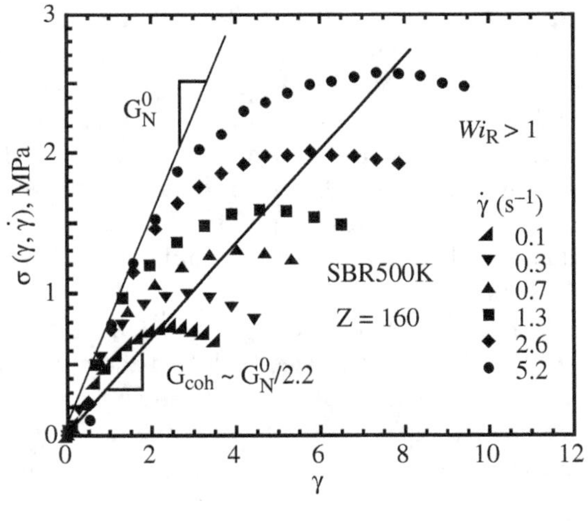

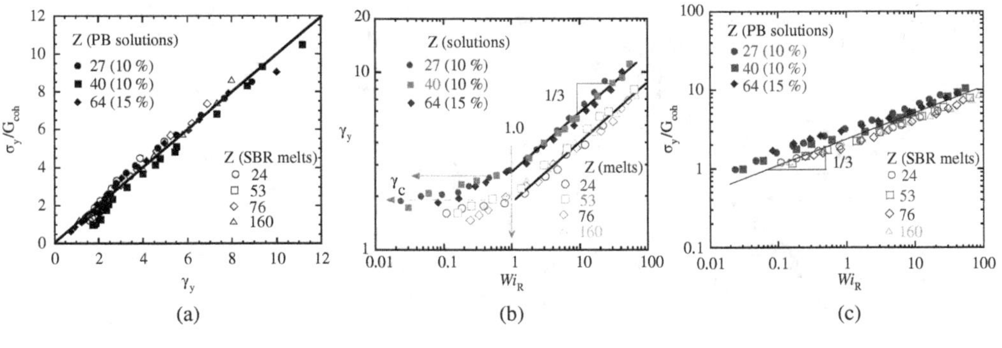

Figure 7.8 (a) Linear dependence of the normalized peak shear stress on the yield strain γ_y for the seven samples listed in Table 7.1. (b) Scaling behavior of γ_y against Wi_R for three entangled PB solutions of different Z (filled symbols) and four SBR melts of different molecular weights (open symbols). (c) Combination of (a) and (b) shows the scaling dependence of the normalized peak stress on Wi_R. *Source:* Data taken from Boukany *et al.* 2009 [24].

The combination of Eqs. (7.4a) and (7.5) shows that σ_y scales as $G_{pl}(Wi_R)^{1/3}$. Figure 7.8(c) collects data at different rates with different degree of chain entanglement from three entangled PB solutions and four entangled SBR melts. Interestingly, the 1/3 power law also seems to apply below $Wi_R = 1$. Here, the separation of the melt data from those of solution has the same origin as the split of the two sets of data in Fig. 7.8(b). Collapse can be made to happen by a horizontal shift by a factor of about 3. Thus, if we choose τ_R to be given by τ/Z in Table 7.1 for the three solutions the open and filled data in Fig. 7.8(b,c) would overlap.

According to the definition $\gamma_y = \dot{\gamma}t_y$, we have the following scaling law for the time t_y, at the yield point, based on the expression of Eq. (7.5).

$$t_y/\tau_R = \gamma_c(Wi_R)^{-2/3}. \tag{7.6}$$

Moreover, Eq. (7.4a) can be rewritten as

$$\sigma_y = G_{coh}\gamma_y = G_{coh}\gamma_c^{3/2}(t_y/\tau_R)^{-1/2}. \tag{7.4c}$$

Indeed, the data in Fig. 7.9 show the character of the overshoot to be consistent with Eq. (7.4c): σ_y shifts to shorter t_y as a power law of inverse square root as Wi_R increases in the elastic deformation regime.

7.2.1.3 Contrast between Two Different Regimes

Figure 7.8(b) reveals that for $Wi_R < 1$ the shear stress maximum is reached at a nearly constant yield strain γ_y around 2. In contrast, in the elastic deformation regime $Wi_R > 1$, there is a significant increase in γ_y with Wi_R, namely, a scaling law with 1/3 exponent seems evident. The Rouse time appears to be an important quantity. From an experimental viewpoint, it is not without ambiguity to identify or evaluate τ_R and Wi_R. Thus, it is uncertain whether or not the lack of overlap in Fig. 7.8(b) between solution and melt data arises from an inability to determine Wi_R accurately. In the original paper,[24] the data in Fig. 7.8(b,c) were presented in a single master curve, while using the same formula $\tau_R = \tau/Z$ for both solutions and melts to estimate τ_R from the terminal relaxation time τ.

The current choices for τ_R are explained in the footnote to Table 7.1. If the choices are correct, the difference between solution and melt data in Fig. 7.8(b) means that solutions are more resistant to yielding via chain disentanglement because γ_y of the solution is larger than γ_y of the melt at any given Wi_R. According to Eq. (2.37), the entanglement strands are significantly longer in a 10% solution than in a melt. Thus, at the same strain, the entropic (retraction) force

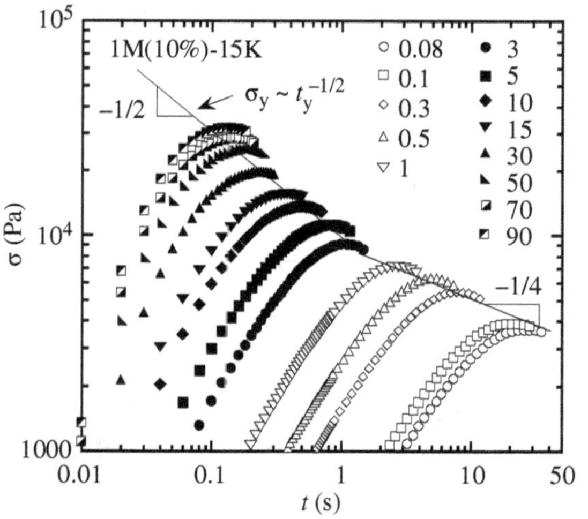

Figure 7.9 Shear stress overshoot at different applied rates, ranging from 0.08 to 90 s^{-1} in a PB1M(10%)-15K solution. *Source:* Data taken from Ravindranath and Wang 2008 [23].

(cf. Eq. (1.53)) in an entanglement strand is weaker in the solution than in the melt. Then an argument (cf. Chapter 16) can be made to suggest that the yielding occurs at smaller strains in melts. On the other hand, can it be coincidental for data on solutions and melts to merge onto a single curve? To answer this question is beyond the scope of this book. Future studies may determine whether or not γ_y at fixed Wi_R decreases as the polymer concentration increases.

7.2.2 Elastic Recoil from Startup Shear: Evidence of Yielding

With the exception of Maxwell and Nguyen's paper,[5] in the literature on nonlinear rheology of entangled polymers, the stress overshoot has never been regarded as a signature for yielding. But yielding is inevitable, for any viscoelastic liquid or solid, in response to a sudden startup deformation, for either $Wi < 1$ or $Wi > 1$. As amply demonstrated in Section 7.2.1, for $Wi > 1$, there is stress overshoot for entangled solutions and melts. The nature of the overshoot can be further examined using a combination of rheological procedures. Figure 7.10 depicts the significance of the stress maximum: After being sheared to any strain less than γ_y, upon removing the external constraints that produced the shear, the sample undergoes elastic recoil and can return to its initial position. Conversely, if the same startup shear persists longer, past the stress overshoot, the elastic recoil will be incomplete. Specific details are provided in the following.

7.2.2.1 Elastic Recoil for $Wi_R > 1$

Elastic recoil experiments can be carried out to probe the nature of the stress response to startup shear using a stress-controlled rheometer. In a cone-plate setup, the plate (or cone) is held fixed while shearing is produced by rotating the cone (or plate). At the moment, the elastic recoil is to begin, the rotating cone (plate) is set to freely rotate without exerting any torque on the sample. In the elastic deformation regime, namely, for $Wi_R > 1$, Fig. 7.11 shows that up to the yield point at $t_y = 0.27$ s, the elastic recovery is complete. A measurable level of strain becomes irrecoverable when shearing continues beyond the global yield point, as shown in diamonds, read from the right-hand side Y-axis. But this irrecoverable strain of less than 1 is quite negligible relative to the elapsed strain that is around 11.5. In other words, only 10% was irrecoverable. The data in Fig. 7.11 also reveal a sign of imperfection. There is sizable inertia that caused the cone to oscillate back and forth around its initial position so that the strain γ turned negative. The oscillation settled down fast enough relative to the reptation time $\tau = 83$ s, so that the elastic recoil is not much affected.

7.2.2.2 Irrecoverable Shear at $Wi_R < 1$

When the same solution is sheared at a low rate satisfying $Wi_R < 1$, measurable irrecoverable strain takes place even before the stress peak is reached, as shown, for example, by the

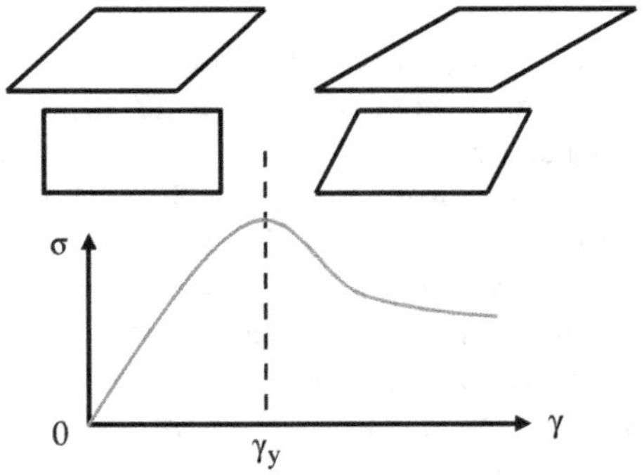

Figure 7.10 Depiction of two elastic recovery tests in which the shear stress is set to zero either before or after reaching the peak stress at γ_y during startup shear. There is then elastic recovery, the sample returning to or toward its original shape. Characteristically, the elastic recovery is nearly complete for $\gamma < \gamma_y$ and incomplete for $\gamma > \gamma_y$.

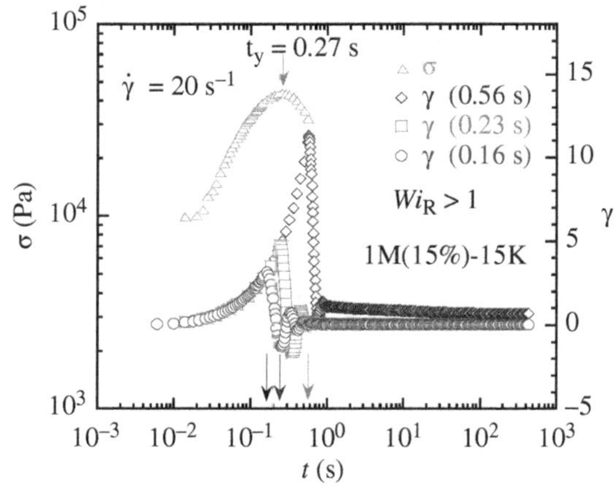

Figure 7.11 Elastic recovery experiments, after a startup shear at $\dot{\gamma} = 20\ s^{-1}$ for a duration of either 0.16 or 0.23 or 0.56 s, presented in double Y-axes. The up-pointing triangles refer to the time-dependent shear stress (left-hand-side Y-axis) as a function of time up to $t = 0.56$ s. The strain $\gamma = \dot{\gamma}t$ (right-hand Y-axis) increases linearly with time, the increase terminating at the three different times when the elastic recoil begins. If elastic recoil begins prior to t_y, there is complete elastic recoil. *Source:* Data taken from Ravindranath and Wang 2008 [23].

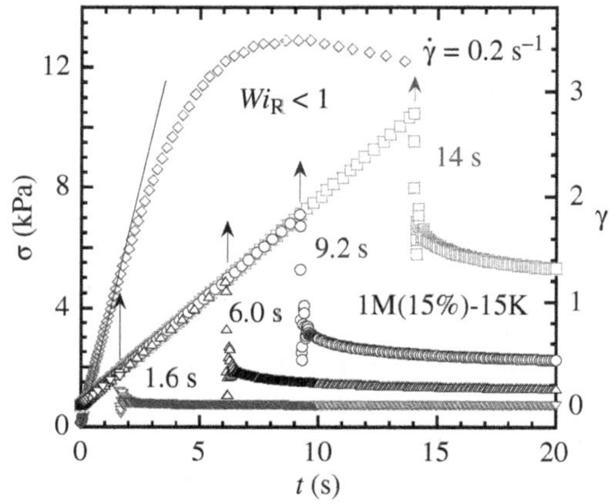

Figure 7.12 Elastic recovery experiments from startup shear, made at a shear rate $0.2\ s^{-1}$ much smaller than the rates used in Fig. 7.11, for durations of 1.6, 6, 9.2, and 14 s, respectively, on the same double Y-axes as used in Fig. 7.11. Here the inertia also takes place to show appearance of overstrain recovery. *Source:* Data taken from Ravindranath and Wang 2008 [23].

up-pointed triangles in Fig. 7.12. Such irrecoverable strain is quite small relative to γ_y, which is close to 2.0. Thus, even for $Wi_R < 1$, the rheological response up to the yield point is essentially solid like. As a result, near-complete elastic recovery can largely occur despite significant disentanglement that may have started at the point of partial yielding.

These elastic recoil experiments illustrate the nature of the overshoot. We call the condition $Wi_R > 1$ the *elastic deformation* regime because the deformation is dominantly elastic up to the overshoot. Even for $Wi_R < 1$, there are appreciable viscoelastic processes during much of the startup shear. For either $Wi_R < 1$ or $Wi_R > 1$, a central task in polymer rheology is to determine why, how, and when affine deformation of the entanglement network ceases and flow commences.

We can be more systematic in the investigation of elastic recovery characteristics that reflect the state of entanglement during startup shear. For example, the degree of elastic recovery from various stages of startup shear can be measured in different regimes ($Wi_R < 1$ or >1), following the generic protocol depicted in Fig. 3.8. Consider an entangled PB solution (1M 5%), as listed in Table 7.2. This solution has little capacity for wall slip in a conventional cone-plate setup where $H \sim 1$ mm because $b_{max}/H \ll 1$. Startup shear at a given rate $\dot{\gamma} = 1.0\ s^{-1}$ produces a familiar stress versus strain curve as shown in Fig. 7.13(a). The arrows indicate the different values of shear strain γ_1 during the startup shear where elastic recovery behavior is examined. Figure 7.13(b) shows 11 such discrete measurements. At the beginning of the elastic recoil, $\gamma(t_2 = 0) = \gamma_1$. Thus, the data in Fig. 7.13(b) should all start at zero. In reality, the rheometer does not have the time

Table 7.2 Characterization of three entangled polybutadiene solutions.

Sample	τ (s)	G_{pl} (Pa)	Z	b_{max} (mm)
1M(5%)-10K	22	1000	18	0.03
2.6M(3%)-9K	190	180	26	0.04
2.6M(3%)-1.8K	770	180	26	

resolution needed to capture initial recoil. Fortunately, the long-time behavior is the focus of interest here. In particular, we are interested in the ultimate recovery, namely, $\gamma(t_2 \to \infty) = \gamma_{2\infty}$. Defining the ultimate recoverable strain as

$$\gamma_r = (\gamma_1 - \gamma_{2\infty}), \tag{7.7a}$$

a full recovery means $\gamma_{2\infty} = 0$, namely, the system returns to the initial position, and $\gamma_r = \gamma_1$.

We can extract $\gamma_{2\infty}$ and γ_r from Fig. 7.13(b). A plot of γ_r against γ_1 appears in Fig. 7.13(c). Several features of Fig. 7.13(c) are noteworthy. First, the linear relation between γ_r and γ_1 roughly lasts until γ_{1y}. Second, there is an overshoot in this relation, mirroring the stress overshoot. The peak of recoverable strain γ_r, denoted by γ_{nwbd}, occurs well after γ_{1y}, where the subscript "nwbd" stands for "network breakdown." γ_{nwbd} should represent the moment that the entanglement network starts to lose global connectivity. Third, γ_r drops to a limiting value $\gamma_{r(ss)}$, signaling the arrival of steady state (SS). It appears that there is no systematic investigation in the literature, exploring how γ_{nwbd} and $\gamma_{r(ss)}$ depend on *Wi* or Wi_R or the degree of entanglement, Z. It is plausible that the normal stress N_1 reaches its peak at γ_{nwbd}. Preliminary yet unpublished data indeed confirm this speculation. This subject should be systematically pursued in the near future to include both $Wi_R \ll 1$, which is not the case in Fig. 7.13, and $Wi_R > 1$. Results for $Wi_R \ll 1$ are scarce in the literature. Based on one set of measurements on the same sample,[30] it seems that the behavior of Fig. 7.13(c) also occurs at smaller rates, as shown in Fig. 7.13(d).

The degree of elastic recovery is presumably related to the structural integrity of the entanglement network, which deteriorates beyond the yield point. As the network continually weakens, it also experiences more straining. Thus, expressing the recoverable strain γ_r in terms of the available driving force, namely, the shear stress $\sigma(\gamma_1)$, as

$$G_{eff}(\gamma_1)\gamma_r = \sigma(\gamma_1) \tag{7.7b}$$

we note that $G_{eff}(\gamma_1)$ is not constant and decreases strongly with the elapsed strain γ_1, as shown in Fig. 7.13(e). The effective modulus G_{eff} keeps dropping beyond γ_y and reaches a minimum at γ_{nwb}. This means that the relationship between G_{eff} and σ is also nonmonotonic as shown in the inset. It is noteworthy, as indicated in the inset, that $G_{eff}(\gamma_{1y})$ coincides with $G_{coh} = \sigma_y/\gamma_{1y} = G_{pl}/2.2$ in Eqs. (7.4a) and (7.4b). Because of the structural evolution during the startup shear, there is no simple linear relationship between γ_r and σ. Instead we have Fig. 7.13(f), showing that the monotonic variation ceases at σ_y, after which more recoverable strain occurs despite the declining driving force.

In terms of the normalized recoverable strain γ_r/γ_1, Fig. 7.14 summarizes the results from six such experiments on two solutions. The data involving three different shear rates collapse on a universal master curve and show a clear trend: Elastic recovery is nearly complete whenever the sample has not reached the yield point defined by the peak of the shear stress at γ_{1y}. Conversely, well beyond the overshoot and thus toward the steady state, much of the elapsed strain is irrecoverable, as shown in Fig. 7.13(c,d).

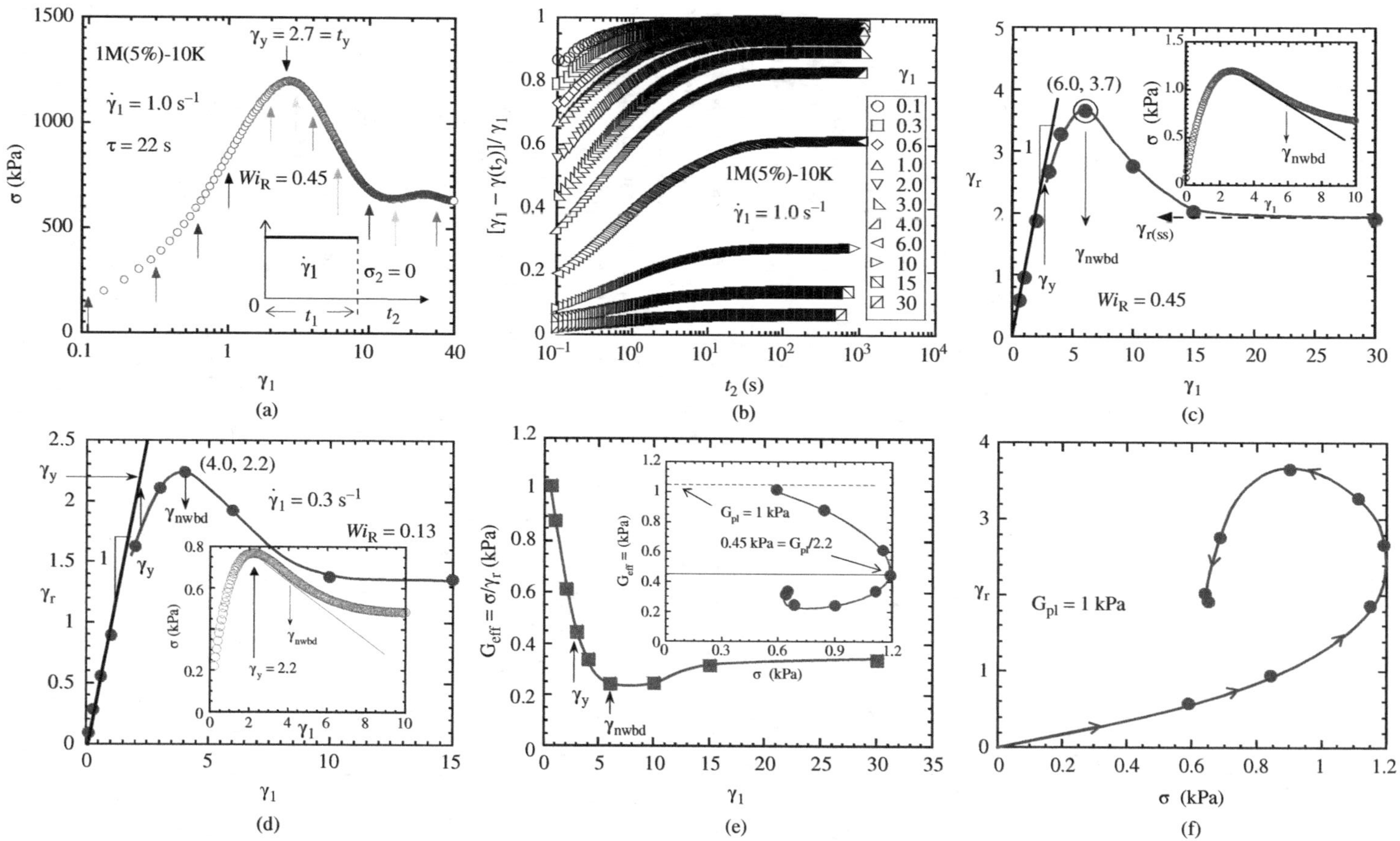

Figure 7.13 (a) Shear stress as a function of time upon startup shear at $\dot{\gamma}_1 = 1\ s^{-1}$ for a PB1M(5%) solution, where the vertical arrows indicate the different stages (i.e., different values of t_1) from which discrete elastic recovery experiments start. (b) Elastic recovery measurements after shearing for the different amounts of strain $\gamma_1 = \dot{\gamma}_1 t_1$ as indicated in (a). (c) Recoverable strain $\gamma_r = \gamma_1 - \gamma(t_2 \gg \tau)$ plotted against the imposed strain γ_1 read from the data in (b). The inset indicates the location of the peak value of γ_r denoted by γ_{nwbd} in the stress versus strain curve. (d) Similar figure to (c) involving a smaller Wi_R of 0.135, based on the data (0.3 s^{-1}) presented in Fig. 7.14. *Source:* Data taken from Wang *et al.* 2012 [30]. (e) Variation of G_{eff} in Eq. (7.7b) as a function of strain γ_1, showing its decrease until the elapsed strain of γ_{nwbd} when it grows again. The inset shows how G_{eff} makes a U-shaped turn with σ. (f) Recoverable strain γ_r versus the driving force, namely, the stress σ at the beginning of the elastic recovery test. *Source:* Data taken from Wang *et al.* 2013 [31].

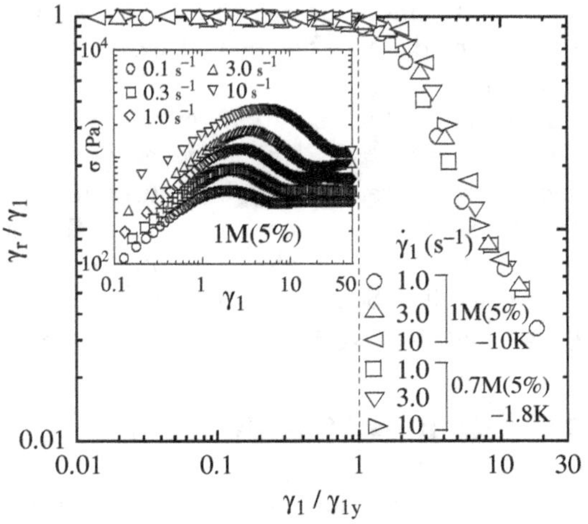

Figure 7.14 Elastic recovery data summarized in the form of the normalized recoverable strain versus the normalized elapsed strain for two PB solutions. The inset shows, in the case of the 1M(5%)-10K solution, the shear stress as a function of the elapsed strain γ_1 for the five discretely applied rates. The 1M(5%)-10K solution involves the same two components as those are listed in Table 12.1 in Chapter 12 from another study, and the 0.7M(5%)-1.8K solution is listed in Table 7.3. *Source:* Data taken from Wang *et al.* 2012 [30].

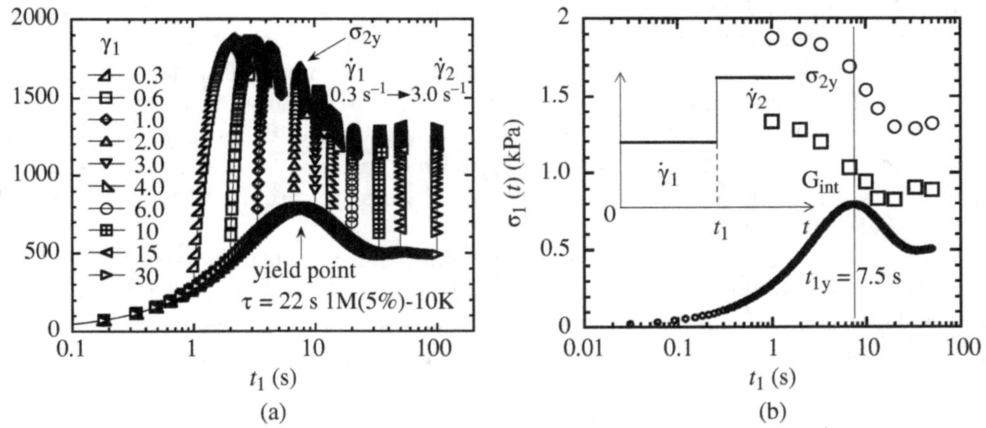

Figure 7.15 (a) Stress overshoot upon the rate switch from 0.3 to 3.0 s^{-1} at different stages during startup shear with $\dot{\gamma}_1 = 0.3$ s^{-1} for the 1M(5%)-10K solution. (b) The peak stress σ_{2y} associated with subsequent rate of 3.0 s^{-1}, along with the stress versus time curve due to the first startup at 0.3 s^{-1}. Also plotted is the initial modulus that characterizes the stress growth upon the rate switch, defined as $G_{int} = (d\sigma_2/d\gamma_2)|t_{2\to 0}$, where $\gamma_2 = \dot{\gamma}_2 t_2$. *Source:* Data taken from Wang *et al.* 2012 [30].

7.2.3 More Evidence of Yielding at Overshoot Based on Rate-Switching Tests

We can interrogate the nature of shear stress overshoot in yet another way. It has already been demonstrated in Section 7.2.1 that the stress overshoot reflects a key characteristic of the entanglement network. In a modern commercial rheometer, it is straightforward to conduct a rate-switching experiment, as discussed in Section 3.4.2, in which the shear rate is changed partway through the experiment. In one such study,[30] a moderately entangled solution was chosen to avoid any edge effect that could affect the interpretation. Figure 7.15(a) describes the typical responses to rate switching. The first startup shear at a low rate $\dot{\gamma}_1 = 0.3$ s^{-1} produces a characteristic overshoot around $t_{1y} = 7.5$ s, namely, $\gamma_{1y} = 2.3$. With rate switching, a second, faster, startup shear replaces the slow shear at various stages. Before the yield point at t_{1y}, the overshoot stress produced by the second shear rate has about the same magnitude as the overshoot stress resulting from the equilibrium sample that experiences the second rate from time zero. The peak stress drops quickly around the overshoot associated with the first startup shear, as shown in Fig. 7.15(b). The stress response to the second rate can also be characterized by the initial modulus G_{int} that measures how rapidly the shear stress grows with the imposed strain immediately after the rate switch. G_{int} is found to decline well before reaching the yield point at 2.3.

Referring to the definition given in Fig. 7.1(b) for *partial yielding*, G_{int} begins to decrease quickly after $t_{1py} = 2$ s, which is where partial yielding starts. In the literature, the decline of the second stress overshoot has been interpreted as a sign of chain disentanglement.[32] Once again, there are not many systematic studies in the literature of the kind discussed in this section. The lack of further studies along this line stems from the common challenge to shear rheometry: edge instability. Successful and reliable investigations require the samples to be essentially free of edge effects, such as those studied in Figs. 7.13–7.15.

7.3 Nature of Steady Shear

Traditionally, characteristic steady shear properties have attracted more attention in the literature. In reality, not much insight, molecular or phenomenological, can be gained without appropriate understanding of the transient response that we have reviewed in the preceding sections. Rubbery elasticity, seen at the onset of a startup shear, obviously cannot be a dominant component after the yielding that is a precursor to the establishment of steady shear. As a side note, we should remark that when the shear stress appears to be nearly constant with increasing shear strain the system may not have reached steady state. Usually, the normal force takes a significantly longer time than the shear stress to approach a constant. Since the normal stress data are often missing in rheometric measurements, we need to be careful in our characterization of steady shear behavior.

In the steady state, there is no longer ongoing elastic deformation, in the sense that the deformation is entirely dissipative. The mechanical work density per unit time $\dot{w} = \sigma^{ss}\dot{\gamma} = \eta\dot{\gamma}^2$ equals the rate of heat generation per unit volume, where η is the steady shear viscosity at the applied shear rate $\dot{\gamma}$. In such a flow state, chains no longer entangle, in the sense that they pass by one another in their deformed states without undergoing any further conformational change in a statistical sense. In other words, all chains have rearranged themselves in such a way that no potential energy barrier exists that could create any additional elastic response.

At a scaling level, we can describe the steady shear stress as follows: For $Wi < 1$, the level of conformational deformation is on the order of Wi because the transition from elastic deformation to flow occurs around τ, for which the elastic deformation lasts a strain of $\dot{\gamma}\tau = Wi$. The steady stress is given by

$$\sigma^{ss} = \eta_0\dot{\gamma} \sim G_{pl}\,Wi, \tag{7.8}$$

where $G_{pl} \sim \eta_0/\tau$. Equation (7.8) gives the Newtonian limit where σ^{ss} is linearly proportional to Wi. When $Wi > 1$, depending on the value of Wi, considerable yielding can take place before steady state is approached. The steady state can be regarded as having a reduced viscosity η and shortened effective relaxation time τ_{eff} so that $Wi_{eff} < 1$. For $Wi > 1$, we write, by analogy to Eq. (7.8), the following expression for the steady stress

$$\sigma^{ss} = \eta\dot{\gamma} = (\eta/\tau_{eff})Wi_{eff}, \tag{7.9}$$

where

$$Wi_{eff} = \dot{\gamma}\tau_{eff}.$$

Phenomenologically, we know that σ is leveling off toward a value comparable to G_{pl}, as shown in Fig. 7.2(b) for relatively monodisperse polymers. This could mean that

$$(\eta/\tau_{eff}) \sim G_{pl}, \tag{7.10a}$$

and $Wi_{eff} \sim 1$, which in turn would imply that the rate dependence of the shear thinning viscosity arises from the adjustment of the effective relaxation time τ_{eff} to $1/\dot{\gamma}$. For a given applied $\dot{\gamma}$ after yielding, the system settles down to a dynamic packing arrangement such that the relaxation time τ_{eff} of the newly formed structure evolves to become shorter than the reciprocal rate $1/\dot{\gamma}$. Therefore, σ is insensitive to $\dot{\gamma}$, as shown, for example, in Fig. 7.2(b), and $\eta(\dot{\gamma}) \sim \tau_{eff} \sim 1/\dot{\gamma}$. This behavior of η produces the limiting stress plateau behavior, with n in Eqs. (7.1) and (7.2) approaching zero. The following experiments support such a simple analysis.

7.3.1 Superposition of Small-Amplitude Oscillatory Shear onto Steady-State Shear

It is a well-known protocol to superimpose a small-amplitude-oscillatory shear (SAOS) upon a steady shear flow.[33] There is a long history of applying this method to learn about the viscoelastic characteristics during steady shear of polymer solutions,[34–41] associative polymers,[42,43] worm-like micellar solutions,[44] and polymer melts.[45–47] This method can indeed be used to gain insight into the steady shear state of an entangled polymer solution. Various entangled polymer solutions can be used for this study, including polymers that cannot undergo homogeneous shear. It appears that all entangled polymers, whenever it is possible to shear them into steady state, show the features sketched in Fig. 7.16.

On the time scale fixed by the underlying shear rate $\dot{\gamma}$, the chains are constantly being displaced past each other. When $\omega = \dot{\gamma}$ we have $G''(\dot{\gamma}) > G'(\dot{\gamma})$, as indicated in Fig. 7.16. More interestingly, if we examine the viscoelastic structure on time scales much shorter than $1/\dot{\gamma}$, namely, use SAOS with $\omega \gg \dot{\gamma}$, we see that the system in steady shear is solid like, with $G' > G''$, suggesting that on time scales much shorter than the characteristic flow time $1/\dot{\gamma}$, chains remain well entangled. Figure 7.16 summarizes what to expect from SAOS of materials undergoing steady shear. In the following paragraphs, we examine some actual examples.

To reliably and successfully implement this methodology, it is necessary for the steady state to be established without such experimental difficulties as edge instabilities. Many entangled solutions can be made to suit this purpose, including some DNA and PB solutions. The basic properties of entangled DNA and PB solutions are listed in Table 7.3. Figure 7.17(a) shows

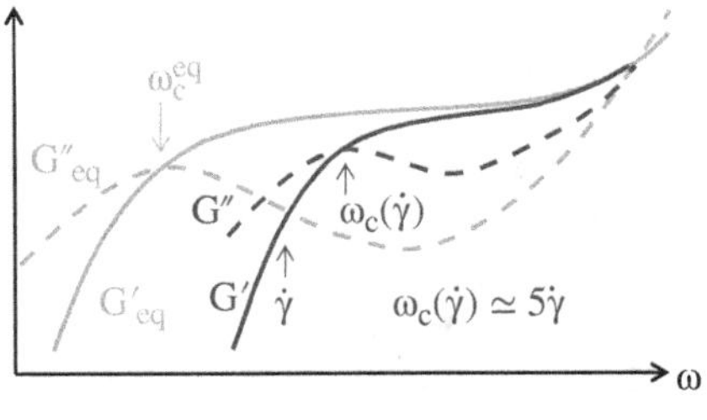

Figure 7.16 Sketches of the storage and loss moduli G'_{eq} and G''_{eq} for a quiescent entangled polymer solution (blue lines) as well as G' and G'' extracted from SAOS superimposed onto steady shear at rate $\dot{\gamma}$ (red lines). *(See color plate section for the color representation of this figure.)*

Table 7.3 Characterization of six entangled DNA and polybutadiene solutions.

Entangled solutions	G_{pl} (Pa)	G_c (Pa)	$Z(\phi)$	τ (s)
0.5 wt% DNA (50M)/glycerol	58	1.4	24	512
1.1 wt% DNA (50M)/glycerol	32	5.3	60	780
2.2 wt% DNA (50M)/glycerol	169	26.1	155	1590
0.7M(5%)PB-1.8K	1766	324	13	29.8
0.7M(10%)PB-9K	5490	1530	27	17.7
1M(10%)PB-15K	6100	1570	40	65.5

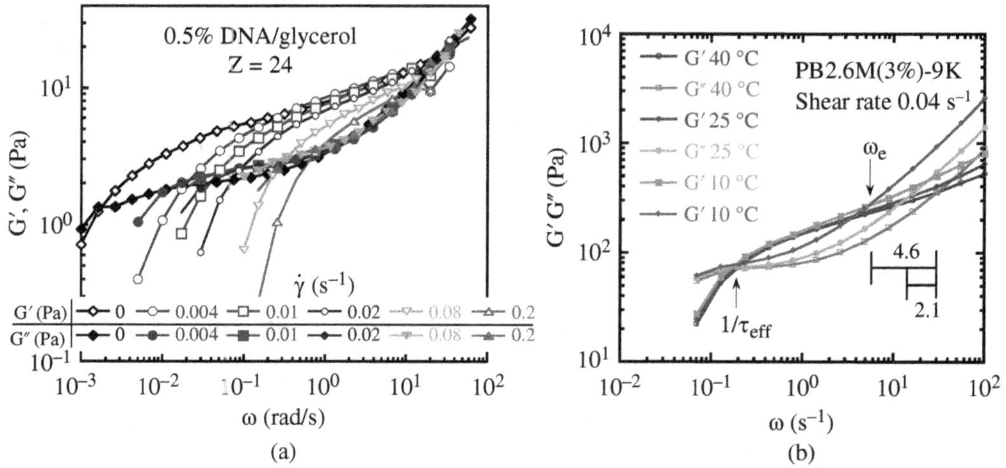

Figure 7.17 (a) Frequency dependences of G′ and G″ from superimposed SAOS of a 0.5% DNA solution in steady shear at different steady shear rates ranging from 0 to 0.2 s^{-1}. *Source:* Data taken from Boukany and Wang 2009 [48]. (b) G′ and G″ curves from imposed SAOS of a 3% entangled PB solution, showing the low-frequency crossover at a common value of $1/\tau_{eff}$ at three different temperatures of 10, 25, and 40 °C. *Source:* Data taken from Boukany and Wang 2009 [48]. As indicated in Table 7.4, the relaxation time $\tau_e \sim 1/\omega_e$ shifts with temperature as follows: from 40 to 10 °C, τ_e increases by a factor of 4.6 and from 40 to 25 °C, τ_e increases by a factor of 2.1.

Table 7.4 Temperature dependence of terminal relaxation rate ω_c of two entangled polymer solutions.

	ω_c(rad/s), PB2.6M(3%)-9K	ω_c(rad/s), PB2.6M(3%)-1.8K
10 °C	2.4×10^{-3}	2.8×10^{-4}
25 °C	5.3×10^{-3}	1.3×10^{-3}
40 °C	1.1×10^{-2}	3.6×10^{-3}

results from an entangled 0.5% DNA solution with Z = 24 entanglements per chain. More information is available in Ref. [48]. Although the sample is rather polydisperse, it still serves to show basic characteristics similar to the illustrative sketch of Fig. 7.16. A second example, based on the PB solution listed in Table 7.2, shows that the effective relaxation time τ_{eff} is regulated by the underlying shear and does not vary with temperature. According to the information in Table 7.4, the quiescent chain dynamics speed up by a factor of 4.6 when the temperature increases from 10 to 40 °C. This polymer solution demonstrates that the effective chain terminal dynamics are controlled by the underlying steady shear and thus do not slow down upon a temperature decrease from 40 to 10 °C. Specifically, Fig. 7.17(b) shows that G′ and G″ cross at the same frequency independent of the experimental temperature. On the other hand, the second (larger) cross-over frequency does shift to smaller frequencies when the temperature is increased, indicating the local segmental chain dynamics are unaffected by the underlying steady shear and still follow the temperature dependence observed with SAOS on a quiescent sample.

The dependence of $\tau_{eff} = 1/\omega_c$ on the underlying *Wi* is shown in Fig. 7.18 for several entangled DNA and PB solutions. Specifically, Fig. 7.18 reveals

$$\tau/\tau_{eff} = 5Wi \tag{7.10b}$$

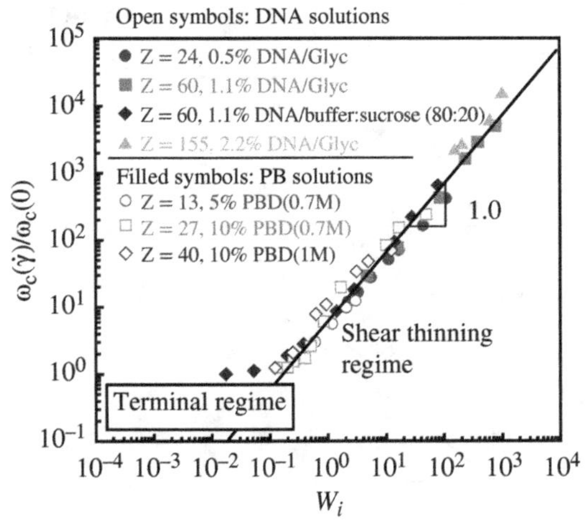

Figure 7.18 Master curves showing that under steady shear the crossover frequency ω_c of both entangled DNA and PBD solutions increases linearly with increasing *Wi*. *Source:* Data taken from Boukany and Wang 2009 [48].

or

$$\tau_{\text{eff}} = 1/\omega_c \sim 1/(5\dot{\gamma}), \tag{7.10c}$$

namely, $Wi_{\text{eff}} \sim 0.2$. Thus, a system undergoing steady shear may be viewed as undergoing terminal flow. The reciprocal relationship between τ_{eff} and $\dot{\gamma}$ as seen in Eq. (7.10c) is consistent with the analysis at the beginning of Section 7.3. Equation (7.10c) is a concrete example of stress plateau and extreme shear thinning: $\sigma^{ss} \sim \eta\dot{\gamma} \sim G_{pl}$ and $\eta = G_{pl}\tau_{\text{eff}} \sim 1/\dot{\gamma}$.

Traditionally, the shear field in a standard rheometric device is assumed to be spatially homogeneous. However, the features (e.g., Eq. (7.10c)) discussed above also show up if the steady state involves stable shear banding.[48] In the presence of inhomogeneous shear, this steady-shear-superimposed-SAOS protocol is still valuable although a more detailed analysis is required.

7.3.2 Two Other Methods to Probe Steady Shear

The nature of steady flow can be further elucidated using rate switching. There are two protocols, as shown in Fig. 7.19(a)–(b), involving either a rate switch after the steady shear has been established with the first rate or a step strain imposed on steady shear. A simple rate switching experiment, following the protocol of Fig. 7.19(a), is already rather revealing, as shown in Fig. 7.20. Even though the applied rate just increases by 11%, namely, from 0.04 to 0.045 s^{-1}, a weak stress maximum shows up. This implies that after the small increment in the applied rate, the system sees a potential barrier "erected," so that some elastic deformation must occur before the barrier is overcome to permit a new steady flow state to be established. When the system is probed on a time scale, just 11% shorter, the entanglement network is no longer fluidic but behaves weakly elastically. Note also that the strain at the stress peak is relatively constant, less than two, reminding us of the behavior observed for $Wi_R < 1$, as reviewed in Fig. 7.8(b). How has the first shear at 0.04 s^{-1} altered the chain dynamics? Can we define an effective Rouse time in the presence of steady shear? Why is there stress overshoot when the second rate is much less than five times the first rate, namely, how can the data in Fig. 7.20 be understood to be consistent with Eq. (7.10c)? What is the new time scale related to the underlying shear rate? Is there still a regime described by the 1/3 scaling law? These interesting questions warrant systematic investigations in the future.

The idea behind the protocol in Fig. 7.19(b) is to produce an excess stress by a superimposed (small) step strain and then measure the relaxation of the induced excess stress $[\sigma - \sigma^{ss}(\dot{\gamma})]$ as

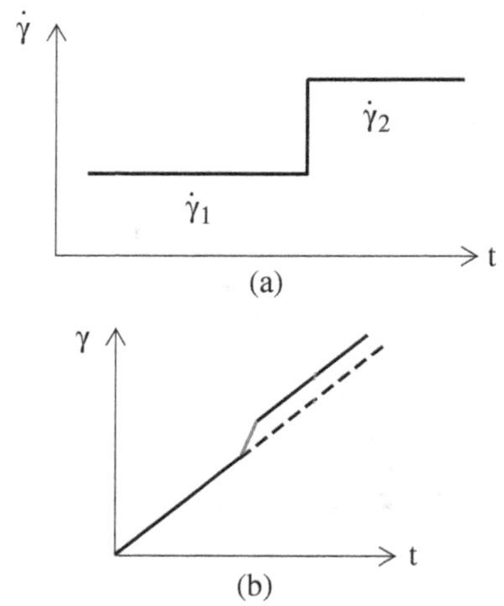

Figure 7.19 Two protocols that are used to probe the nature of steady shear: (a) rate switching, which has been applied in Fig. 7.15(a,b); (b) superimposing a small step strain on the background of steady shear.

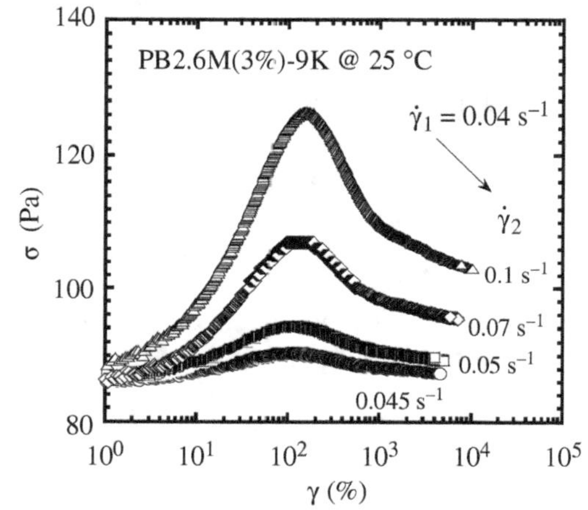

Figure 7.20 Stress change as a function of elapsed strain upon switching from steady-state shear at $\dot{\gamma} = 0.04\,s^{-1}$ to steady-state shear at 0.045, 0.05, 0.07, and $0.1\,s^{-1}$, respectively. *Source:* Data taken from Li and Wang 2010 [49].

a function of the underlying rate $\dot{\gamma}$. Table 7.4 lists two 3% PB solutions and shows the basic information such as the terminal relaxation rate ω_c from small amplitude oscillatory shear measurements. Here, the 1.8K solution is more sluggish and has a stronger temperature dependence of its solution dynamics because of the high vinyl content in the PB-1.8K solvent from Aldrich (category number 20,043-3).

Figure 7.21 shows that as $\dot{\gamma}$ varies by a factor of 10 the excess stress also relaxes on a time scale that is proportional to $1/\dot{\gamma}$. Apparently, the underlying shear flow simply brings the step-strained chains out of their deformed states at a rate proportional to $\dot{\gamma}$. It is straightforward to confirm that the relaxation dynamics are indeed dictated by the kinematics of the imposed continuous shear and not by any thermal diffusion. For example, we can apply the same superimposed step strain test at several different temperatures. Figure 7.22 shows that the same relaxation dynamics take place at these temperatures: the stresses all drop toward zero around 25 s, which is the reciprocal of the underlying rate of $0.04\,s^{-1}$. This lack of temperature dependence is consistent with Fig. 7.17(b). The spread of data at the long times, for example, after 20 s, may be related to molecular diffusion, which is faster at 40 °C than at 10 °C by a factor of 4.6.

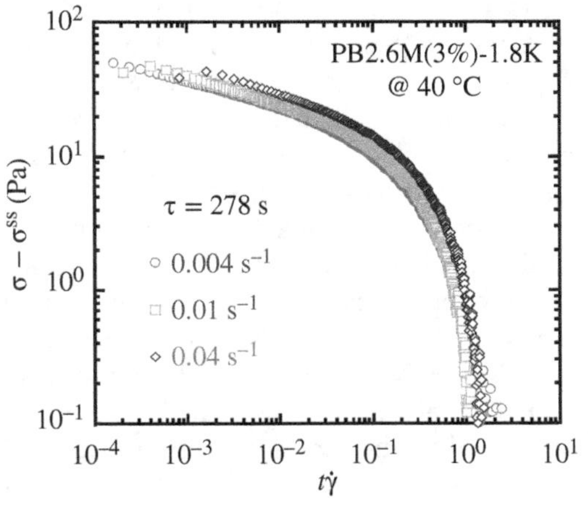

Figure 7.21 Stress relaxation due to the superimposed small strain of magnitude 0.1 (produced at 1 s^{-1}) as a function of the rescaled time, showing that the relaxation is dictated by the underlying shear rate (0.004, 0.01, and 0.04 s^{-1}). *Source:* Data taken from Li and Wang 2010 [49].

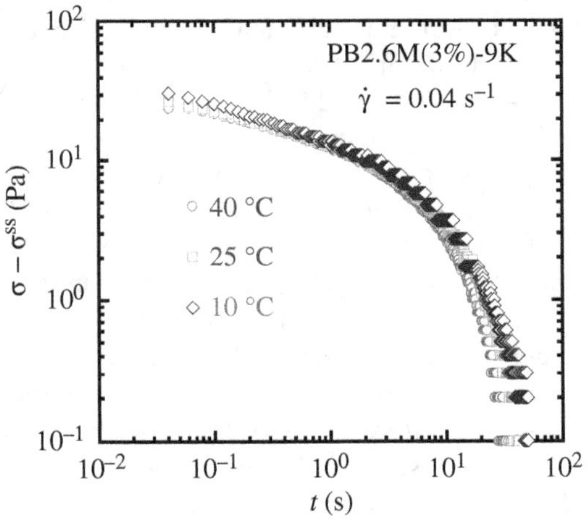

Figure 7.22 Stress relaxation from the superimposed small strain (0.1), at three different temperatures, in steady shear at 0.04 s^{-1}. *Source:* Data taken from Li and Wang 2010 [49].

7.4 From Terminal Flow to Fast Flow under Creep: Entanglement–Disentanglement Transition

In yield-stress materials, creep at a critical level of shear stress can produce a transition from a solid-like state to a state of considerable flow; cf. Section 4.2. Examples of yield-stress materials include ketchup, tooth paste, yogurt, and cheese. To examine this transition-like character, stress-controlled shear in principle can be applied to determine a threshold stress at which the system undergoes "yielding." In practice, the lack of sufficient adhesion between samples and shear surfaces makes it difficult to perform simple shear tests. Often a vane-like apparatus is employed.

Highly entangled polymers are analogous to yield-stress materials in that they display a yielding character in the form of an entanglement–disentanglement transition during startup creep, namely, at a constant shear stress.[12,50] This transition is the main subject of this section. Entanglement–disentanglement transition is not so different from the stick–slip transition, described in Section 6.2, which also takes place under controlled wall stress. We focus on the characteristics of the phenomenon and defer more rigorous discussion of disentanglement to Chapter 16. Since entangled polymeric liquids have not traditionally been treated as yield-stress materials, the convention is to examine their nonlinear rheological responses using rate-controlled tests. However, Fig. 7.2(a,b) shows that extreme shear thinning very much resembles the rheological behavior of yield-stress materials, namely, a sharp drop in the viscosity at a threshold stress.

How does an entangled polymeric liquid transform from its initial state of high viscosity to a final state of low viscosity? The rate-controlled experiments reviewed in the preceding sections do not offer sufficient insight into shear thinning. In contrast, startup deformation at a constant stress directly reveals to how shear thinning takes place. If the applied stress σ is small enough, namely, well below the stress plateau shown in Fig. 2.7, the system should initially deform like a solid. After a period comparable to the terminal relaxation time τ, shear flow takes place at a rate less than the reciprocal terminal relaxation time $1/\tau$: Writing the molecular deformation at τ as $\gamma \sim \dot{\gamma}\tau$, we have $\sigma \sim G_{pl}\gamma = G_p\dot{\gamma}\tau \sim \eta\dot{\gamma}$ with $\eta \sim G_p\tau$. If $\sigma \ll G_{pl}$, we have $\gamma \sim \tau\dot{\gamma} = Wi \ll 1$ or $\dot{\gamma} \ll 1/\tau$. Conversely, when $\sigma > G_{pl}$, the macroscopic elastic deformation is expected to exceed 100%. Such large deformation could cause a transition to flow during the creep. Applying a shear stress comparable to G_{pl} makes it impossible for the entanglement network to maintain its equilibrium state of chain entanglement. The consequence of overcoming the cohesive strength of the transient solid is that shear flow can take place at a far greater rate than $1/\tau$.

After the first creep tests[50] on entangled polymer solutions were reported in the early 2000s, two publications[51,52] suggested that edge instability should be blamed for the observed *entanglement–disentanglement transition (EDT)*. Clearly, the idea of disentanglement has profound implications, perhaps too radical when it was first reported. The question is whether and how an entangled polymer solution can be sheared much faster at the same shear stress level. Strong shear thinning or stress plateau behavior as shown in Fig. 7.2 is an undeniable experimental fact. Figure 7.23 shows that an entanglement–disentanglement transition must take place if a stress plateau emerges from rate-controlled measurements, starting with a highly entangled state with viscosity η_i and ending with a highly disentangled state with viscosity η_1.

Let us consider imposing a constant shear stress σ_1 at time zero. This stress will result in a shear rate $\dot{\gamma}_1$ in steady state with a corresponding viscosity $\eta_1 = \sigma_1/\dot{\gamma}_1$, as indicated in Fig. 7.23. How does the system arrive at this final state of much reduced viscosity from η_i close to the zero-shear Newtonian viscosity η_0? Does the system initially undergo shearing at $\dot{\gamma}_i$ to exhibit a viscosity at the level of η_i? How does the system transform from its initial fully entangled state to the final state (σ_1, $\dot{\gamma}_1$) with viscosity η_1 during creep at stress σ_1? It seems inevitable that

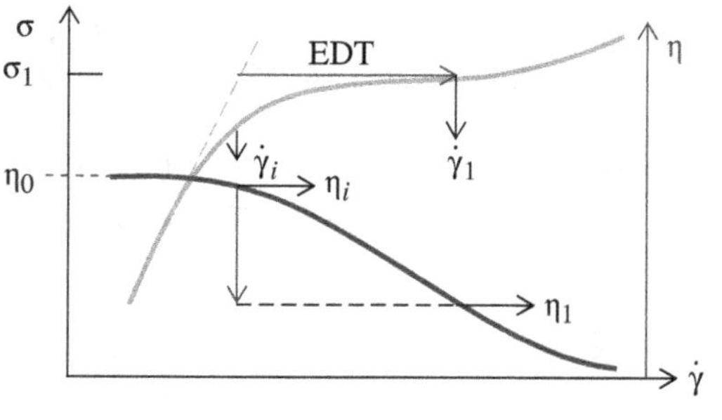

Figure 7.23 Characteristics of a "flow curve," namely, steady shear stress σ versus bulk shear rate $\dot{\gamma}$, showing a stress plateau around σ_1 (left axis) as well as the corresponding steady shear viscosity η (right-hand axis) as a function of $\dot{\gamma}$, displaying shear thinning. Upon creep at stress level σ_1, over time the viscosity of an entangled polymer must undergo a change in viscosity from η_0 to η_1. This transition is an entanglement–disentanglement transition because the initial large viscosity η_i is understood to originate from polymer entanglement, so the decline to the much reduced viscosity η_1 must imply loss of chain entanglement, namely, disentanglement.

the system would speed up, namely, the shear rate would increase in such a startup creep test. Such a change over time corresponds to a temporal decrease of the apparent viscosity at a fixed stress. According to the results of Section 7.3, during creep at σ_1, the effective relaxation time $\tau(\dot{\gamma})$ should decrease from η_i/σ_1 to a final value of η_1/σ_1 as the shear rate $\dot{\gamma}$ changes from $\dot{\gamma}_i$ to $\dot{\gamma}_1$. Since the large relaxation time stems from chain entanglement, it is reasonable to interpret the reduced relaxation time as disentanglement. As long as a stress plateau is present, which the past literature data have uniformly confirmed based on rate-controlled startup shear, we anticipate an *entanglement–disentanglement transition* upon creep at a sufficiently large stress, for example, a level comparable to the stress plateau.

The EDT phenomenon, illustrated in Fig. 7.24, shows how the apparent shear rate rises over time at different applied shear stresses. Figure 7.24 is clearly another way to envision yielding; the system transforms gradually and smoothly from a state with a long relaxation time and a large viscosity to a flow state with a short relaxation time and a small viscosity. Since the EDT could take hundreds of strain units to complete, the observation is prone to experimental difficulties such as edge instabilities. For example, as demonstrated by Inn *et al.*,[51] the influence of edge fracture can be exaggerated by allowing the edge effect to accumulate in order to cast doubt on the existence of the EDT. This line of work was continued[52] without any experimental improvement to prevent sample loss from affecting the rheometric measurements. Facing such uncertainty, a subsequent study[12] used a custom-designed cone-partitioned plate (CPP) assembly as shown in Fig. 5.2. The cone-partitioned plate allows one to show that the EDT is a natural response of strongly shear thinning, monodisperse entangled polymer solutions. The data in Fig. 7.24 are insulated from any edge instability because the CPP decouples any influence of edge failure from the rheometric measurements.

The meaning of EDT is much more straightforward when creep produces homogeneous shear free of edge fracture. The rest of this section focuses on this case. Figure 7.25(a) shows homogeneous entanglement–disentanglement transition, expressed in terms of a "normalized" shear rate, for a weakly entangled polymer solution made of 5 wt.% polybutadiene of $M_w = 1000$ kg/mol in 95% oligomeric PB of 10 kg/mol (see also Table 7.2).

This solution can be examined using startup shear in the rate-controlled mode. The response is the familiar stress overshoot, as shown in Fig. 7.25(b). Such a rate-controlled experiment can hardly show how shear thinning takes place. In contrast to rate-controlled shear, the creep mode is more illustrative of the disentanglement process. Because the applied shear stress is fixed, any change in the nominal shear rate is a direct reflection of the transient shear viscosity. Replotting Fig. 7.25(a) as Fig. 7.25(c) and realizing that $\dot{\gamma}/\sigma$ is simply the reciprocal viscosity,

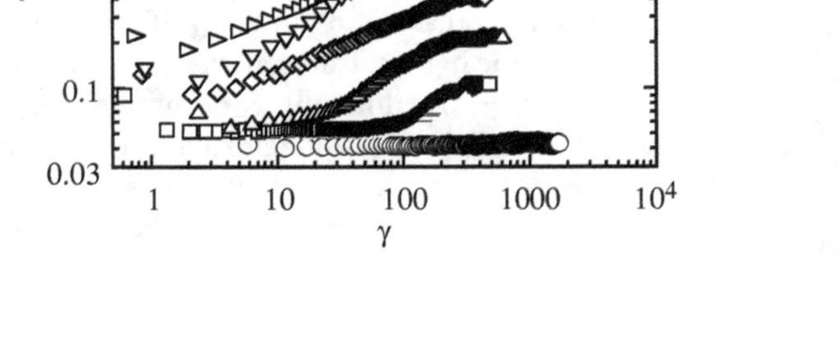

Figure 7.24 Nominal (apparent) shear rate $\dot{\gamma}_{app}$ versus the elapsed strain γ at different applied shear stresses, based on the cone-partitioned plate device (Fig. 5.2). Figure 9.9 in Chapter 9 shows that at the three times indicated by the vertical arrows the velocity profile is not linear, showing apparent wall slip at 101 and 308 s and shear banding at 350 s. Therefore, the bulk shear rate is different from $\dot{\gamma}_{app}$, forcing us not to use $\dot{\gamma}$ as the Y-axis label. *Source:* Data taken from Ravindranath and Wang 2008 [12].

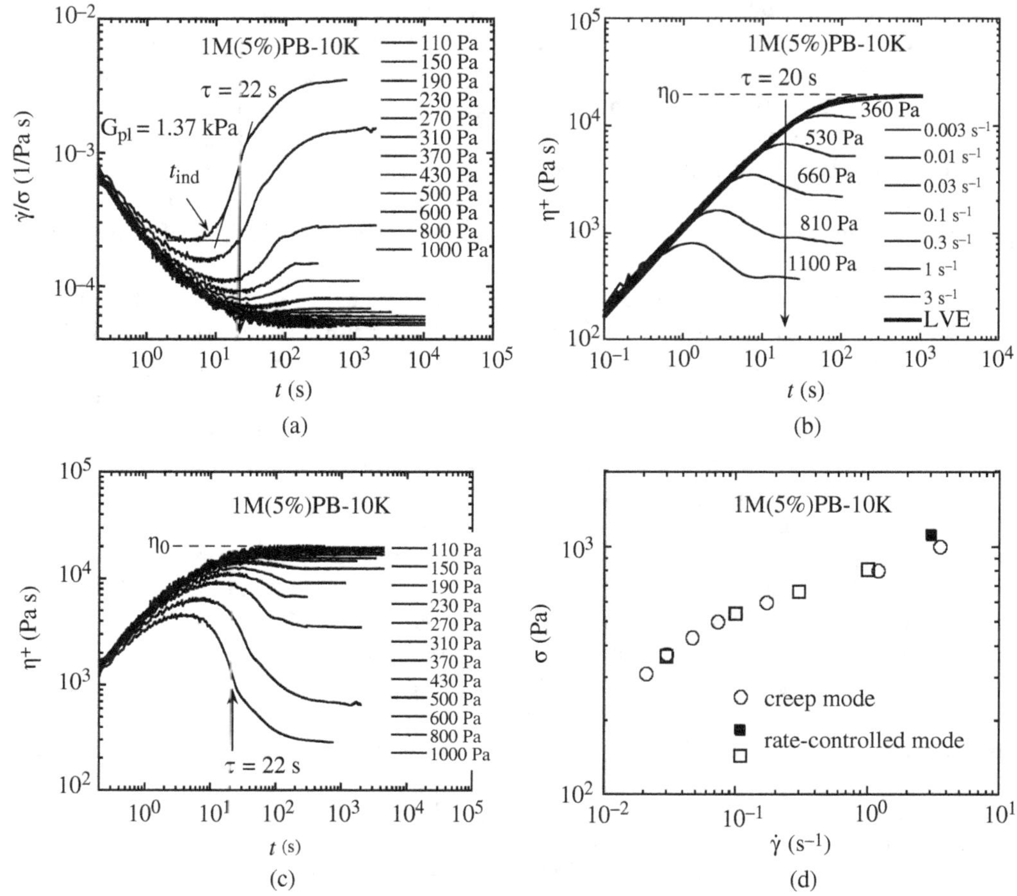

Figure 7.25 (a) Startup shear of PB1M(5%)-10K in creep mode, namely, at different applied stresses, in the absence of any shear strain localization, showing entanglement–disentanglement transition like behavior, represented as $\dot{\gamma}/\sigma$ versus *t*, which allows all the curves to collapse at short times when the responses are elastic. (b) Corresponding more conventional probe of nonlinear rheological responses based on rate-controlled startup shear in terms of the transient viscosity $\sigma(t)/\dot{\gamma}$ where the steady shear stresses are also indicated. (c) Inverse of the data in (a), namely, transient viscosity in creep mode, showing stronger overshoot. (d) Steady shear stress versus steady rate in both creep- and rate-controlled modes. *Source:* Data taken from Ge, Zhu and Wang 2017 [53].

namely, $\dot{\gamma}/\sigma = 1/\eta^+(t)$ we show in Fig. 7.25(c) a different normalized representation of the EDT. Figure 7.25(b,c) shows an overshoot in η^+, signifying yielding, namely, a transition from the initial state of entanglement to a state of flow.

Looking at the nonmonotonicity in Fig. 7.25(b,c), we also notice some important differences. First, the peaks of $\eta^+(t)$ are much closer to the zero-shear transient viscosity envelope in the creep mode of Fig. 7.25(c) than in the rate-controlled mode of Fig. 7.25(b), so we witness a far greater decrease in η^+ in Fig. 7.25(c) than in Fig. 7.25(b). Second, in Fig. 7.25(c), the peak represents the highest transient viscosity that can be measured before the EDT. The meaning of η^+ in Fig. 7.25(b) is different. For example, the peak of η^+ reflects the stress overshoot, showing the highest level of elastic deformation. η^+ defined in rate-controlled shear hardly represents a viscosity before the eventual flow state is established, although it has the dimension of viscosity. Third, in the steady state, the creep and rate-controlled modes should give identical flow curves. Figure 7.25(d) shows that the two modes are indeed in agreement except for the last (filled) square, involving the highest rate of 3 s^{-1}. The difference between the filled square and circle plausibly indicates that the controlled-rate test has not reached the steady state.

7.5 Yielding in Startup Uniaxial Extension

In the shear rheology literature, we heard from Maxwell[28] the idea of *yielding*. In contrast, the long history of extensional melt rheology has rarely concerned itself with yielding. However, any mode of startup deformation of viscoelastic liquids must inevitably result in yielding. The concept of yielding is not specific to shear deformation. What character signifies yielding upon startup uniaxial extension? Historically, the extensional rheology of polymeric solutions and melts has been expressed in terms of the Cauchy stress σ_E, also known as true extensional stress. The traditional perception of polymer solutions or melts is that they are viscoelastic liquids. As liquids, they have short memory. Therefore, only the present state of the liquid is important to quantify, and σ_E represents the present state. This logic dictates that the state of uniaxial extension at any moment be depicted in terms of the true stress $\sigma_E(t)$. After all, in any constitutive description of melt rheology, only the Cauchy stress is a materials variable. Thus, traditionally only σ_E is regarded as the meaningful variable to describe the stress state in uniaxial extension. In contrast to the shear stress σ that is defined for a fixed shearing area and may display overshoot during startup deformation, σ_E involves an ever-shrinking cross-sectional area and is usually only an increasing function of strain, until the sample breaks up, which is a subject treated in chapter 11.

The analysis developed in the recent literature[25,54] and discussed in the following sections shows that the signature of yielding appears in the characteristics of the engineering stress $\sigma_{engr} = \sigma_E/\lambda$. Here, the stretching ratio λ represents the areal shrinkage:

$$A(t)/A_0 = \lambda(t),$$

with A and A_0 being the present and initial cross-sectional areas, respectively. The *engineering stress* σ_{engr}, also known as *Piola–Kirchhoff stress,* shows a maximum as a function of λ, both in the terminal flow regime $Wi < 1$ and in the nonlinear regimes that include viscoelastic ($Wi > 1$ and $Wi_R < 1$) and elastic ($Wi_R > 1$) deformation regimes. We will discuss in Section 7.5.3 why the peak in σ_{engr} signifies yielding during melt stretching.

7.5.1 Myth with Considère Criterion

There is considerable confusion about what the Considère criterion[55] means and when it can be applied. The Considère criterion states that uniform uniaxial extension ceases after the tensile force peaks, at a stretching ratio λ_{max}. This statement originates from the following observation by Considère. When observing tensile deformation of soft iron and steel bars, Considère saw the onset of necking when the tensile force attained a maximum. The Considère criterion has been used to "predict" necking during startup uniaxial extension of entangled melts.[9] Such applications of the Considère criterion to melt rheology[56–59] assume that necking must occur under the Considère condition $d\sigma_{engr}/d\lambda = 0$, namely, $d\sigma_E/d\lambda = \sigma_E/\lambda$. In the polymer engineering literature, the Considère construction is frequently presented.[59–61] However, it has not been agreed upon whether the criterion applies for viscoelastic materials whose mechanical response is strain rate dependent.

Actually, the dimensional stability in tensile extension of viscoelastic materials is a far more complicated problem than implied by the Considère criterion. The stability problem has attracted extensive studies based on various linearized and nonlinear stability analyses.[62–64] In general, strain localization and specimen failure in tensile extension of polymer melts cannot be characterized by the Considère criterion. For example, entangled binary mixtures of SBR melts have been shown to display a pronounced peak in σ_{engr} but homogeneous extension continued far beyond the point where $d\sigma_{engr}/d\lambda = 0$.[65] In such systems, the decrease in σ_{engr} can take place homogeneously, namely, at every value of position z along the stretching direction. Whether and when tensile strain localization occurs depends on the specific microscopic physics, not on a simple characteristic of the stress versus strain curve.

At different levels of investigation, the effect could be mistaken as the cause. Such confusion is the situation with the application of the Considère criterion to extensional rheology of entangled melts. The nonmonotonicity of σ_{engr} is an effect at one (microscopic) level and has been taken as the cause for strain localization at another (macroscopic) level by those who applied the Considère criterion to predict necking in melt extension. Thus, in startup uniaxial extension of entangled melts, the causality becomes reversed when one tries to associate the emergence of a peak in σ_{engr} with the onset of tensile strain localization. At the deeper level, namely the molecular (microscopic) level, we need to ask what causes σ_{engr} to decline and why as well as when tensile strain localization may take place beyond the yield point (which is) at the engineering stress maximum.

The automatic application of the Considère criterion to depict tensile response of entangled polymers is also a good example illustrating the distinct objectives of two different fields: rheology of polymeric liquids and fluid mechanics of polymeric liquids. A polymer rheologist's mission is to determine constitutive behavior from microscopic physics. One task in fluid mechanics is to explain based on reliable constitutive equations and stability analyses how tensile extension of polymer melts may become unstable. If one is not given realistic constitutive relationships, he cannot possibly do a satisfactory job. In passing, we note that ductile polymer solids including semicrystalline polymers and ductile glassy polymers display a yield point, namely, tensile force maximum, during cold drawing. Unlike melt extension, tensile drawing typically produces stable necking. In both melt[55] and glassy[66,67] states, yielding is the cause of the force decline, and necking is the result of localized yielding. The force maximum is a consequence of yielding rather than the cause for necking.

In summary, the statement that the emergence of a maximum in the tensile force during extension causes necking reverses cause and effect. We defer further discussion of the relation between nonmonotonicity of σ_{engr} and strain localization to Chapter 11. This section focuses on the behavior of extensional deformation prior to any visible inhomogeneity.

7.5.2 Tensile Force (Engineering Stress) versus True Stress

With rare exceptions, the literature[68] reports rheological behavior of polymer melts in uniaxial extension in terms of the true tensile stress σ_E. This tradition is consistent with the conventional wisdom that only the current stress state is relevant to describe liquids. Since polymer melts are regarded as liquids, their rheological responses should be expressed in terms of σ_E. But an entangled polymer is hardly a liquid at the onset of startup continuous extension. To appreciate the need to break away from the tradition, namely, to motivate the use of the engineering stress σ_{engr} in rheological studies of extension of entangled melts, let us recall the stress formula for cross-linked rubbers (Eq. (1.52)).

In the presence of cross-linking, σ_{engr} is in principle linearly proportional to λ, until the finite extensibility limit is approached, i.e., when non-Gaussian effects occur. Such a linear relation reflects the success of the rubber elasticity theory summarized by Eq. (1.59). This formula should also describe the initial stress response of an entangled melt to startup extension, so long as the chains remain entangled. When disentanglement occurs, the areal density $\psi(\lambda)$ of entanglement strands starts to decline. At this point, σ_{engr} is expected to not increase as strongly as given by Eqs. (1.58)–(1.59). Upon sufficient disentanglement, namely, when $\psi(\lambda)$ is smaller than its initial value at $\lambda = 1$ by a factor of $(\lambda - 1/\lambda^2)^{-1}$, σ_{engr} declines. Thus, the stress response in terms of σ_{engr} can reveal how the entanglement network weakens and disintegrates (i.e., yields) during melt stretching. We note in passing that the sign for yielding is straightforward in simple shear: Since the shearing area does not change, any decrease in the shear stress with increasing strain is a sign of the failure of the entanglement network.

7.5.3 Tensile Force Maximum: A Signature of Yielding in Extension

Extensional rheological behavior can be described in two different ways, as discussed in the preceding section. By analogy with shear rheology, we look for an effective way to detect the transition during startup extension from elastic deformation to flow. The preceding paragraph asserts the advantage of quantifying the response of entangled melts to startup extension in terms of the engineering stress σ_{engr}. Indeed, σ_{engr} reveals the earliest sign of yielding. There are two reasons to choose σ_{engr} as the pertinent variable. First, when chain entanglement fails during startup extension, σ_{engr} ceases to grow with λ and eventually stops increasing. Second, if steady uniaxial extensional flow could be obtained for an entangled polymer melt, the Cauchy stress σ_E would approach a constant $\sigma_{E(ss)}$, while $\sigma_{engr(ss)} = \sigma_{E(ss)}/\lambda$ would decline as $1/\lambda$ because the cross-sectional area shrinks as $1/\lambda$. As a result, σ_{engr} must display a peak to signal the transition from elastic extension to flow.

For clarity and comparison, we sketch out what must happen during startup deformation, comparing simple shear with uniaxial extension, in terms of the two limits corresponding to the initial and final states. Figures 7.26(a)–(b) depict these limits for shear and extension, respectively, in the form of stress versus strain curves. In Figs. 7.26(a)–(b), the regions where the two regimes (elastic deformation and flow, represented by the thick lines) meet are central features for polymer rheology to determine. The intersection is the crossover where an entangled polymeric fluid turns from elastic to viscous, namely, where the entangled fluid undergoes yielding. As already discussed in Section 7.1, even when $Wi < 1$, yielding must occur. For simple shear, the two thick lines connect smoothly without an overshoot, corresponding to voluntary yielding. For extension, a maximum in σ_{engr} emerges even for $Wi < 1$. This overshoot in σ_{engr} can be regarded as a signature of the voluntary yielding. For well-entangled melts and $Wi > 1$, the regime of steady extensional flow (Fig. 7.26(b)) only exists hypothetically; the regime is therefore denoted by the dashed line because it can rarely be attained in experiment. After the initial elastic extension, σ_{engr} should pass through a maximum (Fig. 7.26(b)) as the system evolves toward a flow state. As will be discussed in Chapter 11, for well-entangled melts at high shear rates ($Wi > 1$), tensile strain localization intervenes, uniform melt extension cannot access flow.

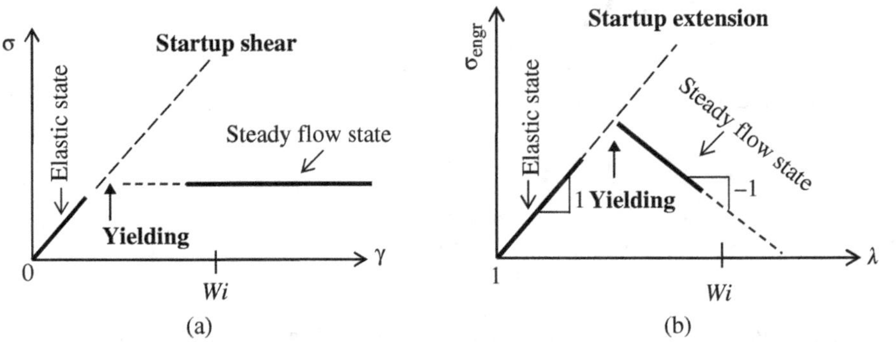

Figure 7.26 Depiction of the limiting responses of entangled polymeric fluids upon startup: (a) simple shear and (b) uniaxial extension, showing initial elastic deformation and eventual flow. Here the illustration does not include the phenomenon of melt rupture observed at sufficiently high rates (cf. Section 11.3).

At even larger rates, σ_{engr} monotonically increases until melt rupture takes place. We return to the lack of flow in continuous melt extension in Chapters 8 and 11.

7.5.3.1 Terminal Flow ($Wi < 1$)

In rheology, the condition $Wi \ll 1$ is often called the terminal flow regime. This phrase "terminal flow" is misleading; it has left some rheologists with the impression that for $Wi < 1$ flow would occur from the onset of either shear or extension. As sketched in Figs. 7.26(a)–(b), the initial deformation is actually elastic, with the stress growing with strain. Using a relatively "fast" melt, a polymer whose reptation time is relatively short, it is straightforward to probe the terminal flow regime on convenient experimental time scales. For example, a fairly monodisperse styrene–butadiene copolymer (SBR100K) with $M_w = 100$ kg/mol has a reptation time, or the so-called *terminal relaxation time* of $\tau = 25$ s as listed in Table 7.1. Figure 7.27 shows the polymer's response to startup extension at rates $\dot{\varepsilon} = 0.004$, 0.008, and 0.016 s^{-1}. Here the Hencky strain ε is used to represent the degree of extension. The engineering stress maxima show the emergence of voluntary yielding at $Wi < 1$. Moreover, the extension with $Wi \ll 1$ approaches steady flow without specimen breakup. These low-rate extension tests show steady flow behavior because chain disentanglement hardly occurs for $Wi \ll 1$.

In this linear response regime, we can apply linear models, such as the Maxwell model, to characterize the rheological response. For $Wi < 1$, the Maxwell model may provide an adequate phenomenological description and can be used to describe the normalized time- or strain-dependent engineering stress as

$$\sigma_{engr}(\varepsilon)/E = Wi \exp(-\varepsilon)[1 - \exp(-\varepsilon/Wi)], \tag{7.11}$$

where E is the tensile plateau modulus, related to the shear plateau modulus G_{pl} as $E = 3G_{pl}$. We obtained Eq. (7.11) by rewriting Eq. (1.23b) for extension in terms of Hencky strain $\varepsilon = \dot{\varepsilon}t$ instead of the time t, and substituting in the Weissenberg number $Wi = \dot{\varepsilon}\tau$ and $E = \eta_E/\tau$. Here, the true stress σ is the Cauchy stress $\sigma_E = \sigma_{engr} \exp(\varepsilon)$. Figure 7.27 demonstrates a good agreement between the Maxwell model and the experimental data and confirms the appearance of a

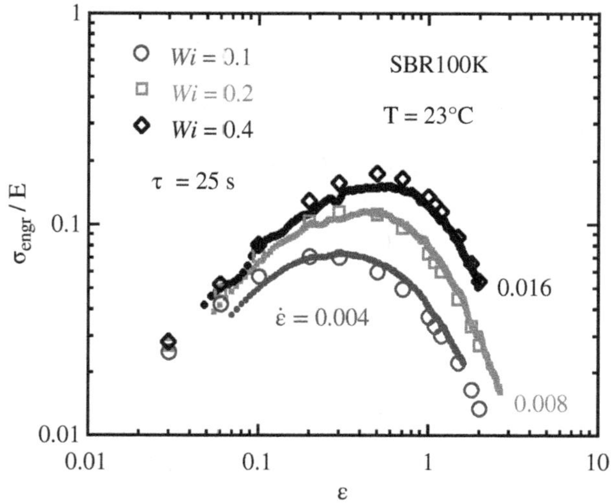

Figure 7.27 Normalized engineering stress σ_{engr} as a function of Hencky strain ε for three Hencky rates of 0.004, 0.008, and 0.016 s^{-1}, where the open symbols are calculated using Eq. (7.11). *Source:* Data taken from Wang and Wang 2008 [54].

pronounced peak in σ_{engr}. According to Eq. (7.11), the position of the peak is given by

$$\varepsilon_y = Wi[ln(1 + Wi) - ln(Wi)]. \tag{7.12}$$

So far as the Maxwell model is concerned, Eq. (7.12) applies for all values of *Wi*. In the limit $Wi \gg 1$, Eq. (7.12) shows $\varepsilon_y(Wi \gg 1) = 1$, Eq. (7.11) becomes $\sigma_{engr}/E = \varepsilon \exp(-\varepsilon)$, and the Cauchy stress σ_E becomes $\sigma_E = E\varepsilon = E\,ln(\lambda)$. This Maxwell formula contrasts strongly with the rubber elasticity formula $G(\lambda^2 - 1/\lambda)$, which is to be expected from fast extension of an entangled melt. These two formulas agree in the limit $\varepsilon \ll 1$, where $\lambda = \exp(\varepsilon) = 1 + \varepsilon + O(\varepsilon^2)$. Thus, the Maxwell model gives an adequate description of extension for $Wi \ll 1$ but describes fast uniaxial extension of an entangled melt rather poorly; the rubber elasticity formula may only hold to a strain as high as or larger than $\varepsilon = 1$ in fast extension.

7.5.3.2 Yielding Evidenced by Decline in σ_{engr}

Just as with simple shear, in response to startup uniaxial extension, entangled melts undergo macroscopic yielding. The initial signature for yielding is a downward deviation of the engineering stress from the ideal rubber elasticity formula (Eq. (1.52)), as shown in Fig. 7.28. The affine deformation limit of Eq. (1.52) is indicated by the straight line. As the applied extension rate increases, the styrene–butadiene rubber (SBR) does not deviate from the limiting elastic deformation until a larger strain is attained, similar to the trend seen in Fig. 7.3 for simple shear. For $Wi > 1$, the tensile force, or its normalized value, the engineering stress declines after the initial rubber-like increase, signaling a gradual loss of elastic deformation. This nonmonotonic behavior of σ_{engr} fits the definition of yielding, as explained in Section 7.5.2. In contrast, during startup uniaxial extension, the Cauchy stress σ_E is not as effective at revealing the state of the entanglement network. If the extension ever reaches steady state, namely, a state of extensional flow, then we may wish to express the stress response in terms of σ_E at long times. However, during transient response to startup extension, σ_E disguises the symptom of yielding: The tensile force may already be declining, but σ_E can still increase monotonically when the decreasing load is more than compensated by the decreasing cross-section. So σ_E keeps increasing as long as the tensile force does not drop as quickly as $1/\lambda$.

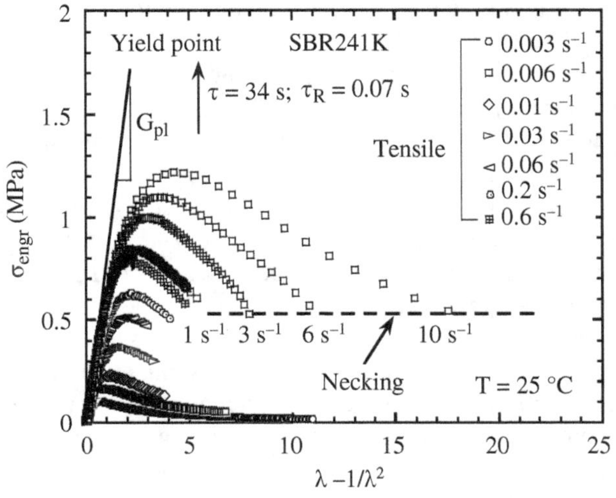

Figure 7.28 Engineering stress σ_{engr} as a function of strain, expressed as $(\lambda - 1/\lambda^2)$, for a wide range of Hencky rate from 0.003 to 10 s^{-1}. The final data points in each curve represent the onset of visible tensile strain localization. *Source:* Data taken from Zhu and Wang 2013 [69].

In order to demonstrate that σ_{engr} gives a more useful representation of the rheological response to startup extension, we need to understand what the decline of σ_{engr} means and what is causing it to decline. In the absence of any *in situ* molecular-level characterization by either experiment or molecular dynamics simulation, we can only speculate microscopic processes based on macroscopic rheometric measurements. It is helpful and convenient to rely on insights from the successful rubber elasticity theory, and the idea of treating entangled melts as transient Gaussian chain networks, as sketched in Fig. 2.9. Specifically, for non-cross-linked entangled polymer melts, we expect the initial stress versus strain relation to approximately follow Eq. (1.52), namely, we expect the initial extension to be affine. For further analysis, it is instructive to express σ_{engr} in terms of the structure of the entanglement network, by analogy with the representation given in Section 1.3.3. We first introduce the concept of the *engineering* areal number density ψ of entanglement strands, which is defined to be the total number of entanglement strands in the total cross section whose normal vector is along the stretching direction, divided by the cross-sectional area in the undeformed state. We then invoke the notion of an intrachain retraction force, denoted $f_{retract}$, for each entanglement strand. The engineering stress σ_{engr} can be expressed as the product of these two quantities

$$\sigma_{engr} = \psi f_{retract}. \tag{7.13}$$

Equation (7.13) is analogous to the formula Eq. (1.59) for the rubber elasticity of a cross-linked Gaussian chain network. Since an entangled polymer melt is not cross-linked, additional concepts are required to describe the response of the entanglement network to melt extension. Specifically, we need to determine whether, when, how, and why the network may fall apart at high strains. In this chapter, we only address "when and how" in a phenomenological sense. We try to answer the question of "why" in Chapter 16. Initially, in the limit of affine deformation, where the rubber elasticity formula Eq. (1.52) holds, we have

$$f_{retract}^{aff} = f_{ent}(\lambda - 1/\lambda^2) \tag{7.14}$$

so that

$$\sigma_{engr}^{aff} = \psi_{ent} f_{retract}^{aff} = G_N^0(\lambda - 1/\lambda^2), \tag{7.15}$$

where the second equality follows from Eq. (2.30c) for G_N^0, with the entanglement force f_{ent} defined in Eq. (2.30d), and ψ_{ent} is given by Eq. (2.31).

We may learn about the structural change of the network through the measurement of $\sigma_{engr.}$ A partial breakdown of the entanglement network leads to a decrease in ψ from its value ψ_{ent} in the undeformed state. Moreover, as the number of entanglements per chain decreases, the remaining entanglement strands become longer than l_{ent} of the original entanglement strands. For example, in Fig. 7.28, the downward deviation of σ_{engr} from Eq. (7.15) is consistent with the interpretation that ψ in Eq. (7.13) has started to decrease from its equilibrium value ψ_{ent}. Furthermore, the magnitude of $f_{retract}$ is no longer determined by the equilibrium prefactor f_{ent} because the surviving strands are longer on average with $l'_{ent} > l_{ent}$. It is necessary to indicate that many or perhaps most of these surviving strands are not the original strands: these new strands are longer because, for example, two consecutive strands from the same chain could have combined into a new strand upon destruction of the middle entanglement point. It is simple to visualize this conversion by borrowing the transient network picture for entanglement in Fig. 2.9. Eventually, the balance tilts: As a function of λ, the loss of active entanglement strands overwhelms the contribution of the surviving entanglement strands to the tensile force. In a zeroth-order picture, we may speculate as illustrated in Fig. 7.29: The extension from (a) to (b) is affine deformation, so that the stress versus strain relation is given by Eq. (7.15). Figure 7.29(c) qualitatively depicts the state of entanglement well after the yield point, showing a significant

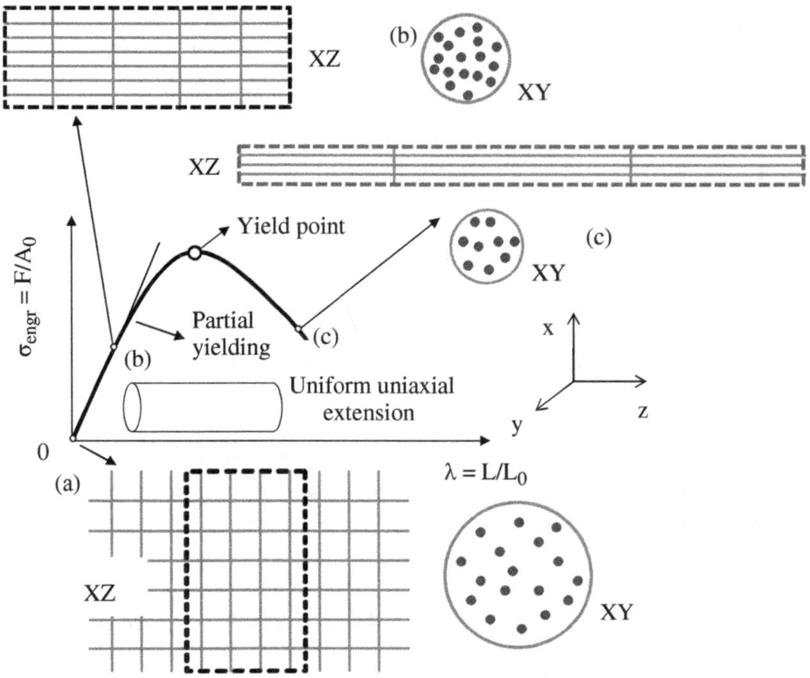

Figure 7.29 Depiction of fast startup uniaxial extension of entangled melts at different stages before and after the yield point (the engineering stress maximum), up to the point where the extension is still homogeneous. Well before the yield point, a non-cross-linked melt acts like a permanent network as depicted. Thus, the number of extended strands per unit cross-sectional area increases from state (a) to (b) at the expense of the non-load-bearing strands such that the mass density remains constant. Clearly, during the affine-like extension, there is geometric condensation (from (a) to (b)) of the active entanglement strands (EntS), namely, ψ increases in the plane whose normal is along the direction of extension.[70] A detailed mathematical account of the geometric condensation concept is provided in Appendix 8.A in Chapter 8. There emerges partial yielding when the stress response softens, namely, no longer increases according to Eq. (1.52). As depicted, the total number of EntS decreases upon reaching the yield point. Further loss of EntS across the sample's cross section accompanies the smooth decline of σ_{engr}. This "dissolution" of the entanglement network eventually reaches a point where uniform extension cannot be sustained, and the sample nucleates a weak segment where uneven extension occurs due to the localized yielding of the entanglement network. The specimen eventually fails. More discussion on this subject is given in Chapter 11.

loss of entanglement. Specifically, in Fig. 7.29(c), ψ is about half of ψ_{ent}, and the surviving entanglement strands have become longer.

In general, it appears necessary to introduce at least two parameters α (<1) and μ (>1) to characterize chain disentanglement and the gradual disappearance of the entanglement network. Upon yielding, we suppose that the average molecular weight of entanglement strands, $M_e' \sim (l'_{ent})^2$, increases from the equilibrium value M_e by factor of

$$\mu = M_e'/M_e > 1. \tag{7.16}$$

Separately, we introduce α to denote the decrease in the engineering areal density of entanglement strands from the equilibrium value ψ_{ent}

$$\alpha = \psi/\psi_{ent} < 1. \tag{7.17}$$

As long as the new longer strands can continue to be treated as Gaussian chains, we can proceed to describe the stress versus strain relation in the presence of yielding.

In terms of α and μ, we can keep track of the evolution of the engineering stress σ_{engr} during startup continuous extension and generalize Eq. (7.15) beyond the affine deformation limit. We explicitly rewrite Eq. (7.13) as

$$\sigma_{engr} = \psi f_{retract} = \left[\frac{\alpha(\lambda)}{\sqrt{\mu(\lambda)}}\right] G_N^0(\lambda - 1/\lambda^2), \tag{7.18}$$

where the second equality follows from rewriting $f_{retract}$ in a form different from Eq. (7.14). Specifically, due to the increased strand size from $l_{ent} \sim (M_e)^{1/2}$ to $l'_{ent} \sim (M'_e)^{1/2}$, namely,

$$l'_{ent} = l_{ent}\sqrt{\mu}, \tag{7.19}$$

we write the retraction force $f_{retract}$ associated with the residual strands as

$$f_{retract} = (k_B T/l'_{ent})(\lambda - 1/\lambda^2) = f^{aff}_{retract}/\sqrt{\mu}, \tag{7.20}$$

which upon insertion of Eq. (7.20) into the first equality of Eq. (7.18) leads to the second equality of Eq. (7.18). Equation (7.20) states that an entanglement strand has a smaller retraction force by a factor of $\mu^{-1/2}$ because the longer strand size l'_{ent} appears in the denominator of the prefactor in Eq. (7.20). Obviously, $\alpha(\lambda)$ and $\mu(\lambda)$ are sensitively dependent on degree of extension, λ, as well as the Hencky rate $\dot{\varepsilon}$. In the derivation of Eq. (7.18), we assumed that the surviving entanglement strands still undergo affine deformation, namely, Eq. (1.52) holds for the surviving entanglement network.

We assert that σ_{engr} sensitively reflects the structural changes in the entanglement network and is an indirect yet effective indicator of molecular processes, namely, chain disentanglement. As long as the decrease in ψ is compensated by the growth of $f_{retract}$ with λ in the surviving entanglement strands, σ_{engr} still increases with λ. The peak of σ_{engr} represents a balance between the remaining longer strands extending farther and the number of active strands diminishing. The overshoot in σ_{engr}, a macroscopic behavior, actually indicates something explicit is happening at the molecular level. Disentanglement causes $(\alpha/\sqrt{\mu})$ to decrease. Eventually, the factor $(\alpha/\sqrt{\mu})$ in Eq. (7.18) decreases faster than $(\lambda - 1/\lambda^2)$ increases.

7.5.3.3 Maxwell-Like Response and Scaling for $Wi_R > 1$

The maximum in σ_{engr} is prominent in Fig. 7.28. We have suggested that this maximum is a sign of yielding. Characteristics of this signature are seen in Fig. 7.30. Also seen in Fig. 7.30 is the Maxwell description of Eq. (7.12), covering a range from $Wi \ll 1$ to $Wi \gg 1$. Up to the onset of

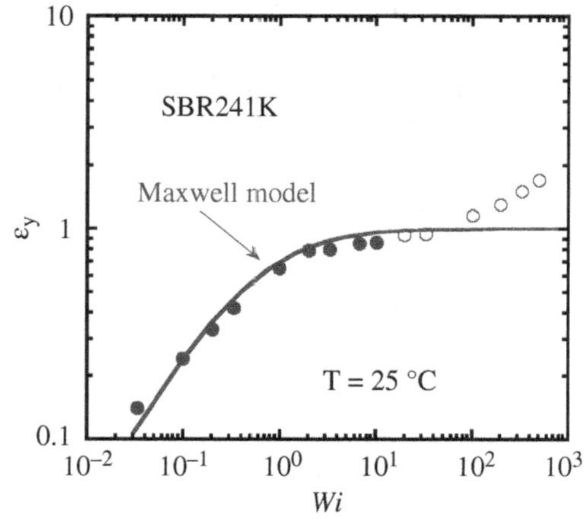

Figure 7.30 Yield Hencky strain ε_y, which is the strain at the peak of σ_{engr}, read from Fig. 7.28, as a function of Wi. The continuous curve is given by the Maxwell model in Eq. (7.12).

a scaling regime, where ε_y starts to grow with the applied rate, the agreement between the data and the Maxwell model is interesting to note.

To determine whether the nascent scaling behavior is universal or not, we analyze the literature data from Ref. [69], based on five styrene–butadiene rubbers having different microstructures and degrees of entanglement. Tables 7.5 and 7.6 list the basic microstructure and property information for these SBR melts. Figure 7.31 agrees with the report[71] that there can be a scaling regime where

$$\varepsilon_y = Wi^{1/3}, \quad \text{for } Wi_R > 1. \tag{7.21}$$

This scaling form resembles Eq. (7.5). Here Wi_R is defined in Eq. (7.3). It remains unknown why startup shear and extension both have a scaling regime with the same exponent 1/3 for the shear yield strain γ_y and for the extensional Hencky yield strain ε_y. In contrast to Fig. 7.8(b),

Table 7.5 Microstructures of five SBR melts.

Sample	Styrene (%)	Butadiene (%)	Vinyl (1,2-BD) (%)
179K	20.0	80.0	65.2
241K	21.0	79.9	39.9
250K	20.0	80.0	54.7
250K*	20.0	80.0	65.0
325K	20.0	80.0	52.6
616K	21.9	78.1	37.5

Table 7.6 Molecular characteristics of five SBR melts.

Sample	M_n (kg/mol)	M_w/M_n	G_N^0 (MPa)	M_e (kg/mol)	Z	τ (s)	τ_R (s)
179K	179	1.14	0.69	3.2	64	75	0.4
241K	241	1.10	0.82	2.7	98	34	0.07
250K	250	1.21	0.77	2.9	105	220	0.6
325K	325	1.36	0.76	2.9	152	714	1.0
616K	616	1.12	0.88	2.5	276	920	0.34

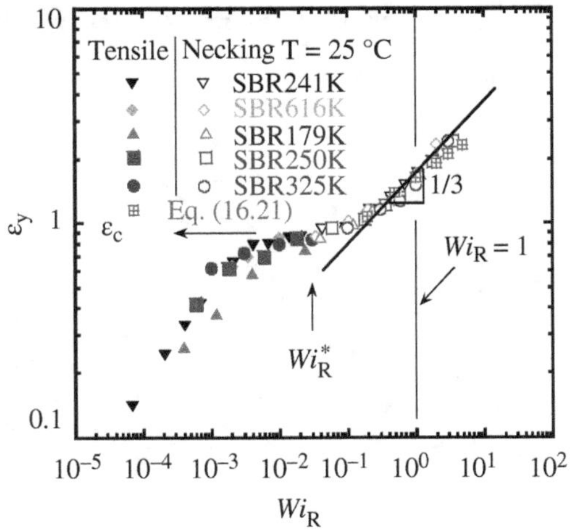

Figure 7.31 Master curve of ε_y versus Wi_R, where the extension tests at smaller Wi_R in filled symbols end up in tensile decohesion and tests at larger Wi_R in open symbols eventually break up through necking-like tensile strain localization. These extensional failures are the subject of Chapter 11. The borderline between tensile failure and shear-yielding-induced necking is denoted Wi_R^*. Around $Wi_R = 1$ and beyond, there appears to be a scaling regime with exponent 1/3, as given by Eq. (7.21), reminiscent of the scaling behavior seen in Fig. 7.8(b). The plus-containing squares are from Eq. (16.21). *Source:* Data taken from Zhu and Wang 2013 [69].

Fig. 7.31 shows that the onset of 1/3 power-law scaling occurs "earlier," namely, for $Wi_R < 1$. There are two plausible explanations: (a) There were errors in the evaluation of the Rouse time τ_R for these melts – τ_R was evaluated according to Appendix 7.A.2 instead of the simple scaling $\tau_R = \tau/Z$ used in Fig. 7.8(b), which will yield significantly larger values for Wi_R for the onset of the 1/3 scaling. (b) Uniaxial extension is more effective than simple shear at causing molecular stretching. Although an accurate determination of τ_R is required to explore possible scaling laws, the uncertainty about the choice of methods to evaluate τ_R from experiment does not prevent us from noting the transition in Fig. 7.31 from a regime where $\varepsilon_y \sim \varepsilon_c$ hardly varies with Wi_R, to another regime where ε_y increases measurably with Wi_R.

It is worth noting that the rheological response to extension, as shown in Fig. 7.30, is essentially Maxwell-like, even for $Wi \gg 1$. In Figs. 7.31 and 7.32, the filled symbols denote the Maxwell-like regime, where the specimen breaks shortly after the peak of the engineering stress, in a rupture-like manner that we call "tensile decohesion" in Chapter 11. The open symbols represent startup uniaxial extension at larger Hencky rates that produces necking-like failure and their scaling behavior. By analogy with startup shear discussed in Section 7.2.1.2, this scaling regime is here described as the "elastic deformation" regime. We defer to Chapter 11 a detailed discussion of different types of extensional failure.

We also defer to Chapter 11 a detailed discussion of a third regime, located beyond the scaling regime at even larger shear rates. At such rates, corresponding to $Wi_R \gg 1$, for example, $Wi_R \sim 10$, the yielding is only partial and σ_{engr} keeps increasing monotonically until the specimen undergoes window-glass-like rupture. Such rupture reveals an intriguing difference between extension and shear deformation beyond the scaling regime, namely that entanglement responds differently to these two different modes of deformation. We will discuss this difference further in Sections 8.3 and 11.3.

Finally, within the explored range of Hencky rates, we note a further similarity between shear and extension. Figure 7.32 shows that the peak value of the engineering stress approximately depends on the stretching ratio λ as

$$\sigma_{engr(y)} = G_{coh(ext)}[\lambda_y - 1/(\lambda_y)^2], \tag{7.22a}$$

where the cohesive modulus $G_{coh(ext)}$ is

$$G_{coh(ext)} = G_N^0/3, \tag{7.22b}$$

Here the subscript "coh(ext)" stands for "cohesion in extension." Equations (7.22a) and (7.22b) are analogous to Eqs. (7.4a) and (7.4b) for startup shear.

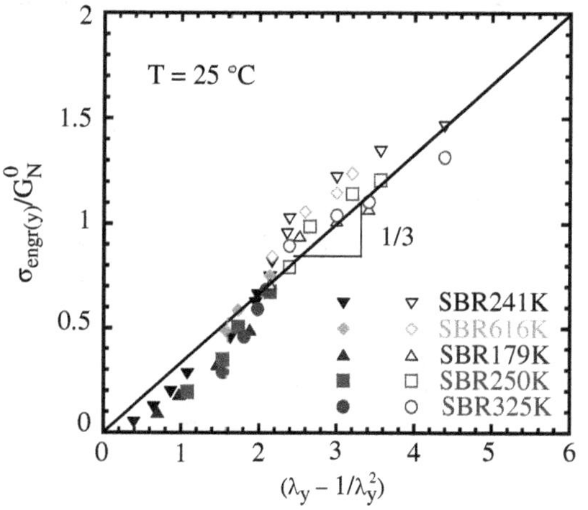

Figure 7.32 Normalized yield engineering stress $\sigma_{engr(y)}$ as a function of the yield strain, expressed as $(\lambda_y - 1/\lambda_y^2)$, where the open and filled symbols have the same meanings as in Fig. 7.31. *Source:* Data taken from Zhu and Wang 2013 [69].

7.5.3.4 Elastic Recoil

To further demonstrate the meaning and significance of the peak in σ_{engr}, we resort to the method of elastic recoil. Let us consider a series of interrupted startup extension tests, stretching an entangled melt to different stages, either before or after the emergence of the stress peak, and then releasing the stretched specimen so that it is free to retract. When the release occurs at $Wi \gg 1$ and before an entangled melt is extended past the peak of engineering stress, the strain is nearly completely recoverable. A lack of full elastic recoil is most obvious when the sample is stretched to $\lambda > \lambda_y$, as schematically shown in Fig. 7.33.

Elastic recoil and strain recovery experiments have been routinely carried out to determine the nature of the deformation state.[72] For the SBR melt whose terminal behavior appears in Fig. 7.27, we can quantify its ability to undergo elastic recoil after startup extension as a function of the amount of melt stretching at different applied Hencky rates, ranging from $Wi < 1$ to $Wi_R > 1$. When $Wi \sim 1$, a great deal of chain relaxation and nonaffine deformation has already taken place before reaching the Maxwell peak given by Eq. (7.12). Conversely, for $Wi \gg 1$, the entanglement network has not fallen apart at the peak of σ_{engr}. In one set of literature experiments,[54] the SBR100K melt was capillary extruded to prepare cylindrical samples to study the elastic recoil behavior. The recoverable Hencky strain can be defined as

$$\varepsilon_{rec} = ln[L(0)/L_{\infty}] = ln[L(0)/L_0] - ln(L_{\infty}/L_0) = \varepsilon - 2ln(r_0/r_f), \quad (7.24)$$

where L_0 is the initial filament length, L(0) is the length at the beginning of the recovery test when the specimen is set stress free, L_{∞} is the length long after the release, and r_0 and r_{∞} are the initial and final radii of the specimen. This expression shows that ε_{rec} can be readily determined by measuring the specimen's radius.

Figure 7.34 shows that for $Wi \gg 1$, $r_f \cong r_0$, namely, we have 100% strain recovery with $\varepsilon_{rec} = \varepsilon$, as long as $\varepsilon \leq \varepsilon_y$ (i.e., up to the force maximum). Beyond ε_y, complete strain recovery is no longer possible as shown by the filled symbols. If Wi is reduced to 10, the recovery is incomplete even before reaching ε_y, as shown by the squares. Partial recovery means that $\varepsilon_r \leq \varepsilon$, so that the data points in Fig. 7.34 stay below the diagonal line. At moderate values of Wi, namely, rates of 0.8 s^{-1} and smaller, ε_y is below unity, as shown in Figs. 7.34 and 7.35, which is consistent with Fig. 7.30.

In another set[31] of elastic recovery experiments that are worth mentioning, the authors used stripe-shaped specimens instead of cylindrical filaments. In this case, the sample was marked by a ruler, and its length or sublength was measured before and after the elastic recoil test. Figure 7.36 shows, after startup extension with $Wi = 34$, the full elastic recovery before the maximum in σ_{engr} that appears at $\lambda_y \cong 3$. The stress versus strain plots in the inset indicate that Cauchy stress σ_E monotonically increases with λ while σ_{engr} has a maximum. The elastic recoil

Figure 7.33 Depiction of two elastic recovery tests involving fast startup uniaxial extension of entangled melts and setting the tensile stress to zero either before or after reaching the peak engineering stress at λ_y. Characteristically, the elastic recovery is nearly complete for $\lambda < \lambda_y$ and incomplete for $\lambda > \lambda_y$.

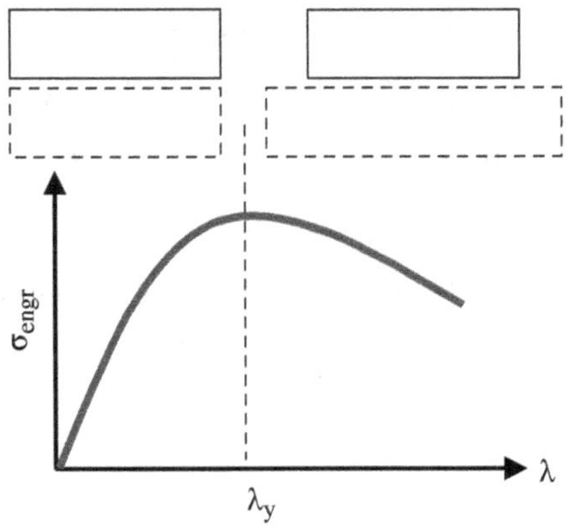

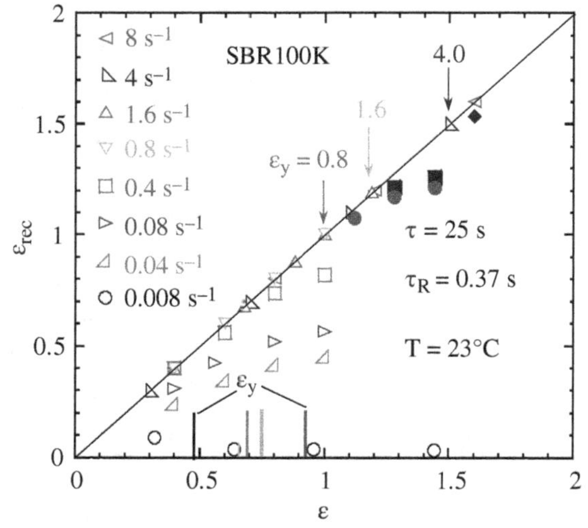

Figure 7.34 Recoverable strain ε_{rec} versus the elapsed strain ε, for eight different extension rates, at different stages of uniaxial extension. Here the vertical arrows indicate the locations (i.e., ε_y) of the yield point at three extension rates 0.8, 1.6, and 4.0 s^{-1} that satisfies $Wi_R > 1$. Beyond the yield points at the respective extension rates, the recovered strains are denoted with filled symbols. *Source:* Data taken from Wang and Wang 2008 [54].

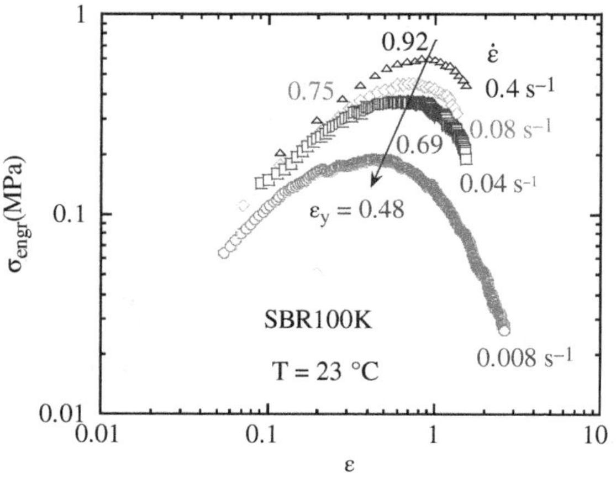

Figure 7.35 Engineering stress σ_{engr} as a function of Hencky strain ε for Hencky rates 0.4, 0.08, 0.04, and 0.008 s^{-1}. These rates produce peak stresses at Hencky strains of $\varepsilon_y = 0.92, 0.75, 0.69$, and 0.48, respectively. *Source:* Data taken from Wang and Wang 2008 [54].

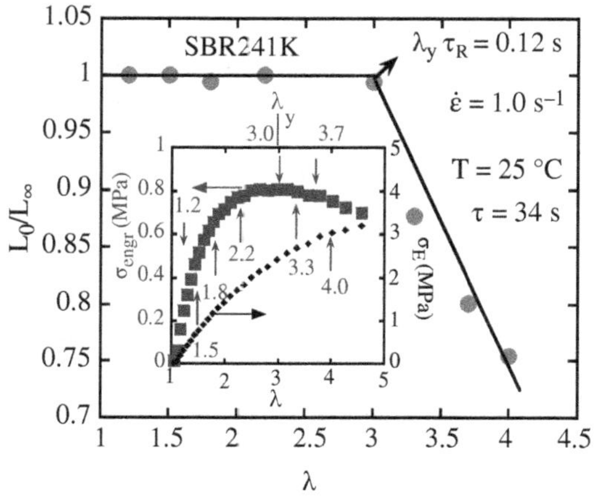

Figure 7.36 Recoverable strain, defined by the ratio of initial (L_0) to final length (L_∞) of the specimen, from different stages in startup uniaxial extension of an SBR241K melt at a Hencky rate of 1 s^{-1}. The inset shows engineering (large squares) and Cauchy (small diamonds) stresses σ_{engr} and σ_E as functions of the stretching ratio λ. The eight vertical arrows indicate the strains at which the extended specimen is released to commence elastic recovery. *Source:* Data taken from Wang *et al.* 2013 [31].

data once again indicate that the maximum in σ_{engr} has specific physical meanings. σ_{engr} shows the massive yielding beyond the maximum that makes it impossible to have full elastic recoil. Systematic studies of melt extension to determine the degree of elastic recoil as a function of the strain for various applied Hencky rates have yet to be carried out.

7.6 Conclusion

This chapter focused on transient nonlinear responses of entangled polymers to startup shear and extension. This core chapter discussed the concept of yielding to show that elastic deformation does not persist indefinitely. In either startup shear or extension, the sample must transition from an initial elastic deformation toward a state of irrecoverable deformation, as shown in Fig. 7.26(a) for simple shear and Fig. 26(b) for uniaxial extension. Therefore, yielding is inevitable from a phenomenological standpoint. We have suggested that macroscopic yielding is due to yielding of the entanglement network, namely, macroscopic yielding originates from microscopic yielding. The yielding was demonstrated to obey scaling law for $Wi_R > 1$. We have invoked the concepts of entanglement and disentanglement in a phenomenological sense. For example, we accept that the high shear viscosity of an entangled polymer liquid stems from chain entanglement, while correspondingly the reduced steady shear viscosity is an indication of disentanglement. Our exploitation of the concept of yielding allows us to provide a unified framework to treat nonlinear rheological responses to both shear and extension. Our identification of the point of maximum tensile force (or engineering stress) as marking the appearance of extensional yielding is novel. It has never been suggested in the literature prior to 2007.[11,71] Nevertheless, the elastic recovery experiments support our simple interpretation of shear stress and engineering stress maxima as a global yield point.

7.A Experimental Estimates of Rouse Relaxation Time

The experimental data in this chapter, especially Fig. 7.8(b), show that the Rouse time τ_R is an important parameter. Here we describe four methods to evaluate τ_R based on experiment.

7.A.1 From Self-Diffusion

We can estimate τ_R from self-diffusion data. In the asymptotic limit of $Z \gg 1$, we have $\tau_d \sim \eta_0$ approaching scaling behavior with exponent 3 (Fig. 2.3) and self-diffusion constant D_s approaching scaling behavior with exponent −2 (Fig. 2.16). In this limit, Eq. (2.47) is valid so that we combine Eqs. (2.40) and (2.47) and write

$$L^2/D_R = R^2/D_{ts} \tag{7.A.1}$$

where $D_{ts} \sim N^{-2}$ is the self-diffusion constant extrapolated from the scaling regime of $D_s \sim M^{-2}$ and is shown, for example, in Fig. 2.17(a) to come from trace diffusion measurements. Then τ_R is given by Eqs. (2.22) and (7.A.1)

$$\tau_R = R^2/D_R = (l_{ent})^2/D_{ts}. \tag{7.A.2}$$

where the entanglement spacing l_{ent} (tube diameter) is readily available for many common melts.[73] Thus, τ_R can be estimated from diffusion measurements.

7.A.2 From Zero-Shear Viscosity

To make sure τ_R scales as M^2, Osaki proposed[74] to estimate τ_R from the zero-shear viscosity η_0 as

$$\tau_R = \frac{6M\eta_0}{\pi^2 cRT}\left(\frac{M_c}{M}\right)^{a-1} \tag{7.A.3}$$

where the scaling behavior of η_0 is assumed to show a power law $\eta_0 \sim M^a$. M_c is usually taken as $2M_e$, where M_e can be obtained from the elastic plateau modulus as shown in Eq. (2.30a).

7.A.3 From Reptation (Terminal Relaxation) Time τ_d

The simplest approach to estimating τ_R is to use literally the relation in Eq. (2.43), namely, we simply have

$$\tau_R = \tau_d/3Z. \tag{7.A.4}$$

For entangled 1,4-polybutadiene solutions, Eq. (7.A.4) gives τ_R that is numerically close to the estimate according to Eq. (7.A.3).

7.A.4 From Second Crossover Frequency ~ $1/\tau_e$

When small-amplitude oscillatory shear data are available to clearly identify τ_e as sketched in Fig. 2.7, it should be reasonable to evaluate Rouse time as

$$\tau_R = \tau_e Z^2, \tag{7.A.5}$$

provided that $Z = M/M_e$ is also known.

References

1 Lodge A. Elastic liquids. Academic press, London and New York; 1964. pp. 118–20.
2 Astarita G, Marrucci G. Principles of non-Newtonian fluid mechanics. McGraw-Hill Companies, New York; 1974.
3 Bird RB, Armstrong RC, Hassager O. Dynamics of polymeric liquids. Volume 1: fluid mechanics. Wiley-Interscience, New York; 1987.
4 Larson RG. Constitutive equations for polymer melts and solutions. Butterworths, Boston; 1988.
5 Maxwell B, Nguyen M. Measurement of the elastic properties of polymer melts. Polym Eng Sci. 1979;19(16):1140–50.
6 Doi M, Edwards S. Dynamics of concentrated polymer systems. Part 1.—Brownian motion in the equilibrium state. J Chem Soc Faraday Trans 2 Mol Chem Phys. 1978;74:1789–801.
7 Doi M, Edwards S. Dynamics of concentrated polymer systems. Part 2.—Molecular motion under flow. J Chem Soc Faraday Trans 2 Mol Chem Phys. 1978;74:1802–17.
8 Doi M, Edwards S. Dynamics of concentrated polymer systems. Part 3.—The constitutive equation. J Chem Soc Faraday Trans 2 Mol Chem Phys. 1978;74:1818–32.
9 Doi M, Edwards S. Dynamics of concentrated polymer systems. Part 4.—Rheological properties. J Chem Soc Faraday Trans 2 Mol Chem Phys. 1979;75:38–54.

10 Tapadia P, Wang S-Q. Nonlinear flow behavior of entangled polymer solutions: yieldlike entanglement–disentanglement transition. Macromolecules. 2004;37(24):9083–95.
11 Wang S-Q, Ravindranath S, Wang Y, Boukany P. New theoretical considerations in polymer rheology: elastic breakdown of chain entanglement network. J Chem Phys. 2007;127(6):064903.
12 Ravindranath S, Wang S-Q. Steady state measurements in stress plateau region of entangled polymer solutions: controlled-rate and controlled-stress modes. J Rheol. 2008;52(4):957–80.
13 Huppler J, Macdonald I, Ashare E, Spriggs T, Bird R, Holmes L. Rheological properties of three solutions. Part II. Relaxation and growth of shear and normal stresses. Trans Soc Rheol. 1967;11(2):181–204.
14 Laun H. Description of the non-linear shear behaviour of a low density polyethylene melt by means of an experimentally determined strain dependent memory function. Rheol Acta. 1978;17(1):1–15.
15 Menezes E, Graessley W. Nonlinear rheological behavior of polymer systems for several shear-flow histories. J Polym Sci Part B Polym Phys. 1982;20(10):1817–33.
16 Pearson DS, Kiss AD, Fetters LJ, Doi M. Flow-induced birefringence of concentrated polyisoprene solutions. J Rheol. 1989;33(3):517–35.
17 Osaki K, Inoue T, Isomura T. Stress overshoot of polymer solutions at high rates of shear. J Polym Sci Part B Polym Phys. 2000;38(14):1917–25.
18 Osaki K, Inoue T, Isomura T. Stress overshoot of polymer solutions at high rates of shear; polystyrene with bimodal molecular weight distribution. J Polym Sci Part B Polym Phys. 2000;38(15):2043–50.
19 Osaki K, Inoue T, Uematsu T. Stress overshoot of polymer solutions at high rates of shear: semidilute polystyrene solutions with and without chain entanglement. J Polym Sci Part B Polym Phys. 2000;38(24):3271–6.
20 Pattamaprom C, Larson RG. Constraint release effects in monodisperse and bidisperse polystyrenes in fast transient shearing flows. Macromolecules. 2001;34(15):5229–37.
21 Shaw MT. Introduction to polymer rheology. Wiley, New York; 2012.
22 Phillies GDJ. Phenomenology of polymer solution dynamics. Cambridge University Press, Cambridge; 2011.
23 Ravindranath S, Wang S-Q. Universal scaling characteristics of stress overshoot in startup shear of entangled polymer solutions. J Rheol. 2008;52(3):681–95.
24 Boukany PE, Wang S-Q, Wang X. Universal scaling behavior in startup shear of entangled linear polymer melts. J Rheol. 2009;53(3):617–29.
25 Wang S-Q, Ravindranath S, Wang Y, Boukany P. New theoretical considerations in polymer rheology: elastic breakdown of chain entanglement network. J Chem Phys. 2007;127(6):064903–14.
26 Wang Y, Wang S-Q. Exploring stress overshoot phenomenon upon startup deformation of entangled linear polymeric liquids. J Rheol. 2009;53:1389.
27 Matsuoka S. Relaxation phenomena in polymers. Hanser, New York; 1992.
28 Maxwell B. The application of melt elasticity measurements to polymer processing. Polymer Eng Sci. 1986;26(20):1405–9.
29 Snijkers F, Vlassopoulos D. Cone-partitioned-plate geometry for the ARES rheometer with temperature control. J Rheol. 2011;55(6):1167–86.
30 Wang Y, Li X, Zhu X, Wang S-Q. Characterizing state of chain entanglement in entangled polymer solutions during and after large shear deformation. Macromolecules. 2012;45(5):2514–21.

31 Wang S-Q, Wang Y, Cheng S, Li X, Zhu X, Sun H. New experiments for improved theoretical description of nonlinear rheology of entangled polymers. Macromolecules. 2013;46(8):3147–59.
32 Robertson C, Warren S, Plazek D, Roland C. Reentanglement kinetics in sheared polybutadiene solutions. Macromolecules. 2004;37(26):10018–22.
33 Ferry JD. Viscoelastic properties of polymers. John Wiley & Sons, New York; 1980.
34 Booij HC. Influence of superimposed steady shear flow on the dynamic properties of non-Newtonian fluids. Rheol Acta. 1966;5(3):215–21.
35 Booij HC. Influence of superimposed steady shear flow on the dynamic properties of non-Newtonian fluids. Rheol Acta. 1966;5(3):222–7.
36 Tanner RI, Simmons JM. Combined simple and sinusoidal shearing in elastic liquids. Chem Eng Sci. 1967;22(12):1803–15.
37 Macdonald IF. Parallel superposition of simple shearing and small amplitude oscillatory motions. Trans Soc Rheol. 1973;17(3):537–55.
38 Laufer Z, Jalink HL, Staverman AJ. Dynamic properties of some polymer solutions subjected to a steady shear superimposed on an oscillatory shear flow. Rheol Acta. 1975;14(7):641–9.
39 Laufer Z, Jalink HL, Staverman AJ. Dynamic properties of some polymer solutions subjected to a steady shear superimposed on an oscillatory shear flow. Rheol Acta. 1975;14(7):650–5.
40 Powell RL, Schwarz WH. Rheological properties of aqueous poly(ethylene oxide) solutions in parallel superposed flows. Trans Soc Rheol. 1975;19(4):617–43.
41 Isayev AI, Wong CM. Parallel superposition of small- and large-amplitude oscillations upon steady shear flow of polymer fluids. J Polym Sci Part B Polym Phys. 1988;26(11):2303–27.
42 Tirtaatmadja V, Tam KC, Jenkins RD. Superposition of oscillations on steady shear flow as a technique for investigating the structure of associative polymers. Macromolecules. 1997;30(5):1426–33.
43 Mewis J, Kaffashi B, Vermant J, Butera RJ. Determining relaxation modes in flowing associative polymers using superposition flows. Macromolecules. 2001;34(5):1376–83.
44 Anderson VJ, Pearson JRA, Sherwood JD. Oscillation superimposed on steady shearing: measurements and predictions for wormlike micellar solutions. J Rheol. 2006;50(5):771–96.
45 Kataoka T, Ueda S. Influence of superimposed steady shear flow on the dynamic properties of polyethylene melts. J Polym Sci Part B Polym Phys. 1969;7(3):475–81.
46 de L. Costello BAL. Papers presented at the polymer melt rheology conference parallel superposition rheology of polyethylene as a function of temperature. J Nonnewton Fluid Mech. 1997;68(2):303–9.
47 Somma E, Valentino O, Titomanlio G, Ianniruberto G. Parallel superposition in entangled polydisperse polymer melts: experiments and theory. J Rheol. 2007;51(5):987–1005.
48 Boukany PE, Wang S-Q. Nature of steady flow in entangled fluids revealed by superimposed small amplitude oscillatory shear. J Rheol. 2009;53:1425.
49 Li X, Wang S-Q. Studying steady shear flow characteristics of entangled polymer solutions with parallel mechanical superposition. Macromolecules. 2010;43(13):5904–8.
50 Tapadia P, Wang S-Q. Nonlinear flow behavior of entangled polymer solutions: yieldlike entanglement–disentanglement transition. Macromolecules. 2004;37(24):9083–95.
51 Inn YW, Wissbrun KF, Denn MM. Effect of edge fracture on constant torque rheometry of entangled polymer solutions. Macromolecules. 2005;38(22):9385–8.
52 Sui C, McKenna GB. Instability of entangled polymers in cone and plate rheometry. Rheol Acta. 2007;46(6):877–88.
53 Ge S, Zhu X, Wang S-Q. Watching shear thinning in creep: Entanglement–disentanglement transition. Polymer. 2017;125:254–264.

54 Wang Y, Wang S-Q. From elastic deformation to terminal flow of a monodisperse entangled melt in uniaxial extension. J Rheol. 2008;52:1275–90.
55 Considère A. Mémoire sur l'emploidu fer et de l'acier dans les constructions Ann des Ponts et Chaussées. 1885;9(6):574.
56 Cogswell F, Moore D. A comparison between simple shear, elongation, and equal biaxial extension deformations. Polymer Eng Sci. 1974;14(8):573–6.
57 Pearson GH, Connelly RW. The use of extensional rheometry to establish operating parameters for stretching processes. J Appl Polym Sci. 1982;27(3):969–81.
58 Malkin AY, Petrie C. Some conditions for rupture of polymer liquids in extension. J Rheol. 1997;41(1):1–25.
59 McKinley GH, Hassager O. The Considere condition and rapid stretching of linear and branched polymer melts. J Rheol. 1999;43(5):1195–212.
60 Vincent P. The necking and cold-drawing of rigid plastics. Polymer 1960;1:7–19.
61 Hillmansen S, Haward RN. Adiabatic failure in polyethylene. Polymer. 2001;42(22):9301–12.
62 Hart E. Theory of the tensile test. Acta Metall. 1967;15(2):351–5.
63 Ghosh AK. Strain localization in the diffuse neck in sheet metal. Metall Trans. 1974;5(7):1607–16.
64 Hutchinson J, Neale K. Influence of strain-rate sensitivity on necking under uniaxial tension. Acta Metall. 1977;25(8):839–46.
65 Wang Y, Cheng S, Wang S-Q. Basic characteristics of uniaxial extension rheology: comparing monodisperse and bidisperse polymer melts. J Rheol. 2011;55:1247–70.
66 Wang S-Q, Cheng S, Lin P, Li X. A phenomenological molecular model for yielding and brittle-ductile transition of polymer glasses. J Chem Phys. 2014;141(9):094905.
67 Wang S-Q, Cheng S. Experiments-inspired molecular modeling of yielding and failure of polymer glasses under large deformation. Chapter 12 in Polymer Glasses, edited by Roth C, CRC press, Boca Raton. 2017.
68 Laun H, Münstedt H. Comparison of the elongational behaviour of a polyethylene melt at constant stress and constant strain rate. Rheol Acta. 1976;15(10):517–24.
69 Zhu X, Wang S-Q. Mechanisms for different failure modes in startup uniaxial extension: tensile (rupture-like) failure and necking. J Rheol. 2013;57(1):223–48.
70 Liu G, Sun H, Rangou S, Ntetsikas K, Avgeropoulos A, Wang S-Q. Studying the origin of "strain hardening": basic difference between extension and shear. J Rheol. 2012;57(1):89–104.
71 Wang Y, Boukany P, Wang S-Q, Wang X. Elastic breakup in uniaxial extension of entangled polymer melts. Phys Rev Lett. 2007;99(23):237801.
72 Meissner J. Elongation behavior of polyethylene melts. Rheol Acta. 1971;10:230–42.
73 Fetters L, Lohse D, Richter D, Witten T, Zirkel A. Connection between polymer molecular weight, density, chain dimensions, and melt viscoelastic properties. Macromolecules. 1994;27(17):4639–47.
74 Osaki K, Inoue T, Uematsu T, Yamashita Y. Evaluation methods of the longest Rouse relaxation time of an entangled polymer in a semidilute solution. J Polym Sci Part B Polym Phys. 2001;39(14):1704–12.

8

Strain Hardening in Extension

"Strain hardening" depicts the response of certain polymer melts to startup uniaxial extension. Following the convention in the literature, the transient response of polymer melts is expressed in terms of the time-dependent extensional viscosity

$$\eta_E^+(t) = \sigma_E(t)/\dot{\varepsilon}, \tag{8.1}$$

obtained from rheometric measurements of the Cauchy stress $\sigma_E(t)$ at a given Hencky rate $\dot{\varepsilon}$. Analogous to the normalization method used to express transient stress responses to startup shear, as shown in Fig. 8.1, it is conventional to compare $\eta_E^+(t)$ to its zero-rate limit η_{E0}^+, as shown in Fig. 8.2. If η_E^+ rises above the zero-rate envelope $\eta_{E0}^+(t)$, as shown in Fig. 8.2, then the sample is said to show "strain hardening." As illustrated in Fig. 8.1, the transient shear viscosity $\eta^+(t)$ observed in startup continuous shear typically stays below the zero-rate envelope $\eta_0^+(t)$, a phenomenon described in the literature as strain softening. Polymer melts with long-chain branching (LCB), such as low-density polyethylene (LDPE), usually display strain softening during startup simple shear, but display "strain hardening" in startup uniaxial extension.[1,2] A common explanation in the literature is that strain hardening is caused by chain stretching.[3] At present, there remains considerable confusion about the nature of this "strain hardening" despite a recent effort[4] to clarify the misunderstanding. Here, the behavior illustrated in Fig. 8.2 is different from true tensile strain hardening defined in Section 8.3. In this chapter, we will show that the "strain hardening" simply reflects a lack of complete yielding at high strains and a wrong choice of the zero-rate envelope as the reference. For this reason, we have used quotation marks to remind the reader that the true strain hardening is rather different from the characteristics sketched in Fig. 8.2. Since long-chain branching can delay complete yielding and thus allow the Cauchy stress to build up before the onset of tensile strain localization, melts with long-chain branching show strong "strain hardening" while melts made of linear chains hardly display such "strain hardening."

8.1 Conceptual Pictures

We begin with an important discussion of the effects of uniaxial extension on the structure of the entanglement network. While the details are presented in Appendix 8.A, the key results of the analysis are used here to advance our discussion. According to the packing model, reviewed in Chapter 2, the network density, for example, given in terms of the areal number density of entanglement strands ψ_{ent} Eq. (2.31), is rather sparse. Specifically, since each strand has a molecular cross-sectional area s given in Eq. (2.38a), the fraction ξ of an XY cross section

Nonlinear Polymer Rheology: Macroscopic Phenomenology and Molecular Foundation, First Edition. Shi-Qing Wang.

Companion Website: www.wiley.com/go/nonlinearpolymerrheology

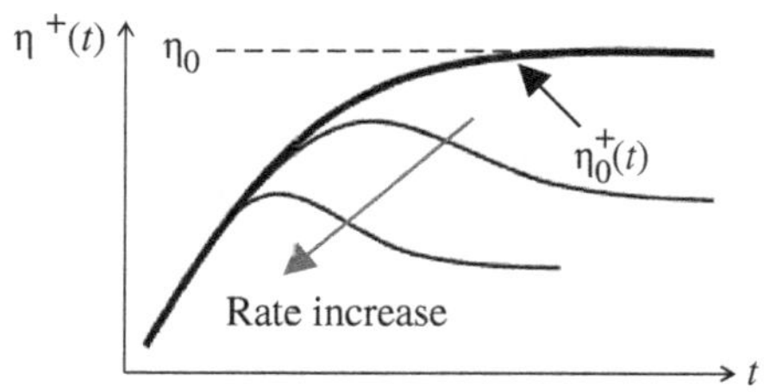

Figure 8.1 Sketch of transient shear viscosity $\eta^+ = \sigma(t)/\dot{\gamma}$ at different rates where the thick curve expresses the zero-shear limit.

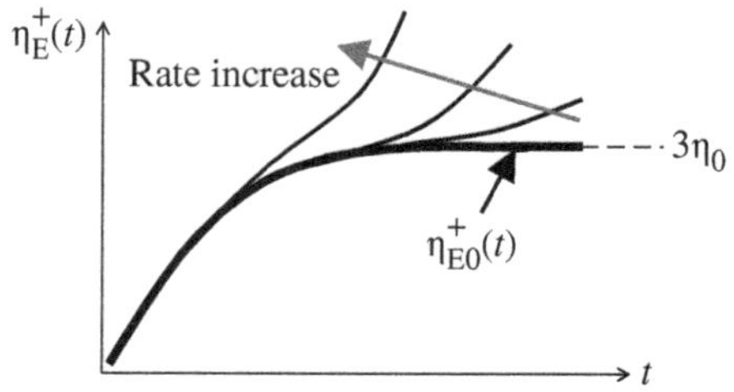

Figure 8.2 Sketch of transient extensional viscosity $\eta_E^+ = \sigma_E(t)/\dot{\varepsilon}$ at different rates where the thick curve expresses the zero-rate limit.

occupied by the entanglement strands is given by $\xi_i = s\psi_{ent} = (N_e)^{-1/2}$. Let us consider affine uniaxial extension of an entangled melt. As the cross-sectional area shrinks as $1/\lambda$, affine uniaxial extension drives the areal number density of entanglement strands to grow in linear proportion to λ. Consequently $\xi = \lambda\xi_i$ and $\psi' = \lambda\psi_{ent}$. Namely, in the affine uniaxial extension limit, entanglement strands are squeezed into an ever-decreasing cross-sectional area. Such a geometric condensation[4] of the entanglement strands reaches a maximum when the entanglement strand is fully extended at λ^*_{eq} (Eq. (8.A.1)), corresponding to $\xi = 1$ and the areal number density changing from the equilibrium value $\psi_{ent} = 1/pl_{ent}$ to $\lambda^*_{eq}\psi_{ent} = 1/s = 1/pl_K$. The detailed analysis is given in Appendix 8.A.

Our zeroth-order explanation of "strain hardening" in melt extension comes from the realization of the basic kinematic difference between shear and extension. In particular, simple shear does not produce any geometric condensation. If the applied shear satisfies $Wi \ll 1$, there is no overshoot. Under this condition, yielding is not forced to occur, and the entanglement network is essentially intact. Consequently, the areal number density of entanglement strands is unchanged and remains the same at ψ_{ent} when viewed in the shear plane XZ, as depicted in (I) of Fig. 8.3. For $Wi > 1$, there can be affine deformation before yielding. But even the affine deformation does not increase the areal number density ψ_{ent} of entanglement strands because the cross-sectional area under load remains constant instead of shrinking. Yielding inevitably occurs so that the areal number density actually decreases as shown in (II) of Fig. 8.3. The

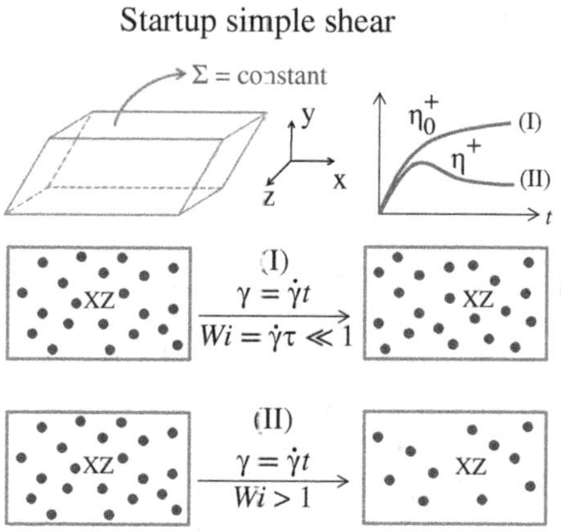

Figure 8.3 States of entanglement in the low-rate ($Wi \ll 1$) regime I and high-rate ($Wi > 1$) regime II, respectively, before and after yielding. Analogous to Eq. (1.55), the shear stress σ is determined by the product of the areal number density ψ of entanglement strands (dots) and the retraction force $f_{retract}$ borne by each strand. In regime II, shear stress decline takes place as the total number of active entanglement strands decreases, leading to the reduced ψ.

normalization of the shear stress by the applied rate, that is, the transient shear viscosity

$$\eta^+(t) = \sigma(t)/\dot{\gamma} \tag{8.2}$$

is conventionally employed to provide a normalized account of the stress response over time t. The shear stress never increases as strongly as linearly with the shear strain $\gamma = \dot{\gamma}t$ because of the yielding signified by the stress overshoot. As shown in Section 7.2, at higher shear rates, the initial linear growth of the shear stress with the strain, characterized by G_{pl}, ceases at shorter times. Thus, for $Wi > 1$, $\eta^+(t) \sim G_{pl}t$ (coming from $\sigma \sim G_{pl}\gamma$) begins to break down at times earlier than τ, while for $Wi \ll 1$, $\eta^+(t)$ grows to $\eta_0 \sim G_{pl}\tau$ at $t \sim \tau$. Namely, for $Wi > 1$, the transient viscosity η^+ involves a time scale significantly shorter than τ, therefore, in Fig. 8.3, curve II stays below curve I.

The strain softening behavior of Fig. 8.1 can be understood in terms of the scaling behavior described in Section 7.2.1. The peak in the transient shear viscosity

$$\eta_y^+ = \sigma_y/\dot{\gamma} = G_{coh}\tau_R(\gamma_y/Wi_R) \tag{8.3}$$

decreases with the shear rate as $\eta_y^+ \sim (Wi_R)^{-2/3}$ because $\gamma_y \sim Wi_R{}^{1/3}$ as indicated by Eq. (7.5). By comparison, on the surface, the opposite trend in apparent viscosity η_y^+ with increasing Hencky extensional rate, depicted as the upward deviation in Fig. 8.2, seems a little hard to understand. The upward deviation is termed "strain hardening," acknowledging a behavior opposite to the strain softening observed in shear. Section 8.2 explores in more detail the origin of strain hardening, distinguishing true strain hardening from the so-called strain hardening seen in Fig. 8.2.

Uniaxial extension actually starts out "normally": There is no geometric condensation for $Wi \ll 1$ because the extension is not affine. Such slow extension leaves the structure of the entanglement network intact, nearly identical to the equilibrium structure. As the cross-sectional area shrinks, the total number of entanglement strands across the cross section decreases in proportion as shown in (I) of Fig. 8.4. Specifically, in the limit of $Wi \ll 1$, uniaxial extension cannot produce geometric condensation because the molecular diffusion renews the entanglement network on time scales much shorter than the time it takes to extend the network. The case of startup extension with $Wi > 1$ is different. Affine extension in entangled melts produces geometric condensation of the entanglement network as shown in Appendix 8.A. Therefore, we cannot use the zero-rate state produced with $Wi \ll 1$ as the reference, namely, we cannot follow the representation that is valid for simple shear without recognizing the kinematic difference between the two different modes of deformation. As shown in (II)

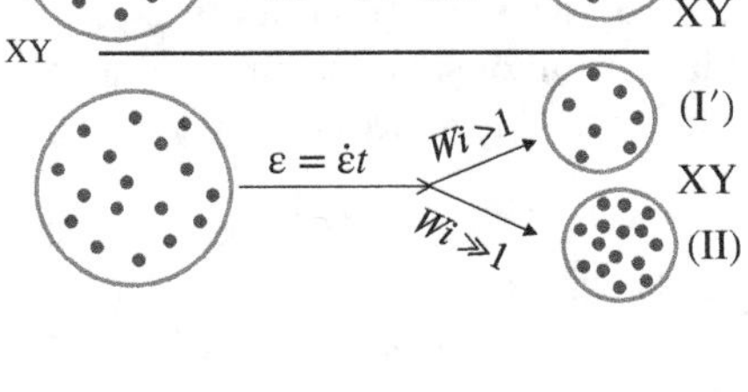

Figure 8.4 States of entanglement in regimes I, I′, and II for $Wi \ll 1$, $Wi > 1$, and $Wi \gg 1$, respectively. Unlike simple shear, there can be a larger areal number density of entanglement strands than is found in an equilibrium system because the loading area, that is, the XY cross section, continuously shrinks during uniaxial extension, as shown in the figure for regime II under the condition of $Wi \gg 1$.

of Fig. 8.4, the entanglement network experiences geometric condensation. Relative to the equilibrium or zero-rate state, the geometric condensation of the entanglement strands leads to the enhanced stress response so that transient extensional viscosity

$$\eta_E^+(t) = \sigma_E/\dot{\varepsilon} \tag{8.4}$$

rises above the envelope of $\eta_{E0}^+(t)$, as indicated in Fig. 8.4. For a larger *Wi*, the upward deviation starts at a shorter time. Even in the presence of yielding via disentanglement, so long as the loss of entanglement strands is not as fast as the shrinkage of the cross-sectional area, the entanglement strand density actually increases. Correspondingly, the response measured in terms of η_E^+ would not drop below the envelope of η_{E0}^+. Specifically, the geometric condensation effect,[4] first depicted in the book in Fig. 7.29, tends to compensate for the continuing disentanglement. Consequently, for entangled *linear chain* polymers, actual areal number density of entanglement strands can remain close to the initial level, shown as (I′) in Fig. 8.4 for $Wi > 1$. In contrast, without the geometric condensation effect, the loss of entanglement reduces the areal number density as shown in case (II) in Fig. 8.3 for simple shear. When $Wi \gg 1$, the geometric condensation can be so strong, as depicted by case II in Fig. 8.4, the lost entanglement is more than compensated by the shrinkage of the cross-sectional area. Therefore, the "strain hardening" takes place because the reference curve (zero-rate) does not contain any effect of geometric condensation, as shown in case (I) in Fig. 8.4.

The phrase "strain hardening" was invented because of the insufficient understanding to reconcile the difference in the responses between shear and extension. It reflects a failure to recognize the kinematic difference. The notion of "strain hardening" stems from an incorrect choice of a reference point. The incorrect representation was made because of the failure to recognize that the stress responses upon startup extension are different, depending on whether *Wi* is much smaller or larger than unity. For simple shear, η_0^+ is a valid reference for η^+ because the sheared area is constant at all stages during startup shear. For uniaxial extension, η_{E0}^+ is a poor reference state because there is no geometric condensation for $Wi \ll 1$, but there is for $Wi \gg 1$. Thus, unlike the case of simple shear, η_{E0}^+ and η_E^+ should not be compared.

Polymers with long-chain branching and entangled polymer mixtures exhibit the most pronounced "strain hardening" for reasons presented in the following section. Melts of linear chains suffer severe yielding and disentanglement at moderately high strain rates. Consequently, little geometric condensation can take place. Therefore, they show little "strain hardening" until $Wi_R \gg 1$, when true strain hardening takes place, as covered in Section 8.3.

8.2 Origin of "Strain Hardening"

8.2.1 Simple Illustration of Geometric Condensation Effect

Whenever yielding is delayed at a large stretching ratio, or yielding is insufficient, the tensile force does not drop sharply, and there will be geometric condensation of entanglement strands in any cross section perpendicular to the direction of extension, as discussed at the start of the previous section and detailed in Appendix 8.A. Consequently, the "normalized" (by rate) Cauchy stress η_E^+ could increase until it was larger than the zero-rate data η_{E0}^+. A mathematical illustration of such "strain hardening" can be straightforwardly demonstrated. For $Wi \ll 1$, η_{E0}^+ can be readily expressed as

$$\eta_{E0}^+ = \eta_0[1 - \exp(-t/\tau)]. \tag{8.5}$$

This formula from the Maxwell model describes the transient behavior all the way to steady flow rather well as shown in Fig. 7.27. For $Wi > 1$, the initial elastic deformation can be approximated by the formula of rubber elasticity from Eq. (1.52), that is,

$$\eta_E^+ = (\eta_0/3Wi)\{\exp[2Wi(t/\tau)] - \exp[-Wi(t/\tau)]\}. \tag{8.6}$$

Plotting Eqs. (8.5) and (8.6) with respect to t/τ in Fig. 8.5, we see that η_E^+ rises above η_{E0}^+ if such elastic extension lasts longer than $\varepsilon = 1$, which arrives at 0.1 and 0.01 of t/τ for $Wi = 10$ and 100, respectively.

Thus, whether "strain hardening" occurs or not merely depends on for how long the rubbery-like elastic stretching can persist before yielding terminates the affine extension. In the affine extension limit, the tensile stress σ_E grows quadratically with the stretching ratio λ. The characteristics of "strain hardening" are closely related to when and how yielding takes place.

8.2.2 "Strain Hardening" of Polymer Melts with Long-Chain Branching and Solutions

8.2.2.1 Melts with LCB

In contrast to linear chains, polymers with long-chain branching resist chain disentanglement, especially in extension. Consequently, yielding in long-chain branched polymers is often partial; the final collapse of the entanglement network is effectively postponed. Such behavior is favorable for stable film blowing and fiber spinning. Low-density polyethylene (LDPE) is a famous example to show "strain hardening" and good processability, while polymer melts of linear chains, for example, high-density polyethylene (HDPE) and linear low-density polyethylene (LLDPE), hardly show "strain hardening" in melt extension. For linear chains, accompanying the shrinking cross section is the loss of chain entanglement due to yielding. Consequently, it is difficult to attain geometric condensation unless the applied Hencky rate is sufficiently high, as discussed in Section 8.3.

In LDPE, "strain hardening" is prevalent, as shown in Fig. 8.6(a) that presents characteristic features of LDPE in response to startup uniaxial extension in the traditional form, that is, transient viscosity as a function of elapsed time t. The "strain hardening" is much simpler to understand if we avoid plotting the data in terms of the transient viscosity. Replotting the data in Fig. 8.6(a) in terms of engineering stress versus strain, as shown in Fig. 8.6(b), we see that σ_{engr} hardly decreases over an extensive range of the stretching ratio λ. This lack of decrease implies that the LDPE resists a transition to flow after the onset of yielding. The presence of such a high level of σ_{engr} indicates that the entanglement network is still connected thanks to the long-chain branching. Because yielding is only partial, there is significant chain stretching that prevents σ_{engr} from plunging. But, the chain stretching is not the cause of the "strain hardening" – chain stretching may instead be understood to be a consequence of the resistance

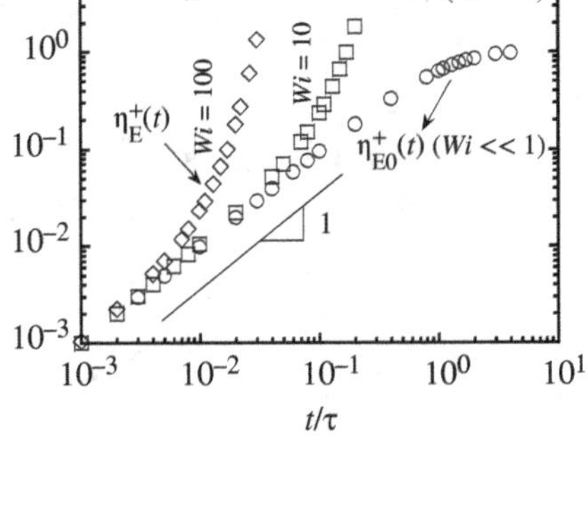

Figure 8.5 Transient viscosity $\eta_E^+(t)$ at a vanishing effective rate ($Wi \ll 1$) according to the Maxwell model of Eq. (8.5) (circles) and at $Wi = 10$ (diamonds) and 100 (squares), respectively, according to Eq. (8.6) for rubber elasticity, resembling Fig. 8.2. The diamonds and squares are also plotted against the Hencky strain on the upper X axes.

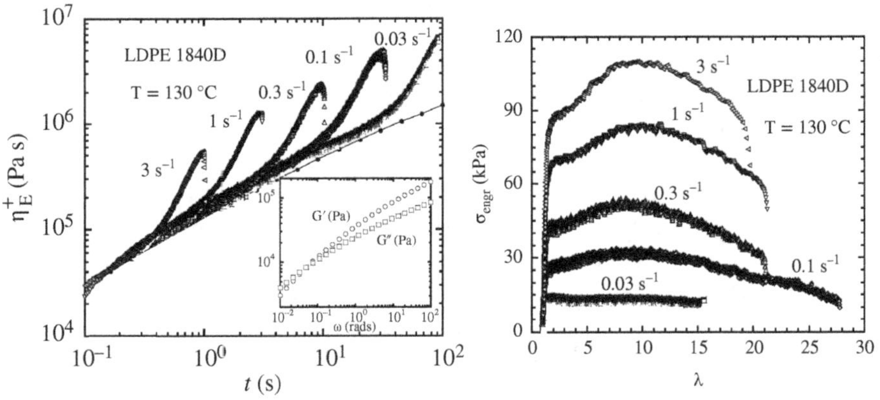

Figure 8.6 (a) Transient viscosity of a low-density polyethylene (LDPE) melt at 130 °C and Hencky rates ranging from 0.03 to 3 s^{-1} from startup uniaxial extension tests, based on an SER apparatus on ARES-G2, where the line with dots is the reference representing the zero-rate envelope. The inset shows the storage and loss moduli of this LDPE. (b) Engineering stress versus stretching ratio, replotted from the identical data in (a). *Source*: Unpublished results by Gengxin Liu at University of Akron.

to yielding. Thus, this confusion concerning "strain hardening" reflects a reversal in causality. Long-chain branching resists yielding and therefore causes such "strain hardening": Because there is significant long-chain branching in LDPE, the sample avoids premature strain localization that would lead to specimen breakup, making it possible to take advantage of geometric condensation.

Our second example is based on a dendritic polyisoprene melt with well-defined architecture. In contrast to the response of melts of linear polymers, as shown in Fig. 7.28, including a sharp engineering stress overshoot, the dendritic melt responds to uniform extension with a nearly constant engineering stress level. As seen in Fig. 8.7, the engineering stress σ_{engr} does not rapidly decline beyond the yield point. Instead, σ_{engr} remains at the yield stress level, even at the two lowest Hencky rates of 0.03 and 0.1 s^{-1}, similar to the behavior of LDPE. If we identify the crossover frequency ω_c as the frequency where $G'(\omega_c) = G''(\omega_c)$ and identify ω_c as the inverse of the terminal relaxation time τ, *Wi* at 0.03 s^{-1} is actually around 0.1. However, as shown in Fig. 8.7, the dendritic PI does not behave like a linear entangled melt that can be described by

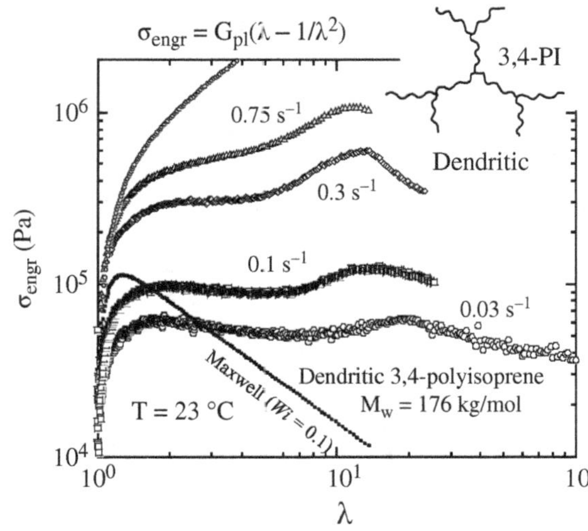

Figure 8.7 Engineering stress versus strain (stretching ratio λ) of a dendritic 3,4-polyisoprene melt, at four Hencky rates from 0.03 to 0.75 s^{-1}, where the top curve is the plot of the rubber elasticity formula. Also plotted is a curve according to the Maxwell model given in Eq. (7.11), with $Wi = 0.1$ and $G_{pl} = 0.53$ MPa, which apparently cannot depict the rheological response well after the yield point. *Source*: Data taken from Liu *et al*. 2013 [4].

the Maxwell ($Wi = 0.1$) curve (cf. Fig. 7.27). The Maxwell curve shows that σ_{engr} sharply declines with increasing λ at $\lambda \sim 1.5$. The lack of the engineering stress decline in Fig. 8.7, in contrast to the sharp drop in Fig. 7.28 for linear chain melts, indicates that the long-chain branching may have significantly deterred breakup of the entanglement network, causing the transient viscosity to increase rapidly with increasing extension due to geometric condensation.

8.2.2.2 Entangled Solutions of Linear Chains

Relative to melts of linear chains, entangled solutions of linear chains may resist yielding and collapse of the entanglement network more effectively. The reason may be that the entanglement strands are longer in solutions, for example, $l_{ent}(\phi) = l_{ent}(\phi = 1)\phi^{-0.6\sim0.65}$ according to Eq. (2.37a), where ϕ is the volume fraction of the high-molecular-weight polymer. Thus, at a given stretching ratio λ, the longer strands in solution are farther away from the finite extensibility limit and can more effectively avoid buildup of high retraction force. The weaker retraction force delays the onset of yielding to larger values of λ. Figure 8.8 shows four sets of data on the transient viscosity η^+ of a styrene–butadiene rubber along with its three binary mixtures. In this representation, the SBR1M melt hardly shows any sign of upward deviation from the scaling limit $\eta_0^+ \sim t$, which is the zero-rate envelope. In contrast, SBR1M(40%) shows a significant upward departure, which is a sign of the "strain hardening." The lower the concentration, the more resistant to yielding the mixture appears to be.

To see the "strain hardening" more clearly in SBR1M(40%), we can add the reference curve, that is, the data at the lowest Hencky rate of 0.03 s^{-1} as shown in Fig. 8.9. The curve for 0.3 s^{-1} is significantly above the data for 0.03 s^{-1}. Specimen breakup, macroscopic material separation, terminates the homogeneous extension, preventing further growth of η^+ for 0.3 s^{-1} beyond the values shown in Fig. 8.9.

In filament stretching rheometry (FSR),[6] uniaxial extension is not globally homogeneous. Specifically, in the filament stretching device, the tensile extension is carried out by adjusting the velocity of the filaments' two ends to ensure that the middle section of the specimen shrinks exponentially, in accordance with the prescribed Hencky rate. The use of FSR for extensional rheological studies will be discussed in more detail in the Introduction of Chapter 11 and in Sections 11.4 and 11.A. For now, it suffices to note that the filament stretching rheometer can produce pronounced "strain hardening" in linear polymer solutions because an FSR test does not terminate even when the thinnest section tends to draw faster. Thus, even melts of linear

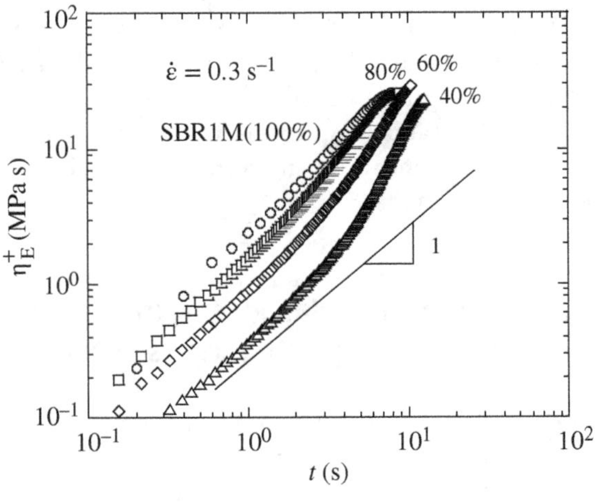

Figure 8.8 Transient viscosity at Hencky rate 0.3 s^{-1} of SBR1M along with its binary mixtures with SBR20K at volume fractions of $\phi = 0.8$, 0.6, and 0.4, respectively, showing "strain hardening" becoming stronger with decreasing ϕ. All four samples have comparable Rouse time. But the reptation time smoothly decreases from 11,000 to 4,100 s with lowering ϕ at room temperature. *Source*: Data taken from Wang *et al.* 2011 [5].

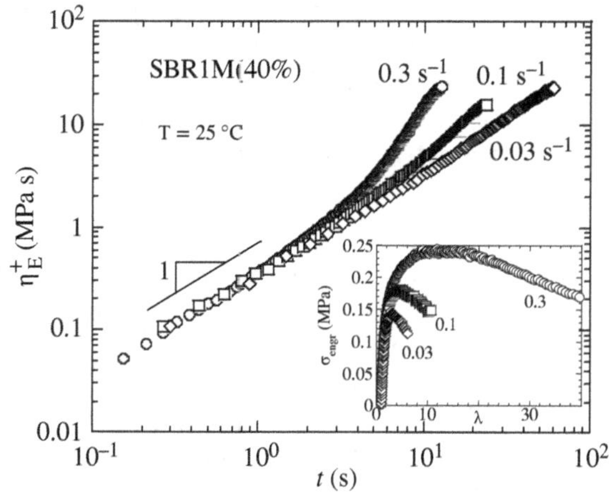

Figure 8.9 Transient viscosity of SBR1M(40%) at three rates 0.03, 0.1, and 0.3 s^{-1} as well as the corresponding representation in terms of the engineering stress σ_{engr} versus strain (λ) in the inset. *Source*: Data taken from Wang *et al.* 2011 [5].

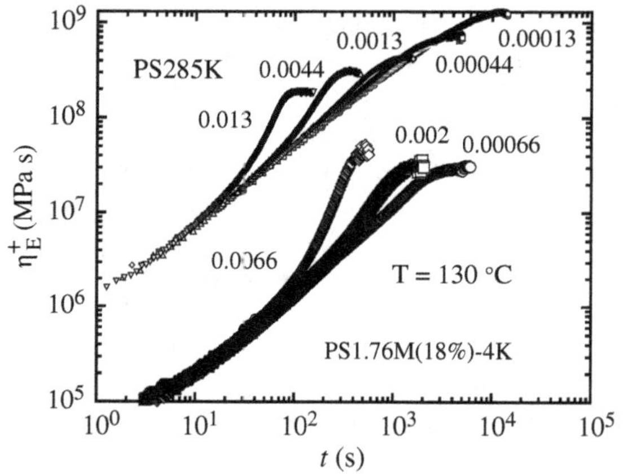

Figure 8.10 Transient extensional viscosity versus time of PS285K melt and PS1.76M(18%)-4K mixture, obtained from filament stretching rheometric measurements at 130 °C, at the equivalent rates defined as the product of Hencky rate and a characteristic relaxation time similar to τ_e. *Source*: Data taken from Huang *et al.* 2015 [7].

chains show mild "strain hardening" in the FSR. Consistent with Fig. 8.8, a polystyrene solution shows stronger "strain hardening" than a PS melt at comparable effective shear rates, as shown in Fig. 8.10. Section 17.5 further contrasts melts with solutions in uniaxial extension. In Section 17.5, we hope to clarify some significant confusion in the literature that stems from this difference in the extensional responses between entangled melts and solutions.

So far, we have emphasized that "strain hardening" in extension versus "strain softening" in shear partially reflects a kinematic effect; uniaxial extension produces geometric condensation while simple shear does not do so. It is also plausible that an entanglement network can better resist yielding in uniaxial extension than in simple shear. Especially for polymer melts with long-chain branching, the onset of strain localization is significantly delayed in uniaxial extension; consequently, the engineering stress does not diminish so that geometric condensation effect builds up.

8.3 True Strain Hardening in Uniaxial Extension: Non-Gaussian Stretching from Finite Extensibility

When a Gaussian chain network is stretched to the extent that the intrachain force in an entanglement strand grows more strongly than the rubber elasticity formula given in Eq. (7.14), the engineering stress σ_{engr} can increase with λ more rapidly than linearly. We would call such behavior true strain hardening because the sample becomes mechanically stiffer with increasing extension. In the following text, we discuss when true strain hardening takes place.

Entangled linear polymer melts can be stretched so fast in uniaxial extension that they lose their ability to fully yield through chain disentanglement. When Wi_R is well above unity, the yielding of well-entangled melts, even of linear chains, is only partial, in the sense that "strain softening," that is, downward deviation of the stress from the rubber elasticity formula of Eq. (1.52), ceases with increasing extension. Strain softening is replaced by a monotonic increase in the engineering stress until the point of melt rupture. Such behavior was extensively studied by Vinogradov *et al.*[8–13] and has since been routinely reported.[4,5,14–16] At high strain rates, chain disentanglement fails to spread along the chain. After losing some entanglements, the remaining entanglements become stronger, perhaps due to the tightened topological constraints. The surviving entanglement strands can extend until the limit of finite extensibility. When a strand is no longer a Gaussian coil, the retraction force within it grows with the stretching ratio much more rapidly than linearly. Such non-Gaussian stretching is the cause of true strain hardening, where σ_{engr} increases more than linearly with λ.

True strain hardening, that is, the upward stress take-off above the curve given by Eq. (1.52), cannot be readily observed in melt extension of "fast" melts such as linear polyethylene, because commercial instruments are rather limited in the highest uniaxial extension rates that they can produce. Linear high-3,4-content polyisoprene and styrene-butadiene rubber (SBR) with large vinyl content are convenient candidates for studying true strain hardening because their molecular relaxation times are quite long at room temperature. In these systems, the equilibrium entanglement strand is usually rather short, in the sense that in the affine stretching limit the strand would become fully stretched at a stretching ratio of about 4.[5] However, the surviving entanglement strands are much longer than the original equilibrium strands and therefore can withstand much greater stretching before reaching the non-Gaussian regime. Detailed depiction of this limit is deferred to Section 11.2.2. Here it suffices to show how, in startup uniaxial extension of linear SBR melts, true strain hardening arises for $Wi_R > 1$. For example, at Hencky rates of 10 and 15 s^{-1}, the engineering stress σ_{engr} turns upward after the initial partial yielding, as shown in Fig. 8.11(a). So the melt extension at 10 and 15 s^{-1} shows true strain hardening.

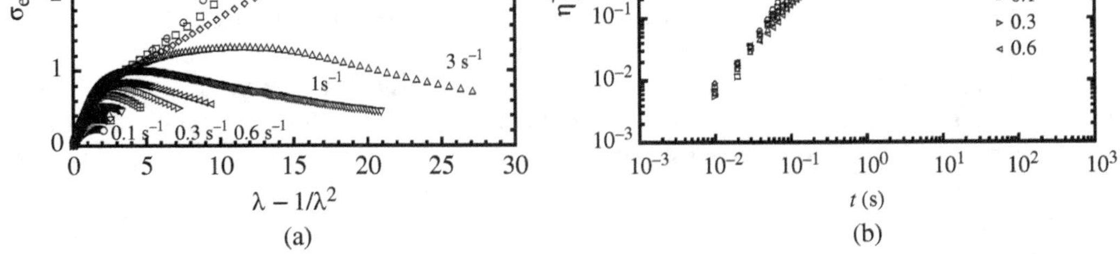

Figure 8.11 (a) Engineering stress σ_{engr} versus strain, expressed as $(\lambda - 1/\lambda^2)$, for a wide range of Hencky rate from 0.001 to 15 s^{-1}, covering tensile failure and melt rupture across the spectrum of different responses (cf. Chapter 11). The last data points in each curve represent the onset of visible tensile strain localization. *Source*: Data taken from Zhu *et al.*, 2013 [16]. (b) Transient extensional viscosity as a function of time, based on the data in (a).

When represented in terms of the transient viscosity, the data stay above the zero-rate envelope even for a Hencky rate of $3\,s^{-1}$, as shown in Fig. 8.11(b). The data at $3\,s^{-1}$ satisfies the conventional characterization of "strain hardening." Such false strain hardening takes place as long as the yielding does not cause σ_{engr} to drop as rapidly as $1/\lambda$.

8.4 Different Responses of Entanglement to Startup Extension and Shear

The contrast between Figs. 8.1 and 8.2 has caused a significant puzzlement in workers who did not entirely appreciate kinematic difference between simple shear and uniaxial extension. How do we compare the two modes of deformation on an equal footing? If we wish to have the stress response reveal something useful about the state of entanglement, we should choose to examine the nonlinear rheological response in terms of the engineering stress σ_{engr} because this representation involves no geometric condensation. As mentioned in Section 7.5.2 and shown in Fig. 7.29, σ_{engr} removes the confusion created by the persistent use of Cauchy stress σ_E to represent the rheological state. If yielding takes place because of chain disentanglement, σ_{engr} shows a decline while σ_E may not do so.

On the other hand, at sufficiently high Hencky rates, for example, $Wi_R \sim 10$, startup extension of entangled melts produces a monotonic increase in σ_{engr} with growing λ, as shown in Fig. 8.11(a), as an exception to the depiction shown in Fig. 7.26(b). In comparison, at the same $Wi_R \sim 10$, startup shear only produces yielding and shear softening in the form of a shear stress overshoot. This contrast between extension and shear is due to the kinematic difference between shear and extension in the following sense: Given the same Wi_R, for the same amount of elapsed time t and in the affine deformation limit, the same strain $Wi_R(t/\tau_R)$ leads to greatly different degrees of molecular deformation. For example, at $\varepsilon = \gamma = 3$, the sample is extended by a factor of 20 in uniaxial stretching; the sample is only displaced three times its thickness in simple shear. Apparently, entanglement can get locked up in startup extension but can escape in startup shear.[17]

8.5 Conclusion

In conclusion, we have devoted a chapter to the subject of the so-called "strain hardening" because of the considerable confusion in the literature about the phenomenon. This false strain hardening is not caused by chain stretching, and the important difference between simple and uniaxial extension is kinematic in origin. Often, a total collapse of the entanglement network is delayed in extension relative to shear. With the postponed breakdown of the entanglement network, entangled melts show increasing Cauchy stress even though the tensile force no longer grows with further extension. Here geometric condensation is a key concept that has been overlooked in the literature until recently.[4] Because this notion of geometric condensation is so important and has been applied to discuss molecular mechanism of mechanical ductility in polymer glasses,[18,19] we have made a thorough analysis in Appendix 8.A. This analysis is placed in an appendix only because it is somewhat technical not because it is any less important than any other content in this chapter.

8.A Conceptual and Mathematical Accounts of Geometric Condensation

Regarding an entanglement strand as a Gaussian coil, we can readily determine, in the affine deformation limit, the effect of uniaxial extension on the structure of the entanglement network whose equilibrium state has been described in Section 2.2.3.1 in terms of the packing model.

Let us consider uniaxial extension of an entangled polymer melt, under which the entanglement strand extends in the affine deformation as shown in Fig. 8.A1(a). Within the space occupied by the pervaded volume of an entanglement strands, namely, in $\Omega_e \sim (l_{ent})^3$, there are universally Q_e such strands. According to Eq. (2.28), $Q_e \sim l_{ent}/p$. Before extension, the areal number density of entanglement strands is ψ_{ent}, which has been derived in Eq. (2.31). Under affine deformation, geometric condensation takes place, as shown in Fig. 8.A1(b), as the cross-sectional area in the XY-plane shrinks reciprocally with the imposed stretching ratio λ. Thus, the areal number density ψ' increases linearly with λ, as illustrated in Fig. 8.A1(b). The strand is fully straightened at the Kuhn segment level to a length of $L_e = N_e l_K$ when λ approaches

$$\lambda^*_{eq} = N_e l_K / l_{ent} = \sqrt{N_e} = l_{ent}/l_K. \tag{8.A1}$$

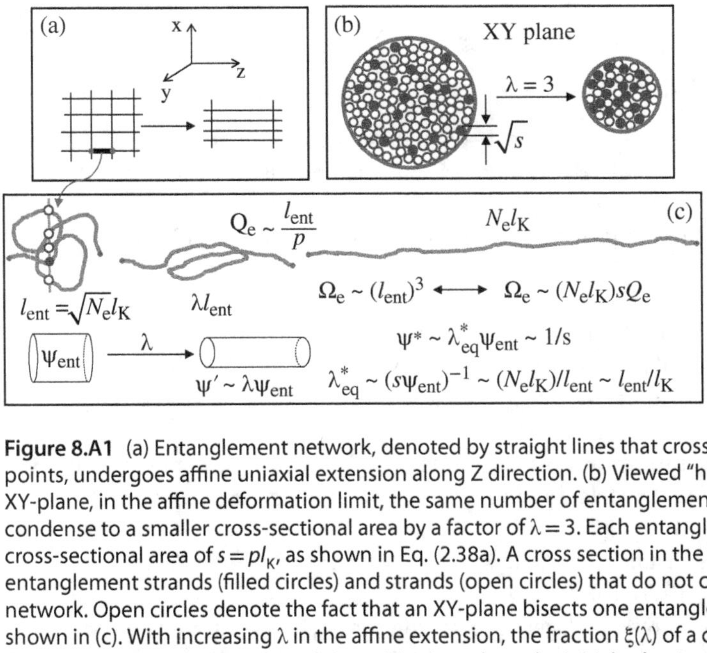

Figure 8.A1 (a) Entanglement network, denoted by straight lines that cross one another at the entanglement points, undergoes affine uniaxial extension along Z direction. (b) Viewed "head-on" along Z-axis in the XY-plane, in the affine deformation limit, the same number of entanglement strands (filled circles) is seen to condense to a smaller cross-sectional area by a factor of $\lambda = 3$. Each entanglement strand has a molecular cross-sectional area of $s = pl_K$, as shown in Eq. (2.38a). A cross section in the XY-plane is occupied by both entanglement strands (filled circles) and strands (open circles) that do not contribute to the entanglement network. Open circles denote the fact that an XY-plane bisects one entanglement strand multiple times as shown in (c). With increasing λ in the affine extension, the fraction $\xi(\lambda)$ of a cross-sectional area occupied by the filled circles increases linearly with λ as $\xi(\lambda) = \lambda\xi_i$, where the initial value ξ_i at $\lambda = 1$ is given by $\xi_i = s\psi_{ent}$, with ψ_{ent} given by Eq. (2.31) and s by Eq. (2.38a). The inverse of this fraction is exactly the number q of times an entanglement strand folds back and forth within its physical volume across an area of $\pi(l_{ent})^2$ according to Eq. (2.39a): $q(n_e) = 1/\xi_i$. (c) In the undeformed state, the Gaussian strand has a coil size l_{ent} with N_e Kuhn segments (Eq. 2.29). From the coiled state with an areal number density ψ_{ent} of entanglement strands, in the affine extension limit, an entanglement strand stretches until it is fully extended at λ^*_{eq} defined in Eq. (8.A1). In this limit of full extension of the entanglement strand, $\xi(\lambda^*_{eq}) = \lambda^*_{eq}\xi_i = 1$ so that the cross section is all occupied by entanglement strands. The mass density conservation means the pervaded volume Ω_e at $\lambda = 1$ should be the same as the volume given by piling up these straightened strands (of length $N_e l_K$ and cross-sectional area s) at λ^*_{eq}, namely, $(N_e l_K) s Q_e = N_e l_K \times p l_K (\Omega_e/v_e) \equiv \Omega_e$ because the physical volume of an entanglement strand is $v_e \equiv (l_{ent})^2 p$, where use is made of Eqs. (2.29) and (2.38a).

From the coil to the straightened state, the cross-sectional area decreases by a factor of λ^*_{eq}, but the number of strands in the cross section remains unchanged in the limit of affine deformation or in the limit of no chain disentanglement. Instead of passing through the cross section multiple times, for example, five or three times as shown in Fig. 8.A1(c), an entanglement strand only passes once in the limit of full extension. In this limit, by definition, on average, there should be one strand per area of s. Thus, the area density should naturally be given by $1/s$, which is indeed the case because

$$\psi^* \sim \lambda^*_{eq}\psi_{ent} \sim \sqrt{N_e}/pl_{ent} = 1/pl_K = 1/s, \tag{8.A2}$$

where use is made of Eq. (2.31) for ψ_{ent}. The present discussion demonstrates that Gaussian coils are highly ineffective in building a molecular network. To fully exploit the molecular strength, it is desirable to produce geometric condensation as shown in Fig. 8.A1(b,c). Fiber spinning and many other industrial processes utilize this principle. When affine deformation ceases, the surviving entanglement strands are longer with $N_e' > N_e$, the preceding analysis will need to be made in terms of the new parameter N_e'. For example, we rewrite Eq. (8.A1) as $\lambda^* = \sqrt{N_e'} > \lambda^*_{eq}$ in Eq. (11.5).

References

1 Meissner J. Modifications of the Weissenberg rheogoniometer for measurement of transient rheological properties of molten polyethylene under shear. Comparison with tensile data. J Appl Polym Sci. 1972;16(11):2877–99.

2 Münstedt H, Laun H. Elongational properties and molecular structure of polyethylene melts. Rheol Acta. 1981;20(3):211–21.

3 McLeish TCB, Larson RG. Molecular constitutive equations for a class of branched polymers: the pom-pom polymer. J Rheol. 1998;42(1):81–110.

4 Liu G, Sun H, Rangou S, Ntetsikas K, Avgeropoulos A, Wang S-Q. Studying the origin of "strain hardening": basic difference between extension and shear. J Rheol. 2013;57(1):89–104.

5 Wang Y, Wang S-Q. Salient features in uniaxial extension of polymer melts and solutions: progressive loss of entanglements, yielding, non-Gaussian stretching, and rupture. Macromolecules. 2011;44(13):5427–35.

6 Tirtaatmadja V, Sridhar T. A filament stretching device for measurement of extensional viscosity. J Rheol. 1993;37(6):1081–102.

7 Huang Q, Hengeller L, Alvarez NJ, Hassager O. Bridging the gap between polymer melts and solutions in extensional rheology. Macromolecules. 2015;48(12):4158–63.

8 Vinogradov G, Yanovskii YG, Borisenkova E, Yarlykov B, Berezhnaya G. Viscoelastic properties and flow of narrow distribution polybutadienes and polyisoprenes. J Polym Sci Part B Polym Phys. 1972;10(6):1061–84.

9 Vinogradov G. Viscoelasticity and fracture phenomenon in uniaxial extension of high-molecular linear polymers. Rheol Acta. 1975;14(10):942–54.

10 Vinogradov G, Malkin AY, Volosevitch V, editors. Some fundamental problems in viscoelastic behavior of polymers in shear and extension. J Appl Polym Sci Appl Polym Symp. 1975;27:47–59.

11 Vinogradov G, Volosevitch V, Shatalov V, Yudin V. Flow, high-elastic (recoverable) deformation, and rupture of uncured high molecular weight linear polymers in uniaxial extension. J Polym Sci, Polym Phys Ed. 1975;13(9):1721–35.

12 Vinogradov G. Ultimate regimes of deformation of linear flexible chain fluid polymers. Polymer. 1977;18(12):1275–85.
13 Vinogradov G, Dreval V, Borisenkova E, Kurbanaliev M, Shalganova V. Uniaxial extension of linear flexible-chain polymers in an extremely broad range of stresses and strain rates. Rheol Acta. 1981;20(5):433–42.
14 Wang Y, Wang S-Q. Rupture in rapid uniaxial extension of linear entangled melts. Rheol Acta. 2010;49(11–12):1179–85.
15 Wang Y, Cheng S, Wang S-Q. Basic characteristics of uniaxial extension rheology: Comparing monodisperse and bidisperse polymer melts. J Rheol 2011;55:1247–70.
16 Zhu X, Wang S-Q. Mechanisms for different failure modes in startup uniaxial extension: tensile (rupture-like) failure and necking. J Rheol. 2013;57(1):223–48.
17 Sun H, Wang S-Q. Shear and extensional rheology of entangled polymer melts: similarities and differences. Sci. China-Chem. 2012;55(5):779–86.
18 Wang S-Q, Cheng S, Lin P, Li, X. A phenomenological molecular model for yielding and brittle-ductile transition of polymer glasses. J Chem Phys 2014; 141(9):094905.
19 Wang S-Q, Cheng S. Experiments-inspired molecular modeling of yielding and failure of polymer glasses under large deformation. Chapter 12 in Polymer Glasses, edited by Roth C, CRC press, Boca Raton, 2017.

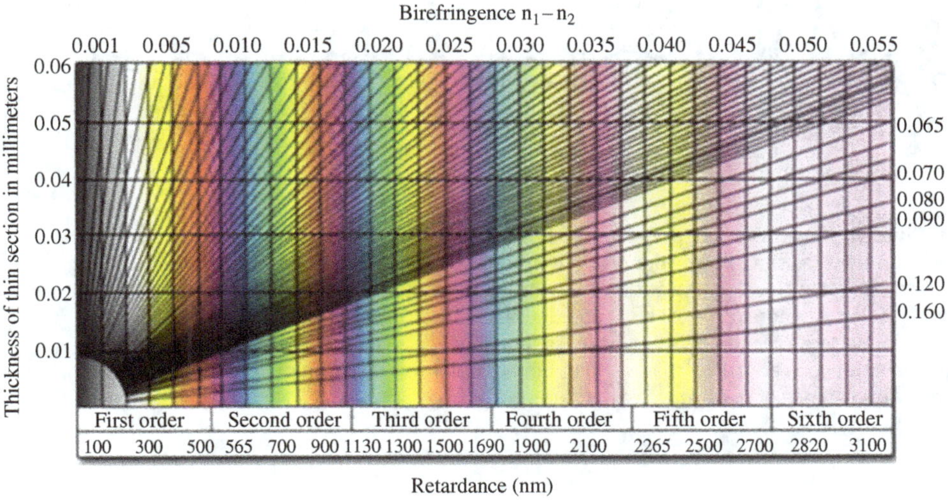

Figure 3.7 Michel-Levy birefringence chart from http://www.olympusmicro.com/primer/techniques/polarized/michel.html. *Source:* Reproduced with permission of Olympus America Inc.

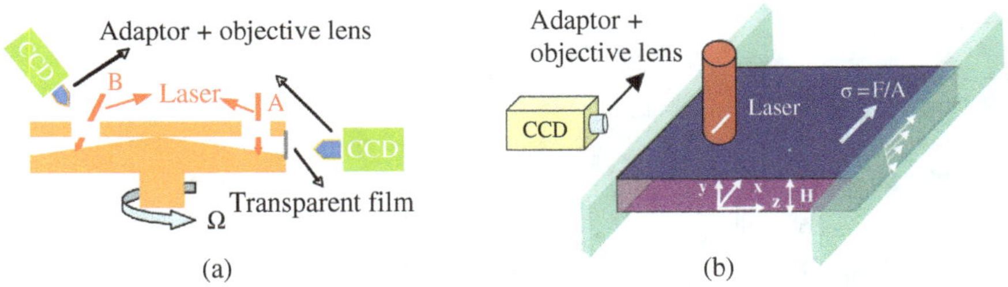

Figure 4.4 (a) Illustration of a cone-plate apparatus coupled to particle-tracking velocimetry (PTV). *Source:* Wang 2007 [4]. Reproduced with permission of Wiley. (b) PTV setup for a sliding-plate shear cell. *Source:* Boukany and Wang 2007 [8]. Reproduced with permission of American Institute of Physics.

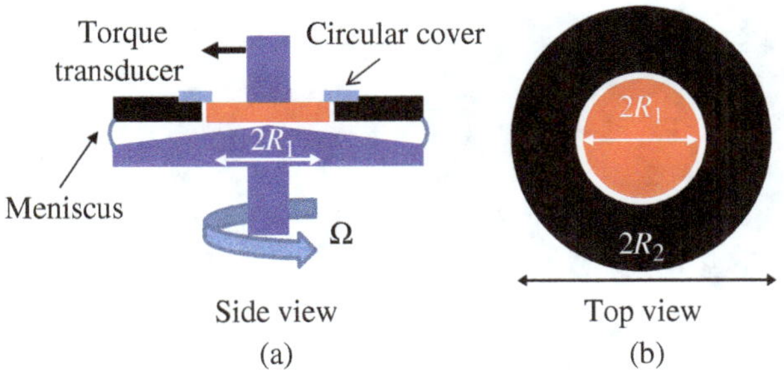

Figure 5.2 Cone-partitioned plate (CPP) device is made of a large cone (of radius R_2) against a partitioned plate that is made of a smaller disk of radius R_1 and a ring of width approximately equal to $(R_2 - R_1)$. (a) Side view showing the rotating cone, the stationary upper plate of radius R_1 and a ring of inner radius $R_1 + \Delta$ and outer radius R_2 where Δ is the gap distance between the disk and ring and (b) top view of the partitioned plate.

Nonlinear Polymer Rheology: Macroscopic Phenomenology and Molecular Foundation, First Edition. Shi-Qing Wang.

Companion Website: www.wiley.com/go/nonlinearpolymerrheology

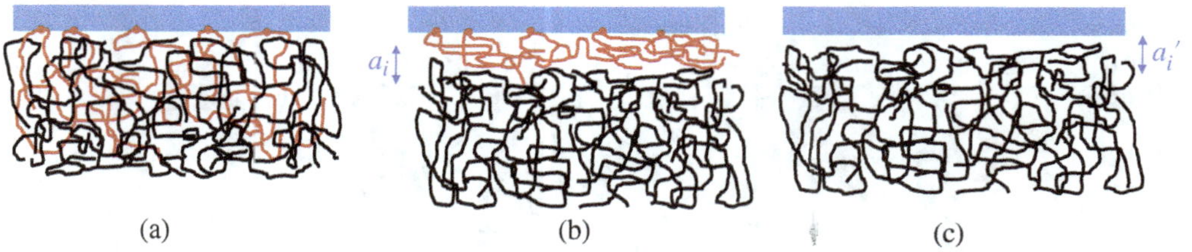

Figure 6.1 Different states at various polymer/wall interfaces: (a) in quiescence with strong polymer adsorption or under weak shear (corresponding either to slow shear rate or to low shear stress) where the depiction is somewhat misleading: the adsorbed chains are actually still inter-penetrating into the unbound chains (see Fig. 16.6 for a more realistic illustration of the interfacial layer) and unbound chains are disentangling in the sense that there is significant sliding between adsorbed and free chains; (b) full disentanglement between adsorbed chains and unbound (bulk) chains during complete wall slip; (c) chain desorption in the presence of shear at weakly adsorbing interfaces.

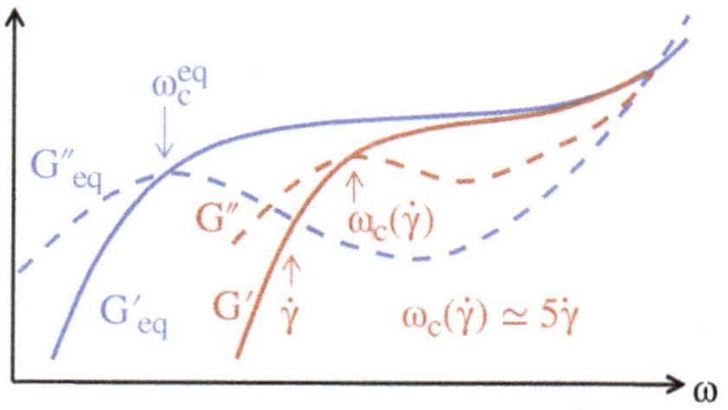

Figure 7.16 Sketches of the storage and loss moduli G'_{eq} and G''_{eq} for a quiescent entangled polymer solution (blue lines) as well as G′ and G″ extracted from SAOS superimposed onto steady shear at rate $\dot\gamma$ (red lines).

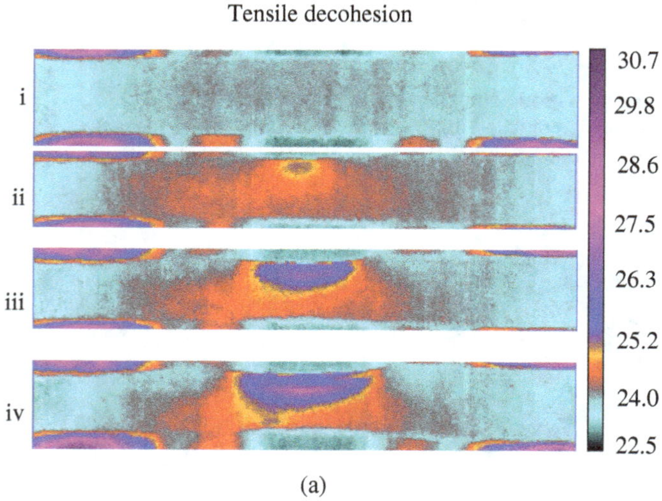

Figure 11.5 (a) Infrared snapshots of an extending specimen (aqua) in a Sentmanat extensional rheometer (SER), taken with an IR thermal imaging camera (FLIR SC325), during startup uniaxial extension of SBR616K at room temperature, at a Hencky rate of 0.06 s^{-1}. Here the color codes on the right-hand side indicate the actual sample temperatures at four different stages. The specimen's length in the SER, which is the length of the photos, is around 1.3 cm. Unpublished observations by Panpan Lin at University of Akron.

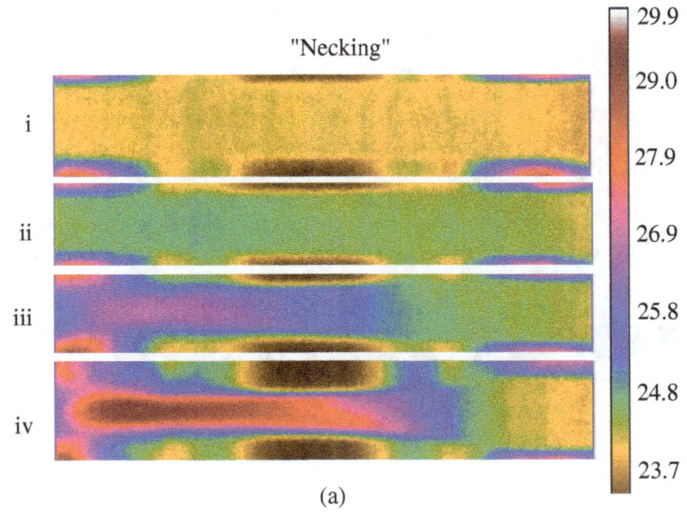

Figure 11.8 (a) Snapshots from the IR thermal imaging camera (FLIR SC325) during startup uniaxial extension of SBR616K at room temperature, at a Hencky rate of 0.6 s^{-1}, where the color codes on the right-hand side indicate the actual sample temperatures at four different stages. The photo length is the sample length, about 1.3 cm. The specimen is initially orange in color. Unpublished observations by Panpan Lin at University of Akron.

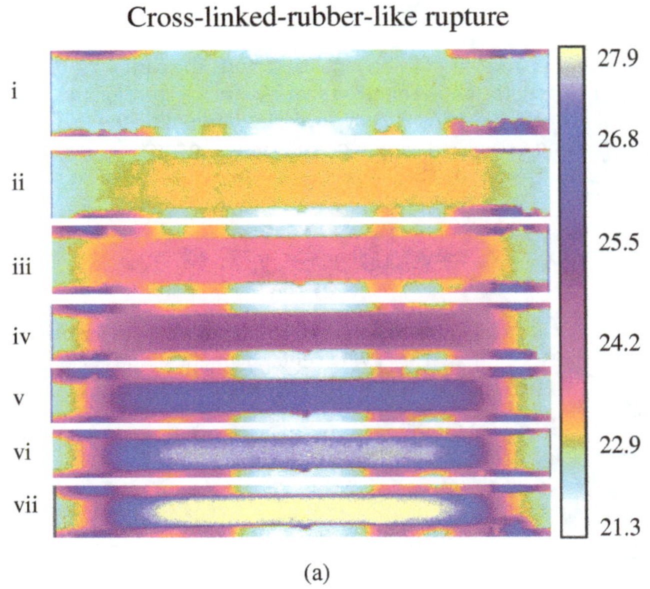

Figure 11.13 (a) Snapshots (initially aqua) from an IR thermal imaging camera during startup uniaxial extension of SBR153K at 22 °C at a Hencky rate of 0.3 s^{-1}, where the color codes on the right-hand side indicate the actual sample temperatures at seven different stages. *Source:* Reproduced from unpublished observations by Panpan Lin at University of Akron.

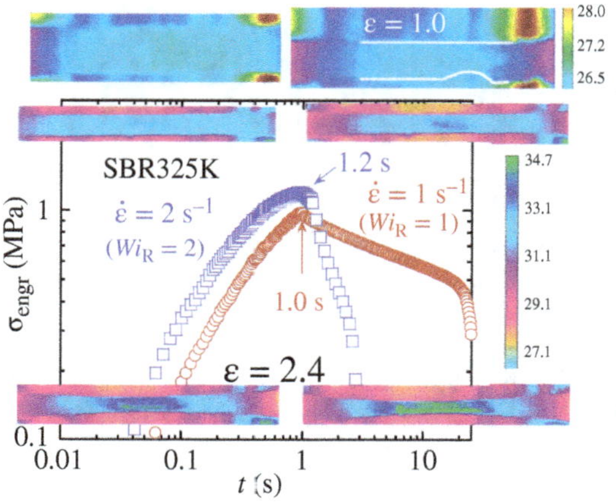

Figure 13.9 Tensile decohesion (from step extension produced at 1 s^{-1} for 1 s) or necking (at 2 s^{-1} for 1.2 s) due to shear yielding after step extension of SB325K. The ambient temperature is 26 °C. The first two color photos at the top show the temperature profiles, respectively, during step extension (ε = 1.0) and after the emergence of tensile decohesion where temperature rise is insignificant, and the white lines indicate the edges of the specimen. The rest of four color photos show the specimen immediately after the step extension of ε = 2.4 (the first two) and onset of necking moments later (the last two), showing a temperature rise to as high as 35 °C. *Source:* Data taken from Sun *et al.* 2015 [6].

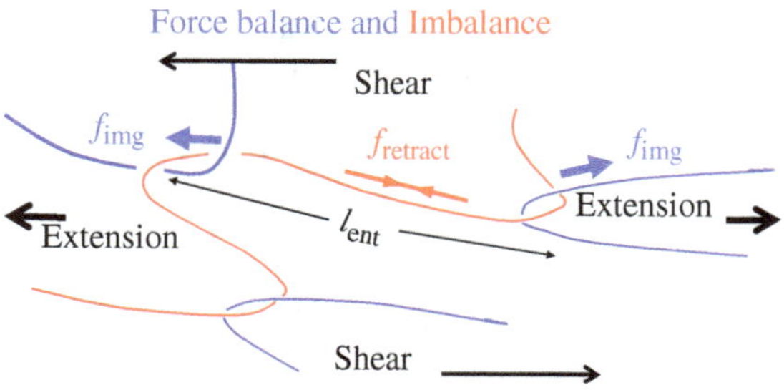

Figure 16.2 A cartoon to depict an entangled polymer in terms of a network-like picture. The cartoon shows that the red strand between two entanglement points gets strained either in shear to γ or in extension to λ. The elastic retraction force $f_{retract}$ ($\sim\gamma f_{ent}$ for simple shear and $\sim(\lambda - 1/\lambda^2)f_{ent}$ for uniaxial extension) originates from the molecular deformation, the deformation occurring because the intermolecular gripping force f_{img} prevents chain sliding at the entanglement points. *Source:* Wang *et al.* 2013 [4]. Reproduced with permission of American Chemical Society.

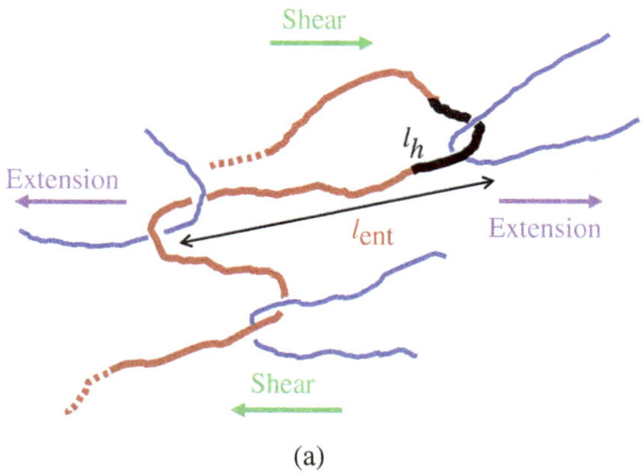

Figure 16.4 (a) Sketch of large chain deformation for $Wi_R > 1$ when entanglement strands of length l_{ent} arise from pairs of hairpins whose characteristic size is l_h.

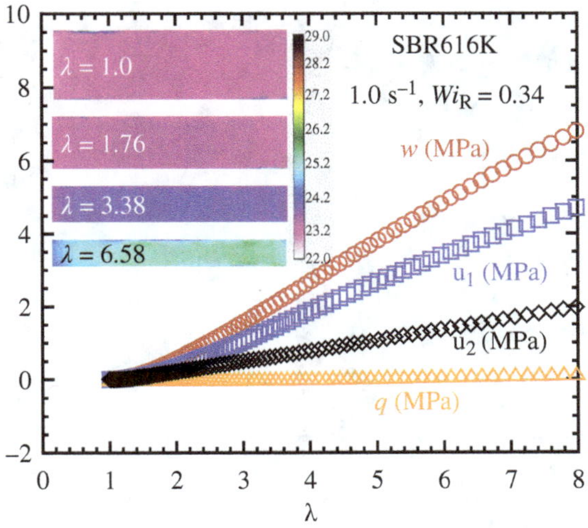

Figure 17.3 Mechanical work density *w* (circles) and internal stored energy density u_2 (diamonds) as well as u_1 and *q* as a function of stretching ratio λ. Since *q* is negligibly low relative to *w*, the test is essentially adiabatic. The inset shows the IR images at the different stages of extension and corresponding temperature rise. *Source:* Data taken from Lin *et al.* 2017 [6].

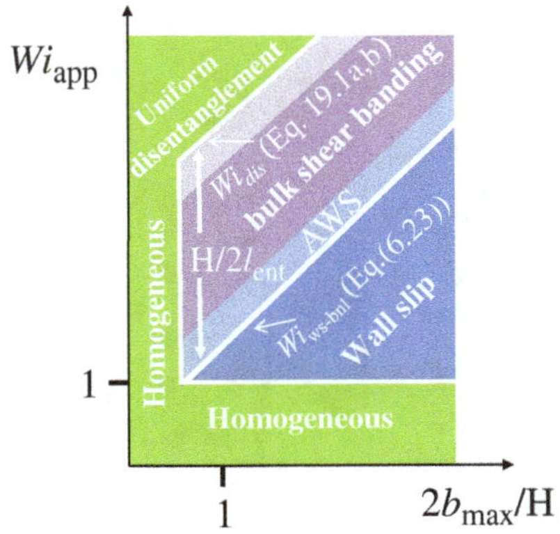

Figure 19.1 Diagram in the phase space of Wi_{app} by $2b_{max}$/H on double log scales to show, at a fixed value of H in steady state, the different rheological responses in terms of the state of shear strain field, reproduced from Ref. [3]. Above $Wi_{ws\text{-}bnl}$, Weissenberg number $Wi > 1$. During high shear, disentanglement could start from the polymer/wall interfaces, producing apparent wall slip (AWS), as shown in the lighter shaded region, instead of any spatially resolvable shear banding in the bulk, which occurs only when Wi_{app} is significantly larger than $Wi_{ws\text{-}bnl}$. A lighter shade is also used to indicate that bulk banding may be replaced by homogeneous shear well before reaching the limit of Wi_{dis} (Eq. (19.1a and b)). Shear homogeneity is restored well below Wi_{dis} because of the reason (shear thinning) presented in Section 9.1.2.5. *Source:* Wang *et al.*, 2011 [3]. Reproduced with permission of American Chemical Society.

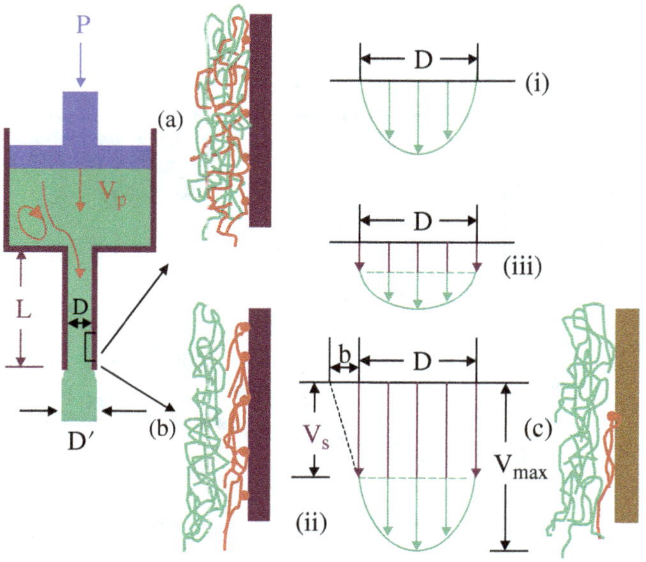

Figure 19.3 Depiction of capillary extrusion (showing vortex formation with a loop), interfacial conditions and corresponding velocity profiles: (a) stick boundary condition due to adequate chain adsorption shown by the anchors (small dots) at the wall; (b) slip boundary condition due to the interfacial disentanglement where adsorbed chains pull out of their engagement with the unbound chains; (c) slip due to polymer desorption on a treated wall. Upon stick–slip transition, the velocity profile changes from (i) to (ii) at a constant pressure. An unstable interfacial state occurs when the velocity profile is given by (iii) where the wall shear stress is lower than in (i) and (ii). The transient state (iii) occurs in the piston-speed-controlled mode of extrusion, that is, constant rate extrusion, typical of industrial operations.

9

Shear Banding in Startup and Oscillatory Shear: Particle-Tracking Velocimetry

In the nonlinear response of many viscoelastic materials, shear banding is common. Different systems are subject to shear banding via different microscopic mechanisms. Strain localization can occur if a system can access two states of vastly contrasting mechanical properties at a comparable shear stress. Upon startup deformation, the initial mechanical response should be solidlike on time scales much shorter than the terminal relaxation time τ. Fast large deformations produce yielding. Different materials have different specific mechanisms for yielding; the yield may not always be spatially homogeneous. For example, wormlike micellar solutions are known to show shear banding.[1–11] At sufficiently high shear rates, shear banding involves a very sharp fast band, reflecting the ability of micelles to undergo scission.[12] Entangled synthetic polymers do not have the option to undergo chain scission because shear deformation typically does not produce high enough stresses. Until 2006, we had no systematic evidence for shear banding in polymer solutions or melts. Thus, no textbook or monograph discussed the phenomenology of shear banding in entangled polymers. This chapter discusses various circumstances under which shear banding takes place in highly entangled polymer solutions. Whenever shear banding takes place, it challenges conventional rheometry. Traditionally, rheometric measurements are meaningfully carried out when the experimentalist can produce homogeneous shear.

If shear strain localization (shear banding) creates troubles for traditional rheometry, why are we searching for it? Polymer melts encounter difficulties such as extrudate distortion during processing. The search for the molecular origin of melt fracture[13–17] propelled us to explore the possibility of shear banding in simple-shear apparatuses. Since only well-entangled melts at high extrusion speed show extrudate distortion, we focus on strongly entangled polymers sheared at Wi_{app} (Eq. (6.5)) $\gg 1$. Whenever there is strain localization or wall slip (an extreme form of strain localization), one must acknowledge that the rheometer can only impose and measure Wi_{app}, not Wi, the latter being termed in this book as the bulk Weissenberg number. Whenever $Wi_{app} \neq Wi$, we use Wi_{app} to describe the imposed deformation condition, because Wi cannot be controlled by the experimenter, and is in fact not measured unless the strain field is determined by means such as the particle-tracking velocimetric method (cf. Section 4.4).

Although shear banding can occur in both viscoelastic solids and liquids, only liquids can possibly display *stable* shear banding during continuous shear. For sufficiently entangled polymers, the most severe form of shear strain localization is wall slip, a subject that has been thoroughly discussed in Chapter 6. A large correction to the textbook assumption of shear homogeneity occurs when $b/\mathrm{H} > 1$.

In this chapter, we specify in a phenomenological way the conditions under which bulk shear banding has been observed during startup shear and during large-amplitude oscillatory shear. It has been shown in Section 6.4.1 that the condition $Wi \gg 1$ requires Wi_{app} to exceed the threshold value of $Wi_{ws\text{-}bnl}$ given in Eq. (6.23). Given in Eq. (4.4), V_s^* must be small enough for the

Nonlinear Polymer Rheology: Macroscopic Phenomenology and Molecular Foundation, First Edition. Shi-Qing Wang.

Companion Website: www.wiley.com/go/nonlinearpolymerrheology

low-speed particle-tracking velocimetric method to properly characterize the shearing field. Thus, we need to have systems that have rather long relaxation times. The systems should not undergo massive wall slip, so that the condition $Wi_{app} \gg Wi_{ws\text{-}bnl}$ can be readily attained.

Our understanding of wall slip behavior provides an effective strategy: According to Eq. (6.13), the maximum extrapolation length b_{max} can be manipulated, for a given degree of polymer entanglement, by adjusting η_s. At a given volume fraction ϕ of the parent polymer of high molecular weight M, different solvents can be chosen. The solvent can be a low-molar-mass hydrocarbon oil. It can also be a low-molecular weight oligomer of the parent polymer so that the solution is actually a binary mixture. Throughout this book, we will call such mixtures "solutions." At a given ϕ, the number of entanglements per chain $Z = M/M_e(\phi)$ is fixed. The zero-shear solution viscosity η does not change linearly with η_s of a polymeric solvent. Thus, the desired value of b_{max} can be achieved by the choice of M_s to effect η_s in Eq. (6.13). Upon complete wall slip, the viscosity η_i in the slip layer can be regarded as that of a binary mixture of $(1-\phi)M_s + \phi M_e$, namely, a binary mixture containing a volume fraction $(1-\phi)$ of short-chain solvent molecular weight M_s, plus a volume fraction ϕ of the disentangled long chains of the solute, having an equivalent molecular weight M_e. In the case of $\phi \ll 1$, we can approximate η_i in Eq. (6.3) to be $\eta(M_s) = \eta_s$. The polybutadiene solutions listed in Table 9.1 were all designed and characterized in such a way that most of these solutions are well entangled to display shear banding but cannot undergo significant wall slip.

Rheometric measurements as a convention apply Wi_{app}, assume no-slip boundary condition and a uniform deformation field, and measure the resulting stress, as shown in Figures 1.1 and 1.2 for the case of simple shear. If the resulting stress is appreciably lower than the elastic plateau modulus G_{pl}, which is the case when $Wi_{app} < 1$, then the simple shear is indeed homogeneous. An equivalent way to produce homogeneous response is to apply a constant shear stress σ that is significantly lower than the plateau modulus G_{pl}. When Wi_{app} is much larger than unity or the creep involves a shear stress level comparable to G_{pl}, there is no *a priori* guarantee of shear homogeneity. We could make the false argument presented in Section 4.1.1.

Table 9.1 Characterization of various entangled PB solutions with and without shear banding.

Samples	Sources	References	M_w (kg/mol)	M_w/M_n	Z	τ (s)
0.7M(5%)-1.8K	Bridgestone/Aldrich	18	750	1.02	13	40
0.7M(10%)-9K	Bridgestone/Aldrich	18	1050	1.05	27	16
1M(10%)-9K	UAkron/Goodyear	18	1050	1.05	40	63
1M(15%)-9K	UAkron/Goodyear	18	1050	1.05	64	65
1.8M(10%)-9K	Goodyear/Goodyear	18	1860	1.19	70	71
1.8M(15%)-15K	Goodyear/Bridgestone	18	1860	1.19	119	250
0.7M(5%)-4K	Bridgestone/Goodyear	19	750	1.02	13	3
1M(10%)-1.5K	UAkron/Aldrich	19	1050	1.05	40	16
1M(10%)-5K	UAkron/Bridgestone	19	1050	1.05	40	25
1M(10%)-15K	UAkron/Bridgestone	19	1050	1.05	40	71
1.6M(10%)-5K	P9114/Bridgestone	20	1616	1.20	63	63
1.6M(13%)-5K	P9114/Bridgestone	20	1616	1.20	86	100
1.4M(14%)-8K	UAkron/Bridgestone	20	1400	1.66	90	125
1M(15%)-15K	UAkron/Bridgestone	21	1050	1.05	64	88
1.2M(15%)-1.5K	PBd1914/Aldrich	21	1243	1.13	80	11

However, shear homogeneity may not be a reality. For decades, the traditional characterization was based on the convenient hypothesis that homogeneous deformation and flow occur both during startup deformation and during and after stepwise strain of any magnitude. Chapters 9–11 examine whether this hypothesis is valid for startup deformation. Part 3 investigates strain localization after stepwise deformation. This chapter mainly treats the subject of shear banding during startup shear in Section 9.1. Here we carefully discuss in Section 9.1.2 the various experimental factors that affect the responses of entangled polybutadiene solutions. It is emphasized by comparison of data in Section 9.1.3 with data in Section 9.1.4 that the criterion for measurable shear banding depends on the system's dimension, for example, the thickness H of the sample in simple shear. Section 9.2 shows how to avoid wall slip so that shear banding can be observed. Section 9.3 briefly indicates that strain localization readily forms in large-amplitude oscillatory shear.

9.1 Shear Banding After Overshoot in Startup Shear

Most rheometric studies of nonlinear responses used entangled polystyrene (PS)[22–26] and polybutadiene (PB)[18,19,27–30] solutions. Existing commercial rheometers are not suitable for studying the nonlinear rheological behavior of well-entangled polymer melts because both the torque and the normal force readily exceed the limits of the instruments. The high normal and shear stresses arise from a high melt plateau modulus G_{pl}. The lowest G_{pl} among common polymers is 0.2 MPa for polystyrene; a few exotic species such as hydrogenated polystyrene have smaller G_{pl} values. For other melts such as 1,4-polybutadiene with $G_{pl} = 1$ MPa, wall slip and edge fracture are the symptomatic responses, making a reliable rheometric study nearly impossible for these melts.

Reliable, systematic nonlinear rheometric data from commercial rotational rheometers requires us to make entangled polymer solutions whose plateau moduli $G_{pl}(\phi)$ drop quickly with ϕ, as seen in Eq. (2.36c). Given the discussion in Sections 3.1.2.1 and 3.1.2.2, it is clear that ϕ of 1,4-PB solutions must be low, namely, well below $\phi = 0.2$ to avoid overload for torque and normal force in commercial rheometers. The solutions also need to be sufficiently entangled. If $b_{max}/H \ll 1$, as shown in Section 9.1.2.4, shear banding cannot occur. Conventional rheometric setups typically have $H \sim 1$ mm. Thus, solutions need to have b_{max} in an ideal range, just below a millimeter but larger than 0.01 mm. For a PB solution at $\phi = 0.1$, sufficient entanglement and appreciable b_{max} are realized by using parent polymers of ultrahigh molecular weight, namely, ca. 10^6 g/mol. For a 10% polystyrene solution, it is desirable to have molecular weight of about 10^7 g/mol to have the number of entanglements per chain $Z = 40$ in order to observe shear banding.

9.1.1 Brief Historical Background

In the absence of any experimental indication to contradict the shear homogeneity assumption, a major theoretical revision[31,32] of the original tube model of Doi and Edwards[33] was made to remove its seemingly incorrect speculation that the shear stress versus rate relation is nonmonotonic. Convective constraint release (CCR) was proposed[32] as a path to restoring a *monotonic* stress versus rate relation. Convective constraint release brought the tube theory into agreement with the available experimental information. However, because of this *ad hoc* remedy, the tube theory[34] can no longer be regarded as being able to predict the basic nonlinear rheological behavior of entangled polymers.

Until 2005[35,29] we had no idea whether or not entangled (synthetic) polymer solutions and melts show shear strain localization. There was one brief report of shear banding based on a polyacrylamide solution.[36] This publication was not viewed as being generically significant because the authors regarded shear banding as a specific chemical effect, stating "the origin of shear banding in (relatively polydisperse) polyacrylamide solutions is associated with hydrogen-bond breakage above a critical shear rate."

Thus, the first report of shear banding in entangled polybutadiene solutions[29,37,38] came as a surprise. The revised tube theory[34] did not anticipate shear banding during startup shear of entangled polymers. To reconcile experiment with tube theory, Hu *et al.* suggested[39] that shear banding would not survive in steady state. For insufficiently entangled PB solutions, shear banding was indeed transient; it disappeared in the steady state. Separate PTV studies also showed that shear banding did not persist to steady state if the PB solutions were insufficiently entangled[18] or if the solvent[19] was of a sufficiently large molecular weight. The samples 0.7M(10%)-9K and 1M(10%)-15K listed in Table 9.1 display, respectively, each of these behaviors. PTV observations confirm[18,19] that shear banding persists to steady state for sufficiently large values of Z. PTV studies confirm[18] that shear strain localization does not occur if $b_{max}/H \ll 1$. Only an adequately large Z ensures that b_{max}/H is not too small to create shear strain localization. A large number of entangled polybutadiene solutions have been examined with PTV.[18–21,39–41] As listed in Table 9.1, many samples were designed based on the consideration that b_{max}/H should be moderate, close to unity, not too far below and not too high above unity (to avoid wall slip).

Since entangled DNA solutions can be sheared in commercial rotational rheometers without creating edge effects, PTV observations have also been carried out to explore the universal nature of shear banding in well-entangled DNA solutions. Tables 9.2 and 9.3 present basic characteristics of two pairs of DNA solutions. The first pair, described in Table 9.2, contains two concentrations using a single solvent. The second pair, shown in Table 9.3, has a single DNA concentration prepared with two different solvents.

9.1.2 Relevant Factors

9.1.2.1 Sample Requirements: Well Entangled, with Long Reptation Time and Low Polydispersity

To obtain a solution that shows shear banding, the sample should be well entangled so that there is a wide separation of time scales between the terminal (reptation) relaxation time and other molecular relaxation mechanisms such as Rouse dynamics. Correspondingly, the solution should have b_{max} that is nonnegligible relative to H. As discussed in Section 4.4.1.2,

Table 9.2 Characterization of DNA solutions at two concentrations based on glycerol.

c (mg/ml)	G_{pl} (Pa)	$M_e(c)$ (g/mol)	$Z(c) = M_w/M_e(c)$	τ (s)	$\tau_R = \tau/3Z$ (s)	$a = [6R_g^2/Z(c)]^{1/2}$ (μm)	η (Pa s)	$b_{max} = (\eta/\eta_s)a$ (mm)
11.0	32	0.84×10^6	60	780	4.3	0.205	7800	2.1
22.0	169	0.32×10^6	156	1590	3.4	0.127	58450	10

Table 9.3 Characteristics of two additional entangled DNA solutions at c = 22 mg/ml.

Solvent	$\eta_0 \times 10^3$ (Pa s)	η_s (Pa s)	b_{max} (mm)	τ (s)	G_{pl} (Pa)	V_s^* (mm/s)
Water	5.8	0.001	1176 (737)	56	166	21 (13)
Ethylene glycol	13	0.023	139 (71)	145	173	0.96 (0.49)

the applied speed of the shearing plate should be below 5 mm/s in order to apply low-speed particle-tracking velocimetry. To have $Wi_{app} > 50$, we need the reptation (terminal relaxation) time of the solution to be at least as long as 10 s. A nominal shear rate V/H around $\dot{\gamma} \sim 1\,s^{-1}$ is ideal. Both a sufficiently high molecular weight of the parent polymer and a sluggish solvent are required to prepare a suitable solution.

9.1.2.2 Controlling Slip Velocity

When $b_{max}/H \gg 1$, Wi_{app} must be much larger than unity before the condition $Wi > 1$ can take place in the bulk, as explained in Section 6.4.1. To reach the condition of bulk nonlinear response, the shearing surface must have a speed V in excess of $2V_s^*$ as indicated in Eq. (6.24a). Section 6.2.3 lists the limiting slip velocity V_s^* for a number of polymer melts. So far, little corresponding information is available for entangled polymer solutions. In solutions, V_s^* will depend on ϕ as well as the solvent η_s.

It is instructive to compare V_s^* between an entangled polymer melt and a binary mixture that is based on the same polymer. We take a ratio of Eq. (6.10)–(6.9a) and have

$$\frac{V_s^*(\phi)}{V_s^*(\phi = 1)} = \left(\frac{l_{ent}(\phi)}{l_{ent}(\phi = 1)}\right)\left[\frac{G_{pl}(\phi)}{G_{pl}(\phi = 1)}\right]\left(\frac{\eta_e}{\eta_s}\right) = \left(\frac{\eta_e}{\eta_s}\right)\phi^{1.6-1.65}, \quad (9.1)$$

where use is made of $\eta_e \sim G_{pl}(\phi = 1)\tau_e$ in the first equality. The second equality follows from the scaling relations in Eqs. (2.36c) and (2.37a). Both ϕ and the viscosity ratio (η_e/η_s) affect the ratio of V_s^* of solutions to V_s^* of the melt. If the solution is made of a very inviscid small-molecular organic liquid so that $\eta_e/\eta_s \gg 1$, we can have a larger slip velocity in solution than is encountered in the melt. Conversely, if the solvent is the same polymer with $M_s \sim M_e$, the ratio in Eq. (9.1) simplifies to $\phi^{1.6-1.65} < 1$. For a binary mixture with $M_s \gg M_e$, we have

$$\frac{V_s^*(\phi)}{V_s^*(\phi = 1)} = \left(\frac{M_e}{M_s}\right)^{3.4}\phi^{1.6-1.65}, \quad (9.2)$$

which can be much smaller than unity as commented in the following paragraph unless the low-molecular-weight component also undergoes disentanglement. We note that Eq. (9.2) is closely related to Eq. (6.18c).

Since $V_s^*(\phi = 1)$ is about 10 mm/s for 1,4-PB melts at room temperature, while the limiting speed for PTV observations is $V^* = 5$ mm/s in Eq. (4.4), and since the displacement of the shear surface must attain a speed V in excess of $2V_s^*$ before the bulk moves into the nonlinear response regime (cf. Section 6.4.1), use of binary mixtures with $M_s > M_e$ is convenient to reduce V_s^*. Consistently with this recommendation, all PTV reports of shear banding in 1,4-PB solutions involved marginally or moderately entangled 1,4-PB as the solvent. For example, Tables 6.1 and 6.2, respectively, show $V_s^* = 10$ mm/s for 1,4-polybutadiene melts and $V_s^* = 0.167$ mm/s and 0.006 μm/s for two 1,4-PB solutions, respectively. Apart from the choice of M_s that can regulate $V_s^*(\phi)$ as demonstrated in Section 6.4.2.2, there is another simple reason why $V_s^*(\phi)$ of mixtures is usually smaller than that of the corresponding polymer melt: According to Eq. (6.1a), it is the shear stress level that determines V_s. The stress level is determined by and comparable to $G_{pl}(\phi)$. At the low volume concentration of 10%, the stress level is lower by a factor of about 200 relative to that for melts, bringing down V_s^*. Equation (9.1) accounts for these factors.

9.1.2.3 Edge Effects

When the bulk of an entangled polymer enters the nonlinear response regime $Wi > 1$, the meniscus instability can take place as indicated in Section 5.2. Such edge effects have made it challenging to study the nonlinear rheological behavior of well-entangled polymers. In a conventional rheometric apparatus, for example, a cone-plate device, the only way to avoid edge fracture on rheometric measurements is to terminate startup shear experiments before such effects accumulate to an unacceptable level. A recent study most clearly disclosed the dilemma.[26] Several approaches have been taken to minimize, suppress, or remove this intrinsic experimental difficulty. Before describing these different ways to cope with edge fracture, we emphasize that step strain tests never really suffer from this complication because the edge failure, a finite size effect, is accumulative and thus can be avoided at a shear strain of, for example, 2. Chapter 12 treats rheological responses after large stepwise shear.

Edge fracture is known to be severe when significant stresses are produced during startup shear. When polymer chains are sufficiently long, solutions can have significant entanglement at rather low values of ϕ. Nonlinear rheological behavior can then be probed at fairly low stresses because the elastic plateau modulus G_{pl} decreases rapidly with ϕ, as shown in Eq. (2.36c). Both wormlike micellar solutions[12] and entangled aqueous DNA solutions[42–45] have been studied with PTV to show shear banding without any edge effects. A 1,4-PB solution at $\phi = 3.5\%$ is edge fracture free in the range of *Wi* that can produce shear banding. To reach a chain entanglement Z of 50, we need the molecular weight of the PB to reach 4000 kg/mol. Such a system has recently been studied (M.C. Wang, unpublished work at University of Akron).

For PTV measurements, edge fracture can be effectively removed by Scheme A in Figure 4.4(a). Here, the meniscus is wrapped around by a flexible (transparent) plastic sheet. A proper installation of the film involves complete wetting of the sheet by the sample, leaving no gap or bubbles trapped between the sample and the film. The introduction of such a stationary boundary at the meniscus affects the accuracy of the rheometric measurements to a tolerable degree. On the other hand, the installment makes it more convenient to carry out PTV observations. The location of the PTV observation can be 3–4 mm inward, away from the fixed film, which is an amply long distance from the stationary boundary, judging from the PTV measurements: Upon a startup shear, the velocity field is seen to immediately attain linearity and remains so at all times, either when $Wi_{app} < 1$ or for insufficiently entangled polymer solutions. Figure 9.1(a,b)

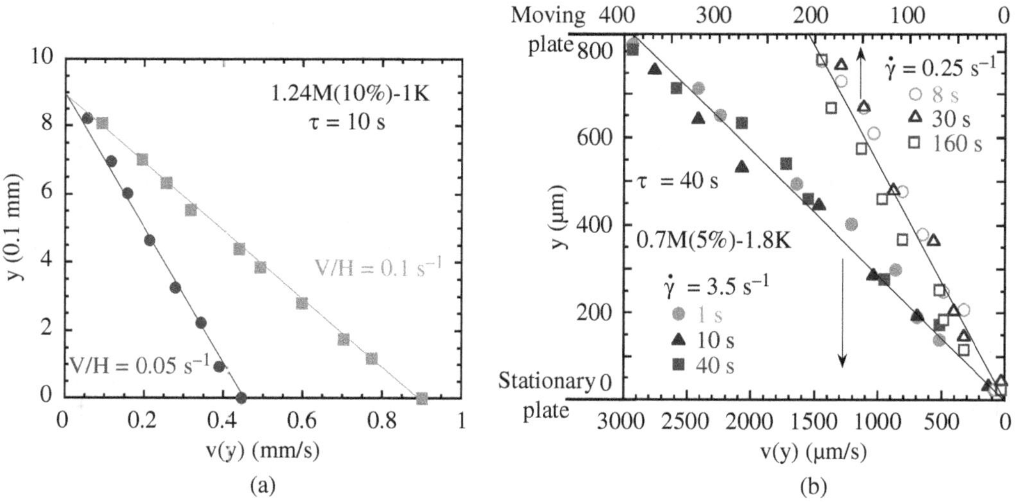

Figure 9.1 (a) Particle-tracking velocimetric measurements of velocity profiles showing the velocity v(y) at each height y along the gap distance, at $Wi_{app} = 0.5$ and 1, respectively, of a 10% polybutadiene (PB) solution, using data from Fig. 2(b) in Ref. [29]. In Ref. [29] both axes were mislabeled, in incorrect units of centimeters and centimeters per second. (b) Velocity profiles at different times at rather high Wi_{app} of 10 and 140 of a 5% PB solution, using data from Fig. 1(b) in Ref. [18]. With the solution viscosity $\eta_0 = 21{,}062$ Pa s and $\eta_s = 97$ Pa s, the slip length b_{max} of PB0.7M(5%)-1.8K is estimated to be $(21{,}062/97)l_{ent}(\phi = 1)0.05^{-0.6} < 0.01$ mm where $l_{ent}(\phi = 1)$ is ca. 3.7 nm. *Source*: Data from Tapadia and Wang 2006 [29].

shows two examples. On the other hand, when well-entangled PB solutions are subjected to startup shear with $Wi_{app} \gg 1$, strong deviation from the linear profile does take place after the stress overshoot, to be discussed in Section 9.1.3. For step-strain and LAOS tests, often no wrapping with film was employed when Scheme B of Figure 4.4(a) was implemented.

Edge instabilities can affect rotational rheometric measurements because the largest contribution to the torque comes from the portion of the sample at large values of the radius, as seen in Eqs. (3.1) and (3.2). Edge instabilities reduce the accuracy of creep tests with a rotational rheometer. To remove the effect of edge failure on rheometric measurements, we have to insulate the measurements from the edge effects. The cone-partitioned plate device is very effective in removing edge effects, as discussed in Section 5.2. For PTV studies of well-entangled solutions, cone-partitioned plate is an ideal setup because the point of PTV observation can be very far away from the rim. There is clear evidence from Figure 5.3 that the cone-partitioned plate can eliminate the effect of edge fracture on rheometric measurements.

Finally, edge effects can be minimized by reducing the gap distance H of a shear cell, for example, by making measurements using parallel disks. In a parallel-disk assembly, when H is reduced from 1 mm to 50 μm while keeping R the same, the aspect ratio is increased by a factor of 20, making it far more difficult for any edge effect to propagate radially far enough inward and affect the strain field many millimeters away from the rim. Such a strategy has been adopted to demonstrate the inherent nature of shear strain localization, as shown in Section 9.1.4.

9.1.2.4 Absence of Shear Banding for $b/H \ll 1$

We should distinguish two different types of shear strain localization, namely transient shear banding and persistent shear banding. With either type of banding, different states of entanglement coexist at the same shear stress. Transient shear banding may be initiated by localized yielding, initially arising from one monolayer of chain disentanglement. Whether monolayer disentanglement produces a measurable discontinuity in the velocity field depends on the magnitude of the "internal slip." When localized yielding occurs shortly after the stress overshoot, there could be a discontinuity of magnitude V_s in the velocity field, as shown in Figure 9.2. Here, we cannot quantify the magnitude of the internal slip in terms of the extrapolation length b of Eq. (6.2) through the shear viscosity: The expression $\sigma = \eta\dot{\gamma}$ is only valid in the flow state and does not apply in the transient elastic-dominant stage of startup shear, for example, at the yield point defined by the peaking of the shear stress σ_y.

Let us evaluate V_s associated with this layer of internal slip relative to the imposed V. We may do this by realizing that (a) V_s depends on the stress level σ and (b) σ is upper-bounded by the yield stress $\sigma_y \sim \gamma_y G_{coh}$ (cf. Eq. (7.4a)). The ratio V_s/V at σ_y is a measure of the correction to the shear field and has a form analogous to Eq. (6.19), given $G_{coh} = G_{pl}/2.2$,

$$V_s/V \sim (b_0/H)(\gamma_y/2.2Wi_{app}), \quad (9.3)$$

where use is made of Eq. (7.4b) and b_0 is given in Eq. (6.20). When $V_s/V \ll 1$, the correction to the velocity field is negligible so that the stress and strain states are unaffected by the first emergent layer of internal slip. In the absence of measurable shear strain localization, all other

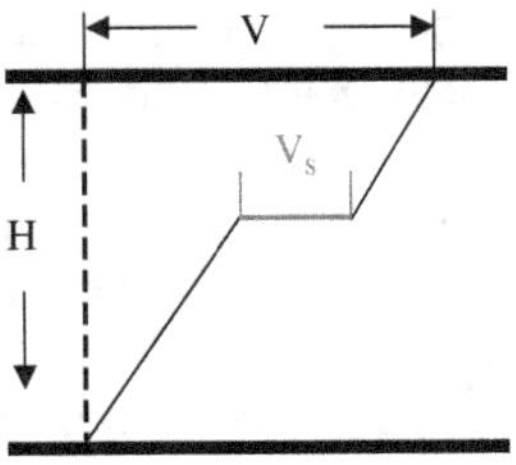

Figure 9.2 Strain field in simple shear in the presence of "internal slip," depicted by a discontinuity of magnitude V_s in the velocity file, so that the local shear rates deviate from the nominal rate or averaged rate of V/H.

layers meet the same condition for "internal slip" through chain disentanglement, and disentanglement can be expected to occur in all the layers simultaneously. Therefore, if $V_s/V \ll 1$, we can expect the yielding to take place homogeneously so that shear banding does not occur.

According to Eq. (9.3), at moderate values for Wi_{app}, the ratio b_0/H determines whether shear strain localization due to one layer of disentanglement would be measurable or not. For insufficient entangled polymers sheared in a sizable gap distance H, the condition $b_{max}/H \ll 1$ is valid and guarantees $V_s/V \ll 1$. In this case, yielding can take place homogeneously, and no shear banding occurs during startup shear. Figure 9.1(b) shows an example of how this ratio dictates the strain response: the maximum value of the slip length, b_{max}, is below 0.01 mm and H is ca. 1 mm. Consequently, homogeneous shear prevails in Fig. 9.1(b). The discussion in this section can be regarded as an extension of the extrapolation-length concept to describe shear strain localization. The argument based on Eq. (9.3) explicitly introduces the system dimension H as a pertinent parameter, implying that when shear banding appears at H shear banding may not occur when the gap distance increases to 30H. The phenomenon is system-size dependent.

9.1.2.5 Disappearance of Shear Banding at High Shear Rates

At high values of Wi_{app} even for well-entangled polymers, the ratio in Eq. (9.3) can become diminishingly small although $b_{max}/H \gg 1$. Specifically, we have, rewriting Eq. (9.3)

$$V_s/V \sim (b_0/H)(Wi_{app})^{-1} \tag{9.4a}$$

if we are in the viscoelastic regime with $\gamma_y \sim 2$, or

$$V_s/V \sim (b_0/H)(Wi_{app})^{-2/3} \tag{9.4b}$$

in the elastic deformation (scaling) regime where $\gamma_y \sim (Wi_{app})^{1/3}$ (cf. Eq. (7.5) and Fig. 7.8(b)). In any case, $V_s/V \ll 1$ is satisfied and shear banding is avoided when $Wi_{app} \gg b_{max}/H > b_0/H$, or $Wi_{app} \gg (b_{max}/H)^{3/2}$. Since shear banding requires $Wi_{app} > Wi_{ws\text{-}bnl}$, as discussed in Section 6.4.1, the range of Wi_{app} in which shear banding takes place is not so large. Equation (9.4) thus provides a rationale for why shear banding disappears at sufficiently high shear rates, as shown in Figs. 9.5 and 9.8(c). The preceding argument is not a molecular theory and is based on having one monolayer of disentanglement in the bulk. The condition is relaxed when multiple monolayers suffer "internal slip."

When shear banding is absent, the shear stress still decreases with the elapsing strain as the entanglement network uniformly collapses. Shear banding is therefore not a necessary condition for the stress overshoot. Conversely, the stress overshoot is not a sufficient condition for shear banding.

9.1.2.6 Avoiding Shear Banding with Rate Ramp-Up

In the absence of any appreciable wall slip, startup shear with $Wi_{app} \gg 1$ means that many strain units of shear can be imposed in a time much shorter than the reptation (terminal relaxation) time τ. How can the entanglement network remain intact if there is a great deal of imposed shear during a time period too short for chain diffusion to adjust the entanglement structure? In fact, the entanglement network cannot sustain its initial elastic response indefinitely; it eventually yields, as discussed in Chapter 7. The macroscopic yielding revealed by the emergence of the shear stress overshoot stems from the microscopic breakdown of the entanglement network. This microscopic breakdown may not take place in a spatially homogeneous manner. For well-entangled polymers, the shear banding resulting from localized yielding is apparently stable.

The nature of the shear banding state can be investigated by advancing to $Wi_{app} \gg 1$ slowly. Specifically, the applied shear rate can be gradually ramped up to a value much larger than the equilibrium relaxation rate $1/\tau$, starting with a value of the shear rate that is smaller than $1/\tau$. Using either entangled DNA[45] or PB solutions,[20] it is shown in the literature that by avoiding sudden startup shear and gradually increasing the applied rate over many hours these samples respond homogeneously from the beginning to the end. Such behavior raises a serious difficulty for modeling shear banding. Apparently, the state of shear banding is not the only possible response for a given Wi_{app}. Since a nonmonotonic constitutive relation between shear stress and rate necessarily leads to shear banding, the results of these rate ramp-up experiments indicate that the physics behind the shear band formation is not constitutive nonmonotonicity, which is a consequence of the Doi–Edwards tube model in the absence of convective constraint release.[32] A more explicit discussion on rate ramp-up and corresponding theoretical implications appear in Sections 15.5 and 16.1.3.4, respectively.

9.1.3 Shear Banding in Conventional Rheometric Devices

To investigate bulk constitutive behavior, it is necessary to know and understand fully polymer wall slip behavior. Chapter 6 has provided all the essential background. The analysis in Chapter 6 clearly shows that well-entangled melts inevitably undergo massive wall slip. If a well-entangled melt can be made to adhere to the shearing surfaces, then it can be driven to the point of strong shear banding (see Section 12.2.2 for an example).

In this chapter, we focus on particle-tracking velocimetric observations of entangled polymer solutions based on rotational rheometric devices. To minimize wall slip, inviscid solvents are avoided. The analysis in Chapter 6 shows that wall slip can be effectively suppressed using a polymeric solvent, namely, a moderate-molecular-weight oligomer, molecular weight M_s, having the same chemical structure as the parent polymer. A short-chain melt chosen to be the solvent may be entangled, namely, $M_s > M_e$. Most PTV observations of shear banding have been based on binary mixtures of 1,4-polybutadiene (PB). We call these "solutions" because of the extremely wide separation of time scales in chain dynamics between the long and short chains.

9.1.3.1 Shear Banding in Entangled DNA Solutions

The strategy of selecting viscous solvents works more generally than for the polybutadiene solutions. For example, entangled DNA solutions can be prepared with 1–2 wt% DNA and solvents such as water, ethylene glycol, and glycerol. As shown in Fig. 6.15, the aqueous deoxyribonucleic acid (DNA) solutions display massive wall slip because of the large b_{max} listed in Table 9.3. By using a viscous solvent such as glycerol, $\eta_s = 0.7$ Pa s, a viscosity that is greater than that of water by a factor of 700, wall slip can be sufficiently suppressed (cf. Eq. 6.13 and Table 9.2). At shear rates similar to those involved in Fig. 6.15(a,b), bulk shear banding appears, as seen in Fig. 9.3(b). However, in discussing shear banding during startup shear, one must always first present the stress versus time plot, as seen in Fig. 9.3(a). Note that the shear field is homogeneous even at the stress overshoot (cf. Fig. 9.3(b)). One first needs to use the particle-tracking velocimetry to confirm that the velocity field is linear before the shear stress peaks. Such information can be treated as a calibration of the PTV technique.

There are no experimental complications when studying these well-entangled DNA solutions: Edge effects are negligible because the solutions are extremely soft, showing an elastic plateau modulus G_{pl} of 32 Pa at a DNA concentration of $c = 11$ mg/ml (Table 9.2). Consequently, the entanglement–disentanglement transition can take place at an applied shear stress of 22 Pa, as shown in Fig. 9.4. The inset shows PTV tracking of the development of shear strain field within the sample gap as a function of time. With enough suppression of wall slip, the bulk of the sample is able to undergo an entanglement–disentanglement transition (cf. Section 7.4). PTV measurements reveal that in the $c = 11$ mg/ml solution the rise of shear rate at a given applied shear stress cannot occur uniformly during creep. When

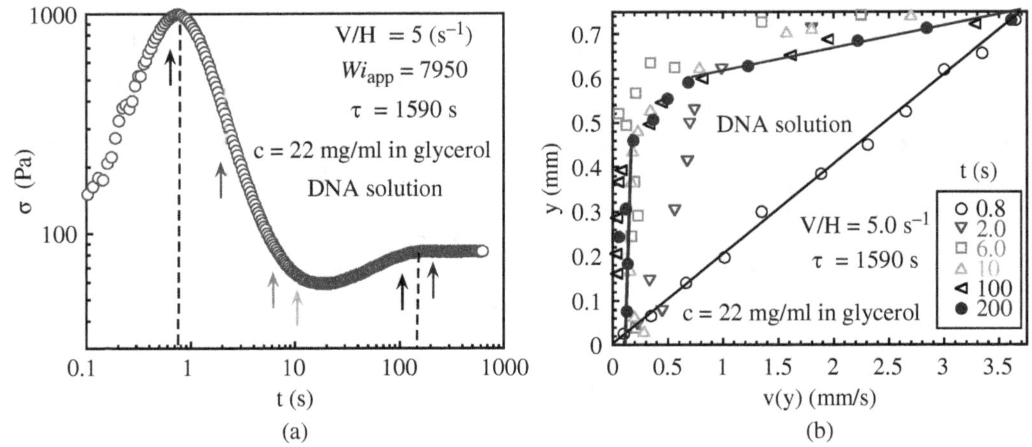

Figure 9.3 (a) Time-dependent shear stress at V/H = 5.0 s^{-1}, showing stress overshoot of a well-entangled DNA solution. (b) The corresponding velocity profiles at different times. *Source*: Data taken from Boukany and Wang 2009 [42].

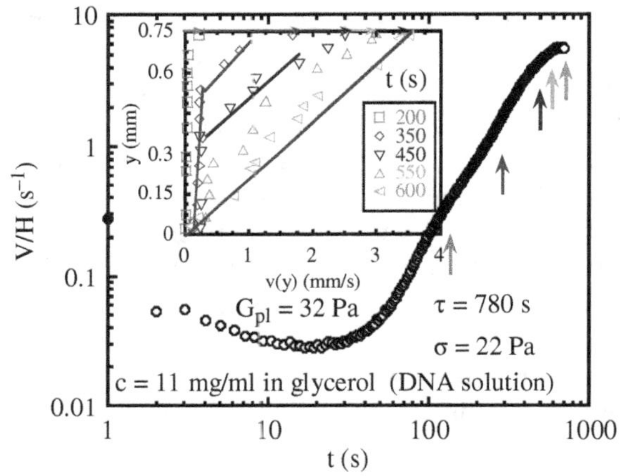

Figure 9.4 Entanglement–disentanglement transition (EDT) of a well-entangled DNA solution at an applied shear stress of 22 Pa, where the inset shows the PTV measurements, revealing the time evolution of the actual velocity profiles. *Source*: Boukany and Wang 2009 [42]. Reproduced with permission of American Institute of Physics.

the entanglement–disentanglement transition has developed fully, the sample achieves almost homogeneous shear at a shear rate of 4.6 s^{-1}, corresponding to $Wi_{app} = 3640$. Like the entangled polybutadiene solutions discussed below, DNA solutions tend to recover shear homogeneity at sufficiently high shear rates. This recovery is not unique to the creep mode. Figure 9.5 shows that at an applied rate of 4 s^{-1} shear homogeneity is restored consistent with the creep data in the inset of Fig. 9.4. Despite an even larger shear rate and correspondingly much larger Wi_{app}, a more concentrated DNA solution still undergoes shear banding at 5 s^{-1} as shown in Fig. 9.3(b). This 22 mg/ml DNA has more than twice as many entanglements per chain as does the 11 mg/ml solution and a significantly larger b_{max}.

9.1.3.2 Transient and Steady Shear Banding of Entangled 1,4-Polybutadiene Solutions

Most PTV data in the literature are based on carefully designed 1,4-polybutadiene solutions that typically involve 1,4-polybutadiene of low molecular weight as the solvent. The strong tendency of 1,4-polybutadiene melts to undergo massive wall slip is bypassed by studying the binary mixtures. For the PTV method to be applicable on commercial rotational rheometers, a polybutadiene solution needs to be sufficiently low in concentration. Mixing short chains into

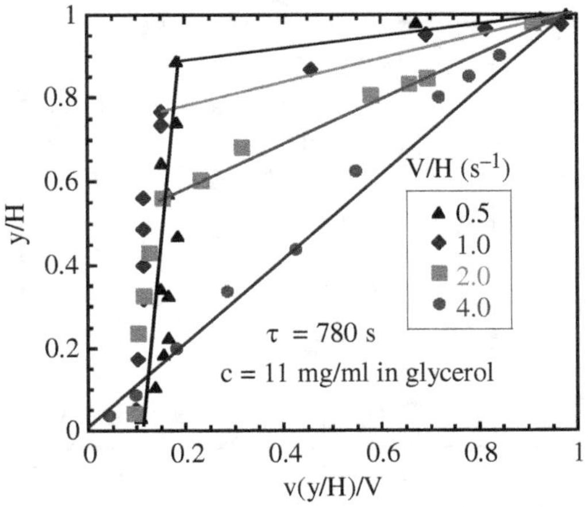

Figure 9.5 Normalized velocity profiles, where H is the gap distance and V is the velocity of the moving surface, showing how the system transforms from shear banding to the approximate shear homogeneity as V/H increases. The particle-tracking velocimetric video[1] shows that the shear banding seems to prevail for 2 s^{-1} at long times, namely, after 350 s. *Source*: Data taken from Boukany and Wang 2009 [42].

long chains achieves both effects: lowering the stress level because of the strong concentration dependence of the plateau modulus $G_{pl}(\phi)$ (Eqs. 2.36a–2.36c), and reducing the tendency to undergo wall slip. The first 15 solutions in Table 9.1 were made in this way to satisfy the conditions for minimal wall slip and strong shear banding in some of these solutions. The search for shear banding is not only an academic exercise, even though shear banding is a profoundly interesting phenomenon. Shear strain localization phenomena are actually prevalent and universal for entangled polymers. They are at the heart of frequently encountered industrial polymer processing difficulties. We return to industrial applications in Chapter 19.

Out of the many solutions listed in Table 9.1, we will only discuss a few. We review particle-tracking velocimetric (PTV) measurements of a moderately entangled PB solution, namely, 0.7M(10%)-9K. Figure 9.6 shows that it responds to startup shear at $Wi_{app} = 16$ with a stress overshoot depicted in the inset, followed by *transient* shear banding. However, the solution recovers shear homogeneity at the long times. Such behavior is typical of weakly entangled PB solutions.[39]

Next we examine a slightly more entangled PB solution 1M(10%)-9K whose linear viscoelastic characteristics are shown in Fig. 9.7. Its maximum extrapolation length b_{max} appears in Table 9.4. 1M(10%)-9K is an ideal solution for PTV measurements because the solution

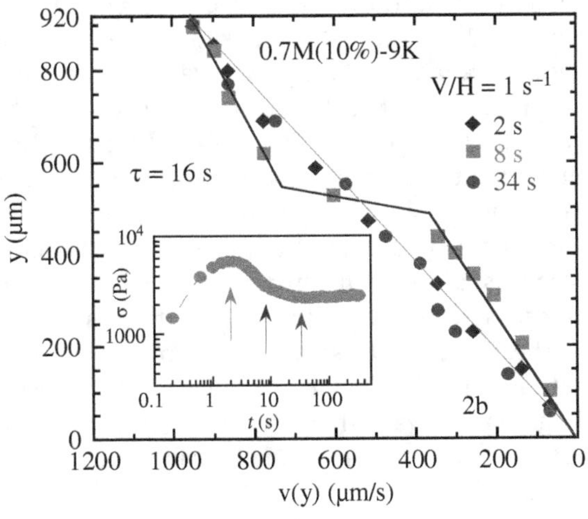

Figure 9.6 Velocity profiles of startup shear at V/H = 1 s^{-1} at various stages indicated by the vertical arrows in the inset, which shows the transient stress response in the form of an overshoot. *Source*: Data taken from Ravindranath et al. 2008 [18].

1 Video 9.1: 1% DNA solution is in steady shear at the rate of 2 (1/s) at room temperature, made at University of Akron.

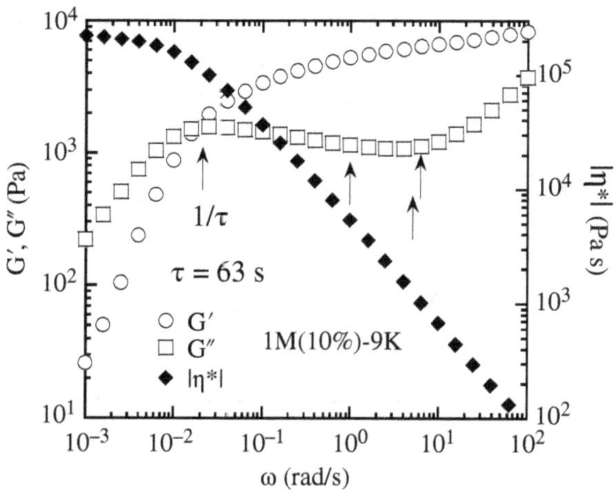

Figure 9.7 Storage and loss moduli G′ and G″ from small amplitude oscillatory shear of 1M(10%)-9K solution, revealing a terminal relaxation time of $\tau = 63$ s. *Source:* Ravindranath *et al.* 2008 [18]. Reproduced with permission of American Chemical Society.

Table 9.4 Basic rheological and wall slip characters of four 1M(10%) PB solutions.

Solvent	$\eta_0 \times 10^3$ (Pa s)	η_s (Pa s)	b_{max} (mm)	τ (s)	G_{pl} (Pa)	V_s^* (mm/s)
1.5K[a)]	50	0.7	1.9 (1.07)	16	5500	0.12 (0.07)[b)]
5K[a)]	96	1.8	1.3 (0.82)	25	6300	0.052 (0.033)[b)]
9K[c)]	230	10	0.53 (0.35)	63	5600	0.0084 (0.0055)[b)]
15K[a)]	250	36	0.18 (0.104)	71	~6000	0.0025 (0.0015)[b)]

a) Ref. [19].
b) The values in the parentheses are obtained using Eq. (6.13) for b_{max} and Eq. (6.11) for V_s^*. The discrepancy occurs because there is a numerical prefactor relating η_0 to $G_{pl}\tau$.
c) Ref. [18].

dynamics are sufficiently slow. It has a conveniently long terminal relaxation time, so that a high value of Wi_{app} can be applied, far exceeding $Wi_{ws\text{-}bnl}$ of Eq. (6.23). Section 6.4.1 showed that in the presence of wall slip the bulk will be sheared at $Wi > 1$ only when $Wi_{app} > Wi_{ws\text{-}bnl}$. Figure 9.7 indicates the three frequencies that are numerically equal to the three shear rates of 1, 5, and 6 s^{-1}, for which the corresponding Wi_{app} values are 63, 315, and 378, all much larger than $Wi_{ws\text{-}bnl}$. Consequently, the condition for shear banding is met. In contrast to 0.7M(10%)-9K solution, in this solution, shear banding appears shortly after the stress overshoot and persists into the steady state, where the shear stress has leveled off. Figure 9.8(a) and its inset, respectively, show the *in situ* velocity profiles from PTV measurements and corresponding stress–strain relationship. Unfortunately, these figures did not report the time-dependent normal stress N_1, so we cannot judge whether the normal force has attained steady state. For an impression of such PTV observations, see Video 9.2 on 1M(15%)-9K during startup shear at apparent rate 1.3 s^{-1}.[2]

The 1M(10%)-9K solution continues to show shear banding when subject to a larger shear rate at $V/H = 5\ s^{-1}$ as shown in Fig. 9.8(b). However, if the shear rate is instead set to the slightly larger rate of 6 s^{-1}, the solution begins to return at long times to shear homogeneity. At a shear rate 6 s^{-1}, it becomes difficult for the system to sustain two different states of shear at the same

2 Video 9.2: PTV movie capturing startup shear of 1M(15%)-9K solution at 1.3 (1/s) at room temperature, made at University of Akron.

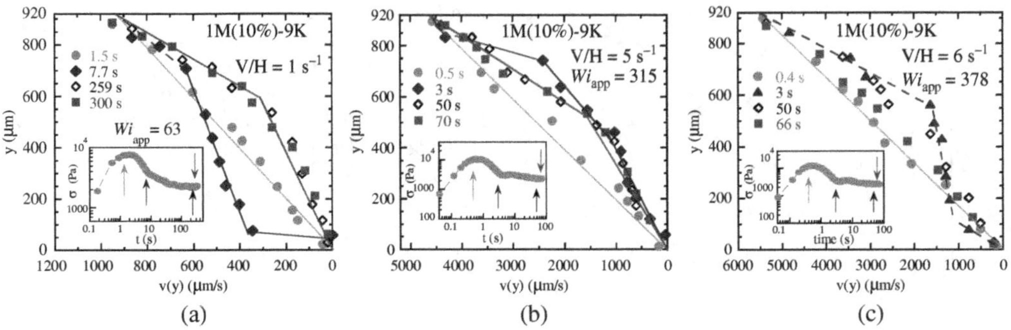

Figure 9.8 Velocity profiles at different times at three respective apparent rates at V/H = (a) 1 s^{-1}, (b) 5 s^{-1}, and (c) 6 s^{-1}. Corresponding rheological data are given in the insets. *Source*: Data taken from Ravindranath *et al.* 2008 [18].

time in a stable manner. Apart from the argument in Section 9.1.2.5, we have not developed a first-principles theoretical understanding that predicts when shear banding is not the prevailing response for well-entangled polymers. We can make the following observations: The local shear rate $\dot{\gamma}_1$ in the slow band has already reached a very high value, in excess of 2 s^{-1}, as shown in Fig. 9.8(b). According to the discussion near Eq. (7.10c), the local state should satisfy $\tau_{\mathrm{eff}}\dot{\gamma}_1 \sim 1/5$. Thus, in the slow band, the relaxation time has shortened by a factor of $\tau/\tau_{\mathrm{eff}} \sim 5(\tau\dot{\gamma}_1)$, where $\tau = 63$ s and $\dot{\gamma}_1 > 2\,\mathrm{s}^{-1}$. Equivalently, translating this inferred change in the relaxation time into a change in the number of entanglements per chain, thereby replacing Z with Z_{eff}, we find the ratio of Z_{eff} to the original Z to be

$$Z_{\mathrm{eff}}/Z = (\tau_{\mathrm{eff}}/\tau)^{1/3.4}, \tag{9.5}$$

where we have taken an empirical relation $\tau \sim Z^{3.4}$ instead of invoking Eq. (2.45a). Inserting $\tau/\tau_{\mathrm{eff}} > 630$ into Eq. (9.5), we get

$$Z_{\mathrm{eff}} < Z/6.7 = 6. \tag{9.6}$$

Thus, the effective degree of entanglement is rather weak even in the slow shear band. It is therefore perhaps not surprising that the system loses its ability to maintain shear banding in steady shear when the shear rate is made larger than 5 s^{-1}. The return of shear homogeneity, as hinted by Fig. 9.8(c), is therefore inevitable.

Finally, based on a different polybutadiene solution, namely, 1M(10%)-1.5K, during the entanglement–disentanglement transition produced at a constant shear stress of 2400 Pa, shear banding and apparent wall slip emerge, as shown in Fig. 9.9. This result is consistent with results on the entangled DNA solution in Fig. 9.4. For this solution, the limiting slip velocity $V_s^* \sim b_{\max}/\tau$ can be estimated from the value of $b_{\max} = (\eta/\eta_s)l_{\mathrm{ent}}(\phi = 0.1)$ and $\tau = 16$ s. According to the information available in Ref. [19] V_s^* is on the order of 0.1 mm/s. However, at $t = 308$ s the data show in Fig. 9.9 that slip-like shear banding has a velocity discontinuity in excess of 1 mm/s. This "slip velocity" is much larger than V_s^*, revealing a sharp shear banding at the sample/wall interface. At $t = 350$ s, the EDT evolves unevenly, forming two visible layers having different shear rates in addition to the two fast bands at the interfaces. Thus, although the creep mode gives the system the freedom to attain any velocity at the movable plate, the system still prefers to split into multiple coexisting layers that have different shear rates. The presence of shear banding is somewhat unexpected in creep where the nominal shear rate is freely adjustable variable and shows that homogeneity is not automatically preferred in the presence of sufficient entanglement even in creep.

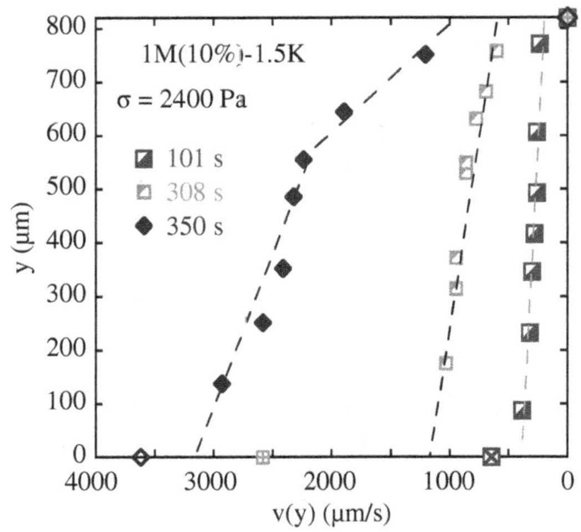

Figure 9.9 PTV determination of velocity profiles at various stages during creep at 2400 Pa corresponding to the time-dependent rise of the apparent shear rate in Fig. 7.21, where the symbols at y = 0 and y = H are values of the speeds at the two walls. *Source*: Data taken from Ravindranath and Wang 2008 [19].

9.1.4 From Wall Slip to Shear Banding in Small Gap Distance

According to the analysis in Chapter 6, the correction to the bulk velocity field is measured in terms of b/H and thus can be much greater in systems having a much smaller value of H. Several reasons motivate simultaneous PTV and rheometric observations for a gap distance H as small as 50 μm. In particular, edge effects cannot propagate into the point of PTV observation, which in our system was 3 mm from the rim, as shown in Fig. 4.3. Such a reduction from the conventional gap size of 1 mm introduces a huge "aspect ratio" of 3 mm (distance to the edge) to 0.05 mm (the gap distance), which is 60.

Changing H by a factor of 20 also provides a way to verify our theoretical description of the relationship between wall slip and bulk shear banding. The key parameter is $Wi_{wb\text{-}bnl}$ of Eq. (6.23). When $b_{max}/H \gg 1$, the apparent Weissenberg number Wi_{app} needs to exceed $Wi_{ws\text{-}bnl} \sim 2b_{max}/H$, so that the bulk can be sheared with $Wi > 1$. Table 6.2 shows two PB solutions of the same concentration with different values of b_{max} and therefore $2b_{max}/H$. The first sample, 1M(13%)-1.5K, only shows wall slip at H = 50 μm, at least within the PTV-accessible shear-rate range, as shown in Section 6.4.2.2.

In this chapter, we focus on the second solution, 1M(13%)-10K, which has a massive correction from wall slip as seen in Fig. 6.24(b), despite having a modest b_{max} of 0.4 mm. The large correction from wall slip arises from the use of such a small gap distance. However, as predicted in Section 6.4.1, because this solution has a much larger solvent viscosity the sample can readily display shear banding even at H = 50 μm. Figure 9.10(a) shows the shear stress responses of this second solution at three different apparent rates, namely 1.5, 3, and 6 s^{-1}. Follow the transient evolution of the strain field. Figure 9.10(b,c) presents the PTV measurements of the velocity profiles in the quasi-steady state, at the moments indicated by the vertical arrows in Fig. 9.10(a). Video 9.3 provides an example of the PTV movie at the rate 5 s^{-1}.[3]

A number of comments can be made. First, there is apparent wall slip. In particular, the PTV reveals an apparent "slip velocity" at the top surface, on the order of 30 μm/s in Fig. 9.10(b) and 80 μm/s in Fig. 9.10(c). These slip velocities considerably exceed the limiting slip velocity $V_s^* = 6.2$ μm/s from Table 6.2. Presumably, in this system, there is an unresolvable, thin surface layer of rapid band, which we should call the "surface slip band" to distinguish it from the bulk shear banding and the ideal wall slip. Second, there are two or three macroscopic bands having

3 Video 9.3: PTV movie observing steady shear of 1M(13%)-10K solution at 5 (1/s) at room temperature, scanning different XZ planes over time, made by P. Boukany at OSU. To avoid edge effect the gap distance is only 50 micron.

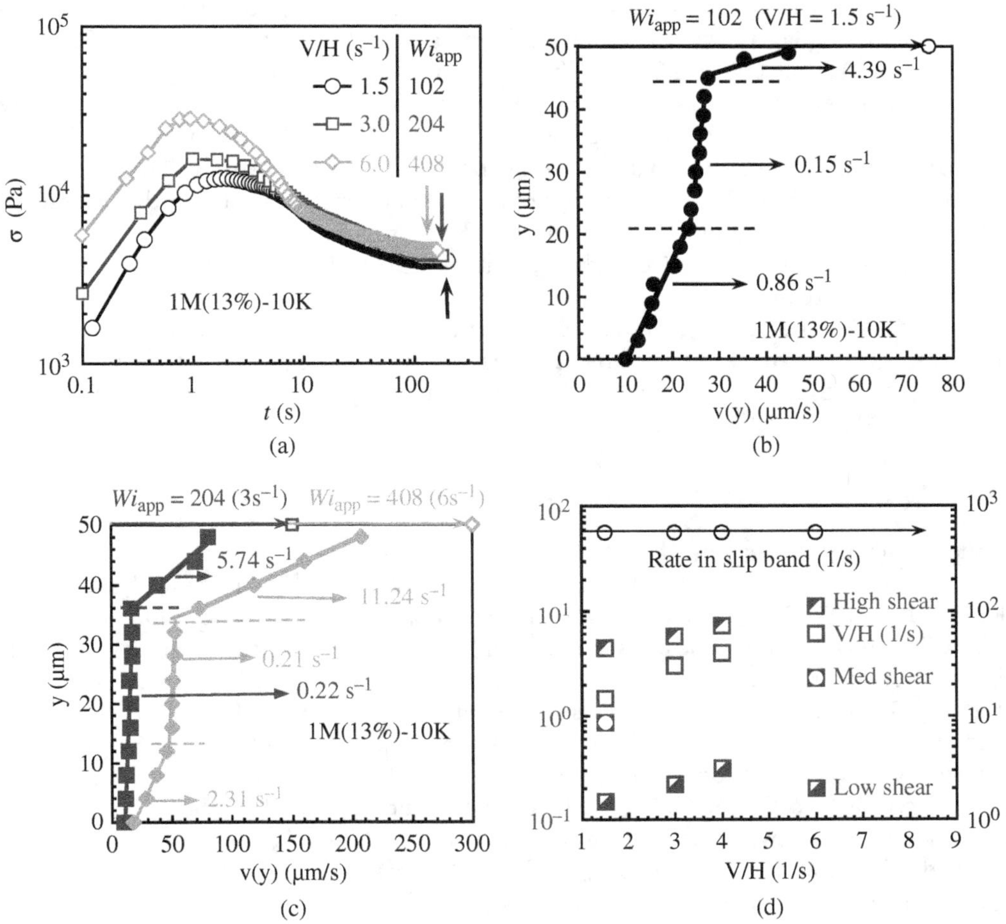

Figure 9.10 (a) Stress responses to startup shear of PB1M(13%)-10K solution at nominal rates of 1.5, 3, and 6 s^{-1}. (b) Velocity profile at V/H = 1.5 s^{-1}, which shows three bands of rates different from 1.5 s^{-1}. (c) Velocity profiles at the other two rates of 3 and 6 s^{-1}. (d) Different local shear rates at the four different applied nominal rates (open squares). Here, the open circles represent the level of rate in the slip band at the moving plate plotted against the right-hand-side Y axis. *Source*: Boukany *et al.* [47] http://pubs.rsc.org/-/content/articlehtml/2015/sm/c5sm01429h. Used under CC BY 3.0 https://creativecommons.org/licenses/by/3.0/.

different local shear rates. In Fig. 9.10(d), we plot these local shear rates against the apparent shear rate V/H. The largest and smallest local shear rates were nearly two decades apart. The shear rate in the surface slip band is much larger than that in the largest shear macroscopic band. Clearly, a wide range of different local shear rates can coexist during startup shear of an entangled PB solution.

9.2 Overcoming Wall Slip during Startup Shear

9.2.1 Strategy Based on Choice of Solvent Viscosity

All the successful PTV observations of bulk shear banding require a proper understanding of the influence of wall slip on the intended rheometric measurements. Shear banding cannot take place when $b_{max}/H \ll 1$. A sufficient degree of polymer entanglement is prerequisite for a sizable $b_{max} \sim H$. For $b_{max}/H \gg 1$, wall slip remains dominant until the maximum wall slip occurs in the stress plateau when V_s attains its limiting value V_s^*. For the bulk to be effectively sheared, Eq. (6.24a) must be satisfied. As discussed in Chapter 4, for a low-speed PTV method to provide accurate determination of the shear field in a gap distance of about 1 mm, the velocity of the shearing surface should be bounded by $V^* = 5–10$ mm/s, as indicated in Eq. (4.4). Thus, if a polymer has $V_s^* > V^*$ at the stationary wall, it can be challenging to determine the velocity field in the gap. For polymer melts, Eq. (6.9a) shows that we have little control over the magnitude of its V_s^* at a given temperature. For entangled solutions, significant control can be exerted to affect the value of V_s^* by the choice of the solvent viscosity η_s in Eq. (6.10) and lowering the concentration ϕ to reduce G_{pl}.

Although V_s^* does not depend on the degree of entanglement or molecular weight (MW) as shown in Eq. (6.10), stronger entanglement and larger value of the terminal relaxation time τ do broaden the rate range for wall slip. Wall slip begins at $Wi_{app} > 1$ and remains dominant until $Wi_{ws\text{-}bnl} \sim (2V_s^*/H)\tau$, which grows as strongly with the molecular weight as τ does. Actually, in steady shear, whenever $b/H > 1$, namely,

$$V_s/H > \dot{\gamma} = Wi/\tau \tag{9.7}$$

wall slip is a significant response, by definition.

There are two examples in the literature to demonstrate the strategy to control the magnitude of wall slip. First, we already showed in Section 6.3.2 that a water solution of DNA at $c = 22$ mg/ml shows nothing else but slip in the PTV observation. The massive wall slip shown in Fig. 6.15 stems from the use of the inviscous water as the solvent, leading to a large slip velocity V_s^*. Table 9.3 shows that the massive magnitude of wall slip in the water solution can be effectively reduced by use of ethylene glycol. When η_s changes by a factor of 23 from water to ethylene glycol the solution viscosity η and relaxation time τ did not change by the same amount. According to Eq. (6.10), V_s^* can be estimated directly based on the value of η_s and plateau modulus G_{pl} as well as the slip layer thickness $l_{ent}(\phi)$. The values are listed in Table 9.3. Also listed in Table 9.3 are the values of b_{max} taken as $V_s^*\tau$ as shown in Eq. (6.11). Equivalently, if b_{max} is evaluated according to Eq. (6.13) and V_s^* is in return obtained from b_{max} using Eq. (6.11), we have the numbers in parentheses in Table 9.3. The difference arises from the numerical simplification that takes the prefactor as unity in $\eta \sim G_{pl}\tau$.

Because the solvent viscosity is increased by a factor of 23 upon switching from water to ethylene glycol, V_s^* decreases by a similar amount, as predicted by Eq. (6.10). In contrast to Fig. 6.15, a V/H of $5\,s^{-1}$ is now enough to assure $Wi \gg 1$ and to produce shear banding, as shown in Fig. 9.11(a). Also presented is the accompanying stress versus time measurement in Fig. 9.11(b). At $V/H = 5\,s^{-1}$, in comparison to $V = 5.0\,(s^{-1}) \times 1\,mm = 5$ mm/s, $V_s^* = 0.96$ mm/s (Table 9.3) of the DNA–ethylene glycol solution is only a small fraction of V. Thus, the bulk must be sheared at an average rate of $(5–2 \times 0.96)$ (mm/s)/1 mm = $3.1\,s^{-1}$, corresponding to $Wi = 450$, which is very large. Consequently, there is a very pronounced stress overshoot. It is interesting to note that if we further boost the solvent viscosity using glycerol, wall slip is completely suppressed as shown in the preceding Fig. 9.3(b).

In the second example, we demonstrate that with a sufficient reduction of V_s^* by employing solvents of increasingly high viscosity, even shear banding can be made to disappear (cf. Fig. 9.15(a)). Table 9.4 lists four 1,4-polybutadiene solutions, all at 10 volume percent ($\phi = 0.1$), the solutes being 1,4-polybutadiene of four different molecular weights 1.5, 5, 9, and 15 kg/mol, respectively. Here we estimate V_s^* using Eq. (6.10) and b_{max} using (Eq. (6.11)) with $l_{ent}(\phi) = \phi^{-0.6} l_{ent}(\phi = 1) = 15$ nm, given $l_{ent}(\phi = 1)$ is 3.75 nm.[46] Clearly, 1M(10%)-1.5K solution has a large enough b_{max} to show slip-like behavior. Figure 9.12 shows apparent wall slip in the sense that the shear band at the bottom surface is of a thickness too thin to be resolved by the

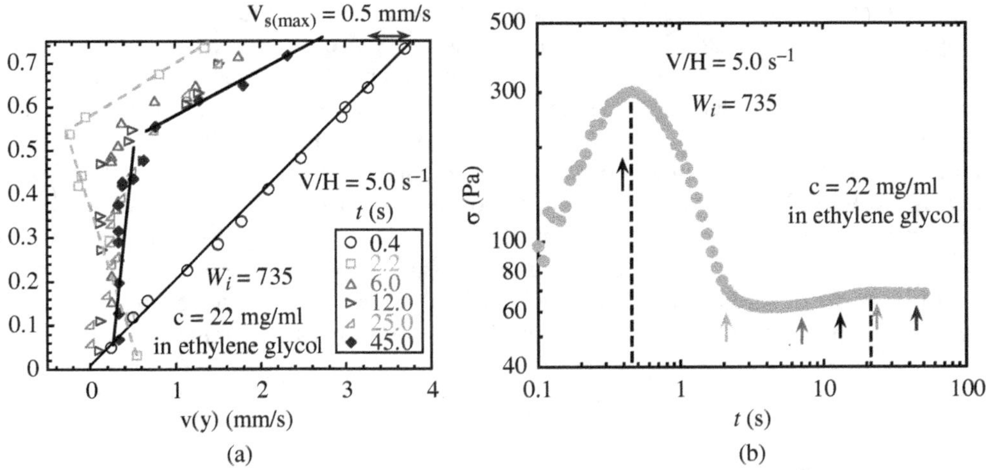

Figure 9.11 (a) Velocity profiles of an entangle DNA solution at different stages during startup shear at an apparent rate of 5 s^{-1}. (b) The corresponding stress versus time curve where the vertical arrows indicate the stages for the PTV observations in (a). *Source*: Data taken from Boukany and Wang 2009 [43].

PTV measurement. Since we have no reason to distrust the estimated V_s^* in Table 9.4, the "slip velocity" on the order of (800–200) μm/s = 0.6 mm/s is clearly much too large in comparison to the estimated V_s^* in the range 0.07–0.12 mm/s. In passing, it is interesting to also estimate V_s^* for 1M(10%)-1.5K from Eq. (9.2). Reading the value for $V_s^*(1) = 10$ mm/s from Table 6.1, we have $V_s^*(0.1) = V_s^*(1)(1.5/1.5)^{3.4}(0.1)^{1.6} \sim 0.25$ mm/s, taking $M_e = 1.5$ kg/mol for 1,4-PB melts. This estimate is different from that list in Table 9.4 only by a factor of 2. Thus, we confirm that the two independent estimates of V_s^* are self-consistent, as expected.

The apparent wall slip even shows up in 1M(10%)-5K solution, as shown in Fig. 9.13. This solution should not exhibit measurable wall slip, yet both circles and squares show a "slip velocity"

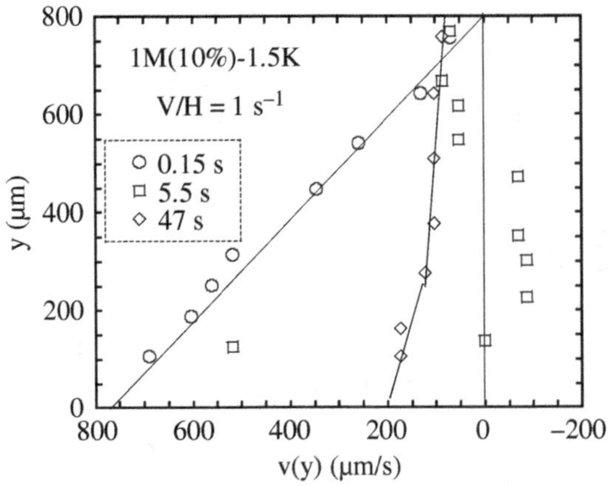

Figure 9.12 Velocity profiles at three different moments during startup shear at an apparent rate of 1 s^{-1}, showing apparent wall slip. *Source*: Data taken from Ravindranath and Wang 2008 [19].

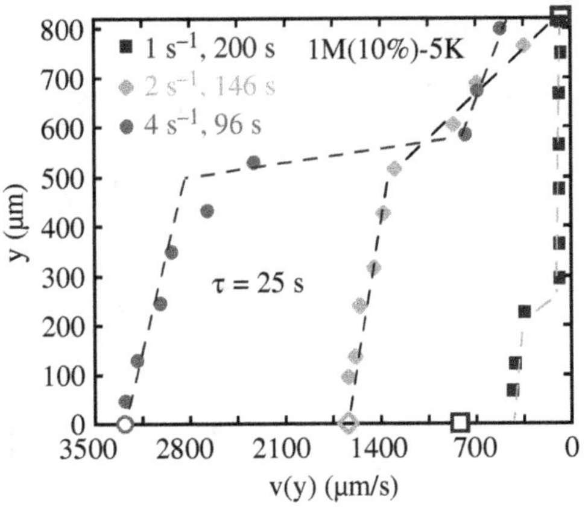

Figure 9.13 Long-time velocity profiles at three different shear rates during startup shear of PB1M(10%)-5K, showing shear banding and apparent wall slip. Here the open symbols denote the speeds of the walls in the absence of wall slip. *Source*: Data taken from Ravindranath and Wang 2008 [19].

of approximately 0.4–0.5 mm/s. The data seem to show that a thin ultrafast surface shear band can appear in entangled solutions with appreciable b_{max} either at the shearing or at the stationary surfaces. Perhaps, once full chain disentanglement takes place at the interface(s), the disentanglement can spread from the surfaces into the bulk. Elsewhere in the bulk, the shear bands do not reach such an extreme level of high shear rate. Currently, there is no good understanding concerning the origin or mechanism for how surface shear bands form. If the slip magnitude is further reduced by using PB9K as the solvent, the surface shear essentially disappears, as shown in Figure 9.14(a,b). Finally, when PB15K is chosen as the solvent, further reducing V_s^* and b_{max}, as shown in Table 9.4, approximate shear homogeneity returns, as demonstrated (Fig. 9.15(a)) by the particle-tracking velocimetric observations. During startup creep, at different stages, shear homogeneity also prevails in this solution, as shown in Fig. 9.15(b).

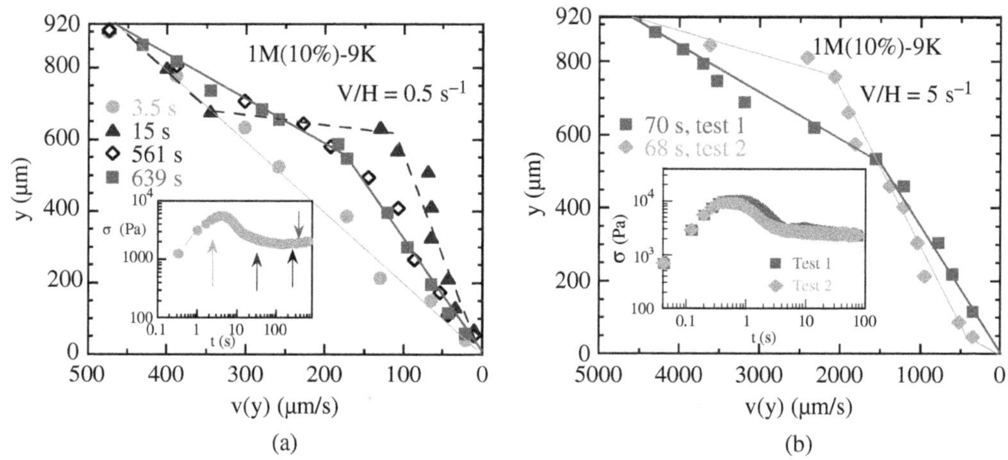

Figure 9.14 (a) Velocity profiles at four different moments during startup shear at 0.5 s^{-1} along with the corresponding stress versus time plot in the inset. (b) Long-time velocity profiles at V/H = 5 s^{-1} from two different loadings along with the stress versus time plot in the inset. *Source*: Data taken from Ravindranath et al. 2008 [18].

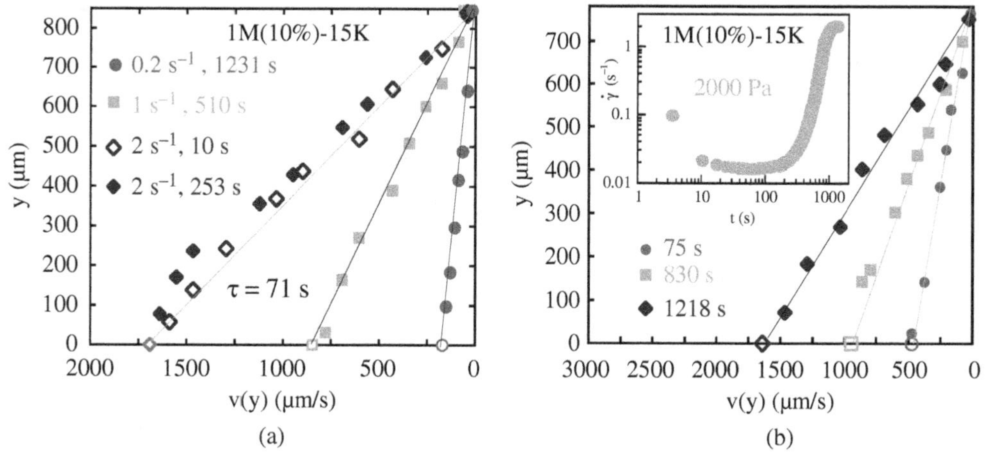

Figure 9.15 (a) Linear velocity profiles for PB1M(10%)-15K upon startup shear at apparent rates 0.2, 1, and 2 s^{-1}. (b) Linear velocity profiles upon startup creep at 2000 Pa at different stages of the same solution where the inset shows the entanglement–disentanglement transition, namely, the rise of the shear rate over time during the creep. *Source*: Data taken from Ravindranath and Wang 2008 [19].

9.2.2 Negligible Slip Correction at High Wi_{app}

To overcome the influence of wall slip, namely, for wall slip to be a small correction to the shear strain field, we need the shearing velocity $V = v(y = H)$ to be much greater than V_s^*. At a larger molecular weight, the reptation time or terminal relaxation time τ is longer, thus, $(V_s^*/H)\tau$ is greater, namely, $Wi_{ws\text{-}bnl}$ of Eq. (6.23) is larger. With increasing Wi_{app}, the transition from wall slip and bulk nonlinear deformation occurs at $Wi_{app} = Wi_{ws\text{-}bnl} \sim (2V_s^*/H)\tau$ that varies linearly with τ (V_s^* being molecular weight independent). To make wall slip unimportant, we must either reduce V_s^* (as demonstrated in Section 9.2.1) or we must apply sufficiently large $V \gg V_s^*$. Under the condition $V_s^* < V$, $Wi > 1$ occurs at all stages of startup shear. As illustrated in Fig. 6.26(a), when Wi_{app} increases beyond $Wi_{ws\text{-}bnl}$, b starts to drop below b_{max}, eventually leading to diminished wall-slip corrections when Wi_{app} increases sufficiently. From Eq. (6.28) and Fig. 6.27, $Wi = Wi_{app} - Wi_{ws\text{-}bnl}$ for $Wi_{app} > Wi_{ws\text{-}bnl}$. In this regime, according to Eq. (6.27), the condition $b/H \ll 1$ is met when $(Wi_{app})^{(1-n)} \gg b_{max}/H \sim Wi_{ws\text{-}bnl}$. In this limit, the slip correction can be evaluated in terms of the velocity ratio, obtained by inserting $b(Wi)$ of Eq. (6.27) into Eq. (6.4a),

$$V_s/V \sim (b_{max}/H)Wi^{n-1}. \tag{9.8}$$

If $b_{max}/H \sim 1$, $V_s/V \ll 1$ is realized for $Wi_{app} \gg Wi_{ws\text{-}bnl}$, which leads to $Wi \gg 1$. As shown in Table 9.4, most polybutadiene solutions studied in the literature (cf. Table 9.1) belong to the class having $b_{max}/H \sim 1$. Consequently with clean metallic shear surfaces, wall slip is not a complicating factor so long as $Wi_{app} \gg 1$. When $b_{max}/H \gg 1$, as shown in Tables 9.2 and 9.3 for entangled DNA solutions and in Table 6.2 for entangled PB solutions in a low-gap shear cell, significant wall slip is dominant unless $Wi_{app} \gg Wi_{ws\text{-}bnl}$, a circumstance shown in Figs. 9.3(a,b), 9.4, 9.5, and 9.10(a–d).

9.2.3 Summary on Shear Banding

Several characteristics of shear banding are apparent from multiple particle-tracking velocimetric (PTV) studies. (a) A high enough value of Z is only a necessary condition for shear banding; it is not a sufficient condition. Even when Z is large, we can still make b_{max}/H small by having a large η_s in Eq. (6.10) or a sufficiently large value for H. The 1M(10%)-15K solution in Table 9.4 used a 15 kg/mol PB as the solvent. Its zero-shear viscosity η_s(15K) is 20 times the viscosity η_s(5K) of 1,4-polybutadiene with M = 5 kg/mol. Consequently, b_{max} of 1M(10%)-15K is reduced to ca. 0.14 mm, making b_{max}/H so small that shear banding is no longer seen for Wi_{app} ranging from 14 to 71, as shown in Fig. 9.15(a,b). In contrast, the 1M(10%)-5K solution with the same Z = 40 readily undergoes shear banding. Comparison between Figs. 9.13 and 9.15(a) demonstrates that b_{max}/H is a pertinent parameter for linear polymers. (b) The character of the shear field changes with the applied shear rate that determines the value of Wi_{app}. There can be more than two bands across the gap. Local shear rates in the bands depend on Wi_{app}, as shown in Fig. 9.14(a,b). (c) Different loadings produce different shear bands, suggesting that the band structures are not deterministic, as shown by the comparison between Figs. 9.8(b) and 9.14(b) involving three different loadings. (d) Shear banding also occurs during startup shear in stress-controlled mode, as shown in Figs. 9.4 and 9.9. (e) When b_{max}/H is equal to or larger than unity, shear banding often takes place at the shearing surface(s) and is spatially too thin to be resolved by PTV. Such surface-shear bands look like apparent wall slip, as shown in Fig. 9.12. (f) In theory, the shear banding profile should be S shaped, namely, symmetric with respect to the middle plane of the sheared sample. More often than not, there is a lack of such symmetry, perhaps because the sample loading or device geometry is not perfectly symmetric or because only one of the two surfaces moves. We usually see two bands, with the fast band occurring next to the moving surface. Unlike the cone-plate assembly, in the parallel-disk, parallel-sliding plate geometries, sometimes there are three bands situated somewhat symmetrically. We suspect that a stationary film wrapped around the meniscus in the particle-tracking velocimetric setup could also introduce symmetry breaking. When simple shear is generated with a counter-rotating parallel-disk shear cell, the symmetry is often restored as shown in the movie in Video 9.4.[4] (g) PTV has its limitations. The present PTV technique typically records the velocity field over a small area, $2 \times 1.5\ \mathrm{mm}^2$. Measurements on such a small area are adequate provided that the strain field is invariant along the shearing direction and independent of the position along the vorticity direction.

Little is known about whether and how entangled polymers with long-chain branching would suppress or delay shear banding. It is desirable for future studies to systematically investigate the effect of long-chain branching on wall slip and shear banding. Morever, it remains unknown whether the PB solutions are most prone to shear banding, relative to entangled solutions based on polyisoprene, polystyrene, etc.

9.3 Nonlinearity and Shear Banding in Large-Amplitude Oscillatory Shear

Large-amplitude oscillatory shear (LAOS) is a widely used method to probe nonlinear responses of viscoelastic materials including entangled polymeric liquids. LAOS refers to experiments in which the amplitude of oscillatory shear is so large that the rheological responses may be nonlinear and change with time. Equation (1.11) is inadequate to describe LAOS experiments. LAOS experiments are typically interpreted based on the assumption that the sample experiences uniform strain at all times during each cycle of oscillatory shear. Moreover, most past studies have focused on steady-state behavior and ignored the evolution of the system from its initial state to a final state.

9.3.1 Strain Softening

Strain softening occurs when the steady-state peak stress decreases with increasing strain amplitude. Strain softening is typically obtained by programming a strain sweep test in a commercial rheometer. A more informative way to explore this behavior is to apply a sinusoidal strain and monitor the stress as a function of time. Steady-state peak stress versus the strain amplitude γ_0 can be obtained by discrete experiments at different values of γ_0. From each experiment, it is useful to collect the raw stress and strain data. A commercial rheometer can usually export such data, allowing explicit measurement of $\sigma(t)$ as a function of time. When γ_0 exceeds a critical value (ca. 100% for entangled polymers), the stress peak may decrease over time, typically accompanied by an increase in $\tan\delta$. When strain softening is observed, it is important to question whether the sample is experiencing homogeneous shear. Until recently,[48] such a concern did not exist in most experimental and theoretical studies. Many rheological studies involving LAOS still overlook the possibility of inhomogeneous strain across the sample thickness.

---

4 Video 9.4: PTV movie capturing startup shear of DNA-glycerol solution at concentration of 11 mg/mL and a shear rate of 0.2 (1/s) using a specially designed counter-rotating device fit onto MCR 301 rotational rheometer at room temperature, made at University of Akron.

9.3.2 Wave Distortion

If the stress response is observed to be nonsinusoidal, then nonlinearity has obviously emerged. Unlike small-amplitude oscillatory shear (SAOS), this nonlinearity in LAOS plausibly originates from changes in the underlying microstructure. Structural changes can develop over many cycles in time. In the long time limit, the final state may or may not be stable. When the state of viscoelasticity changes during a cycle due to emergence of unstable structures, nonsinusoidal stress should arise. Conversely, does a sinusoidal stress response necessarily imply that the LAOS is homogeneous? In general, we do not know the answer. To quantify the wave distortion, one can carry out Fourier transform analysis.[49,50]

9.3.3 Shear Banding

Nonlinear responses in LAOS do not have to be accompanied by either wave distortion or shear inhomogeneity. When the structural rearrangement manages to take place without any localization, a homogeneous strain field should prevail. Strain softening in LAOS can take place in the absence of shear banding when the entangled polymers are very weakly entangled. Further investigation remains to be carried out concerning the threshold value of Z (the number of entanglements per chain) above which shear banding develops under LAOS over time.

In general, rheological characteristics of LAOS may not be used as rheological fingerprints. Moreover, we cannot perform meaningful analysis of the wave distortion if shear banding takes place. Furthermore, no features of LAOS can be used to discern whether shear banding has taken place. Thus, it is necessary to carry out PTV observations before making any attempt to characterize the stress response. In LAOS, the state of entanglement can evolve over a long period of time, for example, longer than the terminal relaxation time. Consequently, shear banding occurs more readily under LAOS than during startup continuous shear. Any sample that shows shear banding in startup shear should also undergo strain localization in LAOS.

PB1M(15%)-9K solutions show strong shear banding, as reported in the literature.[18] Thus, it is straightforward to show how shear banding occurs in such a well-entangled polymer solution. LAOS for entangled polymer solutions occurs when γ_0 is comparable to unity and the oscillation frequency ω exceeds the crossover frequency ω_c. To determine the parameters for LAOS, we can apply a sufficiently high γ_0 and $\omega > \omega_c$ as shown in Fig. 9.16. Clearly, the amplitude of $\gamma_0 = 70\%$ is hardly high enough to qualify as a condition for LAOS. On the other hand, 100%

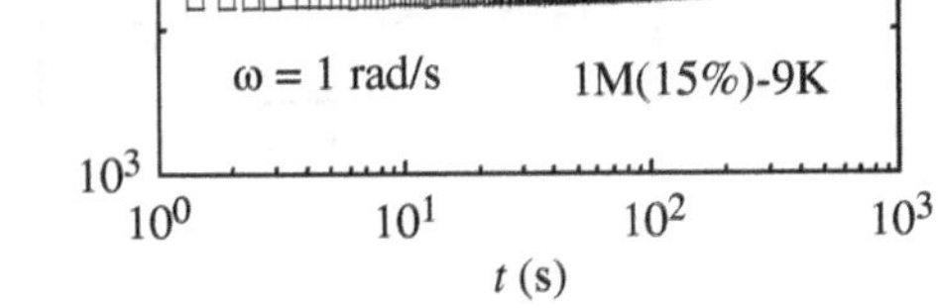

Figure 9.16 Apparent storage and loss moduli G′ and G″ of 1M(15%)-9K solution at room temperature from LAOS at strain amplitude $\gamma_0 = 70\%$ and 100% and frequency $\omega = 1$ rad/s, based on 25 mm parallel-plate disks where γ_0 is taken as the strain at a radial distance of 4 mm from the edge. *Source*: Data taken from Ravindranath and Wang 2008 [48].

is high enough to force the solution to change its viscoelastic characteristics significantly over time. However, the declining G′ and corresponding rising G″ do not reveal whether such "strain softening" has taken place homogeneously or not. Raw data in terms of the waveforms are also inadequate to signal any emergence of shear banding. Conversely, the characteristics of the stress wave would lose their inherent meanings if shear banding is present. It is pertinent to have information about the strain field using an effective technique (e.g., PTV). For example, Fig. 9.17 in combination with Fig. 9.16 shows that the initial deterioration of viscoelasticity (the state of entanglement) occurs homogeneously and the corresponding Lissajous plot (not shown here but available in Ref. [48]) remains a symmetric (nondistorted) ellipse. Three shear bands eventually emerge in time. Because γ_0 is relatively low, a long time is required before shear banding can form. At larger γ_0 shear banding can occur much earlier in time than observed for a smaller γ_0. In LAOS, the peak shear rate is determined by the product of ω and γ_0. Figure 9.18 shows the strain field for LAOS at frequencies 0.25, 0.5, and 1 rad/s at the same amplitude $\gamma_0 = 400\%$.[5] The velocity field in LAOS is typically quasi-symmetric in the presence of shear banding, namely, the fastest band is found in the middle of the sample, with slower bands on each side. Shear banding during startup continuous shear usually lacks any symmetry, for example, the fast band is often found near the shearing surface as shown in Fig. 9.8(a–c).

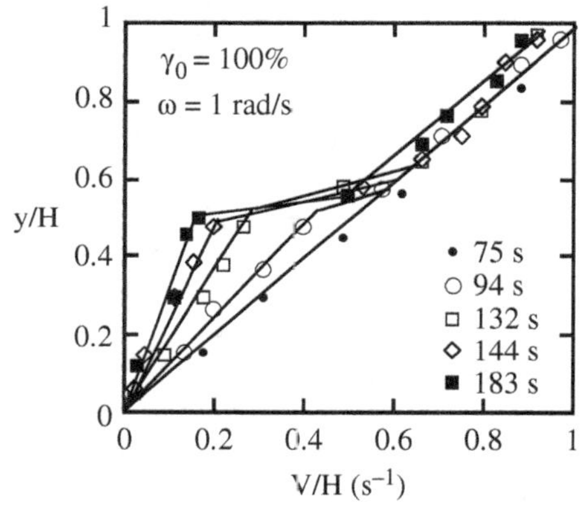

Figure 9.17 Velocity profiles at the instant of maximum plate speed for different times at a radial distance of 4 mm from the edge in 25 mm parallel-plate disks. The sample at this radial distance experienced 100% strain amplitude and the applied frequency was 1 rad/s. *Source*: Data taken from Ravindranath and Wang 2008 [48].

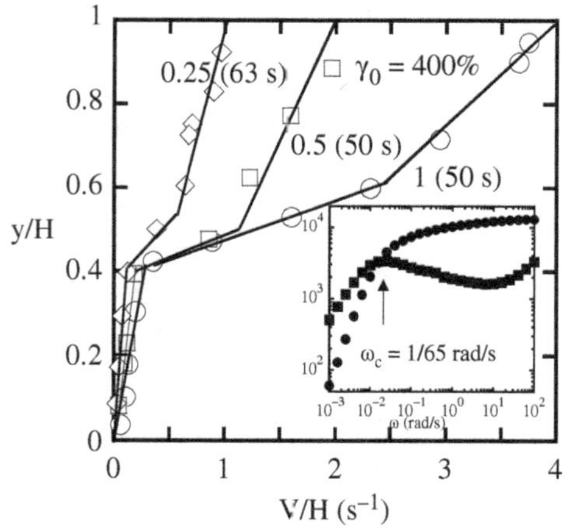

Figure 9.18 Velocity profiles at three different frequencies of 0.25, 0.5, and 1 rad/s at 63, 50, and 50 s from the start of the LAOS, respectively. *Source*: Data taken from Ravindranath and Wang 2008 [48].

5 Video 9.5: LAOS of 1M(15%)-9K solution at amplitude 400% and frequency 1 rad/s, at room temperature, made at University of Akron.

In passing, we should mention that nonlinear response to oscillatory shear can also be produced in the creep mode. In other words, we can apply a sinusoidal stress wave and monitor the strain response. In such a large-stress oscillatory shear (LSOS) mode, nonlinear response occurs when the peak stress and peak strain are not linearly related. In LSOS, the "strain softening" will take place in the sense that over time the strain amplitude γ_0 would increase at a given $\sigma_0 sin(\omega t)$. For entangled polymers, LSOS can be more interesting than LAOS to probe how the state of entanglement evolves under high stress. In fact, we can expect entanglement–disentanglement transition-like responses during large-stress oscillatory shear[51]. In other words, there could be a threshold beyond which γ_0 at long times grows sharply over a narrow range of σ_0. Few systematic studies have been carried out using large-stress oscillatory shear, especially studies that also performed PTV. Hopefully, future investigations will consider this interesting topic.

In conclusion, like the subjects of wall slip, yielding, "strain hardening" treated in Chapters 6 through 8 respectively, this chapter on particle-tracking velocimetric characterization of the shear field during startup shear and large-amplitude oscillatory shear is also unique in a similar sense that the topic has not been studied in available books on polymer rheology. The role of particle-tracking velocimetry to reveal shear inhomogeneity cannot be understated: The discovery of bulk shear banding allows the notion of yielding to be proposed and confirmed. Conversely, this chapter shows how shear inhomogeneity can be made to diminish by reducing the magnitude of the extrapolation length b relative to the gap distance H. Since shear strain localization ruins the rheology tradition (based on characterization of the stress response to a presumed deformation field), it would be desirable that shear banding would not occur under all circumstances. Aiming to show that shear banding should not occur, Li *et al.*[52] applied weak shear conditions and claimed to observe no shear banding, without marching deeply into the regime of shear banding. They were misled to assume and speculate that shear banding was caused by edge fracture. Consequently, they avoided the condition of high shear required to produce shear banding yet claimed that shear banding was due to edge failure. If they did not produce and witness edge failure, then on what basis[53] did they speculate about a correlation between shear banding and edge effects? As pointed out in Ref. [54], their main sample was mischaracterized and was insufficiently entangled to produce shear banding. Ref. [55] summarized the responses to Ref. [53], which was a rebuttal to Ref. [54].

References

1 Callaghan P, Cates M, Rofe C, Smeulders J. A study of the "spurt effect" in wormlike micelles using nuclear magnetic resonance microscopy. J Phys II. 1996;6(3):375–393.

2 Mair R, Callaghan P. Observation of shear banding in worm-like micelles by NMR velocity imaging. Europhys Lett). 1996;36(9):719.

3 Britton MM, Callaghan PT. Two-phase shear band structures at uniform stress. Phys Rev Lett. 1997;78(26):4930.

4 Britton M, Mair R, Lambert R, Callaghan P. Transition to shear banding in pipe and Couette flow of wormlike micellar solutions. J Rheol. 1999;43(4):897–909.

5 Fischer E, Callaghan P. Is a birefringence band a shear band? Europhys Lett. 2000;50(6):803.

6 Salmon J-B, Colin A, Manneville S, Molino F. Velocity profiles in shear-banding wormlike micelles. Phys Rev Lett. 2003;90(22):228303.

7 Salmon J-B, Manneville S, Colin A. Shear banding in a lyotropic lamellar phase. I. Time-averaged velocity profiles. Phys Rev E. 2003;68(5):051503.

8 Bécu L, Manneville S, Colin A. Spatiotemporal dynamics of wormlike micelles under shear. Phys Rev Lett. 2004;93(1):018301.
9 Hu Y, Lips A. Kinetics and mechanism of shear banding in an entangled micellar solution. J Rheol. 2005;49(5):1001–1027.
10 Lopez-Gonzalez M, Holmes W, Callaghan P. Rheo-NMR phenomena of wormlike micelles. Soft Matter. 2006;2(10):855–869.
11 Gurnon AK, Lopez-Barron C, Wasbrough MJ, Porcar L, Wagner NJ. Spatially resolved concentration and segmental flow alignment in a shear-banding solution of polymer-like micelles. ACS Macro Lett. 2014;3(3):276–280.
12 Boukany PE, Wang S-Q. Use of particle-tracking velocimetry and flow birefringence to study nonlinear flow behavior of entangled wormlike micellar solution: from wall slip, bulk disentanglement to chain scission. Macromolecules. 2008;41(4):1455–1464.
13 Tordella J. Fracture in the extrusion of amorphous polymers through capillaries. J Appl Phys. 1956;27(5):454–458.
14 Tordella JP. Unstable flow of molten polymers: a second site of melt fracture. J Appl Polym Sci. 1963;7(1):215–229.
15 Tordella JP. Unstable flow of molten polymers. In: Eirich FR, editor. Rheology: theory and applications 5. Academic Press, New York; 1969. pp. 57–92.
16 Piau J, El Kissi N, Tremblay B. Influence of upstream instabilities and wall slip on melt fracture and sharkskin phenomena during silicones extrusion through orifice dies. J Nonnewton Fluid Mech. 1990;34(2):145–180.
17 Mackley M, Rutgers R, Gilbert D. Surface instabilities during the extrusion of linear low density polyethylene. J Nonnewton Fluid Mech. 1998;76(1):281–297.
18 Ravindranath S, Wang S-Q, Olechnowicz M, Quirk RP. Banding in simple steady shear of entangled polymer solutions. Macromolecules. 2008;41(7):2663–2670.
19 Ravindranath S, Wang S-Q. Steady state measurements in stress plateau region of entangled polymer solutions: controlled-rate and controlled-stress modes. J Rheol. 2008;52(4):957–980.
20 Cheng S, Wang S-Q. Is shear banding a metastable property of well-entangled polymer solutions? J Rheol. 2012;56(6):1413–1428.
21 Hu YT. Steady-state shear banding in entangled polymers? J Rheol. 2010;54(6):1307–1323.
22 Menezes E, Graessley W. Study of the nonlinear response of a polymer solution to various uniaxial shear flow histories. Rheol Acta. 1980;19(1):38–50.
23 Osaki K, Inoue T, Isomura T. Stress overshoot of polymer solutions at high rates of shear. J Polym Sci Part B Polym Phys. 2000;38(14):1917–1925.
24 Osaki K, Inoue T, Isomura T. Stress overshoot of polymer solutions at high rates of shear; polystyrene with bimodal molecular weight distribution. J Polym Sci Part B Polym Phys. 2000;38(15):2043–2050.
25 Osaki K, Inoue T, Uematsu T. Stress overshoot of polymer solutions at high rates of shear: semidilute polystyrene solutions with and without chain entanglement. J Polym Sci Part B Polym Phys. 2000;38(24):3271–3276.
26 Pattamaprom C, Larson RG. Constraint release effects in monodisperse and bidisperse polystyrenes in fast transient shearing flows. Macromolecules. 2001;34(15):5229–5237.
27 Menezes E, Graessley W. Nonlinear rheological behavior of polymer systems for several shear-flow histories. J Polym Sci Polym Phys Ed. 1982;20(10):1817–1833.
28 Tapadia P, Wang S-Q. Nonlinear flow behavior of entangled polymer solutions: yieldlike entanglement–disentanglement transition. Macromolecules. 2004;37(24):9083–9095.
29 Tapadia P, Wang S-Q. Direct visualization of continuous simple shear in non-Newtonian polymeric fluids. Phys Rev Lett. 2006;96(1):016001.

30 Wang Y, Li X, Zhu X, Wang S-Q. Characterizing state of chain entanglement in entangled polymer solutions during and after large shear deformation. Macromolecules. 2012;45(5):2514–2521.
31 Cates M, McLeish T, Marrucci G. The rheology of entangled polymers at very high shear rates. Europhys Lett. 1993;21(4):451.
32 Marrucci G. Dynamics of entanglements: a nonlinear model consistent with the Cox-Merz rule. J Non-Newtonian Fluid Mech. 1996;62(2–3):279–289.
33 Doi M, Edwards S. Dynamics of concentrated polymer systems. Part 4.—rheological properties. J Chem Soc Faraday Trans 2 Mol Chem Phys. 1979;75:38–54.
34 Graham RS, Likhtman AE, McLeish TCB, Milner ST. Microscopic theory of linear, entangled polymer chains under rapid deformation including chain stretch and convective constraint release. J Rheol. 2003;47(5):1171–1200.
35 Preliminary results disclosed in oral presentations at Annual Meeting of the Society of Rheology, October 2005, Portland, OR, and Annual Meeting of Division of Fluid Dynamics of the American Physical Society, Chicago, November 2005.
36 Callaghan P, Gil A. Rheo-NMR of semidilute polyacrylamide in water. Macromolecules. 2000;33(11):4116–4124.
37 Tapadia P, Ravindranath S, Wang S-Q. Banding in entangled polymer fluids under oscillatory shearing. Phys Rev Lett. 2006;96(19):196001.
38 Wang S-Q, Ravindranath S, Boukany P, Olechnowicz M, Quirk RP, Halasa A, et al. Nonquiescent relaxation in entangled polymer liquids after step shear. Phys Rev Lett. 2006;97(18):187801. A refined explanation for the observed elastic yielding was subsequently proposed. J Chem Phys 2007;127:064903.
39 Hu YT, Wilen L, Philips A, Lips A. Is the constitutive relation for entangled polymers monotonic? J Rheol. 2007;51(2):275–295.
40 Hu YT. Response to: CPP vs circular couette. J Rheol. 2012;56(4):683–686.
41 Ravindranath S, Wang Y, Boukany P, Li X. Letter to the editor: cone partitioned plate (CPP) vs circular couette. J Rheol. 2012;56(4):675–681.
42 Boukany PE, Wang S-Q. Shear banding or not in entangled DNA solutions depending on the level of entanglement. J Rheol. 2009;53(1):73–83.
43 Boukany PE, Wang S-Q. Exploring the transition from wall slip to bulk shearing banding in well-entangled DNA solutions. Soft Matter. 2009;5(4):780–789.
44 Boukany PE, Hu YT, Wang S-Q. Observations of wall slip and shear banding in an entangled DNA solution. Macromolecules. 2008;41(7):2644–2650.
45 Boukany PE, Wang S-Q. Shear banding or not in entangled DNA solutions. Macromolecules. 2010;43(17):6950–6952.
46 Fetters L, Lohse D, Richter D, Witten T, Zirkel A. Connection between polymer molecular weight, density, chain dimensions, and melt viscoelastic properties. Macromolecules. 1994;27(17):4639–4647.
47 Boukany PE, Wang S-Q, Ravindranath S, Lee LJ. Shear banding in entangled polymers in the micron scale gap: a confocal-rheoscopic study. Soft Matter. 2015;11(41):8058–8068.
48 Ravindranath S, Wang S-Q. Large amplitude oscillatory shear behavior of entangled polymer solutions: particle tracking velocimetric investigation. J Rheol 2008;52:341.
49 Wilhelm M. Fourier-transform rheology. Macromol Mater Eng. 2002;287(2):83–105.
50 Li X, Wang S-Q, Wang X. Nonlinearity in large amplitude oscillatory shear (LAOS) of different viscoelastic materials. J Rheol. 2009;53(5):1255–1274.
51 Ge S, Zhu X, Wang S-Q. Watching shear thinning in creep: Entanglement–disentanglement transition. Polymer. 2017; 125:254–264.

52 Li YF, Hu M, McKenna GB, Dimitriou CJ, McKinley GH, Mick RM, et al. Flow field visualization of entangled polybutadiene solutions under nonlinear viscoelastic flow conditions. J Rheol. 2013;57(5):1411–1428.
53 Li YF, Hu M, McKenna GB, Dimitriou CJ, McKinley GH, Mick RM, et al. Response to: sufficiently entangled polymers do show shear strain localization at high enough Weissenberg numbers". J Rheol. 2014;58(4):1071–1082.
54 Wang S-Q, Liu G, Cheng S, Boukany PE, Wang Y, Li X. Letter to the Editor: sufficiently entangled polymers do show shear strain localization at high enough Weissenberg numbers. J Rheol. 2014;58(4):1059–1069.
55 Wang SQ. Challenging conditions to observe shear banding in highly entangled polybutadiene solutions Rheol Open Access 2017;1:104.

10

Strain Localization in Extrusion, Squeezing Planar Extension: PTV Observations

The preceding chapter illustrated strain localization in simple shear geometry, in which the shear stress is the same at all points along the velocity gradient direction (Y axis). For well-entangled polybutadiene solutions, particle-tracking velocimetric observations reveal that different shear rates can coexist and persist over hundreds of strain units during startup shear. If the physics behind shear banding is yielding of the entanglement network through chain disentanglement, as suggested in Chapter 16, then strain localization could also be significant in more complex geometries, such as those encountered in industrial polymer processing. This chapter describes shear strain localization in less well-defined configurations, such as the convergence of a polymer melt into a narrow die from a large reservoir (barrel) during extrusion and squeezing.

10.1 Capillary Rheometry in Rate-Controlled Mode

Capillary rheometry is a useful way to study the nonlinear rheological characteristics of entangled polymer melts. Such a pressure-driven shearing mode is actually more relevant to polymer processing and therefore can reveal features of direct industrial relevance. Although pressure-controlled capillary rheometry is the right apparatus to delineate the nature of stick–slip transition (SST), as demonstrated in Chapter 6, in processing the extrusion is usually carried out at constant volumetric throughput by a single-screw or twin-screw extruder. Lab-scale capillary extrusion is usually produced by displacing a piston in the barrel at a constant speed. When well-entangled melts with pronounced wall slip are being studied, interesting transient behavior shows up. Intriguing yielding and strain localization events can take place when entangled melts are forced from their resting state in the barrel into a narrow capillary die or slit-like channel. The breakdown of the entanglement network can occur at the die entry in a spatially extensive region. The deformation field in extrusion processing can be much more complicated than the deformation field generated in a boundary-driven simple shear apparatus. For example, the strain field can be unsteady even though the driving conditions, such as the piston speed, are constant over time. Section 10.2 describes the die entry instability based on a combination of capillary rheometry and particle-tracking velocimetry.

10.1.1 Steady-State Characteristics

A steady-state relationship between the capillary volumetric flow rate Q_f and the pressure P at the die entry is a common characterization in the capillary extrusion of entangled polymer melts. For sufficiently well-entangled melts, wall slip and the finite compressibility of the melt make this relationship between Q_f and P much more interesting to describe; the relationship is not a simple monotonic flow curve. Under controlled pressure, a measurable stick–slip

Nonlinear Polymer Rheology: Macroscopic Phenomenology and Molecular Foundation, First Edition. Shi-Qing Wang.

Companion Website: www.wiley.com/go/nonlinearpolymerrheology

transition appears if the extrapolation length b_{max} is larger than the die diameter D, as shown in Section 6.2.1. The stick–slip transition produces a discontinuity in the Q_f versus P plot, as well as hysteresis, as shown in Fig. 6.4.

In rate-controlled extrusion, the piston speed V_p prescribes a volumetric throughput according to

$$Q_b = (\pi D_b^2/4)V_p, \tag{10.1}$$

where D_b is the diameter of the cylindrical barrel. There can be pressure oscillation at intermediate values of Q_b, when the melt/wall boundary condition oscillates back and forth between the stick and slip. The volumetric flow rate Q_f that describes the output exiting the die can be different from Q_b because there can be melt compression or decompression. In general, there are four distinctly different regimes I through IV, as shown in Fig. 10.1. According to the description given in Fig. 10.1, pressure oscillation occurs in regime II, which corresponds to the discontinuity in Q_f observed from pressure-controlled capillary rheometry. Figure 10.1 contains mostly steady-state data and also shows a peculiar regime III where the steady-state pressure is lower than some of the pressures found in regime I. The behavior in regime III independently confirms that the wall slip behavior is hysteretic. In regime IV, wall slip is a stable boundary condition. Here, a larger Q_f produces a larger shear rate in the bulk of the capillary die. From the data in Fig. 10.1, we can approximately estimate the limiting slip velocity attained during the stick–slip transition at the end of the regime III. At this point, the die output $Q_{slip} \sim 1\ mm^3/s$. From $Q_{slip} \sim (\pi/4)D^2V_s^*$ and D = 1 mm, we obtain $V_s^* \sim 4/\pi \approx 1.3$ mm/s, which is consistent with the value in Table 6.1.

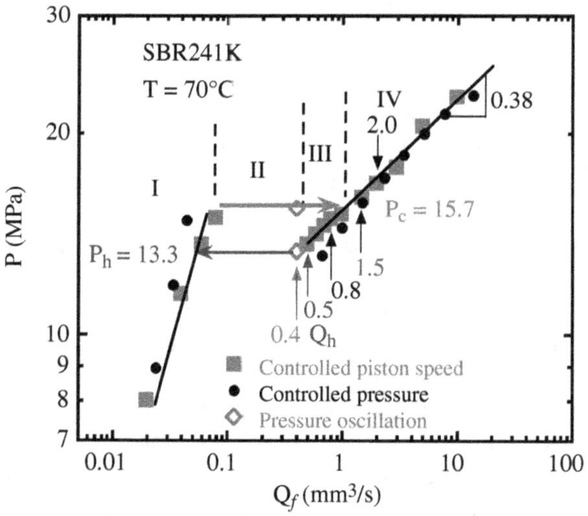

Figure 10.1 Pressure P at the die entry versus volumetric flow rate Q_f from the die in steady state, calculated from constant piston speed and constant pressure, for SBR241K melt whose characteristics are provided in Tables 7.5 and 7.6. At 70 °C, the terminal polymer dynamics are faster than at room temperature by a factor of 24. For the no-slip branch under constant Q_f, a die of L = 32.0 mm and D = 2.0 mm is used while the rest of the constant-Q_f data is based on a die of L = 16.0 mm and D = 1.0 mm. The controlled-pressure data (circles) were obtained from a Monsanto capillary rheometer while the controlled-speed data (squares and diamonds) were obtained from a Rosand capillary rheometer. *Source:* Reproduced from unpublished research results based on the doctoral thesis of Xiangyang Zhu.

10.1.2 Transient Behavior

10.1.2.1 Pressure Oscillation and Hysteresis

Unlike pressure-controlled capillary rheometry where the steady state is almost immediately established, the response to the application of a constant piston speed V_p can be slow. Due to the finite compressibility of the melt, the barrel pressure P gradually builds up. There is no simple, linear relationship between P and the die volumetric flow rate Q_f. Initially, Q_b is larger than $Q_f(P)$, so that the melt in the barrel undergoes compression, and the driving pressure P builds up. In the absence of wall slip, which is the case at low Q_b, P increases in time monotonically until reaching saturation when $Q_b = Q_f$.

For polymer melts capable of significant wall slip, the behavior is more interesting. Figure 10.2 shows an example of capillary rheometry in the mode of constant piston speed, that is, constant Q_b. Here the pressure at the die entry, or more precisely the shear stress at the capillary wall $\sigma = DP/4L$, is plotted against the strain at the wall, defined as $\gamma_w = \dot{\gamma}_w t$, where $\dot{\gamma}_w = (32/\pi)(Q_b/D^3)$. First, in regime II, pressure oscillation is the dominant response, involving boundary conditions that are alternating between stick and slip. For example, at the end of the regime II, with $Q_b = 0.4\,mm^3/s$, a strong pressure oscillation is seen in Fig. 10.2. After P increases to the critical level, corresponding in this system to a wall stress $\sigma_c = 0.245$ MPa, wall slip takes place so that Q_f exceeds Q_b. The melt in the barrel starts to decompress, and correspondingly P decreases. But $Q_b = 0.4\,mm^3/s$ is not large enough to sustain a critical level of capillary flow and to keep up P above P_h. Therefore, the system drops out of the hysteretic loop. Upon returning to a no-slip boundary condition (after sufficient decompression), Q_f suddenly drops to a much smaller value, so that $Q_f < Q_b$ occurs again. Now P starts to increase again due to the melt compression in the barrel. The cycle can repeat perpetually if the barrel is infinitely large. The period of the cycle depends on Q_b, the die diameter D, and the amount of melt in the barrel.[1] The pressure oscillation quickens when there is less material in the barrel; the period is shorter when a larger die diameter D is used. Beyond regime II, the cusp in P still exists for Q_b between 0.5 and 1.0 mm^3/s, for the same reason as it exists at $Q_b = 0.4\,mm^3/s$, namely there is a mismatch between Q_b and Q_f. However, pressure oscillation ceases in regime III because $Q_b > Q_h$ ensures that P is above P_h, despite considerable wall slip, with $V_{s(h)} < V_{s(max)}$. In other words, in regime III, the driving throughput Q_b is sufficiently high to ensure that the wall stress due to bulk capillary flow is higher than σ_h.

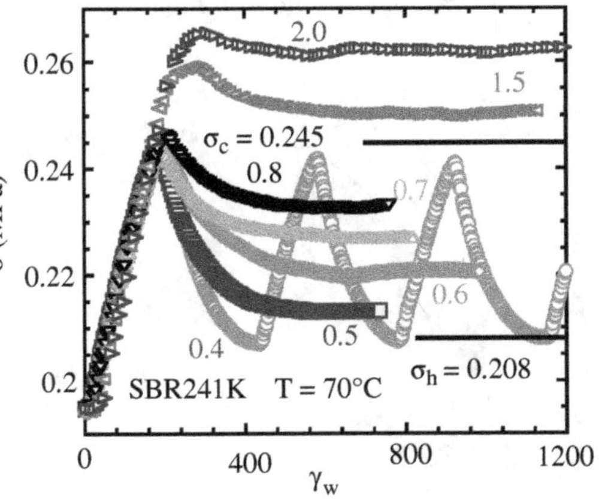

Figure 10.2 Pressure responses expressed in terms of wall shear stress $\sigma = PD/4L$, where P is the piston pressure, at seven different applied volumetric throughputs, determined by the piston speed according to Eq. (10.1), $Q_b = 0.4$, 0.5. 0.6, 0.7, 0.8, 1.5, and 2.0 mm^3/s respectively. *Source:* Reproduced from unpublished research results based on the doctoral thesis of Xiangyang Zhu.

10.1.2.2 Input vs. Throughput, Entry Pressure Loss and Yielding

Conventional rate-controlled capillary rheometry uses the piston speed V_p to specify the volumetric flow rate Q_f. This interpretation is valid in the steady state. To describe the transient behavior associated with melt compression/decompression in the barrel, one must measure Q_f directly from the die output. For example, the input Q_b of Eq. (10.1) is initially larger than Q_f, causing melt compression to take place. As long as $Q_b > Q_f$, P increases. To clearly characterize the transient behavior, we need an independent *in situ* measurement of Q_f. Because there is melt compression and decompression, the mass flow rate, that is, weight of the extrudate per unit time, may not be a good measure of Q_f. The flow rate Q_f can be measured from the flow field inside the die using particle-tracking velocimetry. Specifically, let us consider a slit die with gap $H = 0.8$ mm, width $w = 8$ mm, and length $L_{slit} = 30$ mm. The slit die is attached to the barrel through an adaptor with a rectangular cross section of 8 mm by 4 mm, as shown in Fig. 10.3. Sending a sheet of laser light whose normal is along Y axis, we can record the illuminated YZ plane at point B, using a CCD camera that is focused through an optical window and is looking into the gap. From such PTV observations, the velocity field in the slit die can be determined directly.

According to the PTV observations of SBR241K, the velocity profile is plug-like, that is, the maximum velocity V_{max} inside the slit is close to the slip velocity V_s at the wall. Therefore, the *in situ* measurement of V_s as a function of γ_w as shown in Fig. 10.4 can be made to evaluate $Q_f(P) = wHV_s(\gamma_w)$. Although the input, that is, the driving rate $Q_b = 4\ mm^3/s$, is a constant, the output $Q_f(P)$ is far from constant over a period that is as long as more than 2 min. The PTV data indicates that Q_f smoothly increases due to the rising pressure P. The maximum in P indicates the end of barrel melt compression when the condition of $Q_b > Q_f$ no longer holds. We note that Q_f is not synchronized with P as indicated in Fig. 10.4. At the moment indicated by the vertical line around $\gamma_w = 200$, the corresponding V_s around 0.4 mm/s is considerably smaller than the steady-state value $V_{ss} = 0.625\ mm/s = Q_b/wH$, although P has reached the same level as its value in steady state. If we label P_{die} as the pressure available to produce the channel flow, then P_{die} at this point is clearly lower than P_{die} in steady state. The so-called entrance pressure loss simply reflects this pressure difference

$$P_{entrance} = P - P_{die}, \tag{10.2a}$$

where in principle P_{die} can be evaluated from the characteristics of the strain field inside the channel. In other words, as a portion of the total pressure, $P_{entrance}$ is "consumed" to transform the resting entangled melt into a "flowable state" at the die entry. Before reaching steady

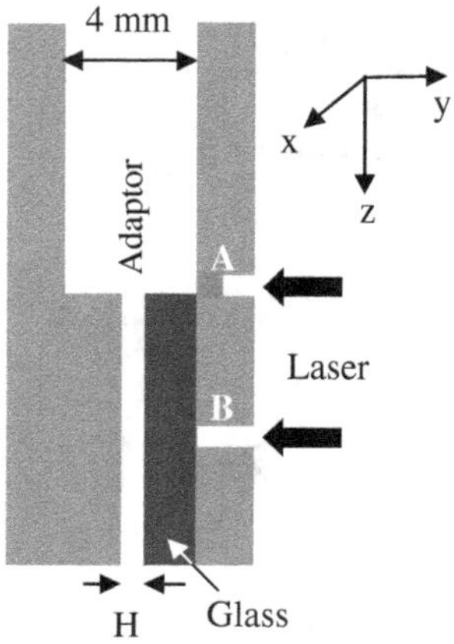

Figure 10.3 Schematic illustration of particle-tracking velocimetric setup for observations at the die entry A and die inland B from a slit die.

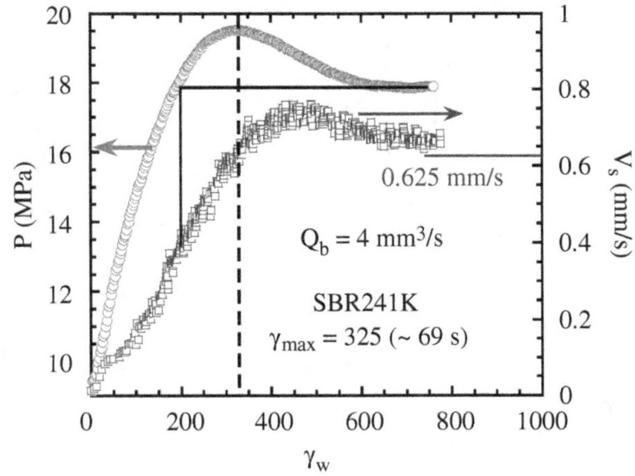

Figure 10.4 Piston pressure P (on the left-hand-side Y axis) and slip velocity in the slit die (on the right-hand-side Y axis) according to PTV measurements, against $\gamma_w = \dot{\gamma}_w t$, which is proportional to the elapsed time, at constant throughput $Q_b = 4\ mm^3/s$, based on a Rosand capillary rheometer. The onset at $\gamma_w = 0$ is defined as the moment when the piston pressure P has built up to a level around 9 MPa. P reaches a maximum at $\gamma_{max} = 325$, that is, $t = 69$ s. At around $\gamma_w = 200$, P reaches the same level as its steady-state value, around 18 MPa. Also indicated is the steady-state slip velocity of $V_{s(ss)} = 0.625\ mm/s = Q_b/wH$.

state, $P_{entrance}$ evolves in time, giving rise to the desynchronization between P and Q_f as shown in Fig. 10.4.

Since Fig. 10.4 depicts the metastable wall slip in the regime III,[1] the relationship between Q_f and P_{die} can be read from the steady-flow curve in Fig. 10.1 as $P_{die} \sim Q_f^{0.38}$. Therefore, we can approximately estimate this pressure loss as follows. Taking $P_h = 13.3$ MPa and $Q_f = 0.5\ mm^3/s$ as the first data point in the scaling regime III from Fig. 10.1, the relationship between the wall stress and rate can be readily established using the formula for Newtonian fluids in Sections 3.1.3.1 and 3.1.3.2:

$$\sigma = \sigma_h(\dot{\gamma}_w/5.1)^{0.38}, \tag{10.2b}$$

where $\sigma_h = (DP_h/4L_{capil}) = 0.21$ MPa and the shear rate at σ_h is $Q_f(32/\pi D^3) = 0.5(32/\pi) = 5.1\ s^{-1}$. If we simply assume Eq. (10.2b) to hold true for the slit die, we can estimate $P_{die(slit)}$ as

$$P_{die(slit)} = (2L_{slit}/H)\sigma = 15.75(\dot{\gamma}_w/5.1)^{0.38}, \tag{10.2c}$$

where $\dot{\gamma}_w = 6Q_f/wH^2$ with $Q_f = V_s wH$, that is, $\dot{\gamma}_w = 6V_s/H$. By plugging the values of V_s from Fig. 10.4 into $\dot{\gamma}_w = 6V_s/H$ and reading the data for $P(\gamma_w)$ from Fig. 10.4, $P_{entrance}$ of Eq. (10.2a) can be evaluated from Eq. (10.2c) as a function γ_w, as shown in Fig. 10.5.

Figure 10.5 reveals interesting information. The most striking feature is the overshoot of $P_{entrance}$, reminiscent of the stress overshoot in startup deformation discussed in Chapter 7. Indeed, the application of Q_b to produce the extrusion is analogous to a startup deformation test. At negligibly low P and similarly low Q_f, $P_{entrance}$ is also negligibly low, as shown by the dashed lines. The resting melt resists more against being pushed into the die as Q_f increases in the initial elastic regime until a maximum degree of elastic deformation takes place at the die entrance. The decline of $P_{entrance}$ signifies *yielding* that shares the same physics as that responsible for the stress overshoot in startup shear and extension. Presumably, the yielding is realized through chain disentanglement so that the "fight" to rush into the die from the entangled resting state subdues. It is this sense that the conventional simple shear and uniaxial extension tests provide insights into other geometrically less-well-defined forms of deformation such as extrusion. At the present, there have been no further investigations to determine how the peak of $P_{entrance}(Q_b)$ varies with Q_b.

---

1 At the same wall shear rate $\dot{\gamma}_w$, the volumetric flow rate of the slit die (with H = 0.8 mm and w = 8 mm) is higher than that of the capillary die (with D = 1 mm) by a factor of $(wH^2/\pi D^3)(32/6) = 8.7$. Thus, the power law behavior with exponent 0.38 should occur in the range of Q_f between $0.4 \times 8.7 = 3.5$ and $1.0 \times 8.7 = 8.7\ mm^3/s$ according to Fig. 10.1 that shows the range of Q_f for the capillary die is between 0.4 and 1.0 mm³/s.

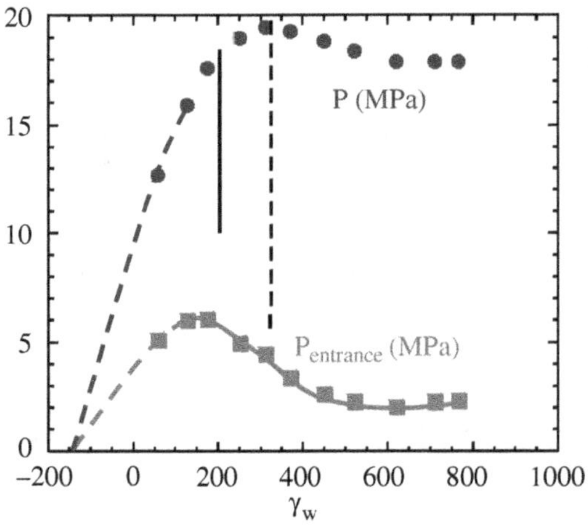

Figure 10.5 Entrance pressure loss $P_{entrance}$ and total pressure P versus γ_w, both showing a maximum of different origins, where the dashed lines are drawn to "fill in" the missing data prior to the emergence of the visible wall slip. On the dashed lines, the extrusion involves the no-slip branch, characterized by high piston pressure at negligible Q_f as shown in Fig. 10.1. As in Fig. 10.4, γ_w is defined to be zero when $V_s = 0$. Thus, during the period of no-slip, γ_w is negative.

10.2 Instabilities at Die Entry

10.2.1 Vortex Formation vs. Shear Banding

When highly viscoelastic fluids enter a narrow die from a widely open reservoir as in the case of capillary extrusion, they may respond differently at the die entry region. For example, it is well known that LDPE shows vortex formation in the barrel while HDPE does not. Traditionally, constitutive models have been applied to carry out finite-element calculations on the strain field at the die entry.[2–7] The parameters in the constitutive equations are constant, independent of both space and time. Thus, the simulated instabilities at the die entry reflect the mathematical properties of the underlying nonlinear constitutive equations. One characteristic instability is vortex formation around the die entry. Remarkably, certain constitutive relationships can produce vortices. However, the finite-element calculations do not explain the entry instabilities in terms of molecular-scale structural changes.

DNA solutions at concentrations of 0.1% and 0.5% readily[8] develop vortex formation, while more concentrated DNA solutions of 1% concentration with stronger viscoelasticity avoid vortex formation and show shear banding over a wide range of flow rates, as shown in Fig. 10.6. More interestingly, the 0.7% DNA buffer solution that shows no vortex formation can be made to display strong vortices by replacing the 40% buffer with sucrose, as shown in the photos in Fig. 10.7. The only difference between the two solutions is that the sucrose-containing 0.7% DNA solution involves a much more viscous (by a factor of 8) solvent. Conventionally, all the parameters needed in a constitutive equation are encoded in such information as the storage and loss moduli from SAOS measurements. As shown in Ref. [8], the two 0.7% solutions have the same relaxation spectrum, except for a small difference in the terminal relaxation time τ, by a factor of less than 2. Since the existing constitutive models do not have the extrapolation length as a parameter, a conventional finite-element calculation is not expected to predict shear banding in the buffer solution and vortex formation in the other solution. The particle-tracking velocimetric observations shown in Figs. 10.6 and 10.7 have a natural explanation: fluids having a sufficiently large value of *b* yield through strain localization (shear banding) and avoid a vortex-type response. Specifically, *b* decreases from 36 to 1.4 mm when the DNA concentration changes from 1% to 0.5% and changes from 8.1 to 1.5 mm when the solvent contains 40% sucrose instead of 100% water.

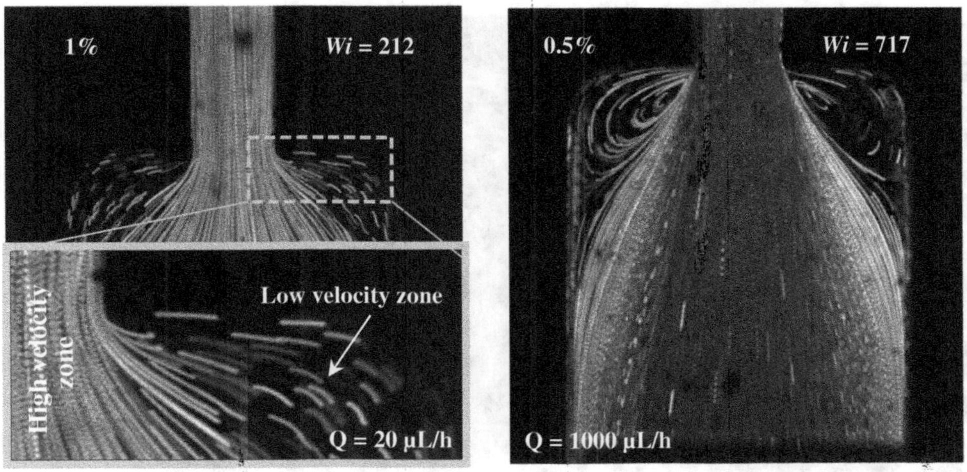

Figure 10.6 Streak images of PTV particles in the convergence region of a microfluidic device, showing (a) 1% DNA solution at an effective $Wi = 212$ evaluated from $Q_f = 20$ μL/h and (b) 0.5% DNA solution at $Wi = 717$ ($Q_f = 1000$ μL/h). In (a), there is a zone where the stream lines involve high velocity near the entry, with the two corners relatively stagnant, as shown in the magnified image. In contrast, the DNA solution of lower concentration is incapable of developing shear banding and cannot avoid vortex formation. *Source:* Hemminger *et al.* 2010 [8]. Reproduced with permission of Elsevier.

10.2.2 Stagnation at Corners and Internal Slip

In contrast to LDPE that cannot display a sizable *b*, molten HDPE (made of linear chains) can readily undergo localized yielding (internal slip) and thus does not usually show vortex formation at the die entry. To apply the PTV technique to reveal the different entry flow instabilities of entangled melts, it is convenient to use a rubbery polymer melt such as styrene butadiene rubber (SBR) and polybutadiene. Die entry instability is known to produce extrudate distortion. As the state of chain entanglement varies in a chaotic manner both in space and time upon entry into the die at a high throughput, no steady state can be established during the capillary flow. The exiting sample does not have time to erase the memory of the "initial" state produced at the die entry. Thus, a varying degree of relaxation and corresponding die swell take place in a quasi-chaotic manner. It can be readily demonstrated that the severe extrudate distortion, often known as gross melt fracture, originates from events at the die entry. We can straightforwardly demonstrate its origin using a pressure-controlled capillary rheometer.

After loading a sufficient amount of an SBR melt into the barrel, we apply pressure to first fill the capillary die. Then the piston pressure is reduced to zero to allow the melt in the die and the barrel to rest and relax for a sufficiently long time. Subsequently, a pressure higher than the critical pressure for the stick-slip transition is applied. Initially, the extrudate first comes out smoothly with little die swell; at later times, we encounter nearly chaotic extrudate distortion and significant die swell, as shown by the photo in Fig. 10.8 as well as the movie[2] that shows gross melt fracture. The first segment of the extrudate is smooth because this part of the extrudate was the prefilled melt that has undergone complete relaxation in the die. The subsequent extrudate is freshly extruded from the barrel, having passed through the die entry but having had no time to relax from its deformation history associated with having been in the barrel and then experiencing the forced rush-in at the die entry: At a volumetric flow rate of $Q_f \sim 12\ \mathrm{mm}^3/\mathrm{s}$, the melt passes through a die of $L = 10$ mm in a rather short residence time $t_{resid} \sim L/(Q_f/D^2) \sim D^2$ s,

2 Video 10.1: Extrusion showing smooth (from the sample in the prefilled die) and gross distortion of the extrusion.

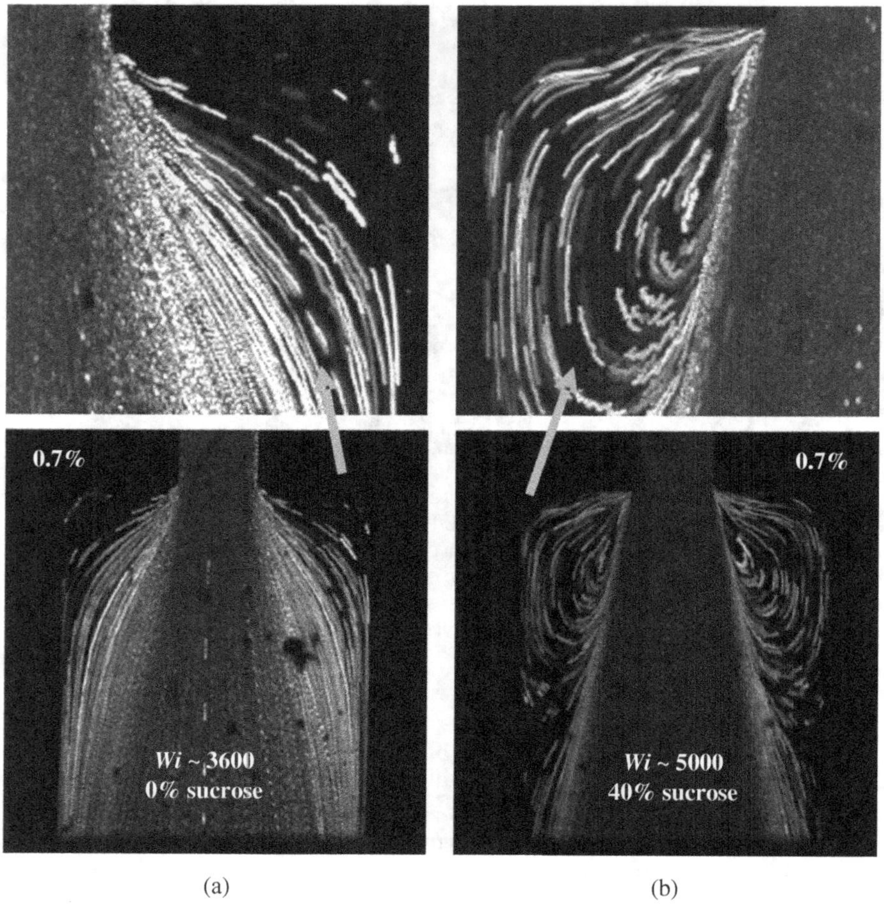

Figure 10.7 Streak images of PTV particles in the convergence region of a microfluidic device, showing (a) 0.7% water-based DNA solution at an effective $Wi = 3000$ ($Q_f = 800\ \mu L/h$) and (b) 0.7% DNA solution based on 40% sucrose and 60% water at $Wi = 5000$ ($Q_f = 800\ \mu L/h$), showing strain localization (shear banding) and vortex flow, respectively. The blow-up are the two images on the first row. *Source:* Hemminger *et al.* (2010) [8]. Reproduced with permission of Elsevier.

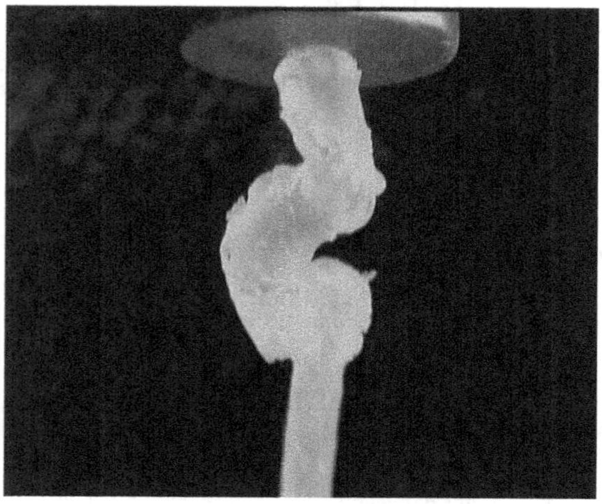

Figure 10.8 Snapshot of an extrudate prepared by prefilling the die, followed by resting before application of a high piston pressure to drive the capillary extrusion well beyond the stick–slip transition.

where D is in the unit of millimeter. For D = 1–2 mm, t_{resid} is a few seconds while $\tau = 820$ s for SBR536 at room temperature.[9] Thus, the highly deformed polymer chains after entry do not have enough time to fully relax and erase their deformation history before arriving at the die exit. The observed strong extrudate distortion originates from the "chaotic" entry of the SBR melt into the capillary die.

To go beyond this convincing but indirect inference that the gross melt fracture originates from entry instabilities, "flow" visualization can be carried out to show how the sample flows from the barrel into the die. One straightforward method is particle-tracking velocimetry, as illustrated in Fig. 10.3, which can monitor the streamlines in real time. For example, at $Q_f = 12\ mm^3/s$, the PTV movie shows the instant velocity field in a 2 mm by 1.5 mm field of view at the die entry, as shown in Fig. 10.9. Although the stress field is expected to vary smoothly in space, there emerges a discontinuity in the velocity field. Specifically, there is a "stagnant" corner where the PTV particles hardly move. Since the reptation time of SBR536K is so long, a shear rate corresponding to $Wi > 1$ can be immeasurably low. Thus, the stagnation does not imply vanishing stress. Without a detailed analysis, it is unclear how much larger the shear rate is in the high velocity zone. However, one thing is clear from Fig. 10.9: the velocity discontinuity, as shown in Fig. 10.10, indicates the presence of a very narrow band with a high local shear rate. This internal "slip" indicates that massive chain disentanglement has presumably taken place. Because the internal slippage did not occur on a fixed wall, the location of the interface is not spatially fixed. The instant profiles shown in Figs. 10.9 and 10.10 are actually unsteady, resulting in the chaotic distortion of the extrudate as shown in Fig. 10.8. In closing, it is worth mentioning that the irregular extrudate is of a much larger diameter compared to the initial smooth extrudate (cf. Fig. 10.8), suggesting that chain deformation experienced at the die entry significantly influences the extent of die swell. Two videos are provided to show the strain localization and temporal and spatial variations of the entry deformation.[3]

Based on the preceding results in Sections 10.2.1 and 10.2.2, we can speculate that long-chain branching could suppress internal slip and cause vortex formation at the die entry. As indicated in Chapter 8, long-chain branching tends to delay the onset of a catastrophic breakdown of the

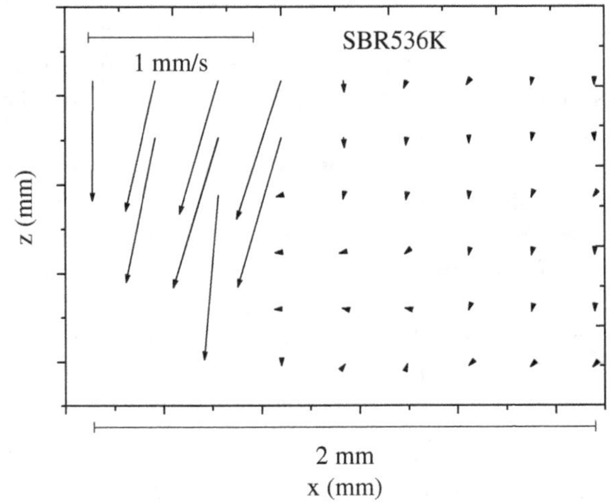

Figure 10.9 Deformation field during the "steady" extrusion revealed by PTV measurements in the entry region for SBR536K using a slit die shown in Fig. 10.3 mounted onto a Rosand capillary rheometer. The volumetric throughput $Q_b = 12\ mm^3/s$ and P is around 20 MPa. *Source:* Zhu *et al.* (2013) [9]. Reproduced with permission of American Institute of Physics.

3 Video 10.2: PTV observation at die entry above the stick-slip transition for (a) linear polybutadiene melt (M = 200 kg/mol) and (b) linear styrene-butadiene rubber (M = 616 kg/mol), at room temperature.

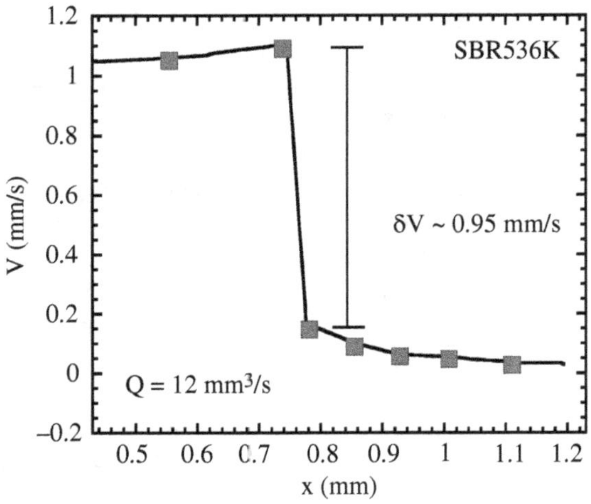

Figure 10.10 Velocity profile for SBR536K according to the particle-tracking velocimetric observation, corresponding to Fig. 10.9. *Source:* Data taken from Zhu *et al.* (2013) [9].

entanglement network in bulk. In the absence of a total failure of the network, a polymer with significant long-chain branching may negotiate to satisfy the no-slip boundary condition and develop vorticities at the die entry. Such expectation is of course consistent with the observation of vortex formation in LDPE.

10.3 Squeezing Deformation

Let us consider the one-dimensional squeezing depicted in Fig. 5.4. In the absence of lubrication at sample/wall interfaces, and assuming a no-slip boundary condition, the behavior of a Newtonian fluid has been quantitatively illustrated in Section 5.3.2. Although the analysis in Section 5.3.2 is based on the simplest constitutive relation, namely Newton's laws, the analysis raises several interesting questions. Even though the squeezing is only along the Z axis, significant shear stress σ_{zx} can arise in the sample (except for the middle layer where $z = 0$) because of the no-slip boundary condition, as shown in Eq. (5.21). For a Newtonian fluid, the magnitude of σ_{zx} increases with the speed with which the thickness of the system decreases. Equation (5.21) gives us the insight that the shear stress would be zero at the center of the apparatus, that is, at $x = 0$, even for a generic viscoelastic fluid. Thus, even well-entangled polymer melts should not undergo wall slip at the center. But there is a basic difference: being Newtonian, all the characteristics, for example, velocity field and shear stress, are not explicit functions of time. In contrast, an entangled polymeric liquid can yield during the squeezing. For example, chain disentanglement may take place to produce an overshoot in both the shear stress σ_{zx} and resisting pressure $P(x)$ as the system goes beyond the initial elastic response.

It is impossible to solve the equation of motion, Eq. (5.5), in the absence of a full constitutive equation. However, analogous to capillary extrusion, specifically to Eqs. (3.7a) and (3.13), a generic relation between the pressure gradient and shear stress can be established by integrating Eq. (5.5) to obtain

$$\sigma_{zx} = -z\frac{\partial P'}{\partial x} \tag{10.3a}$$

which is valid for any viscoelastic material. In general, we expect Eq. (5.10) to hold true for P'. Therefore, based on Eq. (5.10), we can rewrite Eq. (10.3a) as

$$\sigma_{zx}(x, z) = 3\overline{P}(2x/L)(2z/L), \tag{10.3b}$$

where $\overline{P}$ is the average pressure, given by integration of $P'(x)$ over x from $-L/2$ to $L/2$. $\overline{P}$ plus the normal force σ_{zz} is equal to the total force F_z divided by the total area wL. For simplicity, here we treat the normal force (produced by chain deformation) to be negligible. The shear stress is bilinearly proportional to x and z, attaining maximum $\sigma_{edge(w)}$ at $x = L/2$ and $z = \pm h$, given by

$$\sigma_{edge(w)} = P_{max}(4h/L), \tag{10.4}$$

which resembles Eq. (3.7b) for capillary rheometry, where $P_{max}(x = 0) = (3/2)\overline{P}$.

In Section 5.3.2, an example was worked out in detail on the basis that the sample is a Newtonian fluid. In this section, we examine how a pure elastic body responds during squeezing. Such an exercise is relevant to squeezing an entangled melt because such a sample must initially respond as an elastic solid. We again make use of Eqs. (5.6)–(5.8) to solve for the velocity profiles. To do so, we first write the Hooke's law [different from Eq. (A25) in Ref. [10] that seems to have a typographical error because of the inconsistent dimensions] as

$$\sigma_{zx} = -G_{pl}\frac{\partial X(x, z)}{\partial z} \tag{10.5a}$$

so that its time derivative is given by

$$\dot{\sigma}_{zx} = -G_{pl}\frac{\partial v_x}{\partial z}. \tag{10.5b}$$

Now we can take a time derivative on both sides of Eq. (5.5) and integrate the resulting equation by inserting the differentiation of P' from Eq. (5.10) as well as Eq. (10.5b). Since v_x is given by Eq. (5.7), we have under the no-slip boundary condition, that is, $v_x(x, z = \pm h) = 0$,

$$f(z) = (\dot{B}/G_{pl})(h^2 - z^2), \tag{10.6}$$

where $\dot{B}$ is the time derivative of B in Eq. (5.10). The functional form of Eq. (10.6) prescribes how both $v_x(x, z)$ and $v_z(z)$ vary with the coordinates x and z, through Eqs. (5.7) and (5.8). During squeezing specified in Eq. (5.14), the velocity profiles can be explicitly described in terms of C given in Eq. (5.15b) and

$$\dot{B} = (3/2)G_{pl}(\dot{h}/h^3). \tag{10.7}$$

By replacing B/η in Eqs. (5.13a) and (5.13b) with $(\dot{B}/G_{pl})$ given in Eq. (10.7), we confirm[4] that the velocity profiles $v_x(x, z)$ and $v_z(z)$ have respective expressions identical to Eqs. (5.16) and (5.17).

To derive the relation between the squeezing displacement and the resulting pressure P', Eq. (10.7) can be integrated to show

$$B = (3/2)(G_{pl}/2h^2)[1 - (h/h_0)^2], \tag{10.8}$$

where h_0 denotes the sample thickness before squeezing. Upon inserting Eq. (10.8) into Eq. (5.10), we obtain the formula for the position-dependent pressure as

$$\begin{aligned} P' &= (3/2)(G_{pl}/2)[(L/2h)^2 - (x/h)^2][1 - (h/h_0)^2] \\ &= (3/2)G_{pl}(\Delta z/2h_0)[(L/2h)^2 - (x/h)^2](1 - \Delta z/4h_0) \end{aligned} \tag{10.9}$$

4 This occurs because the constitutive laws for Newtonian fluids and Hookean solids are both linear. More generally, we should recall by analogy that during simple shear, the velocity field would be the same as $v_x(y) = \dot{\gamma}y$ regardless of whether the material under shear is an elastic solid or viscous liquid, provided that there is no shear strain localization. Precisely because of this expectation, it turned out to be less straightforward to recognize that there is little flow, that is, little irrecoverable deformation, when a viscoelastic fluid initially undergoes startup shear. As discussed in detail in Chapter 7, when an entangled polymeric fluid is subjected to startup deformation, before yielding, the deformation is actually largely elastic. The mere existence of a velocity field does not reveal the physical nature of the deformation. The presence of a velocity gradient merely means that there is a finite rate of deformation.

for affine deformation, where the total displacement is defined by

$$\Delta z = 2\dot{h}t = |v_z(z = h)|t = 2(h_0 - h). \tag{10.10}$$

In contrast to Eq. (5.18) for Newtonian fluids where P′ has an initial magnitude (at $t = 0$) proportional to the rate ($\dot{h}/h$), P′ for Hookean solids is zero at $t = 0$ at h_0 and $\Delta z = 0$, as expected from deformation of solids. Newtonian liquids and Hookean solids also share a common characteristic: the pressure is the highest at the center (x = 0), greater than the shear stress level of $G_{pl}(\Delta z/2h_0)$ around $x = h$ by an amplifying factor $(L/2h)^2$. The averaged pressure $\overline{P}$ is related to the squeezing displacement as

$$\overline{P} = (G_{pl}/2)(L/2h_0)^2[(h_0/h)^2 - 1]. \tag{10.11}$$

According to Eq. (10.11), when significant squeezing is elastically applied in the absence of wall slip, the resulting pressure can well exceed G_{pl}.

Finally, we can evaluate the shear stress σ_{zx} by inserting Eq. (10.11) into Eq. (10.3b) and integrating over z to obtain

$$\sigma_{zx} = (3/2)G_{pl}(xz/h_0^{\,2})[(h_0/h)^2 - 1] = G_{pl}\gamma(x, z), \tag{10.12}$$

where the shear strain

$$\gamma = (3/2)(xz/h_0^{\,2})[(h_0/h)^2 - 1] \tag{10.13}$$

can also be obtained from time integration of the shear rate in Eq. (5.21). As stated in the beginning of Section 10.3, upon yielding, the pressure P′ and shear stress σ_{zx} are expected to decline. Moreover, for well-entangled melts, shear strain localization could take place due to plausible localized yielding. In particular, wall slip may first take place when the shear stress grows sufficiently large. Thus, it is desirable to apply the particle-tracking velocimetric (PTV) method to study how the velocity field may deviate from the prescribed profile upon yielding.

Preliminary and unpublished PTV observation shows that shear banding emerges at the late stage of squeezing as shown by a PTV movie that[5] recorded the strain field as a function of time at $x \sim 0$. According to Eq. (10.12) at $h/h_0 \sim 1/2$, shear stress can reach a level comparable to G_{pl} at $x \sim h$, which is within the field of view covered by PTV. Such a high level of stress is expected to produce shear banding as well as nonquiescent relaxation upon cessation in any well-entangled melt. Further systematic PTV observations should be carried out, along with the time-dependent stress response, that is, *in situ* measurements of the resulting pressure.

A particle-tracking velocimetric investigation can also be carried out in the mode of controlled-pressure instead of controlled displacement. Equations (10.3b) and (10.4) show that at a given applied F_z, the wall shear stress $\sigma_{zx}(x, z = \pm h)$, proportional to the available $\overline{P} = F_z/wL - \sigma_{zz}(\gamma)$, varies with the sample thickness $2h$. When $\overline{P}$ is sufficiently high so that σ_{zx} is on the order of the plateau modulus G_{pl} at $z = \pm h(t)$ and $x \sim L/2$, the creep should produce an entanglement–disentanglement transition similar to that observed in simple shear configuration in Section 7.4. Such a transition should start from interfacial slip and plausibly followed by shear banding. Under a constant force, the squeezing is somewhat analogous to pressure-controlled capillary extrusion. However, unlike pressure-driven capillary extrusion, squeezing under controlled pressure is not shear deformation at a constant shear stress. The squeezing is more complicated than simple shear. For example, at a constant force, the entanglement–disentanglement transition is not sustainable because of the reducing thickness: according to Eq. (10.4), the available shear stress decreases linearly with h.

5 Video 10.3: Speed-controlled squeeze of a 1,4-polybutadiene melt ($M_w = 600$ kg/mol) at room temperature, made by X. Y. Zhu at University of Akron.

The attached movie[6] shows pressure-controlled squeezing that turns the initial elastic deformation into an entanglement–disentanglement transition (when the squeezing speeds up), followed by a "halt" when the maximum wall shear stress at $x = L/2$ is eventually not even high enough to produce wall slip. The system returns to a fully entangled state and terminal flow.

10.4 Planar Extension

By introducing lubrication between the sample and the squeezing walls to remove the no-slip boundary condition, the apparatus depicted in Fig. 5.4 can be used to produce planar extensional deformation. When the lubrication works ideally, the sample stretches along the X axis as its thickness decreases along the Z axis, so that planar extension can be realized.

Under large fast deformation, an entangled melt tends to lose its structural integrity as it transforms from the initial elastic deformation to the eventual state of flow. In uniaxial extension, the yielding leads to tensile strain localization, to be discussed in Chapter 11. The planar extension may allow us to study yielding through chain disentanglement without the structural failure.

In summary, shear strain localization of highly entangled polymer melts also takes place in geometries other than simple shear. Specifically, the shear banding formed at the die entry is shown to be the origin of gross extrudate distortion. Highly entangled melts with narrow molecular weight distribution show more severe shear banding because of their intrinsically larger magnitude of the extrapolation length b_{max}. Conversely, modestly entangled melts are less likely to undergo the shear strain localization at the die entry and are expected to be free of gross extrudate distortion. Systematic studies should be carried out in the future to quantify this statement. Squeezing deformation should be recognized as an important protocol to interrogate strong nonlinear response of well-entangled melts to large deformation because it has the advantage of being free of complication from edge fracture.

References

1 Pearson JR. Mechanics of polymer processing. Springer Science & Business Media, New York; 1985.

2 Mackley M, Moore I. Experimental velocity distribution measurements of high density polyethylene flowing into and within a slit. J Nonnewton Fluid Mech. 1986;21(3):337–358.

3 Feigl K, Öttinger H. The flow of a LDPE melt through an axisymmetric contraction: a numerical study and comparison to experimental results. J Rheol. 1994;38(4):847–874.

4 Béraudo C, Fortin A, Coupez T, Demay Y, Vergnes B, Agassant J. A finite element method for computing the flow of multi-mode viscoelastic fluids: comparison with experiments. J Nonnewton Fluid Mech. 1998;75(1):1–23.

5 Clemeur N, Rutgers R, Debbaut B. Numerical simulation of abrupt contraction flows using the Double Convected Pom–Pom model. J Nonnewton Fluid Mech. 2004;117(2):193–209.

6 Video 10.4: Squeezing at constant pressure $\overline{P} = 4.3$ MPa of SBR250K whose characteristics are listed in Tables 7.5 and 7.6. With $h_0 = 2$ mm, the maximum shear stress at the edge is $\sigma_{edge(w)} = 1.3$ MPa $> G_{pl} = 0.77$ MPa. The instant elastic squeeze is soon followed by a fast decrease of the gap distance because of yielding via chain disentanglement. Video (a) was made with a camera placed near the center at $x = 0$, video (b) was made with a camera placed at $x = 6$–8 mm. Both obtained by Dr. Xianggang Li at University of Akron.

6 Olley P, Wagner M. A modification of the convective constraint release mechanism in the molecular stress function model giving enhanced vortex growth. J Nonnewton Fluid Mech. 2006;135(1):68–81.
7 Karapetsas G, Mitsoulis E. Some experiences with the slip boundary condition in viscous and viscoelastic flows. J Nonnewton Fluid Mech 2013;198:96–108.
8 Hemminger OL, Boukany PE, Wang S-Q, Lee L. Flow pattern and molecular visualization of DNA solutions through a 4: 1 planar micro-contraction. J Nonnewton Fluid Mech. 2010;165(23):1613–1624.
9 Zhu X, Yang W, Wang S-Q. Exploring shear yielding and strain localization at the die entry during extrusion of entangled melts. J Rheol. 2013;57(1):349–364.
10 Laun HM, Rady M, Hassager O. Analytical solutions for squeeze flow with partial wall slip. J Nonnewton Fluid Mech. 1999;81(1–2):1–15.

11

Strain Localization and Failure during Startup Uniaxial Extension

When entangled polymeric liquids under a given shear stress can access two states having vastly different rheological properties, strain localization may occur. The preceding two chapters described several cases in which shear banding arose from localized yielding during ongoing deformation. During extension, strain localization can also be expected to take place, following localized yielding. At long times, well-entangled polymer melts may not undergo homogeneous uniaxial extension after yielding. In uniaxial extension, Chapter 7 shows that the engineering stress varies nonmonotonically with the extension ratio λ for a wide range of Hencky rates, from $Wi = 1$ to $10(\tau_d/\tau_R)$, namely, $Wi_R \sim 10$.

When an initially homogeneous entangled melt of macroscopic size is subjected to global uniaxial extension, experiment reveals that the specimen undergoes a variety of instabilities that terminate uniform stretching. Vinogradov and coworkers extensively studied polybutadiene and polyisoprene melts in uniaxial extension and observed a "fluidic to elastic" transition as a function of the applied Hencky rate.[1–3] The magnitude of the applied Hencky rate determines which type of specimen failure takes place. There are three ways in which nonuniform extension can occur. Our description coincides with the classification suggested by Malkin and Petrie,[4] and subsequently elaborated by Malkin and coworkers,[5] as shown in Fig. 11.1. Experimentally, entangled polymer melts behave similarly to the behavior summarized in Fig. 11.1. In particular, the terminal data points of each curve in Fig. 7.28 reveal the Hencky strain ε_{inhm} at the onset of inhomogeneous extension for a series of applied Hencky strain rates, where the subscript "inhm" stands for the onset of inhomogeneous extension. In Fig. 7.28, these end points trace a U shape (oriented horizontally) similar to the U shape displayed by the regimes II and III of Fig. 11.1.

Data in Fig. 7.28 did not reach regime IV. In regime IV, the melt acts as if it were cross-linked. Similar to cross-linked systems, styrene–butadiene rubber (SBR325K in Tables 7.5 and 7.6) suffers sharp rupture in regime IV.[6] For this sluggish melt shown in Fig. 11.2, at room temperature it is only convenient to explore regimes IV, III, and a portion of II.

Regarding regime III, there is a significant discrepancy between the description of Malkin and Petrie in Fig. 11.1 and the quantitative diagram seen in Fig. 11.3, the latter figure being extracted from Figs. 7.28 and 11.2. When Fig. 11.3 divides the responses into regimes according to the natures of their tensile failures, as discussed in detail in Section 11.2 strain localization in regime III is necking-like. In regime III, ε_{inhm} increases monotonically with *Wi*. However, according to Figs. 7.28 and 11.2, the onset of inhomogeneous extension occurs well after the yield point where the engineering stress σ_{engr} peaks. Consequently, the deformation is not "completely elastic" because the melt extension does yield, with σ_{engr} declining over significant strains before the point of specimen failure.... The elastic recoil tests in Fig. 7.36 show that recovery is only partial after the yield point has been passed. Because structural inhomogeneity only sets in well beyond

Nonlinear Polymer Rheology: Macroscopic Phenomenology and Molecular Foundation, First Edition. Shi-Qing Wang.

Companion Website: www.wiley.com/go/nonlinearpolymerrheology

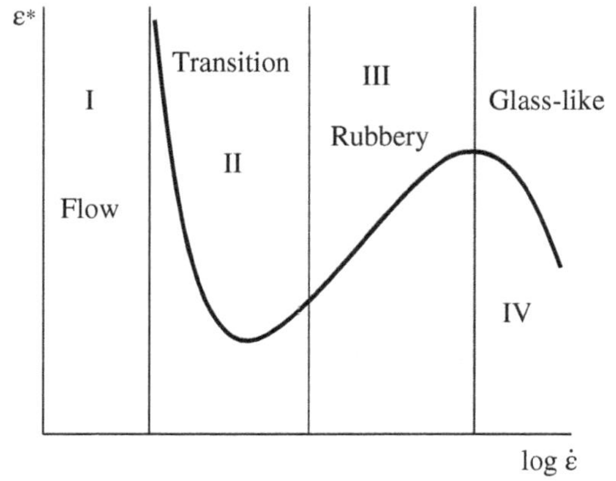

Figure 11.1 Master curve of Malkin and Petrie,[4] schematically illustrating the dependence of the limiting (failure) strain ε^* on Hencky strain rate $\dot{\varepsilon}$, in uniaxial extension of polymer liquids, drawn according to Ref. [4]. Regime I is a flow-dominant domain with $Wi \ll 1$ where there is little ongoing elastic process. Malkin and Petrie labeled regime II as the transition zone, involving a "superposition of elastic deformation and viscous elongational flow." Regime III was termed "rubbery," where "the entire deformation is completely elastic." Finally, regime IV was identified as the "glassy-like" regime, although the reason to use such a term is not obvious.

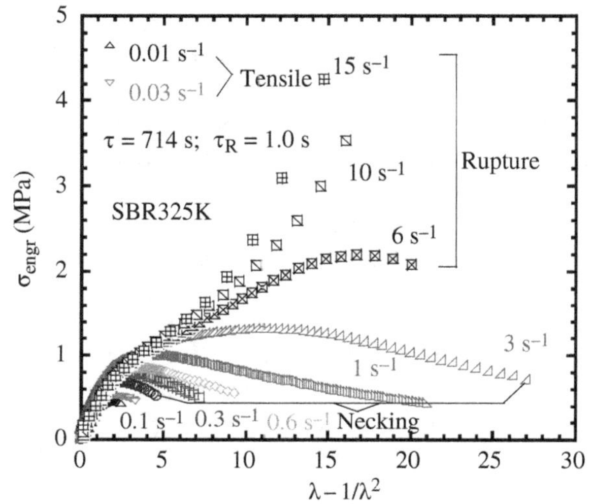

Figure 11.2 Engineering stress versus strain ($\lambda - 1/\lambda^2$) of SBR325K at room temperature. *Source:* Zhu and Wang 2013 [6]. Reproduced with permission of American Institute of Physics.

the yield point, the state is not fully elastic or rubbery, in disagreement with the characterization of Malkin and Petrie. Only regime IV involves "complete elasticity" in the sense that (i) σ_{engr} grows monotonically to the rupture point and (ii) a ruptured specimen is capable of full recovery to its initial dimensions. The stretching in regime IV is non-Gaussian, due to true strain hardening (due to lack of complete yielding), as discussed in Section 8.3. In regime IV, the Hencky strain ε_{inhm} at rupture decreases with increasing extension rate. In short, instead of Fig. 11.1, the uniaxial extension of entangled polymers can be more accurately quantified according to the precise way in which the stretched specimen breaks up: smooth steady flow in regime I, tensile-like failure in regime II, necking in regime III, and rupture in regime IV, as shown in Fig. 11.3.

Before describing the different forms of extensional strain localization, it is necessary to note a different line of research. This line of research employed a filament-stretching rheometry (FSR), in which two plates sandwich a disk-shaped polymer sample, and extension is produced by displacing one plate while holding the other fixed. Instead of strain localization, the investigations claimed to attain "steady state" during continuous extension of entangled melts and solutions.[7,8] Such studies appear to contradict the observations[4,5,9–11] that steady flow cannot be achieved for well-entangled polymer melts unless $Wi \ll 1$. The same workers also applied FSR to study stress relaxation by moving the two ends of the specimen toward each other to avoid[12,13] the elastic breakup observed after significant stepwise extension. This chapter

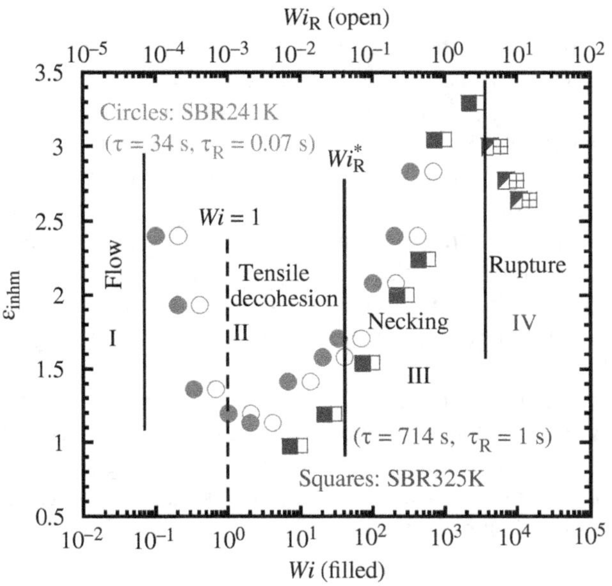

Figure 11.3 "Phase diagram" depicting four regimes from I and IV as a function of either *Wi* in filled and semifilled symbols or Wi_R in open and plus-filled open symbols, for two SBR melts (SBR241K and SBR325K in Tables 7.5 and 7.6) as an example. Specifically, ε_{inhm} denotes the Hencky strain at which homogeneous extension ceases during startup uniaxial extension at the various Hencky rates where the top X axis refers to Wi_R and the bottom X axis shows the value of *Wi*. The borderline between regime II (Tensile decohesion) and regime III (Necking) is marked by Wi_R^* that by definition coincides with the threshold value indicated in Fig. 7.31. *Source:* Zhu and Wang 2013 [6]. Reproduced with permission of American Institute of Physics.

focuses on startup extension while the following chapter studies rheological responses after stepwise extension. In Section 11.4, we discuss the discrepancies between results from the filament-stretching rheometer and results from methods that produce homogeneous extension until the point of strain localization.

Unlike the shear device where the sample is in contact with high heat-capacity shearing plates that insure isothermal tests, and literature melt extension experiments involving immersion of the sample in a circulating liquid,[14–17] the results in this chapter from ambient experiments may not meet the isothermal conditions. Since air convection is ineffective at insuring fast thermal equilibration between the sample under extension and the ambient air, often the specimen's temperature increases during melt extension. Infrared (IR) thermal imaging camera has been used to monitor the sample's temperature in order to report actual rheometric data.[18] As shown in Section 17.1, some fast startup extension tests are actually adiabatic. In the absence of such temperature monitoring, it is still useful to examine the rheological responses of entangled melts to startup extension. This chapter focuses on understanding these responses without *in situ* temperature reading because the conclusions do not change when the corrections due to temperature changes are included.

11.1 Tensile-Like Failure (Decohesion) at Low Rates

The scaling characteristics of the yield point were described in Sections 7.2.1 and 7.5.3, which treat startup shear and extension, respectively. Entangled polymer solutions and melts are clearly able to reach the yield point, signified by a peak in either the shear stress or the engineering stress. At larger extension rates, namely, for $Wi_R > 1$, scaling laws have been identified, showing that the yield strain increases with the applied extension rate. In contrast, for $Wi_R < 1$, the yield strain tends to reach rather small values, namely, $\gamma_y = 2$ for shear and $\varepsilon_y = 1$ for extension. An extensive discussion considering why there are at least two different regimes (with $Wi_R > 1$ and $Wi_R < 1$, respectively) is deferred to Chapter 16, which presents a coherent theoretical discussion of the conceptual foundation for nonlinear polymer rheology.

The scaling behavior associated with the characteristics of yielding, reviewed in Section 7.5.3, gives us helpful information on how to anticipate the appearance of different types of extensional strain localization above the engineering stress maximum. Specifically, Fig. 7.31 shows that the yield strain ε_y is rather insensitive to the applied Hencky rate up to a threshold value Wi_R^*. Yielding is rather different in regime II (where $Wi_R < Wi_R^*$) than it is in the scaling regime III where $Wi_R > Wi_R^*$ and ε_y starts to rise with increasing Wi_R. With $Wi_R < Wi_R^*$, the entangled melt apparently cannot be extended much beyond $\varepsilon = 1$; at larger ε, breakup occurs shortly after the maximum of the engineering stress. In absence of a thorough theoretical description of the polymer entanglement under large deformation, some simple understanding of regime II is obtained by assuming that the entangled melt can be treated as a Gaussian chain network with a mesh size l_{ent} that equals the entanglement spacing.

As discussed in Section 7.5.3.2, we can express the stress response in terms of the "engineering" areal number density ψ of entanglement strands and the intrachain retraction force $f_{retract}$ of each active entanglement strand, as shown in Eq. (7.13). We estimate the magnitude of $f_{retract}$ by comparing Eq. (7.13) with experiment. Figure 11.4 shows examples of stress versus strain curves in regime II, where the largest-strain data points of each curve indicate the termination of uniform stretching. Consistent with Fig. 7.32, in regime II the yield stress does not exceed the melt plateau modulus G_N^0, namely, $\sigma_{engr(y)} < G_N^0$. We thus have

$$\sigma_{engr} = \psi f_{retract(II)} < G_N^0 = \psi_{eq} f_{ent}, \tag{11.1}$$

where use is made of Eq. (2.30c). Eq. (11.1) indicates

$$f_{retract(II)} \sim f_{ent} = k_B T / l_{ent}. \tag{11.2}$$

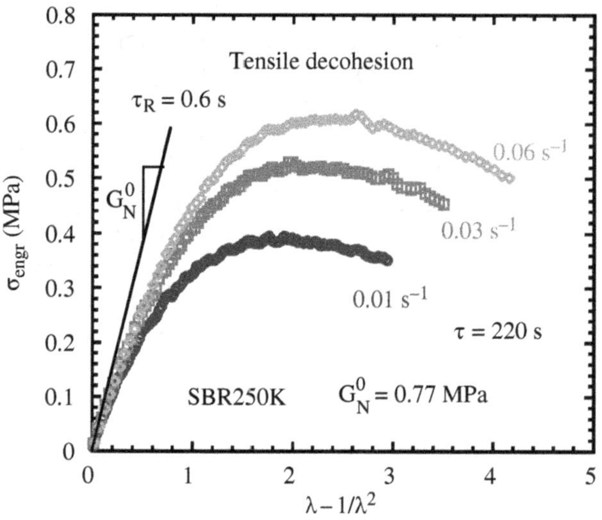

Figure 11.4 Engineering stress versus strain ($\lambda - 1/\lambda^2$) of SBR250K at low Hencky rates (regime II). *Source:* Zhu and Wang 2013 [6]. Reproduced with permission of American Institute of Physics.

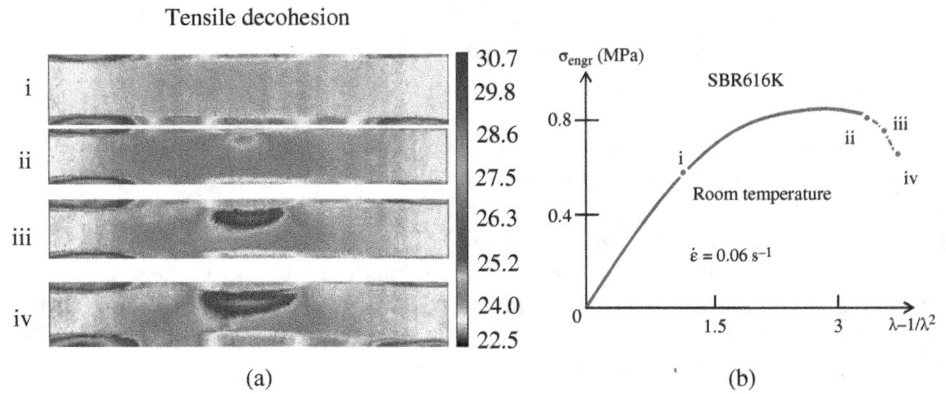

Figure 11.5 (a) Infrared snapshots of an extending specimen (aqua color online) in a Sentmanat extensional rheometer (SER), taken with an IR thermal imaging camera (FLIR SC325), during startup uniaxial extension of SBR616K at room temperature, at a Hencky rate of 0.06 s^{-1}. Here the color codes on the right-hand side indicate the actual sample temperatures at four different stages. The specimen's length in the SER, which is the length of the photos, is around 1.3 cm. Unpublished observations by Panpan Lin at University of Akron. (*See color plate section for the color representation of this figure.*) (b) Schematic curve of engineering stress versus stretching ratio λ, showing the four stages.

If we assume that the inequality in Eq. (11.1) stems from $\psi < \psi_{engr}$, there must be in regime II some loss of entanglement during extension. Our analysis suggests that entangled melts are only able to build a rather moderate level of intrachain retraction force, namely, $f_{retract(II)} \sim f_{ent}$, at modest extension rates. On the other hand, this weak retraction force is sufficient to overcome the cohesion of the entanglement network, provided that we interpret the observed termination of homogeneous extension to arise from structural failure via disentanglement. In other words, we suspect that disentanglement initializes strain localization.

Figure 11.5(a) shows the failure is sharply localized. These are color pictures taken from an IR camera to show the spatial temperature profiles. The rising temperature at the location of strain localization may largely be due to emergent dissipative flow. The characteristic is rupture-like, which we prefer to call tensile decohesion (tensile-like failure). While rupture means elastic failure, Fig. 11.5(a) seems to indicate that the strain localization is highly viscous, initiating at the edge of the specimen where the temperature first rises. Figure 11.5(a) nicely demonstrates the characteristics of strain localization, as the extension continues along the stress versus strain curve shown in Fig. 11.5(b).

11.2 Shear Yielding and Necking-Like Strain Localization at High Rates

11.2.1 Shear Yielding

At larger extension rates we reach regime III, where the rheological behavior and failure mode are rather different. Figures 7.28 and 11.2 show representative stress versus strain curves, which are labeled as "Necking." As indicated in Fig. 7.31, a distinctly different type of failure, namely, necking-like, shows up to replace tensile decohesion as Wi_R increases. Uniaxial extension may perhaps enter the highly stretched state at a smaller Wi_R, namely, for $Wi_R < 1$ than in simple shear: Uniaxial extension may be more effective in causing chain deformation than simple shear. In contrast to regime II, in regime III the styrene–butadiene rubbers continue to stretch well past the yield point at the peak of σ_{engr}. The yield stress also grows larger with increasing extension rate, as shown in Figs. 7.28 and 11.2. Figure 7.31 shows that the yield strain ε_y can grow to a Hencky strain of 2 for $Wi_R > 1$.

In regime III, much larger yield strain than $\varepsilon_c \sim 1$ (cf. Fig. 7.31) and stress can be achieved. We defer further discussion of the origin of much larger yield strain for $Wi_R > 1$ to Chapter 16. Here we use detailed experimental observations to explore the mechanism for shear yielding that leads to unstable "necking."

As an example, consider the behavior of SBR616K (Table 7.5 and 7.6). At a Hencky rate $\dot{\varepsilon} = 1\ s^{-1}$ and $Wi_R = 0.34 < 1$, this melt shows a pronounced maximum in σ_{engr} in the stress versus time relation in Fig. 11.6. Extension past the yield point appears to be homogeneous extension until the last data points of each curve are reached in Fig. 11.6. However, if we closely examine the kinematics of the strain field instead of the overall shape of the specimen, which appears uniform at $t = 2.2$, we see a sign of shear yielding, emerging from the moment marked by the vertical dashed line. In the time window where the Cauchy stress σ_E is nearly constant, which could be mistakenly regarded as attainment of steady flow,[7] a closer examination shows that extension is no longer uniform and globally uniaxial. But the extension of course did not reach steady flow: If the extension during the time interval $t = 1.8$–2.0 was steady flow, *why should the extension subsequently have turned inhomogeneous?*

A comprehensive study[6] used particle-tracking velocimetric observations, made by video recording the ongoing extension of a sheet-like specimen whose surface had been sprayed with light-reflecting particles. Figure 11.7(a) shows the first moment of nonuniform extension during the startup extension. At a later stage, the deformation is no longer extensional, as shown in Fig. 11.7(b). In fact, the deformation is now shear-like, even though the specimen was only subjected to uniaxial extension. Simple shear and extension are both generated by moving two parallel surfaces in opposite directions: We perform uniaxial extension by holding the ends of a specimen and then stretching the sample by displacing the ends in opposite directions without being able to guarantee that the sample will always follow the kinematics of uniaxial extension

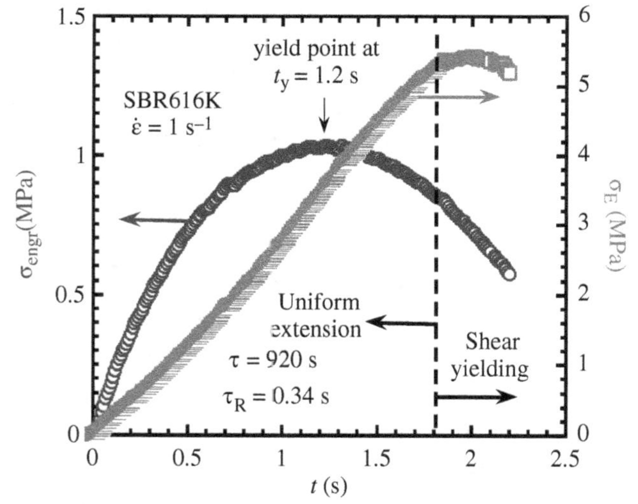

Figure 11.6 Engineering stress σ_{engr} (left-hand-side Y axis) as a function of time upon startup extension of SBR616K at a Hencky rate $1.0\ s^{-1}$ and the time dependence of Cauchy stress σ_E against the right-hand side of Y axis. The extension is further analyzed by examining the local strain field at the late stage, namely, past the vertical dashed line. This analysis is made in Fig. 11.7(a,b). *Source:* Zhu and Wang 2013 [6]. Reproduced with permission of American Institute of Physics.

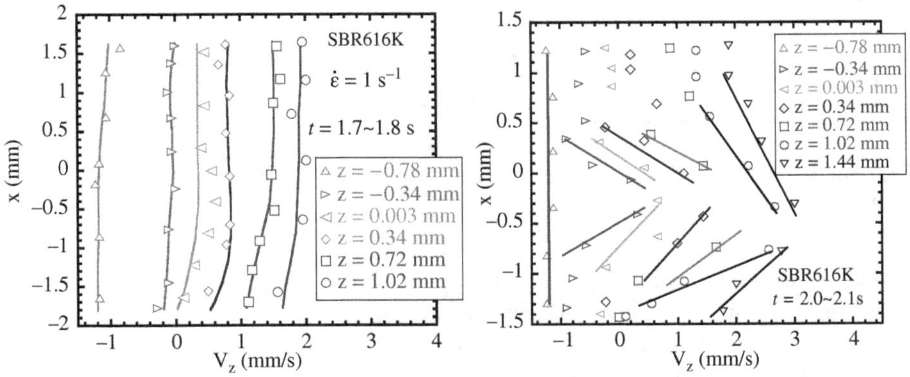

Figure 11.7 The velocity field V_z (in the stretching direction, the Z axis) across the sample width in different Z positions along the specimen's length during a startup extension of Hencky rate of 1.0 s^{-1}, when (a) the field is still uniform because the lines drawn to trace the data are nearly vertical and (b) shear movements are visible, where lines are also drawn to show that the velocities at different values of x along the specimen's width (shown by different symbols) have rather different values. *Source:* Zhu and Wang 2013 [6]. Reproduced with permission of American Institute of Physics.

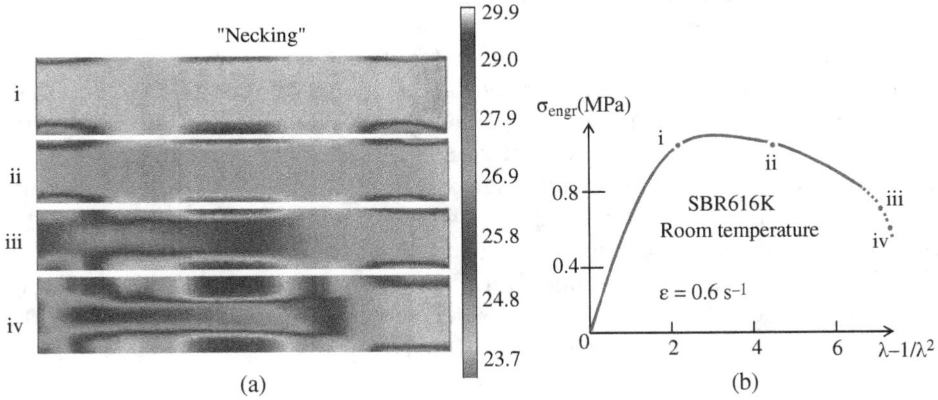

Figure 11.8 (a) Snapshots from the IR thermal imaging camera (FLIR SC325) during startup uniaxial extension of SBR616K at room temperature, at a Hencky rate of 0.6 s^{-1}, where the color codes on the right-hand side indicate the actual sample temperatures at four different stages. The photo length is the sample length, about 1.3 cm. The specimen is initially orange in color (online). (*See color plate section for the color representation of this figure.*) Unpublished observations by Panpan Lin at University of Akron. (b) Schematic curve of engineering stress versus stretching ratio λ, showing the four stages.

everywhere. We see from Fig. 11.7(b) that homogeneous extension cannot prevail in regime III at long times.

The termination of uniform extension has also been captured with an IR. In contrast to Fig. 11.5(a), Fig. 11.8(a) shows an extensive region of the extending specimen where the sample temperature is different from the rest of the specimen. Once shear yielding takes place, leading to nonuniform extension, the stress drops more quickly, as depicted schematically in Fig. 11.8(b).

The behavior depicted in Fig. 11.7(a,b) indicates that shear yielding produces strain localization in the stretched melt. According to Fig. 11.9, the tensile force can project a component at an angle to the extension direction. Correspondingly, we can perceive a shear stress. Depending on the inclining angle θ, this shear stress is related to the Cauchy stress σ_E by

$$\sigma = (\sigma_E/2)sin(2\theta). \tag{11.3}$$

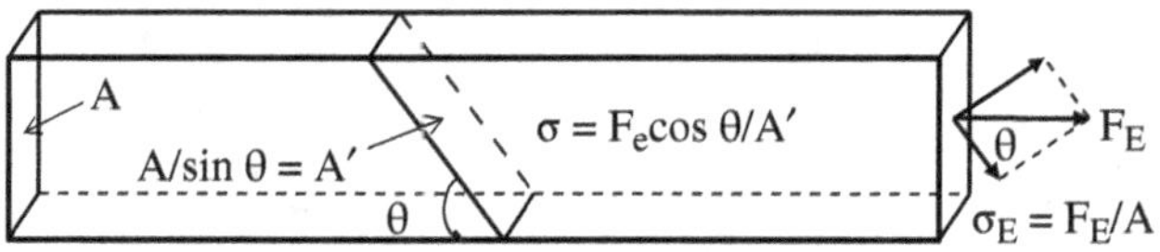

Figure 11.9 The tensile stress can be decomposed along an inclined plane. The maximum shear stress occurs for $\theta = \pi/4$ according to Eq. (11.3).

Then the critical condition for yielding could simply be that the shear stress σ exceeds the cohesive strength of the entanglement network. According to the description in Chapter 7, the minimum cohesive strength is approximately the melt plateau modulus G_N^0. In regime III, the engineering stress does not drop as fast as the cross-sectional area (which goes as $1/\lambda$) decreases, a circumstance that persists until specimen breakup. During homogeneous uniaxial extension, the Cauchy stress $\sigma_E = \sigma_{engr}\lambda$ is a monotonically increasing function of λ. On reaching $\sigma_E = 2G_N^0$, we have from Eq. (11.3)

$$\sigma_{sy} = G_N^0 \text{ at } \theta = \pi/4, \tag{11.4}$$

which may be the minimum condition for shear yielding (sy).

Apparently the stress level estimated by Eq. (11.4) can occur for $Wi_R < 1$. As indicated by the solid symbols in Fig. 7.31, shear yielding begins before the scaling behavior $\varepsilon_y \sim (Wi_R)^{1/3}$ is applicable. Shear yielding could then perhaps take place as soon as the condition of Eq. (11.4) is met, even though $Wi_R < 1$. Thus, the borderline dividing tensile failure and necking is not determined by $Wi_R = 1$ but rather by Eq. (11.4). Indeed, as shown by Fig. 11.3, much of regime III involves the condition $Wi_R < 1$. On the other hand, it is the condition $Wi_R > 1$ that permits entangled melts to stretch to increasingly large yield strains.

There are several issues to discuss with respect to regime III: First, according to Ref. [3], the boundary between regime II and regime III indeed seems to occur where σ_E approaches $2G_N^0$, so that shear stress of critical magnitude given in Eq. (11.4) can cause shear yielding. The master curve in Fig. 11.10 demonstrates convincingly that shear yielding replaces tensile yielding whenever Eq. (11.4) for shear yielding is met. At the larger rates in regime III, apparently the entanglement network does not quickly fall apart beyond the global yielding point at the maximum of σ_{engr}. Inside regime III, the shear stress at the 45° inclined plane always exceeds σ_{sy} of Eq. (11.4).

Second, Fig. 11.6 indicates that (a) the state of deformation after the yield point is not flow, since σ_E continues to rise monotonically and (b) shear yielding occurs in the domain of saturated σ_E. We could speculate that there is still plenty of chain entanglement after the engineering stress maximum. While the decrease of σ_{engr} may imply that more active entanglement strands are being destroyed, it is plausible that some surviving entanglement strands tighten up in presence of strong chain retraction forces. When σ_E exceeds $2G_N^0$, namely, the condition Eq. (11.4) is met, an entangled melt has the choice between shear yielding and tensile yielding. Shear yielding is apparently preferred.

Third, when $Wi_R > 1$, it is not difficult to imagine that, with a reduced importance of local chain relaxation, the chains can tighten up at the entanglement points during extension, making tensile yielding less favorable. In this last portion of regime III, σ_E considerably exceeds the threshold of $2G_N^0$. Failure due to necking through shear yielding is observed to occur at increasingly larger strains. We propose that as the applied extension rate increases the cohesive strength increases above the level set by G_N^0. The data summarized in Figs. 7.8(a) and 7.32 show that the stress maximum can considerably exceed G_N^0.

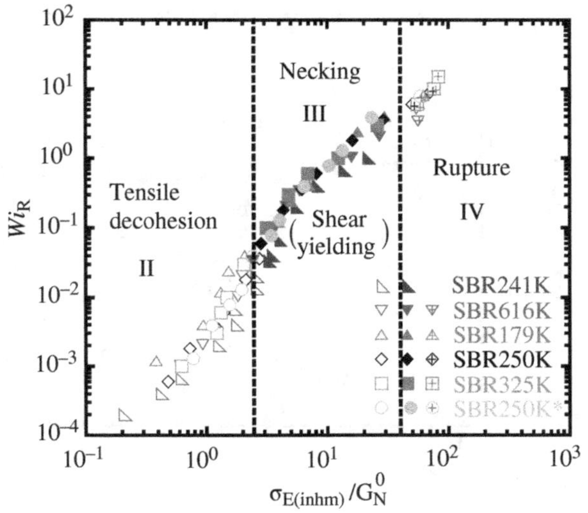

Figure 11.10 Summary of different failure modes in a "phase diagram" defined by Wi_R versus the normalized Cauchy stress at the onset of extensional inhomogeneity. *Source:* Zhu and Wang 2013 [6]. Reproduced with permission of American Institute of Physics.

The present discussion cannot proceed further without invoking a molecular conjecture about how cohesion of the entanglement network is enhanced in the presence of fast deformation with $Wi_R > 1$. We defer further analysis to Chapter 16.

11.2.2 Constant Normalized Engineering Stress at the Onset of Strain Localization

One more feature of regime III is rather noteworthy. Figure 11.3 shows that with increasing Wi the onset of inhomogeneous extension in regime III shifts systematically to larger Hencky strain. The corresponding Cauchy stress $\sigma_{E(inhm)}$ increases as shown in Fig. 11.10. A closer examination of Figs. 7.28 and 11.2 indicates that homogeneous extension terminates when the engineering stress σ_{engr} reaches the same level. Figure 11.11 shows a master curve for the normalized onset stress $\sigma_{engr(inhm)}$ as a function of the normalized Hencky rate (Wi_R). It is perhaps infeasible to explain this intriguing universal condition without discussing the state of chain entanglement just before the shear yielding-induced strain localization. However, an interpretation is desirable about why the onset of necking occurs at a common value of $\sigma_{engr(inhm)}/G_N^0$ over two decades of Hencky rate.

Specifically, according to Eq. (7.18) and Fig. 7.29 and their discussions, macroscopic yielding through molecular disentanglement may be depicted in zeroth order in terms of a reduced engineering areal density of entanglement strands ψ of by the factor α defined in Eq. (7.17) and the reduction of the retraction force $f_{retract}$ of Eq. (7.20) by the factor $\sqrt{\mu}$ given in Eq. (7.16). There is a simple correspondence between the engineering stress σ_{engr} and the two parameters α and μ. The state of yielding is reached because the intermolecular forces created by chain uncrossability cannot match the monotonic increase of the intrachain retraction force $f_{retract}$. A point of force imbalance between interchain and intrachain forces is inevitable, as described in Chapter 16. It is plausible that at the onset of inhomogeneous extension well beyond the yield point $f_{retract(inhm)}$ is comparable to f_{ent} in Eq. (2.30d).

To carry out further analysis, we first recall the condition under which an entanglement strand is fully extended. An equilibrium entanglement strand of N_e Kuhn segments requires a molecular stretching ratio of λ^*_{eq} in Eq. (8.A1) to become straightened. Since values for l_{ent} and l_K are available for many common polymer melts, this limiting ratio λ^*_{eq} can be readily estimated. Upon loss of original entanglements, the surviving entanglement strands are longer,

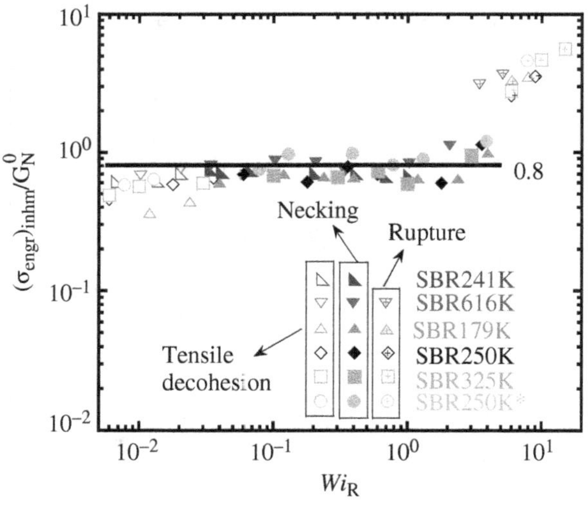

Figure 11.11 Normalized engineering stress at the onset of visible strain localization at different applied rates for six SBR melts, which is constant among these samples of different levels of chain entanglement. *Source:* Zhu and Wang 2013 [6]. Reproduced with permission of American Institute of Physics.

having $N_e'(> N_e)$ Kuhn segments. These new strands are straightened at

$$\lambda^* = \sqrt{N_e'}. \tag{11.5}$$

In terms of Eqs. (8.A1) and (11.5), μ of Eq. (7.16) can be rewritten as

$$\sqrt{\mu} = \lambda^*/\lambda^*_{eq} > 1. \tag{11.6}$$

When there are only two surviving strands per chain, μ approaches its upper bound $\mu_{max} = Z/2$.

The data on regime III in Fig. 11.11, revealing the condition of $\sigma_{engr(inhm)}/G_N^0 \simeq 0.8$ along with Eq. (2.30c), allow us to write Eq. (7.13) as

$$0.8\psi_{eq} f_{ent} \simeq \sigma_{enrg(inhm)} = \psi_{inhm} f_{retract(inhm)}, \tag{11.7}$$

which can be translated into the form

$$\alpha_{inhm}(f_{retract(inhm)}/f_{ent}) \simeq 0.8. \tag{11.8}$$

Since $(\lambda_{inhm})^2 \gg 1$, we can approximate the expression for $f_{retract(inhm)}$ of Eq. (7.20) as

$$f_{retract(inhm)} = f_{ent}\lambda_{inhm}(\mu_{inhm})^{-1/2} = f_0(\lambda_{inhm}/\lambda^*_{inhm}) \tag{11.9}$$

with

$$f_0 = \lambda^*_{eq} f_{ent} = k_B T/l_K, \tag{11.10}$$

which corresponds to the force necessary to orient a Kuhn segment. We conjecture that inhomogeneous extension occurs when the entanglement network loses its structural integrity, corresponding to $\alpha_{inhm} = \psi_{inhm}/\psi_{ent}$ reaching a specific value. Taking α_{inhm} to be 0.4 as an arbitrary choice, Eq. (11.8) yields $f_{retract(inhm)} \sim 2f_{ent}$. Namely, at the onset of inhomogeneous extension well beyond the yield point, $f_{retract(inhm)}$ is comparable in magnitude to the cohesion force f_{ent} defined in Eq. (2.30d). More important, if α_{inhm} is a universal constant (not greatly smaller than unity), independent of the applied rate in regime III, $f_{retract(inhm)}$ is also constant. With a larger value of $\dot{\varepsilon}$, the onset of inhomogeneous extension occurs at a larger stretching ratio λ_{inhm}. But $f_{retract}$ may not be larger because it involves a longer strand. In Eq. (11.9), the ratio $(\lambda_{inhm}/\lambda^*_{inhm})$ stays the same. At $\alpha_{inhm} = 0.4$ and $f_{retract(inhm)} \sim 2f_{ent}$, Eq. (11.9) gives $(\lambda_{inhm}/\lambda^*_{inhm}) = 1/2$, taking $\lambda^*_{eq} = 4$. For example, the value of λ^*_{eq} for these SBR melts is around 4. The condition $\lambda_{inhm} = \frac{\lambda^*_{inhm}}{2}$ says that, for any applied Hencky rate in regime III, at the onset of inhomogeneous extension the surviving strands are stretched half way toward the point of full extension. Thus, the present analysis has provided plausible information about the state of deformation and state of entanglement at the point that necking begins.

In summary, we have parametrized the structural change of the entanglement network during uniaxial extension in terms of the reduction of the areal density of remaining entanglement strands by $\alpha < 1$ and of the increase of the molecular weight between entanglements of the surviving strands by μ. Figure 11.12(a,b) illustrates schematically for two Hencky rates how these two structural parameters change during startup uniaxial extension as functions of the stretching ratio $\lambda = L/L_0$ The stress versus strain curves are also sketched.

11.3 Rupture-Like Breakup: Where Are Yielding and Disentanglement?

At the even larger Hencky rates in regime IV, the rheological response is qualitatively different from the responses in regimes II and III. The transition from regime III to IV, which we call yielding-to-rupture transition (YRT), is termed by Vinogradov and coworkers the "fluid to elastic" transition.[1–5] In regime IV, there is only partial yielding in the following sense: although σ_{engr} deviates downward from the rubber elasticity limit given in Eq. (1.52), it no longer exhibits a peak and only increases monotonically with increasing strain, as shown in Fig. 11.2 (two largest Hencky rates). Uniform extension continues until the specimen suffers rupture. The fracture surfaces are as sharp as those from fracture of a window glass. The IR imaging camera reveals perfectly uniform extension until rupture occurs, as shown in Fig. 11.13(a,b). Figure 11.14 shows simultaneous readings of stress and temperature. According to Table 11.1, at 22 °C, SBR153K has a terminal relaxation time of more than 5000 s. Since the rupture takes place around a Hencky strain $\varepsilon_{rupt} \sim 2.0$, or larger, and $\lambda^*_{eq} \simeq 4.0 < \exp(2)$, the entanglement strands at rupture must be longer than the equilibrium strands that contain N_e Kuhn segments. Some chain disentanglement apparently takes place to produce partial yielding so that the stress versus strain curve shows lower stress than would arise from the affine extension of the equilibrium entanglement network. However, in the later stages of extension, the response

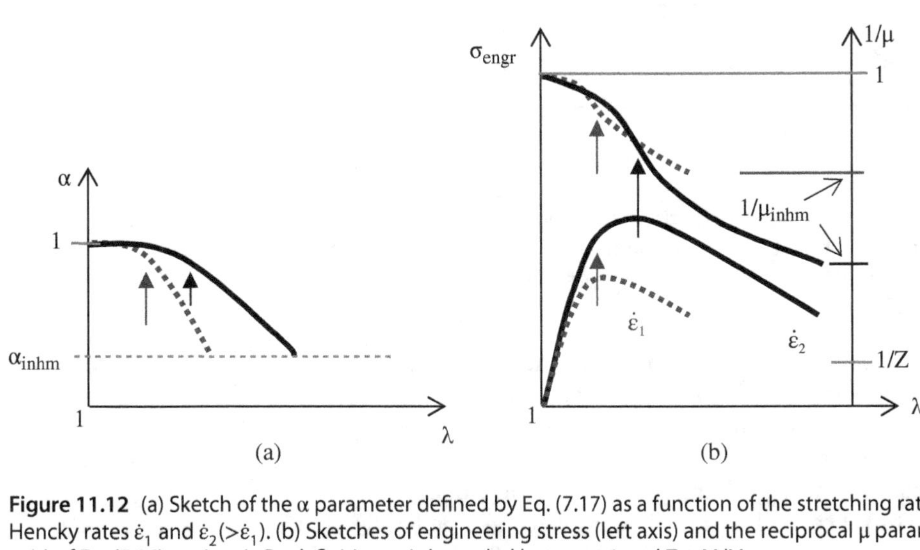

Figure 11.12 (a) Sketch of the α parameter defined by Eq. (7.17) as a function of the stretching ratio λ for Hencky rates $\dot{\varepsilon}_1$ and $\dot{\varepsilon}_2(>\dot{\varepsilon}_1)$. (b) Sketches of engineering stress (left axis) and the reciprocal μ parameter (right axis) of Eq. (7.16) against λ. By definition, μ is bounded between 1 and $Z = M/M_e$.

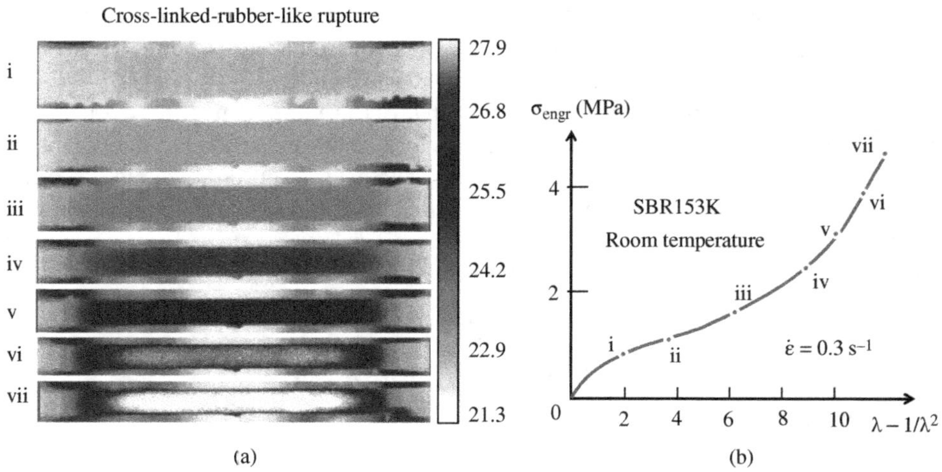

Figure 11.13 (a) Snapshots (initially aqua color online) from an IR thermal imaging camera during startup uniaxial extension of SBR153K at 22 °C at a Hencky rate of 0.3 s^{-1}, where the color codes on the right-hand side indicate the actual sample temperatures at seven different stages. (*See color plate section for the color representation of this figure.*) (b) Schematic curve of engineering stress versus stretching ratio λ, showing the various stages from (i) to (vii). The temperature dependence of the terminal relaxation time for SBR153K is given in Fig. 17.5. *Source:* Reproduced from unpublished observations by Panpan Lin at University of Akron.

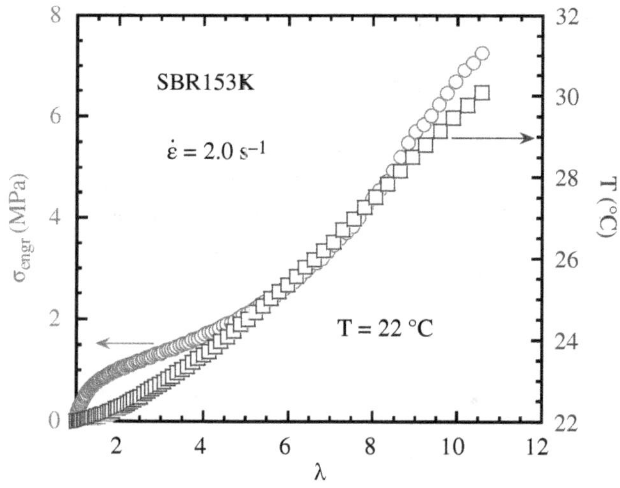

Figure 11.14 Engineering stress σ_{engr} (circles referring to the left-hand-side Y axis) and the corresponding temperature rise (squares referring to the right-hand-side Y axis) as a function of stretching ratio λ for SBR153K at a Hencky rate of 2.0 s^{-1} and T = 22 °C. *Source:* Reproduced from unpublished observations by Panpan Lin at University of Akron.

stiffens, manifesting non-Gaussian stretching of the residual entanglement network. The locking-up, or "freezing in" of the surviving entanglement points also allows the specimen to return to its original dimensions after rupture. Rupture is expected to be triggered by a few chain scission events that unleash massive chain disentanglement. At the same Wi_R, at which extension leads to partial yielding, strain hardening (due to non-Gaussian stretching), and melt rupture, startup shear only results in complete yielding (cf. Section 8.4): In general, the shear stress peaks and declines smoothly (see Chapter 17 for an exception).

Table 11.1 Molecular characteristics of one additional SBR melt at 22 °C.

M_n (kg/mol)	M_w/M_n	τ (s)	τ_R (s)	G_N^0 (MPa)	Styrene (%)	Butadiene (%)	Vinyl (%)
153	1.05	5060	51	0.53	25.6	74.4	70

11.4 Strain Localization Versus Steady Flow: Sentmanat Extensional Rheometry Versus Filament-Stretching Rheometry

It is conventional wisdom that for liquids only the current stress state is a useful quantity for characterizing their rheological behavior. In the past, rheological responses of polymer melts and solutions to startup extensional deformation have been expressed in terms of the Cauchy stress σ_E. Except for studies[1–5] from Russia, literature results from research groups of Sridhar, McKinley, and Hassager might leave one with the impression that steady-state flow can be attained in uniaxial extension. However, the preceding Sections 11.1 through 11.3 amply demonstrate that the opposite is true in a Sentmanat extensional rheometer (SER). Such strain localization implies that steady flow does not take place. As reviewed by Malkin and Petrie,[4] specimen failure is common in melt extension. The viewpoint, namely, that steady flow state is not attainable in extension, is consistent with previous results[1–5] that were obtained using Instron-type extension of entangled polymer melts. To reiterate, at a fixed Hencky rate with $Wi > 1$, startup uniaxial extension of well-entangled melts cannot reach a fully developed flow state before strain localization terminates homogeneous deformation.

Representing data similar to those in Fig. 7.28 in terms of the Cauchy stress σ_E, we can replot the data in regime III from Fig. 7.28 and obtain Fig. 11.15. The final data points in Fig. 11.15 indicate the strains at which shear yielding and subsequent inhomogeneous extension start to take place. These end points fall onto a straight line that goes through the origin, indicating a common engineering stress $\sigma_{engr(inhm)}$, consistent with results discussed in Section 11.2.2. Specifically, Fig. 11.15 shows $\sigma_{E(inhm)}$ to be linearly proportional to λ_{inhm}, implying $\sigma_{engr(inhm)} = \sigma_{E(inhm)}/\lambda_{inhm}$ is constant, independent of the Hencky rate. Here we note that $(\lambda - 1/\lambda^2)$ is numerically close to λ for $\lambda = 5$ or larger.

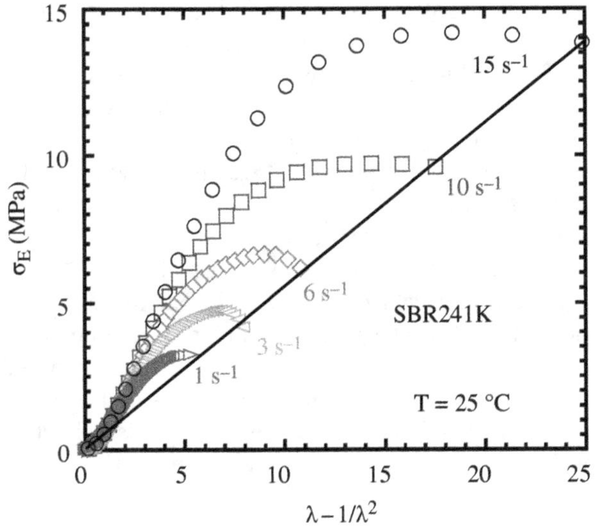

Figure 11.15 True stress σ_E as a function of strain $(\lambda - 1/\lambda^2)$ of SBR241K at T = 25 °C at five different Hencky rates from 1 to 15 s^{-1} in regime III where the last data points are at the onset of visible necking and form a straight line to indicate that $\sigma_{E(inhm)}$ at these points grows approximately linearly with λ. *Source:* Data taken from Zhu and Wang 2013 [6].

The leveling off of σ_E at Hencky rates of 10 and 15 s^{-1} gives the impression that steady state has been reached. If a flow state has been reached, why then would the melt suddenly undergo strain localization? At smaller extension rates, σ_E shows a maximum and then a decline before the specimen breaks up. The data in Fig. 11.15 show a lack of steady state. This conclusion is at odds with classical literature on melt extension of low-density polyethylene (LDPE). Using an Instron-type stretcher, Laun and Münstedt showed not only the level-off of the Cauchy stress but also the saturation of the recoverable strain at Hencky strain above four.[15,17] It remains to be clarified whether long chain-branching plays a role.

Along a different, second line of research, studies based on the FSR[19–21] have claimed to observe steady flow behavior for $Wi > 1$ for entangled melts and solutions,[7,8] consistent with the conclusion of Refs. [15,17]. Uniaxial extension at a constant Hencky rate was only realized at the middle point of the specimen by moving the two plates away in whichever manner to assure that the middle point shrinks exponentially in time.[7] The specimen initially has the shape of an hourglass. The device permits different segments along the specimen length to undergo extension at different stress states with different local strain rates. The so-called "steady-state" properties are not obtained from homogeneous uniaxial extension and therefore cannot be compared with any molecular theories that evaluate rheological properties on the basis of homogeneous extension.

Unlike earlier pioneering measurements from Vinogradov's lab, studies based on the filament-stretching rheometer did not aim at producing homogeneous uniaxial extension. In fact, the measurement scheme disguises strain localization behavior. The tendency to undergo strain localization in the middle segment is suppressed by instantly adjusting the crosshead speed. Consequently, the extension is inherently nonuniform. To reiterate, specimens in filament-stretching rheometers do not undergo homogeneous extension at any time. The application of FSR is even more questionable when it is applied to analyze stress relaxation after stepwise extension. Stress relaxation under extension in FSR will be discussed further in the last paragraph of Section 13.2. In passing, we note that the same authors who claimed to achieve steady extensional flow in FSR suggested that the emergence of a maximum in σ_{engr} should produce necking.[22]

Whenever we cannot generate a homogeneous deformation field, we lose our ability to discern whether at a given speed there is a unique rheological state. The given speed might be the velocity of the shearing surface in simple shear or the relative velocity of the two ends under uniaxial extension. Let us recall the case of shear banding. Over a gap distance on the order of 1 mm, we cannot dictate that shear homogeneity always prevails. For sufficiently entangled polymer solutions and melts, a startup shear with $Wi \gg 1$ can produce long-lived shear banding that survives in a seemingly steady state.[23] We cannot be sure which local shear rate is unique. If there are several of them, then none of them is unique at a given shear stress. Moreover, as discussed in Section 9.2.3, we are not always sure that shear banding is a unique state. Similarly, at $Wi \gg 1$, in uniaxial extension produced by holding the two ends of a macroscopic specimen and moving them apart an exponentially increasing velocity does not guarantee homogeneous extension of an entangled polymer melt.

Experiments performed with a SER reveal various types of strain localization. The phenomenon observed with a filament-stretching rheometer looks deceptively different. In the feedback-loop dictated filament, stretching rheometry, filament failure, or strain localization is replaced by nonuniform stretching. Apart from the issue of whether a fully developed flow state is reached, where the only evidence is a short period of relatively constant Cauchy stress, we note that the so-called "steady state" of the midpoint of the filament is at most one possible state. When global homogeneity is not observed, as is the case when shear banding persists during startup continuous shear, we can no longer assert that a unique rheologically steady state exists. A filament-stretching rheometer is accessing multiple states when it reports a quasi-steady true stress for the middle segment of the specimen: Since the tensile force is the same everywhere along the specimen but the cross-sectional area varies along the length of the specimen, the Cauchy stress and Hencky strain along the length of the sample have values different from their values at the middle. More investigations should be made in the future to inquire about the validity of comparisons of data from filament-stretching rheometry and theoretical calculations.

One conclusion can already be drawn. Filament-stretching rheometry does not produce a globally uniform uniaxial extension. In the traditional sense, the filament-stretching rheometer is not a suitable rheometric device because the Cauchy stress and Hencky rate both vary along the specimen. Filament-stretching rheometry for uniaxial extension is analogous to capillary rheometry for shear deformation. It is not straightforward to explore either the transient extensional rheological behavior or the "steady-state" behavior using these instruments because the strain field cannot be prescribed over an extensive region. When the constitutive behavior of the sample is unknown, capillary rheometry is powerless. This technique cannot be applied to determine any detailed constitutive relationship. It only provides two simple numbers: the throughput Q and the piston pressure or die entry pressure P. If results from filament-stretching rheometry are to be compared with theory, one must perform a fluid mechanical calculation to take into account the spatial variations of the tensile stress, extensional strain, and Hencky rate for the sample at each point between two separating end plates. The evidence for nonuniform extension and absence of any prescription or control of the strain field comes from the decorrelation between the apparent Hencky strain ε_Z and the local Hencky strain ε, as measured at the middle of the specimen. For example, Fig. 11.16 shows this decorrelation and confirms that filament-stretching rheometer does not prescribe homogeneous extension because ε does not linearly vary with ε_Z. Additional discussion is presented in Appendix 11.A to indicate that it is questionable to take the rheological information from the filament-stretching rheometry and interpret it as fully constitutive.

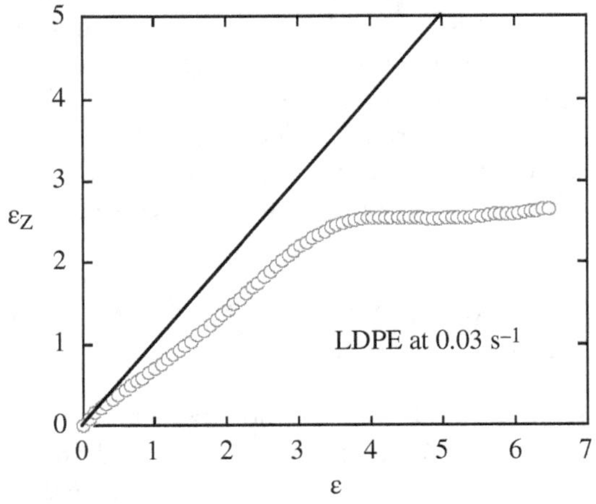

Figure 11.16 Hencky strain according to the length L of the specimen relative to its initial length L_0, given by $\varepsilon_Z = ln(L/L_0)$, as a function of the *local* tensile strain at the middle of specimen that can be defined in terms of the ratio of the specimen's diameter D to the original diameter D_0: $\varepsilon = (1/2)ln(D_0/D)$. The data were based on melt extension of low-density polyethylene melt at Hencky rate 0.03 s^{-1}, using filament-stretching rheometry. *Source:* Data taken from Marín *et al.*, 2013 [24].

11.5 Role of Long-Chain Branching

The discussion so far in this chapter on uniaxial extension has referred to polymer melts of linear chains. The behavior of entangled melts with sufficient long-chain branching is rather different. Quantifying the linear viscoelastic properties of polymers with long-chain branching is theoretically rather challenging.[25] In practice, it is not straightforward to produce such polymers with well-defined branching architecture. Although systematic studies remain to be carried out to identify the role of long-chain branching in the nonlinear response of polymer melts, we have some general knowledge. Specifically, based on the examples of LDPE and dendritic PI, we can conclude that in presence of long-chain branching, the point of macroscopic separation occurs at larger extensional ratio, well beyond the yield point, as shown in Section 8.2.2. Binary mixtures containing a component of ultrahigh molecular weight also show homogeneous extension to high stretching ratio.[11,26] Finally, we note that a systematic description of how tensile failure takes place as a function of the Hencky rate has yet to be completed for entangled melts made of chains with long-chain branching. In particular, it remains unknown whether melts with sufficient long-chain branching can attain steady state during uniform uniaxial extension. More work should be carried out to show whether low-density polyethylene is indeed able to reach a fully developed flow state, as reported in the literature.[14,15,17] Moreover, it remains unknown whether long-chain branching can suppress shear inhomogeneity during startup shear. It is desirable that a future study will be carried out to provide some clues about the effect of long-chain branching on nonlinear responses of entangled polymers to startup shear.

This chapter describes strain localization in uniaxial extension as Chapter 9 shows strain localization in shear. The results of shear banding in Chapter 9 are the inspiration for the studies carried out in this chapter because the physics behind the strain localization is the same: localized yielding and failure of the entanglement network. In shear, the magnitude of extrapolation length b relative to the gap distance H dictates whether or not shear banding takes place. No pertinent parameters have been identified to show how inhomogeneous extension can be prevented. Moreover, for a well-entangled melt, as a function of increasing Hencky rate, there are three different types of extensional strain localization, tensile decohesion, necking-like failure, and melt rupture, respectively. Because of the strain localization behavior, homogeneous flow state cannot be reached. This conclusion contrasts sharply with the literature data based on filament-stretching rheometry that claim to attain steady flow. Appendix 11.A cautions us against any broad interpretation of the data from the filament-stretching rheometry. In particular, it is shown that the configuration of filament-stretching rheometer disguises the key characteristics of nonlinear responses of entangled melts to startup extension: elastic breakdown leading to macroscopic separation of the specimen under extension. More studies have yet to be performed to determine whether various modes of strain localization including melt rupture would disappear when the level of entanglement, quantified by the number of entanglements per chain $Z = M/M_e$, decreases below some threshold value.

11.A Analogy between Capillary Rheometry and Filament-Stretching Rheometry

Section 3.1.3.1 discussed how capillary rheometry can be applied to learn about the shear rheology based on the entry pressure P and throughput Q_f. Even in the absence of a simple scaling law, we can still use Eq. (3.12c) to evaluate the shear rate at the capillary wall. It is necessary to note that expressions such as Eq. (3.12c) treat the capillary flow in the limit of steady state by assuming the rheological properties are not time-dependent and ignoring the die entry effect discussed in Chapter 10. Therefore, capillary rheometry is not a preferred method to study constitutive shear behavior of any viscoelastic materials including materials dealt with in this book, namely, entangled polymer solutions and melts. Let us suppose that in principle we have long enough dies and thus are able to observe steady flow in the capillary die at different values of the pressure gradient dP/dz to reliably calculate the shear stress σ_w at wall according to Eq. (3.7a). Then we can apply a visualization method such as particle-tracking velocimetry and explicitly measure the shear rate $\dot{\gamma}_{w(n\text{-}Newt)}$ at the wall. So we can presumably obtain a curve relating σ_w to $\dot{\gamma}_{w(n\text{-}Newt)}$, which is often known as the flow curve. Thus, by examining the rheological state at the capillary wall, we could be led to conclude that the constitutive relationship in steady shear could be obtained in capillary rheometry.

The filament-stretching rheometry (FSR) can claim to have accomplished the same for uniaxial extension by only examining the rheological behavior at the middle of an extending specimen. Both capillary rheometry and FSR seem rather clever because both claim to have obtained the steady-state rheological characterization. However, because both describe only the rheological state at one location, either at the capillary wall or middle of a long specimen, they are inherently incapable of describing, anticipating, and therefore accounting for states over an extensive spatial region. The relation of σ_w versus $\dot{\gamma}_{w(n\text{-}Newt)}$ from capillary rheometer cannot possibly say anything about whether bulk shear banding can take place or not. Similarly, the FSR with a feedback loop by design avoids the interesting physics of tensile strain localization. Thus, capillary rheometry and filament-stretching rheometry both by default miss the opportunity to learn about and reveal any plausible strain localization behavior. Consequently, they have very limited utility and must not be applied to produce misleading information.

According to the capillary rheometry, the information on shear stress and rate at the wall is available through Eqs. (3.7a) and (3.12c). But it would be a mistake if we start to think that the relation σ_w vs $\dot{\gamma}_{w(n\text{-}Newt)}$ actually provided full constitutive information about steady shear rheology. Obviously this "flow curve" says nothing about whether shear banding would take place in steady state.

In the past, the rheological information from the filament-stretching rheometry has been interpreted to represent the constitutive behavior in melt extension in numerous studies, which are too many to list here. But such information from the filament-stretching rheometry cannot be used to anticipate tensile strain localization, found using Sentmanat Extensional Rheometry. Yet, the tensile strain localization always occurs for entangled melts whenever $Wi > 1$. By analogy, it would be misleading and actually wrong for us to simply invoke capillary rheometry and claim to be able to provide the full shear rheology information based on measuring the shear rate at the capillary wall and shear stress from the pressure gradient, and for us to stop using all the other designs for shear rheometry including the cone-plate device.

In conclusion, in either capillary or filament stretching configuration, we cannot prescribe the strain field without the explicit constitutive law that governs the rheological behavior of the sample under study. The complex strain field actually varies with the imposed external condition and sensitively depends on the material's constitutive behavior. Thus, capillary and filament setups fall short of being good rheometric apparatuses.

References

1 Vinogradov G. Viscoelasticity and fracture phenomenon in uniaxial extension of high-molecular linear polymers. Rheol Acta. 1975;14(10):942–54.

2 Vinogradov G, Malkin AY, Volosevitch V, editors. Some fundamental problems in viscoelastic behavior of polymers in shear and extension. J Appl Polym Sci Appl Polym Symp. 27, 1975.

3 Vinogradov G, Volosevitch V, Shatalov V, Yudin V. Flow, high-elastic (recoverable) deformation, and rupture of uncured high molecular weight linear polymers in uniaxial extension. J Polym Sci Polym Phys Ed. 1975;13(9):1721–35.
4 Malkin AY, Petrie C. Some conditions for rupture of polymer liquids in extension. J Rheol. 1997;41(1):1–25.
5 Malkin AY, Arinstein A, Kulichikhin V. Polymer extension flows and instabilities. Prog Polym Sci. 2014;39(5):959–78.
6 Zhu X, Wang S-Q. Mechanisms for different failure modes in startup uniaxial extension: tensile (rupture-like) failure and necking. J Rheol. 2013;57(1):223–48.
7 Bach A, Rasmussen HK, Hassager O. Extensional viscosity for polymer melts measured in the filament stretching rheometer. J Rheol. 2003;47(2):429–41.
8 Costanzo S, Huang Q, Ianniruberto G, Marrucci G, Hassager O, Vlassopoulos D. Shear and extensional rheology of polystyrene melts and solutions with the same number of entanglements. Macromolecules. 2016;49(10):3925–35.
9 Wang Y, Boukany P, Wang S-Q, Wang X. Elastic breakup in uniaxial extension of entangled polymer melts. Phys Rev Lett. 2007;99(23):237801.
10 Wang Y, Wang S-Q. From elastic deformation to terminal flow of a monodisperse entangled melt in uniaxial extension. J Rheol. 2008;52:1275–90.
11 Wang Y, Wang S-Q. Salient features in uniaxial extension of polymer melts and solutions: progressive loss of entanglements, yielding, non-Gaussian stretching, and rupture. Macromolecules. 2011;44(13):5427–35.
12 Nielsen JK, Rasmussen HK, Hassager O. Stress relaxation of narrow molar mass distribution polystyrene following uniaxial extension. J Rheol. 2008;52(4):885–99.
13 Huang Q, Agostini S, Hengeller L, Shivokhin M, Alvarez NJ, Hutchings LR, et al. Dynamics of star polymers in fast extensional flow and stress relaxation. Macromolecules. 2016;49(17):6694–9.
14 Laun H, Münstedt H. Comparison of the elongational behaviour of a polyethylene melt at constant stress and constant strain rate. Rheol Acta. 1976;15(10):517–24.
15 Laun HM, Münstedt H. Elongational behaviour of a low density polyethylene melt. Rheol Acta. 1978;17(4):415–25.
16 Wagner M, Laun H. Nonlinear shear creep and constrained elastic recovery of a LDPE melt. Rheol Acta. 1978;17(2):138–48.
17 Münstedt H, Laun HM. Elongational behaviour of a low density polyethylene melt. Rheol Acta. 1979;18(4):492–504.
18 Lin P, Wang S-Q. Erratum: nonisothermal condition in past melt extension experiments. J Rheol. 2015;59(5):1329–34.
19 Sridhar T, Tirtaatmadja V, Nguyen DA, Gupta RK. Measurement of extensional viscosity of polymer-solutions. J Nonnewton Fluid Mech. 1991;40(3):271–80.
20 Tirtaatmadja V, Sridhar T. A filament stretching device for measurement of extensional viscosity. J Rheol. 1993;37(6):1081–102.
21 McKinley GH, Sridhar T. Filament-stretching rheometry of complex fluids. Annu Rev Fluid Mech. 2002;34:375–415.
22 McKinley GH, Hassager O. The Considere condition and rapid stretching of linear and branched polymer melts. J Rheol. 1999;43(5):1195–212.
23 Ravindranath S, Wang S-Q, Olechnowicz M, Quirk RP. Banding in simple steady shear of entangled polymer solutions. Macromolecules. 2008;41(7):2663–70.

24 Marín JMR, Huusom JK, Alvarez NJ, Huang Q, Rasmussen HK, Bach A, et al. A control scheme for filament stretching rheometers with application to polymer melts. J Nonnewton Fluid Mech. 2013;194:14–22.
25 McLeish T, Larson R. Molecular constitutive equations for a class of branched polymers: the pom–pom polymer. J Rheol. 1998;42(1):81–110.
26 Wang Y, Cheng S, Wang S-Q. Basic characteristics of uniaxial extension rheology: comparing monodisperse and bidisperse polymer melts. J Rheol. 2011;55:1247–70.

Part III

Decohesion and Elastic Yielding After Large Deformation

Part 3 describes one of the most convenient and useful ways to learn about nonlinear rheological responses of entangled polymers. Specifically, we examine what happens during stress relaxation after a large stepwise shear or extension. Unlike continuous shear, which typically involves a rotational rheometer, stepwise shear (step strain in short) can be made in a slide-plate device (cf. Section 3.1.1) without accumulative edge effects due to meniscus instability. Since stepwise deformation is typically produced "instantly," that is, it takes place on the time scale negligibly short relative to the chain relaxation times, the initial response is fully elastic. Here we demonstrate that upon sufficient elastic deformation, either in simple shear or uniaxial extension, dramatic changes can occur in the structural integrity of the step-strained entangled polymers. Chapters 12 and 13 describe the aftermaths, after a significant amount of shear or extension, respectively, has been applied. Part 3 is closed with Chapter 14 that introduces the concept of finite cohesion in entangled polymers.

Nonlinear Polymer Rheology: Macroscopic Phenomenology and Molecular Foundation, First Edition. Shi-Qing Wang.

Companion Website: www.wiley.com/go/nonlinearpolymerrheology

12

Nonquiescent Stress Relaxation and Elastic Yielding in Stepwise Shear

The protocol of a step strain test has been introduced in Chapter 1 using the example of simple shear, as shown in Fig. 1.1(e). Step shear is not only a valuable method to characterize the quiescent chain dynamics in the linear response regime but also most convenient to apply in order to probe nonlinear rheological responses. After shear cessation, any modern rheometric instruments can record how the resulting shear stress decreases in time, for any imposed shear magnitude γ. For entangled polymer solutions and melts, we observe linear response for $\gamma < 0.6$. For any magnitude of step shear, as long as $\gamma < 0.6$, the stress relaxation occurs in the same quantitative manner, and there is a unique relaxation modulus $G(t)$ as shown in Eq. (1.14) that describes the stress decay. Large step strain is accomplished by displacing the upper shearing surface by distances appreciably greater than the gap distance H, for example, having $X = 2H$ or 3H in Fig. 1.1(a). Stress relaxation measurements involve holding the displaced surface fixed while measuring the force $F(t, \gamma)$ on the either surface. At such a large strain, the functional dependence of $F(t, \gamma)$ on time t is no longer given by a single function $G_{eq}(t)$ but by different functions at different values of γ. Such behavior is inherently nonlinear and commonly observed.[1] In both pre- and post-tube-model eras, step shear had been thought to result in quiescent stress relaxation regardless of how large γ is. Here quiescent relaxation means that after shear cessation, an entangled polymer solution or melt displays no macroscopic motions as the shear stress relaxes in time.

Until 2006, the response of a viscoelastic fluid such as an entangled polymer to step shear was perceived to be *quiescent*: The moving shear plate in a simple shear cell is displaced for a short duration and then brought to stop. Once the shear plate stops moving the sample immediately[1] becomes macroscopically stationary. In 2006, the phenomenon of *nonquiescent relaxation* was discovered:[2,3] Particle-tracking velocimetry (PTV) revealed macroscopic motions in the step-sheared sample during stress relaxation, after the moving shear plate had been brought to a stop, contrary to the commonly held assumption that stress relaxation should always be quiescent. Subsequently, as a prediction, it was demonstrated that step uniaxial extension also results in strain localization.[4,5] For extension, nonquiescent stress relaxation means that the step-extended specimen break up, a subject for Chapter 13.

To explain nonquiescent relaxation after a sizable step deformation,[2,3,5] the concept of elastic yielding was introduced.[6] Actually, the nonquiescent relaxation phenomenon led to the discovery of the *elastic yielding*. Elastic stress built during stepwise deformation can be so high that, after termination of the large deformation, it destroys the entanglement network. Nonquiescent relaxation is the macroscopic manifestation of network breakup, which is largely elastically driven. Elastic yielding is the most radical and novel concept to be introduced in this book.

1 The inertia dies out rather rapidly in a viscous material. The time scale on which a moving sample comes to stop can be estimated as $t_{inert} \sim \rho H^2/\eta \sim 10^{-3}/\eta$ Pa, which is ca. 1 ms for $\eta = 1$ Pa s, where $\rho \sim 10^3$ kg/m^3 and $H \sim 10^{-3}$ m.

Nonlinear Polymer Rheology: Macroscopic Phenomenology and Molecular Foundation, First Edition. Shi-Qing Wang.

Companion Website: www.wiley.com/go/nonlinearpolymerrheology

Unlike some other recently discovered novel phenomena such as shear banding, there was no prior warning from any theory that elastic breakup might exist. As a contrast, consider shear banding, discovered in 2006 for entangled polybutadiene solutions.[7] The original Doi–Edwards tube model[8] of 1979 already suggested shear banding based on its symptom of a constitutive nonmonotonicity. Furthermore, shear banding had been reported by 1996 as a characteristic nonlinear response in wormlike micellar solutions.[9–16] The first observation of shear banding in a polymer solution was actually made[17] by Callaghan and Gil in 2000. No follow-up study took place, and little attention was given to the significance of this report, perhaps because the authors interpreted the banding as arising from specific chemical effects, namely hydrogen bonding in the aqueous polyacrylamide solution, leading to specific associations among polyacrylamide chains.

This chapter focuses on a phenomenological description of nonquiescent relaxation and elastic yielding phenomena observed during relaxation from large step strain. While the notion of elastic yielding was discovered thanks to the observation of nonquiescent relaxation, elastic yielding also takes place during quiescent relaxation. In other words, after stepwise simple shear, breakdown of entanglement network can occur homogeneously. We begin in Section 12.1 by reviewing the conventional description of step strain tests from the literature, then proceed in Section 12.2 to detail the recent PTV observations of macroscopic breakdown after step shear, and end with a discussion of interfacial elastic yielding in Section 12.4 after a short description of quiescent elastic yielding in Section 12.3. When the elastic yielding is only partial for an intermediate strain magnitude, stress relaxation also occurs quiescently, a topic discussed in Section 14.1.

12.1 Strain Softening After Large Step Strain

12.1.1 Phenomenology

A step strain test aims to reveal how chain relaxation takes place after instantaneous deformation. According to the picture of simple shear illustrated in Fig. 1.1, we displace the upper plate for a prescribed distance X, as described in Section 1.1.2, hold both shearing plates fixed at $t \geq 0$ and measure the relaxing shear stress $\sigma(t, \gamma)$. It is conventional to introduce a relaxation modulus $G(t, \gamma)$ that is related to the relaxing shear stress by

$$\sigma(t, \gamma) = G(t, \gamma)\gamma, \tag{12.1}$$

which generalizes Eq. (1.14). By definition, any dependence of $G(t = 0, \gamma)$ on γ indicates nonlinear response. Typically, "strain softening" occurs when the stepwise shear is not produced sufficiently rapidly, and the stress is not linearly proportional to strain. An example is shown in Fig. 12.10(a). Nonlinear responses arise when the relaxation modulus $G(t, \gamma)$ decreases in time differently for different values of γ because γ is beyond a threshold value γ_{LR}.

The linear response regime is characterized by an equilibrium relaxation modulus $G_{eq}(t)$ that captures the time dependence of the relaxing shear stress $\sigma(t)$ over a small range of γ

$$G_{eq}(t) = \sigma(t, \gamma < \gamma_{LR})/\gamma. \tag{12.2a}$$

This expression, identical in form to Eq. (1.14), implicitly assumes that $\sigma(t = 0, \gamma < \gamma_{LR})$ is linearly proportional to γ. A more appropriate criterion for linear response to step strain should be simply specified by a threshold strain γ_{LR}, below which the ratio

$$\sigma(t, \gamma < \gamma_{LR})/\sigma(t = 0, \gamma < \gamma_{LR}) = \overline{G}_{eq}(t) \tag{12.2b}$$

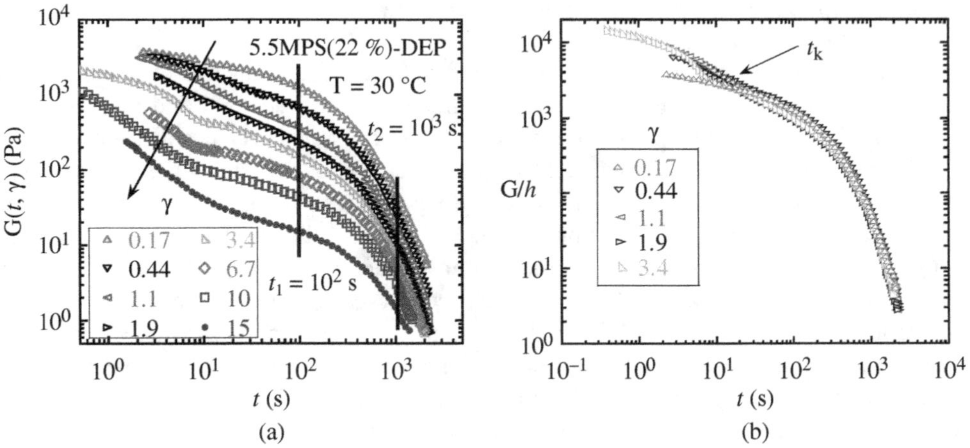

Figure 12.1 (a) Relaxation modulus $G(t, \gamma)$ at 30 °C for different magnitudes of step strain with γ ranging from 0.17 to 15.3. The sample was a 22% polystyrene solution: polystyrene in diethyl phthalate where PS had $M = 5.53 \times 10^6$ g/mol. The figure is based on data in Fig. 3 in Ref. [1], originally from Ref. [18]. The downward deviation from the 0.17 curve (up-pointed triangles) of the 0.44 curve (down-pointed triangles) plausibly reflects some premature wall slip. (b) Data after shifting the data in (a) to match at $t_2 = 10^3$ s. This amounts to defining G/h, where h is given in Eq. (12.4). Here the kink-like relaxation appears at each γ except for $\gamma = 0.17$, at which linear response can be expected. The collapse of data beyond t_K shows that the long-time stress relaxation curves are superimposable. *Source:* Data taken from Osaki 1993 [1], originally from Fukuda *et al.* 1975 [18].

has the same time dependence for all values of $\gamma < \gamma_{LR}$. This definition of a dimensionless relaxation modulus includes cases in which the shear stress σ deviates from linear dependence on γ.

Nearly all experimental data in the literature for step strain have been presented in terms of the relaxation modulus $G(t, \gamma)$, traditionally defined as

$$G(t, \gamma) = \sigma(t, \gamma)/\gamma. \tag{12.3}$$

Figure 12.1(a) is an example. If G decreases with increasing γ for a given time t, as is the case in Fig. 12.1(a), then the sample is said to show *strain softening*. Experimentally, it is often observed for entangled polymers that, after some initial drop, namely, for $t > t_k$, the shear stress $\sigma(t, \gamma)$ relaxes over time in the same manner as $\overline{G}_{eq}(t)$ or $G_{eq}(t)$ does. This feature is known as time-strain separability. Here the subscript "k" means "kink," because there can be an obvious kink in the stress relaxation curve, corresponding to a moment of sharp stress decline at early times. The data in Fig. 12.1(a) display the kinks and approximate time-strain separability.[18] The return to the equilibrium stress relaxation curve at long times indicates that dependence of the stress relaxation on γ at the long times can be captured in terms of a damping function h.[19,20]

It is customary to describe the strain softening by making a further normalization that is claimed to "remove" the common (equilibrium) time dependence, leading to a nominally time-independent damping function $h(\gamma)$

$$h(\gamma) = \frac{G(t > t_k, \gamma)}{G_{eq}(t > t_k)} = \left(\frac{\gamma_1}{\gamma}\right)\left[\frac{\sigma(t > t_k, \gamma)}{\sigma(t > t_k, \gamma_1 < \gamma_{LR})}\right]. \tag{12.4}$$

Here the second equality follows upon use of Eqs. (12.2a) and (12.3). When the stress growth during stepwise shear is weaker than linear in γ, $G(t > t_k, \gamma)$ in Eq. (12.4) for the damping function $h(\gamma)$ has additional γ dependence that is unrelated to the stress relaxation behavior. In other words, h drops below unity and is already a decreasing function of γ if $\sigma(t=0, \gamma)$ does not increase with γ as strongly as linearly.

Operationally speaking, to evaluate the damping function h one measures the stress levels at, say, $t_1 = 10^2$ s or $t_2 = 10^3$ s from Fig. 12.1(a), for the available values of γ and divides these stress values by $G(t, \gamma = 0.17)$ at the same time t_1 or t_2. Because $\gamma = 0.17$ is sufficiently small, $G(t, \gamma = 0.17)$ is to good approximation equal to $G_{eq}(t)$. These ratios give the dependence of $h(\gamma)$ on γ. Using $h(\gamma)$ obtained at 10^3 s, the $G(t, \gamma)$ data in Fig. 12.1(a) can be shifted relative to the equilibrium curve $G_{eq}(t)$ ($\gamma = 0.17$), leading to Fig. 12.1(b). The shifted data in Fig. 12.1(b) is $G(t, \gamma)/h(\gamma)$, first introduced in Ref. [20]. h records how much vertical shift has occurred as a function of γ. As seen in Fig. 12.1(b), the various $G(t, \gamma)$ curves collapse together at long times after the vertical shift. Thus, Fig. 12.1(b) confirms time-strain separability, namely, the stress relaxation for $t > t_k$ approximately follows the same dynamics as observed in the linear response regime. According to particle-tracking velocimetric observations in Section 12.2, there can be significant macroscopic motions after shear cessation so that the onset time for the superimposition (i.e., recovery of equilibrium relaxation dynamics) is often determined by when the quiescent relaxation resumes.

Upon a closer examination of the available stress relaxation data, it is observed that different systems show differing degrees of strain softening. Thus, Osaki[1] categorized the experimental data in the literature into at least three classes according to whether or not the damping function h matches that of the Doi–Edwards tube model,[21] given numerically by

$$h_{DE}(\gamma) \cong 1/(1 + \gamma^2/5). \tag{12.5}$$

Osaki observed[1] that a subset of data falls below the DE curve and called such behavior type C; above the Doi–Edwards curve are labeled as type B by Osaki, as sketched in Fig. 12.2. Data in quantitative agreement with the DE curve are type A. Here it is important to note that any data similar to h_{DE} show a steep decline with γ as $1/\gamma^2$.

Thanks to the existence of time-strain separability, the damping function has been widely recognized as a useful rheological fingerprint to characterize different polymers.[1,22–24] As shown later, this representation of stress relaxation data in terms of the damping function $h(\gamma)$ or the relaxation modulus $G(t, \gamma)$ masks a more revealing way to describe the stress response of entangled polymers to large step strain. Worse, the representation based on h has also made it less convenient to reveal another characteristic of the tube model, concerning its predictions for step strain.

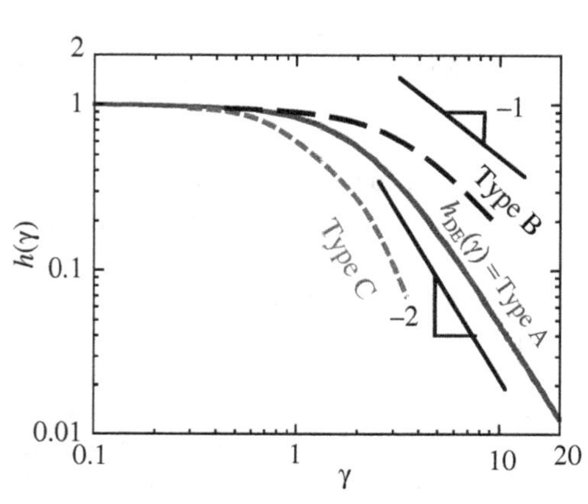

Figure 12.2 Classification of different stress relaxation behaviors according to Osaki,[1] with respect to the predictions of the Doi–Edwards tube model, in terms of characteristics of the damping function h_{DE}.

12.1.2 Tube Model Interpretation

12.1.2.1 Normal Doi–Edwards Behavior

As reviewed by Osaki,[1] stress relaxation experiments had been extensively carried out[18] before the tube model calculation.[19] But it was not straightforward to rationalize the data. Because the tube model could explain the experimentally observed time-strain separability[18,25] and nonlinear stress relaxation behavior, including the severe strain softening indicated by Eq. (12.5), its success was instantly established.[26] To understand the well-known agreement between a subset of the published data and the model calculations, and to pinpoint the misleading implication of this agreement, let us examine the published data in a more meaningful way. Specifically, let us normalize $\sigma(t > t_k, \gamma)$ with $G_{eq}(t)$. By definition, the normalized stress at long times is given by the product of the damping function h and the strain γ

$$\sigma(t > t_k, \gamma)/G_{eq}(t > t_k) = \gamma h(\gamma). \tag{12.6}$$

Instead of Eq. (12.6), most literature data are presented in terms of h, or in terms of the relaxation modulus G along with G_{eq}, which can be combined to reconstruct h using Eq. (12.4).

Equation (12.6) allows us to recover the raw data on the long-time shear stress from the damping function h. Moreover, by inserting the Doi–Edwards damping function Eq. (12.5) into Eq. (12.6), we see that the Doi–Edwards tube model predicts that the relaxing shear stress at $t > t_k \sim \tau_R$ is given by a nonmonotonic orientation function $Q_{xy}(\gamma)$

$$\sigma_{DE}(t > \tau_R, \gamma)/G_{eq}(t) = Q_{xy}(\gamma) = \gamma h_{DE}(\gamma) \cong \gamma/(1 + \gamma^2/5). \tag{12.7}$$

A basic premise of the tube model is that affinely stretched chains would retract around a Rouse time τ_R after the instantaneously imposed step shear, causing the stress relaxation to dip and show a kink at $t_k \sim \tau_R$. The amount of the stress decline is greater for a larger strain magnitude γ. At times larger than τ_R, the residual shear stress is entirely due to the surviving chain orientation. Therefore, the tube model naturally prescribes a separation of time scales: chain retraction at the Rouse time τ_R and recovery of isotropic chain orientation after the reptation time τ_d. Because of the shear alignment, a more oriented chain contributes less to the shear stress for $\gamma > \gamma_{max} = 2.2$, at which $Q_{xy}(\gamma)$ in Eq. (12.7) attains a maximum. Consequently, Fig. 12.3 states that for step strains with $\gamma > 2.2$ the residual shear stress at long times ($t > \tau_R$) is actually smaller at a larger value of γ. Since a significant fraction of all literature data agree quantitatively with the Doi–Edwards curve, such experimental data have received an explicit explanation from the tube model: The phenomenon $\sigma(t, \gamma_2) < \sigma(t, \gamma_1 < \gamma_2)$ is not surprising; it arises from the excessive chain orientation produced by the larger step strain. But it is counterintuitive that a step strain of larger magnitude results in a smaller residual stress. Thus, one

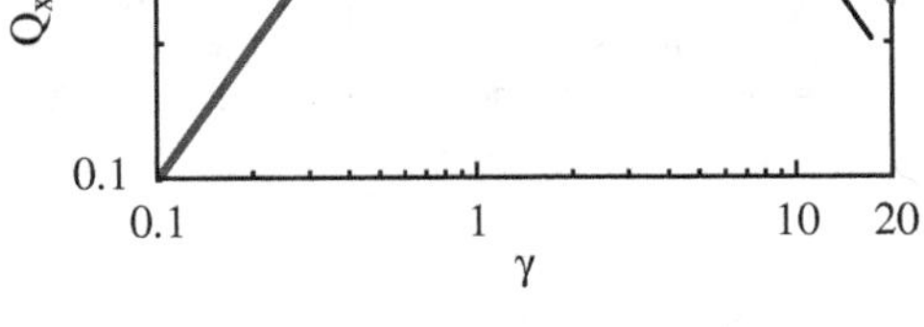

Figure 12.3 Normalized residual stress after $t > t_k \sim \tau_R$ as a function of the magnitude γ of the step strain, predicted by the tube model in terms of the orientation function Q_{xy} of Eq. (12.7).

has to wonder whether the severe shear strain softening depictable by $h_{DE}(\gamma)$ is caused by new physics that is *not* encompassed by the tube model.

12.1.2.2 Type C Ultra-strain-softening

Contradicting the agreement between some experimental step-strain data[25] and the tube model, the type B and C data[1] are threatening the prevailing theoretical understanding. Here we note that the experimental paper containing the data that strongly disagreed with the tube model was actually published 2 years earlier[20] than the data[25] that agreed with the tube model. The agreement was highlighted and the disagreement was unresolved.[21] Let us see how type C disagrees with the tube model. We use the data in Fig. 12.1(a) to calculate $h(\gamma)$ according to its definition in Eq. (12.4) by reading either of the two vertical lines in Fig. 12.1(a) and in Fig. 12.4 plot the derived $h(\gamma)$ against γ. $h(\gamma)$ falls well below the Doi–Edwards curve. Thus, the data in Fig. 12.1(a) actually show type C behavior.

As an alternative to evaluating $h(\gamma)$, we can also examine directly the relaxing shear stress. The representation in terms of relaxation modulus in Fig. 12.1(a) is uninspiring because it does not explicitly reveal any interesting trends. Similarly, Fig. 12.4 conceals the useful information. In place of either Fig. 12.1(a) or 12.4, multiplying the data in Fig. 12.1(a) with the corresponding magnitudes γ, we recover in Fig. 12.5 the actual residual stress data. There are two surprising features. First, even at short times the stress for $\gamma = 15$ (solid dots) is already below that produced at $\gamma = 10$ (squares). After the kink-like fast stress drop (around 10 s), at $t = 100$ s the stress produced with $\gamma = 15$ becomes as small as that produced at $\gamma = 0.17$. Multiplying the data in Fig. 12.4 with γ leads to Fig. 12.6, which can also be obtained by reading the stress values at t_1 and t_2 in Fig. 12.5. Most data points in Fig. 12.6 are well below the Doi–Edwards prediction, and importantly they show nonmonotonic variation with step-strain magnitude γ. In the absence of strain softening, the data should have followed the dashed line, namely, the residual stress should linearly increase with γ. Surprisingly, even for $\gamma = 0.44$, the data is considerably below the ideal trend given by the solid curve, suggesting that the bulk of the sample did not undergo the prescribed magnitude of strain, implicating that complications such as wall slip must have occurred.

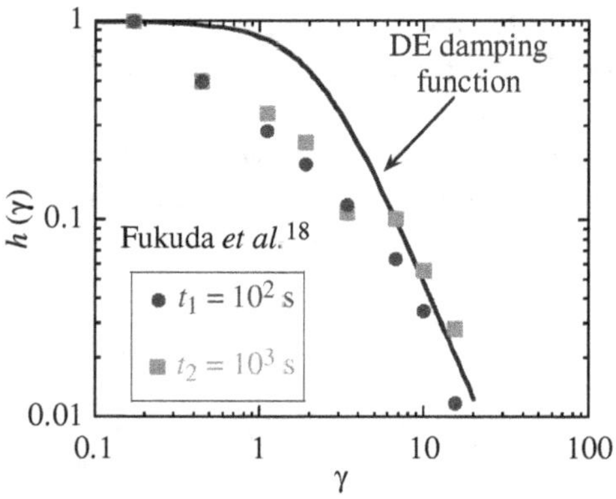

Figure 12.4 Damping function h according to the definition in Eq. (12.4), evaluated from the data in Fig. 12.1(a) at either $t_1 = 10^2$ s or $t_2 = 10^3$ s, where the DE damping function is drawn as a reference.

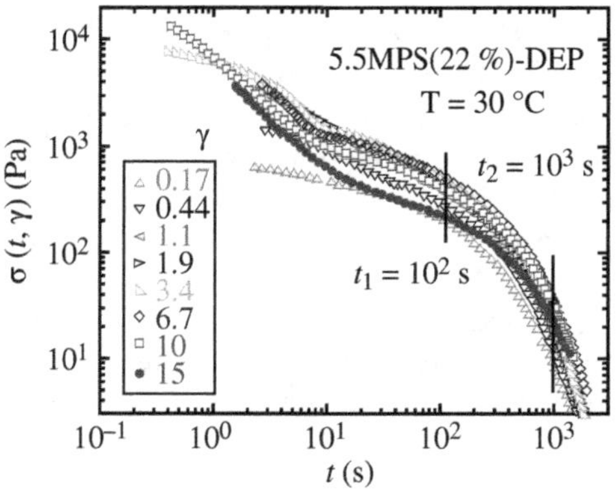

Figure 12.5 Stress relaxation over time for different strain magnitudes, obtained by multiplying the data in Fig. 12.1(a) with γ. At least for this set of data, the time-strain separability is approximately valid except for the data (solid dots) at $\gamma = 15$, which is clearly in some error, showing a time dependence different from the rest of the curves.

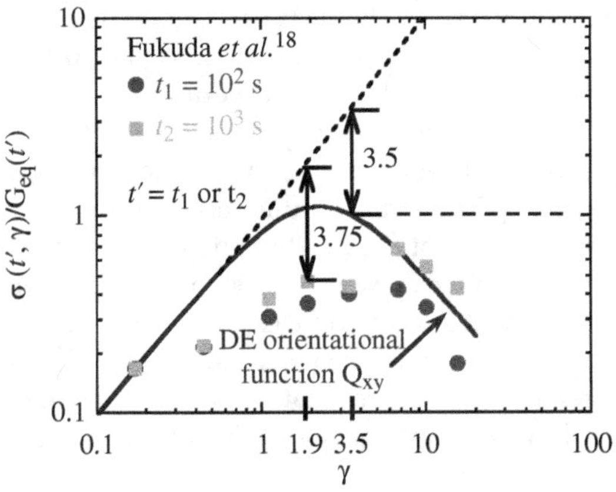

Figure 12.6 Normalized residual stresses, read from Fig. 12.5 at $t_1 = 10^2$ and $t_2 = 10^3$ s, respectively, where the DE orientational function Q_{xy} is also plotted as a reference.

Moreover, at $\gamma = 1.9$ (first vertical line) the actual data (points) differs from the hypothetical non-strain-softening response (dashed line) by a factor of 3.75. The Doi–Edwards curve for the residual stress also shows impressive strain softening. For example, at $\gamma = 3.5$ (second vertical line) the DE curve is smaller than the non-strain-softening prediction by a factor of 3.5.

Ultra-strain-softening did cause the authors of Ref. [18] to look for evidence of wall slip as the origin of the type C behavior. They drew a vertical ink line across the gap on the meniscus before the step strain test. "After the application of strain, the line remained continuous, namely, the line on the sample connected the ends of lines on the cone and on the plate. Moreover, the line on the sample was straight after the experiment." Unfortunately, the authors[18] did not specifically indicate for which strain magnitude such a verification was carried out.

Let us discuss a few more examples of type C behavior, taken from the literature. Figure 12.7(a) shows an extreme case of type C behavior, involving step strain of a 1,4-polybutadiene (PB) melt of $M_w = 350$ kg/mol.[27] For 1,4-PB melts,[28] the melt plateau modulus is $G_N^0 = 10^7$ dyn/cm^2. Thus,

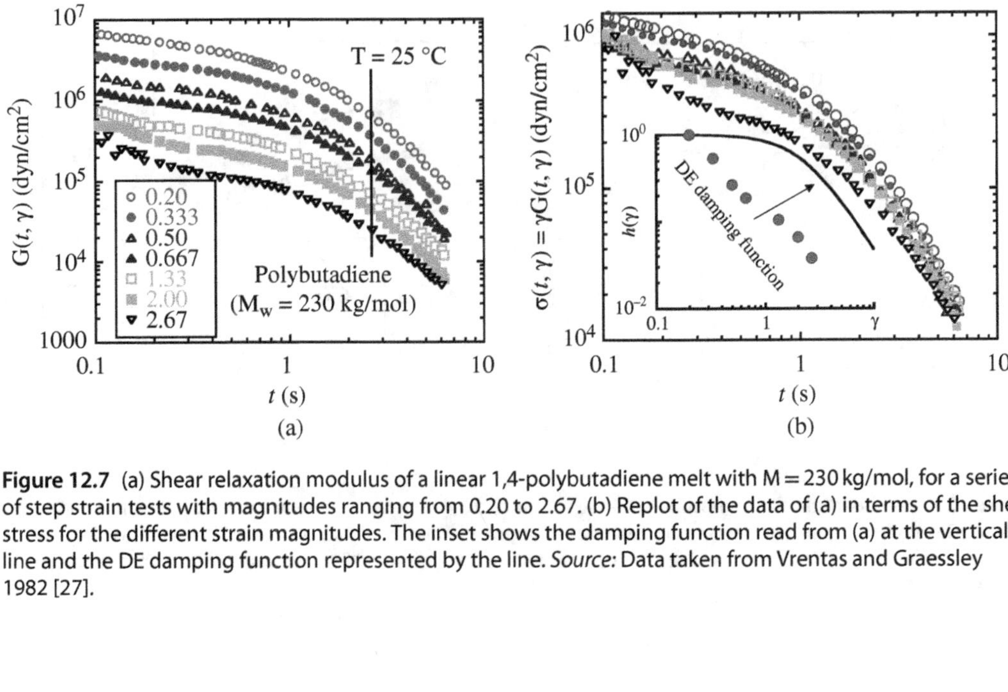

Figure 12.7 (a) Shear relaxation modulus of a linear 1,4-polybutadiene melt with M = 230 kg/mol, for a series of step strain tests with magnitudes ranging from 0.20 to 2.67. (b) Replot of the data of (a) in terms of the shear stress for the different strain magnitudes. The inset shows the damping function read from (a) at the vertical line and the DE damping function represented by the line. *Source:* Data taken from Vrentas and Graessley 1982 [27].

all the curves for the relaxation modulus are below G_N^0. The four lowest curves are below G_N^0 by a factor of 10 or more, namely, $h(\gamma = 1) < 0.1$. As a reference, we note at $h_{DE}(\gamma = 1) = 0.83$. What could cause the relaxation modulus to be smaller than the plateau modulus by one order of magnitude? The nature of this ultra-strain-softening is more readily understood by converting the original data into a plot of the shear stress, as shown in Fig. 12.7(b). The stress is all larger at a smaller γ. The damping function in the inset reveals extreme strain softening. It is most plausible that the bulk of the polybutadiene melt did not undergo shear deformation as described by γ. Instead, massive wall slip took place to prevent the buildup of shear stress in proportion to γ. There must also be some experimental error in this set of data because Fig. 12.7(b) shows that there is a kink between 0.1 and 0.4 s in the down-pointed triangles, while no such feature is present in the curves that involve larger initial stress. Using a method as primitive yet effective as drawing a vertical line on the meniscus and observing its displacement during and after these step strain tests, it should be possible to verify that massive wall slip took place to produce the data in Fig. 12.7(a). If the shearing metallic surfaces are clean and enough contact time is allowed between a comparable PB melt and the walls, strain softening could be weaker than that demonstrated in Fig. 12.7(a,b).

Similar ultra-strain-softening was reported for concentrated PB solutions.[29] By multiplying the original data for $G(t, \gamma)$ in Fig. 12.8(a) by the step-strain magnitude γ, we convert the relaxation modulus into the time-dependent shear stress. Figure 12.8(b) shows somewhat different features from Fig. 12.7(b). First, the initial stress is now in the right order, namely, $\sigma(0, \gamma)$ is larger for larger γ as shown in the inset of Fig. 12.8(b). This recovery of normalcy occurs because premature wall slip is less likely to happen for such a sticky solution. Nevertheless, there is already strain softening, indicated by the severe deviation of $\sigma(0, \gamma)$ from the linear relationship $\sigma = G_{pl}\gamma$. Here $G_{pl}(\phi = 0.21) = 32$ kPa was estimated from $G_{pl}(\phi) = \phi^{2.2}G_{pl}(\phi = 1)$ with $G_{pl}(\phi = 1) = 1$ MPa for 1,4-polybutadiene.[28] Specifically, Fig. 12.8(b) reveals similar problems: By the time of $t = 1$ s, $\sigma(t, \gamma = 3.33)$ drops below the curves of 2.67, 2.00, and 1.33. Such results are similar to the data of polystyrene solution in Fig. 12.5.

Type C relaxation behavior, particularly the data in Ref. [27] (Fig. 12.7(a,b)), prompted the suggestion by Marrucci and Grizzuti[30] that there was an elasticity instability based on the

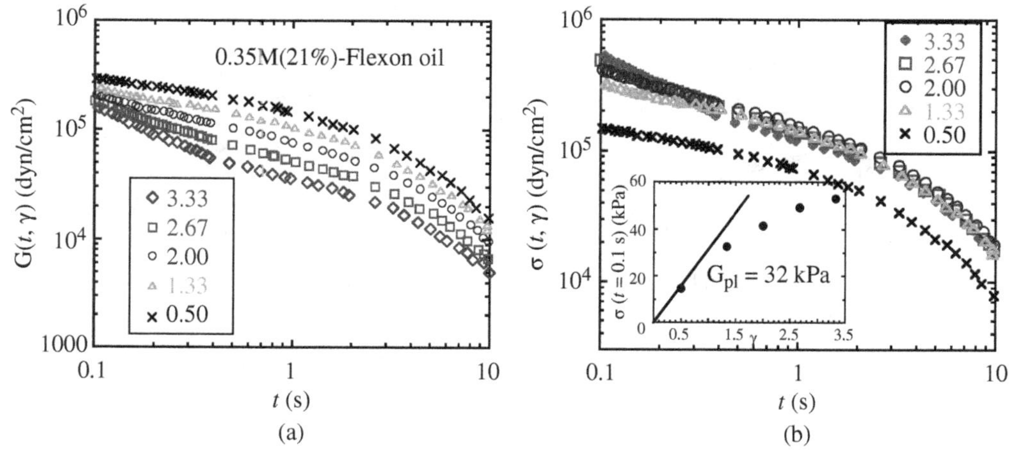

Figure 12.8 (a) Shear relaxation modulus of 21% solution of 1,4-polybutadiene with M = 350 kg/mol and a hydrocarbon oil, for a series of step strain tests with magnitude ranging from 0.50 to 3.33. (b) Replotting the data in (a) in terms of the shear stress for the different strain magnitudes. *Source:* Data taken from Vrentas and Graessley 1981 [29].

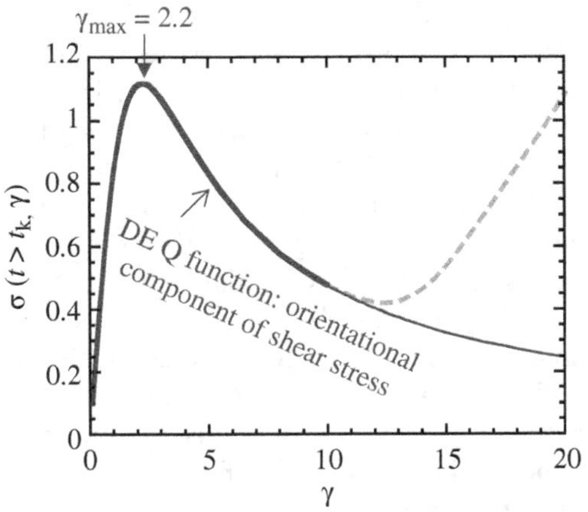

Figure 12.9 Relaxing shear stress at long times as a function of the magnitude of the step strain γ, revised from that in Fig. 12.3 with addition of the dashed curve that hypothetically allows one to argue how strain localization could take place after step strain.

nonmonotonic dependence of $\sigma_{DE}(t > \tau_R, \gamma)$ on γ displayed by the tube model (Fig. 12.3 and Eq. (12.7)). They asserted that if the relaxing system adopts a nonuniform strain field the free energy can be lowered. They did not prescribe a molecular mechanism for the strain localization and did not explain why or how a "deformational phase separation" would produce residual stresses below the Doi–Edwards limit. More importantly, their tube-model-based theoretical study did not consider how a band of larger strain can be stabilized, for example, what entanglement states are present in each of the different shear bands. In other words, for the idea of Marrucci and Grizzuti to work, we have to add new features to Fig. 12.3, so that it will look like Fig. 12.9. However, the dashed curve in Fig. 12.9 is inconsistent with the premise of the tube model: Beyond $\gamma_{max} = 2.2$, the tube model prescribes a simple monotonic decrease of the residual stress is due to the surviving chain orientation. According to the tube model, only the orientational stress survives beyond the Rouse time. There is no known effect that can produce an increasing residual stress $\sigma(t > \tau_R, \gamma)$ with increasing γ (as represented by the dashed curve in Fig. 12.9).

Theoretical difficulty often stems from unclear exploration of causality. The experimentally observed nonmonotonicity sketched by Fig. 12.3 is only an effect and not the cause of any strain localization. Moreover, the remedy provided in Ref. [30] does not explain why the data points in Fig. 12.6 are well below the DE theoretical curve for γ below the peak strain at 2.2, where instability is not expected. In closing, we remark that any experimental data (such as those in Figs. 12.6 and 12.7(b)) below the DE curve may suggest that the sample never experienced as much bulk strain as imposed, implying that wall slip or shear banding could have taken place before shear cessation.

12.2 Particle Tracking Velocimetry Revelation of Localized Elastic Yielding

We should define relaxations types, sketched in Fig. 12.2, in a different manner: Type B relaxations have weaker strain softening: $h(\gamma) \sim \gamma^{-a}$ where $a < 2$. Only type A behavior satisfies the condition $a = 2$. At long times (in the superimposable regime) where the stress relaxation shows a universal time dependence, the (normalized) residual stress is by definition given by $\gamma h(\gamma)$. Data in agreement with the Doi–Edwards damping function for stress relaxation show a long-time stress that varies with γ in a nonmonotonic manner as shown in Eq. (12.7) and Fig. 12.3.

In general, by converting published data such as Fig. 12.1(a) into the raw data for stress (Fig. 12.6), we show that the relationship between the long-time (residual) stress and γ is nonmonotonic. Although the initial stress is larger because of the larger γ, the larger γ is, the smaller the residual stress is at long times. The analysis, made to reveal nonmonotonicity in Fig. 12.3 for data that agree with the Doi–Edwards damping function, prompted the first particle-tracking velocimetric observations of stress relaxation behavior, because the nonmonotonicity suggested that the entanglement network may be breaking apart during stress relaxation.[2] A breakdown of the entanglement network can plausibly be the reason for why the stress declines sharply and exhibits a kink. Perhaps a structural failure to initiate macroscopic recoil can release so much shear stress that the residual stress is smaller at larger γ. It was thus reasonable to speculate that at least type A and C relaxations may involve nonquiescent relaxation. Thus, the "normal" data found to be in numerical agreement with the tube model may actually involve nonquiescent relaxation, given the strong nonmonotonicity shown in Fig. 12.3, in which case their agreement with the tube model calculation would be accidental. Particle-tracking velocimetry (PTV), as described in Section 4.4, can reveal whether quiescence prevails upon shear cessation from large stepwise shear. Nonquiescent relaxation should clearly be a result of localized structural failure.

12.2.1 Nonquiescent Relaxation in Polymer Solutions

There have been a modest number of PTV studies on rheological responses upon step strain. Samples that undergo shear banding during startup shear are obviously suitable candidates for a PTV investigation that aims to look for evidence of localized structural breakdown. If we wish to examine anything but wall slip-like behavior, the capacity to undergo wall slip as measured by b_{max}/H should be minimized. Table 12.1 lists a number of polybutadiene solutions, intentionally made with polymeric solvents to provide a contrast with conventional solutions[29] made with small-molecule solvents. For example, Fig. 12.10(a) involves a 1M(10%)-15K PB solution that is designed to suppress wall slip, while in contrast the 1M10%-oil PB solution is capable of massive wall slip. On one hand, b_{max}/H of 1M(10%)-15K is small enough to prevent shear banding during startup shear for *Wi* from 14 to 140 (see Fig. 9.15(a)). On the other hand, b_{max}/H of 1M(10%)-oil is as large as 5 mm, which encourages wall slip at moderate values of *Wi* for H around 1 mm.

12.2.1.1 Elastic Yielding in Polybutadiene Solutions

Our discussion of elastic yielding and nonquiescent relaxation begins with a sample without significant wall slip so that we can focus on the bulk response. Typically, data on stress relaxation

Table 12.1 Characterization of various entangled PB solutions.

Samples	Source	References	$\eta_0 \times 10^3$ (Pa s)	η_s (Pa s)	Z	τ (s)	b_{max} (mm)
1M(10%)-oil	UAkron/Imperial Oil	3	60.5	0.2	40	18	5.1
1M(10%)-15K	UAkron/Bridgestone	3	250	36	40	67–71	0.12
1M(10%)-1.5K	UAkron/Aldrich	31	50	0.7	64	17	1.2
1M(10%)-10K	UAkron/Bridgestone	31	170	14	64	50	0.2
1M(10%)-46K	UAkron/Goodyear	31	390	260	64	75	0.002

2 Video 12.1: Particle-tracking velocimetric observation of nonquiescent relaxation of 1M(10%) solution described in Ref. [2], made at University of Akron.

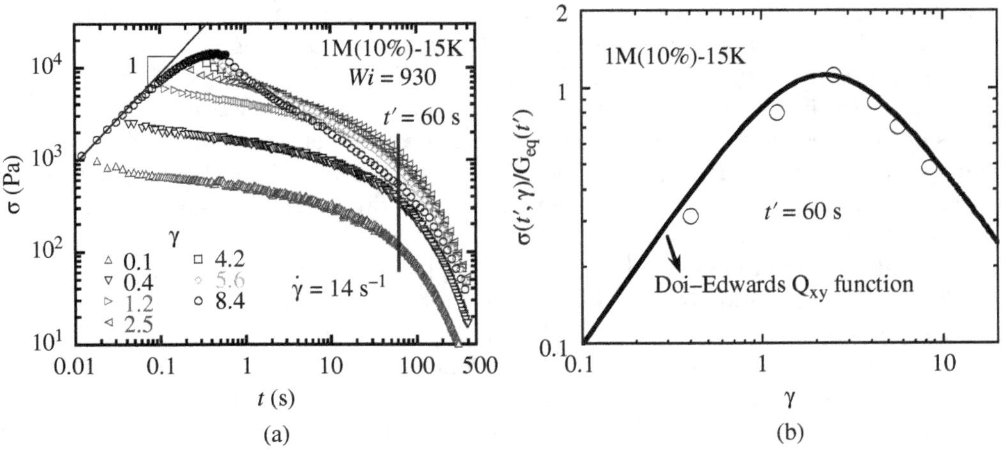

Figure 12.10 (a) Stress growth during startup shear and stress relaxation after termination of a startup shear produced at the rate $14\,s^{-1}$, equivalent to $Wi = 930$, at various levels of magnitude from $\gamma = 0.1$ to 8.4. The last of these magnitudes is already beyond the stress maximum. (b) Normalized residual shear stress at 60 s, based on data at $t' = 60$ s in (a). Also plotted is the Doi–Edwards orientation function Q_{xy}. *Source:* Data taken from Ravindranath and Wang 2007 [3].

are presented in the literature without disclosing any information before shear cessation. Thus, Fig. 12.10(a) is somewhat unconventional. It describes not only how the stress decays in time but also the stress buildup while the stepwise shear is being applied. The textbook definition of step strain specifies the stepwise shear to occur "instantly," namely, it is usually assumed that it takes a vanishingly short time to produce the prescribed value of γ. In reality, we only require duration of step strain to be much shorter than the dominant chain relaxation time such as the Rouse time τ_R. When the step strain is not "instant," we use a more general term of stepwise shear.

Although the stepwise shear in Fig. 12.10(a) involves a rather high rate of $14\,s^{-1}$, corresponding to a Rouse Weissenberg number[3] $Wi_R = \dot{\gamma}\ \tau_R \sim 8.4$, a stress maximum is inevitable and appears at $\gamma_y = 6.3$ ($t_y = 0.45$ s). Since t_y is comparable to τ_R, the step strain is not instant at all. As shown by the circles, there is already softening during the stress buildup, namely, σ deviates downward from a linear increase with increasing γ. This "strain softening" is compounded with the strain softening that occurs after shear cessation. Taking $G_{eq}(t) = G(t, \gamma = 0.1) = \sigma(t, \gamma = 0.1)/0.1$ we directly present in Fig. 12.10(b) the normalized relaxing shear stress as a function of the step strain magnitude γ, reading the stress values at the vertical line in Fig. 12.10(a). Intriguingly, the data in Fig. 12.10(a) agree with the Doi–Edwards orientation function $Q_{xy}(\gamma)$ in Eq. (12.7). As shown in Fig. 12.10(a), the stress does not increase linearly with strain. This downward deviation from a linear stress versus strain relationship also contributes to the decline of the normalized stress with γ. In principle, the observed nonmonotonicity could be due to the use of an insufficiently high shear rate that resulted in "strain softening" before shear cessation. In other words, partial yielding defined in Figs. 7.1(b) and 7.6 may have caused the accelerated stress decline, driving the residual stress in Fig. 12.10(b) at $t' = 60$ s to show a nonmonotonic dependence on the strain magnitude. We should note that some previous literature data might also contain this compounding effect of "strain softening" before shear cessation. Thus, it is desirable for any future study to take the prerelaxation "strain softening"

3 Several methods have been used in the literature to evaluate the Rouse time τ_R, see Section 7.A in Chapter 7. For entangled PB solutions, the simple formula of Eq. (2.43) appears to yield comparable results for τ_R to those from the viscosity-based method presented in Appendix 7.A. Given $Z = 40$ and $\tau_d = 71$ s, we have $\tau_R = 0.6$ s instead of the value given in Ref. [3].

into account. The instrument performance needs to match the materials characteristics; in particular, the duration of the stepwise shear must be sufficiently shorter than τ_R.

Figure 12.10(a) offers a continuous view of the stress behavior, including both stress buildup and relaxation. The initiation of the stepwise shear was at $t = 0$; the time was not reset to zero at the onset of shear cessation. Fixing the origin of the time axis at the beginning of shear does not introduce any significant difference from the conventional analysis. Figure 12.10(b) was generated using measurements at $t' = 60$ s, which is such a long time in comparison to the duration (less than 1 s) of the stepwise shear.

Some literature data are likely to involve either wall slip or shear inhomogeneity. In other words, before shear cessation in Figs. 12.5, 12.7(b), and 12.8(b), it is plausible that there was wall slip, especially at the highest strains. Wall slip makes it possible for $\sigma(t, \gamma_2)$ to be smaller than $\sigma(t, \gamma_1)$ at the onset of stress relaxation, even though $\gamma_2 > \gamma_1$. The inset of Fig. 12.8(b) shows that the initial stress grows more weakly than linearly with γ for $\gamma > 1.5$. These results are signs that either the bulk of the sample did not undergo the nominal level of shear or the rate used to produce the step strain was low, which is unlikely to be the case. It is therefore necessary to analyze the rheometric information and if possible to perform *in situ* particle-tracking velocimetric determination of the strain field during the step shear. The particle-tracking velocimetry confirms for the measurements in Fig. 12.10(a) that step shearing at the rate of $14\,s^{-1}$ is homogeneous and without wall slip and strain localization.

It has been known for decades that stress relaxation is faster at larger step strain. Figure 12.11(a) for the relaxation modulus confirms this knowledge. A fast relaxation can result in nonmonotonicity, as demonstrated by the circles in Fig. 12.10(b). Here *in situ* PTV observations disclose one explanation for the nonmonotonic variation of the residual stress with magnitude γ. There is detectable macroscopic motion already for $\gamma = 1.2$, as shown in Fig. 12.11(b), and very significant motions at $\gamma = 2.5$, after the peak of the stress at $t' = 60$ s versus step strain γ plot in Fig. 12.10(b). By the time $t' = 60$ s, the macroscopic motions have stopped. Quiescent relaxation ensues with increasing time when the stress declines according to $G_{eq}(t) \sim \sigma(t, \gamma = 0.1)$. Based on the PTV phenomenology, the reason for time–strain separability is rather clear: After *healing* of the failing sample, the residual shear stress relaxes in the

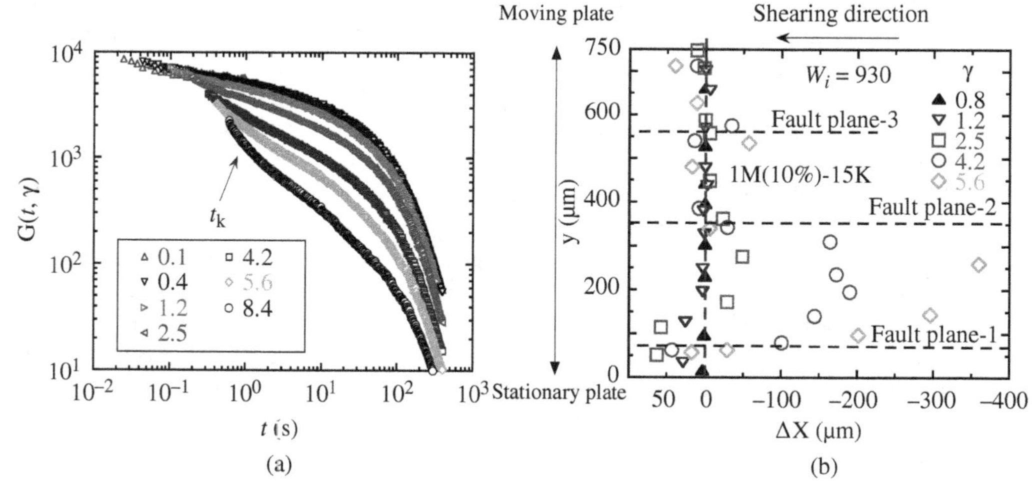

Figure 12.11 (a) Relaxation modulus $G(t, \gamma)$ evaluated from Fig. 12.10(a) according to Eq. (12.3). (b) Particle-tracking velocimetric measurements of macroscopic motions after various discrete step strains, showing the distances ΔX that various tracking particles traveled from their initial positions at shear cessation. *Source:* Data taken from Ravindranath and Wang 2007 [3].

same way as observed in the linear-response regime, through molecular diffusion by means of reptative dynamics, a well-accepted mechanism for stress relaxation in small-magnitude step strain. Therefore, the relaxation dynamics come into agreement with $G_{eq}(t)$ as soon as the macroscopic motions cease.

The macroscopic motions are essentially the internal recoil of the strained sample. During the recoil, the shear stress decreases with time faster than $G_{eq}(t)$ does. The kink-like "relaxation" for the relaxation modulus $G(t, \gamma)$, as shown clearly in Fig. 12.11(a), originates from the macroscopic structural failure that takes place either in the bulk or at the sample/wall interfaces. We note from Fig. 12.11(b) that the magnitude of macroscopic motions after shear cessation grows with γ. These features should be relatively universal for well-entangled polymers.

It is necessary to emphasize that the discovery of nonquiescent relaxation is highly surprising and uncovers a deeper previously unknown level of physics. We term the new effect elastic yielding.[2,6] Unlike yielding during ongoing deformation, here the *yielding* (leading to a structural breakdown) takes place after termination of the external deformation. Elastic yielding is driven by buildup of a significant *elastic* stress. Before this finding, our perception of stress relaxation from step strain of any magnitude is *quiescence*; namely, there should be no macroscopic motion after step strain. Nonquiescent relaxation occurs for γ as low as 1.2, well before the maximum around 2.2 in Fig. 12.10(b). Thus, the nonquiescence has little to do with the nonmonotonic dependence of the residual stress on the magnitude of the step strain. Nonmonotonicity is *not* the cause of strain localization. On the other hand, a structural breakdown in the entanglement network and ensuing macroscopic elastic recoil may be necessary for the nonmonotonicity in Fig. 12.10(b). More discussion of the molecular mechanism for nonquiescent relaxation after large step strain will be presented in Chapter 16.

12.2.1.2 Suppression of Breakup by Reduction in Extrapolation Length b

In parallel to the studies of the role of solvent viscosity in controlling wall slip and strain localization during startup shear, a particle-tracking velocimetric study[31] has also examined the effect of the solvent on the magnitude of nonquiescent relaxation. As shown in Fig. 12.11(b), 1M(10%)-15K solution undergoes considerable movements in the sample interior after shear cessation. At the same concentration of the same polybutadiene, the 1M(10%)-oil solution shows strikingly different behavior. In Section 12.4, we will show that replacing the PB15K solvent with a hydrocarbon oil creates a system capable of significant wall slip. Figures 12.11(b) and 12.26(c) in Section 12.4 make a sharp contrast: As shown in Fig. 12.26(c), the movements are due to the shear-induced wall slip in the 1M(10%)-oil solution while Fig. 12.11(b) depicts internal motions. As seen in Figs. 12.12(a,b), a high-molecular-weight solvent nearly suppresses macroscopic motions after a huge step strain. Although all three solutions exhibit the same level of entanglement and corresponding elasticity, the characteristics of localized elastic yielding are rather different. Evidently, the parameter b_{max} regulates to what degree a step-strained entanglement network will undergo macroscopic displacement after shear. Although the same level of the driving force (the shear stress before shear cessation) is involved, rather different levels of motion are shown in Fig. 12.12(a). Since 1M(10%)-1.5K has undergone a greater amount of movement, its stress relaxation should be faster. Figure 12.12(b) confirms this expectation: the stress of 1M(10%)-1.5K declines faster than that of 1M(10%)-10K over the entire course of the nonquiescent relaxation.

12.2.1.3 Nonquiescent Relaxation in Polystyrene Solutions

A great deal of stress relaxation data is based on studies of entangled polystyrene (PS) solutions. Given that preceding PTV data have drawn a connection between strain softening (faster stress relaxation) and nonquiescent relaxation, it is reasonable to suspect that some past step

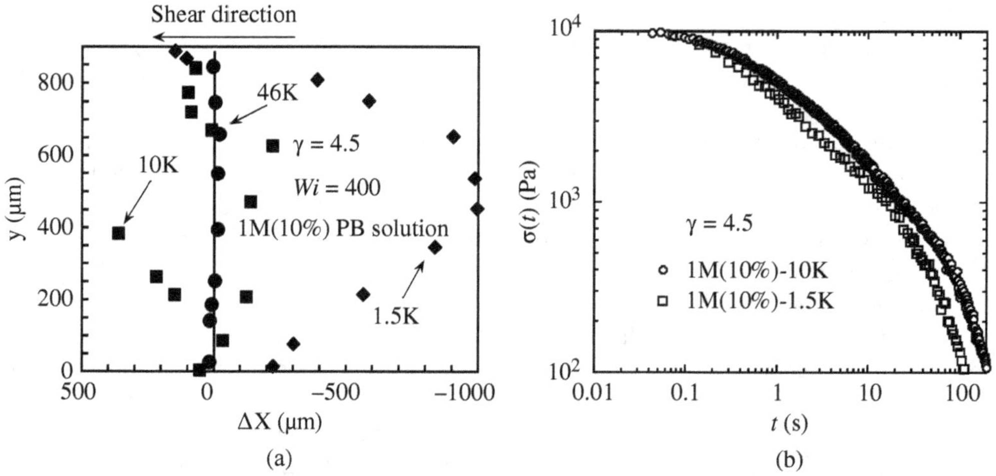

Figure 12.12 (a) Total particle displacement during nonquiescent relaxation from a step strain at $\gamma = 4.5$ produced with $Wi = 400$, at the same 10% concentration based on three different polybutadiene solvents of molecular weight equal to 1.5, 10, and 46 kg/mol, respectively. (b) The corresponding stress relaxation for both 1M(10%)-1.5K and 1M(10%)-10K, after step strain at $\gamma = 4.5$. *Source:* Data taken from Ravindranath *et al.* 2007 [31].

strain experiments on PS solutions show similar behavior. Fukuda *et al.*[18] actually examined the possibility of wall slip and nonquiescence by visual inspection of the free surface on the meniscus and found no evidence. Although they did not specify the magnitude of the step strain for their visual inspection, it can be assumed that they found no sign of movement on the meniscus even at high strains. In the literature, there is only one brief PTV study that reported evidence for nonquiescence in a well-entangled PS solution.[32] Contrary to Fukuda *et al.*,[18] the PTV study found macroscopic motions on the meniscus.[32] Table 12.2 shows the relevant characteristics of this PS solution. Subjecting each sample to several discrete step strains, stress relaxation behavior is measured by conventional rheometry as shown in Fig. 12.13(a). *In situ* PTV observations explain why the relaxing shear stress at $\gamma = 4$ (circles) becomes smaller than that produced at $\gamma = 3$ (squares). In other words, Fig. 12.13(b) confirms that the significant macroscopic motions are of a larger magnitude at a larger value of γ. On bare smooth metallic surfaces, for $\gamma = 2.0$ the observed macroscopic motions appear to originate from the interfaces. However, the maximum recoils evidently occurred in the bulk at large magnitudes of $\gamma = 3.0$ and 4.0.

12.2.1.4 Strain Localization in the Absence of Edge Instability

Step strain is the simplest mode to demonstrate nonlinear rheological responses of entangled polymers. The simplicity comes from the fact that step strain does not involve a large, accumulating amount of strain. A shear strain of a few units, namely, two to three units, is sufficient to illustrate how localized elastic yielding leads to nonquiescent relaxation, as shown in the preceding Sections 12.2.1.1 and 12.2.1.2. Since nonquiescent relaxation is often accompanied by macroscopic movements at the meniscus, as seen in Fig. 7 of Ref. [32], it is necessary to show that the nonquiescence is not due to instability on the free surface of the meniscus. Based on the

Table 12.2 Characterization of one entangled PS solution.

Samples	Source	Z	$\eta_0 \times 10^3$ (Pa s)	η_s (Pa s)	τ (s)	τ_R (s)	b_{max} (mm)
PS20M(7%-TCP)	Polym Stand/Aldrich	50	123	0.23	490	3.7–4.9	3.8

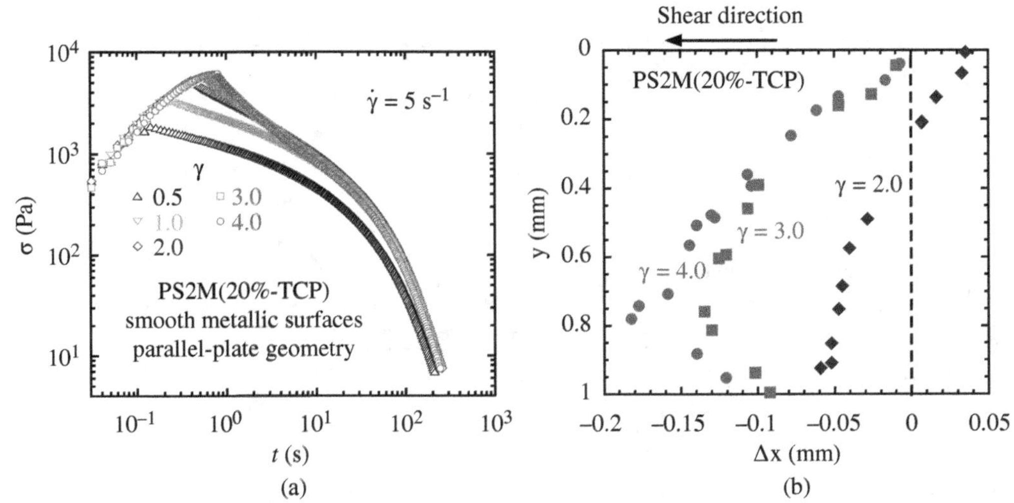

Figure 12.13 (a) Shear stress growth and relaxation of PS2M(20%-TCP) during and after discrete step strain of different magnitudes at shear rate of 5 s^{-1}. (b) Corresponding PTV measurements of the total particle displacements during nonquiescent relaxation for strains 2.0, 3.0, and 4.0. *Source:* Data taken from Liu and Wang 2012 [32].

same principle that guides the design of the cone-partitioned plate rheometer, one can devise an instrument as shown in Fig. 12.14, which is an alternative to a typical cone-partitioned plate device. This fixture is designed to have greatly reduced shear strain at the sample edge. Such a design can ensure that at the rim the shearing condition is in the linear response regime when the sample within the cone-plate of 25 mm diameter has reached the critical condition for non-linear responses, especially nonquiescent relaxation. According to the specifications given in Fig. 12.14, when the step strain has magnitude $\gamma = 2.8$, the step strain at the meniscus is only $\gamma_{rim} \sim 1.0$.

The device in Fig. 12.14 can be used to show that there is little edge effect in step strain tests. For example, Fig. 12.15(a) shows that addition of the protection of the meniscus, achieved by adopting the edge-enlarged cone-partitioned plate of Fig. 12.14, does not alter the rheometric measurement under a large step strain, by comparison with that measured using a conventional cone-plate shear cell.[33] More explicit evidence comes from the simultaneous PTV observations of the sample interior with CCD1 and the meniscus of the enlarged edge with CCD2 as depicted

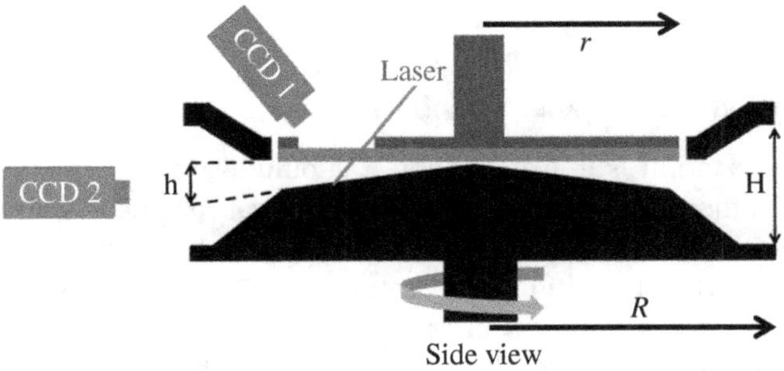

Figure 12.14 Edge-enlarged cone-partitioned plate (CPP) design for rotational rheometers. With cone angle of 4° and R = 20 mm, and inner disk of r = 12.5 mm, h = 1.4 mm, H is chosen to be 6.3 mm so that H/h = 4.5. When the sample in the small gap undergoes a step strain of γ, the shear strain at the meniscus with height H is given by $\gamma_{rim} = \gamma(R/r)(h/H) \sim \gamma/2.8$. *Source:* Data taken from Li and Wang 2010 [33].

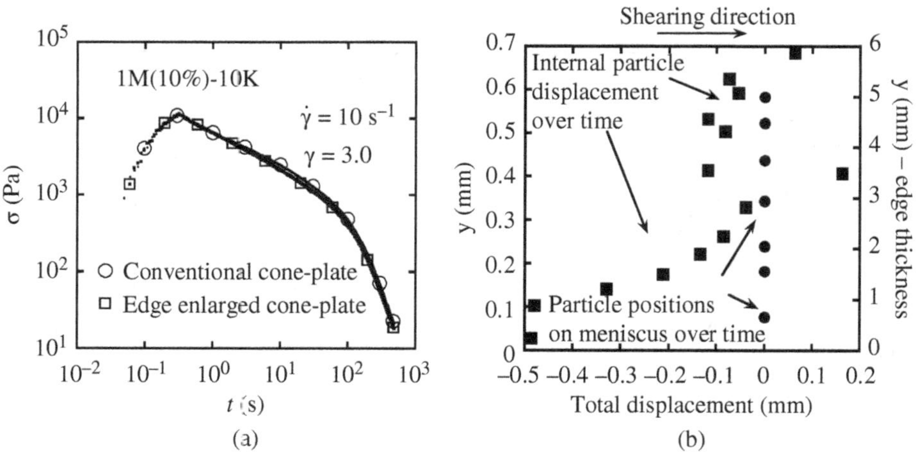

Figure 12.15 (a) Comparison of stress growth and relaxation behavior between data collected with a conventional cone-plate and data obtained with the edge-enlarged CPP. The identical data show (1) the edge enlarged CPP works properly and (2) the meniscus in conventional cone-plate has no effect on the step strain with magnitude as high as $\gamma = 3.0$. (b) PTV observations of particle movements both in the interior and on the meniscus of the edge-enlarged CPP, for a step strain $\gamma = 3.0$ produced with shear rate of 10 s^{-1}. Here the total displacement on the meniscus after shear cessation is found to be undetectable by CCD camera 2 as shown by the filled circles (the right vertical axis). A great deal of motion is revealed by simultaneous PTV observations of the sample interior namely, about 10 mm from the center, where the initial positions right after shear cessation are at 0, and the gap distance is about 0.7 mm, by CCD camera 1 (the left vertical axis) that records particle movements (in squares) about 10 mm from the center where the gap distance is about 0.7 mm. *Source:* Data taken from Li and Wang 2010 [33].

in Fig. 12.14. Figure 12.15(b) shows the detection of significant movements in the bulk after a step shear of magnitude $\gamma = 3.0$. In contrast, the meniscus remains completely stationary. Such data should be taken as indisputable evidence that the observed nonquiescent relaxation is inherent, associated with localized elastic yielding, and is not caused by edge failure.

Solution 1M(10%)-10K is expected to show stress relaxation behavior similar to that in Fig. 12.10(a). Indeed, Fig. 12.16 also shows accelerated stress relaxation phenomenon, in which the relaxing shear stress satisfies $\sigma(t', \gamma_2) < \sigma(t', \gamma_1)$ even though $\gamma_2 > \gamma_1$. Such behavior is usually accompanied by nonquiescent relaxation, as seen in Figs. 12.11(b), 12.12(a), 12.13(b), and 12.15(b). Structural failure causes elastic recoil at the plane(s) of localized yielding and disentanglement, leading to a quick loss of the elastic stress built during the step strain. The macroscopic motions ensue until reentanglement leads to structural repair and healing of the entanglement network.

12.2.2 Nonquiescent Relaxation in Styrene–Butadiene Rubbers

If the physics behind nonquiescent relaxation is universal, then it should also take place in well-entangled melts, provided that sufficiently large shear deformation can be successfully imposed. Wall slip needs to be prevented, especially for melts with $b_{max}/H \gg 1$. For styrene–butadiene rubbers with sufficiently high glass transition temperatures, it is possible to fasten the SBR to metallic shearing plates using conveniently available superglue. It is fortunate that this chemical method works, because physical means such as roughening the shear surfaces is not expected to achieve the same effect. Once good adhesion is ensured, it is straightforward to use a sliding plate apparatus and couple the shear rheometry to a particle-tracking velocimetric setup as illustrated in Fig. 4.4(b). Commercial rotational rheometry is generally not suitable

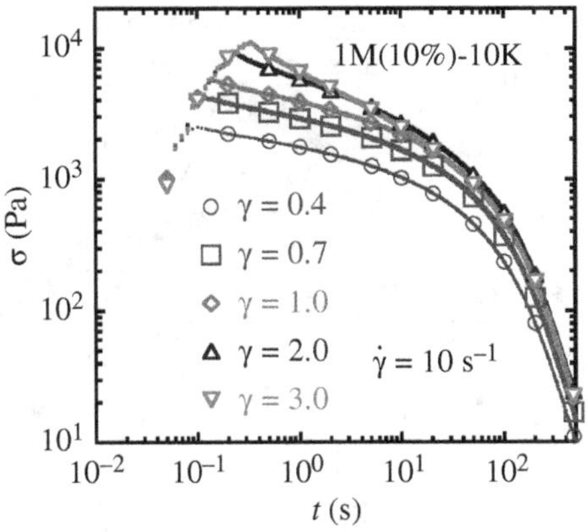

Figure 12.16 Stress growth and relaxation after step strains of various magnitudes produced at a shear rate of 10 s^{-1} in a conventional cone-plate setup. The down-pointed triangles are the same as the data in Fig. 12.15(a). At long times the data points marked by the down-pointing triangles ($\gamma = 3.0$) are smaller than the up-pointing triangles ($\gamma = 2.0$). *Source:* Data taken from Li and Wang 2010 [33].

to study the nonlinear shear rheology of entangled melts: Because of the high shear stress and normal force, the shear cell will need to be exceedingly small. Such a small device exaggerates the effects of edge fracture. Moreover, the small dimensions do not permit sufficient space to set up *in situ* particle-tracking velocimetric determination of the deformation field. It is noted that the superglue cannot prevent the bulk from undergoing localized yielding through chain disentanglement. When such disentanglement occurs, its effect on the strain field in the bulk can also be quantified in terms of b_{max}/H.

12.2.2.1 Induction Time and Molecular Weight Dependence

Particle-tracking velocimetric observations have been carried out to record the aftermath of large step strain in styrene–butadiene rubbers of different molecular weights.[34] Tables 7.1 and 7.5 list the basic characteristics of these styrene–butadiene rubbers. Relative to the nonquiescent stress relaxation behavior of polybutadiene solutions, styrene–butadiene rubbers exhibit more useful characteristics. First, as noted in Fig. 12.17(a), there is no observable motion immediately after the shear cessation while the induction time for nonquiescence in polybutadiene is too short to be captured by the slow-speed (30 fps) video recording. Many seconds later, macroscopic motions set in, building to a climax before settling to quiescence. Specifically, for a strain magnitude of $\gamma = 2.1$ macroscopic motions emerge after an induction period of 15 s, as revealed by the PTV data in Fig. 12.17(b,c). On the other hand, for low shear magnitudes, up to $\gamma = 0.7$ there are no measurable movements. Second, Fig. 12.17(b) shows the severe recoil-like motion: The upper half snaps to the right while the bottom half recoils to the left, causing the kink-like stress drop (diamonds) in Fig. 12.17(a). The drop is so significant that the shear stress (diamonds) approaches the quiescently relaxing stress produced at $\gamma = 0.7$ (circles). The time scale associated with the kink is $t_k = 20$ s, which is considerably longer than the Rouse time. It is worth noting that according to the PTV observations the step strain was produced homogeneously.[34]

The styrene–butadiene rubbers also offer an opportunity to examine the influence of molecular weight, revealing how the degree of chain entanglement Z (i.e., the number of entanglements per chain) affects the induction time. Figure 12.18(a) illustrates the rapid decline in the stress during the period of nonquiescent motion. The inset in Fig. 12.18(a) shows that the onset and duration of the breakup motions scale with the Rouse relaxation time τ_R. Similarly, the elastic recoil after breakup occurs at a larger speed for styrene–butadiene rubber of smaller molecular weight.[34] As shown by a detailed analysis of the video recording from the PTV observations, the induction time needed to build up to the maximum recoiling speed scales with τ_R: As shown in Fig. 12.18(b), the kinetic curves collapse onto one another when the time is rescaled with τ_R. See a movie showing the nonquiescent relaxation during

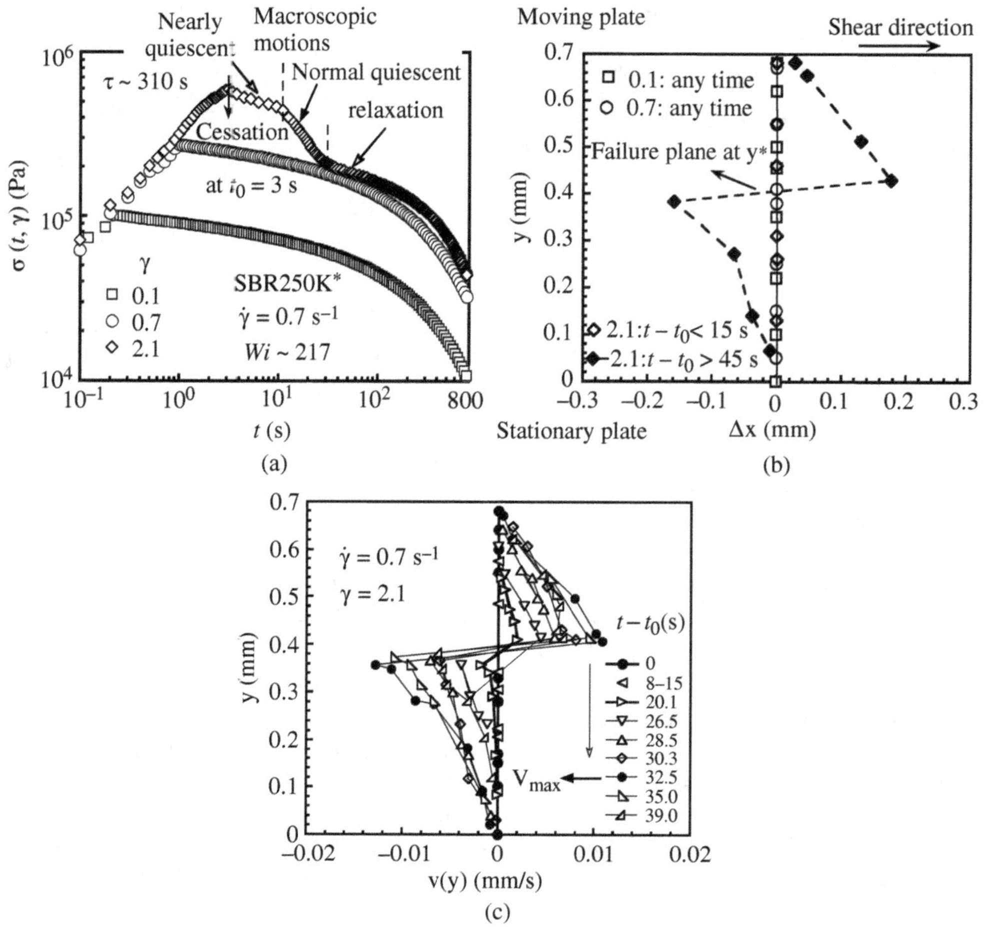

Figure 12.17 (a) Shear stress growth during step strain produced with a rate of 0.7 s^{-1} and relaxation after shear cessation for SBR250K melt, for $\gamma = 0.1$, 0.7, and 2.1. (b) PTV data of the relaxation behavior according to the tests of (a): No macroscopic motion is seen for $\gamma = 0.1$ and 0.7 at any time, and none is seen for $\gamma = 2.1$ during the first 15 s of relaxation. After an induction time of about 15 s, considerable displacements of the tracked particles were observed to occur, as shown by the filled diamonds. After 45 s, namely, for $t - t_0 > 45$ s, the macroscopic motions cease. (c) Depiction of the induction in quiescence and characteristic movements during nonquiescence in terms of the velocity of the tracking particles (horizontal axis) at different positions across the shear cell (vertical axis) at various times. The maximum velocity was observed about 33 s after shear cessation. *Source:* Data taken from Boukany *et al.* 2009 [34].

stress relaxation.[4] Nonquiescent relaxation of these SBR melts has allowed us to observe the consequences of elastic yielding. A finite induction time is required before breakup is observed. So what happens during the initial quiescence? Is there any topological hindrance against chain disentanglement? What is the barrier against disentanglement since step strain at $\gamma = 0.7$ does not produce any macroscopic motions? Does elastic yielding always produce macroscopic motion? These questions motivate further discussion.

The remarkable elastic yielding, revealed by the PTV observations of rupturing-like failure in the bulk after shear cessation of styrene–butadiene rubbers, is obviously impossible to observe

4 Video 12.2: Nonquiescent relaxation after stepwise of 210% of SBR250K*, after 15 s of quiescent induction period, made at University of Akron.

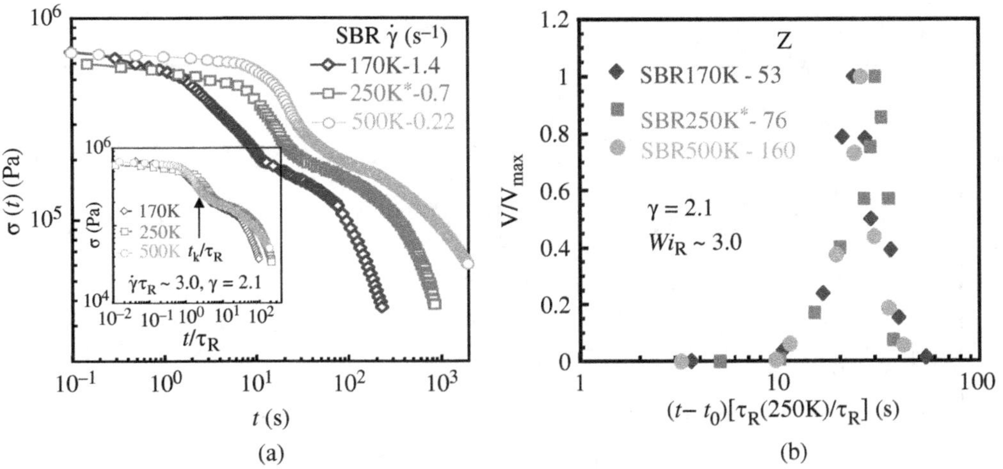

Figure 12.18 (a) Shear stress relaxation of three SBR melts for step strain $\gamma = 2.1$, produced with a Rouse Weissenberg number $Wi_R = \dot{\gamma}\tau_R$ equal to 3.0. The inset indicates the approximate collapse of the time-dependent stress upon normalizing the time t with τ_R. The time at the kink scales with τ_R, as indicated by the arrow. (b) Normalized velocity V/V_{max} at the rupture planes for all three SBR melts as a function of the normalized time $(t - t_0)[\tau_R(250K)/\tau_R]$, revealing the significance of the Rouse relaxation time τ_R to characterize the onset of elastic yielding. *Source:* Data taken from Boukany *et al.* 2009 [34].

unless wall slip is eliminated. In other words, the data in Fig. 12.17(c) contrast sharply with Fig. 6.16(b). It was necessary to apply superglue to fix the melts onto the shearing plates. But are such experimental results practically relevant? Do the PTV observations offer useful insights in polymer processing? Where else do we see "melt fracture" if we regard the data in Fig. 12.17(b,c) as indicating melt fracture? These obvious questions immediately urged us to search in other geometries for strain localization due to breakdown of the entanglement network, including melt entry to an extrusion channel, squeezing, and planar extension, as described earlier in Sections 10.2–10.4.

12.2.2.2 Severe Shear Banding before Shear Cessation and Immediate Breakup

No commercial rheometric apparatus is designed as a suitable device to study startup shear of a well-entangled polymer melt until steady state. On the other hand, using parallel-sliding plate rheometry, coupled with particle-tracking velocimetry, it is possible to shear a melt for up to 10 strain units. At moderate rates, the yield strain γ_y is significantly smaller than 10. Figure 12.19(a) shows two startup shears, imposed at rates of 0.7 and 1.4 s^{-1}, respectively, followed by cessation at $\gamma = 5.0$ and 7.14, respectively. As shown in Fig. 12.19(b) shear banding emerges at $t = 6.7$ just before shear cessation. There is no reason to suggest that this massive shear strain localization would not survive to steady state. On the other hand, it is still practically infeasible to attain steady shear of such well-entangled polymer melts. Thus, we do not know for sure whether shear banding would persist during steady shear of well-entangled polymer melts. Because shear banding has already formed before shear cessation, the melt takes the advantage of the emergent banding to immediately undergo much more severe breakup. For example, the recoiling speed – in Fig. 12.19(c) in excess of 0.1 mm/s is far larger than the recoiling speed seen in Fig. 12.17(c).

12.2.2.3 Rate Dependence of Elastic Breakup

A linear increase of stress with increasing strain is prescribed by the rubber elasticity theory of Eq. (1.50) for a cross-linked Gaussian chain network. The shear banding observed during

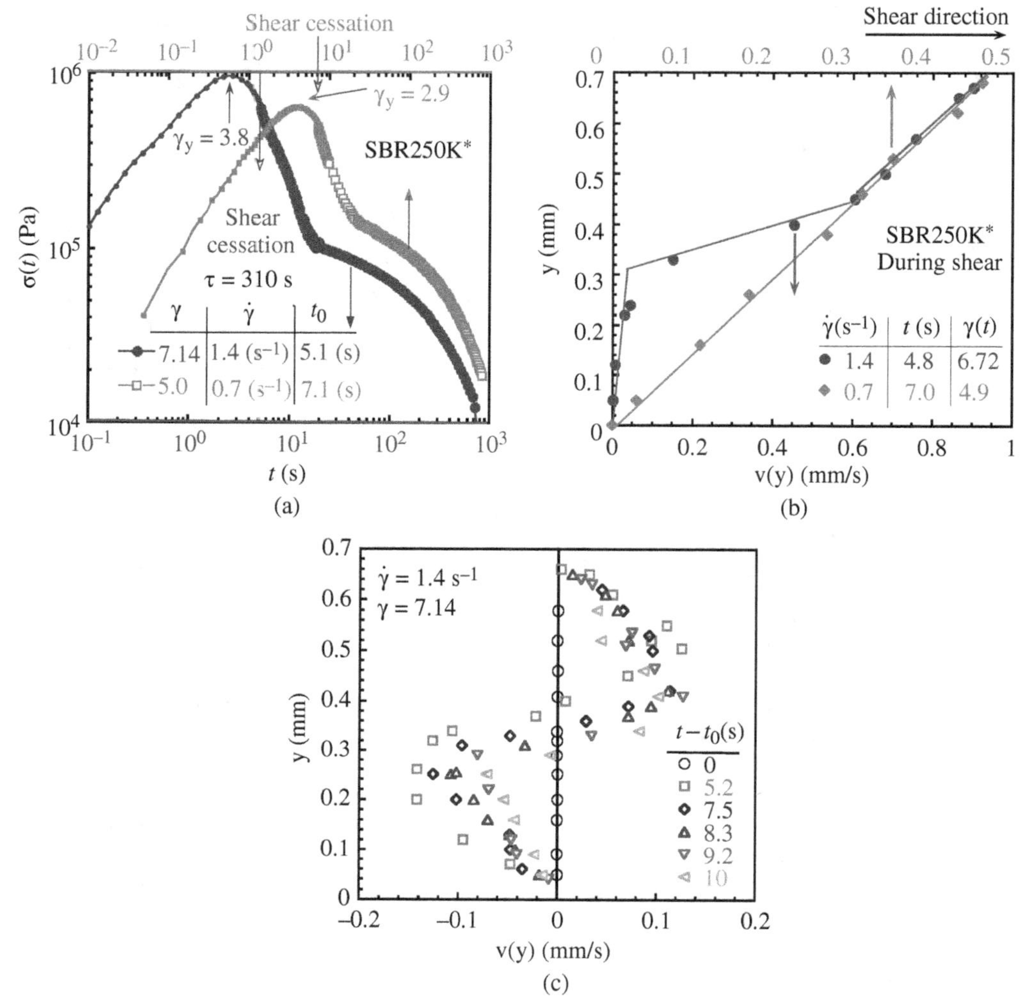

Figure 12.19 (a) Stress responses to startup shear at two rates (thin lines) and stress relaxation after the stress overshoots (circles and squares). (b) PTV observations during startup shear depicted in (a). Homogeneous shear prevails up to $\gamma = 4.9$ at the rate of $0.7\ s^{-1}$, passing the peak stress at $\gamma_y = 2.9$ as well as shear banding at $\gamma = 6.7$. (c) Elastic breakup during stress relaxation from step strain at $\gamma = 7.14$. *Source:* Data taken from Boukany *et al.* 2009 [34].

ongoing shear in Fig. 12.19(b) is plausibly due to a structural breakdown of the entanglement network. Chain disentanglement could have started when the shear stress deviated from its initial linear growth with the strain. In Chapter 7, we called this nonlinear "strain softening" *partial yielding*. By the time we reach γ_y, significant damage to the entanglement network may have occurred. In other words, slowing down of stress growth is a plausible indicator of a loss of balance between further deformation of the surviving entanglement strands and decline in the total number of effective, active entanglement strands.

If this proposed interpretation is correct, then elastic breakup should begin earlier, when shear is terminated closer to γ_y. If shear cessation occurs after γ_y, when the entanglement network is already falling apart, the induction time for macroscopic motions should be even shorter. Figure 12.19(c) shows the "extreme" situation: no induction time was needed to produce nonquiescent relaxation at $\gamma = 7.14$ with $\dot{\gamma} = 1.4\ s^{-1}$.

As discussed in Section 7.2.1.2, the shear stress maxima, produced at different applied rates under the condition of $Wi_R > 1$, correspond to a similar state because it can be viewed to have a common reduced modulus that is smaller than the equilibrium plateau modulus G_{pl} by a factor of 2.2. The closer the entangled polymer is sheared toward the shear stress maximum at γ_y, the weaker the network has become. Since γ_y increases with the imposed shear rate as $\dot{\gamma}^{1/3}$, at a given γ the system is further away from its yield point for a larger applied rate. Figure 12.20(a) demonstrates this statement with the vertical line at $\gamma = 2.1$ that crosses the three stress versus strain curves at shear rates 0.7, 5.7, and $14\,s^{-1}$, respectively. At a strain $\gamma = 2.1$, the sample sheared at $0.7\,s^{-1}$ is much closer to (σ_y, γ_y) than are the samples sheared more rapidly (at 5.7 and $14\,s^{-1}$). If shear cessation occurs at $\gamma = 2.1$ using the three different shear rates, the preceding discussion suggests that the induction time for nonquiescence is shortest for the test involving the $0.7\,s^{-1}$ shear rate.

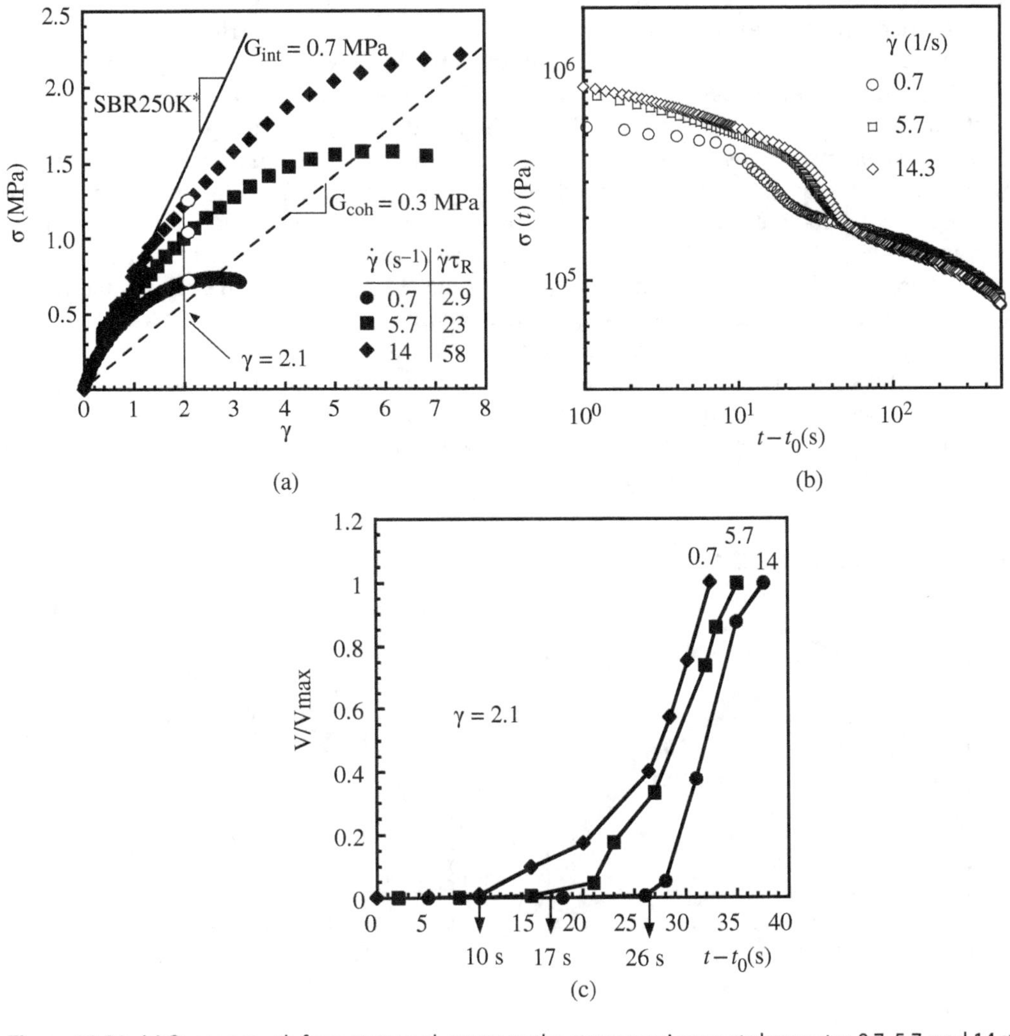

Figure 12.20 (a) Stress growth from startup shear up to the stress maximum at shear rates 0.7, 5.7, and $14\,s^{-1}$. At strain $\gamma = 2.1$, the stress levels are 0.72, 1.04, and 1.27 MPa, respectively. (b) Stress relaxation from the termination of the three startup shear tests in (a) at $\gamma = 2.1$. (c) Particle-tracking velocimetric observations of the stress relaxation in (b) in terms of the induction times and speed of the moving particles. *Source:* Data taken from Boukany *et al.* 2009 [34].

Figure 12.20(b) confirms the prediction: Although the induced stress at $\gamma = 2.1$ is smallest with the slowest shear rate of 0.7 s^{-1}, the system shows the fastest stress relaxation. This may appear counterintuitive, for example, contrary to the expectation based on the stress relaxation from steady shear. Shear relaxation is usually faster when the steady shear stress is larger. Consistent with Fig. 12.20(b) and as shown in Fig. 12.20(c), PTV observations show that upon step strain of magnitude $\gamma = 2.1$ produced with the three discrete rates, an SBR250K melt starts to undergo macroscopic breakup after different amounts of induction time, ranging from 10 to 26 s. Once the macroscopic motions begin, the SBR sheared at the highest rate of 14 s^{-1} takes the shortest time to reach its peak recoiling speed, although it has the longest induction time, 26 s.

12.2.2.4 Unconventional "Step Strain" Produced at $Wi_R < 1$

Elastic yielding takes place because a large step strain produces sufficient chain deformation. A sufficiently large chain retraction force can overcome the cohesive strength of the entanglement network. According to the prevailing theoretical description of nonlinear rheology of entangled polymers, namely, the tube theory,[35] only chain orientation can occur when startup shear is carried out at sufficiently low rates, namely, for $Wi_R \ll 1$. Therefore, by performing nontraditional step strain, namely, interrupting a startup shear imposed at $Wi_R \ll 1$, along with *in situ* particle tracking velocimetric observations, we can discern whether there is nonquiescent relaxation that can only stem from a failure of the strained entanglement network. We assume, as argued in Chapter 16, that sufficient chain deformation drives strong localized elastic yielding. If the tube theory is correct and for $Wi_R \ll 1$ there is only chain orientation, nonquiescent relaxation from a slowly generated step strain should not take place. Figure 12.21(a) shows the stress buildup and relaxation. The relaxation is kink-like, suggestive of an internal breakup. The PTV observation confirms that the relaxation is only quiescent initially, as shown in Fig. 12.21(b).

At later times, nonquiescent relaxation sets in. This structural breakdown presumably indicates that there was enough chain deformation at the end of the stepwise shear to cause decohesion of the entanglement network. In other words, it is less straightforward to imagine a structural breakup after cessation if there is only chain orientation: How can the network

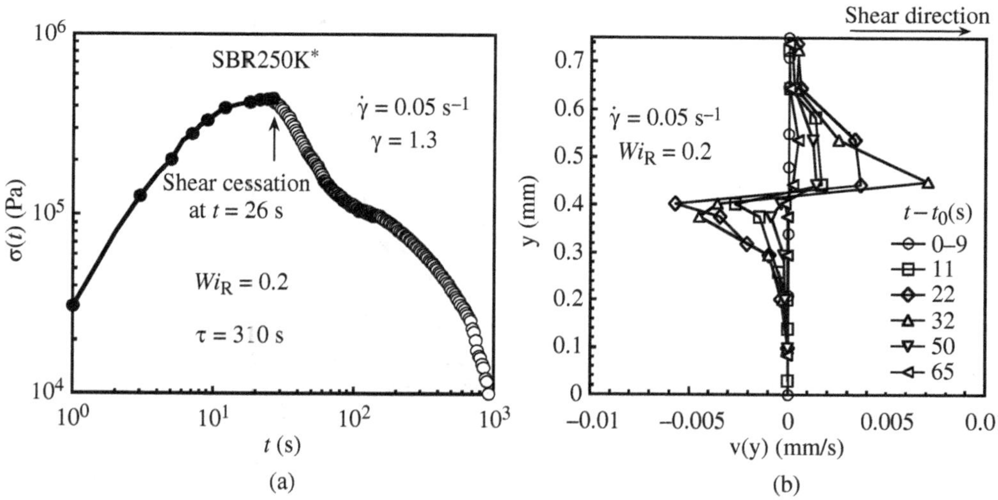

Figure 12.21 (a) Shear stress growth and relaxation of SBR250K produced by a low shear rate of 0.05 s^{-1} corresponding to $Wi_R = \dot\gamma\tau_R \sim 0.21 < 1$. To reach a strain of 1.3 the shearing lasted 26 s. (b) PTV description of the elastic breakup after shear cessation, revealing an induction time of 9 s. *Source:* Data taken from Boukany *et al.* 2009 [34].

disintegrate in absence of chain stretching? Should we not consider chain stretching on scales larger than the entanglement spacing? Thus, experiment like the one in Fig. 12.21(a,b) draws into question the central assumption of the tube model that every chain performs Rouse movements in its tube and therefore cannot undergo chain stretching for $Wi_R < 1$. Given its importance, more systematic studies should be carried out to examine the generality of this result.

Numerical calculations have been performed, based on the tube model, for comparison with the PTV measurements of melt breakup during relaxation.[36–38] Unfortunately, such studies do not explicitly describe disentanglement that could result in a structural collapse of the strained entanglement network. The irreconcilable differences between the tube model description and our alternative conceptual framework will be treated in detail in Chapter 16.

12.3 Quiescent and Uniform Elastic Yielding

12.3.1 General Comments

Entangled polymers in their liquid state are widely known to display considerable rubbery elasticity on macroscopic length scales. Four of the five indications for chain entanglement, namely, elastic recoil, rubbery plateau in creep compliance, the plateau in the relaxation modulus, and the plateau in storage modulus, are all associated with this elasticity. Only the 3.4 power law scaling of the zero-shear viscosity with molecular weight is not associated with elasticity. Elasticity of polymer solutions and melts is already evident in the linear response regime. What does the elasticity do in the regime of large deformation? According to the phenomenology reviewed in Chapter 9 and earlier, the elasticity associated with deformation of the entanglement network, or the growing elastic deformation, is a "time-bomb," embedded in the system to cause catastrophe, namely, structural breakup either during or after deformation.

For decades, the elastic-driven disaster had hidden itself very well. The nonquiescent relaxation after a sizable step strain was not just a surprise when first revealed in 2006.[2] It was completely unanticipated, not to mention that it is also nontrivial to understand its origin. We must resolve the following paradox and explain the counterintuitive nature of the phenomenon. Figure 12.22(a) depicts the "paradox." The system breaks down, showing strain localization, upon a step strain of magnitude γ_{ebd}, while shear homogeneity persists between γ_{ebd} and γ_{sb}

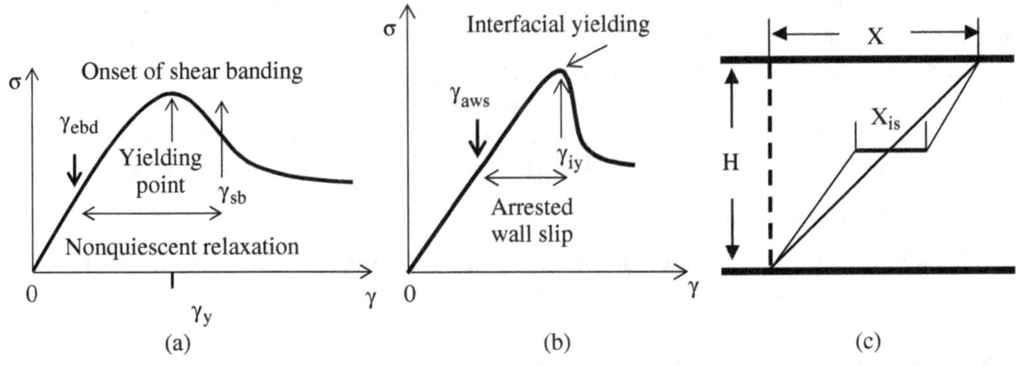

Figure 12.22 (a) Sketch to summarize different PTV observations such as elastic breakdown at the threshold γ_{ebd} during nonquiescent relaxation and shear banding during postyield shearing beyond γ_{sb}. (b) Summary of the relationship between interfacial yielding (iy) and arrested wall slip (aws). (c) Depiction of internal slip involving disentanglement of one monolayer in the bulk, with a crack length of X_{is}, relative to the elapsed displacement of X on the moving surface over a gap distance of H.

during the same startup shear. The subscript "ebd" stands for "elastic breakdown" and "sb" for shear banding. Here γ_{sb} can be significantly larger than γ_{ebd}. Why does ongoing shear delay the emergence of localized yielding and subsequent structural breakup? Conversely, how does the entanglement network undergo failure at a moderate strain after the external deformation has ceased? How does the cessation of external deformation trigger elastic yielding? Our understanding of polymer rheology cannot advance without facing and answering such questions. In parallel, a similar counterintuitive phenomenon exists in the literature for entangled melts capable of significant wall slip. Figure 12.22(b) indicates that interfacial yielding leading to observable wall slip occurs when the growing strain reaches γ_{iy} during startup shear. However, when the same startup shear is terminated at a *smaller* strain $\gamma_{aws} < \gamma_{iy}$, interfacial slip is observed, where the subscript "aws" stands for arrested wall slip, which is the subject of the subsequent Section 12.4. So, why does wall slip show up at $\gamma_{aws} < \gamma_{iy}$ upon shear cessation on one hand, and no wall slip occurs in presence of the ongoing shear till $\gamma_{iy} > \gamma_{aws}$ on the other hand?

12.3.2 Condition for Uniform Yielding and Quiescent Stress Relaxation

According to the PTV observations in Figs. 12.17(b), 12.18(b), and 12.20(c), after a moderate step strain there is an induction period of quiescence. During this quiescent period, the elastic stress does not decrease substantially. Eventually, namely, for $t > t_k$, somewhere in the sample the disentangling network reaches the point of structural disintegration, and the "horse-race" produces a winner, namely, failure of the network at some location y^*. Under sufficient straining, every layer rushes to disentanglement and is therefore in a horse race. This internal "slip" is felt everywhere since the sample is mechanically correlated over the length scale of the sample dimension H, namely, the shear stress is the same at every value of y along the sample thickness direction. The strained sample takes advantage of this internal disentanglement to undergo elastic recoil, releasing the retractive stress. As the stress drops quickly, layers other than the "slip layer" no longer have any chance to undergo the same breakdown that has already occurred at y^*, at least so long as the disastrous "internal slip" is of a sufficient magnitude, namely, capable of producing a sizable correction to the strain field.

The question of whether and how the elastic yielding becomes localized can be further investigated at a quantitative level as follows. Let us consider making a step strain of magnitude $\gamma = X/H$ that is homogeneous. When γ is significantly larger than unity, internal slip can occur at a random position y^*, due to loss of chain disentanglement in the monolayer at y^*. The internal slip (is) produces a correction of magnitude X_{is} to the strain field, as shown in Fig. 12.22(c), where the size X_{is} of the "crack" is determined by the shear stress $\sigma = G_{pl}\,\gamma$. If this correction is significant, it would cause the sample to recoil appreciably, leading to a fast drop in the shear stress. Elastic yielding would localize because with the fast stress decline layers away from y^* no longer meet the condition for chain disentanglement. It is clear from Fig. 12.22(c) that the effect of the recoil on the stress relaxation can be characterized in terms of the ratio

$$X_{is}/X = (b_0/H)(t_{is}/\tau), \tag{12.8}$$

where $X_{is} = V_{is}t_{is}$ with $V_{is} = \sigma/\beta = G_{pl}\gamma/\beta$, $X = \gamma H$, and use is made of Eq. (6.20) for b_0. Here t_{is} is the duration of the internal slip, which ends when the crack heals. Surely, we always have $t_{is}/\tau < 1$ because a great deal of shear stress would have relaxed after a reptation time τ. The correction is significant if X_{is}/X is not negligibly smaller than unity. For X_{is}/X not to be vanishingly small, so that nonquiescent can be expected, b_0/H needs to be on the order of unity. We note that the analysis given in Eq. (12.8) is based on internal slip at one monolayer. In reality, nonquiescent relaxation could readily occur if the "internal slip" involves multiple layers.

Conversely, the stress relaxation is unaffected by any internal slip when $X_{is}/X \ll 1$. In this case, quiescent relaxation can be expected because chain disentanglement can occur uniformly. The logic goes as follows. Since one monolayer of chain disentanglement does not result in any modification of the homogeneous strain field (when $X_{is}/X \ll 1$), the other monolayers in the gap of thickness H can continue to disentangle in the "horse race." The result is uniform yielding through homogeneous disentanglement. According to Eq. (12.8), the condition $b_0/H \ll 1$ guarantees quiescent relaxation since $t_{is}/\tau < 1$: Had the internal slip lasted as long as τ, chain deformation that drives disentanglement would have relaxed. Figure 12.12(a) illustrates for polybutadiene solutions how macroscopic motions can indeed be suppressed by reducing b_{max}/H sufficiently without reducing the level of chain entanglement. Elastic yielding refers to disintegration of the entanglement network and takes place only when the magnitude of the stepwise deformation exceeds a threshold. In conclusion, we emphasize that without the PTV revelation of the breakdown of step-sheared solutions and melts, the concept of elastic yielding would have remained undiscovered. The nonquiescent relaxation is clearly the most revealing among all known rheological phenomena.

12.3.3 Homogeneous Elastic Yielding Probed by Sequential Shearing

The condition $X_{is}/X \ll 1$ can be met by reducing the level of entanglement in polymer solutions. In the absence of macroscopic motions, the stress relaxation can be readily interpreted and compared with the prevailing theoretical description based on the tube model. No experimental technique is available to identify molecular events associated with elastic yielding, namely, strain-induced chain disentanglement. Thus, we have to contemplate the macroscopic consequence of chain disentanglement. Nonquiescent relaxation revealed by PTV observation is a striking macroscopic indication of elastic yielding. According to the preceding discussion, lack of macroscopic motions implies that either disentanglement can take place uniformly in space or disentanglement did not occur. What would be the rheological effects of chain disentanglement?

The advanced rheometric protocol discussed in Section 3.4.2 can be adopted to obtain additional rheometric information on the nature of the elastic yielding. In particular, to examine the consequence of stepwise shear during different stages of the quiescent relaxation, a second startup shear can be applied as sketched in Fig. 12.23. Here t_w is the elapsed time during the relaxation. To generalize the definition of step strain as well as examine the nature of stress relaxation, one study in the literature chose a rather low shear rate $\dot{\gamma} = 0.25\,s^{-1}$, namely, $Wi_R = 0.1$, to impose a stepwise shear with a magnitude of $\gamma = 1.5$. During the various stages (i.e., at different times of t_w) of the stress relaxation from this slowly produced stepwise shear, a startup shear with $\dot{\gamma}_2 = 15\,s^{-1}$ is applied in separate tests.

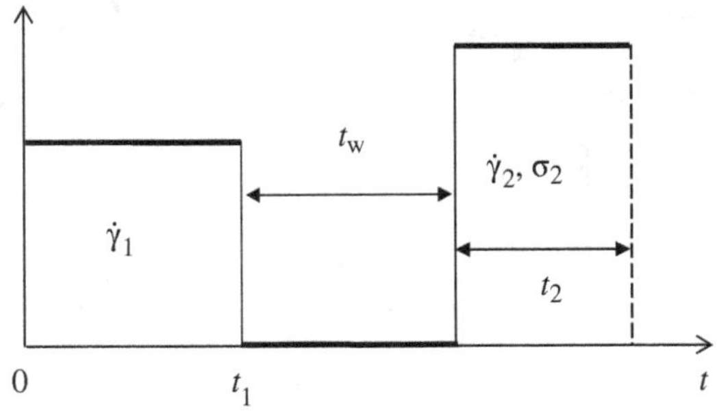

Figure 12.23 Schematic of the protocol reminiscent of Fig. 3.8 represented in Section 3.4.2, which is applied in Fig. 12.24.

Figure 12.24 summarizes such a study.[39] As expected, the squares show that the state of entanglement produces a smaller stress maximum σ_{2y} after the step strain. Such behavior is consistent with the fact that as shown in Fig. 12.25 partial yielding has already taken place, namely the shear stress deviates downward from a linear dependence. More interestingly, the behavior of σ_{2y} is nonmonotonic with respect to the duration of relaxation: During the relaxation at $t_w > 0$, σ_{2y} further declines before rising toward the horizontal line (equilibrium value). The lowest point of σ_{2y} occurs during the relaxation at $t_w = 6$ s. In other words, elastic yielding, namely, the strain-induced chain disentanglement, lasted 6 s before healing (reentanglement) started. Even though there are no detectable motions, we could still delineate the nature of the stress relaxation created by a sufficiently large stepwise shear. Apparently, $\gamma = 1.5$ is sufficiently large to produce elastic yielding, evidenced by the dip in the curve of squares in Fig. 12.24.

To the leading order, we speculate that elastic yielding during relaxation requires sufficient chain deformation. According to the prevailing tube model, in this experiment (Fig. 12.24) there could not be any appreciable chain deformation since the stepwise shear was made at $Wi_R = 0.1$. Consequently, after shear cessation, the pictures embraced by the tube model should predict that the state of chain entanglement will improve monotonically toward equilibrium. With

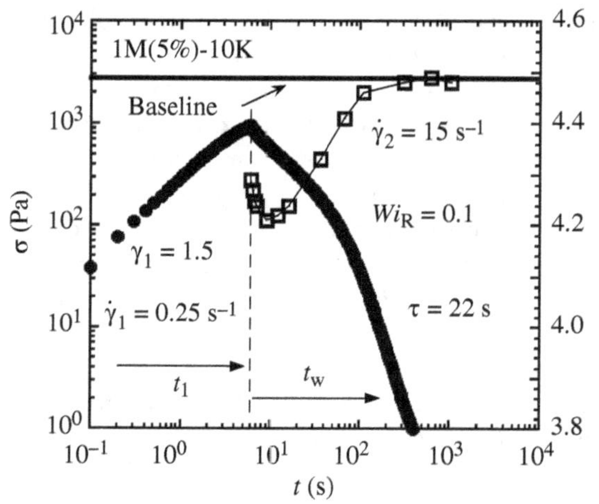

Figure 12.24 Stress growth and relaxation during and after step strain at $\gamma = 1.5$, produced with $Wi_R = 0.1$, as a function of time, plotted against the left-hand-side Y-axis. During the relaxation, a second startup shear with a rate 60 times the first rate is applied after different amounts of waiting time to produce stress overshoot σ_{2y}, which are presented as squares on the right-hand-side Y-axis. *Source:* Data taken from Wang *et al.* 2012 [39].

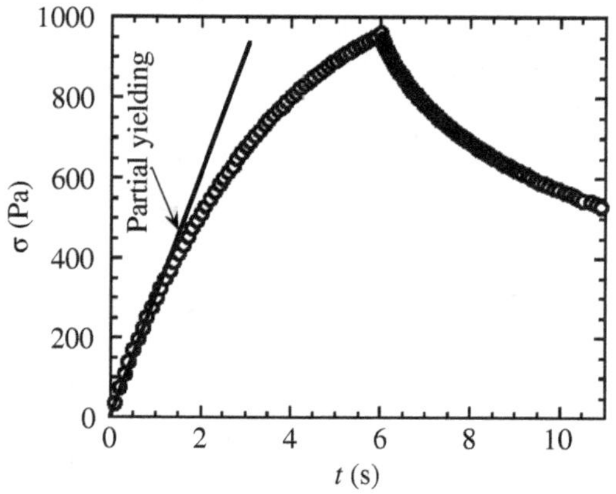

Figure 12.25 Replot of Fig. 12.24 on double-linear scale that shows partial yielding around $t = 2$ s.

improved entanglement, the subsequent startup shear should only result in a monotonic relation between σ_{2y} and the delay time t_w in contradiction to the nonmonotonic behavior shown in Fig. 12.24.

12.4 Arrested Wall Slip: Elastic Yielding at Interfaces

12.4.1 Entangled Solutions

The physics responsible for elastic yielding, either localized (which leads to nonquiescent relaxation) or uniform (which retains quiescent relaxation), can also produce delayed wall slip[40] leading to nonquiescent relaxation. Delayed wall slip is as unexpected and counterintuitive as is the bulk nonquiescent relaxation as described in the preceding Section 12.2.

For example, when a 1M10% polybutadiene solution is made in a conventional way using ditridecyl phthalate as the solvent,[29] stepwise shear produces nonquiescent relaxation because there is apparent failure at the shearing surfaces. Figure 12.26(a) shows the stress buildup during stepwise shear and relaxation after shear cessation for eight discrete magnitudes of the step strain. Although the study did not mention whether or not interfacial slip took place during the startup shear, there cannot be significant wall slip before γ_y. To verify this assertion, in Fig. 12.26(b) we replot the stress versus strain on linear scales. We see that the stress growth follows the straight line $\sigma = G_{pl}\,\gamma$ until $\gamma = 1.2$.

In contrast to the other solutions studied in Ref. [3] to show nonquiescent relaxation due to structural disintegration in the bulk, the 1M(10%)-oil solution allows us to examine whether wall slip is induced by shear cessation. The PTV observations in Fig. 12.26(c) show that severe apparent wall slip occurs upon shear cessation. Given that the maximum slip length b_{max} of this solution is larger than 5 mm, the observed nonquiescence is most plausible due to true wall slip. We call the phenomenon arrested wall slip because it was arrested by the shear cessation, namely, it took place not during shear but after shear.

12.4.2 Entangled Melts

It is interesting to find out how arrested wall slip could occur in entangled melts. We take an example from the literature.[41] Figure 12.27(a) shows the stress buildup and relaxation upon startup shear of a 1,4-polyisoprene melt at $\dot{\gamma} = 2.2\,s^{-1}$. The startup is interrupted discretely at three different stages, corresponding to applied strains γ of 0.22, 1.1, and 2.0, respectively. Unlike the smooth stress relaxation in the linear response regime, in which $\gamma = 0.22$, the stress plunges rapidly after shear cessation at $\gamma = 1.1$ and 2.0, resembling Fig. 12.19(a). Particle tracking velocimetric data in Fig. 12.27(b) indicate that there is little wall slip up to $t = 0.8$ s, namely, $\gamma \sim 1.8$ although severe wall slip does occur at $\gamma = 2.2$ if the startup shear is to continue beyond 2.0.

While no wall slip occurs for gamma up to 1.8, PTV data in Fig. 12.27(c) shows substantial wall slip after termination of the startup shear at $t = 0.5$ s, corresponding to $\gamma = 1.1 < 1.8$. As a reference, step shear with $\gamma = 0.4$ produces complete quiescence while the relaxation from step shear with $\gamma = 2.0$ involves immediate massive sample recoil. Apparently, a strain of 0.4 is insufficient to cause elastic yielding at the polymer/wall interfaces. The significant wall slip at 1.1 and 2.0 is clearly the reason for the sharp stress decline, *in situ* recorded in Fig. 12.27(a). Similar to Fig. 12.18(b), it is possible to identify a short induction time for $\gamma = 1.1$ as shown in

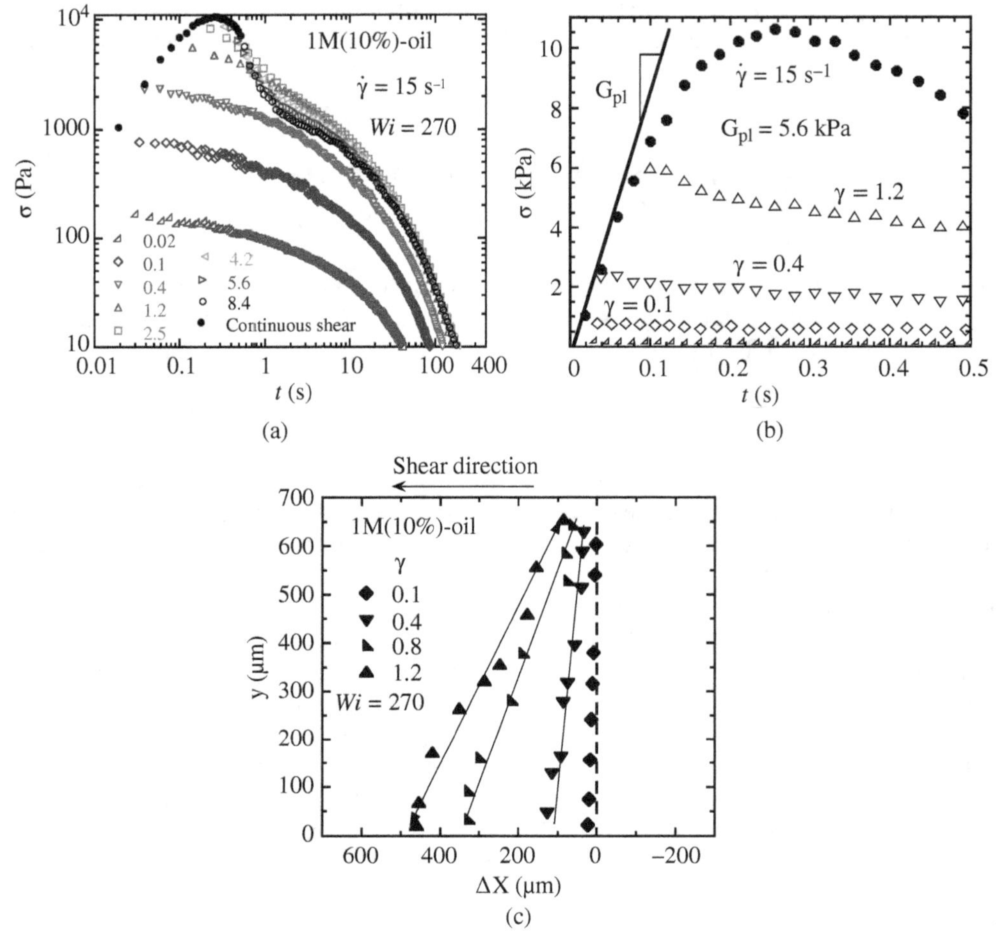

Figure 12.26 (a) Shear stress growth and relaxation during and after discrete step strain of different magnitudes. (b) Stress versus strain and stress relaxation (open symbols) on linear scales to examine the initial stress growth. (c) Particle-tracking velocimetric (PTV) detection of macroscopic motions after various discrete step strains, showing the distances traveled by the tracked particles since the onset of the stress relaxation test; these distances suggest apparent wall slip. *Source:* Data taken from Ravindranath and Wang 2007 [3].

Fig. 12.27(d). In other words, it takes a finite fraction of a second before wall slip shows up, suggesting there is some barrier resisting the interfacial disentanglement.

It is interesting to note that the arrested wall slip demonstrated in Fig. 12.27(a–d) was actually first predicted by authors of Ref. [44] before carrying out the PTV observations for verification. If nonquiescent relaxation can take place due to localized elastic yielding in the interior of a step-strained polymer, the same physics should be in play at the polymer/wall interface. Moreover, in Ref. [41] the study of interfacial elastic yielding after shear cessation also provided an explanation for previously reported "delayed slip."[40] This earlier study was part of the continuing investigations[42,43] on the type C stress relaxation behavior reviewed in Section 12.1. Using a transparent plate as the shearing surface and dispersing micron-sized particles in an entangled PS solution it was reported[40] that the kink-like stress relaxation synchronizes with

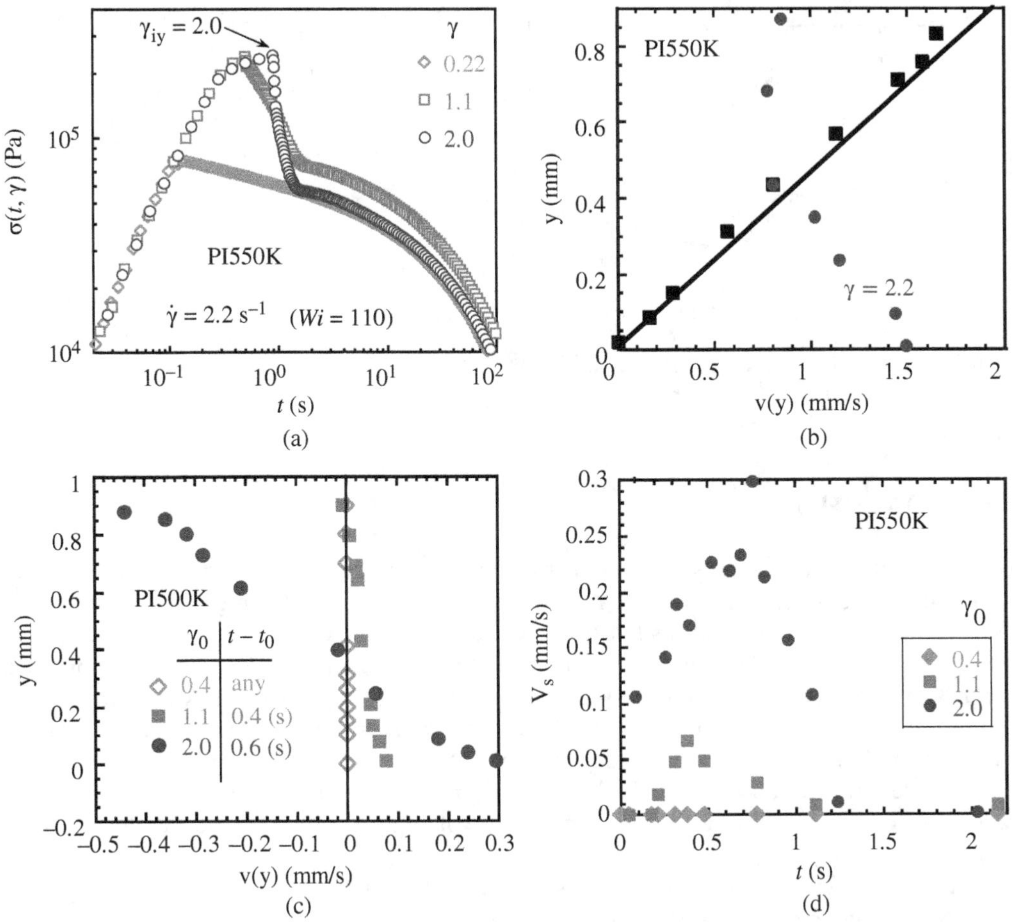

Figure 12.27 (a) Shear stress growth upon startup shear at 2.2 s^{-1} and stress relaxation after a series of discretely applied step strains (0.22, 1.1, and 2) for PI550K melt, which are simply the truncated startup shear. (b) PTV determination of the velocity field during shear, showing no slip condition up to $\gamma = 1.8$ (squares), interfacial failure at $\gamma = 2.2$. (c) Interfacial failure, as observed with PTV when the step strains are 1.1 and 2.0, respectively, expressed in terms of the velocity field shortly after shear cessation at t_0. (d) After step strains 0.4, 1.1, and 2.0, the interface remains intact for $\gamma = 0.4$ and takes a fraction of a second of induction time to develop wall slip at $\gamma = 1.1$, while no resoluble time is required for the PI550K melt to undergo wall slip for $\gamma = 2.0$. *Source:* Data taken from Boukany and Wang 2009 [41].

the fast movement of particles at the sample/wall interface, as shown in Fig. 12.28. This correlation is consistent with the speculation of Osaki[1] that type C behavior might be associated with wall slip. In other words, Fig. 12.28 was strong evidence that many large step-strain stress relaxation tests in the literature may not have involved quiescent relaxation. Thus, among the vast literature[18,20,27,44,45] on step strain of entangled polymers, this pioneering publication[40] has unique merit even though it was not possible for the authors to provide the molecular mechanism for the delayed slip. The so-called "delayed slip" has the same origin as the arrested wall slip observed by our particle-tracking velocimetry, shown in Fig. 12.26(c) for a PB solution, observed in entangled melt in Fig. 12.27(a–d) for a polyisoprene melt.

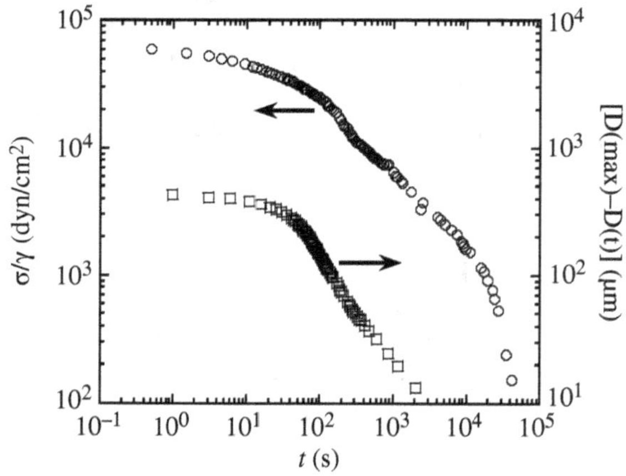

Figure 12.28 Displacement of particles at the shearing surface after a step strain at $\gamma = 2$ (right-hand-side Y-axis) and the corresponding relaxation modulus from the rheometric measurement (left-hand-side Y-axis). *Source:* Data taken from Archer *et al.* 1995 [40].

12.5 Conclusion

This important chapter focuses on nonlinear responses to large step shear strain, in particular, nonquiescent relaxation. Section 12.1 reviewed the pertinent literature that revealed strong strain softening. A simple analysis led to the conclusion that such strain softening plausibly arose from either internal or interfacial failure of the entanglement network. In Section 12.2, examples of nonquiescent relaxation were presented based on particle-tracking velocimetric observations. Section 12.3 discussed the condition for quiescent relaxation in presence of either partial or complete elastic yielding. Finally, arrested interfacial yielding along with nonquiescent relaxation was shown in Section 12.4. The physics for elastic yielding is the same, regardless of where it occurs, in the bulk or at the interface, involving a force imbalance as described in Section 16.3.1.

Given the correlation between nonquiescent relaxation and strong strain softening, it is reasonable to understand why type B relaxation usually occurs in systems involving either broad molecular weight distribution or long-chain branching or moderately entangled polymer solutions and melts where localized elastic yielding and nonquiescent relaxation is suppressed and why type C and B relaxations typically involve linear polymers with strong entanglement and significant extrapolation length that show considerable macroscopic movement after shear cessation due to the localized failure of the entanglement network. It is worth noting that for the same subject of nonlinear stress relaxation treated in this chapter there is a parallel description published just before particle-tracking velocimetric observations had emerged. The contrast between the present chapter and Chapters 10 and 11 of Ref. [46] is starkly alarming.

Since stress relaxation was perceived to be quiescent before the discovery of macroscopic motions by particle-tracking velocimetry, we reviewed ample evidence to show otherwise in step strain of sufficient magnitude. Little attention is devoted to identifying the threshold strain magnitude γ_{ebd} for the localized elastic breakdown. Nevertheless, it is important to point out that γ_{ebd} is just above unity considerably smaller than $\gamma_{max} = 2.2$ in Figs. 12.3 and 12.9. Therefore, the nonquiescent relaxation has nothing to do with the nonmonotonicity, which is the symptom of the tube model. Further discussion about the condition is provided in Section 14.1 in relation to the scenario where the stress relaxation can take place quiescently. The theoretical (scaling) analysis made to determine γ_{ebd} can be found in Section 16.2.3.2.

References

1 Osaki K. On the damping function of shear relaxation modulus for entangled polymers. Rheol Acta. 1993;32(5):429–37.
2 Wang S-Q, Ravindranath S, Boukany P, Olechnowicz M, Quirk RP, Halasa A, et al. Nonquiescent relaxation in entangled polymer liquids after step shear. Phys Rev Lett. 2006;97(18):187801. A refined explanation for the observed elastic yielding was subsequently proposed in J Chem Phys 127, 064903 (2007).
3 Ravindranath S, Wang S-Q. What are the origins of stress relaxation behaviors in step shear of entangled polymer solutions? Macromolecules. 2007;40(22):8031–9.
4 Wang Y, Boukany P, Wang S-Q, Wang X. Elastic breakup in uniaxial extension of entangled polymer melts. Phys Rev Lett. 2007;99(23):237801.
5 Wang Y, Wang S-Q. From elastic deformation to terminal flow of a monodisperse entangled melt in uniaxial extension. J Rheol. 2008;52:1275–90.
6 Wang S-Q, Ravindranath S, Wang Y, Boukany P. New theoretical considerations in polymer rheology: elastic breakdown of chain entanglement network. J Chem Phys. 2007;127(6):064903.
7 Tapadia P, Wang S-Q. Direct visualization of continuous simple shear in non-Newtonian polymeric fluids. Phys Rev Lett. 2006;96(1):016001.
8 Doi M, Edwards S. Dynamics of concentrated polymer systems. Part 4.—Rheological properties. J Chem Soc Faraday Trans 2 Mol Chem Phys. 1979;75:38–54.
9 Callaghan P, Cates M, Rofe C, Smeulders J. A study of the "spurt effect" in wormlike micelles using nuclear magnetic resonance microscopy. J Phys II. 1996;6(3):375–93.
10 Mair R, Callaghan P. Observation of shear banding in worm-like micelles by NMR velocity imaging. Europhys Lett. 1996;36(9):719.
11 Britton MM, Callaghan PT. Two-phase shear band structures at uniform stress. Phys Rev Lett. 1997;78(26):4930.
12 Britton M, Mair R, Lambert R, Callaghan P. Transition to shear banding in pipe and Couette flow of wormlike micellar solutions. J Rheol. 1999;43(4):897–909.
13 Holmes W, Lopez-Gonzalez M, Callaghan P. Fluctuations in shear-banded flow seen by NMR velocimetry. Europhys Lett. 2003;64(2):274.
14 Lopez-Gonzalez M, Holmes W, Callaghan P, Photinos P. Shear banding fluctuations and nematic order in wormlike micelles. Phys Rev Lett. 2004;93(26):268302.
15 Lopez-Gonzalez M, Holmes W, Callaghan P. Rheo-NMR phenomena of wormlike micelles. Soft Matter. 2006;2(10):855–69.
16 Callaghan PT. Rheo NMR and shear banding. Rheol Acta. 2008;47(3):243–55.
17 Callaghan P, Gil A. Rheo-NMR of semidilute polyacrylamide in water. Macromolecules. 2000;33(11):4116–24.
18 Fukuda M, Osaki K, Kurata M. Nonlinear viscoelasticity of polystyrene solutions. I. Strain-dependent relaxation modulus. J Polymer Sci Polymer Phys Ed. 1975;13(8):1563–76.
19 Doi M. Molecular rheology of concentrated polymer systems. I. J Polymer Sci Polymer Phys Ed. 1980;18(5):1005–20.
20 Osaki K, Kurata M. Experimental appraisal of the Doi–Edwards theory for polymer rheology based on the data for polystyrene solutions. Macromolecules. 1980;13(3):671–6.
21 Doi M, Edwards SF. The theory of polymer dynamics, Oxford University Press, New York; 1988.

22 Ianniruberto G, Marrucci G. A simple constitutive equation for entangled polymers with chain stretch. J Rheol. 2001;45(6):1305–18.
23 Rolón-Garrido VH, Wagner MH. The damping function in rheology. Rheol Acta. 2009;48(3):245–84.
24 Ianniruberto G, Marrucci G. Convective constraint release (CCR) revisited. J Rheol. 2014;58(1):89–102.
25 Osaki K, Nishizawa K, Kurata M. Material time constant characterizing the nonlinear viscoelasticity of entangled polymeric systems. Macromolecules. 1982;15(4):1068–71.
26 Graessley WW. Polymeric liquids and networks: dynamics and rheology. Garland Science, New York; 2008.
27 Vrentas CM, Graessley WW. Study of shear-stress relaxation in well characterized polymer liquids. J Rheol. 1982;26(4):359–71.
28 Fetters L, Lohse D, Richter D, Witten T, Zirkel A. Connection between polymer molecular weight, density, chain dimensions, and melt viscoelastic properties. Macromolecules. 1994;27(17):4639–47.
29 Vrentas C, Graessley W. Relaxation of shear and normal stress components in step-strain experiments. J Nonnewton Fluid Mech. 1981;9(3–4):339–55.
30 Marrucci G, Grizzuti N. The free energy function of the Doi–Edwards theory: analysis of the instabilities in stress relaxation. J Rheol. 1983;27(5):433–50.
31 Ravindranath S, Wang S-Q, Olechnowicz M, Chavan V, Quirk R. How polymeric solvents control shear inhomogeneity in large deformations of entangled polymer mixtures. Rheol Acta. 2011;50(2):97–105.
32 Liu G, Wang S-Q. A particle tracking velocimetric study of stress relaxation behavior of entangled polystyrene solutions after stepwise shear. Macromolecules. 2012;45(16):6741–7.
33 Li X, Wang S-Q. Elastic yielding after step shear and during LAOS in the absence of meniscus failure. Rheol Acta. 2010;49(10):985–91.
34 Boukany PE, Wang S-Q, Wang X. Step shear of entangled linear polymer melts: new experimental evidence for elastic yielding. Macromolecules. 2009;42(16):6261–9.
35 Graham RS, Likhtman AE, McLeish TCB, Milner ST. Microscopic theory of linear, entangled polymer chains under rapid deformation including chain stretch and convective constraint release. J Rheol. 2003;47(5):1171–200.
36 Adams J, Olmsted P. Nonmonotonic models are not necessary to obtain shear banding phenomena in entangled polymer solutions. Phys Rev Lett. 2009;102(6):067801. The authors apply a version of the tube model to assert that shear banding can take 40place if there is a stress gradient in the simple shear setup.
37 Wang S-Q. Comment on "nonmonotonic models are not necessary to obtain shear banding phenomena in entangled polymer solutions". Phys Rev Lett. 2009;103(21):219801.
38 Agimelen OS, Olmsted PD. Apparent fracture in polymeric fluids under step shear. Phys Rev Lett. 2013;110(20):204503.
39 Wang Y, Li X, Zhu X, Wang S-Q. Characterizing state of chain entanglement in entangled polymer solutions during and after large shear deformation. Macromolecules. 2012;45(5):2514–21.
40 Archer L, Chen YL, Larson R. Delayed slip after step strains in highly entangled polystyrene solutions. J Rheol. 1995;39(3):519–25.
41 Boukany PE, Wang S-Q. Exploring origins of interfacial yielding and wall slip in entangled linear melts during shear or after shear cessation. Macromolecules. 2009;42(6):2222–8.

42 Larson RG, Khan S, Raju V. Relaxation of stress and birefringence in polymers of high molecular weight. J Rheol. 1988;32(2):145–61.
43 Morrison FA, Larson RG. A study of shear-stress relaxation anomalies in binary mixtures of monodisperse polystyrenes. J Polym Sci Part B Polym Phys 1992;30(9):943–50.
44 Rolon-Garrido VH, Wagner MH. The damping function in rheology. Rheol Acta. 2009;48(3):245–84.
45 Islam MT, Sanchez-Reyes J, Archer LA. Nonlinear rheology of highly entangled polymer liquids: step shear damping function. J Rheol. 2001;45(1):61–82.
46 Dealy JM, Larson RG. Structure and rheology of molten polymers. Hanser, Munich; 2006

13

Elastic Breakup in Stepwise Uniaxial Extension

In well-entangled solutions and melts, large stepwise shear produces nonquiescent relaxation. This relaxation arises from localized elastic yielding either in the bulk or at polymer/wall interfaces. The mechanism for localized elastic yielding should be universal. Therefore, we anticipate nonquiescent relaxations to occur after stepwise uniaxial extension. The phrase "stepwise extension" includes step extension not produced at very high rates. For simplicity, we shall call all stepwise extension "step extension" in this chapter. Conveniently, PTV measurements are not needed in order to examine the possibility of tensile strain localization leading to breakup of a step-extended melt. Long ago, Smith reported[1] that even cross-linked styrene–butadiene rubbers (SBRs) could suffer breakup after step extension. Somehow, it was considered unsurprising to observe delayed rupture of a step-stretched cross-linked rubber. Such delayed breakup occurs on the time scale of tens of hours, unless small cuts are introduced perpendicular to the extension direction, in which case rupture occurs immediately. Because no molecular model was offered to explain the phenomenon, and perhaps because the phenomenon was regarded as something specific to cross-linked rubbers, there was no way for us to anticipate what would happen to a step-stretched non-cross-linked SBR. Since entangled polymer melts are observed to undergo nonquiescent relaxation from step shear,[2] the same melts could be expected to undergo elastic breakup after a large step extension.[3] Such expectation arises because the physics behind elastic yielding in shear and in extension are the same. Chapter 12 described elastic yielding after stepwise shear. This chapter examines the phenomenon produced by stepwise extension.

Literature data reveal that below a threshold (ca. $\varepsilon_{ebd} = 0.6–0.7$) step extension of entangled melts does not produce localized elastic yielding and subsequent breakup.[3] Here the subscript "ebd" stands for "elastic breakdown." Failure of a step-stretched melt occurs when the intrachain retraction force is high enough to overcome the cohesion associated with the entropic penalty for chain disentanglement (as explained in Chapter 16). The driving force for breakup of a melt-stretched sample after termination of extension is understood to be significant chain stretching. This understanding motivates one to study elastic yielding and failure in multiple ways. In particular, according to the phenomenological understanding, the imposed rate is an important variable. As illustrated in Fig. 13.1, the entanglement network extends differently, depending on the Hencky rate: There are three manifestations of extensional strain localization during continuous stretching (cf. Chapter 11); with increasing rate, the response changes from (1) to (2) and eventually to (3). Different forms of breakup might also appear after step extension, depending on how an entangled melt is stretched. Initial reports[3–5] of elastic breakup after step extension did not address whether different forms of the breakup exist. As we shall see later, not only can stretched melts of linear chains break apart, but they can do so in different manners. The onset strain λ_{ebd} may in principle vary with the Hencky rate, unlike the simplifying sketch in Fig. 13.1 that shows a common value. In contrast, melts made of chains with long-chain

Nonlinear Polymer Rheology: Macroscopic Phenomenology and Molecular Foundation, First Edition. Shi-Qing Wang.

Companion Website: www.wiley.com/go/nonlinearpolymerrheology

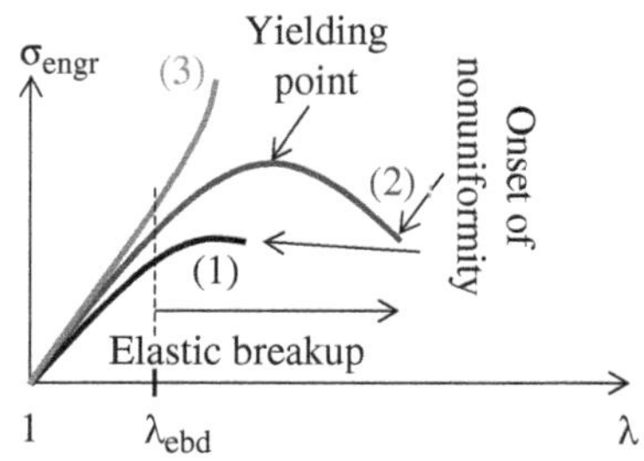

Figure 13.1 Summary of stress versus strain behavior during startup continuous extension of entangled melts in three different regimes: (1) tensile-like failure, (2) shear yielding/unstable necking, (3) rupture as if the melt was chemically cross-linked. There exists a threshold λ_{ebd}: if the startup extension is terminated beyond λ_{ebd}, the relaxing specimen undergoes structural instabilities, making it infeasible to observe quiescent relaxation. Similar to Fig. 12.22(a) for startup shear, this figure intends to indicate that "elastic breakdown" (ebd) can take place well before the yield point is passed.

branching resist the elastic breakdown, as briefly indicated in the following chapter. Chapter 14 also addresses the behaviors observed for intermediate magnitudes of stepwise deformation where only partial yielding takes place and relaxation remains quiescent.

13.1 Rupture-Like Failure during Relaxation at Small Magnitude or Low Extension Rate ($Wi_R < 1$)

The different styrene–butadiene rubbers in Tables 7.1 and 7.4 allow us to access a wide range of Hencky rates. Rates from $Wi > 1$ to $Wi_R > 1$ lie within the accessible window of conventional rheometric instruments. In this section, we discuss the most common tensile strain localization after step extension, involving either a moderate magnitude of the step extension or the Hencky rate that is lower than the Rouse relaxation rate $1/\tau_R$, namely, $Wi_R < 1$.

13.1.1 Small Magnitude ($\varepsilon \sim 1$)

When the step extension is of a modest magnitude, there is a universal mode in which an entangled melt undergoes elastic-driven breakup. For example, suppose we stretch the SBR250K melt to a Hencky strain $\varepsilon = 1$ at the relatively high rate of $3.4\,s^{-1}$, namely, $Wi_R = 2$. As shown in Fig. 13.2, the tensile stress (circles) smoothly decreases until breakup causes the measured stress to plunge toward zero. A close examination of the video recording shows that in first several seconds the relaxation is essentially quiescent. This tensile decohesion takes an induction time $t_{tdc} = 6$ s, where the subscript "tdc" stands for *tensile decohesion*, that being the manner in which the breakdown occurs. The photographs in Fig. 13.2 illustrate the sharply localized "rupture-like" failure. If chain retraction, as envisioned by the tube model, were the mechanism for tensile decohesion, visible breakup should have taken place at the Rouse relaxation time τ_R, which was ~0.6 s. Instead the tensile decohesion time was ~6 s, which is about 10 times the Rouse relaxation time. Video 13.1 shows such failures after step extension.[1]

13.1.2 Low Rates Satisfying $Wi_R < 1$

The second set of data (squares) in Fig. 13.2 involves a step extension of SBR250K produced with an unconventionally low Hencky rate, $0.3\,s^{-1}$. This rate corresponds to $Wi_R = 0.18$ and would effectively produce a level of chain stretching given by $\varepsilon \sim Wi_R = 0.18$ according to the tube model, rather insignificant relative to the threshold $\varepsilon_{ebd} = 0.6 \sim 0.7$ that must be exceeded to have specimen breakup. Despite the low Hencky rate, the termination of the extension at $\varepsilon = 1$ also resulted in breakup around t_{tdc}. In this limit of "no chain stretching", the tube theory cannot identify any molecular mechanism leading to a structural breakdown. If there were not enough chain stretching, what could have caused the tensile decohesion? As discussed in Chapter 16,

1 Video 13.1: Elastic breakup after fast step extension of SBR240K, showing tensile decohesion, made at University of Akron.

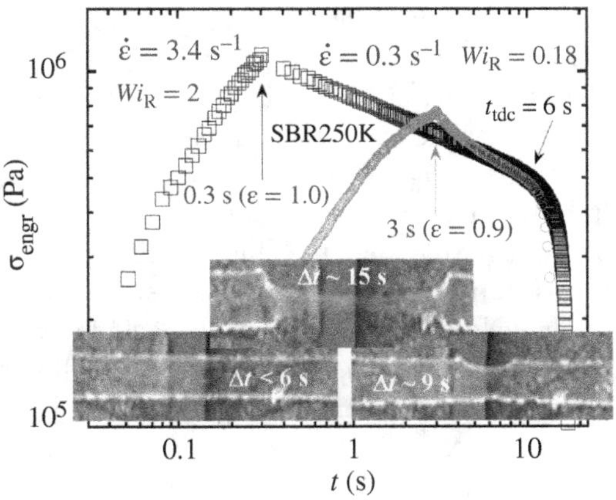

Figure 13.2 Tensile decohesion from step extension of SBR250K, produced at Hencky rates $\dot{\varepsilon} = 0.3$ and $3.4\ s^{-1}$ to two different Hencky strains of 0.9 and 1, respectively, as indicated by the vertical arrows to mark the times of 0.3 and 3 s. These step strains were performed on Physica MCR301 (circles) and TA ARES (squares), respectively. It took a long period of $t_{tdc} = 6$ s before any sign of breakup showed up in both cases. The inset photos show the features of such a tensile decohesion at various times $\Delta t = (t - t_0)$ after termination of extension at t_0 in the example of $\dot{\varepsilon} = 3.4\ s^{-1}$. *Source:* Data taken from Sun *et al.* 2015 [6].

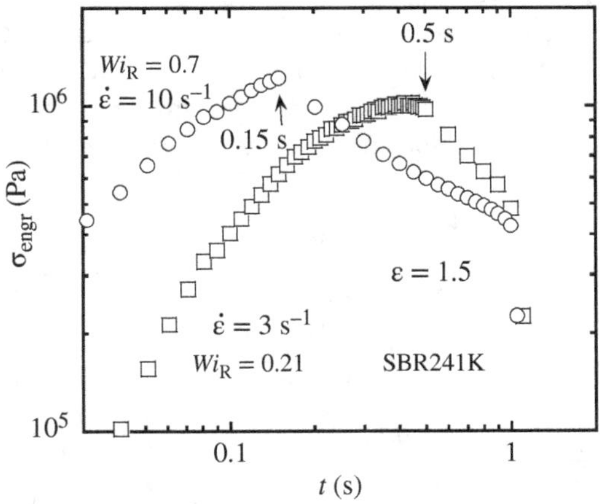

Figure 13.3 Tensile decohesion from two step extension tests ($Wi_R < 1$) of SBR241K, where the conditions are explicitly described, including the rate and strain magnitude. Both experiments satisfy the condition $Wi_R < 1$. When cessation of extension occurs after the stress maximum (circles), the breakup characteristic as tensile decohesion is less distinct. *Source:* Data taken from Sun *et al.* 2015 [6].

even at $Wi_R = 0.18$ there may actually be chain stretching on scales larger than the entanglement spacing because of the entropic barrier. Consequently, the elastic retraction force produced during step extension can be high enough to drive the network to reach a point of decohesion during the stress relaxation.

At $\dot{\varepsilon} = 10\ s^{-1}$, the termination of the startup extension of SBR241K at $\varepsilon = 1.5$ also produces tensile decohesion. The breakup took place immediately. After cessation of the extension, there was little induction time. Relaxation was never quiescent; Fig. 13.3 shows that the tensile stress (squares) decreases to zero within a second. A startup extension at a smaller rate of $3\ s^{-1}$ produces a maximum in the engineering stress before extension was terminated at 0.5 s and a magnitude $\varepsilon = 1.5$. At the smaller shear rate, the tensile decohesion characteristic is less distinct because yielding through disentanglement had already begun at t = 0.4 s when the peak stress was passed. The failure mode remains essentially tensile-decohesion-like as long as the step extension is produced at $Wi_R \ll 1$. This finding might be universal for all well-entangled polymers of linear chains.

13.2 Shear-Yielding-Induced Failure upon Fast Large Step Extension ($Wi_R > 1$)

The analyses of available data in Section 7.5.3 indicate that Wi_R is a pertinent variable to characterize the response of an entanglement network to startup extension. Moreover, the data in

Chapter 11 reveal that an entangled melt can be far more stretched at $Wi_R > 1$ than at $Wi_R < 1$ before reaching the yield point. Thus, it is worthwhile to determine whether or not the failure characteristics after termination of a fast extension to high magnitude could be different from those of tensile decohesion depicted in Section 13.1. Figure 13.4 shows two examples, under the conditions $Wi_R = 6$ and $\varepsilon = 1.5$ (circles) or $Wi_R = 2$ and $\varepsilon = 2$ (squares). The photos in Fig. 13.4 indicate that the failure at high Wi_R (>1) and high magnitude involves a more extensive region of the specimen than the failure in Fig. 13.2. By comparison with the description of Section 11.2 on a second type of failure during startup continuous extension, we identify the failure here as shear-yielding-induced unstable necking. Video 13.2 shows a case of elastic breakup that should be shear-yielding-induced, based on a cylindrical filament (extrudate).[2]

Once the extensional rate is high enough, namely, $Wi_R > 1$, it is the magnitude of step extension that determines whether elastic yielding results in tensile decohesion or alternatively in shear yielding. Based on a fourth sample, SBR616K, Fig. 13.5 shows the contrast of two discrete tests involving the same $Wi_R = 3.4$ but attaining two different strains, namely 1 (circles) and 2 (squares), respectively. A video recording of the elastic breakup indicates two clearly distinct manners in which the failure can occur. To quantify nonquiescent relaxation following 0.2 s of extension at $\dot{\varepsilon} = 10\ s^{-1}$, a trace amount of tracking particles placed on the surface of SBR616K allows the motions generated by the localized elastic yielding to be tracked by a video camera.

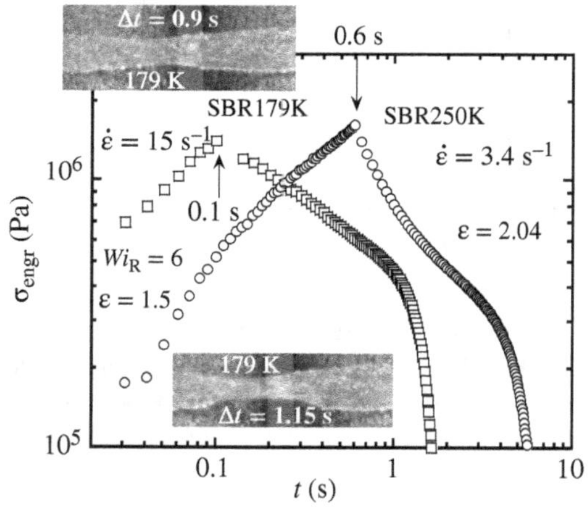

Figure 13.4 Shear yielding and necking of two samples: SBR179K ($Wi_R = 6$, $\varepsilon = 1.5$) and SBR250K ($Wi_R = 2$, $\varepsilon = 2$). The inset photos show that necking (after startup extension with $Wi_R = 6$ at $t = 0.1$ s) involves an extensive region where strain localization takes place. *Source:* Data taken from Sun *et al.* 2015 [6].

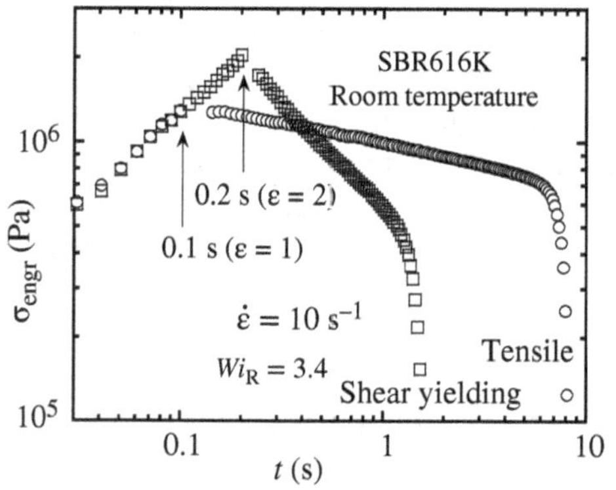

Figure 13.5 Tensile decohesion or necking due to shear yielding, depending on the magnitude of step extension of SBR616K, where the conditions of the two tests are explicitly described, including the rate and strain magnitude. *Source:* Data taken from Sun *et al.* 2015 [6].

2 Video 13.2: Elastic breakup after fast step extension of SBR250K*, showing necking-like failure, made at University of Akron.

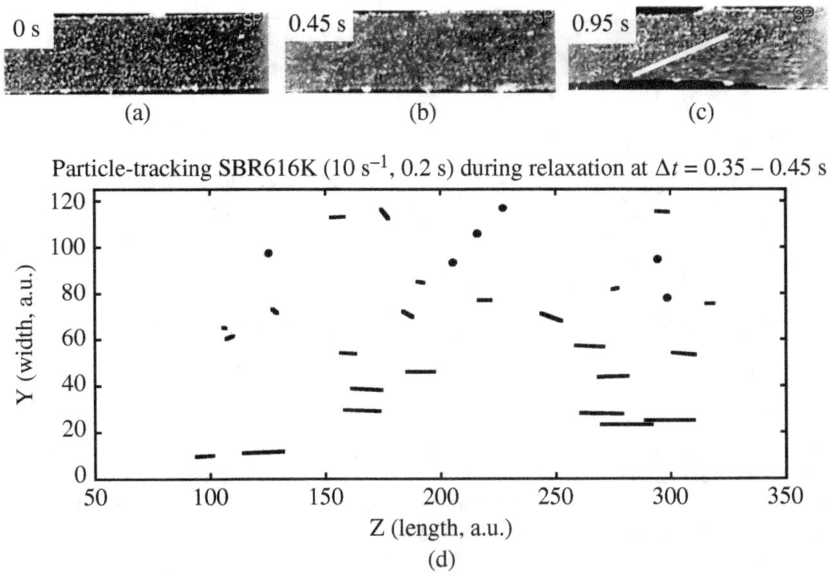

Figure 13.6 (a–c) Three photos showing particle tracking of nonquiescent relaxation after the step extension depicted in Fig. 13.5, 0.2 s after startup extension at 10 s^{-1}. The line in (c) separates the quiescent part of the specimen from the fast retracting part of specimen where the tracking particles show streaks due to the finite exposure time in the video recording. (d) Particle-tracking measurements of nonquiescent relaxation 0.35 s after cessation of startup extension produced at 10 s^{-1} for 0.2 s. Here the horizontal Z axis is the direction of tensile stretching, and the vertical Y axis is along the width direction of the specimen. Between 0.35 and 0.45 s, the movements of traceable particles are measured. The displacements during this period are denoted by the bars and dots to show the amounts of movement at different locations on the relaxing specimen, on some arbitrary scale. There is shear because the amount of displacement at different "heights" is not the same. *Source:* Sun *et al.* 2015 [6]. Reproduced with permission of American Institute of Physics.

Figure 13.6(a–c) displays three photos of the specimen during the relaxation. The specimen at the beginning of the relaxation in Fig. 13.6(a) looks essentially identical to that at $\Delta t = 0.45$ s in Fig. 13.6(b). However, the relaxation from (a) to (b) in Fig. 13.6 was not entirely quiescent. Figure 13.6(c) shows (at $\Delta t = 0.95$ s) visible nonuniform movements of the relaxing sample: The fast moving particles leave bright streaks in the lower right part of the photo, and in contrast the rest of the sample appears nearly quiescent. Figure 13.6(d) indicates motions occurring after completion of the step extension. Specifically, between $\Delta t = 0.35$ and 0.45 s, there are different amounts of movement across the specimen in the width direction, revealing shear deformation that precedes unstable necking.

Based on these observations, we can draw a diagram that summarizes failure behavior due to localized elastic yielding after step extension. In a two-parameter space showing imposed Wi_R versus magnitude ε, we observe in Fig. 13.7(a) that shear yielding is found for $Wi_R > 1$ and $\varepsilon > 1.5$. Quiescent relaxation is found for $\varepsilon < 0.6$–0.7 and all values of Wi_R, everywhere to the left of the vertical dashed line. The remainder of this Wi_R versus ε plot is dominated by a "tensile decohesion" type of failure, that is typically rupture-like. At the upper right corner there is shear-yielding-induced failure: A large value of ε is necessary but not sufficient for shear yielding. At a marginal strain of $\varepsilon = 1.5$, SBR241K shows tensile decohesion and SBR179K displays shear yielding and necking. The former involves $Wi_R = 0.21$ or $0.7 < 1$ while the latter has $Wi_R = 6 > 1$. Conversely, even for $Wi_R > 1$, we have tensile decohesion as long as the magnitude is insufficiently high, namely, below a Hencky strain of 1.5. Correspondingly, we can map the "phase diagram" of Fig. 13.7(a) into a diagram in terms of the Cauchy stress and strain

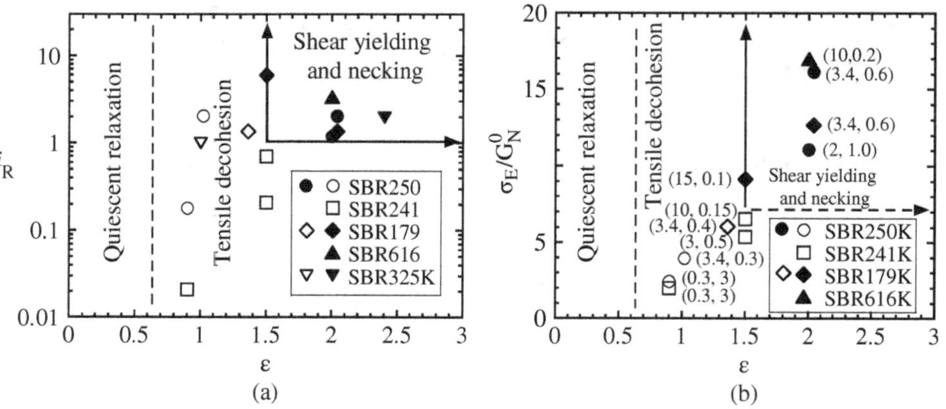

Figure 13.7 "Phase" diagrams mapping the failure behavior during relaxation from step extension at room temperature in terms of either (a) applied Wi_R versus magnitude of step strain ε or (b) normalized stress (at the end of step extension) versus ε, with filled symbols designating necking, and open symbols representing tensile decohesion. In (a), necking only occurs in the upper right quadrant. Tensile decohesion takes place everywhere else to the right of the vertical dashed line, which denotes the threshold magnitude $\varepsilon_{ebd} = ln\,\lambda_{ebd}$ for unstable relaxation. Quiescent relaxation occurs when ε is below a threshold $\varepsilon_{ebd} = 0.6–0.7$. The numbers in the parentheses denote the Hencky rate and the duration in seconds of step extension, respectively. In (a), a solid line defines the border between tensile decohesion and necking because the Rouse Weissenberg number Wi_R appears to control the type of the failure during step extension. *Source:* Sun *et al.* 2015 [6]. Reproduced with permission of American Institute of Physics.

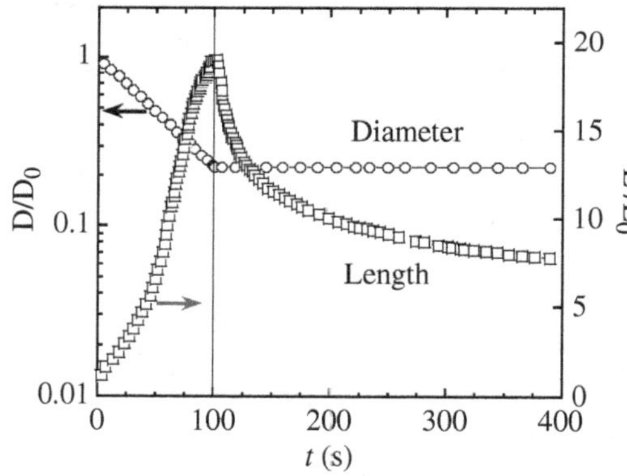

Figure 13.8 Stress relaxation based on filament-stretching rheometry after melt extension of polystyrene at 120 °C for 100 s at 0.03 s^{-1} to a Hencky strain ε = 3. Circles and squares denote the change of the filament diameter D(*t*) and sample length L(*t*), respectively, as a function of time *t*. To keep D constant at the onset of "stress relaxation" at *t* = 100 s, L(*t*) is made to decrease rapidly. If the length L was kept constant, D(*t*) was observed to shrink with time according to Ref. [7], similar to the various observations made in this chapter. *Source:* Data taken from Nielsen *et al.* 2008 [7].

magnitude, as shown in Fig. 13.7(b). The normalized stress level at the end of a step extension does not determine which type of failure takes place, although higher tensile stress typically results from larger Wi_R.

We now consider how filament-stretching rheometry has been used to study the rheological behavior of entangled melts during and after step extension.[7] The preceding sections have shown that after large step extension elastic yielding can terminate a meaningful stress relaxation test; the step-extended specimen starts to break up. In filament-stretching rheometry, a highly stretched polymer melt would of course also suffer from tensile strain localization. To avoid elastic breakup following the imposed extension, instead of holding the ends of the specimen stationary, the specimen was allowed to "shrink" by moving the ends toward each other continuously, so as to prevent the monitored middle section from becoming thinner via elastic breakup. Figure 13.8 illustrates this scenario where the total length was made to shrink in order to keep the diameter of the middle section constant. The interpretation of such a measurement is highly challenging because the time-dependent stress relaxation did not occur under the condition of maintaining the sample in the extension state by fixing the two ends. Except for a very thin middle section, every part of the specimen is externally made to move during "stress relaxation." The protocol is closer to a controlled recoil experiment than to a stress relaxation test. Such data would require new and extended fluid mechanical modeling and should not be directly compared with a rheological model such as the tube model that treats the stress relaxation to be quiescent.

13.3 Nature of Elastic Breakup Probed by Infrared Thermal-Imaging Measurements

After tensile decohesion failure, a specimen can typically recover its initial dimensions prior to the step extension. In contrast, when the breakup is due to shear yielding, extensive portions of the specimen in the "neck" are significantly thinner than the initial dimensions of the specimens. There is clearly a significant amount of flow taking place during shear-yielding-induced breakdown. Figure 13.5 illustrates the contrast. With a stretching magnitude $\varepsilon = 1$, there is barely enough elastic retraction available to overcome the cohesion of the entanglement network associated with the entropic barrier. Consequently, the elastic energy remains stored until the point of tensile yielding of the entanglement network. Here we are borrowing language that will be developed in Chapter 16. Since there is no need to undergo shear yielding, the localization is rather sharp. At $\varepsilon = 2$, however, little induction is required because the elastic force exceeds the entropic cohesion force. Decohesion through tensile breakup is in competition with an easier escape: shear yielding to achieve disentanglement.

It is helpful for us to further verify the difference between tensile decohesion and shear yielding. An IR camera (FLIR SC325) can video record the step extension and subsequent failures during relaxation. Based on a fifth sample, SBR325K from Tables 7.5 and 7.6, Fig. 13.9

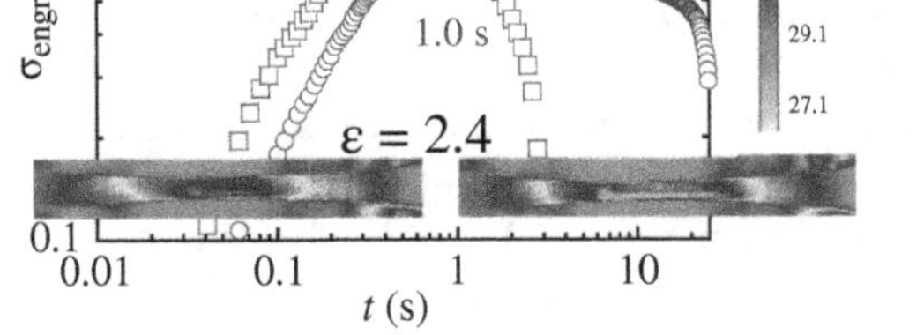

Figure 13.9 Tensile decohesion (from step extension produced at 1 s^{-1} for 1 s) or necking (at 2 s^{-1} for 1.2 s) due to shear yielding after step extension of SB325K. The ambient temperature is 26 °C. The first two color photos at the top show the temperature profiles, respectively, during step extension ($\varepsilon = 1.0$) and after the emergence of tensile decohesion where the temperature rise is insignificant, and the white lines indicate the edges of the specimen. The rest of four color photos show the specimen immediately after the step extension of $\varepsilon = 2.4$ (the first two) and onset of necking moments later (the last two), showing a temperature rise to as high as 35 °C. (*See color plate section for the color representation of this figure.*) *Source:* Data taken from Sun *et al.* 2015 [6].

shows startup extension with Wi_R = either 1 or 2, terminated at two different strains, 1.0 and 2.4, respectively. The first one second of extension is largely elastic, as shown by the stress versus strain curve. Correspondingly, the IR measurement of the specimen's temperature shows that the temperature hardly increases during extension to $\varepsilon = 1$. There is a significant temperature increase of 6 °C (from 26 to 32 °C) during the extension[8] from epsilon = 1 to $\varepsilon = 2.4$. Moreover, shear yielding after a step extension produces an extensive region of uneven temperature. Heat generation is spatially nonuniform, as shown in the four photographs in Fig. 13.9. During necking, the sample's temperature further increases to 35 °C. No external work is being supplied once the step extension has been completed; the heat production arises from the stored elastic energy.

13.4 Primitive Phenomenological Explanations

The map in Fig. 13.7 clearly describes two regimes where different failure characteristics are observed. Tensile decohesion occurs at low magnitude, and alternatively at low extension rates satisfying $Wi_R < 1$. Under these conditions, the entanglement network is only weakly stretched. For $Wi_R < 1$, it may be proposed that the entanglement points are not "tightened up." During relaxation, there are two plausible ways for entanglement strands to recover from the stretched state to the isotropic equilibrium state. One way is molecular diffusion, which is expected to take the reptation time τ_d. Another way is for stretched chains to disengage from the entanglement network. On shorter time scales, such disengagement encounters an entropic barrier, associated with significant conformational changes. In the tensile decohesion regime, with low magnitude or low extension rate, the step extension does not sufficiently deform an entangled polymer to build the retraction force needed for a stretched entanglement strand to drag along surrounding chains during retraction. In the absence of strong chain stretching, fast retraction of entanglement strands meets the entropic barrier. This barrier assures that the system remains intact after step extension, for extensions less than a critical magnitude $\varepsilon \sim 0.6$–0.7. With larger degrees of step extension, after a sufficiently long activation time under a large enough retraction force $f_{retract}$ ($>f_{ent}$) chains can hop over the barrier. Eventually, disentanglement nucleates and causes localized failure of the entanglement network. We call the behavior rupture-like tensile decohesion seen in Figs. 13.2 and 13.9.

At $Wi_R > 1$ stretching of surviving entanglement strands continues even after the yield point. Strain localization is eventually initiated by shear yielding, as discussed in Chapter 11. Apparently, the tightened network finds shear yielding to be the preferred way to undergo disentanglement during ongoing extension. On termination of a fast startup extension at a sufficiently large strain, there is enough elastic retraction force available to enable chain retraction. Fast chain retraction is a collective, multichain process because a retracting chain must drag surrounding chains along. If the retraction were tensile, namely, along the direction of extension, the entanglement network might only tighten further. In contrast, it may be easier for entanglement strands to slide by each other when the motion is at an angle relative to the stretching direction. Finally, we note that there are no systematic investigations to determine how the number Z of entanglements per chain affects the response of melts to large step extensions. It could be proposed that, below a threshold value Z^*, local chain and Rouse relaxation mechanisms can effectively relieve the elastic strain on the entanglement network before strain localization has enough time to develop. It is worthwhile for a future study to identify Z^*.

13.5 Step Squeeze and Planar Extension

The setup of Fig. 5.4 along with particle-tracking velocimetric observations can be employed to contrast responses of entangled polymer melts during relaxation from large step one-dimensional squeezing (cf. Sections 5.3.2 and 10.3) and planar extension (cf. Sections 5.3.3 and 10.4). Squeezing can produce significant shear deformation, so it should be possible to observe recoil-like breakup upon cessation of squeezing, namely, relative displacement between neighboring layers along the Z axis upon termination of squeezing. The configuration of the setup is appealing since it is essentially closed, with the meniscus situated at $x = \pm L/2$ with $L/h \gg 1$. Therefore, any nonquiescent relaxation would be free of any influence from edge instability. In presence of lubrication, the squeezing produces planar extension. Particle-tracking velocimetry can be applied along with the full characterization of nominal stress and strain to study nonlinear responses of entangled melts.

In conclusion, Chapter 13 is the mirror of Chapter 12 and deals with elastic yielding and nonquiescent relaxation after step extension instead of step shear. The same physics dictates the response to large stepwise deformation (shear and extension): Sufficiently large step strain builds large retractive forces in the entanglement strands that can overcome the cohesion of the entanglement network upon withdrawal of the external deformation (see Chapter 16 for explanation). Depending on how the step extension creates the intrachain retractive stress associated with chain deformation, a step-extended melt undergoes two different forms of instability to terminate quiescent relaxation.

References

1 Smith TL, Stedry PJ. Time and temperature dependence of the ultimate properties of an SBR rubber at constant elongations. J Appl Phys. 1960;31(11):1892–8.
2 Boukany PE, Wang S-Q, Wang X. Step shear of entangled linear polymer melts: new experimental evidence for elastic yielding. Macromolecules. 2009;42(16):6261–9.
3 Wang Y, Boukany P, Wang S-Q, Wang X. Elastic breakup in uniaxial extension of entangled polymer melts. Phys Rev Lett. 2007;99(23):237801.
4 Wang Y, Wang S-Q. From elastic deformation to terminal flow of a monodisperse entangled melt in uniaxial extension. J Rheol. 2008;52:1275–90.
5 Wang Y, Cheng S, Wang S-Q. Basic characteristics of uniaxial extension rheology: comparing monodisperse and bidisperse polymer melts. J Rheol. 2011;55:1247–70.
6 Sun H, Lin PP, Liu GX, Ntetsikas K, Misichronis K, Kang N, et al. Failure behavior after stepwise uniaxial extension of entangled polymer melts. J Rheol. 2015;59(3):751–67.
7 Nielsen JK, Rasmussen HK, Hassager O. Stress relaxation of narrow molar mass distribution polystyrene following uniaxial extension. J Rheol. 2008;52(4):885–99.
8 Lin PP, Wang SQ. Erratum: nonisothermal condition in past melt extension experiments. J Rheol. 2015;59(5):1329–34.

14

Finite Cohesion and Role of Chain Architecture

This chapter discusses the concept of finite cohesion associated with chain entanglement. In Section 14.1, we suggest based on experiment that an entangled polymer melt can withstand a critical level of external straining without encountering any structural collapse. In Section 14.2, we show that relative to entangled melts of linear chains long-chain branching significantly increases the threshold strain for elastic breakup after stepwise deformation. Although this chapter is rather short, it serves two important purposes. First, it describes a crucial link between linear response and the highly nonlinear responses associated with strain localization. Second, it is a precursor to Part Four, which presents the theoretical foundation of nonlinear rheology of entangled polymers.

In the preceding two chapters, we have seen the most severe form of elastic yielding after large stepwise shear and extension. A severely strained entanglement network cannot remain intact and undergoes a structural breakdown. The existence of structural breakdown after termination of external deformation is highly unexpected because it contradicts the prevailing perception in the past. It was common knowledge that there should be *quiescence* after cessation of external deformation. Had we not applied the particle-tracking velocimetric method to show nonquiescent relaxation, the concept of elastic yielding would not have been discovered. On the other hand, elastic yielding that produces accelerated stress relaxation can take place homogeneously and therefore quiescently, as discussed in Chapter 12.

In Fig. 14.1, we draw a simple chart to categorize the responses of entangled polymers to stepwise deformation. Specifically, away from the elastic breakup regime, the quiescent relaxation is actually comprised of two regimes. It is necessary to elucidate the physics that determines the boundary at γ_{LR} or ε_{LR} as well as the boundary at γ_{ebd} or ε_{ebd}, where the subscript ebd stands for elastic breakdown via localized disentanglement. This task of exploring the condition γ_{ebd} or ε_{ebd} is deferred to Chapter 16. In this chapter, we examine experiments that identify the upper bound (i.e., γ_{LR} or ε_{LR}) of the linear-response regime and discuss whether the experimental characteristics in the quiescent relaxation domain agree with the conventional theoretical understanding.

The description in Fig. 14.1 does not include one important case: As demonstrated in Section 12.3.2, when $b_{max}/H \ll 1$, even large-magnitude step strain may not produce macroscopic motions. There are at least two cases where this situation can occur: Either the sample is not sufficiently entangled to have a large b_{max}, or the effect of chain disentanglement has been suppressed, namely, b_{max} is made small despite a large degree of entanglement. The former case has yet to receive explicit experimental confirmation, whereas the latter case is presented in Fig. 12.12(a). Specimen breakup via tensile decohesion does not involve shear yielding. Thus, the condition of $b_{max}/H \ll 1$ does not apply to step extension. Experimental information is presently unavailable to determine whether a very weakly entangled melt would undergo breakup after large step extension.

Nonlinear Polymer Rheology: Macroscopic Phenomenology and Molecular Foundation, First Edition. Shi-Qing Wang.

Companion Website: www.wiley.com/go/nonlinearpolymerrheology

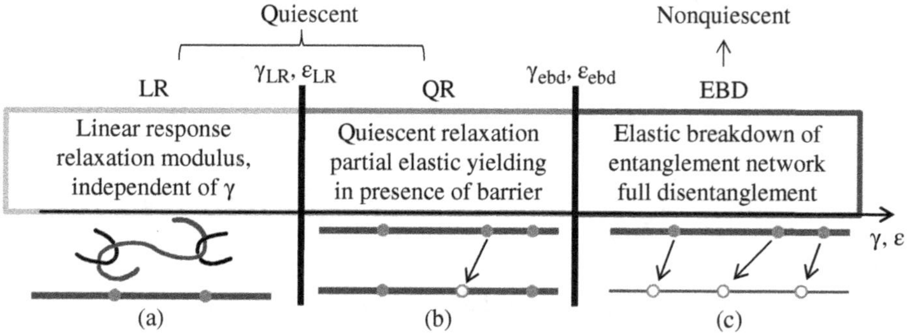

Figure 14.1 Depiction of different states of entanglement after stepwise deformation at increasing levels of strain (γ or ε): (a) linear response (LR) regime where chain entanglement is intact because the magnitude of step shear or step strain is below a threshold; (b) quiescent relaxation (QR) regime at intermediate magnitudes of deformation where partial elastic yielding occurs due to loss of some entanglement strands, for example, the three entanglement points represented by the three dots turning into two plus an open circle that denotes a removed entanglement point; (c) regime of elastic failure of the entanglement network through full disentanglement of whole chains.

14.1 Cohesive Strength of an Entanglement Network

PTV observations[1,2] of nonquiescent stress relaxation from large step strain are rather insightful. They disclose that the entanglement network is of finite strength and can be elastically destroyed. Nonquiescent relaxation indicates the occurrence of localized elastic yielding. On the other extreme, in the linear response (LR) regime, the stress relaxation can be described by an equilibrium relaxation modulus $G_{eq}(t)$ that is independent of the strain magnitude γ. Beyond the threshold for linear response, at $\gamma > \gamma_{LR}$ or $\varepsilon > \varepsilon_{LR}$, the stress relaxation presumably decreases faster than $G_{eq}(t)$ does. However, a rapid stress relaxation may not result in nonquiescence, as long as $\gamma < \gamma_{ebd}$ or $\varepsilon < \varepsilon_{ebd}$. For example, Fig. 12.11(b) shows that there is no macroscopic motion at $\gamma = 0.8$. Therefore, we have $\gamma_{ebd} > 0.8$. Similarly, below some threshold around $\varepsilon_{ebd} = 0.6 \sim 0.7$, stepwise extension of entangled polymer melts does not result in specimen breakup.[3,4]

It may be argued that the intertwining of entangling Gaussian chains naturally produces an entropic barrier. Because of the intimate intermolecular topological constraint, retraction of a deformed chain on length scales *much larger than* the entanglement strand cannot take place on time scales much shorter than the reptation time without paying a conformational entropic penalty. This view is suggested by the experimental data to be presented below. On the theoretical side, the implication of the entropic barrier to ensure a sizable linear response regime is discussed in Section 16.2.3. Unlike nonpolymeric systems, the range of the linear response regime is appreciable as demonstrated here: an entangled polymer can stay within the linear response regime for step strain with $\gamma_{LR} \sim 0.6$. Due to the entropic barrier, the stress relaxation in the quiescent relaxation regime is considerably slower than it is in the elastic breakdown (EBD) regime. We describe experiments that have been carried out to identify γ_{LR} and ε_{LR} as follows. The experimental results will be compared with the conventional theoretical description based on the tube model.

Most literature data on stepwise deformation have focused on whether and how strain softening takes place under rather large levels of deformation. Results were expressed in terms of the damping function of Eq. (12.4). To look for the threshold magnitude γ_{LR} and γ_{ebd} as defined in Fig. 14.1, the stress relaxation should be examined at different values of γ. Let us take an example based on SBR153K, whose storage and loss moduli are given in Fig. 14.2 as

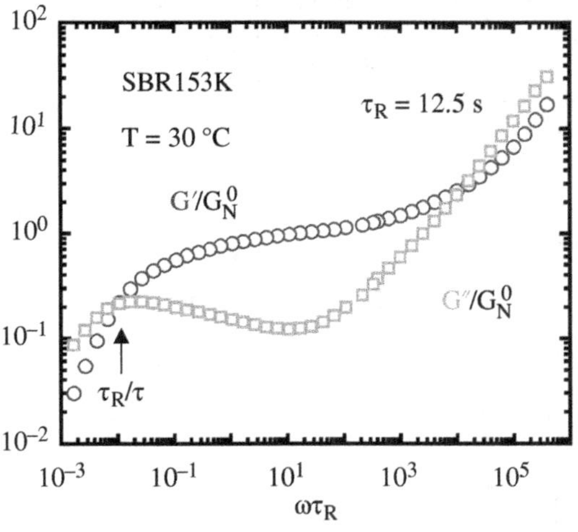

T °(C)	τ (s)	τ_R (s)
0	63,400,000	622,000
10	326,000	3,200
20	13,200	130
30	1,280	12.5
40	202	1.98
50	50.7	0.497
60	14.8	0.145
70	5.90	0.0578
80	2.40	0.0235
90	1.20	0.0118

Figure 14.2 Small-amplitude oscillatory shear (SAOS) data of SBR153K, showing the normalized storage and loss moduli G'/G_N^0 and G''/G_N^0 as functions of the normalized frequency or Rouse Deborah number. Here the elastic plateau modulus G_N^0 is 0.76 MPa (cf. Table 7.6). The accompanying table shows the temperature dependence of the terminal relaxation time τ and Rouse time τ_R. *Source:* Data taken from Sun *et al.* 2013 [5].

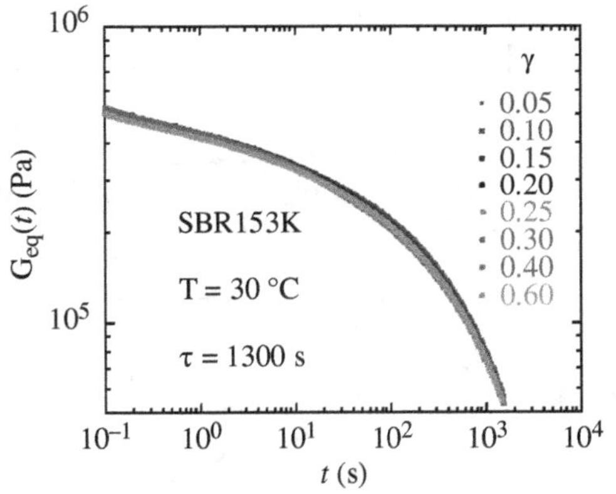

Figure 14.3 Equilibrium relaxation modulus G_{eq} for different magnitudes of step strain from $\gamma = 0.05$ to 0.6, based on SBR153K at T = 30 °C. *Source:* Adapted from Cheng *et al.* 2016 [6].

functions of frequency. Other characteristics of this melt appear in Table 11.1. Figure 14.3 shows the relaxation modulus from step strain in the regime of small magnitude $\gamma \leq 0.6$. Within the experimental error, these relaxation modulus curves are essentially the same for all values of γ. Within the experimental accuracy, the linear response regime can be defined by $\gamma < \gamma_{LR} \sim 0.6$ for this relatively monodisperse SBR153K melt. As seen in Fig. 14.4, at larger γ, G(t) becomes measurably dependent on γ. The data in Fig. 14.3 defines a unique relaxation modulus $G_{eq}(t) = G(t, \gamma = 0.1\text{–}0.6)$, which can be regarded as the equilibrium relaxation modulus. Slight differences show up in Fig. 14.4 involving strain magnitude up to $\gamma = 1.1$, but the relaxation appears quiescent. The behavior seen in Figures 14.3 and 14.4 belongs to the quiescent relaxation regime (cf. Fig. 14.1).

Similarly, relaxation tests from step extension can be carried out to identify the linear response regime for extension. For example, at magnitude of stretching ratio $\lambda = 1.2$, the stress relaxation from step extension produces an identical relaxation modulus as shown in Fig. 14.5. Here up to $\lambda = 1.5$, the relaxation dynamics are indistinguishable as shown in Fig. 14.6 where $G(t, \lambda) = \sigma_{engr}(t, \lambda)/(\lambda - \lambda^{-2})$.

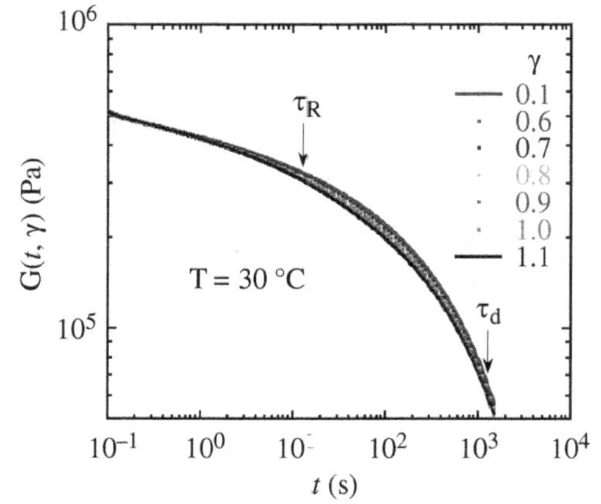

Figure 14.4 Strain-dependent relaxation modulus G(*t*) as a function of time at magnitudes between 0.6 and 1.1, showing a slight downward deviation from G_{eq} represented by that obtained at $\gamma = 0.1$. *Source:* Adapted from Cheng *et al.* 2016 [6].

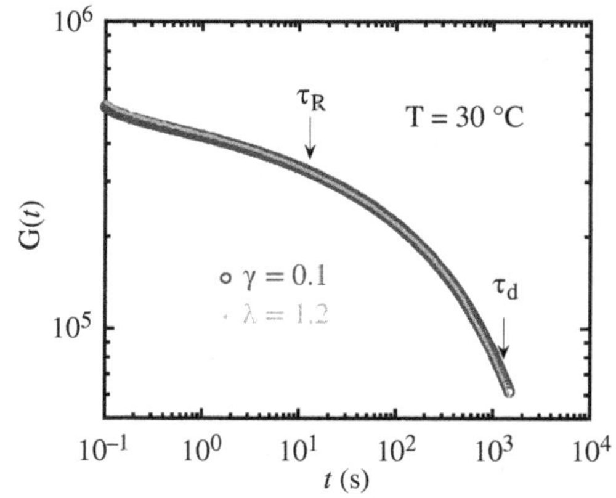

Figure 14.5 Equilibrium relaxation modulus G_{eq} from both small step shear and extension, at $\gamma = 0.1$ and $\lambda = 1.2$, respectively, based on SBR153K at T = 30 °C. *Source:* Adapted from Cheng *et al.* 2016 [6].

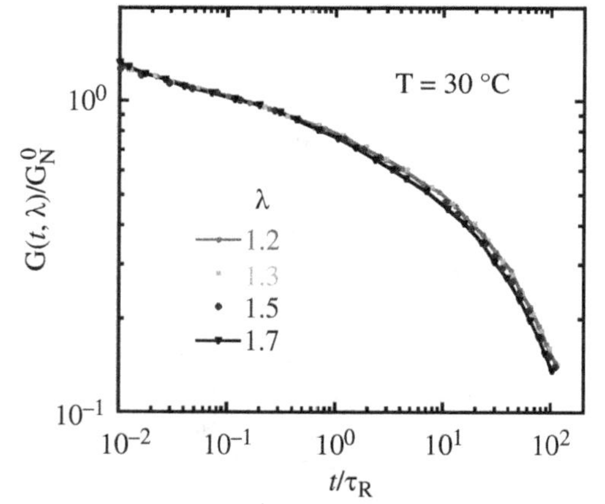

Figure 14.6 Normalized relaxation modulus as a function of time from step extension at stretching ratios $\lambda = 1.2$, 1.3, 1.5, and 1.7, where only the data at 1.7 shows a slight downward deviation from G_{eq} represented by that obtained at $\lambda = 1.2$. *Source:* Adapted from Cheng *et al.* 2016 [6].

According to the tube model,[7,8] the chain stretching (defined on the length scale of tube segment) contribution to the shear stress σ diminishes on the Rouse time scale τ_R due to chain retraction. The residual stress $\sigma(t)$ beyond τ_R is essentially due to the surviving chain orientation on tube segment scales. Since the chain stretching and orientational contributions to σ vary with γ and ε in different ways, the stress relaxation is predicted by the tube model to be different for different values of γ and ε. As a result, in the tube model, there is, in principle, no measurable threshold below which linear response prevails for step strain. In other words, according to the tube theory, the linear response regime is vanishingly narrow, namely, $\gamma_{LR}, \varepsilon_{LR} \ll 1$.

We take the most advanced version of the tube theory, for example, the GLaMM model,[8] and calculate the relaxation modulus to compare with the experimental data from Figs. 14.3 through 14.6. To examine whether there is any qualitative difference between the data and the GLaMM theory,[8] the ratio of the relaxation modulus $G(t, \gamma)$ to $G_{eq}(t)$ is evaluated as a function of either γ or λ at various times.[6] For linear response, this ratio is by definition unity at all times. The open symbols in Figs. 14.7 and 14.8 show that up to $\gamma_{LR} = 0.6$ and $\lambda_{LR} = 1.5$ the stress relaxation shows linear response. The predictions of the tube theory, shown by the filled symbols, reveal a qualitative disagreement with the experimental data. Specifically, as expected from the construct of the tube model, the ratio $G(t, \gamma)/G_{eq}(t)$ decreases smoothly from unity as a function of either γ or λ. This explicit tube-model calculation indicates that the linear response regime is indeed negligibly narrow: No matter how small the magnitude is, there is always the barrier-free chain retraction at τ_R to produce a stress drop whose magnitude increases with the level of step strain.

This difference between the experiment and theory is qualitative. The difference demonstrates the need to consider a different theoretical framework. This issue will be discussed at a far greater length in Chapter 16. At a quantitative level, there are also some noteworthy differences. The experimental data show rather insignificant strain softening, represented by the deviation of the experimental data from the horizontal line $G(t, \gamma)/G_{eq}(t) = 1$ for $\gamma_{LR} > 0.6$,

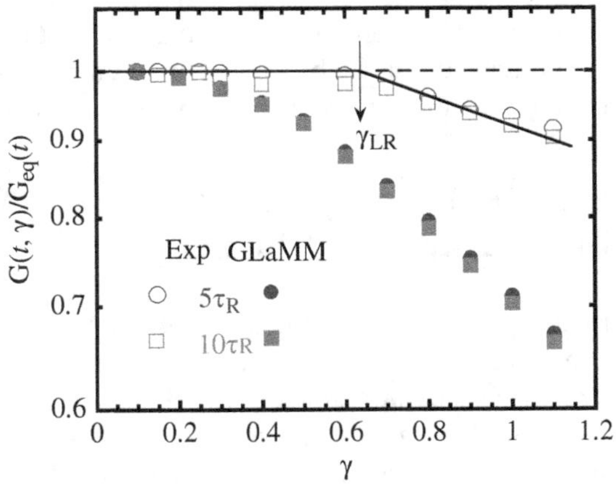

Figure 14.7 Strain dependence of normalized relaxation modulus obtained from step shear experiments (open symbols) and theoretical calculations based on the tube model – GLaMM[8] (filled symbols). The two symbols correspond to $t = 5\tau_R$ and $10\tau_R$, respectively. *Source:* Adapted from Cheng *et al.* 2016 [6].

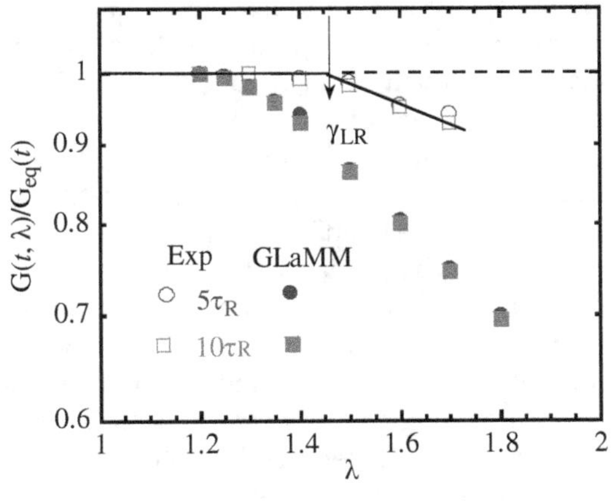

Figure 14.8 Strain dependence of normalized relaxation modulus obtained from step uniaxial extension experiments (open symbols) and theoretical calculations based on the tube model – GLaMM[8] (filled symbols). Here, $G(t, \lambda)$ is obtained from $\sigma_{engr}/(\lambda - 1/\lambda^2)$. The two symbols correspond to $t = 5\tau_R$ and $10\tau_R$, respectively. *Source:* Adapted from Cheng *et al.* 2016 [6].

while the GLaMM calculations show a markedly growing separation not only from the horizontal line but also from the experimental data points. So the experimental data in Figs. 14.7 and 14.8 disagree with the theoretical account in a disproportionate manner. This large discrepancy contradicts the foundation of the tube theory because the barrier-free chain retraction on the Rouse time scale and decoupling of stress into orientation and stretching components at the tube segment level are the central ansatz and starting point of the tube model. The demonstrated existence of sizable γ_{LR} and λ_{LR} (ε_{LR}) indicates that chain deformation and stress arising from chain deformation cannot be characterized on a single length scale of the tube segment. Chain deformation and relaxation takes place on multiple length scales.

We must admit, it is an enormous, perhaps intractable, task to properly account for the intermolecular entanglement effects in a realistic quantitative manner. However, the lack of a quantitative theory to replace the tube model should not be used as an excuse for ignoring the qualitative and quantitative differences seen in Figs. 14.7 and 14.8, and for continuing to use the deGennes–Edwards–Doi tube model that has been constructed to examine chain deformation and evaluate stress only on one length scale corresponding to the entanglement spacing (tube diameter).

As long as the magnitudes γ_{ebd} and λ_{ebd} are not reached, stress relaxation is still quiescent even though it deviates from linear response behavior. Beyond the linear response regime, where $\gamma > \gamma_{LR}$ or $\lambda > \lambda_{LR}$, the stress relaxation can be slightly faster than the linear response expectation because of partial elastic yielding. Since not all entanglement strands are of the same length, there is a simple explanation for the accelerated stress relaxation in the quiescent relaxation regime: Some entanglement strands have been strained to a greater degree and may overcome the entropic barrier, causing mutual sliding at the entanglement points as illustrated in Fig. 14.1. This readjustment at the entanglement strand level does *not* lead to a macroscopic collapse of the entanglement network and corresponding nonquiescence.

14.2 Enhancing the Cohesion Barrier: Long-Chain Branching Hinders Structural Breakup

In the presence of significant long-chain branching, entangled melts possess two different dynamical regimes, associated with the side chains and with the backbone, respectively. This dynamical separation introduces a mechanism to delay structural collapse of the entanglement network. The side branches have much faster dynamics than do the backbone. Thus, after significant extension, for example, after a Hencky strain of 2, side chains can quickly undergo chain retraction and return toward their nondeformed states. As long as the backbones remain entangled with one another, the disentanglement of side chains does not lead to the failure of the entanglement network. At longer times, when the backbones try to undergo disentanglement through chain retraction, thanks to the retracted side chains, the backbones face a much higher entropic barrier than linear chains do. For the backbone to retract, the already relaxed, "reentangled" side chains must undergo conformational changes. This amounts to significantly more entropic loss than retraction of a linear chain. The entropic barrier can be so large that the overall entanglement network survives and the relaxation remains quiescent. Unless the magnitude of step extension is so large that chains are sufficiently deformed to overcome the larger entropic barrier due to long-chain branching, elastic breakdown of the overall entanglement network will not occur and the relaxation will remain quiescent. We present three cases that demonstrate the resistance against elastic breakup of entangled melts containing long-chain branching as follows. Figure 14.9 shows the three chain architectures that have been studied rheologically, with measurements in Figs. 14.10–14.12.

Extensive data in Chapter 13 showed that entangled melts of linear chains always undergo elastic breakup after stepwise extension of magnitude $\varepsilon = 1$ or >1. This information provides a reference point to contrast with the behavior of melts with long-chain branching.

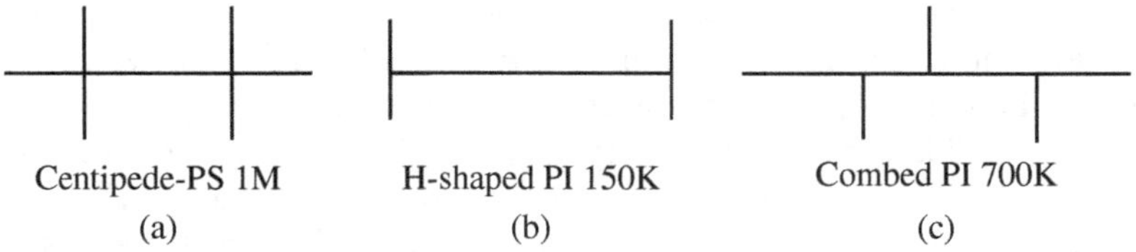

Figure 14.9 (a) Centipede–polystyrene is approximately made of two four-arm stars on average, where each arm is around 125 kg/mol, making the entire chain to be around 10^6 g/mol. (b) H-shaped 3,4-polyisoprene has an average total molecular weight of 210 kg/mol, with a backbone of $M_n = 150$ kg/mol and end-side chains of $M_n = 25$ kg/mol. (c) Comb-like 3,4-polyisoprene is a randomly grafted PI, where each of the three side chains has $M_n = 20$ kg/mol and the backbone has $M_n = 700$ kg/mol.

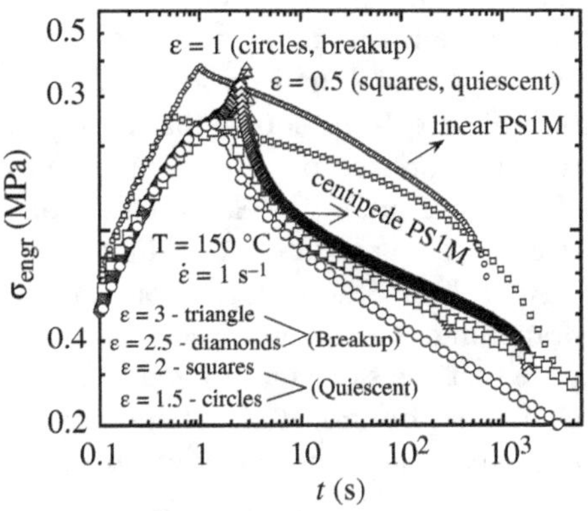

Figure 14.10 Stress growth and relaxation during and after step extension, respectively, carried out at Hencky rates of $1\ s^{-1}$, of both linear PS1M (small symbols) and centipede-PS1M (large symbols). Linear PS1M undergoes quiescent relaxation at $\varepsilon = 0.5$ but breaks at $\varepsilon = 1.0$, whereas centipede-PS1M only undergoes breakup when the step extension is beyond $\varepsilon = 2$. *Source:* Adapted from Liu *et al.* 2013 [9].

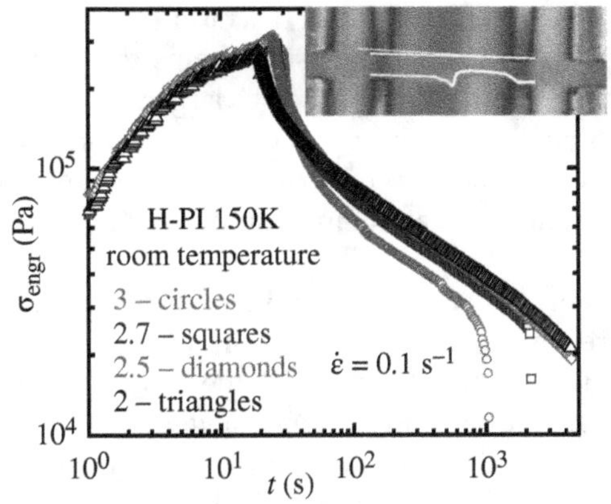

Figure 14.11 Stress growth and relaxation of H-shaped PI150K after step extension of different magnitudes. Breakup occurs only for Hencky strains larger than 2.5. *Source:* Adapted from Sun *et al.*, 2015 [4].

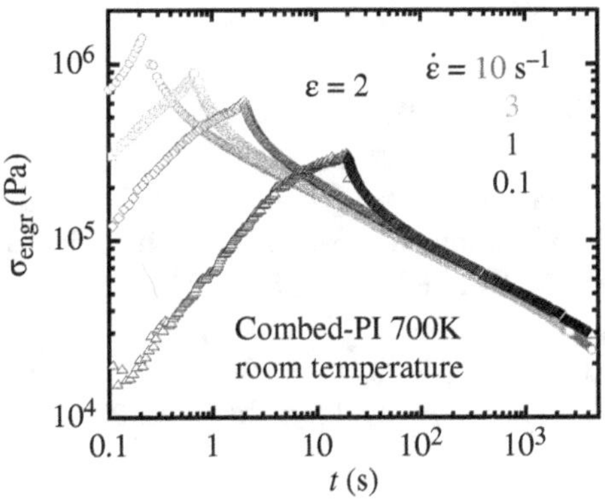

Figure 14.12 No failure after step extension for comb-PI 700K for step Hencky strain of 2, produced at four different Hencky extension rates. *Source:* Adapted from Sun *et al.*, 2015 [4].

Figure 14.10 shows the difference between linear and branched polymers. The small circles represent the stress relaxation behavior of a linear polystyrene with molecular weight of 10^6 g/mol and the larger symbols show quiescent relaxation of a centipede–polystyrene melt up to Hencky strain $\varepsilon = 2.0$. At still larger magnitudes, namely, $\varepsilon = 2.5$ (diamonds) and 3.0 (triangles), the step-extended centipede–PS does undergo elastic failure, causing the tensile stress to vanish. The time required for failure to take place is impressively long. We speculate, based on the stress level at long times, that the entanglement associated with the chain backbone is continuingly weakening during the long "induction."

Different chain architectures that contain sufficient long-chain branching are equally effective in deferring elastic breakup after large stepwise extension. Star-shaped chains with one branching point do not qualify as a long-chain branched polymer because they cannot resist elastic breakdown and actually behave as if they were linear chains. There must be at least two branch points in a polymer before we encounter long-chain-branching behavior. This classification of branching is based on the nonlinear rheological responses of the polymers in question. Figure 14.11 shows that indeed a polymer with minimal long-chain-branching characteristics, for example, an H-shaped polyisoprene, can effectively postpone elastic breakup. The H-polymer suffers the elastic failure only when Hencky strain ε is larger than 2.5, corresponding to a stretching ratio $\lambda = 12$, which is a tremendous degree of extension.

Apart from the criterion of minimal two branch points, to qualify as having long-chain branching, the polymer's side chain obviously needs to be sufficiently long. It is rational to suppose that the side chains need to be long enough to participate in chain entanglement. If side chains are too short to form entanglements, they can slow down the retraction of the backbone but cannot erect an effective entropic barrier against retraction. Let us take the example of Fig. 14.9(c). For polyisoprene, a side chain having $M_n = 20$ kg/mol is apparently sufficiently long to have a long-chain branching effect, judging from the polymer's ability to resist elastic breakup during stress relaxation from large stepwise extension. Specifically, Fig. 14.12 shows that with three such side chains the combed-PI 700K was able to resist any elastic breakup for extension magnitude as large as $\varepsilon = 2$. At present, it remains unknown whether long-chain branching is sufficient to prevent nonquiescent relaxation after large step strain. Hopefully, future studies can be carried out to demonstrate the effects of long-chain branching on shear strain localization, including wall slip, either during startup or after step shear.

In this chapter, we discussed the concept of finite cohesion and showed a sizable linear-response regime. While the effect of long-chain branching to delay or suppress localized yielding of entangled polymers is conveniently explored based on melt extension tests, there has not been any systematic demonstration of whether and how long-chain branching delays the localized breakup of the entanglement network after large step shear, either at the polymer/wall interface or in the bulk. The future work should study whether long-chain branching can increase the threshold strain below which quiescent relaxation prevails.

References

1 Ravindranath S, Wang S-Q. What are the origins of stress relaxation behaviors in step shear of entangled polymer solutions? Macromolecules. 2007;40(22):8031–9.
2 Boukany PE, Wang S-Q, Wang X. Step shear of entangled linear polymer melts: new experimental evidence for elastic yielding. Macromolecules. 2009;42(16):6261–9.
3 Wang Y, Boukany P, Wang S-Q, Wang X. Elastic breakup in uniaxial extension of entangled polymer melts. Phys Rev Lett. 2007;99(23):237801.

4 Sun H, Lin PP, Liu GX, Ntetsikas K, Misichronis K, Kang N, et al. Failure behavior after stepwise uniaxial extension of entangled polymer melts. J Rheol. 2015;59(3):751–67.
5 Sun H, Ntetsikas K, Avgeropoulos A, Wang S-Q. Breakdown of time-temperature equivalence in startup uniaxial extension of entangled polymer melts. Macromolecules. 2013;46(10):4151–9.
6 Cheng S, Lu Y, Liu G, Wang S-Q. Finite cohesion due to chain entanglement in polymer melts. Soft Matter. 2016;12(14):3340–51.
7 Doi M, Edwards SF. The theory of polymer dynamics. Oxford University Press, New York; 1988.
8 Graham RS, Likhtman AE, McLeish TCB, Milner ST. Microscopic theory of linear, entangled polymer chains under rapid deformation including chain stretch and convective constraint release. J Rheol. 2003;47(5):1171–200.
9 Liu GX, Ma HW, Lee H, Xu HD, Cheng S, Sun H, et al. Long-chain branched polymers to prolong homogeneous stretching and to resist melt breakup. Polymer. 2013;54(24):6608–16.

Part IV

Emerging Conceptual Framework and Beyond

Having now reviewed the essential phenomenology on the nonlinear rheology of entangled polymeric liquids, in Part 4 we propose a conceptual framework that explains the phenomenology. On this new molecular foundation, we can build a unifying account of phenomena ranging from shear strain localization (e.g., wall slip and shear banding) during startup shear to elastic breakup after stepwise extension. Besides heuristically explaining how chain entanglements respond to large deformation, we discuss the yielding during startup deformation and the elastic yielding after termination of large external deformation. Chapter 15 briefly describes homogeneous entanglement and disentanglement. Chapter 16 first reviews the standard tube model, indicating that the tube model suffers the inherent difficulty that both chain deformation and accompanying stress are evaluated on a single length scale, that is, the tube segment. Chapter 16, the core of Part 4, then presents the new paradigm for nonlinear polymer rheology in terms of three molecular forces. The chapter goes on to interpret the primary rheological behaviors in terms of the new framework. Beyond the phenomena presented in Chapters 6–15, there are a number of emerging and tentative directions. These topics and directions are summarized in Chapter 17. To reiterate the distinction between old and new treatments of nonlinear rheology of entangled polymers, Chapter 18 discusses the challenges facing the standard paradigm for nonlinear polymer rheology from both theoretical and experimental standpoints, drawing insights from the phenomenology presented in Chapters 6 through 15 and theoretical discussions in Chapter 16. The connection between the present treatment of nonlinear polymer rheology and well-known difficulties in polymer processing is discussed in Chapter 19. We close the book in Chapter 20 with a brief indication of the future directions.

Nonlinear Polymer Rheology: Macroscopic Phenomenology and Molecular Foundation, First Edition. Shi-Qing Wang.

Companion Website: www.wiley.com/go/nonlinearpolymerrheology

15

Homogeneous Entanglement

By "homogeneous deformation," we mean that under shear and extension, the state of polymer entanglement remains spatially uniform. In the linear response regime, the equilibrium state of entanglement is left nearly intact during slow or small-strain deformations. Nonlinear responses involve fast deformations and large strains, that is, for $Wi > 1$ and large values of γ or ε. There is ample evidence that, during startup shear with $Wi \gg 1$, the initial elastic deformation does not last over a period as long as the terminal relaxation time τ, because the yield strain γ_y is much smaller than Wi. Equivalently, the time t_y corresponding to the yield strain is much shorter than τ:

$$t_y = (\gamma_y / Wi)\tau \ll \tau. \tag{15.1}$$

The transition from elastic deformation to flow may not always retain homogeneous deformation, contrary to assumptions in past textbooks on polymer rheology. Until we contemplate seriously the microstructure of viscoelastic materials, here entangled polymeric liquids, we cannot be certain that the state of polymer entanglement can uniformly respond to a fast startup deformation. Because it is plausible that there are inhomogeneous responses, in which during and after a deformation the state of polymer entanglement becomes spatially nonuniform, it is challenging to develop a quantitative theoretical framework of nonlinear responses. It is appropriate to begin our theoretical development with a discussion of homogeneous entanglement.

15.1 What Is Chain Entanglement?

Throughout the book up to this point, we have freely used the phrase "entanglement" without being specific as to what we thought entanglements are or how they behave. We discussed the entanglement concept from various perspectives, including rheological consequences of entanglement (Section 2.2.1), the packing model for the onset of chain entanglement (Section 2.2.3.2), and the entanglement dynamics according to the reptation and tube models (Section 2.3). But what is the microscopic origin of polymer entanglement? How can we describe it in mathematical terms? As a topological question, it seems unknown whether or not effective tools are available from the field of topology in mathematics.[1] On the other hand, polymer physicists have recently attempted to quantify chain entanglement in the context of polymer rheology. This research direction has been pursued notably by Milner and coworkers[2–7] and Likhtman and coworkers.[8,9]

Without understanding where entanglement comes from, we cannot explain how it disappears in response to large deformations. Although there is no rigorous theory that analytically describes the entanglement of linear flexible chains, we can speculate at a conceptual level how entanglements arise. Let us imagine a polymer melt magnified by a factor of a million (from

Nonlinear Polymer Rheology: Macroscopic Phenomenology and Molecular Foundation, First Edition. Shi-Qing Wang.

Companion Website: www.wiley.com/go/nonlinearpolymerrheology

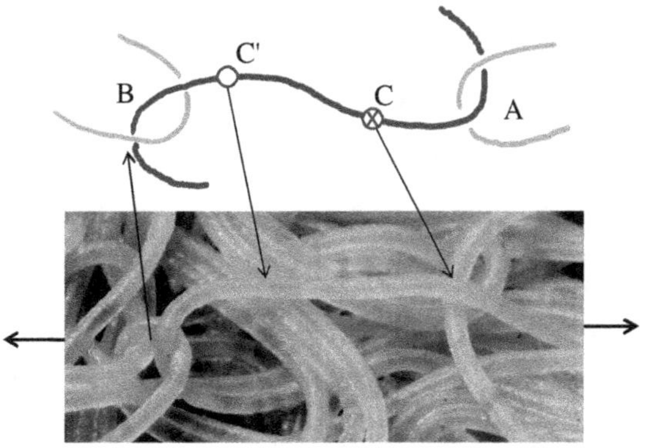

Figure 15.1 Magnified view of Fig. 2.8 showing the mechanically strong entanglements at B and (extrapolated) A due to pairing of hairpins versus the interchain confinements at C and C′.

nanometers to millimeters). The melt may look like the picture in Fig. 2.8. If we zoom in, the picture of a crystal-noodle bowl in Fig. 2.8 looks like Fig. 15.1, which shows the chains are locally intertwined, symbolized by the noodle strands. The topological relationships between the chains are determined by the intersegmental uncrossability. This intermolecular uncrossability is due to the excluded volume effects; chains cannot pass across each other. When chains are short, thermal movements make the uncrossability transient and inconsequential. When chains are sufficiently long and densely packed, they are unable to move in directions transverse to their backbones due to the uncrossability; they stay trapped for a long time until the reptation rearranges the intermolecular packing.

Entanglement points can be defined along a given test chain. These entanglement points are obviously not uniform in character[8,10] and, therefore, are not equivalent. For example, in Fig. 15.1, entanglement C is "weaker": The chain at entanglement point B has less freedom to move transversely than it does at entanglement point C. Thus, at C the test chain can move anywhere but downward toward the perpendicular chains, while the chain at B cannot move anywhere but to the left. Transiently, how much the test chain can move at C depends on how mobile the other surrounding chains are. Similarly, although uncrossability also occurs at C′, this test strand does not stay localized due to C′. A description of entanglement needs to capture the "caging" effect and determine the length scale beyond which the strand is transiently immobilized. For such a complex problem, it would be challenging to do better than hand-waving. For rod-like polymers, Sussman and Schweizer have made a breakthrough[11,12] to account for effects of intermolecular uncrossability on rheological behavior, as briefly mentioned in Section 16.8.

The topological relationship among neighboring chains also depends on the mode and directionality of the external deformation. In presence of extension, as indicated by the two horizontal arrows in Fig. 15.1, there obviously cannot be an affine deformation for the subchain between B and C. The "entanglement" that may be operative at C could prevent free Rouse-like diffusion in quiescence but is clearly irrelevant in fast extension. Thus, the definition of chain entanglement appears to vary from that for linear response to that for nonlinear response. Since the main focus of this book is nonlinear behavior, it is necessary to depict the entanglements that give rise to networking, for example, the type B entanglement in Fig. 15.1. Imagine there is another similar entanglement beyond the field-of-view of the photo, sketched as A. Affine deformation is possible for the strand defined by the pair of entanglement points at A and B in Fig. 15.1. Since a significant population of entanglement strands is present for polymers of sufficiently high molecular weight, entangled polymers can presumably deform the same way that a cross-linked rubber would, even to large strains, as long as the entanglements are not destroyed by the growing intrachain elastic retraction force. As explained in Chapter 16, an entanglement is active and meaningful when the two chains at A (or B) are locked in place and unable to slide freely relative to each other. Disentanglement occurs when there is irreversible mutual sliding at either A or B. Therefore, a correct interpretation of entanglement should be dynamic rather than static. In terms of the crystal-noodle image, if one pulls sufficiently hard with the chopsticks (corresponding to having fast large deformation), the noodles are pulled out of their entanglements with the rest of the dish and toward the eater's mouth. We should note that static picture of the topological packing cannot reveal whether an effective entanglement is present or not and whether it is undergoing disentanglement.

15.2 When, How, and Why Disentanglement Occurs?

The phenomenological task of rheology is to determine mechanical responses and the corresponding deformation fields. For simple shear produced by the displacement of the confining boundaries, homogeneous deformation is usually assumed to prevail at all times, as illustrated in Fig. 1.1(a) and (d). Indeed, commercial shear rheometric instruments are all designed on the basis of this assumption. Past comparisons between theory and experiment are in general predicated on the condition of shear homogeneity.

Homogeneous deformation holds in the linear response regime. But the assumption of homogeneous deformation is questionable for nonlinear responses. This assumption of shear homogeneity still seems *necessary* in order to advance scientific investigations: As shown in Chapters 1 and 4 (e.g., in Section 4.1.1), the state of deformation can be simply controlled only if there is shear homogeneity. When the homogeneity assumption holds, the strain field and the velocity field are known *a priori*, and the mission of rheology reduces to mere measurements of the mechanical responses. The assertion of shear homogeneity is also *convenient*; it allows the macroscopic mechanical characterization to determine intrinsic materials properties. This premise of practical rheology seems valid, according to the reasoning given in Section 4.1.1.

However, for well-entangled polymer solutions or melts, shear banding is sometimes found in either rate- or stress-controlled mode, as shown in Chapter 9. There must therefore be a flaw in the reasoning provided in Section 4.1.1. There is no *a priori* basis for claiming that a one-to-one correspondence between σ and $\dot{\gamma}$ is guaranteed to exist. For entangled polymers, the argument presented in Section 4.1.1 is demonstrably wrong. At least transiently, when shear stress overshoot occurs, there is no one-to-one correspondence.

The shear homogeneity assumption holds so long as polymer entanglement remains uniform on macroscopic scales. Upon startup shear with $Wi \gg 1$, how does the state of entanglement respond? When does a temporary solid with a terminal relaxation time τ cease to undergo affine deformation? When does it begin to flow like a liquid? Must the final liquid flow homogeneously? These are the central questions in entangled polymer rheology that the available textbooks and monographs have avoided. No existing book asks when and how an entangled polymeric fluid yields and stops being deformed affinely. Chapter 16 will explain why and how affine deformation ceases, and how and why yielding takes place as consequence of chain disentanglement.

The next question naturally follows: When can the transition from the fully entangled equilibrium state to a state of disentanglement take place homogeneously? There is plenty of evidence, as reviewed in Chapters 9 and 10, that shear inhomogeneity not only occurs after the stress overshoot in the initial stage of startup shear but can even prevail in prolonged shear. Wall slip through chain disentanglement at the polymer/wall interface, the most extreme form of strain localization on a molecular length scale, is prevalent. The existence of interfacial slip of measurable magnitude explicitly invalidates the shear homogeneity assumption for sufficiently entangled polymers. As demonstrated in Chapter 6, the ratio of the maximum extrapolation length b_{max} to the sample thickness H determines the magnitude of the correction to the shear field. As discussed in Section 9.2.1, monodisperse melts and solutions made with small-molecule organic solvents are most prone to wall slip. Shear inhomogeneity including wall slip can be reduced by optimizing the molecular weight distribution. This chapter discusses other conditions, beyond the obvious condition of linear response, under which shear homogeneity prevails.

15.3 Criterion for Homogeneous Shear

Upon startup deformation, the stress response is initially linear in the deformation, approximately following the ideal rubber elasticity formulas: $\sigma = G_{pl}\gamma$ for shear and $\sigma_{engr} = G_{pl}(\lambda - 1/\lambda^2)$ for extension. If the deformation involves $Wi < 1$, the elastic response naturally ceases when molecular diffusion rearranges the original state of entanglement which occurs on the time scale of the reptation time τ. Such voluntary "disentanglement" smoothly and homogeneously removes the potential barrier against deformation, and thereby allows a uniform flow state to be established.

The following chapter is devoted to a detailed theoretical discussion of how and why disentanglement takes place for $Wi > 1$. The first phenomenological sign of disentanglement is the "strain softening" illustrated in Fig. 15.2, that is, the downward deviation of the stress from the linear dependence of stress on strain. The properties of polymer entanglement, for example, the strength and size of a typical entanglement strand, vary widely. It is plausible that some entanglements vanish well before others do, making the stress response soften. Since the surviving entanglement points still permit further affine-like deformation of the remaining entanglement strands, the stress can still grow. A balance is reached at the stress maximum when the ongoing chain disentanglement, that is, the loss of entanglement strands with increasing strain, cancels the increased retraction force due to increased chain deformation. Beyond the maximum, increased strain does not lead to any increase of the overall stress, even though the surviving entanglement strands can still undergo more deformation as the macroscopic strain further increases.

The continuation of entanglement loss beyond the global yield point at the stress maximum could lead to shear banding. After the yield point, let us suppose one monolayer suffers disentanglement so that this layer can be treated as having a much reduced modulus G_s. The monolayer disentanglement could produce a discontinuity in the strain field as depicted by Fig. 15.3. The magnitude of the "crack" is characterized by a displacement

$$x_s = \gamma_s l_{ent} \tag{15.2}$$

over a thickness of a molecular scale that, we assume, is the entanglement spacing l_{ent}. Whether this failure results in a measurable drop in the shear stress depends on how much it affects the

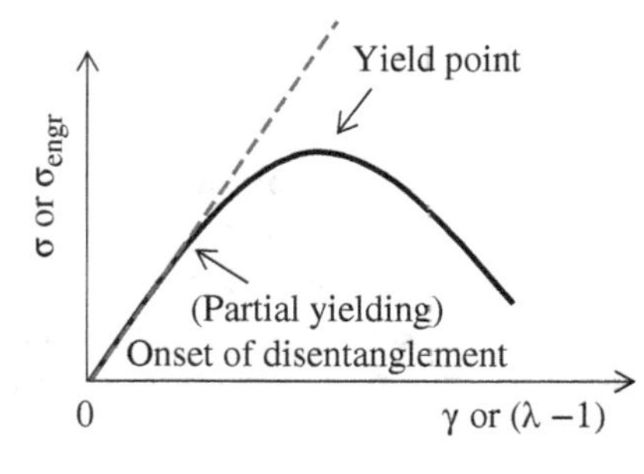

Figure 15.2 Illustration of the stress response to either startup shear or extension, showing partial yielding that plausibly indicates loss of some entanglement strands. According to Maxwell [13, 14], the maximum is a sign of yield.

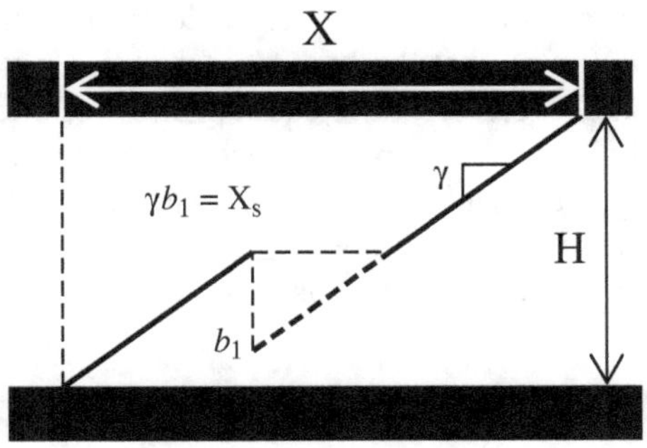

Figure 15.3 Illustration of shear strain localization during continuous shear in terms of the strain field, showing a "crack" (internal slip) of length x_s.

strain field. It is reasonable to measure the "crack" size x_s against the total displacement X, that is, to have the ratio

$$x_s/X = x_s/(\gamma H + x_s) = (b_1/H)/(1 + b_1/H), \tag{15.3}$$

where b_1, defined as an extrapolation length in Fig. 15.3, is given by

$$b_1 = x_s/\gamma = (\gamma_s/\gamma)l_{ent} = (G_{pl}/G_s)l_{ent}. \tag{15.4}$$

Here we have

$$\gamma_s/\gamma = G_{pl}/G_s, \tag{15.5}$$

which is a statement of stress continuity. Equation (15.3) is of the same form as Eq. (6.4a). In terms of b_1, we have $x_s/H = (b_1/H)\gamma = (\gamma_0 - \gamma)$ as an estimate of the normalized "crack" size, where $\gamma_0 = X/H$ is the nominal strain.

If any strain discontinuity takes place in entangled polymeric fluids, its influence is quantified in Eq. (15.3) in terms of b_1. Since G_{pl} does not vary with the degree of chain entanglement, that is, it is independent of the polymer molecular weight M, the expression Eq. (15.4) for b_1 shows little dependence on M, unlike the slip length b of Eq. (6.2), which depends strongly on M. The preceding solid mechanics analysis indicates that the correction due to disentanglement would not depend on the level of polymer entanglement, contrary to the particle-tracking velocimetry (PTV) observations that the level of chain entanglement determines whether shear banding is observable or not. Thus, the preceding analysis to describe a disentangled layer in terms of a finite modulus G_s is incorrect.

One effective way to account for the influence of monolayer disentanglement on the strain field after step strain has been provided in Eq. (12.8). For ongoing shear, the way to quantify the structural discontinuity due to localized chain disentanglement is to evaluate the correction to the velocity field in terms of the slip velocity, as shown in Fig. 9.2. Here we provide an alternative to the discussion surrounding Eq. (9.3). The magnitude of the correction to velocity field due to internal slip is characterized by the ratio

$$V_s/V = V_s/(\dot{\gamma}H + V_s) = (b/H)/(1 + b/H), \tag{15.6}$$

where the extrapolation length b is defined by Eq. (6.2). Equation (15.6) resembles Eq. (6.4a). We note that in the presence of internal slip the local shear rate $\dot{\gamma}$ is reduced from the nominal shear rate V/H by a factor

$$\dot{\gamma}/(V/H) = 1/(1 + b/H) \tag{15.7}$$

where use has been made of Eq. (15.6). Shear homogeneity should prevail when a monolayer of full chain disentanglement does not result in any noticeable change in the velocity field, that is, when the right-hand side of Eq. (15.6) is much smaller than unity, which occurs for $b/H \ll 1$.

In such a case, the shear field is little modified by chain disentanglement in one monolayer, ensuring that the other layers of the sample continue to experience the same shear condition and, therefore, have equal chances to undergo disentanglement. Although Eq. (15.6) generally holds true, it cannot be used to make an *a priori* estimate of V_s/V because b is undetermined. However, if $b_{max}/H \ll 1$, we must have $b/H \ll 1$ so that the bulk shear rate is close to the nominal shear rate according to Eq. (15.7). As described in chapter 6, the value of b_{max} can be theoretically estimated in most cases.

15.4 Constitutive Nonmonotonicity

Experiments indicate that (a) weakly entangled polymer solutions show no shear inhomogeneity during startup (cf. Fig. 9.1(b)), (b) moderately entangled polymer solutions show transient shear inhomogeneity (cf. Fig. 9.6), and (c) only strongly entangled solutions show shear banding in the long time limit (cf. Fig. 9.8(a) and (b)). Historically, there has been no rational understanding of why the state of entanglement does not homogenize across the system thickness in the eventual flow state. If transient shear banding means that there is no constitutive nonmonotonicity in the stress versus rate relationship, why does transient banding take place at all? When shear banding seems to persist forever, as found for well-entangled polymers, is it due to constitutive nonmonotonicity?

Treating shear banding as a continuum phenomenon, steady shear banding has commonly been interpreted as being due to constitutive nonmonotonicity.[15,16] According to the conventional wisdom, for different shear rates to coexist along the velocity gradient direction, the steady stress and shear rate should be related by a nonmonotonic function. Conversely, as argued (incorrectly) in Section 4.1.1, a monotonic relationship between stress and rate should guarantee shear homogeneity. As briefly mentioned in Section 9.1.1, the original tube model of Doi–Edwards prescribes constitutive nonmonotonicity. The Doi–Edwards model speculates, as shown in Fig. 15.4, that there would be two shear bands with local shear rates of $\dot{\gamma}_1$ and $\dot{\gamma}_2$, respectively. The sample across a gap distance H splits into two bands according to

$$V/H = \kappa\dot{\gamma}_1 + (1-\kappa)\dot{\gamma}_2, \tag{15.8}$$

where κ represents the fraction of the gap distance H having a local rate of $\dot{\gamma}_1$ so that the normalized thickness (by H) of the second shear band with $\dot{\gamma}_2$ is $(1-\kappa)$. We first note that the prescription of shear banding by the constitutive nonmonotonicity is incomparable with the observation that the system size plays a pertinent role in determining whether shear band should take place or not, as presented at the end of Section 9.1.2.4.

Moreover, the available data, for example, those in Fig. 15.5, contradict Fig. 15.4 and disagree with Eq. (15.8): The local shear rates in both low and high shear bands vary with the imposed nominal rate V/H and never lock onto two constant values. Moreover, in the presence of shear banding the overall shear stress still weakly increases with Wi_{app}, which is inconsistent with a perfectly flat shear stress plateau. An inevitable feature of the constitutive nonmonotonicity is existence of an absolutely flat stress plateau. According to PTV observations of entangled polymer solutions and melts and uniaxial extensional studies of entangled melts, we have to conclude that the shear banding reflects a structural breakdown of the elastic structure, that is, the entanglement network. In terms of this picture of microscopic yielding, it is more straightforward to understand why transient shear strain localization occurs than to explain why shear banding can persist to steady state.

There are other ways to investigate the nature of shear banding using PTV measurements. For example, if shear banding is somehow due to constitutive nonmonotonicity, as first deduced from the properties of the original Doi–Edwards tube model, it should be present in steady

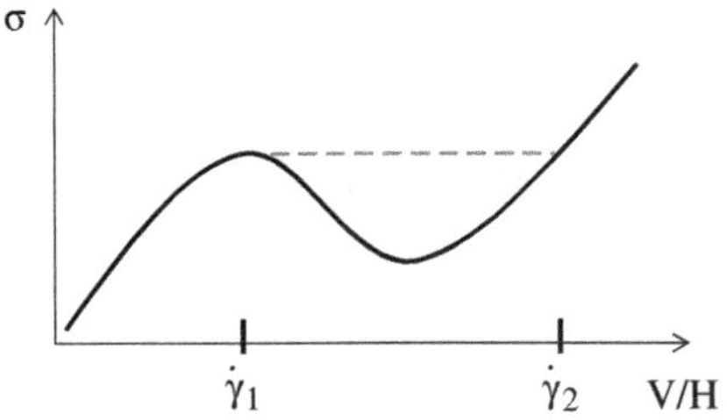

Figure 15.4 Hypothetical nonmonotonicity in shear versus applied rate curve.

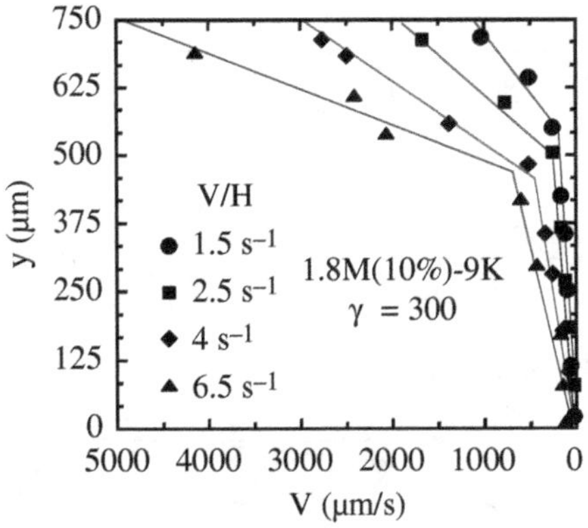

Figure 15.5 PTV observations of shear banding in PB1.8M(10%)-9K, showing different shear rates in each of the fast and slow shear bands at each applied value of V/H, ranging from 1.5 to 6.5 s^{-1}, using data in Ref. [17]. *Source:* Ravindranath *et al.* 2008 [17]. Reproduced with permission of American Chemical Society.

state when V/H falls in the middle of the rate window bounded by $\dot{\gamma}_1$ and $\dot{\gamma}_2$. We examine this scenario in the following section.

15.5 Metastable Nature of Shear Banding

When $Wi < 1$, the entropic potential barrier does not act on time scales comparable to the reptation time τ, because there is molecular diffusion. Elastic deformation is limited to a level of strain given by $Wi < 1$. For $Wi < 1$, elastic retraction forces cannot build up, because it takes times longer than τ to make large deformations. Reptative dynamics renew the entanglement network before the network becomes sufficiently deformed. This trivial observation raises an inspiring question: What happens when an entangled polymer is never given a chance to suffer large elastic deformation?

Instead of applying a sudden startup shear with $Wi \gg 1$, we could begin with $Wi < 1$ and spend a sufficiently long time to ramp up the nominal rate V/H, that is, slowly increase the speed V as depicted by the dashed line in Fig. 15.6(a). An entangled polymer may then uniformly and smoothly adjust its state of entanglement. If V is increasing slowly enough, it should be possible to reach the steady flow state homogeneously, where the effective relaxation time could decrease in proportion to 1/V, that is, to $1/\dot{\gamma}$, as demonstrated in Section 7.3.1. Consequently, the stress response could be monotonic without overshoot as sketched by the dashed curve in Fig. 15.6(b), and the flow field could be specified by a single shear rate instead of shear banding, as illustrated in Fig. 15.6(c).

The protocol depicted in Fig. 15.6(a) recognizes that the conventional concept of $Wi_{app} \gg 1$ refers to an equilibrium sample that has a terminal relaxation time τ. Suppose the external shearing is gradually applied, starting from the condition $Wi < 1$. Enough time is permitted for the system to adjust to the applied shear and reach a steady state before increasing the rate. At each rate and thus in each stage of the slow rate ramp-up, shearing takes place under the weak shear condition. At the end of the rate ramp-up ($t > t_{rru}$), although $Wi_{app} \gg 1$, the

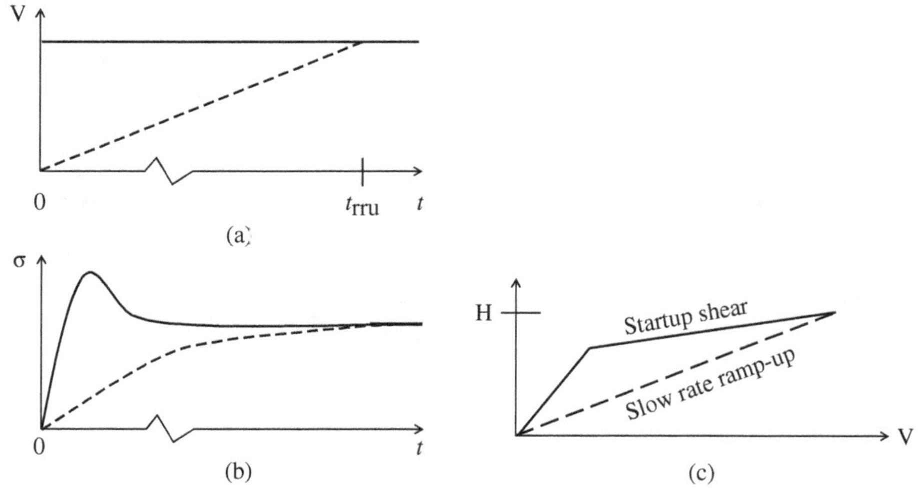

Figure 15.6 (a) Illustration of rate ramp-up in contrast to startup shear where the dashed line shows a gradual ramp-up of the applied speed, eventually reaching the prescribed speed of the moving boundary at time t_{rru} that can be 100 times the terminal relaxation time. (b) Corresponding stress response to the rate ramp-up showing no stress overshoot and characteristic stress overshoot for startup shear with $Wi_{app} \gg 1$. (c) Corresponding velocity profiles of homogeneous shear and shear banding for $Wi_{app} \gg 1$.

effective Weissenberg number, involving an effective relaxation time τ_{eff} that is proportional to the shear rate, is actually always smaller than unity. It is thus predicted that shear banding can be avoided.[18] Specifically, it could be the case that shear banding is only metastable and does not originate from the existence of constitutive nonmonotonicity. If shear banding really originated from nonmonotonicity, shear banding should persist in the steady state, that is, at $t > t_{rru}$, independent of how the condition of $Wi_{app} \gg 1$ was imposed. Because the rate ramp-up protocol involves a great deal of shearing and edge effects can accumulate over time, it is especially challenging to carry out this experiment.

Particle-tracking velocimetric measurements[18] examined the prediction of homogeneous shear, denoted by the dashed line in Fig. 15.6(c). Entangled DNA solutions are ideally suited for this purpose because no edge instability takes place at a stress level as low as several tens of Pascals. The first such experiments were based on a 1.1% DNA solution. As shown in Fig. 15.7(a), when the applied rate increases slowly, reaching $1\,s^{-1}$ after many hours, the shear stress does not exhibit any stress overshoot. Figure 15.7(b) shows that shear banding is essentially eliminated by the slow rate ramp-up while very sharp and pronounced shear banding occurs in response to a sudden startup shear of the same rate.

The data in Fig. 15.7 are based on the same DNA solution as the one studied in Fig. 9.5, which means that shear banding is no longer present in steady state at rates above $4\,s^{-1}$. A second protocol to examine the origin of shear banding, rate quench-down, may then be applied to determine whether shear homogeneity survives at $1\,s^{-1}$ after the sample has been homogeneously sheared at $6\,s^{-1}$. Circles in Fig. 15.7(b) indicate that homogeneous shear prevails at $1\,s^{-1}$. The shear stress is the same as shown by the horizontal line in Fig. 15.7(a) regardless of whether or not there is shear banding and whether the shear involves sudden startup or gradual rate ramp-up.

The conclusion drawn from Fig. 15.7 is rather important. The result appears to rule out constitutive nonmonotonicity as the theoretical explanation for shear banding and challenges the conventional constitutive description of polymeric materials. A typical formulation such as the Kaye–Bernstein–Kearsley–Zapas model[19,20] would relate the stress field and deformation field

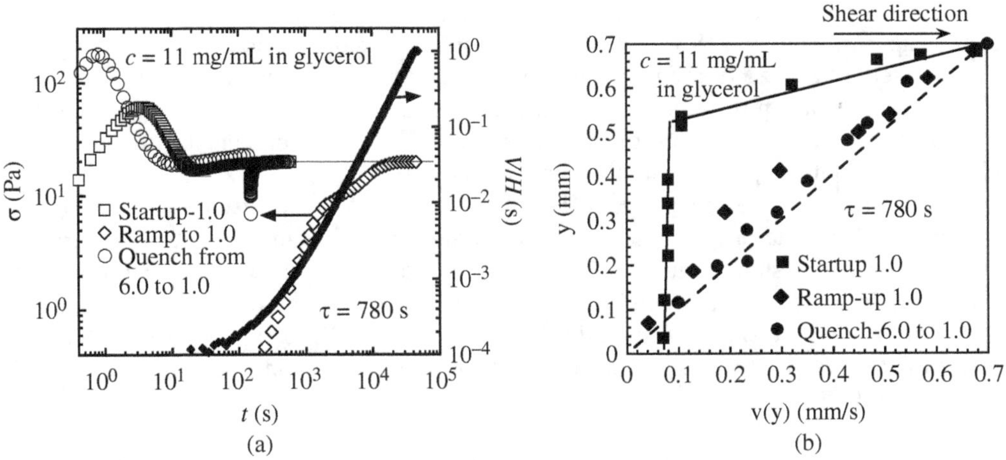

Figure 15.7 (a) Stress responses of an entangled DNA solution (i) to startup shear at an apparent rate of 1 s^{-1} (squares), (ii) upon rate quench to 1 s^{-1} after startup shear at 6 s^{-1} for over 100 s (circles), and (iii) rate ramp-up from 10^{-4} to 1 s^{-1} over a period of 4.6×10^4 s (open diamonds), as a function of time. Also plotted is the time-dependent increase of the applied nominal rate V/H against the right-hand-side *Y* axis. (b) Corresponding velocity profiles: (i) Startup at 1 s^{-1} (squares), (ii) rate quench from 6 to 1 s^{-1} (circles), and (iii) slow rate ramp-up after more than 4×10^4 s (diamonds). *Source:* Boukany *et al.* 2010 [18]. Reproduced with permission of American Chemical Society.

through a *memory* function that involves the *equilibrium* relaxation spectrum. It is known from such formulations that the present rheological state, characterized by the stress $\sigma(t)$ and rate $\dot{\gamma}(t)$, should be uniquely determined by a state at all earlier moments $t' < t - \tau$. The present rate ramp-up experiments challenge such a constitutive description, because shear banding is not a unique state. For the same imposed Wi_{app} that drives shear banding to emerge during startup, shear homogeneity can also be established either by slow rate ramp-up or by rate quench from a homogeneous state produced at a sufficiently large rate.

To verify this conclusion, entangled polymer solutions were subjected to a similar investigation.[21] Given the prolonged shear involved in a rate ramp-up experiment, it is necessary to carry out the inquiry using a cone-partitioned plate assembly (cf. Section 5.2). Figure 15.8(a) summarizes three different ways to impose shear onto a well-entangled 10%

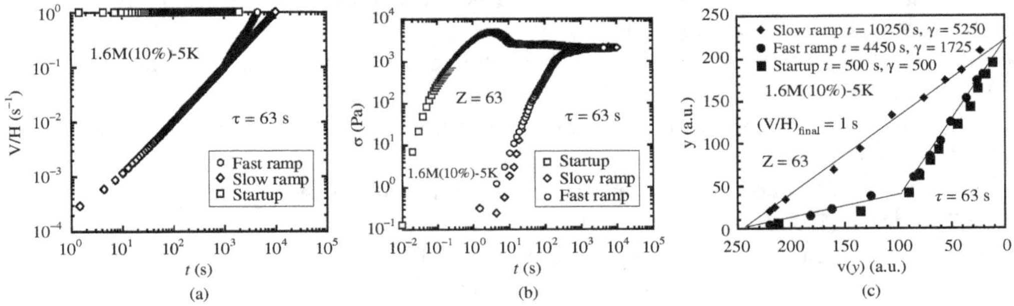

Figure 15.8 (a) Apparent rates as a function of time, involving (i) startup at 1 s^{-1} (squares), (ii) slow rate ramp-up (diamonds) in a time as long as 10^4 s, and (iii) fast rate ramp-up (circles). (b) Corresponding stress responses to (i) startup shear at 1 s^{-1} (squares), (ii) slow ramp-up (diamonds), (iii) fast ramp-up (circles). (c) Corresponding velocity profiles in the respective symbols. All three figures are based on data from Ref. [21]. *Source:* Cheng *et al.* 2012 [21]. Reproduced with permission of American Institute of Physics.

PB solution listed in Table 9.1. Figure 15.8(b), respectively, shows the usual shear stress overshoot upon startup and lack of stress overshoot during slow rate ramp-up. Although the time-dependent stress reaches the same level in the long time limit for the two ramp-up runs, only the slower ramp-up resulted in homogeneous shear as shown in Fig. 15.8(c). Thus, it is much easier for shear banding to show up than to produce shear homogeneity, implying shear banding is more stable than shear homogeneity.

There is also a more recent proposal[22,23] of stress-induced concentration fluctuation as the origin of shear banding. The challenge is to argue that sufficient concentration variation could occur over very short time intervals because nonquiescent relaxation was observed to occur immediately after termination of homogeneous shear. Since strain localization was also observed after stepwise shear of monodisperse styrene–butadiene rubber (SBR) melts that were free of any solvent, the idea of concentration-induced shear banding may not have a wide range of applicability.

In conclusion, under the subject of "homogeneous entanglement," in this chapter we attempted to show (a) when chain disentanglement can take place uniformly in space to ensure shear homogeneity; (b) how available phenomenology on shear banding is inconsistent with the notion of constitutive nonmonotonicity; and (c) how the steady (long-time) shear state can be either inhomogeneous or uniform depending on the strain history under the same final condition.

References

1 Pearl EM. Open problems in topology II: Elsevier; 2011.
2 Qin J, Milner ST. Counting polymer knots to find the entanglement length. Soft Matter. 2011;7(22):10676–93.
3 Bisbee W, Qin J, Milner ST. Finding the tube with isoconfigurational averaging. Macromolecules. 2011;44(22):8972–80.
4 Qin J, Milner ST, Stephanou PS, Mavrantzas VG. Effects of tube persistence length on dynamics of mildly entangled polymers. J Rheol. 2012;56(4):707–23.
5 Qin J, Milner ST. Tube diameter of oriented and stretched polymer melts. Macromolecules. 2013;46(4):1659–72.
6 Qin J, Milner ST. Tubes, topology, and polymer entanglement. Macromolecules. 2014;47(17):6077–85.
7 Cao J, Qin J, Milner ST. Finding entanglement points in simulated polymer melts. Macromolecules. 2015;48(1):99–110.
8 Likhtman AE, Ponmurugan M. Microscopic definition of polymer entanglements. Macromolecules. 2014;47(4):1470–81.
9 Likhtman AE. The tube axis and entanglements in polymer melts. Soft Matter. 2014;10(12):1895–904.
10 Szamel G, Wang T. Long-lived interchain contacts in polymer melts. J Chem Phys. 1997;107(24):10793–8.
11 Sussman DM, Schweizer KS. Microscopic theory of the tube confinement potential for liquids of topologically entangled Rigid Macromolecules. Phys Rev Lett. 2011;107(7):078102.
12 Sussman DM, Schweizer KS. Microscopic theory of topologically entangled fluids of rigid macromolecules. Phys Rev E. 2011;83(6):061501.
13 Maxwell B, Nguyen M. Measurement of the elastic properties of polymer melts. Polym Eng Sci. 1979;19(16):1140–50.

14 Maxwell B. The application of melt elasticity measurements to polymer processing. Polym Eng Sci. 1986;26(20):1405–9.

15 Olmsted P, Radulescu O, Lu C-Y. Johnson–Segalman model with a diffusion term in cylindrical Couette flow. J Rheol. 2000;44(2):257–75.

16 Olmsted PD. Perspectives on shear banding in complex fluids. Rheol Acta. 2008;47(3):283–300.

17 Ravindranath S, Wang S-Q, Olechnowicz M, Quirk RP. Banding in simple steady shear of entangled polymer solutions. Macromolecules. 2008;41(7):2663–70.

18 Boukany PE, Wang S-Q. Shear banding or not in entangled DNA solutions. Macromolecules. 2010;43(17):6950–2.

19 Kaye A. Non-Newtonian flow in incompressible fluids. College of Aeronautics Cranfield; 1962.

20 Bernstein B, Kearsley EA, Zapas LJ. A study of stress relaxation with finite strain. Trans Soc Rheol. 1963;7(1):391–410.

21 Cheng S, Wang S-Q. Is shear banding a metastable property of well-entangled polymer solutions? J Rheol. 2012;56(6):1413–28.

22 Cromer M, Villet MC, Fredrickson GH, Leal LG. Shear banding in polymer solutions. Phys Fluids. 2013;25(5): 051703.

23 Cromer M, Fredrickson GH, Leal LG. A study of shear banding in polymer solutions. Phys Fluids. 2014;26(6): 063101.

16

Molecular Networks as the Conceptual Foundation

In Chapters 6 through 15, we reviewed the key phenomenology on the nonlinear rheological responses of entangled solutions and melts, including shear banding during startup shear and nonquiescent relaxation after step shear, as well as specimen breakup during or after extension. It is clear that entangled polymeric fluids can break down like solids under the condition $Wi \gg 1$. Throughout the book thus far, we have loosely attached new microscopic pictures to these nonlinear rheological phenomena without proper discussion. This practice is hard to avoid because a great deal of the experimental data was collected in response to the emerging picture of nonlinear polymer rheology. Conversely, the experimental observations, many made to test predictions[1] of the emerging molecular network framework,[2] have provided increasing molecular insights. In this chapter, we attempt to present detailed theoretical accounts of the experimental results in Chapters 6 through 15. If chain entanglement initially acts like molecular cross-linking, when during startup deformation does it "dissolve" to terminate the initial elastic rubber-like response? How is the initial elastic deformation replaced by irrecoverable deformation (flow)? Why cannot chain entanglement survive after a sizable stepwise deformation? We are still at the beginning[2,3] of development for a reliable theory of the nonlinear rheology of entangled polymers that can answer these questions from first principles.

The nonlinear rheology of entangled polymers faces first two central questions: (a) How and why does molecular deformation take place? (b) When does affine deformation cease to be true or when does chain deformation cease to increase during startup deformation? Clearly, it may be too simple[4] to treat intermolecular interactions in terms of a smooth tube in which a primitive chain moves like a Rouse chain.

In Section 16.1, we first review the prevailing molecular treatment for polymer rheology, namely, the tube model,[5] in terms of its ansatz and approximations and describe its corresponding pictures. Such a review will help us see its contrast with an alternative framework that qualitatively predicts and explains responses of chain entanglement to fast large deformation. In Section 16.2, we describe the essential ingredients of the alternative conceptual foundation by explaining the origins of three microscopic forces, two of which are entirely foreign to the prevailing theoretical framework (the tube model). In Section 16.3, we show that large step deformation builds high intrachain retraction force to overcome the finite cohesion associated with an entropic barrier. In Section 16.4, we describe our picture of yielding in terms of force imbalance between the driving force, namely, the intermolecular grip force, and the growing intrachain retraction force. In Section 16.5, we show how interfacial yielding occurs by disentanglement of adsorbed chains with unbound chains upon the force imbalance. In Section 16.6, we suggest that long-chain branching should increase the entropic barrier to delay strain localization in both melt extension and shear. In Section 16.7, the occurrence of entanglement–disentanglement transition is interpreted as stress-induced decohesion of the entanglement network that has a finite cohesive strength of entropic origin. Importantly,

Nonlinear Polymer Rheology: Macroscopic Phenomenology and Molecular Foundation, First Edition. Shi-Qing Wang.

Companion Website: www.wiley.com/go/nonlinearpolymerrheology

we indicate in Section 16.8 that Sussman and Schweizer have developed a first self-consistent, force level, dynamic mean field treatment of the nonlinear rheology of entangled polymers. Finally, in Section 16.9, we look forward to future small-angle-neutron scattering studies and molecular-dynamics computer simulation that may provide crucial and differentiating evidence.

Before going any further, we must emphasize that neither the standard theory based on the tube model nor the emerging alternative theoretical framework has received sufficient confirmation based on explicit molecular evidence. Nevertheless, there is a crucial difference between the two models. The recently proposed molecular pictures are rooted in a considerable amount of phenomenology that was simply unavailable at the time when the tube model was invented. When the ability to describe the many-body molecular processes associated with the nonlinear responses is limited, should we throw away essential ingredients of a realistic picture in order to reach a mathematically tractable model? Four decades after the proposal of the tube model, we confront this question for the first time in presence of enormously increased phenomenology. If our objective is to have a basic molecular viewpoint that provides coherent explanations for all known phenomena rather than a mathematical account of quantitative nonlinear rheology that oversimplifies the underlying physics, the choice should be clear.

16.1 Introduction: The Tube Model and its Predictions

The tube model[5–7] is well known in the rheology and polymer physics communities and has been undergoing constant revision to achieve better agreement with experiment. After various improvements,[8,9] the most advanced version of the tube theory was made[12] in 2003. It has currently reached an iconic status. A polymer rheologist's perception of how entanglements respond to large deformation had until recently[4] been dependent on the tube model's conclusions. Despite its extensive applications and impressive popularity, a careful examination[11] of its foundation has rarely been carried out. This introductory section affords us a valuable opportunity to thoroughly explore the structure and construction of the tube model. Except for a brief discussion of the inherent difficulties facing the tube theory,[7,11] there has not been any further systematic critical examination of the foundation of the tube theory by those who routinely apply the model to study nonlinear rheology of entangled polymeric fluids.

In any theoretical description of complex behavior such as the nonlinear responses of entangled polymers, approximations are necessary to simplify the construction and formulation of one's theoretical model. To know what approximations to make, we must have a good idea of what the full physical picture should be. Only then can we be more confident that our simplifications do not throw away the essential physics. In retrospect, if we knew that the nonlinear response of entangled melts to startup shear involves yielding, namely, a transition from elastic affine-like deformation to irrecoverable deformation, as Bryce Maxwell first suggested,[12] would we have proposed a single-chain model such as the tube model to simplify the treatment of the interchain interactions in a smoothed-out manner, namely, representing the presence of entangled surrounding chains by a passive invisible and indestructible tube?

There is little doubt that the tube model is a remarkable theoretical accomplishment in theoretical polymer physics, providing a quantitative description of the linear viscoelasticity of entangled polymer melts. In the linear response regime, the entanglement structure is intact, and the effect of entanglement interactions may be captured by invoking the Edwards tube description.[13] Can one extend the treatment to describe what happens to the entanglement network when it undergoes fast large deformation? Without addressing how the entanglement network may undergo decohesion, Doi and Edwards proceeded to extend the tube model[14–16] into the strongly nonlinear regime[17] and seemed to be able to depict quantitatively many observed rheological phenomena.[10,18] However, the new phenomenology summarized in Chapters 6 through 15 suggests that an alternative theoretical description may be more realistic and useful. In this new description, it is necessary to explore such concepts as the *finite cohesion* of the entanglement network, *elastic yielding* after step deformation, and *yielding* of the entanglement network during ongoing deformation. Before presenting a thorough discourse on this competing theoretical framework, we first dissect the tube model by identifying the simplifications and assumptions that have been applied to construct this model. Other variations (often known as slip-link models)[19–21] have been invented to describe phenomenologically the nonlinear rheology of entangled polymers; these have been briefly reviewed by Graessley.[22] We will omit any discussion of these different theoretical treatments because they are comparable to the tube model in terms of their rigor, for example, they share the same basic deficiency: they have a lack of self-consistency and do not work at the force level.

16.1.1 Basic Starting Points of the Tube Model

Given what we know about entangled polymers, we have to choose how to characterize the interchain interactions. The tube model made its choice nearly four decades ago, inspired by (a) the tube idea of Edwards[13] who treated the environment of a cross-linked rubber in terms of a tube-like constraint to an immersed test chain and by (b) de Gennes' derivation[23] of reptative diffusion of such a test chain in an Edwards tube. The model avoids the task of depicting chain entanglement explicitly. The tube model instead assumes that entanglements are somehow present whenever the concentration is sufficient large and the chain length is sufficiently long.[1] Specifically, the tube model assumes as its principal approximations that (a) interchain interactions merely provide a tube-like constraint to the test chain, (b) affine deformation of the tube prevails, and (c) only intrachain deformation contributes to the stress. Approximation (c) was inspired by the successful classical rubber elasticity theory.[24,25] Thus, the tube model bypasses the challenge of directly answering how affine deformation arises, in the absence of cross-linking. As a result, the tube model does not explain when affine deformation ceases. Finally, the model assumes (d) that after rapid large deformation the test chain undergoes retraction along the primitive path inside its tube on the Rouse time scale so that there is significant conformational recovery toward the equilibrium state. Under (d), the tube model speaks about whether chain deformation is only orientational or stretching, depending on the rate of deformation. The corresponding stress decomposition into stretching and orientation arises because a test chain is assumed to live in a smooth barrier-free tube.

In the tube model made possible by these four approximations, a test chain has restricted transverse degrees of freedom defined by the tube diameter. In the linear response regime, namely, under infinitesimal deformation, the entanglement network is nearly intact and close to equilibrium, making it a good approximation to adopt a smooth tube ansatz. Under large fast deformation, however, it is perhaps not difficult to be convinced that the interchain interactions become more point-like and heterogeneous than they are in quiescence,[26,27] and can be different from the picture of the smooth tube confinement. Moreover, the assumption that polymer coils have barrier-free Rouse motion inside tubes has serious consequences. Experimental evidence to support this ansatz remains scarce. Since assumption (d) is the key premise of the tube model, we first examine the conclusions due to this simplifying assumption in the following section.

1 Entanglement is actually a dynamically emergent effect due to chain connectivity and interchain excluded volume forces. The tube model made no attempt to prescribe or speculate about how the entanglement molecular weight M_e should be related to other molecular characteristics such as the packing length, a concept developed in the packing model (reviewed in Section 2.2.3.1) to explain the emergence of entanglement.

16.1.2 Rouse Chain Retraction

By assuming that a test chain is only transversely restricted and otherwise free to undergo Rouse dynamics inside the tube, the tube model prescribes a two-step stress relaxation and decouples the stress into the two components. Time-strain separability for large step strain is an inherent property of the tube model. For startup deformation in the nonlinear response regime, the smoothed-out simplification of the inherently heterogeneous interchain interactions, namely, assumption (a), is combined with the ansatz of Rouse dynamics in the tube, namely, assumption (d), to ensure that affine deformation lasts only for a time on the order of the Rouse time τ_R. The tube model assumes that a test chain can undergo deformation, enslaved by the affinely deformed tube. Since the test chain obeys Rouse dynamics inside the tube, after affine deformation it retracts to its equilibrium contour length and the coil size shrinks correspondingly. Specifically, upon startup deformation at $t = 0$ chain retracts at a strain of Wi_R corresponding to $t = \tau_R$. For $Wi_R \ll 1$, the tube model claims there is no chain stretching and the stress buildup is due to chain orientation. But of course chain stretching may be present on scales larger than the tube segment. With $Wi_R > 1$, chain retraction leads to stress decline at τ_R. Thus, the strain at the peak of either shear or engineering stress would be comparable to Wi_R, namely, γ_{max} and $\varepsilon_{max} \sim Wi_R$. This expectation from the tube model contradicts the experimentally observed 1/3 scaling law of γ_{max} and $\varepsilon_{max} \sim (Wi_R)^{1/3}$ (cf. Figs. 7.8(b) and 7.31).

16.1.3 Nonmonotonicity due to Rouse Chain Retraction

The tube model is claimed to be the most successful theory in the modern polymer science and is appealing not only because its construction is straightforward but also because it asserts to describe all nonlinear responses.[10,18] To appreciate its success and its iconic status, one has to recall the situation before the tube model era, when only transient network models were available. The transient network models appeared rather *ad hoc* as they postulate a rate-dependent lifetime $\tau(\dot{\gamma})$ to model nonlinear behavior such as shear thinning. With proper adjustment, stress overshoot can be described by a transient network model. However, in the absence of any proper understanding of what the transient junctions represent and how they are removed due to external deformation, the approach of transient network models is rather unsatisfying. In particular, it is not clear that such a phenomenological description has any predictive power.

On the surface, there is nothing *ad hoc* about the tube model's account for the shear stress overshoot. It shows up so naturally that one tends to forget that the tube model is nearly as *ad hoc*, as explained in Section 16.1.3.2. Specifically, it is *ad hoc* to assume that the test chain is Rouse like in a smooth tube. It is an unproven simplification that an imaginary tube can restrict transverse motions and yet permit barrier-free longitudinal motion along the tube. Along with the perceived chain retraction, the coil size decreases considerably, as shown in the case of step strain.[15] Once Rouse chain retraction is taken for granted and mistakenly regarded as being non-*ad hoc*, one starts to equate the tube model to reality. However, on close examination, we find that the tube theory is full of self-inconsistencies. Here, we list five properties of the tube model and indicate experimental observations that appear to contradict these properties.

16.1.3.1 Absence of Linear Response to Step Strain

Because the only pertinent length scale is the tube segment in the tube model, both chain deformation and resulting stress are evaluated at the level of the primitive chain whose repeated unit is the tube segment. Because of the Rouse chain retraction ansatz and the assumption that the retraction force is proportional to the contour length (e.g., Eq. (7.99) in Ref. [5], the response of an entangled melt to step strain is always nonlinear, as shown in Section 14.1. With increasing γ or λ, there would be growing chain stretching contribution $\sigma_{cs}(\gamma)$ or $\sigma_{cs}(\lambda)$ to the total stress. Thus, during the stress relaxation from step strain, since $\sigma_{cs}(\gamma)$ or $\sigma_{cs}(\lambda)$ vanishes at the Rouse time τ_R according to the ansatz (assumption (d)), the relaxing stress shows different time dependence at different strain magnitudes. Consequently, the relaxation modulus $G(t, \gamma)$ or $G(t, \lambda)$ explicitly depends on the magnitude of the step strain, as shown in Figs. 14.7 and 14.8. The tube model concludes that entangled melts would not respond to step strain in a manner that is independent of the step strain magnitude. However, careful experimental determination of $G(t, \gamma)$ and $G(t, \lambda)$ shows that the relaxation modulus has the same time dependence as the equilibrium relaxation modulus $G_{eq}(t)$ as long as $\gamma < 0.6$ and $\lambda < 1.5$. This disagreement between experiment and theory suggests that it is problematic to have decomposition of stress into orientation and stretching and evaluate the relaxing stress only at the tube segment scale. Unfortunately, the next four key features of the tube model are the consequences of the simplification to account for chain deformation and stress on the tube segment scale.

16.1.3.2 Stress Overshoot upon Startup Shear

The original tube model of Doi and Edwards[14–17] has no difficulty in producing shear stress overshoot during startup shear with $Wi > 1$. It actually predicts the overshoot at strain $\gamma_{max} = 2.2$, independent of Wi as long as $Wi < \tau_d/\tau_R$, namely, $Wi_R < 1$. Under this condition, namely, for $1 < Wi < \tau_d/\tau_R$, according to the tube model only the chain orientation contributes to the shear stress. Specifically, σ is given[5] in terms of the orientation function Q_{xy} (cf. Chapter 12)

$$\sigma(\dot{\gamma}, t) = G_{pl} \int_0^t dt' \ \mu(t') \ Q_{xy}(\dot{\gamma} t'). \tag{16.1}$$

Here, G_{pl} is the elastic plateau modulus, μ is a memory function involving the full *equilibrium* relaxation spectrum $\{\tau_p\}$ (Eq. (2.49)), given by $\mu(t') = \partial\psi(t - t')/\partial t'$, where ψ is defined in Eq. (2.48a). $Q_{xy}(\gamma)$ can be numerically approximated by Eq. (12.7). This nonmonotonic function $Q_{xy}(\gamma)$ indicates that when the elapsed molecular strain γ is beyond $\gamma_{max} = \sqrt{5} = 2.24$, further chain orientation results in a lower shear stress. Namely, σ reaches a maximum $\sigma_{max} \cong G_{pl}$ around $\gamma_{max} = 2.2$, as depicted in Fig. 16.1. Remarkably, over a range of Wi just above unity the shear stress overshoot is experimentally observed around a strain of two, in close agreement with the prescription of the tube model. However, σ_{max} usually varies weakly with Wi instead of being a constant. Furthermore, Fig. 7.3 indicates that γ_y, namely, γ_{max}, still does vary appreciably with Wi under the condition of $Wi_R < 1$. Nevertheless, the tube model seems to show the shear stress overshoot as a prediction. The overshoot in the tube model is simply the combined consequence of assumed affine deformation of the tube and postulated free Rouse retraction. If we cannot justify affine tube deformation and Rouse chain retraction, the tube model may not claim that it can describe the stress overshoot for $Wi_R < 1$. In principle, the stress decline may have an alternative origin. For example, it may arise from a breakdown of the entanglement network, as discussed in detail in a subsequent section.

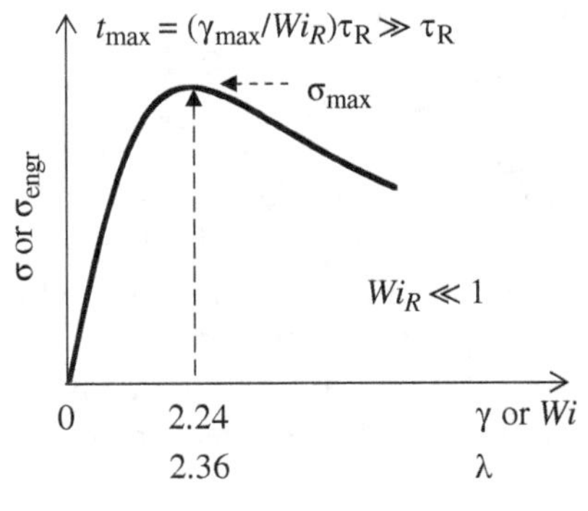

Figure 16.1 Sketch of the tube model prediction, revealing a stress maximum (shear or engineering) upon startup simple shear or uniaxial extension as a function of strain for $Wi_R \ll 1$, on double-linear scales. It is worth noting that $t_{max} \gg \tau_R$ under the condition of $Wi_R \ll 1$.

It is straightforward to observe self-inconsistencies and the *ad hoc* nature of the assumptions in the tube model, arising from the oversimplification that interchain interactions can be represented as a notional tube. For example, at $Wi_R \ll 1$ and $Wi > 1$, the assertion of negligible chain stretching and thus insufficient chain deformation is the assumption and simplification, not a prediction of the tube model. Chain deformation is clearly not affine under this "slow" deformation condition. Yet the tube model maintains that the tube would undergo affine orientation for $Wi_R < 1$ and stress is due to affine chain orientation as shown in Eq. (16.1). In reality how can nonaffine chain deformation produce affine chain orientation?

16.1.3.3 Strain Softening: Damping Function for Stress Relaxation

Stress relaxation after a large step strain is a most straightforward, convenient, and powerful experiment to explore nonlinear response of entangled polymers. Many step strain experiments[28–31] had been carried out before the invention of the tube model. Alarmingly, the data published in 1980 did not agree with Doi's damping function[32] $h(\gamma)$. The widely known agreement between the model and experiment is based on a subsequent experimental study[33] published in 1982. It turns out that only a subset of all available data agrees with the tube model description, as discussed in Chapter 12.[34] Nevertheless, the limited agreement was truly encouraging to the researchers in the field.

It is particularly noteworthy that both experiment and theory exhibit time-strain separability. Both the data and tube model find that, after an earlier decline, the stress relaxation beyond a certain time scale t' ($>t_k$) follows the same time dependence, given by the equilibrium relaxation modulus $G_{eq}(t) = G_{rept}(t)$ of Eq. (2.48a), independent of the strain magnitude. The tube model stipulates that[5]

$$\sigma(\gamma, t > t_k) \cong G_{eq}(t) Q_{xy}(\gamma). \tag{16.2}$$

According to the tube model, after the step strain has caused the tube to shear affinely, chain retraction would occur within the tube around $t_k = \tau_R$ for any value of γ. The residual stress at $t > \tau_R$ is thus entirely due to chain orientation calculated at the tube segment scale. The γ dependence of the normalized shear stress at long times is therefore given by Doi and Edwards[5] the orientation function $Q_{xy}(\gamma)$. The tube model describes the response of an entangled polymeric fluid to step strain as follows: At $t > \tau_R$, namely, after chain retraction, there would still be chain orientation, so some residual stress survives during subsequent relaxation over the reptation time τ_d. The chain orientation relaxes by quiescent reptation dynamics, so that the late-stage stress relaxation obeys $G_{eq}(t) = G_{rept}(t)$.

This theoretical feature of time-strain separability allows one to examine the late-stage stress relaxation by superimposing all stress relaxation curves with a vertical shift according to a universal damping function

$$h_{DE}(\gamma) = Q_{xy}(\gamma)/\gamma = 1/(1 + \gamma^2/5). \tag{16.3}$$

An example of such a shift was presented in Fig. 12.1(b). Indeed, some data[33] can be described by Eqs. (16.2) and (16.3). This quantitative agreement lends strong support for the tube model. The success[33] is widely regarded as a crucial triumph of the tube model.[5] The nonmonotonicity of Q_{xy} in Eq. (16.2), however, seems unphysical, as discussed in Chapter 12. How can the relaxing stress from a step strain having magnitude $\gamma > \gamma_{max} = 2.2$ be actually lower than the relaxing stress from a step strain having $\gamma < \gamma_{max} = 2.2$? Moreover, as surveyed by Osaki[34] and sketched in Fig. 12.2, many data sets do not follow the aforementioned expression for the damping function. Many experimental studies show that the relaxation modulus $G(t, \gamma)$ collapses onto $G_{eq}(t)$ after vertical shifting for $t > t_k$ that are significantly longer[35] than τ_R. When t_k and τ_R differ, the system may have a different mechanism for stress relaxation. In our view, the molecular processes are fundamentally different from anything associated with tube model, as discussed in the following paragraph, inspired by data presented in Chapters 12 and 14.

The stress relaxation behavior obeying Eq. (16.3) is unusual as remarked in Chapter 12 because $\gamma h_{DE} = Q_{xy}$ is already nonmonotonic. As shown in Chapter 12, such nonmonotonicity as an experimental fact suggests that nonquiescent relaxation may have taken place. Besides the types A and B behaviors summarized in Fig. 12.2, some more entangled polymer solutions and melts show type C behavior, in which there is much stronger strain softening than that described by Eqs. (16.2) and (16.3). Many experiments[34–36] have been carried out to show that the damping function is not universal. Interfacial slip is often suspected to cause ultra-strain-softening, namely, the type C behavior.[34] Indeed, some early step strain experiments on polybutadiene melts[37] and solutions[38] may have involved massive wall slip, as mentioned in Chapter 12. In the absence of any direct verification of the assumptions of homogeneous shear and quiescent relaxation, conventional rheometric experiments do not produce more insight. Conversely, the discovery of nonquiescent relaxation by particle-tracking velocimetry raised serious questions about how localized structural failure and consequent macroscopic motions affect the stress relaxation behavior. Chapter 12 systematically investigated an intimate correlation between fast stress relaxation and localized elastic yielding. For example, the appearance of a sharp shear stress decline (kink-like) at t_k often coincides with the onset of massive recoil movements (which rapidly relieves the shear stress), originating from a macroscopically observable structural collapse in the strained sample.

16.1.3.4 Excessive Shear Thinning: The Symptom of Shear Stress Maximum

Let us now turn our attention to the description of steady shear behavior. According to the tube model, under the condition $1 < Wi < \tau_d/\tau_R \sim 3Z$, after the chain retraction around $t = \tau_R$, only chain orientation contributes to the shear stress during further shearing. The maximum attainable chain orientation occurs at τ_d, at which the corresponding strain $\dot{\gamma}\tau_d = Wi$. A larger rate produces a higher molecular strain and greater level of chain orientation. Therefore, the steady shear stress changes with Wi according to[5] $Q_{xy}(Wi)$, which is nonmonotonic (cf. Eq. (12.7). Figure 16.1 sketches this feature, namely, the emergence of a shear stress maximum in the constitutive curve from the original tube model of Doi and Edwards, which implies shear banding. Since shear banding was never systematically observed by experiment, convective constraint release was introduced in an attempt to restore[39] constitutive monotonicity. However, the incorporation of the convective constraint release into the tube model is *ad hoc*. The mere introduction of convective constraint release suggests that the tube model is inherently unable to deal with interchain interactions in a self-consistent manner: The tube model is a single-chain model while convective constraint release is a multiple-chain effect.

Several comments should be made here. (a) According to experiment, the nominal relations between shear stress and applied rate always have a positive slope, namely, the larger the shear rate, the larger the stress. (b) Particle-tracking velocimetric observations of shear banding show that more than two shear rates can coexist across the thickness. (c) Shear banding seems to be only metastable (cf. Section 15.5). The three facts in combination contradict the proposal[40] that there is a nonmonotonic constitutive relationship for shear rheology of entangled polymers, which is a key feature of the original tube model.[5,17] In the absence of constitutive nonmonotonicity, a small shear stress gradient was suggested as a prerequisite for the experimentally observed shear banding.[41,42] Although the tube model has been extensively applied to explain shear strain localization, there is no appropriate structural picture that accompanies its calculations. In reality, conceivably chains constantly slide past one another during steady shear, so the nature of steady stress is necessarily viscous and of interchain origin. At a larger shear rate, chains are more deformed so that the intermolecular interactions, which depend on the chain conformation, are weaker. It remains elusive how states with different shear rates can coexist in a quasi-stable manner.

Although no experiment has ever observed the steady shear stress to decrease with increasing applied shear rate, some simulation study in the literature[43] has shown that a nonmonotonic relation between steady shear stress and rate is possible. If we agree that startup in either rate-controlled mode or creep should produce the same flow curve, namely, the same relationship between steady shear stress and applied rate, then nonmonotonicity can be ruled out: It is clearly improbable that the resulting final shear rate would be smaller at a larger applied stress during creep for nonassociating entangled polymeric fluids.

16.1.3.5 Anticipation of Necking Based on Considère Criterion

Under the conditions $Wi > 1$ and $Wi_R \ll 1$, the tube model of Doi and Edwards also gives an explicit account of the nonlinear response to startup uniaxial *extension*[17]: As the external extension increases in time, chain orientation and resulting Cauchy stress σ_E saturate and level off. The inevitable consequence is a decline in the engineering stress $\sigma_{engr} = \sigma_E/\lambda \sim 1/\lambda$ when the stretching ratio λ satisfies $\lambda > \lambda_{max} = 2.36$ according to the tube model. The original study[17] did not explore any structural meaning of an emergent maximum in σ_{engr}. As expected, σ_{engr} initially increases with λ according to Eq. (1.52). By invoking the Considère criterion,[44] Doi and Edwards concluded that uniaxial extension would be unstable beyond the peak in σ_{engr}. They did not speculate as to the molecular processes leading to tensile strain localization. It is actually common in the literature to apply the so-called Considère criterion to explain any appearance of necking or emergence of nonuniform extension.[45,46] In our understanding, linking a maximum in σ_{engr} to the emergence of necking is to reverse the causality, as emphasized in Section 7.5.1. One should first confront the basic question: What molecular processes produce a decline in σ_{engr}? At the continuum level, the nonmonotonicity was taken as the cause of strain localization per Considère criterion. At the more fundamental, namely, molecular level, the decline of σ_{engr} is merely the effect of chain disentanglement.

Although the tube model never speaks of strain-induced disentanglement, the hypothesized chain retraction at the Rouse time τ_R is admission that affine deformation ceases to continue beyond a strain level of Wi_R. The tube model is built on the *ad hoc* assumption that chain retraction always occurs in startup extension at a strain equal to $\varepsilon_R \sim Wi_R$. For $Wi_R \ll 1$, this assumption means that any structural adjustment finishes at the beginning of the startup extension. If the state of entanglement remains unchanged after Wi_R ($\ll 1$), namely, very soon after startup extension starts, then in the Doi–Edwards model the emergent maximum $\sigma_{engr(max)}$ in σ_{engr} has nothing to do with any structural change of the entanglement network because that change has long since been completed. The Doi–Edwards prescription of a maximum in σ_{engr} is conceptually and quantitatively identical to that of the Maxwell model. Maxwell model gives a simple formula Eq. (7.11) to show $\sigma_{engr(max)}$ by the open symbols in Fig. 7.27. Thus, application of the structural criterion of Considère to interpret the Maxwell-like maximum in σ_{engr} leads to the unreasonable conclusion that the specimen would show tensile strain localization under the conditions $1 \gg Wi_R$ and $Wi > 1$ without any structural breakdown. It is crucial to reiterate that according to the tube model the nonmonotonicity of σ_{engr} versus λ, depicted in Fig. 16.1, has nothing to do with any structural breakdown of the entanglement network. Contrary to the tube model description, in Section 16.4.3 we suggest that the maximum of σ_{engr} indicates the onset of the structural collapse of the entanglement network.

16.1.4 How to Test the Tube Model

The nonmonotonic characteristics in the tube model were interpreted to describe experimental observations. The nonmonotonicity can be traced back to a single conjecture that chain deformation and corresponding stress can be described on a single length scale, namely, the scale of the equilibrium tube segment. In this scenario, all primitive chains are assumed to retract on the Rouse time scale τ_R, and stress is separable into chain orientation and stretching. For highly entangled polymers with a sufficient separation between τ_d and $\tau_{R,}$ the condition $1 < Wi \ll \tau_d/\tau_R$ is feasible to attain experimentally. Under the condition $Wi_R \ll 1$, the shear stress overshoot would emerge from the progressive chain orientation at $\gamma_{max} \sim 2.2$, namely, $t_{max} \sim 2.2/\dot{\gamma} = 2.2\tau_R/Wi_R \gg \tau_R$. The peak shear stress occurs well after τ_R at a strain $Wi_R \ll 1$. Thus, according to the tube model during either startup shear or extension, the shear stress or engineering stress decline beyond the maximum has no structural significance and is not due to disentanglement leading to failure of the entanglement network.

To test this central conclusion of the tube model that there is only chain orientation during startup deformation with $Wi_R < 1$ or the deformed coil size undergoes rapid contraction on global scales a Rouse time after fast step strain, we can in principle resort to molecular dynamics simulations. Further small-angle neutron scattering (SANS) studies can also provide valuable information, beyond the past SANS experiments that have already been carried out to probe the key assumption of the tube model. These two microscopic tools are briefly discussed in Section 16.9.

According to the tube model, there is Rouse chain retraction after affine deformation, causing different levels of stress drop at τ_R for the different magnitudes of step deformation. Thus, in principle, the stress relaxation is always strain dependent, contradicting the fact that entangled melts and solutions can show identical relaxation modulus over a finite range of step strain, as reviewed in Section 14.1. Reference [47] is the only study in the literature that is in direct conflict with the tube model, as admitted by Graham *et al.*[18] More studies should be carried out, especially ones based on well-entangled melts with $Z \sim \tau/\tau_R \gg 1$ that involves (a) fast step uniaxial extension and (b) startup extension with $Wi_R \ll 1$. Step shear tests must demonstrate that they had excellent adhesion of polymer melts to the shear surfaces, while extensional tests are much more straightforward to perform. Superglue may work effectively for shearing styrene–butadiene rubbers with high vinyl content but does not work well for polybutadiene and polyisoprene. Extensional tests do not face the adhesion challenge and are straightforward to perform with Sentmanat-extensional rheometry.

16.2 Essential Ingredients for a New Molecular Model

Entangled polymeric liquids should be regarded as transient molecular networks. Historically, the idea of modeling entangled polymers as transient chain networks was extensively pursued, well before the invention of the tube model.[48] Because the transient network approach seemed rather *ad hoc*, most researchers abandoned it in favor of the tube model. Transient network models were based on classical rubber elasticity theory. Neither did they address where affine deformation comes from nor did they explain adequately how affine deformation ceases. For example, transient network models interpreted shear thinning by allowing the junction lifetime to depend on the applied shear rate in an empirical way.

The tube theory was constructed at a microscopic level, so it does not appear *ad hoc*. However, our dissection of the tube model in Section 16.1 shows that the tube model is also *ad hoc* albeit it is more molecularly based than the earlier transient molecular network models were. The construct of the tube model decouples stress into chain orientation and stretching while the rubber elasticity theory informs us that chain deformation does not separate into such two artificial components. The tube model is built on the assumption that the entanglement structure is always depicted by the equilibrium tube diameter (analogous to cross-linked rubbers that do have fixed network structure), contrary to the fact that entanglement may adjust in response to large deformation. Then the tube model presents its key premise that a test chain performs Rouse dynamics in a barrier-free tube, preventing large chain deformation during startup deformation with $Wi_R < 1$ and making coil size undergo uncontrollable shrinkage after step strain. This chain retraction ansatz was made because it was and still is so hard to describe how entanglement responds to external deformation. In the absence of sufficient phenomenology, it was quite reasonable to model intimate intermolecular interactions in terms of a smooth tube. However, given the recently collected phenomenology on nonlinear rheological behaviors including shear banding and nonquiescent relaxation, we question the construct of the tube model. We wonder whether the tube model has adequately addressed the two core questions: Where does chain deformation come from, and when does affine deformation cease?

We must find a conceptual framework that allows us to describe plausible answers to these two foundational questions. We must develop a more realistic understanding of how chain entanglements respond to fast large deformation and what molecular pictures and processes are responsible for the macroscopic rheological behavior. To build a molecular theory of nonlinear polymer rheology is challenging because it involves bridging a gap of six orders of magnitude in length scales, going from nanometer to millimeter. Indeed, a bowl of "molecular noodles" is too complex to depict explicitly by a quantitative analytical theory of the molecular dynamics under large deformation. Instead of developing a quantitative but clearly unrealistic, oversimplified theory, such as the tube model, we need to start with a qualitative picture, even though it is still currently intractable to formulate the picture into a rigorous mathematical theory. In this section, we present a conceptual treatment of the nonlinear rheological behavior of entangled polymers. This description requires us to discuss three molecular forces, namely, the intrachain entropic force, the intermolecular grip force, and the cohesion (entanglement) force. The latter two forces, which are of intermolecular origin stemming from chain uncrossability, are elusive and difficult to formulate and cannot be characterized within a tube model. In fact, until their introduction[2] in 2007, they had not been discussed in the literature. In this section, we describe these three crucial ingredients in the emerging molecular picture, inspired by the systematic phenomenology, much of which has been presented in Chapters 6–15.

16.2.1 Intrachain Elastic Retraction Force

The intrachain elastic retraction force, originating from the chain deformation, is a familiar concept from the classical rubber elasticity theory. Regarding entangled polymer melts as a network with elastic plateau modulus G_N^0, we recognize that the basic building block is an entanglement strand with an average molecular weight, the entanglement molecular weight M_e, related to G_N^0 according to Eq. (2.30a). In the affine deformation limit, the retraction force $f_{retract}$ in an entanglement strand is linearly related to the macroscopic stress σ. For a startup simple shear or uniaxial extension, we have for the shear stress

$$\sigma = G_N^0 \gamma \tag{16.4}$$

and for the engineering stress

$$\sigma_{engr} = G_N^0(\lambda - 1/\lambda^2). \tag{16.5}$$

In other words, we assume that the classical rubber elasticity formula, Eqs. (1.50) and (1.52), may initially apply. By analogy with the alternative representations of these formulas in Section 1.3.3, we can express the stress in terms of equilibrium areal density of entanglement strands, ψ_{ent} of Eq. (2.31), so that Eqs. (16.4) and (16.5) can be rewritten as

$$\sigma = \psi_{ent} f_{retract(shear)} \tag{16.6}$$

and

$$\sigma_{engr} = \psi_{ent} f_{retract(extension)} \tag{16.7}$$

respectively. Here, the retraction force can be identified as

$$f_{\text{retract(shear)}} = f_{\text{ent}}\gamma \tag{16.8a}$$

or

$$f_{\text{retract(extension)}} = f_{\text{ent}}(\lambda - 1/\lambda^2), \tag{16.8b}$$

where $f_{\text{ent}} = k_B T/l_{\text{ent}}$ is defined in Eq. (2.30d). In rewriting Eqs. (16.4) and (16.5) as Eqs. (16.6) and (16.7), use is made of (2.30c) for the melt plateau modulus G_N^0. Equations (16.8a) and (16.8b) for the intrachain elastic retraction force f_{retract} show that f_{retract} increases linearly with the strain γ in shear and $(\lambda - 1/\lambda^2)$ in extension, respectively, with its magnitude dictated by the entropic force f_{ent}. Such idealized behavior cannot last as the strain is increased because f_{retract} should not increase with γ or λ without bound. Specifically, the affine deformation assumption cannot be valid to arbitrarily large strains. We have to determine when affine deformation ceases. To do so, we must first recognize why affine deformation can take place, as briefly discussed at a conceptual level in Chapter 15 and investigated more in the following section. For entangled solutions, the same formula Eqs. (16.8a-b) hold with f_{ent} given by Eq. (2.35).

16.2.2 Intermolecular Grip Force (IGF)

It is clear from Fig. 15.1 that the initial affine deformation upon startup deformation is due to the intermolecular uncrossability. When the chains are long enough, this excluded volume effect constrains them to reptate during quiescent diffusion in the absence of fast large deformation. Because of the chain uncrossability, the entanglement network can initially deform in an elastic manner, namely, affinely, as though it were cross-linked. For a test (entanglement) strand, depicted in Fig. 16.2, the interchain dragging amounts to[2] having an intermolecular grip force f_{img}. The surrounding blue strands are pulling the red test strand in the middle, causing it to extend. This grip force arises from interchain uncrossability against mutual transverse chain displacement and is dynamic and time dependent. Thus, the grip force's magnitude should depend on the deformation rate, which introduces a time scale to compete the local chain relaxation dynamics. There are two regimes, depending on the applied shear rate. In this section, we only discuss the elastic regime of large *Wi*. Experimental data reviewed in Chapter 7 show that the onset of this regime is marked by $Wi^* = \tau_d/\tau_R$, namely, the Rouse–Weissenberg number $Wi_R = Wi(\tau_R/\tau_d) = 1$. The scaling laws discussed in Sections 7.2.1 and 7.5.3.3 suggest that the grip force f_{img} is strong in both shear and extension, as long as $Wi_R > 1$. As shown in

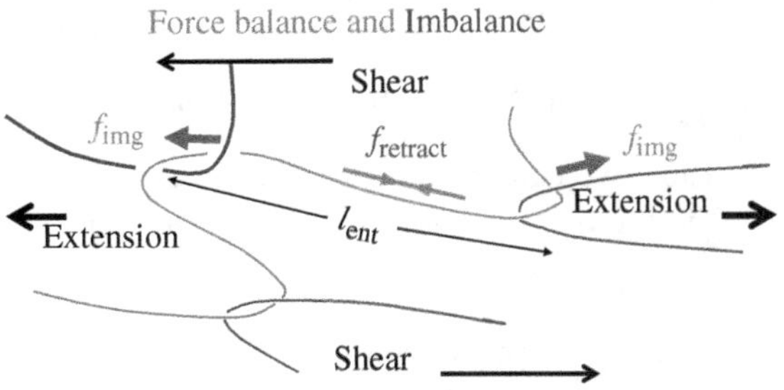

Figure 16.2 A cartoon to depict an entangled polymer in terms of a network-like picture. The cartoon shows that the strand between two entanglement points gets strained either in shear to γ or in extension to λ. The elastic retraction force f_{retract} ($\sim\gamma f_{\text{ent}}$ for simple shear and $\sim(\lambda - 1/\lambda^2)f_{\text{ent}}$ for uniaxial extension) originates from the molecular deformation, the deformation occurring because the intermolecular gripping force f_{img} prevents chain sliding at the entanglement points. *Source:* Wang *et al.* 2013 [4]. Reproduced with permission of American Chemical Society. (*See color plate section for the color representation of this figure.*)

Section 16.4.2, unlike the third force f_{ent} to be introduced in Section 16.2.3, f_{img} is neither viscous nor entropic in origin.

At $Wi_R < 1$, the local chain relaxation is fast enough to render chain uncrossability ineffective in producing strong intermolecular grip. For $Wi_R > 1$, namely, when significant deformation can take place on a time scale shorter than τ_R, the chains surrounding a test strand can effectively deform the test strand. Thanks to this intermolecular gripping, based on which a quantitative analysis is presented in Section 16.4.2, $f_{retract}$ can grow substantially greater than f_{ent}. This means that the elastic deformation can ensue to the strain γ or $(\lambda - 1/\lambda^2)$ much larger than unity per Eq. (16.8a) or (16.8b). As long as $f_{img} > f_{retract}$, the molecular deformation continues to grow. However, not every entanglement strand is the same; there is a Gaussian distribution for the size l_{ent} of entanglement strands. Consequently, some entanglement strands may undergo more deformation than others. Figure 16.2 merely conveys a sense of how to think about the macroscopic deformation of an entanglement network. The chain deformation cannot increase without bound, and $f_{retract}$ cannot grow indefinitely. Mutual sliding between a test strand and its surrounding strands will take place when a molecular force imbalance is reached. The molecular yielding condition is met when $f_{retract}$ reaches f_{img} and chains begin to slide through the entanglement points to allow chain retraction;[2] such retraction causes $f_{retract}$ to decrease with further straining, so that the affine-deformation expressions (Eqs. (16.8a) and (16.8b)) no longer hold.

The stress behaves in a more complicated way because it involves both $f_{retract}$ and the areal density ψ of entanglement strands. To zeroth order, if until the yield point we treat ψ as constant equal to ψ_{ent}, the force imbalance coincides with the onset of stress decline. Most often we have adopted this simplest picture and take the stress decline as indicating that all entanglement strands cease to be further deformed. In a refined picture, actually the surviving strands continue to deform and ψ decreases at the same time. We have also acknowledged that partial yielding occurs before the global yield point; the global yield point is found at the maximum of the shear stress or engineering stress. Using the example of uniaxial extension, a more detailed description of the stress behavior is made in Section 7.5.3.2 by introducing two structural parameters $\alpha(\lambda)$ and $\mu(\lambda)$ in Eqs. (7.17) and (7.16). While Eq. (7.18) illustrates how the engineering stress changes with λ, an equivalent expression can be provided for simple shear as $\sigma = \left[\frac{\alpha_{shear}(\gamma)}{\sqrt{\mu_{shear}(\gamma)}}\right] G_N^0 \gamma$. Here, the maximum in σ, σ_y, emerges when the disentanglement factor before G_N^0 decreases more sharply than $1/\gamma$. Experiments reviewed in Section 7.2.1 show that the yield stress σ_y increases with Wi_R as $(Wi_R)^{1/3}$, as shown Fig. 7.8(b). In Section 16.4.2, we will present an analysis showing how this scaling law may arise at the moment of force imbalance. Unlike the case of uniaxial extension, for simple shear, it can be argued that $\alpha_{shear}(\gamma) = 1/\sqrt{\mu_{shear}(\gamma)}$ and therefore $\left[\frac{\alpha_{shear}(\gamma)}{\sqrt{\mu_{shear}(\gamma)}}\right] = \left[\frac{l_{ent}}{l_{ent}(\gamma)}\right]^2$, where l_{ent} is defined in Eq. (2.29) and $l_{ent}(\gamma)$ is the entanglement spacing at strain γ. Thus, we can further rewrite $\sigma = G(\gamma)\gamma$, with $G(\gamma) = k_B T/p[l_{ent}(\gamma)]^2 = \rho RT/M_e(\gamma)$, where $M_e(\gamma)$ is the average molecular weight of the surviving entanglement strand.

16.2.3 Entanglement (Cohesion) Force Arising from Entropic Barrier: Finite Cohesion

Out of the three forces necessary to depict nonlinear rheological responses of entangled polymers, the third force is least straightforward and most difficult to understand. It stems from an entropic barrier against any rapid changes in chain conformations on global coil size scales. Experiments discussed in Section 14.1 indicate that entangled melts can resist faster relaxation than equilibrium relaxation after a moderate stepwise deformation. Such information reveals

2 Video 16.1: Force imbalance, illustrated by stretching a rubber band between two hands.

that existence of a cohesion barrier, which we show in the following section, is entropic in origin. When polymer chains of high molecular weight mutually intertwine in an entangled polymeric liquid, they are constrained due to interchain uncrossability, giving rise to an entanglement network. Network strands, which are termed entanglement strands, are subchains of an average coil size l_{ent} between two neighboring entanglement points/junctions, such as those shown in Fig. 16.2, which represents a simplified perception of entanglement. Experiments reveal (in Chapters 12 and 14) that elastic yielding occurs only when step strain is beyond a threshold magnitude. Below the threshold, on the order of $\gamma_{ebd} \sim 1.0$ for shear and $\lambda_{ebd} = 1.8$–2.0 for extension, the cohesion of the network can be preserved. Correspondingly, at the threshold strain, the retraction force $f_{retract(coh)}$ is at the level of f_{ent} according to Eqs. (16.8a-b). So it was proposed[2] that there must be a previously unknown counterforce against this level of retraction force. This counterforce is termed interchangeably the cohesion or entanglement force and has a magnitude given by f_{ent} in Eq. (2.30d) or Eq. (2.35). It can balance any retraction force produced by a moderate stepwise deformation by preventing rapid chain contraction on scales larger than l_{ent}. In the following section, a scaling analysis is made to show how such a cohesion arises from entropic penalty against rapid chain retraction on large length scales.

16.2.3.1 Scaling Analysis

Rapid chain retraction on scales well beyond entanglement strand l_{ent} requires a change in chain conformation and thus meets an entropic barrier.[3] Such a barrier may be similar to a Gaussian chain attempting to thread through a narrow ring, as depicted in Fig. 16.3(a). How much force is required to pull a Gaussian chain through the narrow opening with a characteristic scale of D? The problem facing us is equivalent to dragging a subchain having an end-to-end distance R′ into a channel having length L and an opening of size D, as shown in Fig. 16.3(b). In terms of D, the length L is determined by the chain length N′. Specifically, we have the number of blobs given by

$$L/D = N'/g_D, \tag{16.9}$$

where g_D is related to D as $D^2 = g_D l_K^2$ for Gaussian blobs. The entropic penalty or free energy F associated with such a squeeze is given by

$$F(R') \approx k_B T(L/D) = k_B T(R'/D)^2 \tag{16.10a}$$

since the confinement energy per blob is $k_B T$. The potential barrier grows linearly with chain length N′, making it difficult for a subchain of size $R'/D \gg 1$ to thread through the hole in

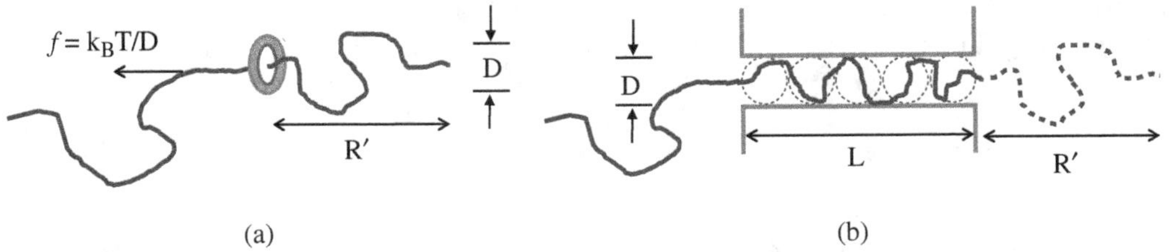

Figure 16.3 (a) Illustration of the conformational entropy penalty for a Gaussian chain of end-to-end distance R′ to thread through a ring-like obstacle of size D. (b) Passage through a channel of size D and length L, equivalent to (a), where the chain of size R′ folds into blobs of size D to go through under a force *f* given in Eq. (16.11).

3 In previous publications (e.g., Refs. [2,4,47]), it was hardly ever specified on which scale the perceived entropic barrier is effective. To be clear, we accept that chain retraction may occur, allowing any stretched *tube segments* to restore their equilibrium length after a step strain. In other words, the contour length measured on the equilibrium tube segment scale could have its equilibrium value after Rouse time during stress relaxation. However, such chain retraction may not take place on larger length scales at Rouse time without encountering barriers of the kind discussed here.

the absence of any external force. Apart from a prefactor, the time $t_D(R')$ required for the chain to diffuse through the channel is inversely proportional to the Boltzmann probability $P \sim \exp(-F/k_BT) = \exp[-\text{const}\cdot(R'/D)^2]$, namely, $t_D(R') \sim \exp[\text{const}\cdot N'(l_K/D)^2]$. This form is reminiscent of the form for the relaxation time of a star polymer[5] that the relaxation time increases exponentially with the arm length N_a. A time-averaged force required to pull the chain through the opening is given by

$$f = dF/dL \approx k_BT/D, \tag{16.11}$$

which is of the same form as Eq. (2.30d) or Eq. (2.35) for f_{ent}.

16.2.3.2 Threshold for decohesion

The length scale examined by the tube model is the equilibrium entanglement spacing l_{ent}. We postulate that a strained chain cannot retract on the scale much greater than $D = l_{ent}$ without encountering an entropic barrier such as the one described by Fig. 16.3 and Eq. (16.10a). Since the coil size R is much greater than the entanglement spacing l_{ent}, chain retraction on the scale of R will meet a high entropic barrier given by

$$F(R) \sim k_BTZ \gg k_BT \quad \text{for} \quad Z \gg 1 \tag{16.10b}$$

This barrier can be overcome after a sufficiently large step strain. The price to pay is entropic change per chain comparable to the potential barrier F(R). A sudden step strain of sufficient magnitude can provide the required level: Let us consider such a step shear of magnitude γ for an entangled melt of elastic modulus G_N^0. The mechanical work done during step strain produces an energy density $\bar{e}$ given by

$$\bar{e} = \int_0^\gamma \sigma d\gamma = \sigma\gamma/2 = G_N^0\gamma^2/2, \quad \text{for shear} \tag{16.12a}$$

$$\bar{e} = \int_1^\lambda \sigma_{engr} d\lambda = G_N^0(\lambda^2/2 + 1/3\lambda - 5/6), \quad \text{for extension} \tag{16.12b}$$

where use is made of the affine deformation expressions, Eqs. (16.4) and (16.5). When every entanglement strand acquires k_BT, each chain with Z strands has reached the condition to overcome the barrier given in Eqs. (16.10b). This condition of each strand having k_BT corresponds to an energy density given by k_BT multiplying the number of entanglement strands per unit volume. This energy density is straightforwardly given by $(\rho/M_e)N_a(k_BT) = G_N^0$, which is the definition for G_N^0 given in Eq. (2.30a). Equating Eqs. (16.12a) and (16.12b) to G_N^0, we arrive at the threshold magnitude γ_{ebd} and ε_{ebd} (defined in Section 14.1) in step strain for complete elastic yielding as

$$\gamma_{ebd} = \sqrt{2} = 1.41 \tag{16.13a}$$

and

$$\varepsilon_{ebd} = \ln\lambda_{ebd} \cong \ln 1.9 = 0.64. \tag{16.13b}$$

Naturally, the critical condition can be weaker than the expressions (16.13a) and (16.13b) because not all chains need to overcome the entropic barrier to cause an elastic breakdown.

Inserting Eqs. (16.13a) and (16.13b) into Eqs. (16.8a) and (16.8b), we see that the retraction force $f_{retract}$ just needs to be on the order of f_{ent} to cause the elastic breakdown. When $f_{retract}$ is much smaller than f_{ent} decohesion cannot take place. This cohesion force f_{ent} is estimated in Eq. (16.11) by equating D to l_{ent}. Since our scaling argument cannot specify the prefactors, the values of γ_{ebd} and λ_{ebd} cannot be determined accurately. Similarly, we only suggest that the cohesion force is of the order of magnitude given by f_{ent}. More important, we note that the existence of this cohesion (entropic) barrier is also the reason why we observe a sizable regime of linear response, as indicated in Fig. 14.1. Based on experimental data in Figs. 14.7 and 14.8, the range of linear response is about $\gamma_{LR} = 0.6$ for simple shear and $\lambda_{LR} = 1.5$ for uniaxial extension.

To be clear, we should note that the entropic barrier in question against large-length scale chain retraction is not to be confused with the tube constraint and does not affect reptative movements. Namely, reptation does not encounter this entropic barrier since the curvilinear diffusion does not perturb the chain conformation.

16.3 Overcoming Finite Cohesion after Step Deformation: Quiescent or Not

In the following Sections 16.3 through 16.7, we briefly show how various rheological phenomena can be interpreted in terms of the new conceptual framework. Although these sections are relatively short, they are central to the book.

Let us first recall Fig. 14.1. Experiment shows that there are three regimes for step deformation in either shear or extension. Depending on the magnitude of the step strain, we can have linear response behavior. Alternatively, elastic yielding can take place homogeneously or in a localized manner that leads to nonquiescent relaxation. Linear response is possible because of the cohesion barrier discussed earlier: Below some threshold magnitude, a step deformation leaves the entanglement network intact so that the stress relaxation behavior is independent of strain. Based on the discussions in Sections 16.2.1 and 16.2.3, a finite range of linear response is expected. At small strains, for $\gamma = 0.1$ or $\lambda = 1.1$, the retraction force $f_{retract}$ in Eqs. (16.8a) and (16.8b) is only 10% of the cohesion force f_{ent} and is thus incapable of causing elastic yielding. We rely on the experimental data to reveal the threshold values γ_{LR} and λ_{LR} and are content to learn that the prefactors in Eqs. (16.8a) and (16.8b) are on the order of unity.

16.3.1 Nonquiescence from Severe Elastic Yielding

Severe force imbalance can take place when a large strain is applied over a negligible amount of time, for example, much shorter than the Rouse time. When a large-magnitude deformation produces a large retraction force $f_{retract}$, as happens with γ or $\lambda \gg 1$, according to Eqs. (16.8a) and (16.8b), the elastic retraction force can overcome the potential barrier, characterized by the entanglement force f_{ent}; after the external deformation stops, only f_{ent} is available to counter $f_{retract}$. For step deformation of magnitude γ or $\lambda \gg 1$, the force imbalance condition readily occurs because $f_{retract} \gg f_{ent}$. In the case of step shear, this condition is necessary but not sufficient for molecular yielding to produce nonquiescent relaxation. As discussed in Section 12.3.2, the ratio b_{max}/H, which quantifies the effect of chain disentanglement, determines whether elastic yielding produces visible macroscopic motions during stress relaxation. For step extension, a critical strain ε_{ebd} appears to be the only criterion. Below this threshold, well-entangled melts can undergo quiescent relaxation after step extension.

16.3.1.1 With $Wi_R > 1$

Conventionally, step deformation tests are performed with the highest displacement rate attainable in the rheometric instrument, involving the so-called stress relaxation mode. Step deformation can be regarded as an interrupted startup deformation: it is never truly instantaneous in experiment. In fact, it is more revealing for us to impose step deformation using a wide range of Wi values. In Section 12.2, from Sections 12.2.1 to 12.2.2.3, and in Chapter 13, we saw nonquiescent relaxation after sizable step shear or extension made at $Wi_R > 1$. In terms of our conceptual picture a molecular-level explanation of these phenomena is readily apparent.

With $Wi_R > 1$, a sizable grip force f_{img} emerges, allowing substantial elastic deformation and corresponding $f_{retract}$ to build up well in excess of f_{ent}. Upon termination of the imposed deformation, the surrounding chains no longer grasp the test strand to displace it in the opposing directions. Only the entropic force f_{ent} is present to fight against the mighty $f_{retract}$ built up during the step deformation. Thus, the condition of force imbalance given by

$$f_{retract} > f_{ent} \tag{16.14}$$

commences immediately after cessation of external deformation. Since $f_{retract}$ is elastic in origin, we expect the condition of Eq. (16.14) to produce *elastic yielding*. In other words, at a scaling level, Eq. (16.14) provides an explanation for why elastic yielding may take place after large step deformation. When such yielding occurs in a localized manner, the elastic yielding reveals itself by producing nonquiescent relaxation.

16.3.1.2 With $Wi_R \ll 1$

In Section 12.2.2.4, stress relaxation was discussed for step strain produced with $Wi_R \ll 1$. According to the tube model, chain can hardly get stretched during such an unconventional stepwise shear test. Consequently, there could not be any driving force to cause the entanglement network to collapse thereby leading to nonquiescent relaxation. A stepwise deformation at $Wi_R \ll 1$ is therefore an interesting experiment to probe the tube model's key assumption of chain retraction. The tube model does not expect the weak shear condition to produce any chain stretching on the equilibrium tube segment scale. But chain stretching can be envisioned to occur on larger scales. Particle-tracking velocimetry along with sliding-plate rheometry reveals that nonquiescent relaxation does take place for $Wi_R = 0.2$ and $\gamma = 1.3$. It is worth noting that at such a small shear rate, the shear stress did not increase linearly with strain even within $\gamma = 1.3$, as shown in Fig. 12.21(a). This considerable "strain softening" is an indication of a lack of affine deformation and occurrence of chain relaxation. Perhaps little grip force is available for $Wi_R \ll 1$.[49] Thus, the way that external straining deforms the entanglement network is elusive in this low-shear-rate limit. We will return to this subject in Section 16.4.1. For now, it suffices to acknowledge that even for a stepwise extension produced at $Wi_R < 1$, chain stretching is expected to occur on scales larger than the equilibrium tube segment so that enough driving force arises to destroy the entanglement network and produce nonquiescent relaxation. The driving force, according to our picture, is the elastic retraction force $f_{retract}$ associated with chain deformation. Apparently, $f_{retract}$ can reach a level sufficiently large even for $Wi_R < 1$ and overcome the cohesion force f_{ent}. Because there is already "strain softening," namely, the much weaker than linear dependence of the shear stress on strain, the entropic force is likely to be already smaller than f_{ent} in Eq. (2.30d), favoring the force imbalance. In summary, the observed failures in Sections 12.2.2.4 and 13.1.2 may be explained as arising from the occurrence of disentanglement through the retraction of deformed chains.

16.3.2 Homogeneous Elastic Yielding: Quiescent Relaxation

As indicated in Fig. 14.1, there is also a regime of quiescent elastic yielding, involving some intermediate magnitude of step strain. For strains between $(\gamma_{LR}, \varepsilon_{LR})$ and $(\gamma_{ebd}, \varepsilon_{ebd})$, the stress relaxation is faster than the stress relaxation in the linear response regime. It is plausible that those strands that suffer more deformation than others first reach the condition of force imbalance, begin to retract, and thus no longer contribute to the relaxing stress. As long as a significant fraction of the entanglement strands survive, the network will not collapse, and stress relaxation can take place without emergence of macroscopic motions. The partial yielding may preserve quiescent relaxation.

Throughout the book, by disentanglement we have meant the loss of some entanglement points and their associated strands because the strands have slided past one another at the entanglement points. Such events occur because some strands inevitably reach the condition of force imbalance earlier than others. If there are a large number of entanglement strands per chain, namely, $Z \gg 1$, partial disentanglement, namely, loss of a fraction of the Z strands per chain, may not produce a structural collapse. This condition of partial yielding is guaranteed by imposing only a moderate level of stepwise deformation.

The molecular yielding can also be homogeneous on macroscopic scales so that the stress relaxation is quiescent, provided that the effect of chain disentanglement in one monolayer does not produce a measurable correction to the strain field. This was experimentally shown in Section 12.2.1.2. It was also theoretically demonstrated in Section 12.3.2 that nearly quiescent relaxation can be expected for sufficiently weakly entangled polymeric liquids despite full elastic yielding.

16.4 Forced Microscopic Yielding during Startup Deformation: Stress Overshoot

The phenomenon of stress overshoot during startup shear, reviewed in detail in Section 7.2, especially in Section 7.2.1, and during startup extension, reviewed in Section 7.5, especially in Section 7.5.3, is a leading characteristic of nonlinear polymer rheology. By definition, stress overshoot is a clear sign that macroscopic yielding has taken place. It literally signals a transition from elastic deformation (where stress grows with strain) toward steady flow (where the stress becomes a constant) as illustrated in Fig. 7.26(a) for simple shear. One central task in the nonlinear rheology of entangled polymeric liquids is to figure out how nonlinear responses such as stress overshoot can be explained in terms of microscopic processes. The macroscopic yielding does seem to imply molecular yielding, namely, yielding of the entanglement network. Depending on the magnitude of the applied shear rate, experiments reveal different features. Since we have no quantitative theory to explain how and why the overshoot coordinates (γ_y, σ_y) show different characteristics in the elastic deformation regime with $Wi_R > 1$ and the viscoelastic regime with $Wi_R < 1$, in the following section we speculate on the plausible pictures and interpret the experimental observations within the framework introduced in Section 16.2.

16.4.1 Chain Disentanglement for $Wi_R < 1$

The experimental observation $\gamma_y \sim 2.0$ hints that the interchain grip force is unimportant for $Wi_R < 1$; it therefore appears that the entropic barrier as introduced in Section 16.2.3 drives chain deformation under the weak deformation condition $Wi_R < 1$. Due to this barrier whose origin is still chain uncrossability, displacement of the two shearing surfaces (cf. Fig. 16.6) or two ends of a specimen in uniaxial extension allows the surface chains to transfer the movement into the bulk. Entanglement strands will initially undergo affine-like deformation. As the entanglement strand deforms, $f_{retract}$ increases. At the beginning $f_{retract}$ is too small to fight against the cohesion barrier. So entanglement strands further deform. As chain further deforms because of the entropic barrier, $f_{retract}$ further increases until it is large enough in comparison to f_{ent}. This point of force imbalance may be simply described as $f_{retract}$ reaching f_{ent}. For $Wi_R < 1$, chain relaxation might also act against the stress buildup. The peak shear stress σ_y slightly increases with the shear rate maybe because there is less effect from chain relaxation, and the yield strain $\gamma_y \sim 2.0$ at the stress maximum is rather insensitive to Wi, as shown in Fig. 7.8(b). In short, the stress overshoot may be approximately understood to occur because of force imbalance between the growing $f_{retract}$ and interchain cohesion force (which is only on the order of f_{ent}). Namely, in our view, the overshoot may be structural in origin and indicates termination of chain deformation and restructuring of the entanglement network. On the other hand, we have no evidence yet in the literature to show that chain deformation is much larger than the tube model prediction and chain overorientation did not occur during startup shear to produce the observed overshoot at $\gamma_y \sim 2.0$.

16.4.2 Molecular Force Imbalance and Scaling for $Wi_R > 1$

The much stronger dependence of σ_y and γ_y on Wi_R for $Wi_R > 1$ indicates that the physics at play is different. It appears that only the condition $Wi_R > 1$ activates f_{img}, thereby allowing a much greater degree of elastic deformation. The experimental data in Section 7.2.1.2 suggest that the time scale defined by the Rouse time appears fundamental. It is not surprising that with increasing Wi_R, the intermolecular uncrossability produces a greater capacity for the entangling chains to grip one another, allowing a larger amount of relative displacement. We envision the stress response in startup deformation with $Wi_R > 1$ to involve continuous events of force balance and imbalance as the original entanglement strand is replaced by new longer entanglement strands: Because coil sizes of entanglement strands has a Gaussian distribution, $f_{retract}$ in a long entanglement strand is weaker. Consequently, such strands may elastically deform considerably further. For a shorter entanglement strand, the moment of force imbalance arrives earlier. The softened stress response, characterized in Chapter 7 as "partial yielding," may in part stem from these earlier local adjustments. Stress overshoot is an indication that the stress response is switching from being elastic-dominant to being viscous-dominant. For this reason, when describing the nonlinear rheology of viscoelastic materials including entangled polymeric fluids, there is merit to referring to viscoelastic materials as *elastoviscous* materials instead of *viscoelastic*, so that we are reminded of what comes first and what ensues subsequently. To the leading order, for $Wi_R > 1$, the stress overshoot is the point of force imbalance between the intrachain $f_{retract}$ and intermolecular grip force f_{img}, which results in chain disentanglement and yielding of the entanglement network. In the remainder of Section 16.4.2, we explore how f_{img} may depend on the value of Wi_R and how γ_y at the point of force imbalance may scale with Wi_R.

The strength of f_{img} is variable (i.e., time and rate sensitive), depending how rapidly the local chain relaxation takes place relative to the imposed shear rate. If we are able to prescribe the characteristics of the grip force such as its strength as a function of time, we may be able to specify the condition of force imbalance that leads to chain disentanglement. It is unclear how to derive a physical model based on first principles for the grip force. However, we can present a rationale for the transition from the elastic to viscoelastic responses (as discussed in Section 7.2.1) by accepting the experimental finding that the transition takes place at $Wi_R = 1$. We can then investigate whether the scaling law with 1/3 exponent appears for $Wi_R > 1$. Without claiming to prove that the picture leading to the scaling form is exclusively correct, in the following paragraphs we explore what scenario can produce a 1/3-power scaling law.

As depicted in Fig. 16.4(a), let us consider the emergence of the grip force at a pair of hairpins due to displacement of entangled chains in opposing directions. We assume that the hairpin has a characteristic segment length $l_h < l_{ent}$ as illustrated. Suppose a force that we call the interchain grip force f_{img} is required to destroy such a hairpin and remove the corresponding entanglement point. By dimensional analysis, the magnitude of f_{img} should be related to l_h by

$$f_{img} \sim k_B T / l_h \tag{16.15}$$

Due to molecular relaxation, the size of the hairpin grows with time because there is a greater relaxation time for a longer hairpin. Thus, f_{img} decreases continuously in time, starting from the onset of the startup shear, because f_{img} is inversely proportional to the hairpin length l_h in Eq. (16.15). When the hairpin involves the entire entanglement strand, the grip has dropped

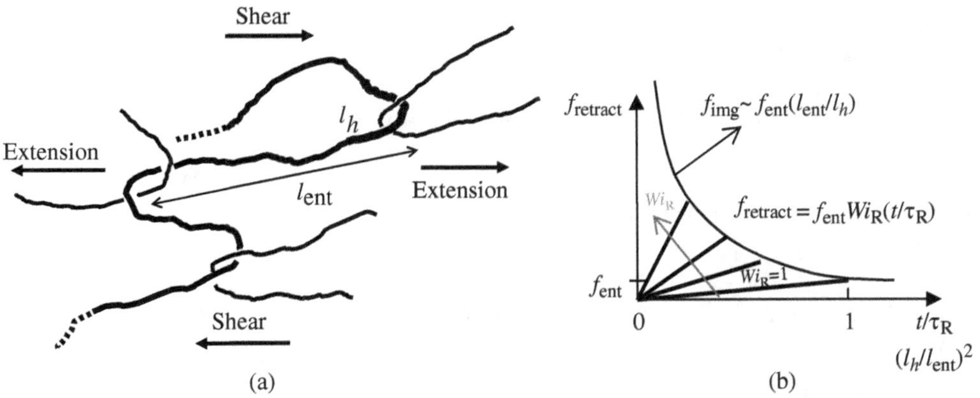

Figure 16.4 (a) Sketch of large chain deformation for $Wi_R > 1$ when entanglement strands of length l_{ent} arise from pairs of hairpins whose characteristic size is l_h. *(See color plate section for the color representation of this figure.)* (b) Depiction of both decreasing f_{img} and increasing retraction force $f_{retract}$ as a function of rescaled time for the case of simple shear where the straight line with the smallest slope represents $f_{retract}$ at $Wi_R = 1$.

to the level of the entanglement force f_{ent}. Namely, when the elapsed time t approaches τ_R, l_h approaches l_{ent}, and $f_{img} = f_{ent}$. Here, our key assumption is that entanglement strands and corresponding hairpins renew on the Rouse time τ_R. When $Wi_R < 1$, f_{img} cannot make $f_{retract}$ grow beyond f_{ent}, because at the elapsed strain $Wi_R < 1$ and time τ_R, f_{img} has already dropped to f_{ent}. Thus, we assume that, when $Wi_R < 1$, the entropic barrier and corresponding entanglement f_{ent} come into play in place of f_{img}. Deformation of entanglement strands occurs through the entropic barrier that provides the network cohesion.

Let us consider the more interesting condition $Wi_R > 1$. We assumed that the hairpin size (l_h), involved to produce the grip force in Eq. (16.15), grows with time t because only the chain segments with longer l_h have not yet relaxed at the longer time. Thus, associated with the segment of length l_h is a particular time scale $\tau_h(l_h)$ on which f_{img} is available to deform chains. To proceed, we need to determine how τ_h changes with l_h. The relaxation time of a segment of length l_h could be argued based on dimensional analysis to be given by

$$\tau_h \sim \zeta_{eff}(N)(l_h)^2/k_B T, \tag{16.16a}$$

where the effective friction coefficient ζ_{eff} represents the slowed-down dynamics of the segment because the hairpin cannot relax independent of the other segments on the chain. Alternatively, we could argue that τ_h should be a fraction of τ_R, namely, proportional to the Rouse time since the equilibrium entanglement strand length l_{ent} (in Eq. (2.29)) has been implicitly assumed in the preceding paragraph to have a lifetime τ_R. Specifically, we postulate

$$\tau_h \sim (l_h/l_{ent})^2 \tau_R. \tag{16.16b}$$

The identification between Eqs. (16.16a) and (16.16b) reveals that the effective friction factor is related to the bare bead friction coefficient by $\zeta_{eff} \sim \zeta N^2/N_e$. Unlike a single chain in dilute solution, the result $\tau_h(l_{ent}) = \tau_R$ stems from chain entanglement.

Initially, f_{img} drives the intrachain retraction force $f_{retract}$ to increase monotonically with time t. Force imbalance occurs when the declining f_{img} and the rising $f_{retract}$ given in Eqs. (16.8a) and (16.8b) are equal, namely, when

$$f_{img}(\tau_{h(y)}) \equiv (l_{ent}/l_{h(y)}) f_{ent} = f_{ent}\dot{\gamma}t_y, \quad \text{for simple shear,} \tag{16.17a}$$

and

$$(l_{ent}/l_{h(y)}) f_{ent} = f_{ent}[\exp(\dot{\varepsilon}t_y) - \exp(-2\dot{\varepsilon}t_y)], \quad \text{for uniaxial extension.} \tag{16.17b}$$

In the case of shear, Eq. (16.17a) shows that

$$l_{\text{ent}}/l_{h(\text{y})} = \gamma_{\text{y}} = Wi_{\text{R}}(t_{\text{y}}/\tau_{\text{R}}). \tag{16.18}$$

The critical scale $l_{h(\text{y})}$ at t_{y} is determined by equating $\tau_h(l_{h(\text{y})})$ in Eq. (16.16b) to t_{y}, which gives

$$(l_{h(\text{y})}/l_{\text{ent}})^2 \sim t_{\text{y}}/\tau_{\text{R}}. \tag{16.19a}$$

The combination of Eqs. (16.18) and (16.19a) gives

$$\gamma_{\text{y}} \sim (Wi_{\text{R}})^{1/3}. \tag{16.20}$$

Equation (16.20) confirms Eq. (7.5) and therefore explains the experimental findings in Fig. 7.8(b) and (c). Similarly, rewriting Eq. (16.19a) as

$$(l_{\text{ent}}/l_{h(\text{y})})^2 \sim Wi_{\text{R}}/\varepsilon_{\text{y}} \tag{16.19b}$$

and substituting Eq. (16.19b) into Eq. (16.17b) yields

$$\varepsilon_{\text{y}}[\exp(\varepsilon_{\text{y}}) - \exp(-2\varepsilon_{\text{y}})]^2 \sim Wi_{\text{R}}. \tag{16.21}$$

In the limit of $\varepsilon_{\text{y}} \ll 1$, Eq. (16.21) is exactly the scaling law

$$\varepsilon_{\text{y}} \sim Wi_{\text{R}}^{1/3}. \tag{16.22}$$

Unfortunately, the yield Hencky strain ε_{y} is typically equal to or larger than unity. Equation (16.21) does not straightforwardly produce Eq. (16.22). To fit the data in Fig. 7.31, we have to introduce a huge prefactor to the right-hand side of Eq. (16.21). The plus-containing squares in Fig. 7.31 are the numerical solution of Eq. (16.21) with 40 as the prefactor; they fit the data approximately but deviate from the 1/3 scaling relation at large values of Wi_{R}.

With increasing Wi_{R}, f_{retract} grows faster so that the point of force imbalance arrives earlier. On the other hand, the grip force is stronger at shorter times and shorter l_h. In the case of simple shear, we have at the point of force imbalance

$$f_{\text{img(y)}} = f_{\text{retract(y)}} = f_{\text{ent}} Wi_{\text{R}}^{1/3}. \tag{16.23}$$

In Fig. 16.4(b), we show for different values of Wi_{R} the trends of f_{retract} and f_{img} as functions of time as well as their intercepts at $f_{\text{img(y)}}$. As illustrated in Fig. 16.4(b), the crossing occurs at shorter times with increasing Wi_{R}. Therefore, equivalently, we can express Eq. (16.15) at the yield point as $f_{\text{img(y)}} = f_{\text{ent}}(t_{\text{y}}/\tau_{\text{R}})^{-1/2}$. This expression is consistent with Fig. 7.9, namely, the yield stress σ_{y} scales with t_{y} at the yield point as $(t_{\text{y}})^{-1/2}$, as indicated in Eq. (7.4c).

It is important to point out that f_{img} does not diverge at short times because the cutoff for l_h is not zero but rather the Kuhn length l_{K} at the bead-spring level. Therefore, f_{img} is bounded by

$$f_{\text{img}}^* \sim f_{\text{ent}}(l_{\text{ent}}/l_{\text{K}}) = \sqrt{N_{\text{e}}}\, f_{\text{ent}} = f_0, \tag{16.24}$$

where f_0 was introduced previously in Eq. (11.10). We have to invoke a more detailed molecular picture if we wish to examine cases where f_{img} is even larger than the expression given in Eq. (16.24). For f_{img} to be larger than f_0, a Kuhn segment would undergo stretching. Realistically, f_{img} can indeed exceed the magnitude given in Eq. (16.24). When fast stretching causes the entanglement to tighten one may encounter non-Gaussian stretching and melt rupture as described in Chapter 11. To reach the magnitude given in Eq. (16.24), namely, for f_{img} to be available despite segmental relaxation, we need the applied rate of startup deformation to exceed a critical value. Since a segment of a hairpin as small as a Kuhn segment has a relaxation time

$$\tau_{\text{K}} \sim (l_{\text{K}}/l_{\text{ent}})^2 \tau_{\text{R}} = \tau_{\text{R}}/N_{\text{e}} \tag{16.25}$$

the elapsed strain within τ_K can exceed unity only if

$$Wi_R(l_K/l_{ent})^2 > 1 \quad \text{or} \quad Wi_R > Wi_R^* = N_e. \tag{16.26}$$

It is plausible that, apart from a prefactor which we cannot determine with the scaling analysis, Eq. (16.26) is related to the condition for the yielding-to-rupture transition in uniaxial melt extension. This condition corresponds to the borderline between regime III (necking) and IV (rupture) depicted in Fig. 11.3. In principle, we can test this prediction for different melts at temperatures sufficiently above T_g to show that the transition occurs at different values of Wi_R because N_e varies according to Fig. 2.10(b) for different polymers. A study of the yielding-to-rupture transition needs to be carried out for different melts to determine if Eq. (16.26) applies to describe the threshold value of Wi_R for the transition.

If the preceding account is correct, then a new theory for chain entanglement needs to show that Rouse time τ_R dictates the dynamics of entanglement strands. Perhaps molecular dynamics simulations, briefly discussed in Section 16.9.1, can be carried out to reveal how the local chain dynamics depend on the model parameters.

Schweizer has developed a separate treatment[49] to show that the scaling behavior seen in Eqs. (16.20) and (16.22) can arise from a different set of considerations. Unlike the present analysis that traces the origin of the grip force to the existence of hairpins due to intermolecular uncrossability, the account of Schweizer associates it with uniform friction experienced by each entangling chain, which is undergoing reptative diffusion even under nonlinear deformation conditions.

The preceding simple scaling argument overlooks more detailed features of the stress overshoot where the force imbalance leads to disentanglement-induced yielding of the entanglement network. We have no more quantitative account of Eqs. (7.4a) and (7.4b) except to note that partial yielding defined in Figs. 7.1(b) and 7.6 occurs before the global yield point at γ_y. Since the entanglement strands are expected to have a broad (plausibly Gaussian) size distribution, the stress level is determined by two competitive processes. As some entanglement strands are removed when they reach force imbalance, the surviving strands continue to undergo affine deformation, so that the stress keeps increasing beyond "partial yielding" in the presence of ongoing disentanglement. The overall stress starts to decrease when further deformation of the residual entanglement strands cannot compensate for the progressive loss of entanglements. At the yield point, according to Eq. (7.4b) the surviving network has a stiffness lower than half of its equilibrium value. The lower modulus $G_{coh} < G_{pl}/2$ occurs because the surviving strands are longer than the equilibrium ones, and the network is effectively less dense. Figure 16.5 presents

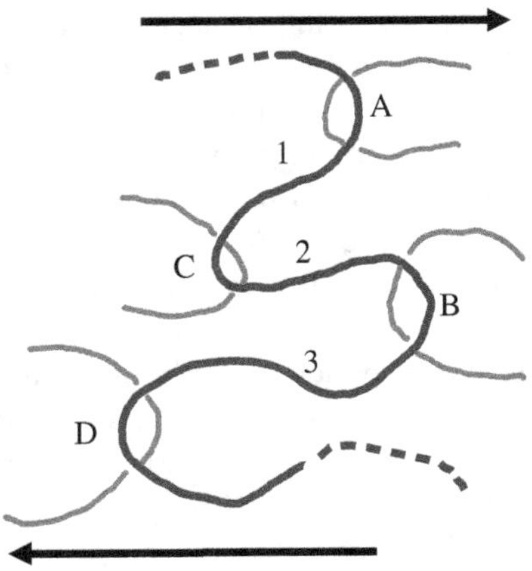

Figure 16.5 Depiction of three strands formed at the hairpins in startup simple shear, where Strand 2 is not undergoing deformation. As a function of time, the retraction force in Strand 1 may increase until the point of force imbalance. A new strand may form between A and B and is longer than the initial stand between A and C.

an image of how chain disentanglement reconstructs the state of entanglement. In reality, chain disentanglement is far more complex than depicted in Fig. 16.5. Nevertheless, Fig. 16.5 may give us a sense of under large deformations why surviving entanglement strands are longer.

16.4.3 Yielding is a Universal Response: Maximum Engineering Stress

It is important to note that the microscopic yielding, namely, chain disentanglement through chain sliding due to force imbalance, terminates affine-like elastic deformation in both startup shear and extension. The signature of yielding is obvious for simple shear, although it has not been a tradition to regard the shear stress overshoot as a yield point despite the suggestion[12] of Maxwell and Nguyen as early as 1979. The year 1979 is also the time when Doi–Edwards published[17] the nonlinear-response aspects of the tube model. Since the concept of yielding was not backed up with a systematic demonstration of its utility and merit, and a microscopic theory based on the reptation model was already available, the Maxwell-Nguyen's notion of yielding never gained adequate acknowledgment in the polymer rheology community. It was the particle-tracking velocimetric observations that made us aware of the necessity to treat an entangled polymer as a network and to contemplate yielding of the network as a leading cause for nonlinear rheological responses.

While it is straightforward to define shear stress overshoot as a sign of yielding, it was more difficult to identify yielding during startup uniaxial extension. If the physics described in Section 16.2 is true, yielding must also take place during startup extension. The force imbalance leads to termination of further chain deformation. But does the areal number density of active entanglement strands decrease in the XY cross section during extension along Z-axis? Because of the geometric condensation effect demonstrated in Appendix 8.A, Cauchy stress may not show a sign of decline when the point of force imbalance is past. Unlike simple shear, in uniaxial extension the loaded area is not constant. Thus, σ_{engr} is explicitly correlated with the structural failure, while the Cauchy stress $\sigma_E = \sigma_{engr}\lambda$ does not. Section 7.5 discussed a new perspective concerning startup extension of entangled polymers. It showed that the peaking of engineering stress σ_{engr} has structural significance: Elastic recoil is complete only when it occurs before $\sigma_{engr(max)}$ is reached, as shown in Figs. 7.34 and 7.36. The overshoot of σ_{engr} and the maximum of the shear stress σ have similar meanings.

16.5 Interfacial Yielding via Disentanglement

There are several theoretical studies of polymer wall slip.[50–55] Most of these studies perceive wall slip by assuming weak chain adsorption and did not explore how and why wall slip takes place. Past theoretical investigations concentrated on describing how the magnitude of slip depends on shear stress or shear rate in the steady state, without describing how and when wall slip arises during startup shear. According to these studies, there cannot be any wall slip if there is complete and strong chain adsorption. Only in the limit of low surface coverage by polymer chains could the models characterize wall slip by assuming the presence of sufficiently large slip velocity V_s at the polymer/wall interface.

Experimentally, at the beginning of startup shear no wall slip takes place in the presence of strong polymer adsorption. Because of the initial no-slip boundary condition, even wall-slip-prone systems undergo significant bulk elastic deformation at the onset of startup shear, as shown in Sections 6.3.2 and 6.4.2.2. The key question is how the boundary condition turns from no-slip to slip, namely, how the interfacial state changes from an entangled to a disentangled state.

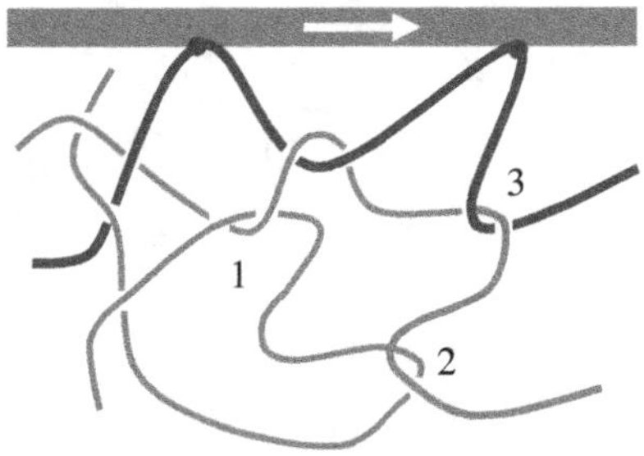

Figure 16.6 Depiction of chain entanglement at a polymer/wall interface due to chain adsorption. Here, the adsorbed chain is the thicker and darker one, anchored at two points on the wall. Upon disentanglement at junction 3 by the adsorbed chain sliding out, entanglements at junctions 1 and 2 may also disappear. Such yielding eventually leads to wall slip that may or may not be observable in macroscopic settings (e.g., with gap distance H), depending on the ratio of the extrapolation length *b* to H.

Parallel to Fig. 16.2 for bulk polymer entanglement, Fig. 16.6 depicts the polymer entanglement at a polymer/wall interface. Upon fast shear, namely, for $Wi_{app} > 1$ or $\gg 1$, the adsorbed (thicker) chains displace to the right, dragging the unbound chains into shear deformation. With increasing strain, interfacial chain disentanglement (mutual chain sliding) eventually occurs at the junctions 1–3, leading to reduced entanglement interactions between the adsorbed chains and the chains in the bulk. The cause of this disentanglement is similar to the one occurring in the bulk, namely force imbalance (cf. Eq. (16.17a)) due to the rising intrachain elastic retraction force. Disentanglement produces interfacial yielding, a transformation from no-slip to slip. Wall slip does not emerge without significant strain, for example, around one strain unit. Experimentally, it is important to understand first this prevalent feature of the nonlinear rheology of entangled polymeric liquids before attempting to study the *bulk* shear rheology of entangled polymers. We therefore started to describe the nonlinear rheological phenomenology by first discussing wall slip in Chapter 6.

The physical origin of wall slip is similar to that of bulk shear banding. Interfacial slip through chain disentanglement usually precedes bulk strain localization and is the extreme formation of shear strain localization. The wall breaks translational invariance because (a) adsorbed chains with slower dynamics are different from the unbound chains and (b) the state of entanglement at the interface may be different from the state in the bulk. Consequently, wall slip is typically the leading form of shear strain localization and is often a precursor to bulk shear banding. Experimentally, we have seen one "exception." Upon application of superglue to fasten styrene–butadiene rubbers to shearing surfaces, the interfacial weakness was eliminated. A stepwise shear of styrene–butadiene rubbers causes the melt to undergo localized elastic yielding in the bulk instead of at interfaces, as shown in Section 12.2.2.

In closing, we comment on the nature of the counterintuitive phenomenon of arrested slip, described in Section 12.4. The physics here is the same as that discussed in Section 16.3.1: the interfacial chain entanglement is overcome by a sufficient buildup of the intrachain retraction force. The molecular mechanism for arrested slip is the same as that depicted by Eq. (16.14), which takes place when the entropic barrier is insufficient to balance the retraction force. Thus, when elastic yielding (cf. Section 16.3.1) takes place at polymer/wall interface after step strain, we may observe arrested wall slip.

16.6 Effect of Long-Chain Branching

Throughout the book, we focus on nonlinear rheological responses of entangled polymers of linear chains. For comparison, the role of long-chain branching is mentioned in several places. (a) In startup extension sufficient long-chain branching delays the final collapse of the entanglement network, preventing the tensile force from dropping sharply after the initial yielding and thus consequently producing a higher transient viscosity than the zero-rate values, namely,

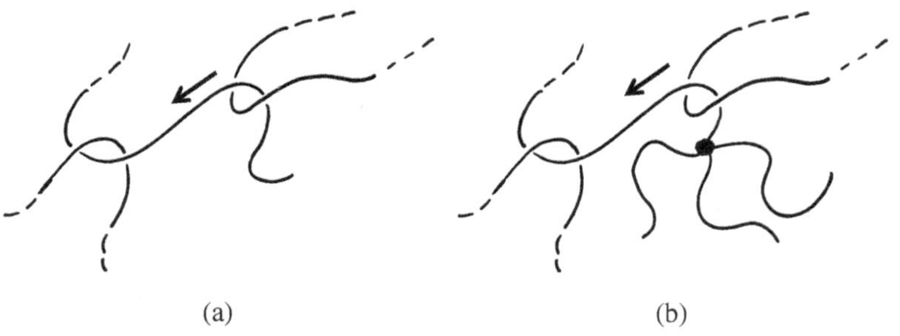

Figure 16.7 Contrast of disentanglement in (a) linear chain and (b) long-chain branched polymers where the arrows indicate the chain sliding driven by the intrachain retraction forces. *Source:* Liu *et al.* 2013 [56]. Reproduced with permission of Elsevier.

the so-called strain hardening, as discussed in Chapter 8. (b) With long-chain branching, vortex formation is preferred over shear banding at the die entry as indicated in Section 10.2.1. (c) Moreover, significant long-chain branching can prevent full elastic yielding after large stepwise extension, thereby preventing the specimen from undergoing macroscopic separation, as described in Section 14.2. An entanglement network made of linear chains is remarkably weaker because sufficiently extended linear chains can readily slide past their entanglement points. Long-chain branching apparently provides extra blockage as illustrated in Fig. 16.7, delaying the yielding of the entanglement network and increasing the cohesion associated with the entropic barrier.

With long-chain branching, the relaxation spectrum is much broader. Side chains, being faster, can first undergo disentanglement. Melts with substantial LCB also behave like a solution, with side chains acting as solvent since their dynamics are much faster. The entanglement network formed by the backbones can be much softer than the entanglement network of a linear-chain melt. During startup deformation or after large step deformation, the side chains disentangle first, well before the backbone network yields through disentanglement. As side chains relax into subcoils, the side-chain contribution to the total entropic barrier is restored. Thus, before the backbone reaches the condition (of force imbalance) to yield through disentanglement, side chains reentangle to increase the entropic barrier against decohesion of the backbone entanglement network. As a consequence, long-chain branching is typically much more resistant to strain localization, as seen in the available cases of melt extension and extrusion. This mechanism suppresses or delays the onset of complete failure of the entanglement network during startup deformation, and promotes homogeneous relaxation after large step deformation, as demonstrated in Chapters 8 and 14, respectively. A more detailed theoretical treatment is needed to address such questions as the dependence of rheological responses on chain architecture. For example, for a polymer of sufficiently high molecular weight, will more branch points be more effective at resisting failures during or after extension? Currently, little is known about correlations between the number of branch points and the ability to resist strain localization. The following situation is equally unsatisfactory: We do not have any systematic experimental data to indicate how long-chain branching affects *shear* strain localization including wall slip and shear banding. Similar to the effect of long-chain branching on tensile strain localization reviewed in Chapters 8 and 14, we can expect long-chain branching to significantly alter the nonlinear responses to simple shear, for example, preventing or postponing the onset of wall slip and shear banding. It is highly desirable to carry out particle-tracking velocimetric observations in the near future based on H-shaped or comb-like rubbery polymer melts, along with linear-chain counterparts.

16.7 Decohesion in Startup Creep: Entanglement–Disentanglement Transition

One of most stunning phenomena in the nonlinear rheology of entangled polymers is their response to startup shear in stress-controlled mode. We call this phenomenon the entanglement–disentanglement transition. The phenomenology was presented in Section 7.4. It effectively and insightfully demonstrates the nature of shear thinning, the drop of shear viscosity by orders of magnitude at shear stresses above some threshold. The large viscosity in polymeric liquids is recognized to be due to strong entanglement. Thus, a drop in viscosity may conveniently be taken to imply loss of entanglement or *disentanglement*.

We can expand the discussion in Section 7.4 and further examine the physics behind the entanglement–disentanglement transition. Based on the concept of finite cohesion, characterized by f_{ent}, depending on the level of the applied shear stress, we anticipate the entanglement–disentanglement transition to take place on different time scales. Straightforwardly, the entanglement structure may determine the cohesion level, which can be characterized by the rubbery plateau modulus G_{pl} as

$$\psi_{ent} f_{ent} = G_{pl}, \tag{16.27}$$

which is Eq. (2.30c). The characteristics of the entanglement–disentanglement transition may depend on whether the applied shear stress is as large as G_{pl} or not. As an example, we reanalyze the entanglement–disentanglement transition studied in Section 7.4, based on a 5% polybutadiene (PB) solution. Corresponding to Fig. 7.25(a)–(d), we supply Fig. 16.8 to show how creep takes place in terms of the shear strain γ. The first feature to notice from Fig. 16.7 is that the shear compliance $J(t)$ collapses to the same line, up to a fraction of τ, for all applied stresses. This feature reveals the initial response to a given applied stress to be fully elastic, satisfying

$$\gamma = \sigma_0 J(t). \tag{16.28}$$

Second, the initial compliance, $J_i = 0.74 \times 10^{-4}\,\text{Pa}^{-1}$, reveals the elastic plateau modulus $G_{pl} = 1/J_i \sim 1.37\,\text{kPa}$, which coincides with $G_{pl} = G_N^0 \phi^{2.2} = 1.37\,\text{kPa}$, given $G_N^0 = 1\,\text{MPa}$ and $\phi = 0.05$. Thus, at $\sigma_0 = 1\,\text{kPa}$, for example, the initial response to the creep should not exceed a strain of $\gamma_i = \sigma_0 J_i$ around 0.74.

Third, as shown in Fig. 7.25(a), at low stresses, for example, at and below 600 Pa, the rise of the responding shear rate does not occur until after the reptation time $\tau = 20$ s. As shown in

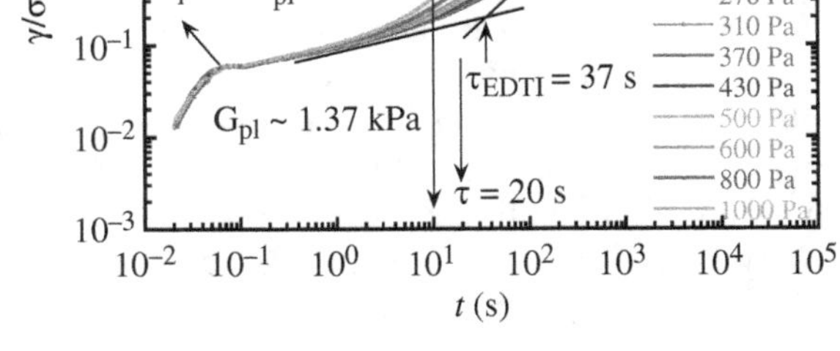

Figure 16.8 Shear strain in units of %, normalized by different applied shear stresses, as functions of time for PB1M(5%)-10K solution. The compliance plateau $J = \gamma/\sigma$ is on the order of 0.001 Pa^{-1}, consistent with the reciprocal of the plateau modulus G_{pl}, which is around 1000 Pa as listed in Table 7.2. *Source:* Data taken from Ge, Zhu and Wang 2017 [57]

Fig. 16.7, creeping starts to pick up speed around $\tau_{EDTI} = 37$ s (>τ) nearly twice as long as τ. We interpret this regime as follows. When applied stress is not high enough to overcome the cohesion of the entanglement network, for example, lower than G_{pl} (Eq. (16.27)) the system has only one option: to wait. With the combined effect of molecular diffusion and stress bias, the state of entanglement gradually evolves over a time scale longer than τ to a new state of entanglement, with a corresponding viscosity smaller than the zero-shear viscosity, as shown in Fig. 7.25(c).

Fourth, at sufficiently high stresses there is another regime. For an applied stress of 1 kPa, by a time of $10\,\text{s} = \tau/2 = \tau_{EDTII}$, the sample has crept to $\gamma = 3$. The system did not wait as long as τ to have its state of entanglement rearranged because the applied stress was large enough to overcome the cohesion forcefully. Regime II is characterized by a time scale $\tau_{EDTII} < \tau$, contrary to the situation in regime I in which the entanglement–disentanglement transition, namely, rise of the shear rate with time, occurs on time scale $\tau_{EDTI} > \tau$.

The entanglement–disentanglement transition is decohesion due to external shear stress. It can take place on time scales shorter or longer than the reptation time. The stress-induced decohesion is forced restructuring of the entanglement structure in response to an imposed stress. Obviously, the most interesting case is regime II, involving an entanglement–disentanglement transition that occurs on time scales shorter than reptation. Being a viscoelastic liquid, in either regime I or II, the system eventually flows as a liquid. It is only a matter of time: the conversion from the elastic to viscous states is faster at a higher applied stress.

16.8 Emerging Microscopic Theory of Sussman and Schweizer

Equilibrium reptation-tube theory[5] is phenomenological in the sense that the anisotropic tube field is postulated. Chain dynamics is not described at the level of forces, and the motion of a tagged chain and its surrounding tube are not treated in a self-consistent manner. For nonlinear rheology, additional assumptions are invoked about what entanglements are and how they respond to strong deformations. The tube is assumed to be elastically unbreakable and essentially unchanged by deformation (except in an affine sense), in analogy with an ideal cross-linked rubber network.

Foundational theoretical progress should address the aforementioned difficult issues in a first-principles-based, bottom-up, self-consistent manner. Szamel met the challenge for the anisotropic equilibrium dynamics of a fluid of needles (infinitely thin rods) based on an ensemble-averaged statistical mechanical approach.[26] The method of Szamel and Schweizer explicitly enforces dynamic uncrossability at the two-chain level.[58] Tube localization and reptation scaling laws for long time transport and relaxation emerge from the microscopic calculation. The key step in this treatment is to construct a *dynamic* mean-field description whereby the motion of a tagged rod is controlled by dynamically emergent tube constraints. These constraints are self-consistently generated by the motion of the other rods that form the confinement tube.

Recently, Sussman and Schweizer qualitatively extended the Szamel approach to construct the full tube confinement field for the transverse motion of rods using statistical mechanical ideas that had been successfully developed and applied for the slow activated dynamics of glass-forming liquids.[59] The crucial new prediction is that the tube is strongly anharmonic, softening as rods displace in the transverse direction beyond a threshold value. Such a theoretical picture is in accord with the experimental observations of highly non-Gaussian displacement distributions in entangled biopolymer solutions.[60] Despite an infinite entropic barrier to transverse motion in equilibrium, a maximum entanglement force keeping rods in their tubes is predicted, which is qualitatively consistent with the finite cohesion conceptual picture described in Section 16.2.3. The rod theory was subsequently generalized to treat entangled chain liquids at the rigid primitive path segment level.[61,62] Predictions for the tube diameter and the full distribution of primitive path step lengths in melts are in good agreement with experiment and computer simulations, respectively.[61]

The new idea of an anharmonic tube confinement field and finite-strength entanglement force has potentially many qualitatively novel consequences for nonlinear rheology. For example, with increasing strain or stress, for entangled needle fluids the tube widens, entanglements are lost, and reptation speeds up.[63] According to the analysis, the entropic barrier for transverse motion becomes finite under deformation, and hence activated hopping transverse to the tube axis emerges as a second relaxation channel that competes with accelerated reptation. Sussman and Schweizer have proceeded to construct initial quantitative theories for the nonlinear rheology of entangled rod solutions subjected to an instantaneous step strain[64] and a startup continuous shear.[65] Here, the tube confinement field is nonlinearly and self-consistently coupled to polymer orientation, reptative motion, and the macroscopic stress–strain response. The limiting concept of "microscopic absolute yielding"[61–63] was proposed, corresponding to the destruction of the entanglement network by an applied force in the nonlinear elastic regime.[66] The specific predictions are inevitably sensitive to numerical prefactors and detailed physical assumptions about precisely how macroscopic stress is transmitted to a force on individual chains. Hence, whether the tube can be literally broken, in contrast to strongly softened, requires future study. All these new behaviors are argued to be qualitatively relevant to polymeric fluids under slow enough nonlinear deformations where contour length equilibration is fast.[66] The predicted weakening of the entanglement network under deformation is similar to the elastic yielding and entanglement–disentanglement transition concepts discussed in this book.

New theoretical connections have also been made between tube softening, the stress overshoot, shear thinning, and the steady state flow curve.[65–67] In these theoretical treatments convective constraint release (CCR) emerges as a natural consequence of a self-consistently determined anharmonic tube field. The stress overshoot is predicted to correspond to an elastic-viscous crossover associated with deformation-induced disentanglement, which is causally related to the emergence of CCR and shear thinning in the nonequilibrium steady state.[65–67] Overall, the new theoretical work has placed our conceptual understanding (depicted in Section 16.2 and applied in Sections 16.3–16.7) on firmer grounds. However, much remains to be done to both explore technical subtleties that depend on quantitative factors, which inevitably enter a microscopic force level theory, and to build a description of the nonlinear rheology of flexible chain melts and solutions, especially when subjected to fast large deformations.

16.9 Further Tests to Reveal the Nature of Responses to Large Deformation

At least two different models exist in the literature, each having merit as a description of the nonlinear rheology of entangled polymers. The tube model is quantitative and has been around for much longer, while our alternative worldview, based on a molecular-networking paradigm, is quite recent and only qualitative, but has been able to make penetrating and insightful predictions[1] that are not shared by the tube theory. The newer view has found theoretical support as reviewed in the preceding section. Can the tube model be further tweaked to incorporate some concepts from the molecular-network paradigm? The foundation of the tube model was constructed in such a way that any additional consideration or improvement, including the previous incorporation of constraint release and convective constraint release, cannot readily be made, and has not been made, self-consistently. If the improvement of the tube model continues to lack self-consistency and predictive power, it may not be fruitful to further revise the model. The difficulty indicates that the two worldviews are incompatible. Besides, given the complexity of entangled polymers as many-body systems, a quantitative agreement between experiment and theory does not necessarily imply the correctness of the tube model.[4] Until experimental means are developed to directly determine molecular processes during large deformation of entangled polymers, it seems necessary to carry out molecular-dynamics-based computer simulations. It would further be desirable to perform new small-angle neutron scattering studies to examine conformation changes during large deformations of polymer melts.

16.9.1 Molecular Dynamics Simulations

With available computing power, it has become feasible to study the molecular processes involved in nonlinear rheology of entangled polymers through molecular dynamics (MD) simulations. MD simulations may play a valuable role in clarifying the molecular mechanisms for nonlinear rheological characteristics.

MD simulation studies should be carried out to address an obvious disagreement between the two models' explanations of the molecular origin of stress overshoot, as found during startup shear and startup extension of entangled melts. Our new picture, inspired by the particle-tracking velocimetric observations of shear banding following overshoot, is that the nonmonotonic change in the stress response arises from chain disentanglement due to force imbalance. The stress decline corresponds to structural breakdown. In contrast, the tube theory prescribes a rather starkly different picture: For $Wi > 1$ but $Wi_R < 1$, nothing structural happens at the maximum of shear stress during startup and at the maximum of engineering stress during startup extension. Passing through the stress maximum each chain continues to increase its orientation with no sign of disentanglement. For $Wi_R < 1$, Gaussian coils simply undergo growing orientation as a function of strain. In simple shear, the orientational stress eventually declines. In uniaxial extension, the Cauchy stress saturates as the molecular orientation levels off, leading to a decline in the engineering stress due to the cross-sectional areal shrinkage. For $Wi_R > 1$, the tube model envisions different dependence of the yielding strain γ_y on Wi_R than revealed by experiment. The alternative paradigm provides a coherent account for the nonlinear rheological responses in both viscoelastic and elastic regimes with $Wi_R < 1$ and $Wi_R > 1$. The two competing pictures cannot be more different and farther apart.

Unfortunately, at the present time, few MD simulations[43,68,69] have been carried out to discern the chain disentanglement process during or after large deformation despite plentiful experimental information in Chapters 6 through 15 that suggests the merit of examining polymer disentanglement by MD simulations. Since MD simulations naturally and systematically include interchain uncrossability, which has been challenging to account for explicitly in analytical

---

4 Here, it is appropriate to cite or paraphrase the statement by E.T. Jaynes: "the principles of scientific inference tell us that a theory is confirmed by observing things which it predicts that are otherwise unexpected; not by observing things that would be expected whatever one's theory..." – *A backward look to the future*, by E.T. Jaynes, p. 261 in *Physics & Probability*, ed. by W.T. Grandy and P.W. Milonni, Cambridge University Press, 1993. This statement not only instructs us on how to understand the merit of a theory but also points out the essence of having a useful theory. The alternative nonmainstream treatment of the subject on nonlinear polymer rheology seems to have precisely demonstrated its quality as a more pertinent theory. It indeed predicts things "that are otherwise unexpected." For example, it predicted arrested wall slip (cf. Section 12.4), which is highly counterintuitive. It anticipated elastic yielding in stepwise melt extension and plausible tensile strain localization during startup melt stretching. These predictions have been summarized in Ref. [1].

theories, they may hold promise to illustrate molecular mechanisms for a host of phenomena, ranging from stress overshoot and elastic yielding to strain hardening.[5]

16.9.2 Small Angle Neutron Scattering Measurements

Postdeformation neutron scattering measurements can be and have been carried out to delineate the conformational response of entangled polymers to startup deformation. Although many experiments have been made in the past, most of them examined the equilibrium structure and dynamics.[70] Nevertheless, there are several publications[71–79] involving deformation of entangled polymers. Because of the poor temporal resolution, small-angle neutron scattering (SANS) measurements involve quenching deformed polymers below their T_g during different stages of ongoing deformation or different stages of relaxation after deformation, and taking the quenched samples to the SANS setup for measurement below T_g.

The early SANS report from Boué and coworkers indicated[73] that "the effect of contraction of the dimensions of the chain is not observed" and this would be in "discrepancy with the tube model, concerning the process of contraction of the deformed chain in its 'tube'." SANS should reveal a transient shrinkage in the coil size due to the perceived chain retraction if the tube model is correct. In particular, the transverse (to the extension direction) dimension of a stretched chain would show a small minimum during the stress relaxation, namely, an initial decrease due to the chain retraction followed by continuous increase toward the equilibrium value. Boué and coworkers were unable to identify this nonmonotonicity, as expected from the Doi–Edwards tube model. The SANS results motivated de Gennes to speculate that there is an absence of chain retraction on the tube segment level.[80]

Following Boué and coworkers, more recently Blanchard *et al.*[77] also searched for a non-monotonic change of the coil dimensions in the direction perpendicular to the tensile stretching direction. Based on a polyisoprene melt with $Z = 58$, they claimed that the chain dimensions did shrink beyond experimental uncertainty during the relaxation, namely, there is a faint level of nonmonotonicity involving the transverse dimension (R_g^{perp}) of labeled chains. The chain transverse dimension decreased by 4% around the Rouse time τ_R. It is worth noting that Boue's experiments had placed an error greater than 5% for this kind of evaluation of R_g. Until the experimental accuracy of SANS measurements improves and the extraction of chain conformation information from the scattering intensity becomes more reliable, it is difficult to draw a convincing conclusion from these opposing results.

Because of the existing uncertainties and discrepancies in the literature on SANS studies of polymer deformation, more studies are highly desirable. The following SANS measurements are clearly feasible and should be pursued. We divide them into two categories, depending whether Wi_R is small or large.

16.9.2.1 Melt Extension at $Wi_R \ll 1$

A stepwise deformation produced with $Wi_R < 1$ is interesting because it would not generate significant chain deformation, according to the ansatz of the tube model. After a stepwise extension of sufficient magnitude, should SANS measurements reveal nonmonotonicity in the chain dimensions during stress relaxation because there is elastic yielding? How could specimen breakup occur, if there is not enough chain deformation during extension with $Wi_R = 0.18$ to cause failure through disentanglement after stepwise extension, as shown by the squares in Fig. 13.2? Can chain orientation produced during stepwise deformation cause chain disentanglement and elastic yielding during stress relaxation? How much chain deformation

5 In our judgment, MD simutions should be restricted to exploring only homogeneous deformation. For example, computer simulation of shear banding in entangled chains (in a simulation box of just tens of nanometers) cannot be used to predict whether or not shear homogeneity prevails in macroscopic experiment.

must be present at $Wi_R < 1$ for elastic yielding to take place, after stepwise shear as shown in Section 12.2.2.4 or extension as shown in Section 13.1.2?

For $Wi_R < 1$ and $Wi > 1$, according to the tube model, it is the saturation of the chain orientation that causes the Cauchy stress to level off at large strains, leading to a maximum in the engineering stress σ_{engr}, as discussed in Section 16.1.3.5. Experiments indeed reveal a peak in σ_{engr} as described in Chapter 11. According to the alternative framework explained in Sections 16.2 and 16.3, the entanglement network undergoes global yielding at $\sigma_{engr(max)}$ due to chain disentanglement. Namely, on the average entanglement strands extend continuously until the yield point at $\sigma_{engr(max)}$ when the force imbalance occurs, leading to disentanglement. This disentanglement is expected to prevent further chain extension, on the average. Could uniaxial melt extension produce any nonmonotonic change in the overall chain dimensions as a function of stretching ratio? Since the surviving entanglement strands will further extend past the yield point, the average conformational changes measured by SANS could be challenging to interpret. Nevertheless, it is meaningful to compare the SANS data with the prediction of the tube model at different stretching ratios.

16.9.2.2 Step Melt Extension With $Wi_R > 1$

It is worth repeating SANS measurements for step extension involving magnitudes as large as $\lambda = 3$. According to rheological tests such as that shown in Fig. 13.5, there may not be any detectable global chain retraction before the sign of decohesion (a tensile-like failure) shows up during the stress relaxation. Up to $\lambda = 3$, would SANS measurements produce evidence for considerable chain contraction, for example, R_g^{perp} displaying a minimum during the relaxation, well beyond the error bars? The circles in Fig. 13.5 show that at $\lambda = 2.7$ the stress relaxation is slow up to $t_{tdc} = 7\,s = 21\,\tau_R$. There should be a much greater nonmonotonicity than there is for $\lambda = 1.7$, which was the case examined by Blanchard *et al.*[77] Thus, the tube model can be more convincingly tested for a step extension of magnitude as large as $\lambda = 3$.

According to the tube model, any sizable step extension would produce a nonlinear rheological response because Rouse chain retraction causes a rapid stress decline about a Rouse time after the onset of stress relaxation, as shown in Fig. 14.8. In contrast, the experiment appears to have identified a threshold for linear response at λ_{LR} around 1.4–1.5, implying that a test chain may not retract much even at a time as long as $20\tau_R$ after such a step extension. SANS measurements should be able to search for evidence of significant chain retraction on all scales, as the tube model assumes. Lack of chain retraction, as suggested in Section 16.1.3.1, would seriously challenge the key ansatz of the tube model and questions the validity to decouple polymer deformation into chain orientation and stretching. All the leading features of the tube model listed in Sections 16.1.3.2–16.3.1.5 hinge on the validity of this foundational construct.

16.10 Conclusion

It has been challenging to infer molecular behavior from macroscopic rheometric measurements of entangled polymers in the regime of nonlinear response. The task of a molecular or microscopic theory is to provide the constitutive relationship, or at least to explain rheological phenomena in terms of molecular processes. How do we build such a theory? What must we do first? We must first collect a sufficient amount of coherent phenomenology. For example, we should understand and demonstrate what shear stress (or engineering stress in uniaxial extension) overshoot means before constructing a molecular theory. Since the stress overshoot indicates macroscopic yielding, we need to identify the corresponding microscopic events. Yielding amounts to a termination of the rubbery-like elastic deformation, so we need to show why and when the initial affine deformation ceases.

We summarize the key contents and conclusions of this chapter as follows. To describe nonlinear responses of entangled polymers, any theoretical treatment must confront the core questions of (a) why molecular (chain) deformation can take place and (b) when the initial elastic deformation is replaced by flow. The tube model addresses these two questions by assuming persistent affine deformation of the confining tube and Rouse dynamics for the test chain in the tube. The alternative picture envisions explicitly that (a) the deformation of a test chain along with growing intrachain retraction force $f_{retract}$ is due to the presence of the surrounding chains and (b) the initial (dominantly elastic) deformation ceases when $f_{retract}$ reaches a level approximately equal to the intermolecular grip force f_{img}. These two competing theories interpret the rheological phenomena in remarkably different ways. For example, according to the molecular-network picture the stress overshoot is due to a structural failure, a breakdown of the entanglement network, due to a molecular force imbalance. In the limit $Wi_R \ll 1$, the tube model predicts no structural adjustment around the stress peak. Similarly, upon stepwise deformation produced with $Wi_R \ll 1$, the tube model should not expect any structural rearrangement during the stress relaxation, while the molecular-network paradigm predicts elastic yielding, namely, disentanglement driven by the elastic force $f_{retract}$. For $Wi_R > 1$, experiment reveals a scaling law $\sigma_y \sim Wi_R^{1/3}$ to characterize the stress overshoot at the yield strain $\gamma_y = \sigma_y/G_{pl} \sim Wi_R^{1/3}$. This law can be rationalized by applying a scaling analysis in the new conceptual framework. The concept of force imbalance does not exist in the tube model. In the tube model, the maximum shear stress occurs because the affinely stretched test chain must retract at τ_R, namely, at a strain of Wi_R. Consequently, according to the tube model σ_y and γ_y would scale linearly with Wi_R, in sharp contradiction to the experimental observation of the 1/3 scaling law. Most important, the tube model assumes barrier-free chain retraction, leading to a significant chain contraction[15] after any magnitude of step strain. Experimental data indicate *lack* of such contraction that would have led to a systematic deviation of the stress relaxation data from the equilibrium relaxation dynamics, which was not observed. If this key assumption is invalidated, which seems to be suggested by the macroscopic experiment, we wonder what is left of the tube model. Microscopic evidence from SANS measurements will be crucial. The first of new SANS studies has just appeared in the literature.[81]

It must be stressed that we have taken an oversimplified view on the nonlinear responses of entanglement to large deformation. This chapter describes our analyses based on rational arguments. The perceived molecular pictures and processes should not be regarded as facts let alone truth. Our approach to the description of nonlinear (entangled) polymer rheology is merely phenomenological, heuristic, conceptual, and often speculative rather than quantitative and numerical. The reader should consult Section 16.8 for the latest theoretical developments that are microscopic, quantitative at the force level and therefore promising and encouraging. To further elaborate on a more sophisticated theoretical description is well beyond the scope of this book. We should just note that despite all the shortcomings of our scaling-level description, we have managed to indicate a coherent physical picture. The predictions[1] confirmed by the experiments are the strongest support for the effectiveness of this simplistic approach.

References

1 Wang S-Q. Nonlinear rheology of entangled polymers at turning point. Soft Matter. 2015;11(8):1454–8.

2 Wang S-Q, Ravindranath S, Wang Y, Boukany P. New theoretical considerations in polymer rheology: elastic breakdown of chain entanglement network. J Chem Phys. 2007;127(6):064903–14.

3 Wang SQ. The tip of iceberg in nonlinear polymer rheology: entangled liquids are "solids". J Polym Sci Part B: Polym Phys. 2008;46(24):2660–5.
4 Wang S-Q, Wang Y, Cheng S, Li X, Zhu X, Sun H. New experiments for improved theoretical description of nonlinear rheology of entangled polymers. Macromolecules. 2013;46(8):3147–59.
5 Doi M, Edwards SF. The theory of polymer dynamics. Oxford University Press, Oxford; 1988.
6 McLeish TCB. Tube theory of entangled polymer dynamics. Adv Phys. 2002;51(6):1379–527.
7 Larson RG. Looking inside the entanglement "tube" using molecular dynamics simulations. J Polym Sci Part B Polym Phys. 2007;45(24):3240–8.
8 Mead D, Larson R, Doi M. A molecular theory for fast flows of entangled polymers. Macromolecules. 1998;31(22):7895–914.
9 Milner ST, McLeish TCB, Likhtman AE. Microscopic theory of convective constraint release. J Rheol. 2001;45(2):539–63.
10 Graham RS, Likhtman AE, McLeish TCB, Milner ST. Microscopic theory of linear, entangled polymer chains under rapid deformation including chain stretch and convective constraint release. J Rheol. 2003;47(5):1171–200.
11 Likhtman AE. Whither tube theory: from believing to measuring. J NonNewton Fluid Mech. 2009;157(3):158–61.
12 Maxwell B, Nguyen M. Measurement of the elastic properties of polymer melts. Polym Eng Sci. 1979;19(16):1140–50.
13 Edwards S. The statistical mechanics of polymerized material. Proc Phys Soc. 1967;92(1):9.
14 Doi M, Edwards S. Dynamics of concentrated polymer systems. Part 1.—Brownian motion in the equilibrium state. J Chem Soc Faraday Trans 2 Mol Chem Phys. 1978;74:1789–801.
15 Doi M, Edwards S. Dynamics of concentrated polymer systems. Part 2.—Molecular motion under flow. J Chem Soc Faraday Trans 2 Mol Chem Phys. 1978;74:1802–17.
16 Doi M, Edwards S. Dynamics of concentrated polymer systems. Part 3.—The constitutive equation. J Chem Soc Faraday Trans 2 Mol Chem Phys. 1978;74:1818–32.
17 Doi M, Edwards S. Dynamics of concentrated polymer systems. Part 4.—Rheological properties. J Chem Soc Faraday Trans 2 Mol Chem Phys. 1979;75:38–54.
18 Graham RS, Henry EP, Olmsted PD. Comment on "new experiments for improved theoretical description of nonlinear rheology of entangled polymers". Macromolecules. 2013;46(24):9849–54.
19 Schieber JD, Neergaard J, Gupta S. A full chain, temporary network model with sliplinks, chain-length fluctuations, chain connectivity and chain stretching. J Rheol. 2003;47(1):213–233.
20 Doi M, Takimoto J-I. Molecular modelling of entanglement. Philos Trans R Soc Lond Ser A. 2003;361(1805):641–52.
21 Masubuchi Y, Takimoto J-I, Koyama K, Ianniruberto G, Marrucci G, Greco F. Brownian simulations of a network of reptating primitive chains. J Chem Phys. 2001;115(9):4387–94.
22 Graessley WW. Polymeric liquids and networks: dynamics and rheology. Garland Science, London; 2008.
23 de Gennes P-G. Reptation of a polymer chain in the presence of fixed obstacles. J Chem Phys. 1971;55:572–9.
24 James HM, Guth E. Theory of the elastic properties of rubber. J Chem Phys. 1943;11(10):455–81.
25 James HM, Guth E. Simple presentation of network theory of rubber, with a discussion of other theories. J Polym Sci. 1949;4(2):153–82.

26 Szamel G, Wang T. Long-lived interchain contacts in polymer melts. J Chem Phys. 1997;107(24):10793–8.
27 Likhtman AE, Ponmurugan M. Microscopic definition of polymer entanglements. Macromolecules. 2014;47(4):1470–81.
28 Einaga Y, Osaki K, Kurata M, Kimura S-I, Tamura M. Stress relaxation of polymer solutions under large strain. Polym J. 1971;2(4):550–2.
29 Einaga Y, Osaki K, Kurata M, Kimura S-I, Yamada N, Tamura M. Stress relaxation of polymer solutions under large strain. Polym J. 1973;5(1):91–6.
30 Fukuda M, Osaki K, Kurata M. Nonlinear viscoelasticity of polystyrene solutions. I. Strain-dependent relaxation modulus. J Polym Sci Part B Polym Phys. 1975;13(8):1563–76.
31 Osaki K, Kurata M. Experimental appraisal of the Doi–Edwards theory for polymer rheology based on the data for polystyrene solutions. Macromolecules. 1980;13(3):671–6.
32 Doi M. Molecular rheology of concentrated polymer systems. I. J Polymer Sci Polymer Phys Ed. 1980;18(5):1005–20.
33 Osaki K, Nishizawa K, Kurata M. Material time constant characterizing the nonlinear viscoelasticity of entangled polymeric systems. Macromolecules. 1982;15(4):1068–71.
34 Osaki K. On the damping function of shear relaxation modulus for entangled polymers. Rheol Acta. 1993;32(5):429–37.
35 Archer LA. Separability criteria for entangled polymer liquids. J Rheol. 1999;43(6):1555–71.
36 Islam MT, Sanchez-Reyes J, Archer LA. Nonlinear rheology of highly entangled polymer liquids: Step shear damping function. J Rheol. 2001;45(1):61–82.
37 Vrentas CM, Graessley WW. Study of shear-stress relaxation in well characterized polymer liquids. J Rheol. 1982;26(4):359–71.
38 Vrentas C, Graessley W. Relaxation of shear and normal stress components in step-strain experiments. J NonNewton Fluid Mech. 1981;9(3–4):339–55.
39 Marrucci G. Dynamics of entanglements: a nonlinear model consistent with the Cox-Merz rule. J NonNewton Fluid Mech. 1996;62(2):279–89.
40 Olmsted PD. Perspectives on shear banding in complex fluids. Rheol Acta. 2008;47(3):283–300.
41 Adams J, Olmsted P. Nonmonotonic models are not necessary to obtain shear banding phenomena in entangled polymer solutions. Phys Rev Lett. 2009;102(6):067801. The authors apply a version of the tube model to assert that shear banding can take place if there is a stress gradient in the simple shear setup.
42 Wang S-Q. Comment on "nonmonotonic models are not necessary to obtain shear banding phenomena in entangled polymer solutions". Phys Rev Lett. 2009;103(21):219801.
43 Mohagheghi M, Khomami B. Molecular processes leading to shear banding in well entangled polymeric melts. ACS Macro Lett. 2015;4(7):684–8.
44 Considère A. Memoirs on the use of iron and steel in construction. Ann des Ponts et Chaussées. 1885;9(6):574.
45 Vincent P. The tough-brittle transition in thermoplastics. Polymer. 1960;1:425–44.
46 McKinley GH, Hassager O. The Considere condition and rapid stretching of linear and branched polymer melts. J Rheol. 1999;43(5):1195–212.
47 Cheng S, Lu Y, Liu G, Wang S-Q. Finite cohesion due to chain entanglement in polymer melts. Soft Matter. 2016;12(14):3340–51.
48 Green M, Tobolsky A. A new approach to the theory of relaxing polymeric media. J Chem Phys. 1946;14(2):80–92.
49 Schweizer KS. 2016, private communication.
50 Brochard F, De Gennes PG. Shear-dependent slippage at a polymer/solid interface. Langmuir. 1992;8(12):3033–7.

51 Adjari A, Brochard-Wyart F, de Gennes P-G, Leibler L, Viovy J-L, Rubinstein M. Slippage of an entangled polymer melt on a grafted surface. Phys A Stat Mech Appl. 1994;204(1–4):17–39.
52 Brochard-Wyart F, Gay C, De Gennes P-G. Slippage of polymer melts on grafted surfaces. Macromolecules. 1996;29(1):377–82.
53 Mhetar V, Archer L. Slip in entangled polymer solutions. Macromolecules. 1998;31(19):6639–49.
54 Joshi YM, Lele AK, Mashelkar R. Molecular model for wall slip: role of convective constraint release. Macromolecules. 2001;34(10):3412–20.
55 Joshi YM, Lele AK. Dynamics of end-tethered chains at high surface coverage. J Rheol. 2002;46(2):427–53.
56 Liu GX, Ma HW, Lee H, Xu HD, Cheng S, Sun H, et al. Long-chain branched polymers to prolong homogeneous stretching and to resist melt breakup. Polymer. 2013;54(24):6608–16.
57 Ge S, Zhu X, Wang S-Q. Watching shear thinning in creep: Entanglement–disentanglement transition. Polymer. 2017;125:254–264.
58 Szamel G, Schweizer KS. Reptation as a dynamic mean-field theory: self and tracer diffusion in a simple model of rodlike polymers. J Chem Phys. 1994;100(4):3127–41.
59 Sussman DM, Schweizer KS. Microscopic theory of the tube confinement potential for liquids of topologically entangled rigid macromolecules. Phys Rev Lett. 2011;107(7):078102.
60 Wang B, Guan J, Anthony SM, Bae SC, Schweizer KS, Granick S. Confining potential when a biopolymer filament reptates. Phys Rev Lett. 2010;104(11):118301.
61 Sussman DM, Schweizer KS. Microscopic theory of entangled polymer melt dynamics: flexible chains as primitive-path random walks and supercoarse grained needles. Phys Rev Lett. 2012;109(16):168306.
62 Sussman DM, Schweizer KS. Entangled polymer chain melts: orientation and deformation dependent tube confinement and interchain entanglement elasticity. J Chem Phys. 2013;139(23):234904.
63 Sussman DM, Schweizer KS. Communication: effects of stress on the tube confinement potential and dynamics of topologically entangled rod fluids. J Chem Phys. 2011;135(13):131104.
64 Sussman DM, Schweizer KS. Microscopic theory of quiescent and deformed topologically entangled rod solutions: general formulation and relaxation after nonlinear step strain. Macromolecules. 2012;45(7):3270–84.
65 Sussman DM, Schweizer KS. Entangled rigid macromolecules under continuous startup shear deformation: consequences of a microscopically anharmonic confining tube. Macromolecules. 2013;46(14):5684–93.
66 Schweizer KS, Sussman DM. A force-level theory of the rheology of entangled rod and chain polymer liquids. I. Tube deformation, microscopic yielding and the nonlinear elastic limit. J Chem Phys. 2016;145:214903.
67 Schweizer KS, Sussman DM. A force-level theory of the rheology of entangled rod and chain polymer liquids. II. Perturbed reptation, stress overshoot, emergent convective constraint release and steady state flow. arXiv:1604.03972.
68 Baig C, Mavrantzas VG, Kroger M. Flow effects on melt structure and entanglement network of linear polymers: results from a nonequilibrium molecular dynamics simulation study of a polyethylene melt in steady shear. Macromolecules. 2010;43(16):6886–902.
69 Cao J, Likhtman AE. Shear banding in molecular dynamics of polymer melts. Phys Rev Lett. 2012;108(2):5.
70 Wischnewski A, Monkenbusch M, Willner L, Richter D, Likhtman A, McLeish T, et al. Molecular observation of contour-length fluctuations limiting topological confinement in polymer melts. Phys Rev Lett. 2002;88(5):058301.

71 Maconnachie A, Allen G, Richards RW. Small-angle neutron scattering from a polymer coil undergoing stress relaxation. Polymer. 1981;22(9):1157–60.
72 Boué F, Nierlich M, Jannink G, Ball R. Polymer coil relaxation in uniaxially strained polystyrene observed by small angle neutron scattering. J Phys. 1982;43(1):137–48.
73 Boué F, Nierlich M, Osaki K, editors. Dynamics of molten polymers on the sub-molecular scale. Application of small-angle neutron scattering to transient relaxation. Faraday Symposia of the Chemical Society; 1983: Royal Society of Chemistry.
74 Boué F. Transient relaxation mechanisms in elongated melts and rubbers investigated by small angle neutron scattering. Polymer physics. Springer, Berlin; 1987. pp. 47–101.
75 Bent J, Hutchings L, Richards R, Gough T, Spares R, Coates PD, et al. Neutron-mapping polymer flow: scattering, flow visualization, and molecular theory. Science. 2003;301(5640):1691–5.
76 Heinrich M, Pyckhout-Hintzen W, Allgaier J, Richter D, Straube E, McLeish T, et al. Small-angle neutron scattering study of the relaxation of a melt of polybutadiene H-polymers following a large step strain. Macromolecules. 2004;37(13):5054–64.
77 Blanchard A, Graham R, Heinrich M, Pyckhout-Hintzen W, Richter D, Likhtman A, et al. Small angle neutron scattering observation of chain retraction after a large step deformation. Phys Rev Lett. 2005;95(16):166001.
78 Graham R, Bent J, Hutchings L, Richards R, Groves D, Embery J, et al. Measuring and predicting the dynamics of linear monodisperse entangled polymers in rapid flow through an abrupt contraction. A small angle neutron scattering study. Macromolecules. 2006;39(7):2700–9.
79 Graham RS, Bent J, Clarke N, Hutchings LR, Richards RW, Gough T, et al. The long-chain dynamics in a model homopolymer blend under strong flow: small-angle neutron scattering and theory. Soft Matter. 2009;5(12):2383–9.
80 de Gennes PG. Molten polymers in strong flows: a nonclassical proposal. MRS Bull. 1991;16(1):20–1.
81 Wang Z, Lam C, Chen W, Wang W, Liu J, Liu Y, et al. Fingerprinting molecular relaxation in deformed polymers.. Phys Rev X.. 2017;7:031003.

17

"Anomalous" Phenomena

We have covered most experimental aspects of nonlinear polymer rheology in Chapters 6–15. We intentionally left out a few more recently emerging topics in the field. According to the conventional wisdom, the behavior to be presented in this chapter would seem to be anomalous. Thus, these new phenomena are highly interesting and appear to further validate the need to develop an alternative theoretical framework for polymer dynamics. Since further systematic studies should be carried out in the near future, we will be brief in the description of these new phenomena. As a necessary background, the following section first examines whether all rheometric measurements in the literature satisfied the isothermal condition or not. This topic has not been sufficiently addressed in the literature. As an extension of Section 17.1, the subsequent analysis in Section 17.2 describes characteristic behavior associated with significant chain stretching and reveals emergence of considerable internal energy during melt stretching to high strains. Section 17.3 explores the breakdown of time–temperature superposition principle in transient nonlinear responses. Section 17.4 shows the striking behavior of strain hardening in startup simple shear. Section 17.5 discusses the alleged differences in the "steady-flow behavior" between entangled polymer solutions and melts. Section 17.6 indicates the occurrence of transient glassy response observed at unconventionally low temperatures that are still considerably above the glass transition temperature.

17.1 Essence of Rheometric Measurements: Isothermal Condition

We begin our discussion with the statement of the first law of thermodynamics. We have the rate of change of the enthalpy h per unit volume to arise from energy transfer due to heat exchange with the surroundings and the mechanical power supplied to a fixed volume[1]

$$\mathrm{d}h/\mathrm{d}t = -\nabla \cdot \mathrm{J}_\mathrm{q} + \mathrm{d}w/\mathrm{d}t, \tag{17.1}$$

where the first term represents heat inflow, as determined by the divergence of the heat flux J_q. Here the mechanical power density is given by

$$\mathrm{d}w/\mathrm{d}t = \sigma\dot{\gamma} \text{ or } \sigma_\mathrm{E}\dot{\varepsilon}, \tag{17.2}$$

for simple shear or uniaxial extension, respectively. By definition, the maximum possible temperature increase due to the mechanical deformation in the absence of any heat inflow, namely, for $\mathrm{J}_\mathrm{q} = 0$, is given by

$$\rho c_\mathrm{p}\frac{\mathrm{dT}}{\mathrm{d}t} = \sigma\dot{\gamma} \text{ or } \sigma_\mathrm{E}\dot{\varepsilon}. \tag{17.3}$$

Nonlinear Polymer Rheology: Macroscopic Phenomenology and Molecular Foundation, First Edition. Shi-Qing Wang.

Companion Website: www.wiley.com/go/nonlinearpolymerrheology

This result is only true *if* any other forms of internal energy are negligibly small. In this case, Eq. (17.3) can be integrated to obtain the maximum possible temperature increase, namely

$$T - T_0 = (1/\rho c_p)\int_0^{\gamma} d\gamma'\sigma(\gamma') \tag{17.4}$$

or

$$T - T_0 = (1/\rho c_p)\int_0^{\varepsilon} d\varepsilon'\sigma_E(\varepsilon') = (1/\rho c_p)\int_1^{\lambda} d\lambda'\sigma_{engr}(\lambda'), \tag{17.5}$$

for shear or extension, respectively, where T_0 is the initial temperature.

In general, any temperature increase reflects an increase in the internal kinetic energy. If we take any volume change to be negligibly small under constant one atmosphere pressure, the enthalpy change is identical to the change of total internal energy u. Denoting the internal potential energy as u_2, we see u_2 as a component that does not vary with the sample temperature. The left-hand side of Eq. (17.1) can then be written as

$$\frac{dh}{dt} = \frac{du}{dt} = \rho c_p \frac{\partial T}{\partial t} + \frac{\partial u_2}{\partial t} \tag{17.6}$$

in the absence of any volume change.

As discussed in Section 2.4, chain dynamics in polymer solutions and melts are temperature dependent. The reason to require an isothermal condition in rheometric measurements is to retain the experimenter's ability to prescribe a given value for *Wi*. According to Eqs. (17.4) and (17.5), there could be a temperature increase in the system, leading to an actual value of *Wi* lower than the prescribed value. Temperature changes are more likely if high stress lasts over a large amount of strain and there is not an adequate heat sink to maintain the sample temperature. Rheological tests can be regarded as acceptable so long as the temperature change is small enough, namely, the mechanical energy involved in the rheological measurements should be small enough so that there can only be a negligible increase in the sample temperature. Otherwise, we must make sure that the heat removal mechanism is effective enough that it prevents an uncontrolled temperature rise. Alternatively, we should specify that the measurement was not carried out at a constant temperature.

To justify a claim of isothermal conditions, we need to evaluate the mechanism for heat removal in various rheometric apparatuses. We note that under extreme conditions, for example, when non-Gaussian chain deformation is prevalent, significant mechanical work could have been spent to increase the internal energy in a form that is temperature insensitive, as discussed in Section 17.2. Moreover, heat production occurs in two different forms. The reversible portion is associated with the conformational entropy reduction, and the irreversible part arises from viscous dissipation. Since different heat-transfer mechanisms are involved for shear and extensional devices, respectively, we carry out further analyses for each class of device separately in the following sections.

17.1.1 Heat Transfer in Simple Shear

From Eq. (17.4) we can readily estimate the maximum temperature rise corresponding to a given stress versus strain curve. For SBR melts, we have the mass density $\rho = 933\,\text{kg/m}^3$ and the specific heat $c_p = 1.9\,\text{kJ/kg K}$. Referring to the data in Fig. 7.7, we see that the total mechanical work up to $\gamma_y \sim 7.0$ for the highest rate of $5.2\,\text{s}^{-1}$ is on the order of $14\,\text{MJ/m}^3$. According to Eq. (17.4), this means that there could be a temperature rise of 7.9 °C if the heat removal were absent and if all the mechanical work turned into internal kinetic energy. Referring to Fig. 17.1, such a temperature increase would speed up the chain dynamics by a factor of 2.5 for SBR500K at room temperature (25 °C). This number is the upper bound.

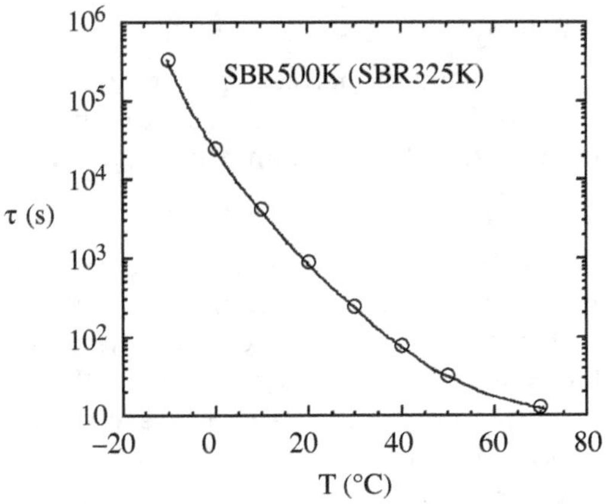

Figure 17.1 Temperature dependence of the terminal relaxation time τ of SBR325K, which contains equivalent information to the WLF (William–Landel–Ferry) shift factor a_T (cf. Eq. (2.53)).

In reality, there is heat conduction in shear rheometric setups; the conduction ensures the isothermal condition, up to a certain limit. We should identify this limit in order to have reliable measurements. Let us consider the geometry of parallel plates for simple shear, separated by a gap distance H. Adopting the coordinate system displayed in Fig. 1.1(a), we rewrite Eq. (17.1) to have

$$\rho c_p \frac{\partial T}{\partial t} = \left(\sigma\dot{\gamma} - \frac{\partial u_2}{\partial t}\right) - \frac{\partial J_q}{\partial y}, \tag{17.7}$$

where use is made of Eq. (17.6), the flux $J_q = \kappa\frac{\partial T}{\partial y}$. The thermal conductivity κ is ca. 0.2 W/mK for polymeric liquids. Equation (17.7) can be further simplified to estimate the temperature rise as a function time for a startup shear with rate $\dot{\gamma}$. In the small gap limit, namely, ignoring any edge effect, the following equation[2] approximately depicts the sample temperature T

$$\rho c_p \frac{dT}{dt} = \left(\sigma\dot{\gamma} - \frac{du_2}{dt}\right) - 4\kappa(T - T_s)/H^2, \tag{17.8}$$

where T_s is the surface temperature of the specimen, and the gap distance H is characteristically around 1 mm.

Since we do not know what fraction of the mechanical energy is spent to increase u_2, we can only estimate an upper bound to the temperature rise by setting u_2 to zero in the absence of any *in situ* temperature measurement. For simple shear of entangled polymeric liquids, the term in Eq. (17.8) involving the time derivative of u_2 may be negligible. In this limit, the general solution of Eq. (17.8) is

$$T - T_s = (1/\rho c_p)\int_0^t \dot{\gamma} ds \;\; \exp(-s/\tau_{tc})\sigma(t - s), \tag{17.9}$$

where the time constant τ_{tc} for thermal conduction characterizes how fast the system reaches thermal equilibration and is given by

$$\tau_{tc} = \rho c_p H^2/4\kappa \cong 1.0\,\text{s}. \tag{17.10}$$

Here the second approximate equality follows from inserting the values for ρ and c_p from the beginning of Section 17.1.1 and taking $H = 10^{-3}$ m and $\kappa = 0.2$ W/mK. In the absence of any heat loss to the conduction, namely, setting κ to zero, Eq. (17.9) reduces to Eq. (17.4). For $t \gg \tau_{tc}$, the solution in Eq. (17.9) is

$$T - T_s \approx \sigma(t)\dot{\gamma}(H^2/4\kappa), \tag{17.11}$$

which in the limit $\frac{du_2}{dt} = 0$ can be obtained by setting the left-hand side of Eq. (17.8) to zero. This is a point of thermal equilibration. In terms of a tolerable temperature rise $(T - T_s)$, Eq. (17.9) or (17.11) dictates the upper limit of the mechanical power density $dw/dt = \sigma\dot{\gamma}$ that can be applied. If the maximum temperature rise must be lower than 1 °C, we need to have

$$\sigma\dot{\gamma} < 0.8\ \text{MPa/s}, \quad \text{for simple shear with } H \sim 1\,\text{mm and } \kappa = 0.2\,\text{W/mK} \qquad (17.12)$$

for the thermal conduction to efficiently prevent temperature increase from exceeding 1 °C. Examining the data in Fig. 7.7, we see that the heat conduction is not sufficiently fast, when there is one degree of temperature difference, at the highest rates: For example, near peak stress, for $\dot{\gamma} = 5.2\,s^{-1}$ dw/dt is larger than $10\,MJ/m^3$ s, and the condition of Eq. (17.12) is not met. The reality is more complicated because some of the mechanical energy may be spent to increase the internal potential energy u_2. Up to $\gamma_y = 7.0$, there is only limited irrecoverable viscous deformation. Thus, the temperature rise should be much less than the upper bound of 7.9 °C, estimated in the beginning paragraph of this section. Since polymer dynamics have strong temperature dependence, for example as shown in Fig. 17.1, in principle high shear rates can be avoided by conducting experiments at lower temperatures. It is useful to note that the same shearing condition, $5.2\,s^{-1}$ at 25 °C, namely, the same $Wi_R = 68$, would, according to Fig. 17.1, involve a shear rate of as low as $0.1\,s^{-1}$ at 0 °C. Correspondingly, dw/dt should also be lower by a factor of 50, namely, around $0.25\,MJ/m^3$ s, which would satisfy Eq. (17.12). In general, for a given Wi, the isothermal condition can be more readily met at lower temperatures where smaller rates are required. We will take advantage of this point in Section 17.3 to examine the temperature dependence of nonlinear melt rheology.

Due to edge effects, shear rheometric apparatuses are usually not employed to study entangled melts. However, data such as those in Fig. 7.7 are not unique. By employing cone-partitioned plate as described in Section 5.2, polystyrene melts have been subjected to high rates of shear, reaching dw/dt in excess of $10\,MJ/m^3$ s, as shown in Fig. 2 of Ref. [3]. Even for entangled solutions, an extremely high shear rate can produce dw/dt in excess of $1\,MJ/m^3$ s, as shown in Fig. 7.4 at $\dot{\gamma} = 90\,s^{-1}$. However, according to Eq. (17.4) the heat accumulation is not sufficient to cause a noticeable temperature rise for 1,4-polybutadiene solutions (examined in Chapters 7, 9, 12, and 14). When Eq. (17.12) is satisfied, no heat can accumulate to raise the sample temperature. When the condition of Eq. (17.12) is not met, it is not necessarily the case that the sample will heat up appreciably. Because the system has a large volumetric heat capacity ρc_p, the temperature will not increase appreciably unless the shear stress is sufficiently large over a wide range of strains. For example, according to Eq. (17.4), no appreciable temperature increase can occur for any of the curves in Fig. 7.4.

In summary, the preceding analysis should always be performed in order to ensure that shear rheometric measurements can meet the desired accuracy. Section 17.1 should be regarded as an integral part of nonlinear polymer rheology studies that involve large stresses and shear rates. Whenever dw/dt in Eq. (17.1) exceeds $0.8\,MJ/m^3$ s, whether a significant temperature rise would take place or not still depends on the total accumulated mechanical work, as shown by Eq. (17.4). The temperature increase during the study can be minimized either by carrying out rheometric measurements at lower temperatures to keep the mechanical power low or by adopting a smaller gap distance H to increase the efficiency of heat conduction.

17.1.2 Heat Transfer in Uniaxial Extension

From Fig. 7.28, we can estimate the accumulated mechanical work density up to the peak engineering stress σ_{engr} at $\lambda_y \sim 5$. At the highest extension rate $15\,s^{-1}$, the integral in Eq. (17.5) produces an energy density larger than $3\,MJ/m^3$. This energy density could result in a temperature rise of up to 1.7 °C. Once again, such a modest temperature increase does not appreciably alter the chain dynamics up to the yield point and thus does not affect the conclusions made in Section 7.5.

In uniaxial extension, air cooling and thermal radiation are the mechanisms that remove heat. The convective heat flux J_q is given by $J_q = \theta(T_s - T_a)$, where the convection heat transfer coefficient θ ranges from 5 for free convection and up to 250 W/m² K for forced convection. For a sheet-like specimen, by analogy with the preceding analysis for simple shear, we have the following equation[4,5] to approximately describe the temperature rise during extension at a Hencky rate of $\dot{\varepsilon}$

$$\rho c_p \frac{dT}{dt} = \left(\sigma_E \dot{\varepsilon} - \frac{du_2}{dt}\right) - 2\theta(T - T_a)/H, \tag{17.13}$$

which follows from insertion of Eq. (17.6) into Eq. (17.1). Here T is an average specimen temperature, slightly higher than[1] T_s, and T_a is the ambient temperature. For easier analysis, from here onward, we approximate T as T_s. For a thinner sample, the air cooling is more effective. The characteristic time τ_{ac} varies with the sample thickness $H(t)$ as

$$\tau_{ac}(t) = \rho c_p H(t)/2\theta. \tag{17.14}$$

When the specimen thickness is on the order of 1 mm, even if θ reaches 100 W/m² K, τ_{ac} is still as long as ca. 10 s. Like the time constant τ_{tc} in Eq. (17.10), τ_{ac} shows the rapidity of heat removal. Using realistic values, we show that the efficiency of heat exchange is much poorer for melt extension in air. Specifically, not much air cooling can occur within τ_{ac}. By immersing the sample in a circulating liquid such as water we can significantly reduce τ_{ac} because θ can jump by more than one decade.

Since air cooling is ineffective, with $\tau_{ac} > 10$ s, any Hencky rate $\dot{\varepsilon} \geq 1\ s^{-1}$ would produce nearly adiabatic melt extension because the test typically terminates due to the tensile strain localization well before reaching $\lambda_{ac} = \exp(\dot{\varepsilon}\ \tau_{ac}) > 2.2 \times 10^4$, as shown in Chapter 11. For example, the onset of necking (regime III) and melt rupture (regime IV) occurs in uniaxial extension of room-temperature SBR melts on a time scale considerably shorter than τ_{ac}, as shown by the data in Fig. 7.28. These experiments were conducted at room temperature without any forced air convection. Therefore, at some extension rates the adiabatic condition was met. Equation (17.5) applies, unless the change in u_2 is nonnegligible, in which case Eq. (17.5) is replaced by

$$T - T_a = (1/\rho c_p) \left[\int_0^{\varepsilon} d\varepsilon' \sigma_E(\varepsilon') - \Delta u_2\right], \tag{17.15}$$

where Δu_2 represents the change in the internal potential energy. Δu_2 could be significant at high extension rates and is definitely not negligible in the rupture regime IV, as shown in Section 17.2. In contrast to simple shear where steady shear at long times is possible, so that it is meaningful to consider the condition $t \gg \tau_{tc}$, here the time of interest can be much shorter than τ_{ac}.

Some further discussion can be made without solving Eq. (17.13). We need $(T - T_a) = 10$ °C to have a cooling rate of 1 MP/s for $\theta = 50$ W/m²K and H = 1 mm. According to Eq. (17.5), such a large temperature increase corresponds to a mechanical work density of 18 MJ/m³ for SBR

1 There is a temperature gradient across the specimen thickness H, on the order of $(T - T_s)/H$ due to the air cooling. Since the heat exchange with the outside can only occur via air cooling at the surfaces, the heat flux due to the thermal conduction cannot exceed the convective heat flux, that is, $0 < \kappa(T - T_s)/H \leq 2\theta(T_s - T_a)$ holds true at all times. This expression can be rewritten as $(T_s - T_a) < (T - T_a) \leq (1 + \alpha)(T_s - T_a)$, where $\alpha = 2hH/\kappa$. In the limit of $\alpha \ll 1$, $T = T_s$. Given $\theta = 20$ W/m² K, H = 0.5 mm, and $\kappa = 0.2$ W/mK, we have $\alpha = 0.1$.

melts. This situation can actually take place well beyond the yield point at the highest rates, as shown in Fig. 7.28. Conversely, air cooling should be effective when

$$\sigma_E\dot{\varepsilon} < 0.1\,\text{MPa/s}, \quad \text{for extension with tolerance of } (T - T_a) \sim 1\,^\circ\text{C}, \tag{17.16}$$

which can be satisfied for $\dot{\varepsilon} \sim 0.1\ \text{s}^{-1}$ with $\sigma_E \sim 1$ MPa.

In summary, with $H_0 = 1$ mm, for several SBR melts uniaxial extension in regime III (necking) is approximately adiabatic. To achieve the isothermal condition, it is necessary to produce a maximum θ using forced convection or immersing the specimen in a circulating liquid and to examine very thin sheet-like specimens at low temperature to lower the actual Hencky rate $\dot{\varepsilon}$. When the time–temperature superposition principle holds, a given effective rate can be reached at a significantly lower $\dot{\varepsilon}$ by carrying out the experiment at a lower temperature. In practice, the temperature rise should be estimated theoretically from Eq. (17.13), using a realistic value for θ and neglecting the du_2/dt term, or should be measured experimentally using, for example, IR thermal imaging as seen in the following section.

17.2 Internal Energy Buildup with and without Non-Gaussian Extension

The discussion of the preceding section allows us to describe the nature of melt extension. The emergence of extensional strain localization described in Chapter 11 clearly indicates that uniform extension does not reach the point of homogeneous flow. Experimental determination of any internal potential energy buildup u_2 during startup extension can provide useful information on a range of questions, for example, (a) whether the state of deformation up to the onset of inhomogeneous extension is still highly elastic, signified by the peak in the engineering stress σ_{engr} and (b) whether the regime IV of melt rupture accumulates considerably more potential energy.

We can integrate Eq. (17.13) over time to evaluate the accumulated mechanical energy density w, the internal energy density u, and the heat loss q due to air cooling. By definition, we have

$$w(\lambda) = \int_1^{\lambda} d\lambda' \sigma_{engr}(\lambda') = u + q. \tag{17.17}$$

The total energy density u is comprised, in principle, of two components, namely,

$$u = u_1 + u_2 = \rho c_p(T_f - T_i) + u_2. \tag{17.18}$$

where T_i and T_f are the initial and final temperatures and c_p is assumed to change little during extension. Using IR thermal imaging to track the specimen's temperature, we can also estimate the heat loss q as

$$q = \int_0^{t} dt'\, 2\theta[T(t') - T_a]/H(t'), \tag{17.19}$$

where T can be measured by an IR thermal imaging camera, T_a is the ambient temperature, and H is the time-dependent thickness of the specimen. Both the first term in Eq. (17.18) and q in Eq. (17.9) can be measured from the time-dependent temperature. Thus, we can find the magnitude of u_2 from an experimental evaluation of Eqs. (17.17)–(17.19). As an example, let us examine melt extension in regime III, namely, a Hencky rate that produces yielding and necking. We find that there is a sizable u_2 even in the case of necking. Rupture typically takes place for Wi_R well above unity, as depicted in Fig. 11.3. Region IV with $Wi_R \gg 1$ is a peculiar regime for two reasons. First, the engineering stress grows monotonically. Second, because of the insufficient yielding, the extending melt eventually ruptures as though it was cross-linked.

Let us first investigate the same data as those presented in Fig. 11.6, for SBR616 at a Hencky rate of 1 s^{-1}. Adding the corresponding temperature reading T(t), we have Fig. 17.2. The various terms in Eqs. (17.17)–(17.19) can be estimated from the data in Fig. 17.2, as shown in Fig. 17.3. It is first confirmed that the measurement is indeed adiabatic; q stays close to zero during this "fast" melt stretching. As a consequence, the sample's temperature increases noticeably, so that a significant portion of the mechanical energy is spent to increase the internal kinetic energy u_1 (squares). More important, the internal energy u_2 is sizable and continues to increase at the onset of inhomogeneous extension, which corresponds to the last data point in Fig. 17.3. These data indicate that a styrene–butadiene rubber under extension is far from a state of fully developed flow before it suffers strain localization. Our conclusion here is consistent with previous interpretations[7,8] concerning why melt stretching does not reach a homogeneous state of flow for $Wi > 1$: When a melt is well entangled, it appears improbable for chain entanglement to "sort out," namely, to disentangle, uniformly in a macroscopic specimen. For well-entangled melts, it would be incorrect to characterize steady-flow properties by taking the last data point of the measurement as being descriptive of a fully developed steady state.

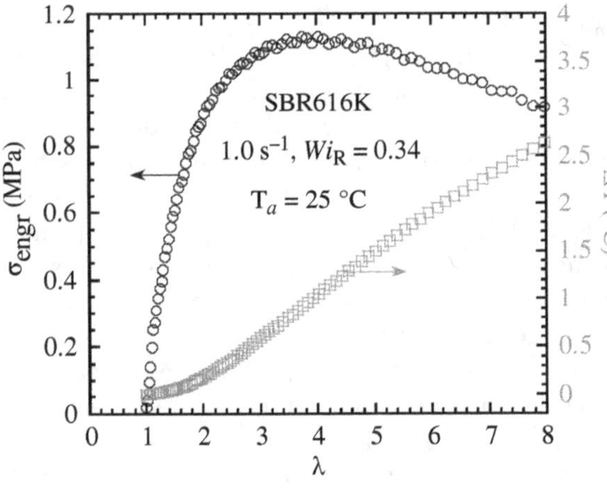

Figure 17.2 Engineering stress σ_{engr} (circles) along with the corresponding temperature rise (squares) as a function of stretching ratio λ at a Hencky rate 1.0 s^{-1} at 25 °C. *Source*: Data taken from Lin *et al.* 2017 [6].

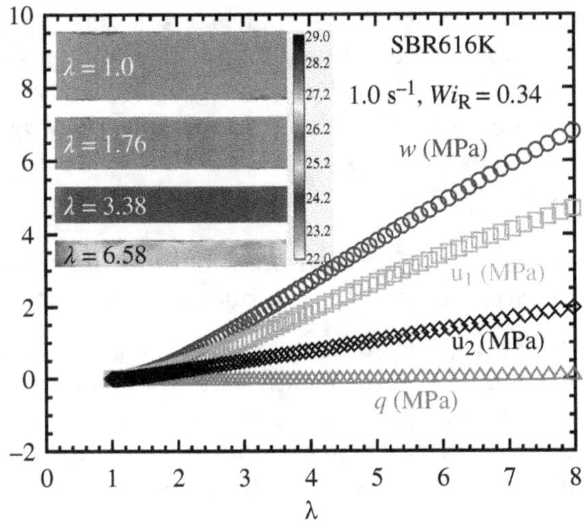

Figure 17.3 Mechanical work density w (circles) and internal stored energy density u_2 (diamonds) as well as u_1 and q as a function of stretching ratio λ. Since q is negligibly low relative to w, the test is essentially adiabatic. The inset shows the IR images at the different stages of extension and corresponding temperature rise. *Source*: Data taken from Lin *et al.* 2017 [6]. (*See color plate section for the color representation of this figure.*)

17.3 Breakdown of Time–Temperature Superposition (TTS) during Transient Response

In polymer rheology, it is convenient and powerful to apply the principle of time–temperature superposition. Time–temperature superposition is usually applied to broaden the range of time scale on which polymer dynamics are probed. The principle of time–temperature superposition is often observed to hold true because the overall (terminal) chain dynamics and segmental dynamics are so far apart in their time scales for well-entangled polymers that we do not typically observe both at a single temperature. When steady shear involves relatively long time scales, which is the case when the Weissenberg number *Wi* is not very large, time–temperature superposition works well.[9,10] *Steady* shear properties do not offer us an opportunity to detect molecular dynamics over a wide range of time scales. To have access to both chain and segmental dynamics, either chain dynamics are made fast by choosing a lower molecular weight polymer or two different techniques are applied.[11] In short, within rheometry, it can be challenging to discover failure of the time–temperature superposition in linear viscoelastic characteristics or steady shear behavior. Plazek applied the creep mode on relatively low-molecular-weight melts, whose terminal chain dynamics lay on time scales not far from the time scales for their segmental dynamics.[12]

Little is known about the validity of the principle of time–temperature superposition regarding transient nonlinear rheological responses of entangled polymers. Failure of time–temperature superposition in nonlinear rheometric measurements would imply that the dynamics controlling nonlinear responses have different temperature dependence from those revealed by small-amplitude oscillatory shear data that defines the terminal relaxation time $\tau(T)$. If time–temperature superposition fails at a fixed Weissenberg number (a product of the deformation rate and τ), the transient stress response may no longer be the same at different temperatures. A breakdown of time–temperature superposition would actually challenge the framework of the bead-spring model on which all molecular models are built because the bead-spring model assumes that the temperature dependences of the relaxation times are the same on all length scales. Specifically, in bead-spring models, the temperature dependence of all macroscopic quantities, such as melt viscosity, explicitly originates from the temperature dependence of the bead friction. In reality, the terminal relaxation dynamics involve significant averages over the local dynamics.[11] Such averaging effects have yet to be adequately built into the theoretical formulations. By examining the nonlinear transient responses of entangled polymer solutions and melts under startup deformation, we can determine whether time–temperature superposition holds for transient nonlinear rheology. If it does not hold, the interchain grip force must be controlled by local chain dynamics that are different from the overall dynamics quantified by the reptation time. Such a scenario is entirely consistent with the analysis presented in Section 16.4.2 on the origins of the grip force and the 1/3 scaling law.

17.3.1 Time–Temperature Superposition in Polystyrene Solutions and Styrene–Butadiene Rubbers: Linear Response

The principle of time–temperature superposition was described in Section 2.4.1. Polymer dynamics are more sluggish at a lower temperature. Empirically, the time–temperature equivalence can be numerically depicted in terms of the WLF shift factor a_T given by Eq. (2.53). Here we present a concrete example to illustrate this principle. We consider an entangled solution of 5 wt% of polystyrene with molecular weight 20,000 kg/mol in tricresyl phosphate (TCP)

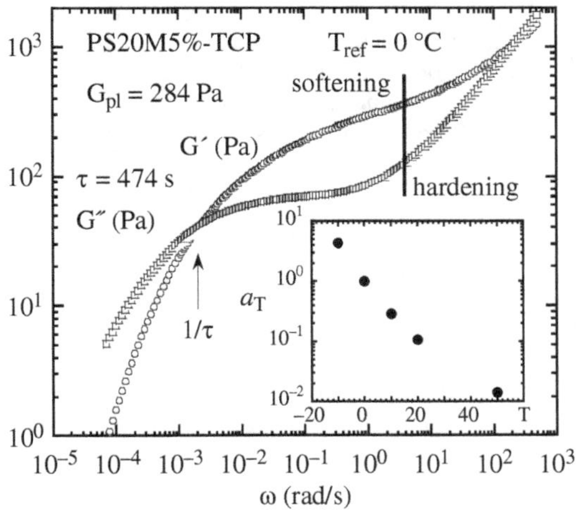

Figure 17.4 Master curves of storage and loss moduli G′ and G″ as a function of oscillation frequency from SAOS measurements at different temperatures from 50 to −10 °C, where the inset shows the WLF shift factor a_T as a function of temperature. This polystyrene solution is examined in startup shear in Figs. 17.6, 17.8, and 17.9 to show a strain softening-to-hardening transition (SHT) at the rate indicated by the vertical line. *Source*: Data taken from Liu and Wang 2016 [13].

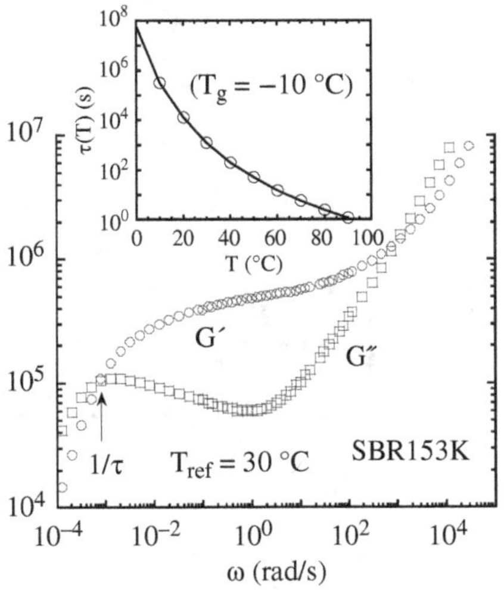

Figure 17.5 Master curves of storage and loss moduli G′ and G″, where the inset shows the WLF shift in terms of the terminal relaxation time τ of SBR153K, which is the plot of the table in Fig. 14.2. *Source*: Data taken from Lin *et al.* 2017 [6].

(Aldrich 1330-78-5). Taking $T_{ref} = 0\,°C$ as a reference temperature, we find by matching G′ and G″ as well as $\tan\delta = G''/G'$, as shown in Fig. 17.4, the WLF shift factor a_T varies with temperature. The temperature dependence of a_T appears in the inset of Fig. 17.4. As a second example, an SBR153K melt is subject to SAOS measurements over a wide range of temperature. The linear viscoelastic characteristics have already been presented in Fig. 14.2, in terms of storage and loss moduli G′ and G″. Because of its high vinyl content, SBR153K exhibits sharply slowed down dynamics below 20 °C, as shown in Fig. 17.5. The temperature dependence of the terminal relaxation time τ contains the same information as the information found in the WLF shift factor a_T.

17.3.2 Failure of Time–Temperature Superposition: Solutions and Melts

17.3.2.1 Entangled Polymer Solutions Undergoing Startup Shear

Entangled solutions containing a polymer having an intrinsically high glass transition temperature T_g reveal a violation of the time–temperature superposition principle. Polystyrene (PS) and poly(methyl methacrylate) (PMMA) are typical examples. The information from the SAOS data shows in Fig. 17.4 how the terminal chain dynamics depend on temperature according to WLF shift factor a_T. The question is whether the temperature dependences of nonlinear rheological responses follow a_T in Fig. 17.4.

For the solution described in Fig. 17.4, the stress responses to startup shear were studied at temperatures 0, −15, and 20 °C. At 0 °C, a shear rate of $4\,s^{-1}$ was applied. According to the SAOS measurements, this shear rate is equivalent to about $37\,s^{-1}$ at 20 °C and ca. $0.23\,s^{-1}$ at −15 °C, respectively. At these shear rates, if TTS holds, the same rheological responses

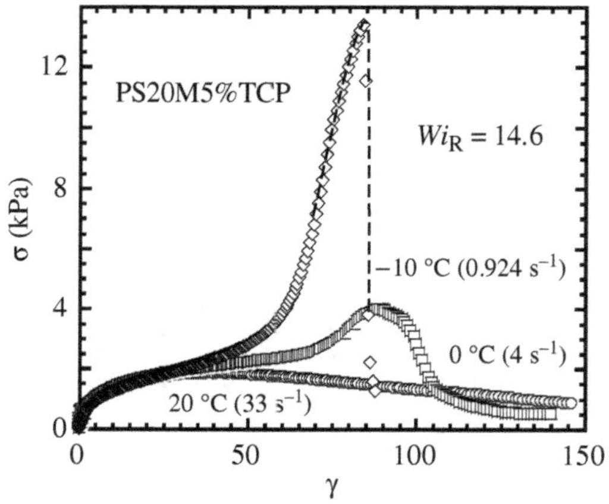

Figure 17.6 Example of TTS breakdown in an entangled polystyrene solution: Shear stress σ versus strain γ upon startup shear of PS20M5%-TCP at −10, 0 to 20 °C, involving the same $Wi_R = 14.6$. *Source:* Data taken from Liu and Wang 2016 [13].

should be found. They were not. Therefore, the data in Fig. 17.6 reveal a clear failure of the time–temperature superposition. The three stress versus strain curves cannot be more different, although they involve the same $Wi_R = 14.6$. At a common Wi_R, the PS solution displays a softening-to-hardening transition (SHT) as the test temperature is varied. The strain softening is characterized by a smooth stress overshoot. Strain hardening is revealed by the divergence of the resulting shear stress, as presented in more detail in Section 17.4.

In closing, we observe that the isothermal condition must be strictly valid during any investigation of time–temperature superposition. At a fixed effective rate, namely, a given value of *Wi*, the measurement at the highest temperature involves a highest rate, pushing the magnitude on the left-hand side of Eq. (17.12) to a higher value. If there is a sufficient temperature increase, the effective *Wi* can be lower than the value that was chosen according to the isothermal condition, resulting in a weaker response and thus a false failure of time–temperature superposition. For Eq. (17.12) to hold, the stress level should be low. For entangled solutions, the stress level is dictated, according to Eq. (2.36c), by the concentration dependence of the plateau modulus G_{pl}. A 5% PS solution is chosen in Fig. 17.6 to ensure that $\sigma\dot{\gamma} \sim 74\,\text{kPa/s} \ll 1600\,\text{kPa/s}$ (Eq. (17.12)).

17.3.2.2 Entangled Polymer Melts during Startup Extension

Before discussing the molecular implication of failure time–temperature superposition in nonlinear transient responses, let us consider a second example, this time involving melt extension.[6] According to the SAOS measurements shown in Fig. 17.5, the temperature dependence of the terminal relaxation time τ, taken to be the reciprocal of the crossover frequency ω_c at which $G' = G''$ and reported in the inset of Fig. 17.5, can be used to choose an effective Hencky rate at different temperatures. If TTS holds, we should have the same stress versus strain curves at different temperatures and a common value of *Wi*. The first example, discussed in the preceding section, provides a couple of useful guidelines. First, any small violation of the TTS may show up in a marked way when a softening-to-hardening transition is examined. Second, for isothermal testing the mechanical power density must be low enough to allow efficient heat transfer.

To ensure thermal equilibration of the sample with the surrounding air during melt extension, the Hencky rate should be sufficiently low. Because of its high vinyl content, SBR154K has sufficiently slow chain dynamics. In melt extension the phenomenon analogous to the softening-to-hardening transition is the yielding-to-rupture transition (cf. Chapter 11).

Figure 17.7 Example of the TTS breakdown based on an entangled melt: Engineering stress σ_{engr} upon startup extension of SBR153K as a function of stretching ratio λ under different conditions involving the same $Wi_R = 3.35$: (a) at a Hencky rate 0.00081 s^{-1} at 9.5 °C (circles); (b) at a Hencky rate 0.006374 s^{-1} at 15 °C (squares); (c) at a Hencky rate 0.0792 s^{-1} in a temperature range between 24.1 and 24.5 °C (diamonds). *Source*: Data taken from Lin et al. 2017 [6].

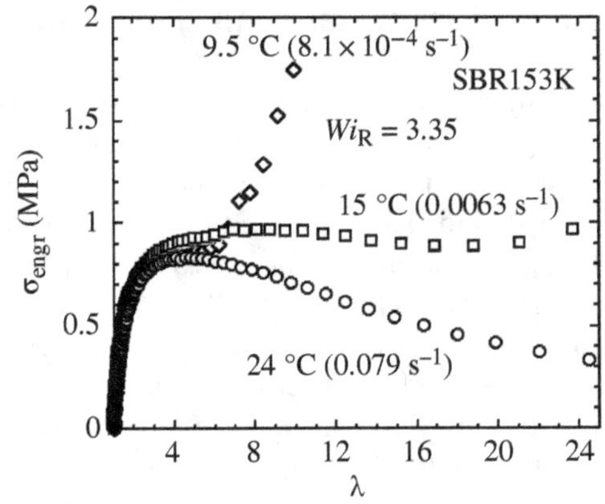

Because of its higher T_g, SBR153K can be studied at room temperature (24 °C) at a rate that is slow enough to achieve isothermal stretching. An infrared thermal-imaging camera can be employed to monitor any temperature rise. Such measurements confirm[6] that there is negligible temperature rise at 24 °C during melt extension tests such as those presented in Fig. 17.7. Figure 17.7 shows differences in the stress versus strain curves during startup extension that indicate a yielding-to-rupture transition at a common Wi_R. At 24 °C, the melt yields, but at 9.5 °C it shows a strong character of true strain hardening evidenced by the diverging stress (leading to melt rupture). The softening-to-hardening transition in Fig. 17.6 for startup shear shows similar behavior; the polystyrene solution undergoes normal yielding indicated by stress overshoot at 20 °C, but at −10 °C the same sample shows shear strain hardening along with a sharp cusp in the stress versus strain curve, which results from shear rupture (cf. Section 17.4).

Since the breakdown of the time–temperature superposition occurs at Hencky rates far below $1/\tau_e$ that represents the relaxation rate of a tube segment or entanglement strand, the failure of the time–temperature superposition is rather unexpected. This TTS failure is rather different from the observed differences in the temperature dependences for terminal relaxation dynamics and for segmental relaxation dynamics in linear viscoelastic characterization. Apparently, the dynamics dictating the nonlinear transient response of entangled melts to uniaxial extension follow a different temperature dependence from that followed by the terminal dynamics. In the language of Sections 16.2 and 16.4, the different temperature dependences for the nonlinear and the terminal dynamics imply that the intermolecular gripping force $f_{img}(T)$ varies with temperature in a manner that cannot be predicted from the SAOS data. We speculate that the relaxation dynamics of the hairpins in Eq. (16.16a) depend on T differently than the reptation time $\tau(T)$ does. As far as the temperature dependence is concerned, Eq. (16.16b) would not hold, in the sense that the temperature dependence of τ_h would be different from that of τ_R. The failure of time–temperature superposition in transient nonlinear responses is a serious challenge to all existing theoretical descriptions of polymer dynamics and may not have an origin similar to the one that produces the observed breakdown of the time–temperature superposition in linear viscoelastic characterization of polymer melts.[12,14–22] Unlike the previous demonstration, which involves a span of time scales to cover both terminal and segmental dynamics, here the local dynamics show up in the nonlinear rheological responses when the applied rate is high enough to access either the softening-to-hardening transition or the yielding-to-rupture transition.

17.4 Strain Hardening in Simple Shear of Some Polymer Solutions

Entangled polymer solutions are known to show stress overshoot during startup shear, as shown in Figs. 7.3 through 7.7. The downward deviation of the shear stress growth from the ideal elastic deformation limit, depicted by $\sigma = G_{pl}\gamma$, may be conveniently called "partial yielding," as indicated in Figs. 7.1(b) and 7.6. All entangled polymer solutions and melts show such behavior before the shear stress reaches its peak. Figure 7.8(b) shows the scaling characteristics of the yield stress as a function of Wi_R. According to Fig. 7.8(b), yielding is the prevailing response, up to $Wi_R \sim 50$, when the yield strain γ_y is greater than 10.

Unlike entangled polybutadiene solutions, on which Chapter 7 is based, entangled polystyrene (PS) solutions exhibit a surprising behavior unknown – until recently.[13] Instead of "strain softening," polystyrene solutions undergo strain hardening. Strain hardening during startup simple shear occurs only when PS solutions are examined at sufficiently low temperatures and high enough shear rates, as shown in Fig. 17.6. The experimental challenge comes from the risk to overload the normal force limit in a typical rotational rheometer. It appears that the earlier rheological studies of entangled PS solutions in the literature did not go to high enough rates and low enough temperatures.[23–25]

The well-studied polystyrene solutions display normal linear viscoelastic characteristics, as seen in Fig. 17.4. At T = 20 °C the solutions show yielding (circles in Fig. 17.6). Specifically, the PS20M(5.5%) solution of Z = 47 entanglements per chain has a reptation time $\tau = 474$ s and a Rouse time $\tau_R = 3.7$ s at 0 °C. Over the same range of Wi_R that has been applied to study polybutadiene solutions, Fig. 17.8 shows strain hardening at $Wi_R = 26$ ($\dot{\gamma} = 7\ s^{-1}$). The shear stress σ keeps increasing with increasing strain and shows a cusp at a strain above 60. The approach to the peak of σ has a concave shape instead of being a convex function of strain. The slope related to the sharp stress increase is five times the elastic plateau modulus G_{pl}. According to Fig. 7.8(b), at $Wi_R = 26$, the yield strain γ_y should be below 8 (independent of Z) if the PS-TCP solution is to behave similarly to the polybutadiene solutions. The PS-TCP solution clearly did not behave like the polybutadiene solutions. The curve at a stress rate of $4\ s^{-1}$ is equally surprising, with σ showing no sign of decrease until a shear strain of 90 units is reached.

To better illustrate the shear softening-to-hardening transition at a critical shear rate, in Fig. 17.9 we plot the first normal stress difference N_1 against the shear stress σ. Marching along the curve, upon reaching σ_{max}, N_1 moves upward for $3\ s^{-1}$ but then downward for 3.5 or $5\ s^{-1}$ as the curves turn "backward," to the left-hand side of the graph, indicating a transition between 3.0 and $3.5\ s^{-1}$. A "normal" U shape involving 1 and $3\ s^{-1}$ is expected from "strain softening": N_1

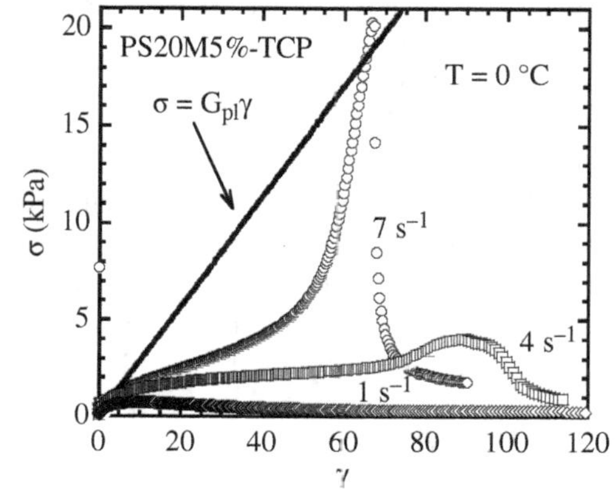

Figure 17.8 Shear stress versus shear strain upon startup shear at three shear rates 1, 4, and $7\ s^{-1}$, respectively, at T = 0 °C, showing a strain softening-to-hardening transition (SHT), where the line represents neo-Hookean rubber elastic response. *Source*: Data taken from Liu and Wang 2016 [13].

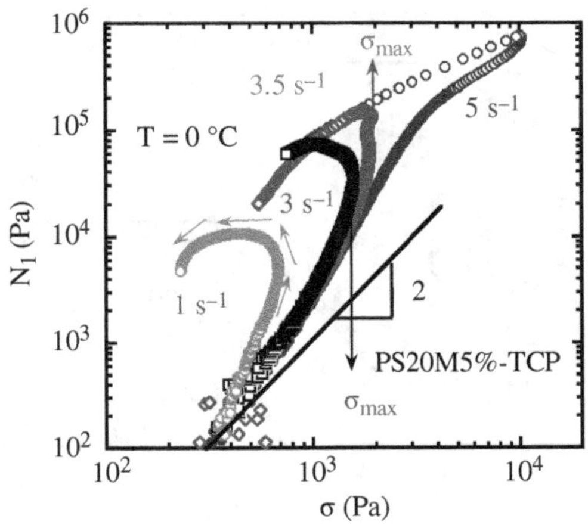

Figure 17.9 Normal stress N_1 versus shear stress σ during startup shear at four shear rates 1, 3, 3.5, and 5 s^{-1}, showing an SHT between 3 and 3.5 s^{-1}. *Source*: Data taken from Liu and Wang 2016 [13].

keeps rising beyond σ_{max}. When shear strain hardening occurs, N_1 ceases to increase beyond the peak of σ.

The cusp in the stress versus strain curve in Fig. 17.8 is a plausible indication of abrupt structural failure in the solution. The characteristic, namely, the cusp, is reminiscent of similar responses observed in aqueous micellar solutions of cetyl trimethyl ammonium bromide (CTAB),[26] as shown in Fig. 17.10. Contrary to shear banding after yielding, rupture occurs at the cusp, as shown by the PTV observations reported in Fig. 17.11, confirming that structural failure is the cause of the steep stress drop. For wormlike micelles, the cusp occurs at a common strain (Fig. 17.10). The rupture in micellar solutions is perhaps not so surprising because the surfactant aggregates are indeed readily breakable. For similar responses in entangled polymer solutions, chain scission likely also takes place. PTV observations can be carried out to identify the nature of the cusp for PS solutions. Preliminary results[13] indeed reveal a breakup profile similar to Fig. 17.11 immediately after the cusp.

Although non-Gaussian stretching leading to true tensile strain hardening has been extensively reported for uniaxial extension, as discussed in Chapter 8, it is not only unknown but also seems implausible that the entanglement mesh could lock up during simple shear. From the discussion in Section 8.4, we would not expect chain entanglement to tighten in simple shear, because the kinematics of shear is rather ineffective in stretching polymer chains, at least

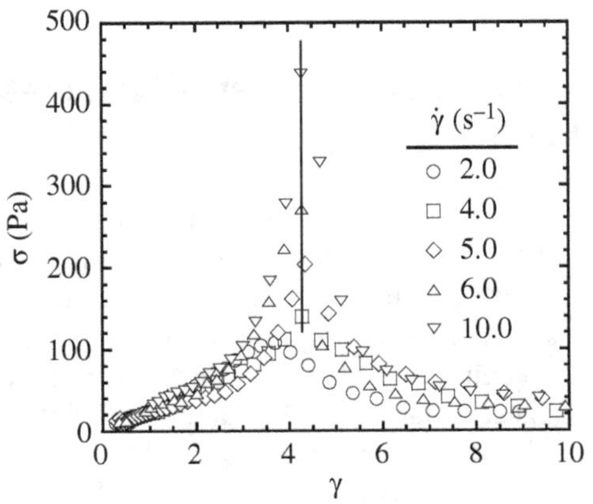

Figure 17.10 Shear stress versus strain at room temperature during startup shear of an entangled wormlike micellar solution – CTAB/NaSal (0.1:0.07mol/l) at five different shear rates, all showing a cusp at a common strain just above γ = 4. *Source*: Data taken from Boukany and Wang 2008 [26].

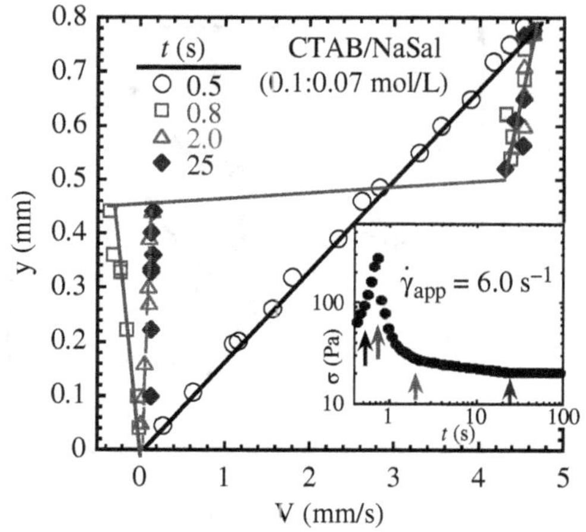

Figure 17.11 PTV observations of homogeneous shear before $t = 0.5$ s and rupture around 0.8 s and strong shear banding at later stages during startup shear at 6 s^{-1}. *Source*: Data taken from Boukany and Wang 2008 [26].

by comparison to uniaxial extension. New physics may also have come into play. When the parent polymer has a high glass transition temperature T_g, so that the experimental temperature is well below T_g, shear strain hardening such as that shown in Figs. 17.8 and 17.9 can take place. Preliminary results indicate that shear hardening can also occur in PMMA-TCP solutions as well as a polystyrene mixture.[13] We expect that shear strain hardening behavior is universal for all entangled solutions containing high T_g polymers that are studied at sufficiently low temperatures. When the testing temperature is sufficiently lower than T_g of the parent polymers, the interchain interactions may turn glassy due to the self-concentration effect[27] so that the grip force becomes sufficiently large to sustain the growing intrachain retraction force until the point of chain scission. Shear hardening thus gives us evidence that entanglement interactions stem from direct interchain contacts.

In passing, it is noteworthy that the same physical effect (leading to the failure of time–temperature superposition in Section 17.3.2.1 and shear strain hardening in this section) has caused the observed difference[28] in polystyrene solutions between the one (solution-4K) made with PS4K as the solvent and the other (solution-1K) made with PS1K as the solvent. Because of the significantly lower glass transition temperature of PS1K, the extensional rheological tests of solution-1K were carried out at significantly lower temperatures. Due to the self-concentration effects, polystyrene chains can have stronger intermolecular grip force in the "solution-1K" because of the slower segmental relaxation at the same effective shear rate. Our simplest explanation is far different from the hypothesis of nematic polymer–polymer interactions.

17.5 Lack of Universal Nonlinear Responses: Solutions versus Melts

Another "anomaly" in the literature of nonlinear polymer rheology is the reported "difference"[29–31] between entangled polymer solutions and melts, concerning their responses to startup uniaxial extension. Specifically, it is reported that the "steady-flow" extensional viscosity shows monotonic extensional thinning for entangled melts,[4] but shows extensional thickening for entangled solutions,[32] as sketched in Fig. 17.12. Such a difference superficially suggests a lack of universality and challenges the conventional wisdom. Various theoretical explanations have been offered.[33–35] However, upon a closer examination, it appears that this mystery is in part related to the question of whether entangled polymer solutions and

Figure 17.12 Depiction of different "steady-state" behaviors of entangled melts and solutions in uniaxial extension in terms of extensional viscosity $\eta_{E(ss)}$.

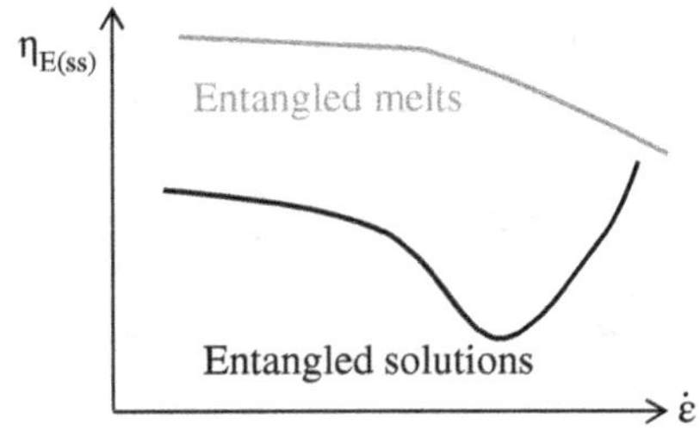

melts can attain uniform steady flow during startup uniaxial extension. We have asserted in Section 11.5 that tensile strain localization always takes place for $Wi > 1$ because there is localized yielding of the entanglement network. For example, the results in Fig. 17.2 indicate that the fully developed flow state is hardly attainable as the diamonds reveal continuation of elastic processes until specimen breakup. But many studies based on the filament stretching rheometry assumed to have access to steady flow.[36–38]

Let us take a look at some data from the literature. Figure 17.13 shows an example of the so-called steady-flow extensional viscosity of polystyrene (PS) melts acquired with filament stretching rheometry.[39] The data seem internally inconsistent, with the two polystyrenes of lower molecular weight showing "thickening," and the two polystyrenes of larger molecular weights displaying only "thinning" behavior. This ambiguity plausibly arises from the protocol to regard the data as coming from steady uniform extensional flow. In a second example, a polystyrene solution is shown in Fig. 17.14 to display the so-called extensional thickening,[32] in contrast to Fig. 17.13. Since *steady-flow* extensional viscosity should *only decrease* with rising Hencky rate, it is plausible that data such as those in Fig. 17.14 (and sketched in Fig. 17.12) do not represent viscosity from a fully developed flow state under the condition of homogeneous extension.

To further demonstrate this difference, the same group made an effort to ensure that a melt and corresponding solution have the same number Z of entanglements per chain.[29] Specifically, Fig. 17.15 compares the reduced "steady-state" viscosity of a PS melt with that of a PS mixture as functions of Wi_R. Two features are noteworthy. First, the melt shows significant extensional thinning, namely, decrease in the dimensionless viscosity with increasing Wi_R while the dimensionless viscosity of the solution stays independent of Wi_R. Second, the reduced viscosity of the solution is larger than that of the melt, at the same Wi_R.

Figure 17.13 "Steady-state" extensional viscosity at 130 °C as a function of Hencky rate for PS50K, PS100K, PS200K, and PS390K. *Source*: Replotted from Nielsen *et al*. 2006 [39].

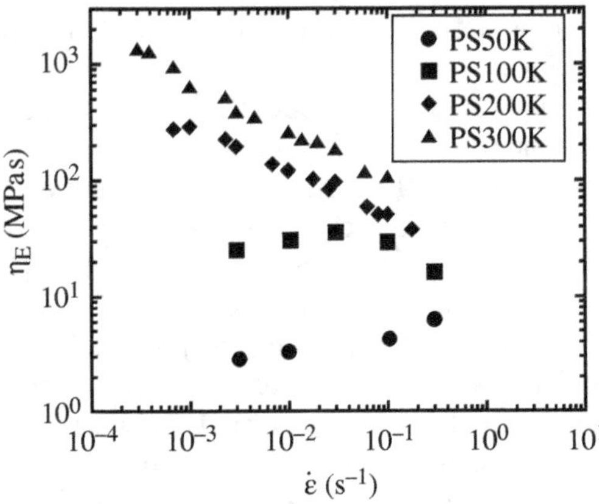

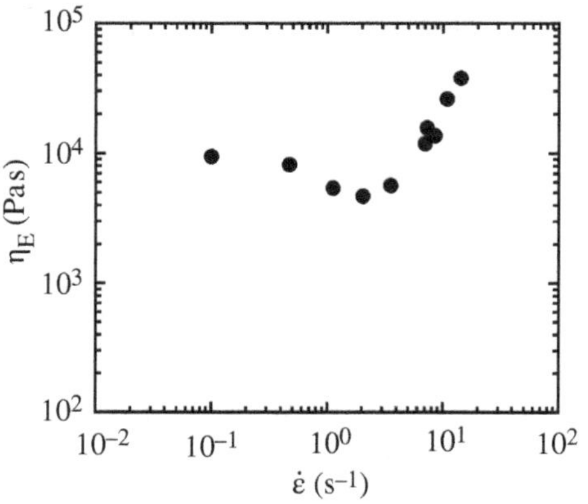

Figure 17.14 "Steady-state" extensional viscosity versus Hencky rate for PS3.9 M(10%) solution. *Source*: Replctted from Bhattacharjee *et al.* 2002 [32].

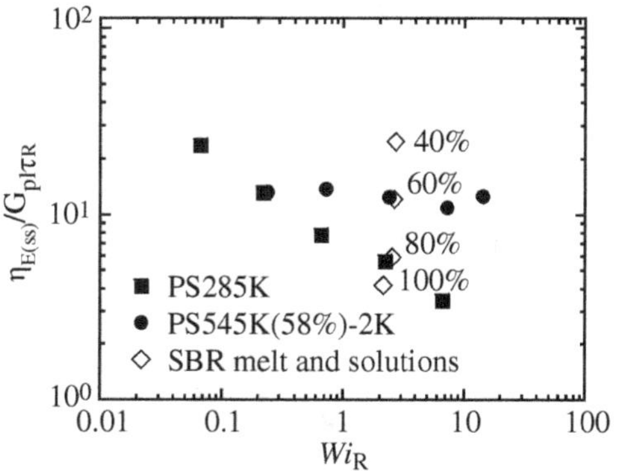

Figure 17.15 "Steady-state" extensional viscosity of pure PS285K and binary mixture PS545K(58%)-2K, based on Figure 12a in Ref. [29] and data in Fig. 17.17(a), which is the last point in each of the four curves in Fig. 17.17(a). *Source*: Data from Huang *et al.* 2013 [29].

In order to explore the origin of these differences, we reproduce the raw data from the literature that produced Fig. 17.15. Instead of the conventional representation of transient viscosity versus time, it is more instructive to plot the engineering stress σ_{engr} against the strain, represented by the melt stretching ratio λ, as shown in Fig. 17.16(a,b). Clearly, the point of yield is appreciably delayed in PS545K(58%)-2K relative to PS285K, both for relatively small and large extension rates. The difference is not due to any correction to the assumed isothermal condition. PS545K(58%)-2K has 31.5 s for its Rouse time, while PS285K's Rouse time is[29] 222 s: To keep the same values of Wi_R, the actual Hencky rate is larger in the extension of PS545K(58%)-2K by a factor of 7. Thus, PS545K(58%)-2K has a larger risk of becoming warmer than the nominal temperature of 130 °C. Consequently, the effective extension rate would have been smaller for PS545K(58%)-2K so that the yield point at the engineering stress maximum would have arrived at a lower strain in PS545K(58%)-2K, contrary to the data in Fig. 17.16.

The difference in extensional rheological responses between entangled melts and solutions seems universal rather than unique. Based on the same parent styrene–butadiene rubber, a series of "solutions" were made using low molar-mass styrene–butadiene-rubber chains as the solvent.[40] Since the unit of building block for the entanglement network is the entanglement spacing, nonlinear transient rheological responses are insensitive to Z when $Z \gg 1$, as shown in Figs. 7.8(b) and 7.31. The high-M_w melt and its mixtures with low-molecular-weight styrene–butadiene rubbers do not have the same Z, their Rouse relaxation times are however comparable. Thus, at a common Hencky rate, Wi_R is similar in value. Fig. 17.17(a) shows a systematic shift of the yield point, defined by the maximum of σ_{engr}, to higher values of λ_y. The notation λ_y was introduced in Fig. 7.33. At similar values of Wi_R, Fig. 17.17(a) reveals characteristics analogous to those in Fig. 17.16(a,b). In all cases, the normalized stress peak takes place at a larger λ_y and is of a larger magnitude for the solution than for the melt. Based on an analysis similar to that provided in Section 11.2.2, it seems that the force-imbalance is deferred in solution relative to melt because the finite extensibility characterized by λ^* is approached at a higher strain in solution.

In passing, we note a logical inconsistency in the literature. If we would follow Doi and Edward[41] and McKinley and Hassager[42] and apply the Considère criterion to suggest that

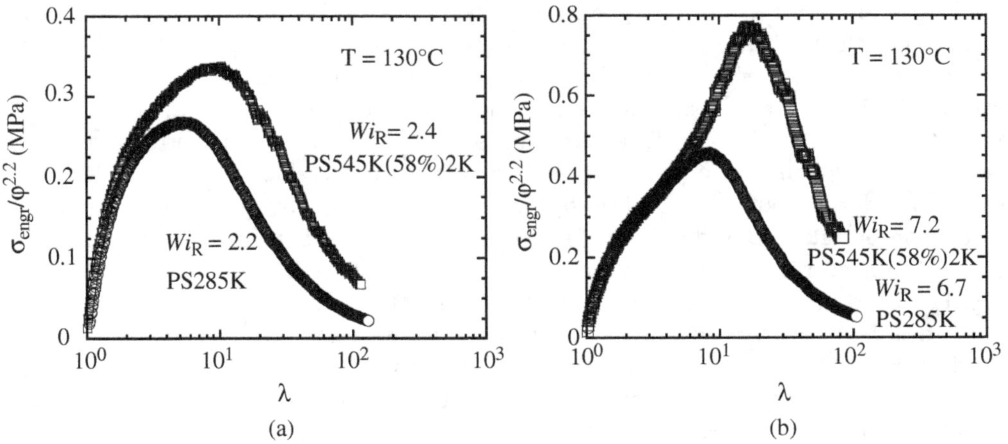

Figure 17.16 Rescaled engineering stress versus stretching ratio of PS285K and PS545K(58%)-2K at comparable Wi_R around (a) 2.2–2.4 and (b) 6.7–7.2. *Source*: Data taken from Huang *et al.* 2013 [29].

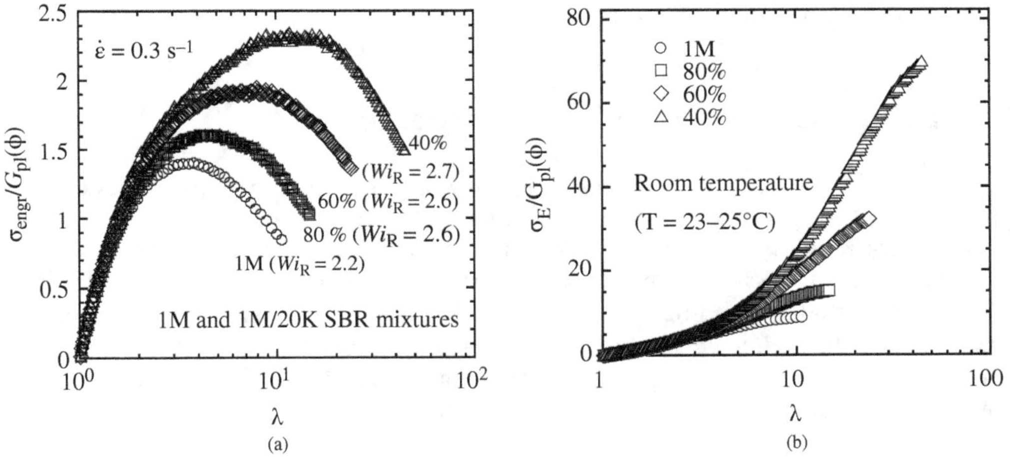

Figure 17.17 (a) Normalized engineering stress as a function of stretching ratio λ for pure SBR1M and its three mixtures with SBR20K. (b) Replot of (a) in terms of the normalized Cauchy stress versus λ. *Source*: Wang and Wang 2011 [40]. Reproduced with permission of American Chemical Society.

there should be "necking," namely, emergence of inhomogeneous extension upon observing a maximum in the engineering stress, then we could not claim that the melt extension would reach steady state. The data used to suggest the discrepancy between solutions and melts in Fig. 17.15 all show peaking of the engineering stress as shown in Fig. 17.16(a) and (b). If "necking" was to occur after the peak, how could the extension continue to a steady-flow state?

Replotting Fig. 17.17(a) as shown in Fig. 17.17(b), we see a far higher normalized Cauchy stress at a lower concentration. The last points on these curves represent the end of homogeneous extension, and the Cauchy stress shows no sign of saturation, so these samples are clearly not approaching steady-flow state. In comparison to simple shear, uniaxial extension is rather special and different: During extension the cross-sectional area keeps shrinking, condensing entanglement strands into a smaller cross-sectional area before yielding. It is this geometric condensation that drives the Cauchy stress σ_E to climb, even though yielding has already begun.

The triangles are much higher than the circles in Fig. 17.17(b) because the 40% solution (triangles) could stretch far more than the melt (circles) before the termination of homogeneous extension due to the emergence of yielding-induced necking. If the ending points of the four curves in Fig. 17.17(b) are mistakenly used to calculate $\sigma_E/\dot{\varepsilon}$ as $\eta_{E(ss)}$, we would have the four diamonds in Fig. 17.15. It is hardly a coincidence that at ca. $Wi_R \sim 2$, the vertical separation between the square and circle (58%) of the PS samples is comparable to the separation of the first (melt) and third (60%) diamonds of the styrene–butadiene rubbers because both involve a similar concentration. But the diamonds here are not steady-state values. The comparison in Fig. 17.15 underscores the origins of the quantitative separation between circles and squares in Fig. 17.15 and the qualitative difference between Figs. 17.13 and 17.14. It is most likely that none of these literature data actually determined the steady-state viscosity. The value of a transient viscosity can vary widely, highly dependent on the state of disentanglement. As pointed out in Section 11.5, under the external condition of globally imposed uniform uniaxial extension, strain localization develops as a consequence of local yielding. The nonuniform extension makes the state of steady flow unattainable.

To clear the confusion about the apparent difference in the rate-dependent "steady-flow" viscosity between entangled solutions and melts, it is necessary and helpful for the same methodology in Section 17.2 to explore the nature of extension in filament stretching rheometer and determine whether u_2 is still growing at the last data point.

17.6 Emergence of Transient Glassy Responses

One of the last frontiers in polymer rheology is to investigate nonlinear responses at deformational rates in the neighborhood of $1/\tau_e$, as shown in Fig. 17.18. So far all the rheological phenomena described in the book involve rates far below the second crossover time scale defined by τ_e. Moreover, all discussions have been based on rheometric and particle-tracking velocimetric measurements at temperatures *well above* the glass transition temperature T_g. For polymer melts, to access the regime circled in Fig. 17.18 using available rheometric instruments, we need to go to sufficiently low temperatures, for example, 120–130 °C for polystyrene. When the experimental temperature is not well above T_g, new physics emerges in this crossover to glass regime.

At $M_w = 10^6$ g/mol, a monodisperse PS has $\tau = 1.6 \times 10^6$ s and $\tau_e = 1$ s at T = 120 °C.[43] Thus, a Hencky rate in the neighborhood of $1/\tau_e$, namely, around 1 s^{-1} conveniently places the experiment in the crossover regime. What stress response should we expect? Should it follow the ideal rubbery elasticity formula of Eq. (1.52) or (7.15)? Figure 17.19 shows something different, namely, an initial solid-like response with a substantially higher elastic modulus $E_{interseg}$. At

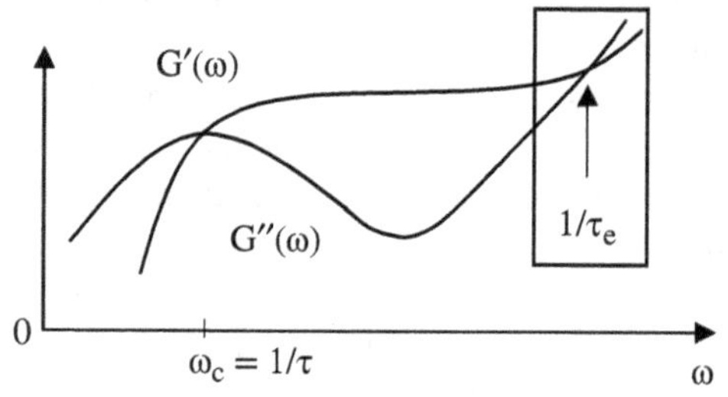

Figure 17.18 Illustration of where new rheological features show up, as indicated by the box, in terms of characteristic G′ and G″ curves from a standard SAOS measurement, when startup shear involves a rate as high as $1/\tau_e$.

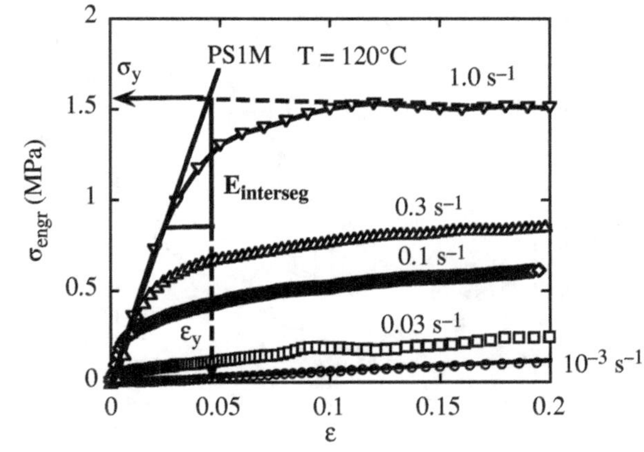

Figure 17.19 Transient elastic response to startup melt extension well above the common rubbery elasticity that emerges upon startup uniaxial extension of PS1M (polystyrene melt with $M \sim 10^6$ g/mol) at various Hencky rates ranging from 10^{-3} to $1\ s^{-1}$. The initial slope $E_{interseg}$ is two orders of magnitude above melt plateau modulus G_N^0. The data at $10^{-3}\ s^{-1}$ shows agreement with the rubber elasticity formula of Eq. (7.15). *Source*: Data taken from Sun *et al.* 2014 [43].

rates ranging from 0.1 (one decade below $1/\tau_e$) to $1\ s^{-1}$, the mechanical response to a startup extension is like a stiff solid with an elastic tensile modulus reaching over 10 MPa. Based on similar experiments at other temperatures, it was concluded[43] that the transient elasticity, which is much stronger than that characterized by G_N^0, occurs only at low enough temperatures and high enough deformation rates. Figure 17.20 summarizes measurements of the observed high transient elasticity. According to the literature,[44–49] a melt shows the characteristics of a glass as its temperature approaches T_g. At temperatures as much as 20–30 °C above T_g, the melt exhibits elastic stiffness far larger than the rubber elasticity indicated by the melt plateau modulus. In earlier work, nonrubber elastic responses were observed only when the experimental temperature was sufficiently close to T_g, as reported by Inoue and coworkers in the 1990s. Figure 17.20 shows that at a given temperature there is a threshold deformation rate below which only rubber elasticity is observed. The initial response is clearly temperature sensitive, as shown by the difference between the three symbols in Fig. 17.20. The magnitude of the elastic rigidity varies markedly with temperature, changing by a factor of almost 10 in the temperature range from 120 to 130 °C. The origin of the transient elasticity remains to be identified quantitatively but should be intermolecular. Indeed, this high elasticity also occurs in unentangled polystyrene

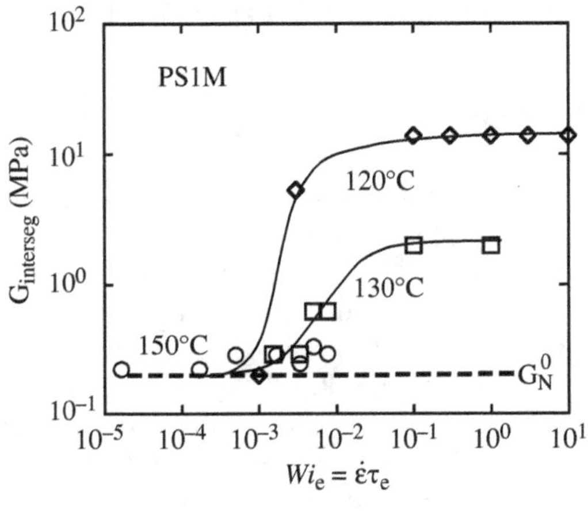

Figure 17.20 Initial modulus $G_{interseg} = E_{interseg}/3$ as a function of the product Wi_e at two temperatures 120 and 130 °C relative to the high-temperature behavior that shows normal rubbery melt elasticity, characterized by G_N^0. *Source*: Data taken from Sun *et al.* 2014 [43].

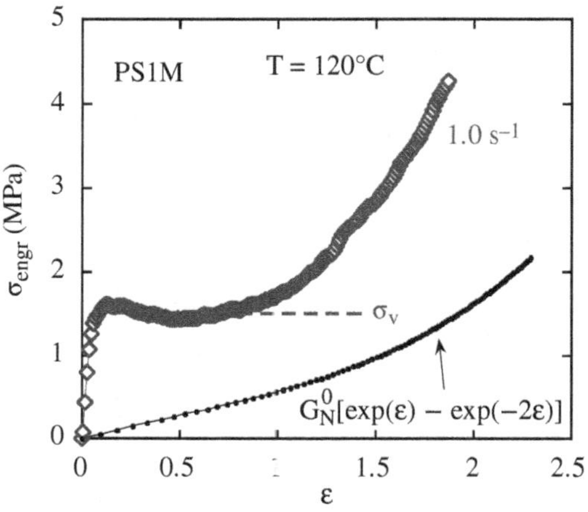

Figure 17.21 Engineering stress versus Hencky strain curve at 1.0 s^{-1} showing both initial "glassy" response and true strain hardening until melt rupture close to $\varepsilon = 2$ in reference to the rubber elasticity curve of $\sigma_{engr} = G_N^0(e^{\varepsilon} - e^{-2\varepsilon})$. *Source*: Data taken from Sun *et al.* 2014 [43].

melts.[43] Moreover, the transient elastic response is universal, independent of the mode of deformation, because it is also found in startup simple shear.[43]

A few more remarks are in order. First, the transient elastic response only lasts for the initial 5% deformation. At larger deformations, the transient structure yields to a state of irrecoverable segmental flow and rubbery elastic deformation. As shown in Fig. 17.19, the yield stress depends on both the applied deformation rate and the temperature.[50] Second, as shown in Fig. 17.21, after yielding, the stress is dominated by interchain friction, given by σ_v. Eventually, the rubber-elastic force produces a stress level comparable to σ_v and takes over subsequently. Since at a Hencky rate of 1 s^{-1}, $Wi_R \gg 1$, the engineering stress monotonically increases, showing non-Gaussian stretching until there is rupture.

In conclusion of this chapter, we point out that for several phenomena there are no available theoretical accounts beyond simple hand-waving speculations. These phenomena should inspire further development of the new theoretical framework and make the tube model further powerless. Actually, each of the five subjects in Sections 17.2–17.6 requires new theoretical considerations beyond Chapter 16. These examples are enumerated here to raise new theoretical questions for future investigation. We expect that further studies in each of these areas will greatly advance the field of polymer rheology research.

References

1 Malvern LE. Introduction to the mechanics of a continuous medium. Prentice-Hall, New Jersey; 1977. See Eq. (5.4.7).
2 Bergman TL, Incropera FP, DeWitt DP, Lavine AS. Fundamentals of heat and mass transfer. John Wiley & Sons, New York; 2011.
3 Schweizer T. Comparing cone-partitioned plate and cone-standard plate shear rheometry of a polystyrene melt. J Rheol. 2003;47(4):1071–85.
4 Bach A, Almdal K, Rasmussen HK, Hassager O. Elongational viscosity of narrow molar mass distribution polystyrene. Macromolecules. 2003;36(14):5174–9.
5 Nielsen JK, Hassager O, Rasmussen HK, McKinley GH. Observing the chain stretch transition in a highly entangled polyisoprene melt using transient extensional rheometry. J Rheol. 2009;53(6):1327–46.

6 Lin P, Zhao Z, Wang Z-G, Wang S-Q. Origin of mechanical stress and rising internal energy during fast uniaxial extension of polymer melts. Polymer 2017;124:68–77.
7 Wang Y, Wang S-Q. From elastic deformation to terminal flow of a monodisperse entangled melt in uniaxial extension. J Rheol. 2008;52:1275–90.
8 Zhu X, Wang S-Q. Mechanisms for different failure modes in startup uniaxial extension: tensile (rupture-like) failure and necking. J Rheol. 2013;57(1):223–48.
9 Ferry JD. Viscoelastic properties of polymers. John Wiley & Sons, New York; 1980.
10 Graessley WW. Polymeric liquids and networks: dynamics and rheology. Garland Science, New York; 2008.
11 Sokolov AP, Schweizer KS. Resolving the mystery of the chain friction mechanism in polymer liquids. Phys Rev Lett. 2009;102(24):248301.
12 Plazek DJ. 1995 Bingham Medal Address: Oh, thermorheological simplicity, wherefore art thou? J Rheol. 1996;40(6):987–1014.
13 Liu G, Wang S-Q. Entangled linear polymer solutions at high shear: from strain softening to hardening. Macromolecules. 2016;49:9647.
14 Plazek DJ. Temperature dependence of the viscoelastic behavior of polystyrene. J Phys Chem. 1965;69(10):3480–7.
15 Plazek DJ, O'Rourke VM. Viscoelastic behavior of low molecular weight polystyrene. J Polym Sci Part B Polym Phys. 1971;9(2):209–43.
16 Plazek DJ, Chelko AJ. Temperature dependence of the steady state recoverable compliance of amorphous polymers. Polymer. 1977;18(1):15–8.
17 Plazek DJ. The temperature dependence of the viscoelastic behavior of poly (vinyl acetate). Polym J. 1980;12(1):43–53.
18 Plazek DJ. The temperature dependence of the viscoelastic softening and terminal dispersions of linear amorphous polymers. J Polym Sci Part B Polym Phys. 1982;20(4):729–42.
19 Ngai K, Plazek D. Relation of internal rotational isomerism barriers to the flow activation energy of entangled polymer melts in the high-temperature Arrhenius region. J Polym Sci Polym Phys Ed 1985;23(10):2159–80.
20 Ngai K, Plazek D. A quantitative explanation of the difference in the temperature dependences of the viscoelastic softening and terminal dispersions of linear amorphous polymers. J Polym Sci Part B Polym Phys. 1986;24(3):619–32.
21 Plazek D, Zheng X, Ngai K. Viscoelastic properties of amorphous polymers. I. Different temperature dependences of segmental relaxation and terminal dispersion. Macromolecules. 1992;25(19):4920–4.
22 Plazek D, Chay I-C, Ngai K, Roland C. Viscoelastic properties of polymers. 4. Thermorheological complexity of the softening dispersion in polyisobutylene. Macromolecules. 1995;28(19):6432–6.
23 Osaki K, Inoue T, Isomura T. Stress overshoot of polymer solutions at high rates of shear. J Polym Sci Part B: Polym Phys. 2000;38(14):1917–25.
24 Osaki K, Inoue T, Isomura T. Stress overshoot of polymer solutions at high rates of shear; polystyrene with bimodal molecular weight distribution. J Polym Sci Part B Polym Phys. 2000;38(15):2043–50.
25 Osaki K, Inoue T, Uematsu T. Stress overshoot of polymer solutions at high rates of shear: semidilute polystyrene solutions with and without chain entanglement. J Polym Sci Part B Polym Phys. 2000;38(24):3271–6.
26 Boukany PE, Wang S-Q. Use of particle-tracking velocimetry and flow birefringence to study nonlinear flow behavior of entangled wormlike micellar solution: from wall slip, bulk disentanglement to chain scission. Macromolecules. 2008;41(4):1455–64.

27 Lodge TP, McLeish TC. Self-concentrations and effective glass transition temperatures in polymer blends. Macromolecules. 2000;33(14):5278–84.
28 Huang Q, Alvarez NJ, Matsumiya Y, Rasmussen HK, Watanabe H, Hassager O. Extensional rheology of entangled polystyrene solutions suggests importance of nematic interactions. ACS Macro Lett. 2013; 2 (8): 741–744.
29 Huang Q, Mednova O, Rasmussen HK, Alvarez NJ, Skov AL, Almdal K, et al. Concentrated polymer solutions are different from melts: role of entanglement molecular weight. Macromolecules. 2013;46(12):5026–35.
30 Rasmussen HK, Huang Q. The missing link between the extensional dynamics of polymer melts and solutions. J Nonnewton Fluid Mech 2014;204:1–6.
31 Huang Q, Hengeller L, Alvarez NJ, Hassager O. Bridging the gap between polymer melts and solutions in extensional rheology. Macromolecules. 2015;48(12):4158–63.
32 Bhattacharjee P, Oberhauser J, McKinley G, Leal L, Sridhar T. Extensional rheometry of entangled solutions. Macromolecules. 2002;35(27):10131–48.
33 Yaoita T, Isaki T, Masubuchi Y, Watanabe H, Ianniruberto G, Marrucci G. Primitive chain network simulation of elongational flows of entangled linear chains: stretch/orientation-induced reduction of monomeric friction. Macromolecules. 2012;45(6):2773–82.
34 Ianniruberto G, Brasiello A, Marrucci G. Simulations of fast shear flows of PS oligomers confirm monomeric friction reduction in fast elongational flows of monodisperse PS melts as indicated by rheooptical data. Macromolecules. 2012;45(19):8058–66.
35 Desai PS, Larson RG. Constitutive model that shows extension thickening for entangled solutions and extension thinning for melts. J Rheol. 2014;58(1):255–79.
36 Ye X, Larson RG, Pattamaprom C, Sridhar T. Extensional properties of monodisperse and bidisperse polystyrene solutions. J Rheol. 2003;47(2):443–68.
37 Rasmussen HK, Nielsen JK, Bach A, Hassager O. Viscosity overshoot in the start-up of uniaxial elongation of low density polyethylene melts. J Rheol. 2005;49(2):369–81.
38 Nguyen DA, Bhattacharjee PK, Sridhar T. Response of an entangled polymer solution to uniaxial and planar deformation. J Rheol. 2015;59(3):821–33.
39 Nielsen JK, Rasmussen HK, Hassager O, McKinley GH. Elongational viscosity of monodisperse and bidisperse polystyrene melts. J Rheol. 2006;50(4):453–76.
40 Wang Y, Wang S-Q. Salient features in uniaxial extension of polymer melts and solutions: progressive loss of entanglements, yielding, non-Gaussian stretching, and rupture. Macromolecules. 2011;44(13):5427–35.
41 Doi M, Edwards S. Dynamics of concentrated polymer systems. Part 4.—Rheological properties. J Chem Soc Faraday Trans 2 Mol Chem Phys. 1979;75:38–54.
42 McKinley GH, Hassager O. The Considere condition and rapid stretching of linear and branched polymer melts. J Rheol. 1999;43(5):1195–212.
43 Sun H, Liu G, Ntetsikas K, Avgeropoulos A, Wang S-Q. Rheology of entangled polymers not far above glass transition temperature: transient elasticity and intersegmental viscous stress. Macromolecules. 2014;47(16):5839–50.
44 Inoue T, Okamoto H, Osaki K. Birefringence of amorphous polymers. 1. Dynamic measurement on polystyrene. Macromolecules. 1991;24(20):5670–5.
45 Inoue T, Hayashihara H, Okamoto H, Osaki K. Birefringence of amorphous polymers. II. Dynamic measurement and relaxation measurement. J Polym Sci Part B Polym Phys. 1992;30(4):409–14.
46 Okamoto H, Inoue T, Osaki K. Birefringence of amorphous polymers. 4. Large deformation of polystyrene near its glass transition temperature. Macromolecules. 1992;25(13):3413–5.

47 Muller R, Pesce J. Stress-optical behaviour near the Tg and melt flow-induced anisotropy in amorphous polymers. Polymer. 1994;35(4):734–9.
48 Inoue T, Ryu D-S, Osaki K. A rheo-optical study on polystyrene under large tensile deformation around the glass transition temperature. Macromolecules. 1998;31(20):6977–83.
49 Inoue T, Osaki K. Role of polymer chain flexibility on the viscoelasticity of amorphous polymers around the glass transition zone. Macromolecules. 1996;29(5):1595–9.
50 Mirigian S, Schweizer KS. Dynamical theory of segmental relaxation and emergent elasticity in supercooled polymer melts. Macromolecules. 2015;48(6):1901–13.

18

Difficulties with Orthodox Paradigms

The reader will have noticed that the ansatz presented here, extracted from the available phenomenology, disagrees with the scheme presented in the tube model that is the convention used by many workers studying nonlinear rheology of polymer melts and entangled solutions. In light of experiment, it is necessary to discuss the differences between standard treatments based on the tube model[1] and the new considerations[2] that explicitly recognize the structural collapse of the entanglement network, i.e., chain disentanglement or disappearance of entanglement strands that may or may not occur in a spatially homogeneous manner, during nonlinear responses.

We should judge the merit of a theoretical description by asking whether or not it permits us to predict behavior that was otherwise unexpected, not by whether it can be used with adequate free parameters to fit experimental data. Hardly any of the description of nonlinear rheology in this book depends on results from the tube model.[1,3]

The concept of yielding was introduced by Maxwell and Nguyen.[4] We rediscovered *yielding* of entanglement network from our analysis of stress overshoot behavior during startup shear. In contrast, the tube model does not prescribe any structural collapse at the overshoot. Rouse chain retraction and resulting convective constrain release[5] can be interpreted as[3] indicating termination of affine deformation. However, the tube model does not explain at a force level why the entanglement network can break up in a nonlinear elastic manner via chain disentanglement. Similarly, when we identified the maximum in the engineering stress as the signature of *yielding* of the network in melt extension,[6] the tube model assigns no structural significance to the maximum. Shear banding[7] and nonquiescent relaxation have been observed experimentally, while the tube model was improved[3,5] to describe homogeneous shear and quiescent relaxation. Homogeneous melt extension was shown to fail to reach steady flow,[6] while the tube model simply assumes that a steady flow state can fully develop. We identified geometric condensation[8] to explain "strain hardening" in melt extension, in the presence of long-chain branching; we recognized the kinematic difference between shear and extension.[9] The tube model did not warn us about choosing the incorrect reference state, that is, zero-rate transient extensional viscosity envelope for comparison with nonlinear response to startup extension. The purpose of this chapter is twofold: to summarize some of the fundamental differences between our model and the tube model and to entice the reader to study the book in detail so that the differences between the models are identified and understood.

18.1 Tube Model Does Not Predict Key Experimental Features

To some, "the tube model is the most successful microscopically based theory for describing the nonlinear flow of entangled polymers."[10] Surely, given the complexity of nonlinear rheological

Nonlinear Polymer Rheology: Macroscopic Phenomenology and Molecular Foundation, First Edition. Shi-Qing Wang.

Companion Website: www.wiley.com/go/nonlinearpolymerrheology

phenomena that the tube model intended to account for, we should greatly appreciate what the tube model has been able to do. However, it may be too early to make such a sweeping assessment and certainly too risky for us to depend on the tube model to provide us the actual molecular pictures. In some instances, the tube model for nonlinear rheology of entangled polymers is like the Maxwell model for phenomenological linear viscoelasticity: There are cases where the molecular picture is hard to identify behind a tube model calculation. For example, Fig. 12.24 shows evidence of elastic yielding after step strain that is produced at $Wi_R = 0.1$. Since the slow shear at $Wi_R = 0.1$ cannot cause much chain deformation, disentanglement should not occur during stress relaxation. The state of entanglement should only improve after shear cessation. Although the tube model cannot describe how the state of entanglement changes during the stress relaxation that leads to the nonmonotonicity, the tube model was nevertheless applied[10] to describe Fig. 12.24. The squares in Fig. 12.24 indicate in our opinion that elastic yielding of the strained network took place to cause the nonmonotonicity.

The objective of molecular rheologists should be to predict and explain why the nonmonotonicity occurs in terms of molecular processes. Otherwise, we would have been content to apply simply the Maxwell model to describe linear viscoelastic characteristics such as the emergence of the terminal relaxation time $\tau = 1/\omega_c$ in Fig. 2.7 without providing a physical picture for how molecular processes determine ω_c in terms of the chain length, as the de Gennes' reptation model achieved to do. Since the Maxwell model can be applied to produce experimental data similar to the solid curves (for G' and G'') in Fig. 2.7, can we claim that the Maxwell model can describe reptation dynamics? If the tube model can be (and has been) worked out to show that a deformed polymer melt can undergo sufficient elastic recovery, as asserted in Ref. [10], can we actually claim that the tube model offers a physical picture for how elastic recovery takes place? Arthur Lodge pointed out that the single-chain-based tube model does not contain the physics required to describe elastic recovery after large step strain:[11] For example, after step extension, the center of mass of each chain is displaced away from the middle point of the sample. Being an isolated chain in a tube where the intrachain retraction force was not perceived to be balanced by the external force during the step extension, each chain can only recoil without moving its center of mass when the external stress is removed. If this was the picture, there should not be any macroscopic recovery. Thus, to claim that the tube model can describe elastic recovery in entangled melts[10] is similar to claiming that the Maxwell model can describe reptative motions in entangled melts.

Moreover, the tube model did not predict many of the key phenomena in the nonlinear rheology of entangled polymeric liquids. The tube model, built on the reptation model of de Gennes and the tube notion of Edwards, successfully simplified the account of interchain interactions for polymer dynamics under small deformation. For large fast deformation, the tube model still assumes that even during chain retraction the tube is infinitely strong, not destructible by elastic forces. The ansatz that the test chain can be treated as a Rouse chain in a tube seems to have been motivated by a desire to produce stress overshoot in startup shear; the assumption is thus *ad hoc*. For nearly four decades, researchers in the field of nonlinear polymer rheology have depended on this standard model to provide them with detailed accounts as well as mental images of polymer chain dynamics during nonlinear response. It is perhaps in this sense, that is, the fact that the tube model has been extensively applied since its invention, that the model has been taken as the most successful model for polymer rheology.

As for equilibrium dynamics and linear rheology, the reptation-tube model seems to have received rather strong phenomenological support. Such an opinion differs from the view expressed by Phillies' monograph[12] on the phenomenology of polymer solution dynamics. Concerning nonlinear polymer rheology, mounting experimental evidence along with the corresponding rational analyses indicates that the tube model is actually unsuccessful. By representing intermolecular interactions with an imaginative tube, the tube model is built on the hypothesis that chain deformation and corresponding stress only need to be depicted on a specific length scale, that is, the equilibrium tube segment. According to the tube model, we only need to speak about whether the tube segments are stretched or only oriented; and the stress can be evaluated in terms of separate contributions associated with tube segment stretching and orientation. The lack of a sizable linear response regime is an inevitable consequence of the tube model framework. The comparison made in Figs. 14.7 and 14.8 crucially challenges the foundation of the tube model. If it is invalid to describe chain deformation and accompanying stress only on the scale of the tube diameter, all five leading features or predictions of the tube model would be untrue, as noted in Section 16.1.3. In the following sections, we summarize a few key examples where our worldview on nonlinear polymer rheology differs from the understanding attributed to the tube model.

18.1.1 Unexpected Failure at $Wi_R \ll 1$

According to the tube model, in the limit of $Wi_R \ll 1$, as illustrated in Fig. 16.1, the shear and engineering stresses show maxima at a time $t_y \sim 2.2/\dot{\gamma}$ (shear) or $t_y \sim ln(2.36)/\dot{\varepsilon} = 0.86/\dot{\varepsilon}$ (extension). The stress drop after t_y ($\gg \tau_R$ because $Wi_R \ll 1$) should have nothing to do with any structural failure of the entanglement network. Experimental observations in Figs. 11.4 and 11.5 starkly contradict this picture from the tube model because at $Wi_R \ll 1$ melts undergo "tensile decohesion," that is, rupture-like sharp macroscopic separation after the peak of σ_{engr}, for example, at melt stretching ratios between 3 and 4, that is, at times much longer than τ_R. According to our new viewpoint, most plausibly, the failure of the entanglement network has occurred through chain disentanglement.

Strongly viscoelastic systems must yield under shear rates that are large relative to their internal relaxation rates or under stresses that are large relative to their elastic plateau moduli. Shear banding clearly reveals the meaning of the stress overshoot: The stress decline is a sign of yielding of the entanglement network.[1] The tube model was not built on the notion of *yielding* of the entanglement network and does not indicate what chain retraction on the Rouse time does to the entanglement. For $Wi_R > 1$, if the shear stress overshoot at strain γ_{max} is due to the chain retraction, the tube model predicts $\gamma_{max} \sim Wi_R$ in contrast to the experimental data showing $\gamma_y \sim (Wi_R)^{1/3}$.

18.1.2 Elastic Yielding Can Lead to Nonquiescent Relaxation

Normally, especially in a lower range of strain magnitude, the larger the magnitude of the step strain the higher the residual shear stress survives at long times. However, a large fraction of the experimental data in the literature on the stress relaxation of entangled polymer solutions reveals an anomalous feature that the relaxing shear stress after an initial quick drop is *smaller* for a *larger* magnitude of step strain, as shown in Fig. 12.6. Even data in perfect agreement with the Doi–Edwards tube model share this paradoxical character, as seen in Fig. 12.10(b). Except for the study of Marrucci and Grizzuti,[13] the literature before 2006 did not confront this reality. In 2006, the nonmonotonic relation between relaxing shear stress at long times and the magnitude of the step strain prompted the first particle-tracking velocimetric observation[14] of nonquiescent relaxation. Here, the question is what causes the nonmonotonicity.

Particle-tracking velocimetry reveals an elastic breakup after large step shear strain.[14] A subsequent study demonstrated a similar elastic failure after large step extension.[15] The key

1 Shear stress overshoot may not be accompanied by shear banding, and shear banding is not a consequence of the overshoot. However, shear banding can accelerate the shear stress decrease after the peak and certainly has allowed us to realize that there can be structural failure during startup shear in well-entangled polymers.

question is what causes the elastic breakup to take place and whether the macroscopic breakup involves a structural breakdown of the entanglement network. Chapter 16 offers a plausible mechanism for elastic failure: The termination of an ongoing deformation induces a force imbalance that leads to chain disentanglement during stress relaxation. The yielding of the entanglement network may not proceed homogeneously. Although the tube model predicts nonmonotonic behavior shown in Fig. 12.3, with which some data agree, it did not inspire any investigation into its structural origin that involves *in situ* characterization of the strain field. More important, since the shear stress increases with the magnitude of the step strain for $\gamma < 2.2$, the tube model predicts no elastic instability for modest step strains with magnitude smaller than 2.2. However, experiment reveals (Figs. 12.11(b) and 12.21(b)) that nonquiescence due to the elastic yielding can take place for strains as small as 1.2–1.3, much smaller than the limiting strain of 2.2. Thus, the observed nonquiescent relaxation for $\gamma < 2.2$ *cannot* be explained by the tube model. Elastic yielding can be the cause of the nonmonotonicity shown in Fig. 12.10(b). However, this nonmonotonicity is not the cause of the elastic yielding, namely, it would be a reversal of causality to identify the observed nonmonotonicity as the cause of the nonquiescence. In fact, the nonmonotonicity is the consequence of the elastic yielding instability.

18.1.3 Meaning of Maximum in Tensile Force (Engineering Stress)

As described in Section 7.5.3 and Chapter 11, except in regime IV, a maximum always appears in the engineering stress σ_{engr}, as found in uniaxial extension of various polymer melts. The Doi–Edwards tube model and its updated version also find a maximum in σ_{engr} because for $Wi_R \ll 1$ the tube model envisions the tensile stress to arise entirely from chain orientation. As the chain orientation saturates, so does the Cauchy stress σ_E. Inevitably, $\sigma_{engr} = \sigma_E/\lambda$ shows a maximum as the cross-sectional area shrinks as $1/\lambda$, λ being the stretching ratio. In the tube model, the maximum of σ_{engr} has no structural meaning because the maximum occurs at $t_y \gg \tau_R$, as noted in the beginning of Section 18.1.1. The ensuing engineering stress decline does not indicate a collapse of the entanglement network according to the tube model. Nevertheless, Doi and Edwards applied[16] the Considère criterion to suggest that there would be instability leading to inhomogeneous extension after a peak in σ_{engr}. But why should there be any instability if the entanglement structure is perceived to be intact during extension? Clearly, the tube model does not contain the physics that explain why and how the entangled melt breaks up, as seen in Fig. 11.5.

In our judgment, the application of the Considère criterion in combination with the tube model reverses the causality.[2] We have already pointed this out in Section 7.5.1. We should ask what causes σ_{engr} to decrease, once melt extension has increased beyond a certain stretching ratio λ, in terms of responsible molecular processes. A nonmonotonic dependence of σ_{engr} on λ does not have to lead to an extension instability. A systematic application of the Considère criterion to infer instability is questionable. Once the microscopic physics behind specimen breakup becomes clear, contrary to conclusions from the Considère criterion, one can demonstrate that a maximum of σ_{engr} in entangled polymer melts does not have to produce an instability. Such a demonstration already exists in the literature. Figure 18.1 shows that, despite pronounced maxima in σ_{engr}, a binary mixture of styrene–butadiene rubbers continues to extend homogeneously at rates equal to and above $2\ s^{-1}$ without extensional instability; the sample eventually yields or

2 "The surface is slippery when wet" is a commonly found warning sign in public space. Indeed, many *smooth* surfaces are slippery when wet. However, sufficiently rough surfaces may not be slippery even when wet. Thus, we must be careful and not generalize by saying "any surface is slippery when wet." More important, the statement that "surface is wet when it is slippery" makes little logical sense because it reverses the causality. A slippery surface does not have to be wet. It could be true that the surface is slippery because it is oily or covered with a thin layer of sands. Thus, there are multiple ways to make a surface slippery other than making it wet with water.

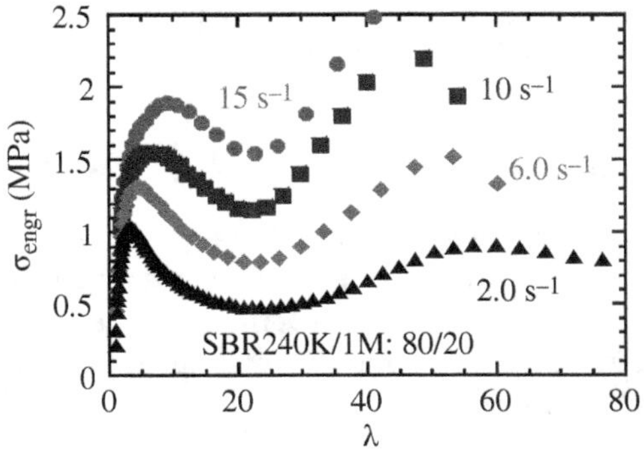

Figure 18.1 Engineering stress σ_{engr} versus stretching ratio λ at various Hencky rates for a binary mixture of two styrene–butadiene rubbers of molecular weights 240 kg/mol and 10^6 g/mol at a composition of 80:20, using data in Ref. [17]. At the rates of 2.0, 6.0, and 10 s^{-1}, the blend twice underwent yielding, corresponding to the two maxima in σ_{engr}. We interpret them to arise from the collapse of the entanglement networks made of SBR240K and SBR1M, respectively. At the largest applied rate of 15 s^{-1}, the entanglement network associated with SBR240K first yields at $\lambda_y \sim 8$, but the second network made of SBR1M continues to extend until melt rupture. *Source:* Wang *et al.*, 2011 [17]. Reproduced with permission of American Institute of Physics.

undergoes melt rupture at a very large stretching ratio $\lambda > 40$, even though the pronounced peak of σ_{engr} occurs at $\lambda < 10$.

In closing, we reiterate that there is a contradiction in the literature. On one hand, the Considère criterion has been applied to infer inhomogeneous extension (necking) beyond any maximum in the engineering stress σ_{engr}. On the other hand, by ignoring the presence of such a maximum, workers in the field claim that in filament stretching rheometry uniaxial extension could reach a state of steady flow (cf. Section 11.4). The reader must wonder why there is necking if the system can attain steady flow.

18.1.4 Other Examples of Causality Reversal

Similar to application of the Considère criterion, the explanation of shear stress overshoot by the tube model in terms of chain overorientation may be also causally incorrect: The stress decline beyond the maximum may be due to chain disentanglement, and the disentanglement is plausibly due to force imbalance (cf. Section 16.4.2). If disentanglement is the cause of the stress overshoot, chain orientation may not continue to grow until its contribution to the shear stress becomes nonmonotonic. With full chain retraction, how can we assume the tube to continue to deform affinely? Without affine deformation of the tube, how could chain show overorientation? If the continuation of chain orientation past the overshoot cannot be validated, the tube model would be unable to prescribe shear stress overshoot as due to chain overorientation.

Another key example of causality reversal is actually the application[18–22] of constitutive nonmonotonicity (where steady shear stress is perceived to decrease with increasing shear rate) to explain formation of shear banding. Such practice does not really address where the nonmonotonicity stems from and what the corresponding molecular picture is. As far as we can clearly indicate, the shear strain localization is due to localized chain disentanglement, that is, inhomogeneous failure of the entanglement network; and the disentanglement is a consequence of force imbalance as explained in Chapter 16. We have no knowledge, *a priori*, whether shear banding is a steady-state characteristic and whether nonmonotonicity can emerge from shear banding. In fact we have no experimental evidence that the nonmonotonicity can take place. Section 15.5 showed that at a given applied nominal shear rate shear banding can not only emerge from a startup shear but also be avoided when either the startup is replaced by a slow rate-ramp-up or when the system is rate-quenched from a homogeneous (higher rate) state. Thus, it makes no sense to invoke constitutive nonmonotonicity as a theoretical explanation of shear banding.

18.1.5 Entanglement–Disentanglement Transition

The tube model did not prompt us to look for the entanglement–disentanglement transition in creep. Although creep has been commonly applied to describe the *linear* viscoelastic characteristics of entangled polymers and used to discover the breakdown of time–temperature superposition,[23] the first creep study[24] to produce nonlinear responses in entangled polybutadiene solutions only appeared as late as 2004. Our interpretation, that the rising nominal shear rate at a given shear stress is evidence of chain *disentanglement,* aroused significant objections.[25,26] However, as subsequent more careful studies showed,[27,28] it is clear that the rising shear rate with increasing time is caused by an entanglement–disentanglement transition (cf. Section 16.7) and is *not* a runaway phenomenon due to sample loss.

18.1.6 Anomalies Are the Norm

The phenomena reviewed in Chapter 17 do not find explanations from the tube model. We list some of them here again.

1) The failure of time–temperature superposition to predict transient nonlinear responses indicates that local chain dynamics, which dictate the affine deformation of entangled polymers, follow a different temperature dependence than does the reptation dynamics (Section 17.3.2). The tube model cannot account for these observations. An intermolecular grip force arising from chain uncrossability is envisioned to vary differently with temperature from the reptation dynamics. The grip force depends on the local relaxation dynamics that are more temperature sensitive than are the averaged terminal chain dynamics.
2) Even at temperatures well above the glass transition temperature, during melt deformation appreciable intrachain internal energy evidently builds up, especially at large shear rates and large strains (Section 17.2). Melt deformation therefore can cause distortion on length scales significantly shorter than the tube segments or entanglement strands. The tube model has not explored and anticipated these issues.
3) Within the tube model it has been challenging to explain the difference in the "steady-state" responses of entangled melts and solutions (Section 17.5). However, since the tube model expects entangled polymeric liquids to have no difficulty attaining steady state in uniaxial extension, it does not help us to clarify the confusion to be discussed in Section 18.2.

The tube theory can indeed produce quantitative results. However, quantitative agreement between a subset of experimental data and a theory cannot be used to prove that a theory is valid. In this book, we often encounter a deeply philosophical point: Is an agreement between macroscopic experiment and the tube model sufficient evidence to assert that the theory is correct and realistic? An excellent example for a negative answer to this question is the seeming success of the tube model description of some stress relaxation data, for example, the agreement shown in Fig. 12.10(b). Without the particle-tracking velocimetric observation in Fig. 12.11(b) to reveal significant nonquiescence, we would have thought that the tube model well explained the data in Fig. 12.10(b).

For many phenomena enumerated above, the tube model is inadequate. Conversely, the idea of yielding does not have to be rejected just because the corresponding molecular picture of chain disentanglement has not matured into a quantitative model. There is the encouraging news that Sussman and Schweizer have laid a foundation for a first self-consistent, force level, dynamic mean field treatment of the nonlinear rheology of entangled polymers.[29–35] Their microscopic approach provides a first self-consistent description of many-body interchain interactions in the nonlinear rheology of entangled polymers.

18.2 Confusion About Local and Global Deformations

Yielding of entangled melts in fast uniaxial extension inevitably leads to the specimen breakup. For $Wi > 1$, such yielding necessarily occurs due to the force imbalance, and steady flow does not take place. However, this assertion supported by our experiments sharply contradicts many studies in the literature that report steady flow in uniaxial extension of entangled polymers. Nearly every researcher in the field of melt extensional rheology, except for the rheologists from Russia[36,37] plus few groups in the United States, conforms to the view that entangled melts can reach a steady flow state before encountering any material instability. However, Vindogradov,[38] Malkin,[39] and others[6,40–42] have shown that during melt extension with $Wi > 1$ entangled melts are incapable of attaining a homogeneous steady flow state. We must resolve this sharp disagreement.

18.2.1 Lack of Steady Flow in Startup Melt Extension

Unlike simple shear, where shear inhomogeneity does not prevent an entangled polymeric liquid from reaching steady flow, uniaxial extension of linear-chain melts always turns inhomogeneous for $Wi > 1$, well before a hypothetical steady flow state is reached. If one relaxes the condition of global homogeneity and adopts filament stretching rheometry,[43] one can under certain circumstances observe some characteristics that resemble steady flow behavior. But an apparent leveling-off of the Cauchy stress σ_E is hardly a reliable indicator of steady flow. Filament stretching rheometric studies do not explain why rupture can sometimes occur,[44] and often take the last data point before filament breakup as steady flow information. To avoid confusion and ambiguity, when these authors claim to report steady flow data, they should always make a caveat that the acclaimed steady state was *not* achieved homogeneously (cf. Fig. 11.16). Clearly, properties identified in filament stretching rheometry cannot be used to reveal nonlinear responses of entangled melts to startup uniaxial extension in a setting that imposes homogeneous melt extension over an extensive spatial range, such as a Sentmanat extensional rheometer does. It is unreliable and unreasonable to identify the last data point before filament breakup as corresponding to steady flow behavior in the bulk, without concern as to the *ad hoc* nature of the claim.

More important, filament stretching rheometry disguises the important phenomenon of tensile strain localization arising from tensile decohesion and shear yielding, as summarized in Chapter 11. At best, the behavior measured by the filament stretching rheometry is merely a small subset of nonlinear responses under the rather peculiar condition of inherent inhomogeneity. As explained in detail in Appendix 11.A, filament stretching rheometry for melt extension is analogous to capillary rheometry for shear: We cannot regard the rheological information from capillary rheometry as constitutive because the relation between the shear stress and shear rate at the die wall does not reveal whether or not shear banding can occur during simple shear across a macroscopic gap. As shown in Chapter 9, shear strain localization does occur upon startup shear. Similarly, the rheological information from filament stretching rheometry should not be taken to represent constitutive behavior of entangled melts under uniaxial extension. For example, this information cannot be used to anticipate whether homogeneous extension can proceed to steady flow in such a configuration as in Sentmanat extensional rheometry. Indeed, as shown in Chapter 11, breakup can occur to terminate initial homogeneous uniaxial extension.

18.2.2 Peculiar Protocol to Observe Stress Relaxation from Step Extension

The use of filament stretching rheometry to study startup extension of entangled polymeric fluids can be misleading. Stress relaxation studies after step extension in filament stretching rheometry[45,46] are not only misleading but also actually problematic. The elastic yielding leading to the specimen breakup after *homogeneous* step extension is avoided in filament stretching rheometry by moving the ends of the filament toward each other during stress relaxation so that the middle section does not change its dimensions (cf. Fig. 13.8). Away from the middle point, the filament is actually in motion. Thus, stress relaxation data from the filament stretching rheometry should not be compared with any theoretical calculation that assumes quiescent relaxation. As demonstrated in Chapter 13, the procedure that keeps the middle section stationary eliminates one's ability to observe and determine the inherent response of entangled melts to step extensions of large magnitude. This response is the interesting and important phenomenon of breakup due to elastic yielding.

18.3 Molecular Network Paradigm

In this section, we contrast our molecular network picture with the essence of the tube model and show where the tube model is deficient in the two most important cases of startup deformation and stepwise deformation. For simplicity, we will only discuss the case of slow startup made with $Wi_R \ll 1$ and instantaneous step strain made with $Wi_R \gg 1$. These are the two most crucial scenarios where a great deal of confusion exists. Admittedly, the discussion in this section is highly speculative, implying that the theoretical description for nonlinear polymer rheology is far from complete. However, the conceptual difference is clear: In the new paradigm, the network failure arises from force imbalance and does not involve such ideas as transverse constraint and lateral cohesion along the chain contour.

18.3.1 Startup Deformation

When the entanglement network is intact, as in equilibrium, the structure may indeed be specified at the level of the equilibrium tube segment. Thus, whenever the deformation is fully elastic (i.e., $Wi_R > 1$), the tube segment can be stretched. However, with $Wi_R < 1$, the chain deformation beyond a strain of Wi_R no longer increases on the equilibrium tube segment scale, as assumed by the tube model through its assumption of the Rouse chain retraction. Because of chain relaxation on different length scales, for example, faster on smaller scales, the mesh size (on which the network experiences elastic deformation) is clearly growing in time. Thus, it should be unreasonable for us to evaluate chain conformation and corresponding stress on the equilibrium tube segment scale past the Rouse time τ_R at strain $Wi_R < 1$. Consequently, the prediction of the stress overshoot (due to the assumed continuing orientation of the equilibrium tube segment) by the tube model is problematic. In our understanding, strands on scales significantly larger than the equilibrium tube segment are still getting deformed past τ_R: The equilibrium tube segments may be partially oriented as a result of the larger scale chain deformation. However, such (equilibrium) tube segment orientation is not affine and therefore may not produce a maximum at γ = ca. 2. Significant chain deformation can occur past the strain of $Wi_R < 1$ because $Wi > 1$. Under this condition chain relaxation is too slow to adjust to the imposed deformation. In other words, strands on larger length scales than equilibrium tube segments may still deform at a strain of unity. The structure of the deformed entanglement network keeps evolving toward larger "mesh sizes" as the time progresses. Therefore, a zeroth order phenomenological picture is one where the mesh size $l_{ent}(\gamma)$ or $l_{ent}(\lambda)$ continuously increases, unlike the tube model assumption of a single length scale.

To further quantify the network picture, we apply the formulation of the rubber elasticity theory. Based on the notion of rubber elasticity, we treat the response to startup deformation of entangled polymeric fluids in terms of a strain-dependent weakening modulus G. This effective elastic modulus is related to the average molecular weight M_e of evolving mesh size or directly to the mesh size $l_{ent}(\gamma)$ or $l_{ent}(\lambda)$ as

$$G(\gamma) = \rho RT/M_e(\gamma) \text{ and } G(\lambda) = \rho RT/M_e(\lambda) \tag{18.1}$$

or

$$G(\gamma) = k_B T/p[l_{ent}(\gamma)]^2 \text{ and } G(\lambda) = k_B T/p[l_{ent}(\lambda)]^2, \tag{18.2}$$

where p is the packing length. The strain dependence of the mesh size can be estimated from Eq. (18.2) in terms of in the strain dependence of the stress, as given by

$$l_{ent}(\gamma) = l_{ent}[\gamma f_p/(l_{ent})^2\sigma(\gamma)]^{1/2} \text{ for startup shear} \tag{18.3}$$

and

$$l_{ent}(\lambda) = l_{ent}[(\lambda - 1/\lambda^2)f_p/(l_{ent})^2\sigma_{engr}(\lambda)]^{1/2} \text{ for uniaxial extension} \tag{18.4}$$

where l_{ent} is the equilibrium entanglement strand, given in Eq. (2.29), i.e., $l_{ent}(\gamma = 0) = l_{ent}(\lambda = 1)$ and $f_p = k_B T/p$. During affine deformation, $l_{ent}(\gamma)$ or $l_{ent}(\lambda)$ stays very close to the equilibrium value l_{ent}. As σ or σ_{engr} starts to deviate from the linear growth with the strain, the mesh size begins to increase. A maximum appears in the stress $\sigma(\gamma)$ or $\sigma_{engr}(\lambda)$ when G decreases faster than $1/\gamma$ or $1/(\lambda - 1/\lambda^2)$. Under the condition of $Wi_R < 1$, origin of the declining G or increasing $l_{ent}(\gamma)$ or $l_{ent}(\lambda)$ is due to (a) chain relaxation that causes molecular deformation to shift to larger length scales as the deformation time increases and to (b) an eventual force imbalance when intrachain retraction force is larger than the cohesion force related to the entropic barrier – the same barrier that produces a finite range of magnitude for linear response to stepwise deformation. We note that this simple molecular-network scenario explicitly disagree with the tube model. In short, the tube model produces stress overshoot by evaluating the (equilibrium) tube segmental affine orientation as the sole contribution to the shear stress. We take a simple standpoint of rubber network elasticity and suggest that the stress overshoot is due to the ever-increasing softening of the elastic response given in terms of $G(\gamma)$ or $G(\lambda)$ due to loss of entanglement. In passing, we should note that since uniaxial extension makes the entanglement network highly anisotropic, it may be too crude to describe the network in terms of one mesh size $l_{ent}(\lambda)$. It may be necessary to resort to the more general formulation Eq. (7.18) for a more realistic description. To the leading order, we can keep just one parameter $l_{ent}(\gamma)$ for simple shear.

18.3.2 Stepwise Deformation

When stepwise deformation is made in the limit of $Wi_R > 1$, chain deformation occurs at the equilibrium tube segment level. The corresponding stress can be described in terms of rubbery deformation of the equilibrium entanglement network, thanks to the presence of intermolecular grip force. Upon termination of external deformation, the intermolecular grip force vanishes. After step strain of a sufficiently large magnitude elastic yielding can take place in the entanglement network due to force imbalance, as discussed in Chapters 14 and 16. When the magnitude is below some threshold (Section 14.1) for which linear response prevails, the relaxation dynamics are dictated by the intact equilibrium network structure. After Rouse time, it may not be valid to regard the equilibrium tube segment as the relevant structural parameter (i.e., the mesh size) for the relaxing entanglement network: The remaining chain deformation exists only in a network of larger mesh size. The mesh size grows with the elapsing time. In this case, the state of stress should obviously not be evaluated on the fixed length of the equilibrium tube segment. Calculations such as the Doi–Edwards damping function would be groundless just as the stress overshoot prediction in the tube model is (as discussed in the preceding section) because the equilibrium tube segment is an irrelevant length scale as the molecular relaxation progresses. In conclusion, the lack of a sizable linear response regime as a key feature of the tube model stems from questionable computation of the macroscopic stress from the relaxing entanglement network on a single length scale of equilibrium tube segments. The origin of this invalidation is the lack of self-consistency in the treatment of intermolecular interactions.

It seems that in the future we have no choice but to take a guarded view toward the tube model and also critically examine the new approach, which is based on the notion from the rubber elasticity theory.

In summary, a microscopic theory for nonlinear rheology of entangled polymers should be based on available phenomenology. Such a theory or model needs to provide molecular pictures for macroscopically observed behavior and demonstrate causal relationships for various phenomena. In the tube model, the core causality questions in nonlinear rheology were not explored in a realistic and adequate way: What causes affine deformation and when the affine deformation ceases? The tube model assumes by construction that chain deformation and corresponding stress can both be described on a single length scale prescribed by the equilibrium tube diameter (or tube segment). As a consequence, stress becomes artificially decomposed into chain orientation and stretching components. In contrast, the alternative semiquantitative framework of molecular network (Sections 16.2–16.7) suggests that interchain uncrossability transmits the external displacements across the sample; the affine deformation ceases upon force imbalance between the growing intrachain retraction force and the intermolecular grip force. Moreover, chain deformation and relaxation are perceived to occur on multiple length scales in an inseparable combination of chain orientation and stretching. Future studies will determine which picture is more realistic.

References

1 Doi M, Edwards SF. The theory of polymer dynamics. Oxford University Press; 1988.

2 Wang S-Q, Ravindranath S, Wang Y, Boukany P. New theoretical considerations in polymer rheology: elastic breakdown of chain entanglement network. J Chem Phys. 2007;127(6):064903.

3 Graham RS, Likhtman AE, McLeish TC, Milner ST. Microscopic theory of linear, entangled polymer chains under rapid deformation including chain stretch and convective constraint release. J Rheol. 2003;47(5):1171–200.

4 Maxwell B, Nguyen M. Measurement of the elastic properties of polymer melts. Polym Eng Sci. 1979;19(16):1140–50.

5 Marrucci G. Dynamics of entanglements: a nonlinear model consistent with the Cox–Merz rule. J Nonnewton Fluid Mech. 1996;62(2):279–89.

6 Wang Y, Wang S-Q. From elastic deformation to terminal flow of a monodisperse entangled melt in uniaxial extension. J Rheol. 2008;52:1275–90.

7 Tapadia P, Wang S-Q. Direct visualization of continuous simple shear in non-Newtonian polymeric fluids. Phys Rev Lett. 2006;96(1):016001.

8 Liu G, Sun H, Rangou S, Ntetsikas K, Avgeropoulos A, Wang S-Q. Studying the origin of "strain hardening": basic difference between extension and shear. J Rheol. 2013;57(1):89–104.

9 Sun H, Wang S-Q. Shear and extensional rheology of entangled polymer melts: similarities and differences. Sci China-Chem. 2012;55(5):779–86.

10 Graham RS, Henry EP, Olmsted PD. Comment on "New experiments for improved theoretical description of nonlinear rheology of entangled polymers". Macromolecules. 2013;46(24):9849–54.

11 Lodge AS. Elastic recovery and polymer–polymer interactions. Rheol Acta. 1989;28(5):351–62.

12 Phillies GDJ. Phenomenology of polymer solution dynamics. Cambridge University Press; 2011.

13 Marrucci G, Grizzuti N. The free energy function of the Doi–Edwards theory: analysis of the instabilities in stress relaxation. J Rheol. 1983;27(5):433–50.

14 (a)Wang SQ, Ravindranath S, Boukany P, Olechnowicz M, Quirk RP, Halasa A, *et al.* Nonquiescent relaxation in entangled polymer liquids after step shear. Phys Rev Lett. 2006;97(18):187801. (b)Wang SQ *et al.* A refined explanation for the observed elastic yielding was subsequently proposed. J Chem Phys. 2007;127:064903.

15 Wang YY, Boukany PY, Wang SQ, Wang XR. Elastic breakup in uniaxial extension of entangled polymer melts. Phys Rev Lett. 2007;99(23):4.

16 Doi M, Edwards S. Dynamics of concentrated polymer systems. Part 4 – Rheological properties. J Chem Soc, Faraday Trans 2: Mol Chem Phys. 1979;75:38–54.

17 Wang Y, Cheng S, Wang SQ. Basic characteristics of uniaxial extension rheology: comparing monodisperse and bidisperse polymer melts. J Rheol. 2011;55:1247–70.

18 Olmsted P, Radulescu O, Lu C-Y. Johnson–Segalman model with a diffusion term in cylindrical Couette flow. J Rheol. 2000;44(2):257–75.

19 Olmsted PD. Perspectives on shear banding in complex fluids. Rheol Acta. 2008;47(3):283–300.

20 Zhou L, Vasquez PA, Cook LP, McKinley GH. Modeling the inhomogeneous response and formation of shear bands in steady and transient flows of entangled liquids. J Rheol. 2008;52(2):591–623.

21 Agimelen OS, Olmsted PD. Apparent fracture in polymeric fluids under step shear. Phys Rev Lett. 2013;110(20):204503.

22 Moorcroft RL, Fielding SM. Shear banding in time-dependent flows of polymers and wormlike micelles. J Rheol. 2014;58(1):103–47.

23 Plazek DJ. 1995 Bingham medal address: Oh, thermorheological simplicity, wherefore art thou? J Rheol. 1996;40(6):987–1014.

24 Tapadia P, Wang SQ. Nonlinear flow behavior of entangled polymer solutions: yieldlike entanglement–disentanglement transition. Macromolecules. 2004;37(24):9083–95.

25 Inn YW, Wissbrun KF, Denn MM. Effect of edge fracture on constant torque rheometry of entangled polymer solutions. Macromolecules. 2005;38(22):9385–8.

26 Sui C, McKenna GB. Instability of entangled polymers in cone and plate rheometry. Rheol Acta. 2007;46(6):877–88.

27 Ravindranath S, Wang S-Q. Steady state measurements in stress plateau region of entangled polymer solutions: controlled-rate and controlled-stress modes. J Rheol. 2008;52(4):957–80.

28 Ge S, Zhu X, Wang S-Q. Watching shear thinning in creep: Entanglement–disentanglement transition. Polymer. 2017;125:254–264.

29 Sussman DM, Schweizer KS. Microscopic theory of the tube confinement potential for liquids of topologically entangled rigid macromolecules. Phys Rev Lett. 2011;107(7):078102.

30 Sussman DM, Schweizer KS. Microscopic theory of topologically entangled fluids of rigid macromolecules. Phys Rev E. 2011;83(6):061501.

31 Sussman DM, Schweizer KS. Communication: effects of stress on the tube confinement potential and dynamics of topologically entangled rod fluids. J Chem Phys. 2011;135(13):131104.
32 Sussman DM, Schweizer KS. Microscopic theory of entangled polymer melt dynamics: flexible chains as primitive-path random walks and supercoarse grained needles. Phys Rev Lett. 2012;109(16):168306.
33 Sussman DM, Schweizer KS. Microscopic theory of quiescent and deformed topologically entangled rod solutions: general formulation and relaxation after nonlinear step strain. Macromolecules. 2012;45(7):3270–84.
34 Sussman DM, Schweizer KS. Entangled rigid macromolecules under continuous startup shear deformation: consequences of a microscopically anharmonic confining tube. Macromolecules. 2013;46(14):5684–93.
35 Sussman DM, Schweizer KS. Entangled polymer chain melts: orientation and deformation dependent tube confinement and interchain entanglement elasticity. J Chem Phys. 2013;139(23):234904.
36 Malkin AY, Arinstein A, Kulichikhin VG. Polymer extension flows and instabilities. Prog Polym Sci. 2014;39(5):959–78.
37 Malkin AY, Petrie CJS. Some conditions for rupture of polymer liquids in extension. J Rheol. 1997;41(1):1–25.
38 Vinogradov GV, Malkin AY, Volosevitch VV, Shatalov VP, Yudin VP. Flow, high-elastic (recoverable) deformation, and rupture of uncured high molecular-weight linear polymers in uniaxial extension. J Polym Sci Part B-Polym Phys. 1975;13(9):1721–35.
39 Malkin AY, Vinogradov GV. Fracture of polymers in the viscous-fluid state under stretching. Vysokomol Soed A. 1985;27(2):227–37.
40 Barroso VC, Andrade RJ, Maia JM. An experimental study on the criteria for failure of polymer melts in uniaxial extension: the test case of a polyisobutylene melt in different deformation regimes. J Rheol. 2010;54(3):605–18.
41 Andrade RJ, Maia JM. A study on the flow, failure, and rupture mechanisms of low-density polyethylene in controlled-stress uniaxial extensional flow. J Rheol. 2011;55(5):925–37.
42 Zhu X, Wang S-Q. Mechanisms for different failure modes in startup uniaxial extension: tensile (rupture-like) failure and necking. J Rheol. 2013;57(1):223–48.
43 McKinley GH, Sridhar T. Filament-stretching rheometry of complex fluids. Annu Rev Fluid Mech 2002;34:375–415.
44 Huang Q, Alvarez NJ, Matsumiya Y, Rasmussen HK, Watanabe H, Hassager O. Extensional rheology of entangled polystyrene solutions suggests importance of nematic interactions. ACS Macro Lett. 2013;2(8):741–4.
45 Nielsen JK, Rasmussen HK, Hassager O. Stress relaxation of narrow molar mass distribution polystyrene following uniaxial extension. J Rheol. 2008;52(4):885–99.
46 Huang Q, Agostini S, Hengeller L, Shivokhin M, Alvarez NJ, Hutchings LR, *et al.* Dynamics of star polymers in fast extensional flow and stress relaxation. Macromolecules. 2016;49(17):6694–9.

19

Strain Localization and Fluid Mechanics of Entangled Polymers

Rheology is conventionally established on the premise that homogeneous deformation prevails in both simple shear and uniaxial extension setups. Whenever this is true, the task of rheology is reduced to rheometric measurements of either the stress response or displacement of the boundaries under the applied stress, and there is no need to set up particle-tracking velocimetry to monitor the shear field.

Well-entangled polymer melts and solutions break this untested rule of rheology. Severe strain localization takes place in startup shear and in extension, as described in detail in Chapters 9 through 13. The nominal stress versus strain (and time) relationship no longer applies to every point in the fluid. Whenever strain localization takes place, we are facing a fluid mechanical problem whose solution requires us to solve the underlying constitutive equations. Thus, the loss of strain homogeneity presents a formidable dilemma. The inhomogeneous deformation spells the end of rheology, but where does the fluid mechanics begin, in absence of a reliable constitutive law, which is supposed to come from rheology? The task of establishing a rheological description by experiment has to take several steps. The first step is to determine when strain localization may take place. The second step involves identifying the experimental parameters that control strain localization in shear and in extension. Whether shear banding occurs or not depends on the ratio b_{max}/H as analyzed in Section 9.1.2.4. For extension, it remains to be established when and how or whether extensional strain localization still occurs when the molecular weight is lowered toward the entanglement molecular weight M_e. The third step is to investigate why strain localization, such as shear banding, takes place. Currently, we have a qualitative but not quantitative understanding of why homogeneous deformation is replaced by inhomogeneous deformation for well-entangled polymers.

We are far from reaching the pivotal goal of polymer rheology, which is to describe constitutive behavior in an analytical and predictive manner.[1] The tube model is a great and celebrated attempt. But the tube model cannot be extended to treat strongly nonlinear rheological behavior, because its foundation was not built to account for nonlinear responses of entanglement interactions to large deformations. The alternative perspective of emphasizing point-like molecular networking is fruitful but only qualitative. Despite its qualitative nature, our recently achieved conceptual understanding as summarized in this book gives a basis for speculating about the molecular origins of various instabilities in polymer processing. In this chapter, we briefly discuss the impact of the emerging knowledge of polymer rheology on our understanding of polymer processing behavior under practical engineering conditions. We begin with a recap of the leading forms of strain localization in simple shear.

Nonlinear Polymer Rheology: Macroscopic Phenomenology and Molecular Foundation, First Edition. Shi-Qing Wang.

Companion Website: www.wiley.com/go/nonlinearpolymerrheology

19.1 Relationship between Wall Slip and Banding: A Rheological-State Diagram

As an integral part of the description of polymer wall slip, we have already discussed in Section 6.4.1 when the bulk enters the regime of nonlinear responses. This condition is reached when $Wi_{app} > Wi_{ws\text{-}bnl}$ of Eq. (6.23). For a given ratio b_{max}/H, as Wi_{app} varies, the system goes through several regimes: (a) Newtonian shear for $Wi_{app} < 1$, (b) wall slip regime, (c) apparent slip, (d) bulk shear banding, and (e) uniform disentanglement. The last regime, uniform disentanglement, occurs when full disentanglement is reached in the bulk. If we attain full disentanglement at the upper end of the stress plateau, the shear rate can be estimated to be $\dot{\gamma}_{dis} \sim G_{pl}/\eta_e = 1/\tau_e$ for melts and $\dot{\gamma}_{dis} \sim G_{pl}(\phi)/\eta_s$ for solutions. Correspondingly, Wi_{app} reaches

$$Wi_{dis} = \tau/\tau_e = b_{max}/l_{ent} \text{ for melts} \quad (19.1a)$$

or

$$Wi_{dis} = (\eta_0/\eta_s) = b_{max}/l_{ent}(\phi) \text{ for solutions,} \quad (19.1b)$$

where the second equalities follow from Eqs. (6.12) and (6.13), respectively, along with the relationships $\eta_0/\eta_e = \tau/\tau_e$ for melts and $\eta_0 = G_{pl}(\phi)\tau$ for solutions. The literature data indicate, as shown in Chapter 9, that the restoration to homogeneous shear typically occurs at $Wi \ll Wi_{dis}$. For example, shear homogeneity is recovered at $Wi_{app} = 378$ (cf. Fig. 9.8(c)) while Wi_{dis} for the solution is around 23,000, according to the viscosity values available in Ref. [2]. Section 9.1.2.5 presents a scaling analysis to show how the condition to avoid shear banding is met without Wi_{app} reaching Wi_{dis}.

Conversely, when b_{max}/H is significantly reduced, as seen in Table 6.2, where the maximum extrapolation length b_{max} is reduced by the use of a polymeric solvent, without changing Wi_{app}, the system changes from wall slip (Fig. 6.23(b)) to bulk shear banding (Fig. 9.10(b) and (c)). The ratio b_{max}/H can also be varied by changing H. For example, at a small gap distance of $H = 50\ \mu m$, wall slip becomes much more significant, as shown in Section 6.4.2.2.

In summary, the borderline between wall slip (ws) and bulk nonlinear (bnl) response, as well as the limit of full disentanglement, can be used to construct a diagram of rheological states in the "phase space" of Wi_{app} versus b_{max}/H, as shown in Fig. 19.1. This figure is a roadmap for well-entangled polymer solutions and melts under steady shear that involves natural physical adsorption between the polymer and shearing walls. It is slightly modified from a similar diagram in the literature[3] to avoid the confusion brought up by Li *et al.*[4–7] The borderline depicted by Eq. (6.23) is derived from a scaling argument and can only be used as a qualitative guide. For $Wi_{app} > Wi_{ws\text{-}bnl}$, there may be shear banding at the wall, showing apparent wall slip, when the band width is too thin to be spatially resolved. We note that true wall slip involves only a monolayer of chain disentanglement having a thickness similar to the entanglement spacing l_{ent}. Thus, we need to have $Wi_{app} \gg Wi_{ws\text{-}bnl}$ before bulk shear banding is detected. Moreover, shear banding may coexist with wall slip or apparent wall slip in the so-called "bulk shear banding" regime. As described in Chapter 9, the typical polybutadiene solutions studied in the past[8] had moderate values of b_{max}, that is, $2b_{max}/H \sim 1$, so that wall slip is not severe and can readily be suppressed. For such samples, $Wi_{ws\text{-}bnl}$ is around unity. Therefore, particle-tracking velocimetric observations can be carried out at $Wi_{app} \gg 1$, that is, $Wi_{app} \gg Wi_{ws\text{-}bnl}$, to enter deep into the bulk shear banding regime. Since the initial PTV studies[2,7] aimed to demonstrate the emergence of bulk shear banding, slip-like behavior at smaller Wi_{app} was not the focus of these studies.

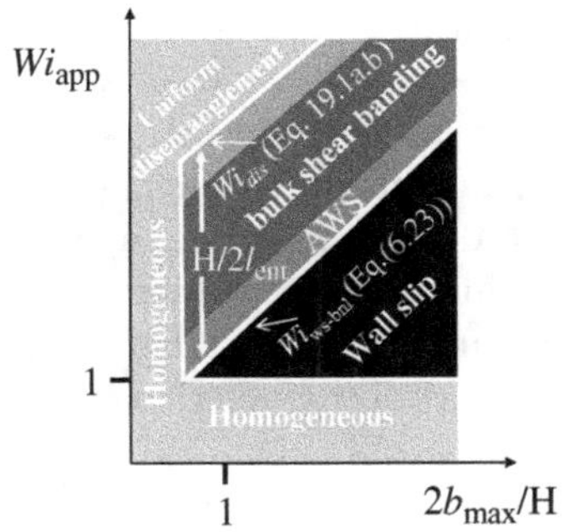

Figure 19.1 Diagram in the phase space of Wi_{app} by $2b_{max}$/H on double log scales to show, at a fixed value of H in steady state, the different rheological responses in terms of the state of shear strain field, reproduced from Ref. [3]. Above $Wi_{ws\text{-}bnl}$, Weissenberg number $Wi > 1$. During high shear, disentanglement could start from the polymer/wall interfaces, producing apparent wall slip (AWS), as shown in the lighter shaded region, instead of any spatially resolvable shear banding in the bulk, which occurs only when Wi_{app} is significantly larger than $Wi_{ws\text{-}bnl}$. A lighter shade is also used to indicate that bulk banding may be replaced by homogeneous shear well before reaching the limit of Wi_{dis} (Eq. (19.1a and b)). Shear homogeneity is restored well below Wi_{dis} because of the reason (shear thinning) presented in Section 9.1.2.5. *Source:* Wang *et al.*, 2011 [3]. Reproduced with permission of American Chemical Society. (*See color plate section for the color representation of this figure.*)

19.2 Modeling of Entangled Polymeric Liquids by Continuum Fluid Mechanics

It is the responsibility of polymer rheology to provide a reliable and realistic constitutive equation. The fluid mechanical phenomena, such as shear banding and vortex formation in the die entry region, can nevertheless reveal something constitutive. It is in this sense that PTV observations of various strain fields, summarized in Sections 10.2–10.4, can be also viewed as a rheological study. Moreover, such PTV investigations help us identify the necessary ingredients that should be included in a realistic constitutive equation. Ultimately, an explanation for the formation of elastic instabilities at die entry, as briefly described in Section 10.2, is a task that belongs to the fluid mechanics of entangled polymers. According to the discussions in Chapter 10, the state of chain entanglement plausibly varies in space, and the effect of disentanglement on the strain field may be measured in terms of b_{max}/D, where D is a characteristic macroscopic dimension of the apparatus (e.g., the die diameter). A constitutive description of the fluid mechanical behavior of entangled polymers must therefore encompass the concept of the extrapolation length b. Separately, we must explain how b varies in space, depending on the local rheological state. A theory for the needed structure–property relations is currently intractable since the desired constitutive equations have yet to be established. The constitutive relationship that we need to find must describe how the state of chain entanglement varies with the local stress and strain fields. At the present, it seems challenging if not infeasible to obtain a realistic and analytical constitutive description of the fluid mechanical behavior of well-entangled polymers. The knowledge presented in the book is merely a modest beginning in our efforts to relate the rheological features to some polymer processing behavior.

So far no constitutive computation of the fluid mechanical behavior of entangled polymeric liquids has incorporated the physics that describes structural variations and the structural influence on the stress versus strain relationship. Parameters in existing constitutive models typically do not vary with space and time. On the other hand, the examples of shear strain localization in Chapter 10 clearly suggest that the state of chain entanglement may be spatially discontinuous. The entanglement network appears intact in one region and undergoes massive collapse in an adjacent region, as shown in Figs. 10.9 and 10.10. In this example, the internal slip, corresponding to the emergence of a very narrow shear band of high shear rate, most plausibly indicates localized chain disentanglement, that is, structural failure in the entanglement network. Certain nonlinear constitutive equations may show mathematical characteristics that resemble the internal slip. But there would be no molecular basis.

While constitutive computations may not be reliable, a great deal can be learned or predicted based on our current theoretical understanding, concerning nonlinear constitutive responses of entangled polymers under large deformation in complex geometries. Figures 10.6 and 10.7 as well Figs. 10.9 and 10.10 provide examples. These experiments were designed to demonstrate and verify the anticipated outcomes. For example, vortex formation is replaced by shear banding at the die entry when the polymer concentration is sufficiently increased or when the solvent viscosity is greatly reduced. In either case, when the extrapolation length b_{max} significantly increases, the solutions have a sufficiently large capacity to undergo internal slip so that vortex formation is avoided. Conversely, upon incorporating sucrose into aqueous DNA solutions to reduce b_{max}, the sample from the reservoir tries to satisfy the no-slip boundary condition at the wall perpendicular to the die flow direction by circulating at the corner, as shown in both Figs. 10.6 and 10.7.

In any currently available constitutive model,[9] b is not a parameter. For example, at a given polymer concentration, when b is substantially altered using a different solvent, the relaxation spectrum obtained from SAOS looks nearly the same. On changing the solvent, there would not be any significant change of the input parameters for any constitutive model. However, we have seen in Fig. 10.7 that the behavior at the die entry switches from shear banding to vortex formation and vice versa, depending on the solvent viscosity. When the chain disentanglement allows massive internal slip to take place (as is also the case in Figs. 10.9 and 10.10 for a polymer melt), the system can avoid vortex formation. It is presently unclear how finite-element calculations can be carried out in a constitutive description of entry instabilities where the disentanglement physics plays an explicit role, and strain localization occurs due to spatially nonuniform yielding of the microstructure.

Finally, in the absence of a reliable constitutive model, we emphasize that in the simplest geometry, that is, simple shear, we can and have made predictions about the fluid mechanics of entangled polymer solutions, in the sense described in Fig. 19.1. On changing the gap distance H, by estimating the intrinsic slip length b_{max}, we can determine whether the response is wall-slip dominant or involves strong shear ($Wi \gg 1$) in the bulk, and whether or not shear banding will emerge. When the level of entanglement is varied, we can also estimate the ratio b_{max}/H, and use this ratio to anticipate characteristic fluid mechanical responses. In general, a realistic fluid mechanical description, based on constitutive information as well as the state of deformation, should also reveal how the structure, for example, state of entanglement, varies in space, along with describing the local stress and strain fields. Despite decades of research, we have not acquired such a description. Consequently, a great deal of new theoretical results have yet to be obtained.

19.3 Challenges in Polymer Processing

There are many instabilities in polymer processing[10] that are bottlenecks limiting attainable production rates and affecting end properties. In general, we expect that most common difficulties met in polymer processing may find solutions from this book after further research and development (R&D). Here we only choose a few pertinent cases where the results presented in this book can be applied to provide a basic understanding.

19.3.1 Extrudate Distortions

Extrusion is a dominant form of polymer processing. For example, pipe extrusion from thermoplastics, tread extrusion from nonvulcanized rubbers, injection molding, film blowing, and fiber spinning, all involve melt extrusion. The simplest feature in extrusion is die swell: the extrudate has a greater dimension than the die diameter. Melts experience severe deformation upon entering a narrow die from the large reservoir and may not relax during die flow. Consequently, significant melt relaxation takes place upon exiting the die. The condition of the die exit also affects the degree of extrude swell.[11] Reliable prediction of die swell requires a realistic fluid mechanical calculation of the entry and exit effects based on sensible constitutive equations. This is a challenging problem since proven constitutive equations are unavailable for well-entangled polymer melts. Then there is the more complicated phenomenon of extrusion instability associated with appearance of irregular extrudate. Such instabilities are collectively known as extrudate distortion or melt fracture. Since linear low-density polyethylene has inherently lower polydispersity, it shows more severe extrudate surface roughness (sharkskin), while low-density polyethylene is free of sharkskin, plausibly due to the long-chain branching that suppresses strain localization, as shown in Sections 8.2.2.1 and 14.2. We need to develop a deeper molecular-level understanding of such behavior to improve the extrusion efficiency and reduce processing costs.

19.3.1.1 Sharkskin Melt Fracture (Due to Exit Boundary Discontinuity)

An extrudate is perfectly smooth when melt extrusion either involves polymers of sufficiently low molecular weight (thus only weakly entangled) or is carried out at a sufficiently small rate. The first type of extrudate distortion is surface roughness, often known as sharkskin melt fracture[12,13] that typically can be suppressed by broadening the molecular weight distribution. To exaggerate the appearance of sharkskin, as shown in Fig. 19.2(a), a metallocene-catalyzed high-density polyethylene resin is used to demonstrate the severity of the instability. The corresponding interfacial condition and velocity profile are expected to look like those in Fig. 19.3(a) and (i). The sharkskin can be made to go away when the die wall at the exit is surface-treated to induce polymer desorption and wall slip. Just the end portion of the die wall (of length comparable to, say, three times the die diameter D) needs to be treated to permit wall slip. Thus, the sharkskin is clearly related to the boundary discontinuity: The sample is first in contact with the die wall and abruptly exits the die to become the extrudate.

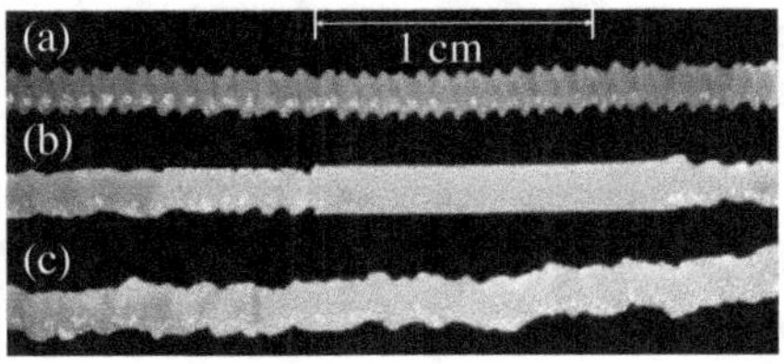

Figure 19.2 Photographs of three extrudates collected under different (piston-speed-controlled) extrusion conditions using a metallocene-catalyzed high-density polyethylene, displaying (a) strong quasi-periodic surface distortion (sharkskin) before the stick–slip transition (SST); (b) alternating smooth and rough sections during pressure (stress) oscillation, involving exactly one cycle of oscillation when the applied throughput is in between the limits set by the SST; (c) gross extrudate distortion above the SST.

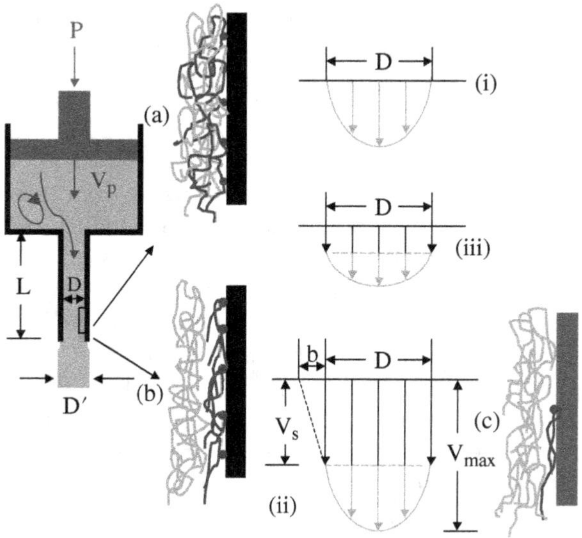

Figure 19.3 Depiction of capillary extrusion (showing vortex formation with a loop), interfacial conditions and corresponding velocity profiles: (a) stick boundary condition due to adequate chain adsorption shown by the anchors (small dots) at the wall; (b) slip boundary condition due to the interfacial disentanglement where adsorbed chains pull out of their engagement with the unbound chains; (c) slip due to polymer desorption on a treated wall. Upon stick–slip transition, the velocity profile changes from (i) to (ii) at a constant pressure. An unstable interfacial state occurs when the velocity profile is given by (iii) where the wall shear stress is lower than in (i) and (ii). The transient state (iii) occurs in the piston-speed-controlled mode of extrusion, that is, constant rate extrusion, typical of industrial operations. (*See color plate section for the color representation of this figure.*)

It was once speculated[14] that the sharkskin reflects an oscillation between no-slip and slip at the die wall near the exit, at a flow rate that is not sufficiently large to cause global wall slip throughout the die. This incorrect interpretation was contradicted by subsequent studies that either visualized the sharkskin formation at the die exit[15] or examined the boundary condition just inside the die wall next to the exit.[16] Twenty years after an oversimplified and incorrect interpretation[17] of sharkskin was given, we realized today that sharkskin represents something we did not really understand in the 1990s: strain localization. It is plausible that the boundary discontinuity at the die exit induces a structural failure, that is, a breakdown of the chain entanglement network. The periodicity in the surface "roughness" arises because, over an extensive region at the die exit, any failure such as shear banding or tensile decohesion results in reduced local strain and stress, so that the melt heals via chain re-entanglement. The healing eventually restores again the condition for the cohesive failure. The cycle repeats. This cyclic behavior is our latest view of sharkskin formation, although we do not have a quantitative depiction of the behavior.

According to experiment, the amplitude of sharkskin increases continuously with the extrusion speed or rate, starting from an undetectable level at low throughputs. The amplitude of the sharkskin also increases with increasing polymer entanglement, as determined by the polymer molecular weight. In particular, the surface roughness diminishes as the molecular weight decreases. It remains unclear how sharkskin behavior is regulated by the extrapolation length b over a relevant length scale, such as the die diameter D: Because the amplitude of the surface roughness seems to scale with D, sharkskin formation may not be simply dictated by the ratio b/D. Tensile strain localization also seems at play. Since the boundary discontinuity increases with D, melt extension, as a key component of the strain field at the exit, is larger for a larger D.

Because the sharkskin instability decreases in magnitude at lower molecular weight, it is perhaps true that at a sufficiently low level of entanglement tensile strain localization no longer occurs. A future study of melt extension as a function of the number Z of entanglements per chain could contribute to a better understanding of the mechanism for sharkskin formation.

19.3.1.2 Gross (Melt Fracture) Extrudate Distortions Due to Entry Instability

With increasing piston speed, the throughput Q_f can be in the middle of the rate window defined by the stick–slip transition that has been depicted in Sections 6.2 and 10.1.2. Since the hysteresis loop is narrow, neither the stick [illustrated by Fig. 19.3(a) and (i)] nor the slip [Fig. 19.3(b) and Fig. 19.3(iii)] boundary condition can satisfy the imposed Q_f. The boundary oscillates between stick and slip, leading to a cyclic pattern in the extrudate, as shown in Fig. 19.2(b). The boundary condition becomes stable-slip when the throughput is above the stick–slip transition, as illustrated in Fig. 19.3(b) and (ii). Upon the stick–slip transition, the melt in the barrel rushes into the die at a high throughout. The sample could experience an entry instability similar to that depicted in Figs. 10.8–10.10. The strongly distorted extrudate in Fig. 19.2(c) is evidence for the unstable entry of the HDPE melt into the die. This photo is similar to that in Fig. 10.8. We are now ready to describe the photo in Fig. 19.2(b). Starting from the left, when capillary extrusion experiences strong wall slip at a high output with a velocity profile depicted by Fig. 19.3(ii), a highly distorted extrudate forms, due to the entry instability. Since the transient output is larger than the value determined by the piston speed, the sample decompresses in the barrel, and the capillary die flow changes from Fig. 19.3(ii) to (iii). Correspondingly, the wall stress, which is proportional to the pressure at the die entry, eventually drops below σ_h, that is, below the hysteresis loop. Because the extrusion returns to the nonslip branch of the flow curve, as shown by Fig. 19.3(i), decompression ceases. Subsequently, melt compression begins again in the barrel because the output from the die, at the rate given by Fig. 19.3(i), is less than the input rate imposed by the piston speed. During this period of slow capillary flow, the HDPE melt has time to relax in the capillary die after passing through the entry instability. With a sufficiently long residence time in the die, the sample loses its memory of the entry instability before exiting. The extrudate is therefore less grossly distorted, but sharkskin reappears. The melt compression terminates when enough entry pressure builds up to drive the capillary extrusion to undergo a stick–slip transition again, changing the flow from Fig. 19.3(i) to (ii). The sample already relaxed in the die comes out smoothly as shown in the photo, similar to the feature in Fig. 10.8, following by gross distortion of fresh extrudate after the extrusion empties the relaxed sample in the capillary die. The exiting melt shows gross distortion because it has freshly undergone entry instability and has no time to relax in the die.

To show explicitly the correlation between the gross melt fracture in Fig. 19.2(c) and the entry instability, we need to apply a visualization method. Although such a task is entirely feasible for polyethylene, it is more convenient to apply room-temperature particle-tracking velocimetric observations and examine similar phenomena in a styrene–butadiene rubber as described in Section 10.2. Since the physics to produce gross melt fracture is the same in high-density polyethylene as in styrene–butadiene rubber, it suffices to demonstrate the strain localization at the die entry with styrene–butadiene rubbers. With rubber polymers, the particle-tracking velocimetric measurements can be conveniently carried out at room temperature.

19.3.1.3 Another Example Showing Pressure Oscillation and Stick–Slip Transition

Capillary rheometry of a monodisperse 1,4-polybutadiene ($M_w = 244$ kg/mol) reveals a stick–slip transition in pressure-controlled mode and pressure oscillation in piston-speed-controlled mode. Figure 19.4 shows pressure oscillation during the capillary extrusion, based on a capillary die of diameter $D = 1$ mm and length $L = 15$ mm, connected to a cylindrical barrel with diameter $D_b = 12$ mm on a Goettfert capillary rheometer (Rheo-Tester 1000). The same

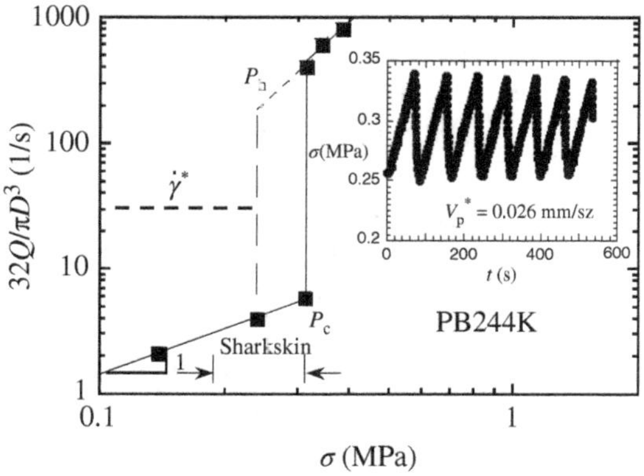

Figure 19.4 Capillary extrusion of PB244K melt (Fig. 6.7) in terms of the normalized die output (shear rate) versus shear stress at the die wall, in either pressure-controlled mode to reveal the stick–slip transition or piston-speed-controlled mode to observe pressure oscillation in the inset. *Source*: Unpublished data obtained by J. Barone at Case Western Reserve University.

sample shows a stick–slip transition, according to results from a Monsanto automatic capillary rheometer. In Figure 19.4, the hysteresis loop is depicted by the dashed line along the upper branch, obtained by lowering the applied pressure on the slip branch, and by the vertical dashed line at P_h, where the flow returns to the lower no-slip branch. For 1,4-polybutadiene of sufficiently high molecular weight, sharkskin typically starts to appear at $\sigma = 0.2$ MPa; the sharkskin grows in amplitude until the stick–slip transition takes place. A constant piston speed $V_p^* = 0.026$ mm/s corresponds to $\dot{\gamma}^* = 30\,s^{-1}$ in the spurt window (between 6 and $400\,s^{-1}$ in shear rate). During melt compression in the barrel the wall shear stress σ builds up to exceed σ_c and causes a disentanglement transition from Figure 19.3(a) to (b). The flow field switches from Figure 19.3(i) to (ii). The initial surge in the output ceases when the wall stress drops below a threshold. Below the critical stress, the adsorbed chains return to their entangled state (as seen in Fig. 19.3(a)), restoring the no-slip boundary condition of Figure 19.3(i). Then compression in the barrel starts again, and $\sigma = PD/4L$ builds up again, leading to the next cycle of the oscillation, which persists as shown by the inset of Fig. 19.4. The period of the cycle shrinks as the amount of sample in the barrel decreases, because at a fixed piston speed it takes a shorter time to compress the melt when there is less in the barrel. Capillary rheometry under controlled pressure can readily reveal sharkskin below the stick–slip transition. Video 19.1 shows capillary extrusion with increasing pressure until the stick–slip transition.[1]

19.3.2 Optimal Extrusion Conditions

For highly entangled melts such as high-density polyethylene, linear low-density polyethylene, and various rubber polymers including polybutadiene, polyisoprene, and styrene–butadiene rubber, extrusion confronts the difficulties summarized in the preceding Section 19.3.1. Even when gross melt fracture does not occur, we must still deal with surface roughness (sharkskin). Such surface roughness makes film blowing unstable. Millions of dollars have been spent to incorporate polymer processing additives (PPA) into polyethylene resins, to induce wall slip by polymer desorption so that sharkskin is removed during extrusion of polyethylene resins. By having wall slip at the die exit, the velocity discontinuity is reduced so that there is weaker melt extension at the exit.

High polydispersity in the molecular weight distribution tends to minimize the amplitude of surface roughness. Since the polydispersity is also known to make shear strain localization less severe,[18] such a correlation seems sensible. Moreover, if the polydispersity reduces tensile strain

1 Video 19.1: Pressure-controlled capillary extrusion of a linear polybutadiene melt – pressure is gradually increased until a stick–slip transition shows up.

localization, we have another explanation for the diminished surface distortion (sharkskin). Commercial low-density polyethylene (LDPE) hardly shows sharkskin. Plausibly long-chain branching in LDPE is an important factor to suppress sharkskin because we have observed the effect of long-chain branching to increase melt cohesion in extension. Since most of the discussion in this book on long-chain branching, in Sections 8.2.2, 11.5, and 14.2, is concerned with melt extension, systematic studies are desirable on how long-chain branching modifies nonlinear responses to simple shear (e.g., whether it suppresses wall slip), in capillary extrusion as well as other forms of deformation including squeezing.

Another way to remove sharkskin is to extrude from the slip branch. More often than not, gross melt fracture would occur because of the greatly increased throughput. The extrudate distortion would be overcome if the entry instability could be made to diminish above the stick–slip transition. The level of polymer entanglement affects the entry instability. With a sufficiently broad molecular weight distribution, for materials such as high-density polyethylene the extrudate is no longer grossly distorted.[19,20] As mentioned in Section 6.2.1.2, the smooth extrudate made it possible to show that the spurt behavior arises from an interfacial stick–slip transition, and *not* from a constitutive instability. Although no systematic studies have been carried out, a preliminary investigation correlated polydispersity with the severity of shear banding.[18] It is plausible that polydispersity suppresses entry instability by reducing the maximum extrapolation length b_{max} relative to the die diameter D. Future work should determine whether long-chain branching is effective at suppressing entry instability during capillary extrusion. Limiting ourselves to the available information on LDPE, we only know that LDPE shows waving extrudate distortion instead of irregular gross distortion at sufficiently large rates.[21]

19.3.3 Melt Strength

In the downstream of polymer melt extrusion, the concept of melt strength is often invoked as a measure of the drawdown tensile force. The melt strength depends on the drawing speed and the basic rheological characteristics.[22] To lowest order, the melt extension rheology presented in Chapters 7 and 10 is directly pertinent to a characterization of the melt strength. For example, we may learn about the melt strength from extension data such as those in Fig. 11.2. We can examine how the tensile force, which is proportional to the engineering stress σ_{engr}, changes with the extension rate at different stretching ratios λ, which can be taken to be approximately equal to the drawdown ratio. Figure 19.5 shows that after a moderate increase with the Hencky rate $\dot{\varepsilon}$ the tensile force saturates and becomes independent of $\dot{\varepsilon}$ at $\lambda = 4.6$. At small rates such as $\dot{\varepsilon} = 1\ s^{-1}$ because of yielding, σ_{engr} is smaller at a larger λ as shown in Fig. 19.5. At sufficiently

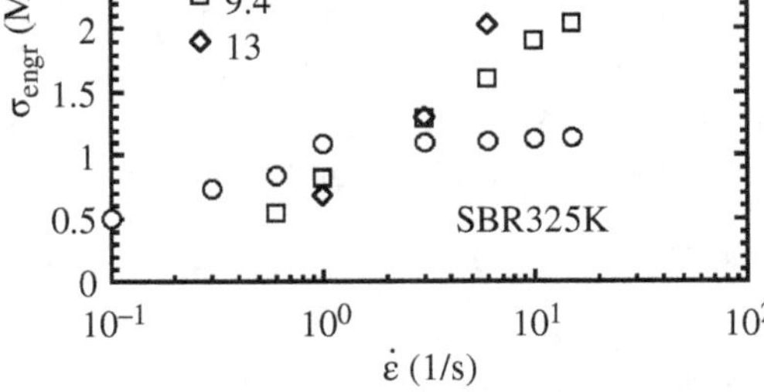

Figure 19.5 "Melt strength" in terms of the normalized tensile force as a function of Hencky rate at three different stretching ratios according to the stress responses of SBR325 to startup uniaxial extension shown in Fig. 11.2.

large extension rates, the melt extension undergoes only partial yielding. σ_{engr} grows with λ monotonically until melt rupture, as indicated in Chapter 11. Consequently, at a large extension ratio of 13, the tensile force (diamonds) rises sharply with $\dot{\varepsilon}$, as shown in Fig. 19.5.

One purpose of this book is to provide a sound basis for phenomenological nonlinear polymer rheology that can be used to discuss and search for solutions to many polymer processing problems. In this chapter, we have only briefly indicated how the end of rheometry is related to the beginning of fluid mechanics for entangled polymeric liquids and shown some examples relating the available knowledge and understanding to well-known processing phenomena.[2] The aim here is to invite and encourage the practitioners in the field to search for new ways to improve polymer processing through a realistic molecular-level understanding of the controlling physics, which has been thoroughly and comprehensively discussed in this book.

References

1 The recent review by Larson and coworkers shows how little we have been able to do to model rheological behavior of polymer solutions and melts: Larson RG, Desai PS. Modeling the rheology of polymer melts and solutions. Annual Review of Fluid Mechanics. 2015;47:47–65.
2 Ravindranath S, Wang SQ, Olechnowicz M, Quirk RP. Banding in simple steady shear of entangled polymer solutions. Macromolecules. 2008;41(7):2663–2670.
3 Wang SQ, Ravindranath S, Boukany P. Homogeneous shear, wall slip, and shear banding of entangled polymeric liquids in simple-shear rheometry: a roadmap of nonlinear rheology. Macromolecules. 2011;44(2):183–190.
4 Li YF, Hu M, McKenna GB, Dimitriou CJ, McKinley GH, Mick RM, *et al.* Flow field visualization of entangled polybutadiene solutions under nonlinear viscoelastic flow conditions. J Rheol. 2013;57(5):1411–1428.
5 Wang SQ, Liu G, Cheng S, Boukany PE, Wang Y, Li X. Letter to the editor: sufficiently entangled polymers do show shear strain localization at high enough Weissenberg numbers. J Rheol. 2014;58(4):1059–1069.
6 Li YF, Hu M, McKenna GB, Dimitriou CJ, McKinley GH, Mick RM, et al. Response to: sufficiently entangled polymers do show shear strain localization at high enough Weissenberg numbers. J Rheol. 2014;58(4):1071–1082.
7 Wang SQ. Meet challenging conditions to observe shear banding in highly entangled polybutadiene solutions Rheology: Open Access. 2016.
8 Ravindranath S, Wang SQ. Steady state measurements in stress plateau region of entangled polymer solutions: controlled-rate and controlled-stress modes. J Rheol. 2008;52(4):957–980.
9 Larson RG. Constitutive equations for polymer melts and solutions. Butterworths, Boston; 1988.
10 Hatzikiriakos SG, Migler KB. Polymer processing instabilities: control and understanding: CRC Press; New York, 2004.
11 Yang XP, Wang SQ, Chai C. Extrudate swell behavior of polyethylenes: capillary flow, wall slip, entry/exit effects and low-temperature anomalies. J Rheol. 1998;42(5):1075–1094.

2 For a technical presentation discussing Nonlinear Polymer Rheology as related to polymer processing, see a recorded lecture at https://www.youtube.com/watch?v=7aw06yn7DbU

12 Piau J, El Kissi N, Tremblay B. Influence of upstream instabilities and wall slip on melt fracture and sharkskin phenomena during silicones extrusion through orifice dies. J Nonnewton Fluid Mech. 1990;34(2):145–180.

13 Mackley M, Rutgers R, Gilbert D. Surface instabilities during the extrusion of linear low density polyethylene. J Nonnewton Fluid Mech. 1998;76(1):281–297.

14 Wang SQ, Drda PA, Inn YW. Exploring molecular origins of sharkskin, partial slip, and slope change in flow curves of linear low density polyethylene. J Rheol. 1996;40(5):875–898.

15 Inn YW, Fischer RJ, Shaw MT. Visual observation of development of sharkskin melt fracture in polybutadiene extrusion. Rheol Acta. 1998;37(6):573–582.

16 Migler KB, Son Y, Qiao F, Flynn K. Extensional deformation, cohesive failure, and boundary conditions during sharkskin melt fracture. J Rheol. 2002;46(2):383–400.

17 Barone J, Plucktaveesak N, Wang S. Interfacial molecular instability mechanism for sharkskin phenomenon in capillary extrusion of linear polyethylenes. J Rheol. 1998;42(4):813–832.

18 Boukany PE, Wang SQ. A correlation between velocity profile and molecular weight distribution in sheared entangled polymer solutions. J Rheol. 2007;51(2):217–233.

19 Wang SQ, Drda PA. Superfluid-like stick-slip transition in capillary flow of linear polyethylene melts. 1. General features. Macromolecules. 1996;29(7):2627–2632.

20 Wang SQ, Drda PA. Stick-slip transition in capillary flow of polyethylene. 2. Molecular weight dependence and low-temperature anomaly. Macromolecules. 1996;29(11):4115–4119.

21 Attalla G, Bertinotti F. Comparison between a linear and a branched low-density polyethylene. J Appl Polym Sci. 1983;28(11):3503–3511.

22 Wagner M, Bernnat A, Schulze V. The rheology of the rheotens test. J Rheol. 1998;42(4):917–928.

20

Conclusion

This book has discussed all essential aspects of nonlinear polymer rheology, including shear and extensional deformations and some other deformations. In most cases, responses of the entanglement network to startup shear and extension are the same, beginning with elastic deformation and transforming with growing strain toward a state of flow. Breaking from the traditional approach to nonlinear rheology of entangled polymers, the discourse is experimentally based, with a great deal of attention given to describing transient responses. The transient time-dependent phenomenology is rich and full of interesting physics. It just so happens that most practical applications in industrial polymer processing do not attain steady state.

From the transient responses, we have learned how to answer the two central questions for the nonlinear rheology of entangled polymers: (a) how macroscopic deformation produces chain-conformational changes and (b) when, how, and why molecular deformation ceases to be elastic at large strains. To address these questions, we must understand what polymer entanglement is and how it responds to macroscopic external deformation. As shown here, the two central characteristics of entanglement are that it is *dynamic* and *collective*, rather than being a static, single-body property. These features make chain entanglement nearly intractable to depict analytically. The famous attempt to describe entanglement in the presence of external deformations is the tube model.[1] The tube model description for nonlinear rheology was built at a time[2] when the coherent phenomenology had yet to be established and the concept of yielding[3] had not had a chance to be discussed and debated in the rheology community. In the absence of the laborious experimental investigations needed to establish a comprehensive phenomenology and a subsequent close scrutiny of this phenomenology, it would be unlikely that anyone could have constructed a reliable and realistic molecular picture of the nonlinear rheology of entangled polymers. Simplifications that are made to make a model tractable must not throw away essential physics. Without sufficient conceptual interpretation of the emerging phenomenology, it is challenging to figure out what assumptions and approximations can be made. One purpose of this book is to present the phenomenology, along with a preliminary understanding of the existing phenomena, in hope that such information can be used to guide the development of a new molecular theory.

The field of nonlinear rheology of entangled polymeric liquids has witnessed significant progress in the past decade. In this book, we have aimed to provide a comprehensive description of how entangled solutions and melts respond to fast large deformations, based on the available information in the literature. In light of strain localization as observed with the particle-tracking velocimetry in simple shear, we inferred that macroscopic yielding originates from the yielding and collapse of the chain entanglement network. To test alternative pictures to the tube model, other more complicated and less conventional experimental arrangements such as squeezing and extrusion have been employed to study localized disentanglement. Moreover, in shear homogeneous yielding via chain disentanglement can also take place when

Nonlinear Polymer Rheology: Macroscopic Phenomenology and Molecular Foundation, First Edition. Shi-Qing Wang.

Companion Website: www.wiley.com/go/nonlinearpolymerrheology

the ratio of the extrapolation length b to the characteristic dimension of the apparatus is much smaller than unity. Even when homogeneity prevails in simple shear, there can still be strong stress overshoot upon startup shear. Thus, stress overshoot does not need to involve shear banding. In general, shear stress overshoot does not need to involve chain disentanglement because overshoot is not unique to entangled polymers. However, the reverse may be true. For well-entangled polymeric fluids, the stress decline during startup shear is due to yielding of the molecular entanglement network through chain disentanglement. In our opinion, for $Wi \gg 1$ deformation-induced failure of the entanglement network is ubiquitous regardless of whether or not strain localization takes place.

20.1 Theoretical Challenges

In any field of scientific inquiry, there are two steps needed toward the establishment of a theoretical framework. The first step is to organize the observed phenomenology into a coherent framework. The second step is to propose an analytical model that accounts for the phenomenology. For the nonlinear rheology of entangled polymers, this first step has been taken. As described in Chapter 16, a conceptual framework has emerged from the wealth of available experimental information. This first step allows us to envision the physics required to answer the elementary questions presented in the second paragraph of this chapter.

On the other hand, the theoretical challenges facing us are profound and demanding. We need a mathematical model that depicts entanglement due to the interchain uncrossability, its response to large deformation, and rheological characteristics in terms of chain entanglement, all in a self-consistent manner. Perhaps we need to divide this enormous task into small pieces, tackling one piece at a time. Finally, such a model will need to be falsifiable.

One key concept to be examined is finite cohesion, as discussed in Section 16.2.3. As a "solid," an entanglement network has a finite cohesion strength. A finite mechanical stress or strain is needed to destroy it. Can we go beyond the toy picture in Fig. 16.3 to establish a theory based on multiple chain interactions, where the entropic barrier concept is rigorously derived? Exploration of this idea will help us clarify the physics of elastic yielding during stepwise deformation and will supply the mechanism for yielding during startup deformation at low shear rates. Sussman and Schweizer have produced promising results in this direction.[4]

Can the characteristics of the intermolecular grip force be investigated by constructing a molecular model? When entangling chains are displaced in opposite directions, how does chain uncrossability produce the grip force due to a deformation field? Conversely, in absence of the grip force, namely, when $1 < Wi \ll \tau/\tau_R$, what could drive the entanglement network to undergo elastic deformation? If entanglements should be treated as point-like interactions, Fig. 20.1 illustrates how chain uncrossability allows external deformations to stretch entangled chains. Here, the dots represent surrounding chains, crossing the XY-plane, that the test chain

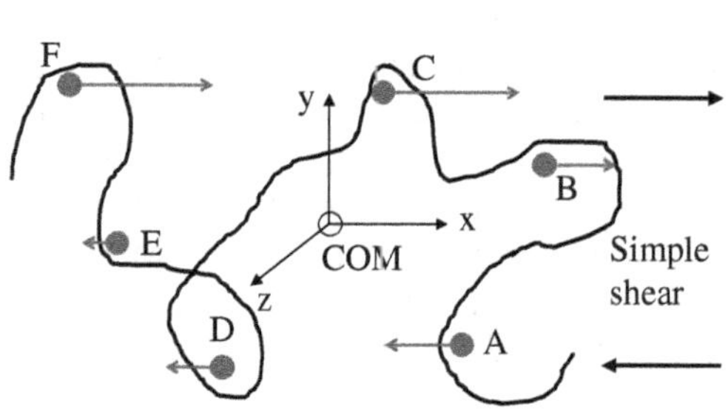

Figure 20.1 Depiction of chain uncrossability, leading to potential entanglement interactions where the dots represent chains surrounding the test chain in simple shear, with the center of mass (COM) of the test chain as the origin of the Cartesian coordinate system.

cannot cross. The effect of shear on the test chain is mimicked by the affine movements of the dots. Conceivably, the strands (subchains) AB and CD first undergo shear deformation. Then the strand FE joins in. There is no entanglement strand for the subchain between chain C and B at the beginning. The surrounding chain D is special. After forming an entanglement strand CD with chain C, another strand, namely, FD may eventually become effective. It is instructive to note that if the shearing direction is reversed, a different set of entanglement strands would come into being. CB may become a strand, CD may also turn into an entanglement strand, and BA and FE are no longer entanglement strands.

It remains to be determined whether polymer entanglement must be treated in such a discontinuous, heterogeneous, and point-like manner in order to capture the physics of entangled polymers under large deformation. In the case of liquids of entangled rods, there are no hairpins to define the point-like interactions. Thus, a self-consistent mean-field approach seems more strongly justifiable, as demonstrated by Sussman and Schweizer.[4–10] The tube theory describes the effect of other chains on a test (flexible) chain also in a mean-field smoothed-out fashion, by considering tube-shaped confinement surrounding the test chain. By assuming the tube to be elastically unbreakable per harmonic confinement, the tube model does not account for interchain interactions in a self-consistent manner. Consequently, a great deal of past work has been made to validate the tube theory, not by examining critically its premises, but by showing how it can be improved to provide calculations that agree with experimental observations. The force-level microscopic theory of Sussman and Schweizer[4–10] provides a self-consistent treatment of chain uncrossability, rigorously for rigid rod-like chains and for primitive-path random walks through coarse-graining. Inevitably, this theory also focuses on the behavior of one test chain in a mean field. Since a mathematical depiction of point-like entanglement interactions is currently intractable, the self-consistent mean-field theory of Sussman and Schweizer is a most promising candidate for nonlinear entangled-polymer rheology. In the spirit of smoothing out heterogeneous interchain interactions, and based on a set of physical considerations that are different from the simple-minded scaling arguments, Schweizer[11] also derived the experimentally observed power scaling law represented in Fig. 7.8(b,c) and summarized it in Eq. (7.5). Although the relationship between the two different approaches, namely, between those presented in Section 16.4.2 and in Ref. [11] has yet to be explored, the recent theoretical efforts[4–10] are truly encouraging.

At large strains in the elastic deformation regime (i.e., for $Wi_R > 1$), it is plausible that hairpins are the true junctions of the entanglement network and the average size l_{hs} of a strand between two hairpins is larger than that of the size l_{ent} of an (equilibrium) entanglement strand. In the packing model depiction of entanglement, two pairs of hairpins are not needed to define an entanglement strand. In the literature we have shown[12] that while l_{ent} scales linearly with the packing length p, the chain flexibility influences l_{hs}. Specifically, l_{hs} is proportional to l_c of Eq. (2.38b) and linearly scales with the Kuhn length l_K. In Ref. [12], the subchain size l_{hs} between two adjacent hairs is denoted by l_{LBS}, where LBS stands for "load-bearing strand." For a given polymer, the ratio l_K/p is a just a numerical constant. Therefore, any difference between l_{ent} and l_{hs} would not change the scaling analysis in Section 16.4.2. However, the difference between l_{ent} and l_{hs} has a specific implication: It has been well established that the initial slope of the stress versus strain curve is proportional to the melt elastic plateau modulus G_{pl}, which according to Eqs. (2.30c) and (2.36a) is related to the structure of the entanglement network. It is conceivable that only the hairpin junctions in the entanglement network survive at large strains to cause growing chain deformation. During large deformations, if hairpin strands are the building blocks of the transformed network, we should observe a different melt modulus from the elastic plateau modulus G_{pl}. It is plausible that the emergence of "partial yielding" indicated in Figs. 7.1(b) and 7.6 is related to this structural shift from the entanglement network of areal density ψ_{ent} in Eq. (2.31) to the chain network[12] made of strands between hairpins with areal density $\psi_{hp} \sim 1/pl_K < \psi_{ent}$.

By construction the tube model decouples the mechanical stress into chain orientation and stretching components at the tube segment scale and overlooks chain deformation at other length scales. These inherent features cause the tube model to contradict the fact that stress relaxation is independent of the magnitude of step strain up to $\gamma_{LR} = 0.6$ and $\lambda_{LR} = 1.5$ (Section 14.1). Chain deformation should not be considered at a single length scale, namely, the size of the equilibrium entanglement strands (tube segments). Moreover, we should not insist that during or after deformation the stress can be separated into one component associated with stretching of the tube segments and another with the orientation of the tube segments. At various stages chain deformation on other length scales may determine the macroscopic stress.

For $Wi_R \ll 1$, although chain stretching is negligible on the tube segment scale, chains in well-entangled polymer liquids do undergo deformation on larger length scales. In general, during startup deformation, the entanglement network initially ($t \sim \tau_R$) deforms on the scale of tube segments. Subsequently, if chain relaxation is not severe, namely, if Wi is much larger than unity, the length scale on which chain deformation occurs may keep growing with increasing strain, as discussed in Section 18.3.1. Conversely, if entanglements survive at large strains, namely, when the yield strain is large for $Wi_R > 1$, chain deformation can spread to smaller length scales than the tube segment scale before the yield point.

Stress relaxation from fast step strain similarly involves multiple length scales. The macroscopic behavior differs, depending on whether the magnitude of step strain exceeds γ_{ebd} (ε_{ebd}) or is below γ_{LR} (λ_{LR}) (cf. Chapter 14). During the elastic yielding (Chapters 12 and 13) retraction force is high enough to break the cohesion of the entanglement network due to the force imbalance. For step strain with magnitude smaller than γ_{LR} (λ_{LR}), chain deformation initially exists on the tube segment scale if the step strain is produced with $Wi_R \gg 1$. Then over time, chain deformation on smaller scales first vanishes. Generally, as a function of time the length scale on which chain remains deformed grows larger; the stress associated with such chain deformation diminishes, as stated in Section 18.3.2. Linear response for step strain is possible because this cascade of chain relaxation on different length scales at different times occurs independent of the magnitude of the step strain. Future theoretical efforts need to capture the essence of changes on multiple length and time scales during startup deformation and after stepwise deformation.

Molecular dynamics simulations may play a role in our search for a tractable molecular model that would capture the essence of entanglement interactions, and their responses to external deformation. More studies should be carried out to clarify the origin of stress overshoot upon startup shear and extension with $1 < Wi \ll \tau/\tau_R$. Such studies should also be aimed to look for evidence or lack of evidence for restructuring of the entanglement network after the stress peak. We note that a past study[13] did not systematically examine how chain deformation occurs on different length scales. Here, it is important to emphasize that, according to the tube theory, the peak in the shear stress or the engineering stress stems from progressive chain orientation on the equilibrium tube segment scale and is not related to any structural rearrangement of the entanglement network. New MD simulations should be able to determine whether this is true or false. Moreover, it is difficult for the tube theory to provide a molecular picture for why elastic yielding can take place after stepwise deformation as shown by the squares in Fig. 13.2 after stepwise extension and by Fig. 12.21(a)-(b) after stepwise shear. These experiments were made at such a low rate that there is only chain orientation according to the tube model. We look forward to future MD simulations to demonstrate how chain deformation occurs on larger length scales and why a step-strained entanglement network undergoes elastic yielding.

20.2 Experimental Difficulties

Most macroscopic rheological experiments have been carried out using conventional rheometric instruments. Further collection of valuable phenomenology requires us to overcome various inherent experimental difficulties. Currently, hardly any data are available on startup shear of well-entangled polymer melts. The reason for this unsatisfactory situation is threefold. First, commercial rotational rheometers do not have sufficient torque and normal force tolerance. Second, in a small cone-plate setup, the effects of edge instability are far more severe for polymer melts because of the high melt plateau modulus G_N^0. Third, it is difficult to set up particle-tracking velocimetry for a small cone-plate apparatus. If we resort to noncommercial, custom-made rotational rheometers, the cone-partitioned plate setup in Fig. 5.2 could be straightforwardly employed to isolate the edge effects from rheometric measurements. In principle, $(R_2 - R_1)$ can always be chosen to be sufficiently large, even larger than R_1, to ensure that simple shear within the cone-plate of radius R_1 remains decorrelated from any edge instability at R_2. For a cone-partitioned plate to accommodate particle-tracking velocimetric observations, R_1 needs to be at least 10 mm. For entangled melts, even for the one with the lowest G_N^0, for example, polystyrene, a new rotational rheometer has to withstand a normal (axial) force as high as 500 N, if not even higher. After meeting all the hardware requirements (e.g., the range of the torque transducer and power of the step motor), we have to confront two additional complications. First, for well-entangled melts with $b_{max}/H \gg 1$, it is challenging to suppress wall slip; in presence of massive wall slip, bulk shear banding will not show up until a very large value of Wi_R in excess of $2b_{max}/H \gg 1$ (cf. Section 6.4.1). Thus, the bulk rheological behavior of entangled melts is challenging to characterize in simple shear: Unlike the situation for solutions, there is not any effective means to suppress wall slip other than employing rough shearing surfaces or superglue. Second, given the high stress levels in entangled melts, the isothermal condition shown in Eq. (17.12) can be more difficult to meet. For example, the shear rate must stay below $1\ s^{-1}$. All these difficulties in making a startup shear experiment on a well-entangled polymer melt underscore the merit of other experiments on polymer melts that have been included in this book, such as uniaxial extension (Section 7.5, Chapters 8, 11, and 13), melt extrusion (Section 10.2), squeezing deformation (Section 10.3), and planar extension (Section 10.4). Since only preliminary results are available on particle-tracking velocimetric observations of die entry, squeezing, and planar extension, much more is expected in the future. The limited data indicate that localized yielding leading to shear banding is also prevalent in more complex geometries than simple shear during large deformation of well-entangled melts.

In case we can successfully build a proper simple shear device with particle-tracking velocimetry capability, several tests await. First, it remains unknown whether shear banding appears in entangled melts that are brought to the condition of $Wi_{app} \gg 1$ by slow rate ramp-up, which is a protocol that has been applied to study entangled solutions in Section 15.5. To establish the phenomenology, it is necessary to understand the relationship between wall slip and bulk shear banding as summarized in Section 19.1. If the melt is too strongly entangled, wall slip prevails because $b_{max}/H \gg 1$. Bulk shear banding would only appear if Wi greatly exceeds $2b_{max}/H$. The condition of $Wi \gg 2b_{max}/H$ can always be accessed if the polymer is amorphous and the time–temperature superposition principle can be applied to slow down chain dynamics, so that the applied rate is small enough to guarantee isothermal condition.

Second, the transient rheological response of melts to startup extension suggests that time–temperature superposition principles break down. The kinematic difference between shear and extension, as mentioned in Section 8.4, reveals that the grip force drives elastic deformation of entangled polymers more efficiently in extension than in shear. Should time–temperature superposition (TTS) remain valid during startup shear of melts? The answer would further demonstrate when the difference between shear and extension must be taken into account. The answer will help clarify whether the observed TTS failure in polymer solutions also occurs in entangled melts under startup shear.

Third, since certain polymer solutions can display strain hardening in startup simple shear, as shown in Section 17.4, it is intriguing to find out whether a pure polymer melt undergoes strain hardening at sufficiently large shear rates. The available observation of a shear strain-softening-to-hardening transition in solutions suggests the temperature needs to be sufficiently close to the glass transition temperature T_g of the parent polymer. Would the transition occur in a melt at the same Weissenberg number when the experimental temperature is varied? This question is related to the preceding question regarding the time–temperature superposition. Clearly, a properly designed shear cell for a wide range of temperatures that supports particle-tracking velocimetry would afford us many new opportunities to study shear rheology of well-entangled polymer melts.

Most of the available evidence for shear inhomogeneity in startup shear or after large stepwise shear is based on studies of polybutadiene and DNA solutions. There are insufficient studies of solutions of other polymers, such as polystyrene and polyisoprene. To determine if the shear banding behaviors are universal, future studies should provide parallel data on entangled solutions of other polymers. Since styrene–butadiene rubbers have been observed to display similar strain localization both during and after startup shear (Chapter 12), in die entry region and squeezing (Chapter 10), we do not have any reason to doubt about the universality.

It would be worthwhile to further clarify the difference between Sentmanat extensional rheometry and filament stretching rheometry. Phenomenologically, the differences are clear. In Sentmanat extensional rheometry, there is no sustainable homogeneous deformation upon startup fast extension, and consequently no fully developed flow state can be reached for monodisperse entangled melts. Measurements in literature have suggested that a polymer melt or solution that suffers strain localization in the Sentmanat extensional rheometry can reach steady-flow state in filament stretching rheometry. Appendix 11.A provided an important analysis, showing that filament stretching rheometry disguises the important physics in melt extension rheology, just as capillary rheometry fails to be a valid protocol for studying bulk shear rheology of entangled polymeric liquids. When strain localization is severe, as is the case for well-entangled melts in regime II (cf. Section 11.1), even filament stretching rheometry cannot avoid tensile instability arising from localized yielding of the entanglement network.[14] With filament stretching rheometry, users have frequently claimed that the last data point before the onset of specimen failure represents the steady-state property. After specimen failure, the sample snaps back. Careful examination should be made to study the degree of elastic recoil associated with the retracted pieces. Users of filament stretching rheometry should further answer a simple question: If the steady-flow state has arrived, so that melt extension can continue without macroscopic separation, why did the specimen suffer a breakdown?

This book has attempted to provide an effective framework for understanding nonlinear polymer rheology, from which one may address challenges in polymer processing. The task of translating the material in this book to develop useful guidelines for processing polymeric products is a major undertaking left to the next generation of practitioners to complete. Fulfilling this task would bring our understanding to fruition.

Finally, we emphasize the following. While the alternative conceptual framework has yet to be developed into an analytical model that supplies quantitative descriptions of nonlinear rheological responses, the framework does satisfactorily rationalize the observed phenomenology and predict many phenomena[15] that were unknown in the past and counterintuitive according to the conventional wisdom. Rather than taking the present interpretations for granted, the reader is encouraged to come up with his or her own explanations for the various phenomena detailed in this book. In examining the available phenomenology, the reader may also benefit from forgetting what he or she has learned from the tube model. Let us suppose that there were no molecular-level interpretations of the phenomenology. How would we proceed to make some sense out of the various experiments? The reader is in control of this mental exercise. Intellectual honesty shall guide where she or he will go.

References

1 Doi M, Edwards SF. The theory of polymer dynamics. Oxford University Press, New York; 1988.
2 Doi M, Edwards S. Dynamics of concentrated polymer systems. Part 4.—Rheological properties. J Chem Soc Faraday Trans 2Mol Chem Phys. 1979;75:38–54.
3 Maxwell B, Nguyen M. Measurement of the elastic properties of polymer melts. Polym Eng Sci. 1979;19(16):1140–50.
4 Sussman DM, Schweizer KS. Microscopic theory of quiescent and deformed topologically entangled rod solutions: general formulation and relaxation after nonlinear step strain. Macromolecules. 2012;45(7):3270–84.
5 Sussman DM, Schweizer KS. Communication: effects of stress on the tube confinement potential and dynamics of topologically entangled rod fluids. J Chem Phys. 2011;135(13):131104.
6 Sussman DM, Schweizer KS. Microscopic theory of the tube confinement potential for liquids of topologically entangled rigid macromolecules. Phys Rev Lett. 2011;107(7):078102.
7 Sussman DM, Schweizer KS. Microscopic theory of topologically entangled fluids of rigid macromolecules. Phys Rev E. 2011;83(6):061501.
8 Sussman DM, Schweizer KS. Microscopic theory of entangled polymer melt dynamics: flexible chains as primitive-path random walks and supercoarse grained needles. Phys Rev Lett. 2012;109(16):168306.
9 Sussman DM, Schweizer KS. Entangled polymer chain melts: orientation and deformation dependent tube confinement and interchain entanglement elasticity. J Chem Phys. 2013;139(23):234904.
10 Sussman DM, Schweizer KS. Entangled rigid macromolecules under continuous startup shear deformation: consequences of a microscopically anharmonic confining tube. Macromolecules. 2013;46(14):5684–93.
11 Schweizer KS. 2016. unpublished.
12 Wang S-Q, Cheng S, Lin P, Li X. A phenomenological molecular model for yielding and brittle-ductile transition of polymer glasses. J Chem Phys. 2014;141(9):094905.
13 Cao J, Likhtman AE. Simulating startup shear of entangled polymer melts. ACS Macro Lett. 2015;4(12):1376–81.
14 Nielsen JK, Hassager O, Rasmussen HK, McKinley GH. Observing the chain stretch transition in a highly entangled polyisoprene melt using transient extensional rheometry. J Rheol. 2009;53(6):1327–46.
15 Wang S-Q. Nonlinear rheology of entangled polymers at turning point. Soft Matter. 2015;11(8):1454–8.

Symbols and Acronyms

ρ	Mass density
ϕ	Polymer concentration in terms of volume fraction
ρ_p	Number density of polymer chains
k_B	Boltzmann constant, 1.38×10^{-23} J/K
N_a	Avogadro's constant, 6.022×10^{23} mol^{-1}
T	Absolute temperature in Kelvin
R	Gas constant, 8.3 J/(K mol) or radius characterizing the dimension of rheometric apparatus
a_T	Williams–Landel–Ferry (WLF) shift factor
C_1, C_{10}	WLF constants (Section 2.4.1)
C_2, C_{20}	WLF constants
T_g	Glass transition temperature
M	Molecular weight
M_e	Entanglement molecular weight
R	End-to-end distance vector of a flexible polymer chain
R	Averaged end-to-end distance of a flexible polymer chain
l_K	Kuhn (segment) length
l_{ent}	Entanglement spacing
$l_{ent}(\phi)$	Entanglement spacing of solution at concentration ϕ
l_h	Characteristic length of a hairpin segment (Section 16.4.2)
N	Number of Kuhn segments per chain
N_e	Number of Kuhn segments per entanglement strand
Z	Number of entanglements per chain (degree of entanglement) (Section 2.3.1)
V_{perv}	Pervaded volume of a Gaussian chain (Section 2.2.3.1)
ν_M	Physical volume of a chain of molecular weight M
p	Packing length (Sections 2.1.4.4 and 2.2.3.1)
s	Molecular cross-sectional area, proportional to p (Section 2.2.3.2)
ψ_{ent}	Areal number density of entanglement strands (Section 2.2.3.1)
ψ_{hp}	Areal number density of hairpins that are junctions to define effective strands (Section 20.1)
ψ^*	Limiting areal number density of entanglement strands (appendix of Chapter 8)
G	Shear modulus
G_{pl}	Elastic plateau modulus
G_N^0	Melt plateau modulus
$G_{eq}(t)$	Relaxation modulus for linear response

Nonlinear Polymer Rheology: Macroscopic Phenomenology and Molecular Foundation, First Edition. Shi-Qing Wang.

Companion Website: www.wiley.com/go/nonlinearpolymerrheology

$G(\gamma, t)$	Relaxation modulus for nonlinear response
G'	Storage modulus
G''	Loss modulus
G_{coh}	Cohesive modulus comparable to the plateau modulus G_{pl} (Section 7.2.1)
J	Compliance (Section 2.2.1)
τ_0	Segmental relaxation time, associated with a Kuhn segment or bead in the bead-spring model
τ	Terminal relaxation time, Maxwell time, reptation time
τ_d	Disengagement (reptation) time
τ_{eff}	Effective relaxation time during steady shear
$\tau(\phi)$	Longest relaxation time of entangled solution at ϕ
τ_e	Entanglement confinement time
τ_R	Rouse relaxation time (Section 2.1.4)
τ_h	Local chain relaxation time associated with a hairpin (Section 16.4.2)
τ_K	Relaxation time of a Kuhn segment in entangled melts
D	Diffusion constant (Section 2.1.4)
D_R	Diffusion coefficient for a Rouse chain (Section 2.1.5)
D_Z	Diffusion coefficient for a Zimm chain
D_s	Self-diffusion coefficient (Section 2.3.1)
η	Shear viscosity, $\eta = G\tau$ from the Maxwell model
η_E	Extensional viscosity
η_0	Zero-shear melt viscosity
$\eta_0(\phi)$	Zero-shear solution viscosity at polymer concentration ϕ
η_i	Viscosity at the polymer/wall interface
η_e	Zero-shear viscosity of a Rouse melt with molecular weight M_e
η_s	Solvent viscosity
$\eta^+(t)$	Transient shear viscosity
$\eta_E^+(t)$	Transient extensional viscosity
γ	Shear strain or magnitude of step strain
γ_y	Yield strain, at which the shear stress attains its maximum (Chapter 7)
γ_r	Recoverable strain (Section 7.2.2)
γ_{LR}	Maximum shear strain for linear response (Section 12.1.1)
γ_{ebd}	Threshold strain for elastic breakdown (nonquiescent relaxation) (Chapter 14)
t_y	Elapsed time to the yield point at peak shear (or engineering) stress
$\dot{\gamma}$	Shear rate in the bulk
$\dot{\gamma}_{app}$	Apparent shear rate or nominal shear rate
$\dot{\gamma}_{dis}$	Shear rate to reach homogeneous fully disentangled state
$\dot{\gamma}_c$	Onset shear rate for shear thinning, equal to $1/\tau$
$\dot{\gamma}_w$	Shear rate at die wall
λ	Stretching ratio
λ^*_{eq}	Maximum stretching ratio in affine extension limit when an equilibrium entanglement strand is fully straightened [Eq. (8.A1)]
λ^*	Limiting stretching ratio in non-affine extension when a surviving entanglement strand is fully straightened [Eq. (11.5)]
ε	Hencky strain equal to $ln\,\lambda$ (Section 3.2)
ε_y	Yield Hencky strain, at which the engineering stress peaks
ε_{rec}	Recoverable Hencky strain

ε_{LR}	Maximum Hencky strain for linear response, $ln\,\lambda_{LR}$ (Chapter 14)
ε_{ebd}	Threshold Hencky strain for elastic breakdown (Chapters 13 and 14)
$\dot{\varepsilon}$	Hencky strain rate
Wi	Weissenberg number, that is, product of bulk deformation rate and terminal relaxation (reptation) time τ (Section 1.2.7)
Wi_R	Rouse–Weissenberg number, that is, product of bulk deformation rate and Rouse relaxation time τ_R
Wi_{app}	Nominal Wi involving the average deformation rate, for example, V/H for a shear cell of gap distance H where one plate is displaced at V while holding the other fixed (Section 6.1.1), equal to $\dot{\gamma}_{app}\tau$
Wi_c	Critical Weissenberg number $\dot{\gamma}_c\tau = 1$
$Wi_{ws\text{-}bnl}$	Threshold value of Wi that divides the wall-slip dominant regime and regime of bulk nonlinear response
Wi_{dis}	Weissenberg number according to $\dot{\gamma}_{dis}\tau$ (Section 19.1)
σ	Shear stress
σ^{ss}	Steady-state shear stress
σ_w	Shear stress at die wall
σ_c	Critical or plateau shear stress for either wall slip or onset of shear thinning
σ_h	Lower bound shear stress for complete wall slip in the hysteresis loop
σ_y	Yield stress, that is, peak of shear stress during fast startup shear, or critical stress in yield-stress materials
σ_E	Extensional Cauchy (true) stress
σ_{engr}	Engineering stress, related to true stress σ_E as σ_E/λ
$\sigma_{engr(y)}$	Yield (engineering) stress in uniaxial extension, that is, the peak of σ_{engr}
a_i	Interfacial slip layer thickness (Section 6.1.1)
V_s	Slip velocity
V_s^*	Limiting slip velocity upon full disentanglement (Section 6.1.3)
b, b_{max}	Extrapolation length, and its maximum value, for melts and solutions
b_0	Variable extrapolation length (Sections 6.3.1 and 9.1.2.4)
β	Interfacial friction coefficient, dependent on the state of interfacial entanglement
β_{min}	Minimum of β in absence of any interfacial chain entanglement
$h(\gamma)$	Damping function (Section 12.1.1)
h_{DE}	Doi–Edwards damping function (Section 12.1.1)
Q_{xy}	Orientation function (Sections 12.1.2 and 16.1.3)
$f_{retract}$	Retractive intrachain force due to external deformation (Section 16.2)
f_{ent}	Entanglement force on length scale of l_{ent} (Section 16.2)
f_{img}	Intermolecular grip force (Section 16.2)
f_0	Segmental force (Sections 11.2.2 and 16.4.2)
e	Enthalpy density
u	Total internal energy density (Chapter 17)
u_1	One component of internal energy sensitive to temperature
u_2	Second component of internal energy insensitive to temperature
q	Heat density
w	Mechanical work density
c_p	Specific heat
TTS	Time–temperature superposition
PTV	Particle-tracking velocimetry

SST	Stick–slip transition
EDT	Entanglement–disentanglement transition
PMC	Polymer mode coupling
SER	Sentmanat extensional rheometry
FSR	Filament stretching rheometry
SBR	Styrene–butadiene rubber
HDPE	High-density polyethylene
LDPE	Low-density polyethylene
PS	Polystyrene
PMMA	Poly(methyl methacrylate)
1,4-PB	Polybutadiene
1,4-PI	Polyisoprene
3,4-PI	Polyisoprene
DNA	Deoxyribonucleic acid

Subject Index

Nonlinear Polymer Rheology: Macroscopic Phenomenology and Molecular Foundation, First Edition. Shi-Qing Wang.

Companion Website: www.wiley.com/go/nonlinearpolymerrheology

www.ingramcontent.com/pod-product-compliance
Lightning Source LLC
Chambersburg PA
CBHW081721120726
47787CB00006B/601
* 9 7 8 0 4 7 0 9 4 6 9 8 5 *